Springer Handbook of Enzymes Volume 34

Dietmar Schomburg and
Ida Schomburg (Eds.)

Springer Handbook of Enzymes

Volume 34
Class 2 · Transferases VII
EC 2.5.1.31–2.6.1.57

coedited by Antje Chang

Second Edition

 Springer

Professor DIETMAR SCHOMBURG
e-mail: d.schomburg@uni-koeln.de

Dr. IDA SCHOMBURG
e-mail: i.schomburg@uni-koeln.de

Dr. ANTJE CHANG
e-mail: a.chang@uni-koeln.de

University to Cologne
Institute for Biochemistry
Zülpicher Strasse 47
50674 Cologne
Germany

Library of Congress Control Number: 2006939794

ISBN-10 3-540-36509-5   2nd Edition Springer Berlin Heidelberg New York
ISBN-13 978-3-540-36509-9   2nd Edition Springer Berlin Heidelberg New York

The first edition was published as Volume 13 (ISBN 3-540-62608-5) of the "Enzyme Handbook".

Springer is a part of Springer Science+Business Media
springer.com
© Springer-Verlag Berlin Heidelberg 2007
Printed in Germany

Cover design: Erich Kirchner, Heidelberg
Typesetting: medionet AG, Berlin

Printed on acid-free paper   2/3141m-5 4 3 2 1 0

Attention all Users
of the "Springer Handbook of Enzymes"

Information on this handbook can be found on the internet at
http://www.springer.com
choosing "Chemistry" and then "Reference Works".

A complete list of all enzyme entries either as an alphabetical Name Index or as
the EC-Number Index is available at the above mentioned URL. You can
download and print them free of charge.

A complete list of all synonyms (> 25,000 entries) used for the enzymes is
available in print form (ISBN 3-540-41830-X).

Save 15 %

We recommend a standing order for the series to ensure you automatically
receive all volumes and all supplements and save 15 % on the list price.

Preface

Today, as the full information about the genome is becoming available for a rapidly increasing number of organisms and transcriptome and proteome analyses are beginning to provide us with a much wider image of protein regulation and function, it is obvious that there are limitations to our ability to access functional data for the gene products – the proteins and, in particular, for enzymes. Those data are inherently very difficult to collect, interpret and standardize as they are widely distributed among journals from different fields and are often subject to experimental conditions. Nevertheless a systematic collection is essential for our interpretation of genome information and more so for applications of this knowledge in the fields of medicine, agriculture, etc. Progress on enzyme immobilisation, enzyme production, enzyme inhibition, coenzyme regeneration and enzyme engineering has opened up fascinating new fields for the potential application of enzymes in a wide range of different areas.

The development of the enzyme data information system BRENDA was started in 1987 at the German National Research Centre for Biotechnology in Braunschweig (GBF) and is now continuing at the University at Cologne, Institute of Biochemistry. The present book "Springer Handbook of Enzymes" represents the printed version of this data bank. The information system has been developed into a full metabolic database.

The enzymes in this Handbook are arranged according to the Enzyme Commission list of enzymes. Some 4,000 "different" enzymes are covered. Frequently enzymes with very different properties are included under the same EC-number. Although we intend to give a representative overview on the characteristics and variability of each enzyme, the Handbook is not a compendium. The reader will have to go to the primary literature for more detailed information. Naturally it is not possible to cover all the numerous literature references for each enzyme (for some enzymes up to 40,000) if the data representation is to be concise as is intended.

It should be mentioned here that the data have been extracted from the literature and critically evaluated by qualified scientists. On the other hand, the original authors' nomenclature for enzyme forms and subunits is retained. In order to keep the tables concise, redundant information is avoided as far as possible (e.g. if K_m values are measured in the presence of an obvious cosubstrate, only the name of the cosubstrate is given in parentheses as a commentary without reference to its specific role).

The authors are grateful to the following biologists and chemists for invaluable help in the compilation of data: Cornelia Munaretto and Dr. Antje Chang.

Cologne
Winter 2006

Dietmar Schomburg, Ida Schomburg

List of Abbreviations

A	adenine
Ac	acetyl
ADP	adenosine 5'-diphosphate
Ala	alanine
All	allose
Alt	altrose
AMP	adenosine 5'-monophosphate
Ara	arabinose
Arg	arginine
Asn	asparagine
Asp	aspartic acid
ATP	adenosine 5'-triphosphate
Bicine	N,N'-bis(2-hydroxyethyl)glycine
C	cytosine
cal	calorie
CDP	cytidine 5'-diphosphate
CDTA	trans-1,2-diaminocyclohexane-N,N,N,N-tetraacetic acid
CMP	cytidine 5'-monophosphate
CoA	coenzyme A
CTP	cytidine 5'-triphosphate
Cys	cysteine
d	deoxy-
D-	(and L-) prefixes indicating configuration
DFP	diisopropyl fluorophosphate
DNA	deoxyribonucleic acid
DPN	diphosphopyridinium nucleotide (now NAD^+)
DTNB	5,5'-dithiobis(2-nitrobenzoate)
DTT	dithiothreitol (i.e. Cleland's reagent)
EC	number of enzyme in Enzyme Commission's system
E. coli	Escherichia coli
EDTA	ethylene diaminetetraacetate
EGTA	ethylene glycol bis(-aminoethyl ether) tetraacetate
ER	endoplasmic reticulum
Et	ethyl
EXAFS	extended X-ray absorption fine structure
FAD	flavin-adenine dinucleotide
FMN	flavin mononucleotide (riboflavin 5'-monophosphate)
Fru	fructose
Fuc	fucose
G	guanine
Gal	galactose

GDP	guanosine 5'-diphosphate
Glc	glucose
GlcN	glucosamine
GlcNAc	N-acetylglucosamine
Gln	glutamine
Glu	glutamic acid
Gly	glycine
GMP	guanosine 5'-monophosphate
GSH	glutathione
GSSG	oxidized glutathione
GTP	guanosine 5'-triphosphate
Gul	gulose
h	hour
H4	tetrahydro
HEPES	4-(2-hydroxyethyl)-1-piperazineethane sulfonic acid
His	histidine
HPLC	high performance liquid chromatography
Hyl	hydroxylysine
Hyp	hydroxyproline
IAA	iodoacetamide
IC 50	50% inhibitory concentration
Ig	immunoglobulin
Ile	isoleucine
Ido	idose
IDP	inosine 5'-diphosphate
IMP	inosine 5'-monophosphate
ITP	inosine 5'-triphosphate
K_m	Michaelis constant
L-	(and D-) prefixes indicating configuration
Leu	leucine
Lys	lysine
Lyx	lyxose
M	mol/l
mM	millimol/l
m-	*meta-*
Man	mannose
MES	2-(N-morpholino)ethane sulfonate
Met	methionine
min	minute
MOPS	3-(N-morpholino)propane sulfonate
Mur	muramic acid
MW	molecular weight
NAD^+	nicotinamide-adenine dinucleotide
NADH	reduced NAD
$NADP^+$	NAD phosphate
NADPH	reduced NADP
NAD(P)H	indicates either NADH or NADPH

NBS	N-bromosuccinimide
NDP	nucleoside 5'-diphosphate
NEM	N-ethylmaleimide
Neu	neuraminic acid
NMN	nicotinamide mononucleotide
NMP	nucleoside 5'-monophosphate
NTP	nucleoside 5'-triphosphate
o-	ortho-
Orn	ornithine
p-	para-
PBS	phosphate-buffered saline
PCMB	p-chloromercuribenzoate
PEP	phosphoenolpyruvate
pH	$-\log 10[H^+]$
Ph	phenyl
Phe	phenylalanine
PHMB	p-hydroxymercuribenzoate
PIXE	proton-induced X-ray emission
PMSF	phenylmethane-sulfonylfluoride
p-NPP	p-nitrophenyl phosphate
Pro	proline
Q_{10}	factor for the change in reaction rate for a $10°C$ temperature increase
Rha	rhamnose
Rib	ribose
RNA	ribonucleic acid
mRNA	messenger RNA
rRNA	ribosomal RNA
tRNA	transfer RNA
Sar	N-methylglycine (sarcosine)
SDS-PAGE	sodium dodecyl sulfate polyacrylamide gel electrophoresis
Ser	serine
T	thymine
t_H	time for half-completion of reaction
Tal	talose
TDP	thymidine 5'-diphosphate
TEA	triethanolamine
Thr	threonine
TLCK	N^{α}-p-tosyl-L-lysine chloromethyl ketone
T_m	melting temperature
TMP	thymidine 5'-monophosphate
Tos-	tosyl- (p-toluenesulfonyl-)
TPN	triphosphopyridinium nucleotide (now $NADP^+$)
Tris	tris(hydroxymethyl)-aminomethane
Trp	tryptophan
TTP	thymidine 5'-triphosphate
Tyr	tyrosine
U	uridine

U/mg	μmol/(mg*min)
UDP	uridine 5'-diphosphate
UMP	uridine 5'-monophosphate
UTP	uridine 5'-triphosphate
Val	valine
Xaa	symbol for an amino acid of unknown constitution in peptide formula
XAS	X-ray absorption spectroscopy
Xyl	xylose

List of Deleted and Transferred Enzymes

Since its foundation in 1956 the Nomenclature Committee of the International Union of Biochemistry and Molecular Biology (NC-IUBMB) has continually revised and updated the list of enzymes. Entries for new enzymes have been added, others have been deleted completely, or transferred to another EC number in the original class or to different EC classes, catalyzing other types of chemical reactions. The old numbers have not been allotted to new enzymes; instead the place has been left vacant or cross-references given to the changes in nomenclature.

Deleted and Transferred Enzymes
For EC class 2.5.1.31–2.6.1.57 these changes are:

Recommended name	Old EC number	Alteration
leukotriene-C4 synthase	2.5.1.37	transferred to EC 4.4.1.20
aristolochene synthase	2.5.1.40	transferred to EC 4.2.3.9
D-aspartate transaminase	2.6.1.10	included with EC 2.6.1.21
tyrosine-pyruvate transaminase	2.6.1.20	deleted
thyroxine transaminase	2.6.1.25	included with EC 2.6.1.24
glutamate synthase	2.6.1.53	transferred to EC 1.4.1.13

Index of Recommended Enzyme Names

Description of Data Fields

All information except the nomenclature of the enzymes (which is based on the recommendations of the Nomenclature Committee of IUBMB (International Union of Biochemistry and Molecular Biology) and IUPAC (International Union of Pure and Applied Chemistry) is extracted from original literature (or reviews for very well characterized enzymes). The quality and reliability of the data depends on the method of determination, and for older literature on the techniques available at that time. This is especially true for the fields *Molecular Weight* and *Subunits*.

The general structure of the fields is: **Information – Organism – Commentary – Literature**

The information can be found in the form of numerical values (temperature, pH, K_m etc.) or as text (cofactors, inhibitors etc.).

Sometimes data are classified as *Additional Information*. Here you may find data that cannot be recalculated to the units required for a field or also general information being valid for all values. For example, for *Inhibitors*, *Additional Information* may contain a list of compounds that are not inhibitory.

The detailed structure and contents of each field is described below. If one of these fields is missing for a particular enzyme, this means that for this field, no data are available.

1 Nomenclature

EC number
The number is as given by the IUBMB, classes of enzymes and subclasses defined according to the reaction catalyzed.

Systematic name
This is the name as given by the IUBMB/IUPAC Nomenclature Committee

Recommended name
This is the name as given by the IUBMB/IUPAC Nomenclature Committee

Synonyms
Synonyms which are found in other databases or in the literature, abbreviations, names of commercially available products. If identical names are frequently used for different enzymes, these will be mentioned here, cross references are given. If another EC number has been included in this entry, it is mentioned here.

CAS registry number

The majority of enzymes have a single chemical abstract (CAS) number. Some have no number at all, some have two or more numbers. Sometimes two enzymes share a common number. When this occurs, it is mentioned in the commentary.

2 Source Organism

For listing organisms their systematic name is preferred. If these are not mentioned in the literature, the names from the respective literature are used. For example if an enzyme from yeast is described without being specified further, yeast will be the entry. This field defines the code numbers for the organisms in which the enzyme with the respective EC number is found. These code numbers (form <_>) are displayed together with each entry in all fields of BRENDA where organism-specific information is given.

3 Reaction and Specificity

Catalyzed reaction

The reaction as defined by the IUBMB. The commentary gives information on the mechanism, the stereochemistry, or on thermodynamic data of the reaction.

Reaction type

According to the enzyme class a type can be attributed. These can be oxidation, reduction, elimination, addition, or a name (e.g. Knorr reaction)

Natural substrates and products

These are substrates and products which are metabolized in vivo. A natural substrate is only given if it is mentioned in the literature. The commentary gives information on the pathways for which this enzyme is important. If the enzyme is induced by a specific compound or growth conditions, this will be included in the commentary. In *Additional information* you will find comments on the metabolic role, sometimes only assumptions can be found in the references or the natural substrates are unknown.

In the listings, each natural substrate (indicated by a bold S) is followed by its respective product (indicated by a bold P). Products are given with organisms and references included only if the respective authors were able to demonstrate the formation of the specific product. If only the disappearance of the substrate was observed, the product is included without organisms of references. In cases with unclear product formation only a ? as a dummy is given.

Substrates and products

All natural or synthetic substrates are listed (not in stoichiometric quantities). The commentary gives information on the reversibility of the reaction, on isomers accepted as substrates and it compares the efficiency of substrates. If a specific substrate is accepted by only one of several isozymes, this will be stated here.

The field *Additional Information* summarizes compounds that are not accepted as substrates or general comments which are valid for all substrates.

In the listings, each substrate (indicated by a bold **S**) is followed by its respective product (indicated by a bold **P**). Products are given with organisms and references included if the respective authors demonstrated the formation of the specific product. If only the disappearance of the substrate was observed, the product will be included without organisms or references. In cases with unclear product formation only a **?** as a dummy is given.

Inhibitors

Compounds found to be inhibitory are listed. The commentary may explain experimental conditions, the concentration yielding a specific degree of inhibition or the inhibition constant. If a substance is activating at a specific concentration but inhibiting at a higher or lower value, the commentary will explain this.

Cofactors, prosthetic groups

This field contains cofactors which participate in the reaction but are not bound to the enzyme, and prosthetic groups being tightly bound. The commentary explains the function or, if known, the stereochemistry, or whether the cofactor can be replaced by a similar compound with higher or lower efficiency.

Activating Compounds

This field lists compounds with a positive effect on the activity. The enzyme may be inactive in the absence of certain compounds or may require activating molecules like sulfhydryl compounds, chelating agents, or lipids. If a substance is activating at a specific concentration but inhibiting at a higher or lower value, the commentary will explain this.

Metals, ions

This field lists all metals or ions that have activating effects. The commentary explains the role each of the cited metal has, being either bound e.g. as Fe-S centers or being required in solution. If an ion plays a dual role, activating at a certain concentration but inhibiting at a higher or lower concentration, this will be given in the commentary.

Turnover number (min^{-1})

The k_{cat} is given in the unit min^{-1}. The commentary lists the names of the substrates, sometimes with information on the reaction conditions or the type of reaction if the enzyme is capable of catalyzing different reactions with a single substrate. For cases where it is impossible to give the turnover number in the defined unit (e.g., substrates without a defined molecular weight, or an undefined amount of protein) this is summarized in *Additional Information*.

Specific activity (U/mg)

The unit is micromol/minute/milligram of protein. The commentary may contain information on specific assay conditions or if another than the natural substrate was used in the assay. Entries in *Additional Information* are included if the units of the activity are missing in the literature or are not calculable to the obligatory unit. Information on literature with a detailed description of the assay method may also be found.

K_m-Value (mM)

The unit is mM. Each value is connected to a substrate name. The commentary gives, if available, information on specific reaction condition, isozymes or presence of activators. The references for values which cannot be expressed in mM (e.g. for macromolecular, not precisely defined substrates) are given in *Additional Information*. In this field we also cite literature with detailed kinetic analyses.

K_i-Value (mM)

The unit of the inhibition constant is mM. Each value is connected to an inhibitor name. The commentary gives, if available, the type of inhibition (e.g. competitive, non-competitive) and the reaction conditions (pH-value and the temperature). Values which cannot be expressed in the requested unit and references for detailed inhibition studies are summerized under *Additional information*.

pH-Optimum

The value is given to one decimal place. The commentary may contain information on specific assay conditions, such as temperature, presence of activators or if this optimum is valid for only one of several isozymes. If the enzyme has a second optimum, this will be mentioned here.

pH-Range

Mostly given as a range e.g. 4.0–7.0 with an added commentary explaining the activity in this range. Sometimes, not a range but a single value indicating the upper or lower limit of enzyme activity is given. In this case, the commentary is obligatory.

Temperature optimum (°C)

Sometimes, if no temperature optimum is found in the literature, the temperature of the assay is given instead. This is always mentioned in the commentary.

Temperature range (°C)

This is the range over which the enzyme is active. The commentary may give the percentage of activity at the outer limits. Also commentaries on specific assay conditions, additives etc.

4 Enzyme Structure

Molecular weight

This field gives the molecular weight of the holoenzyme. For monomeric enzymes it is identical to the value given for subunits. As the accuracy depends on the method of determination this is given in the commentary if provided in the literature. Some enzymes are only active as multienzyme complexes for which the names and/or EC numbers of all participating enzymes are given in the commentary.

Subunits

The tertiary structure of the active species is described. The enzyme can be active as a monomer a dimer, trimer and so on. The stoichiometry of subunit composition is given. Some enzymes can be active in more than one state of complexation with differing effectivities. The analytical method is included.

Posttranslational modifications

The main entries in this field may be proteolytic modification, or side-chain modification, or no modification. The commentary will give details of the modifications e.g.:
- proteolytic modification <1> (<1>, propeptide Name) [1];
- side-chain modification <2> (<2>, N-glycosylated, 12% mannose) [2];
- no modification [3]

5 Isolation / Preparation / Mutation / Application

Source / tissue

For multicellular organisms, the tissue used for isolation of the enzyme or the tissue in which the enzyme is present is given. Cell-lines may also be a source of enzymes.

Localization

The subcellular localization is described. Typical entries are: cytoplasm, nucleus, extracellular, membrane.

Purification

The field consists of an organism and a reference. Only references with a detailed description of the purification procedure are cited.

Renaturation

Commentary on denaturant or renaturation procedure.

Crystallization

The literature is cited which describes the procedure of crystallization, or the X-ray structure.

Cloning
Lists of organisms and references, sometimes a commentary about expression or gene structure.

Engineering
The properties of modified proteins are described.

Application
Actual or possible applications in the fields of pharmacology, medicine, synthesis, analysis, agriculture, nutrition are described.

6 Stability

pH-Stability
This field can either give a range in which the enzyme is stable or a single value. In the latter case the commentary is obligatory and explains the conditions and stability at this value.

Temperature stability
This field can either give a range in which the enzyme is stable or a single value. In the latter case the commentary is obligatory and explains the conditions and stability at this value.

Oxidation stability
Stability in the presence of oxidizing agents, e.g. O_2, H_2O_2, especially important for enzymes which are only active under anaerobic conditions.

Organic solvent stability
The stability in the presence of organic solvents is described.

General stability information
This field summarizes general information on stability, e.g., increased stability of immobilized enzymes, stabilization by SH-reagents, detergents, glycerol or albumins etc.

Storage stability
Storage conditions and reported stability or loss of activity during storage.

References
Authors, Title, Journal, Volume, Pages, Year.

Di-trans,poly-cis-decaprenylcistransferase 2.5.1.31

1 Nomenclature

EC number
2.5.1.31

Systematic name
di-trans,poly-cis-decaprenyl-diphosphate:isopentenyl-diphosphate undecaprenylcistransferase

Recommended name
di-trans,poly-cis-decaprenylcistransferase

Synonyms
C_{55}-OO synthetase
C_{55}PP synthetase
CPDS
DDPPs
UPP synthetase
UPS
Z-prenyl diphosphate synthase
bactoprenyl-diphosphate synthase
cis,polyprenyl diphosphate synthase
dehydrodolichyl diphosphate synthase
di-trans,poly-cis-undecaprenyl-diphosphate synthase
synthetase, undecaprenyl pyrophosphate
undecaprenyl diphosphate synthase
undecaprenyl diphosphate synthetase
undecaprenyl pyrophosphate synthetase
undecaprenyl-diphosphate synthase

CAS registry number
52350-87-5

2 Source Organism

<1> *Lactobacillus plantarum* (ATCC 8014 [1, 3]) [1, 3, 5, 6, 8, 11, 12]
<2> *Escherichia coli* [7, 10, 13, 16, 21, 22, 23]
<3> *Bacillus subtilis* [2, 3]
<4> *Micrococcus luteus* (B-P 26 [4, 15, 16, 17, 19, 20]) [3, 4, 9, 15, 16, 17, 19, 20]
<5> *Salmonella newington* [3]
<6> *Saccharomyces cerevisiae* [14]
<7> *Sulfolobus acidocaldarius* [18]

3 Reaction and Specificity

Catalyzed reaction

di-trans,poly-cis-decaprenyl diphosphate + isopentenyl diphosphate = di-phosphate + di-trans,poly-cis-undecaprenyl diphosphate

Reaction type

alkenyl group transfer

Natural substrates and products

S isopentenyl diphosphate + geranylgeranyl diphosphate <7> (<7>, the enzyme is involved in the biosynthesis of the glycosyl carrier lipid [18]) [18]

P ?

S Additional information <1, 3, 4, 6> (<1, 3, 4>, involved in the biosynthesis of long-chain polyprenyl diphosphate required as a carbohydrate carrier in the biosynthesis of a variety of bacterial cell envelope components [3]; <6>, the synthesized long-chain dehydrodolichyl diphosphate serves as a precursor of glycosyl carrier in glycoprotein biosynthesis in eukaryotes [14]; <4>, the product undecaprenyl diphosphate is indispensable for the biosynthesis of bacterial cell walls [17]) [3, 14, 17]

P ?

Substrates and products

S isopentenyl diphosphate + (E,E)-(2-diazo-3-trifluoropropionyloxy)geranyl diphosphate <1, 2> (<1>, 47% of the activity with trans,trans-farnesyl diphosphate [6]) (Reversibility: ? <1,2> [6,7]) [6, 7]

P (E,E)-(2-diazo-3-trifluoropropionyloxy)polyprenyl diphosphate + diphosphate

S isopentenyl diphosphate + (S)-farnesyl thiodiphosphate <2> (<2>, weak activity, 10000000fold lower than with farnesyl diphosphate [22]) (Reversibility: ? <2> [22]) [22]

P ?

S isopentenyl diphosphate + Δ^3-isopentenyl diphosphate <4> (Reversibility: ? <4> [9]) [9]

P ?

S isopentenyl diphosphate + dimethylallyl diphosphate <1> (<1>, 2% of the activity with farnesyl diphosphate [8]) (Reversibility: ? <1> [8]) [8]

P ?

S isopentenyl diphosphate + farnesyl diphosphate <1-7> (<1, 2, 4>, trans,-trans-farnesyl diphosphate [1, 3, 7, 9, 11, 17]; <1, 4>, cis,trans-farnesyl diphosphate [9, 11]; <3>, (E,E)-farnesyl diphosphate or (Z,E)-farnesyl diphosphate [2]; <1>, trans,trans-farnesyl diphosphate and cis,trans-farnesyl diphosphate [3]; <3>, (Z,E,E)-farnesylgeranyl diphosphate or (Z,Z,E,E)-farnesylgeranyl diphosphate [2]; <1>, farnesyl diphosphate with mixed stereochemistry or trans,trans stereochemistry [8]; <4>, (Z,E,E)-farnesyl diphosphate [19]; <4>, (E,E)-farnesyl diphosphate [20]) (Reversibility: ? <1-7> [1-23]) [1-23]

P C_{55}-polyprenyl diphosphate + C_{50}-polyprenyl diphosphate + diphosphate
<1, 2, 3, 4> (<1>, C_{55} polyprenyl diphosphate with isoprene residues with
cis stereochemistry [1]; <3>, C50-polyprenyl diphosphate and C_{55}-poly-
prenyl diphosphate with Z,E mixed stereochemistry [2]; <3>, formation
of C_{50}-polyprenyl diphosphate and C_{55}-polyprenyl diphosphate in the ratio of 1:4 from (E,E)-farnesyl diphosphate and in the ratio of 4:1 from
(Z,E)-farnesyl diphosphate [2]; <1>, major products are undecaprenol
and probably a longer chain polyprenol [5]; <3>, C_{50}-polyprenyl diphosphate and C_{55}-polyprenyl diphosphate in the ratio of 3:1 from (Z,E,E,E)-
farnesylgeranyl diphosphate and in the ratio 3:2 from (Z,Z,E,E)-farnesyl-
geranyl diphosphate [2]; <1>, C_{55}-polyprenyl diphosphate is the major
product [8]; <2>, with 0.02% Triton X-100 the main product is the C_{55}-
compound, C_{50}-compound is the minor product. With 0.1% Triton X-100
the C_{50}-polyprenyl diphosphate is the main product, the C_{55}-polyprenyl
diphosphate and the C_{45}-polyprenyl diphosphate are minor products.
With 0.5% Triton X-100 the C_{50}-polyprenyl diphosphate compound and
the C_{45}-compound are the main products, the C_{55}-compound is not
formed [10]; <1>, C_{55}-compound is formed as major product from trans,-
trans-farnesyl diphosphate, the C_{50}-compound is formed as the major
product from cis,trans-farnesyl diphosphate [11]; <2>, product distribution under various reaction conditions. In the presence of excess farnesyl
diphosphate the intermediates C_{25}-C_{50} accumulate [13]; <6>, the C_{55} poly-
mer is the major product in presence of Triton X-100, without Triton C_{55}-
C_{75} polyprenyl diphosphates are generated [14]; <4>, E193Q, R197S,
E201Q, R203S, D221A, and D226A show product distribution pattern similar to that of the wild-type enzyme, which produces C_{55} prenyl diphosphate as its major product as well as minor amounts of some intermediates having shorter prenyl chains and C_{60} product. The major products of
the mutant enzymes S74A and E216Q contain C_{35} and C_{50} prenyl chain
length. Mutant enzyme F73A produces a mixture of C_{30} and C_{35} products
[19]; <2>, the major product of the wild-type enzyme, and mutant enzymes H103A, V103A, V105A and I62A in presence of Triton is C_{55}-poly-
prenyl diphosphate. Mutant enzyme L137A produces C_{55}-polyprenyl diphosphate, C_{60}-polyprenyl diphosphate and C_{65}-polyprenyl diphosphate
in the ration 55:41:4 [21]) [1, 2, 5, 8, 10, 11, 13, 14, 16, 19, 21]

S isopentenyl diphosphate + geranyl diphosphate <1, 3, 4> (<1>, about 20%
of the activity with farnesyl diphosphate [8]) (Reversibility: ? <1, 3, 4> [2,
3, 8, 9, 11]) [2, 3, 8, 9, 11]

P C_{55}-polyprenyl diphosphate + C_{50}-polyprenyl diphosphate + diphosphate
<1> (<1>, C_{55}-polyprenyl diphosphate is the major product [8]; <1>, the
C_{50}-compound is the major product [11]) [8, 11]

S isopentenyl diphosphate + geranylgeranyl diphosphate <1, 3, 4, 7> (<3>,
(E,E,E)-geranylgeranyl diphosphate or (Z,E,E)-geranylgeranyl diphosphate
[2]; <1>, trans,trans,trans-geranylgeranyl diphosphate [3]; <1, 4>, trans,-
trans,trans-geranylgeranyl diphosphate and cis,trans,trans-geranylgeranyl
diphosphate [9, 11]; <7>, the enzyme prefers geranylgeranyl diphosphate
to farnesyl diphosphate [18]; <4>, (Z,E,E)-geranylgeranyl diphosphate

[19, 20]) (Reversibility: ? <1, 3, 4, 7> [2, 3, 8, 9, 11, 15, 18, 19, 20]) [2, 3, 8, 9, 11, 15, 18, 19, 20]

P C_{55}-polyprenyl diphosphate + C_{50}-polyprenyl diphosphate + diphosphate <1, 3, 7> (<3>, C_{50}-prenyl diphosphate and C_{55}-prenyl diphosphate in the ratio of 2:3 from (E,E,E)-geranyl diphosphate and in the ratio 1:3 from (Z,E,E)-geranylgeranyl diphosphate [2]; <1>, the C_{55}-compound is formed as major product [11]; <7>, when the amount of isopentenyl diphosphate is 20times larger than that of geranylgeranyl diphosphate the enzyme mainly produces C_{50} and C_{55} polyprenyl diphosphates along with much smaller amounts of C_{60}, in a molar ratio, $C_{50}/C_{55}/C_{60}$ of 3.5:5:1.5 [18]) [2, 11, 18]

S isopentenyl diphosphate + geranylneryl diphosphate <1> (<1>, cis,trans,-trans-geranylneryl diphosphate [3]) (Reversibility: ? <1> [3]) [3]

P ?

S isopentenyl diphosphate + neryl diphosphate <1, 3, 4> (Reversibility: ? <1, 3, 4> [2, 3, 9, 11]) [2, 3, 9, 11]

P C_{55}-polyprenyl diphosphate + C_{50}-polyprenyl diphosphate + diphosphate <3> (<3>, C_{45} prenyl diphosphate, C_{50} prenyl diphosphate and C_{55} prenyl diphosphate in the ratio of 2:10:1 [2]; <1>, the C_{50}-compund is the major product [11]) [2, 11]

S Additional information <4> (<4>, reactivities of the allylic substrates increase with chain length: C_{10}, C_{15}, C_{20}. Each new isoprene unit added has a cis-configuration [9]) [9]

P ?

Inhibitors

(S)-farnesyl thiodiphosphate <2> [22]

EDTA <1, 4> [5, 19]

SO_4^{2-} <4> (<4>, competitive inhibitor with respect to farnesyl diphosphate [20]) [20]

Triton X-100 <1, 4> (<4>, marked decrease of activity when a small amount of detergent is added, activity gradually increases as further detergent is added, being fully restored when the concentration reaches 2% [4]; <1>, inhibition at concentration above 1% [8]) [4, 8]

butanedione <1> [1, 3]

cyclohexanedione <1> [1, 3]

iodoacetate <1> [3]

phenylglyoxal <1> [1, 3]

Additional information <1> (<1>, no inhibition by phosphate [3,5]) [3, 5]

Activating compounds

lysophosphatidylglycerol <4> (<4>, weak effect in activating lipid-depleted enzyme [4]) [4]

Triton X-100 <1, 2, 4, 7> (<1>, activates [5]; <4>, stimulates [9]; <1>, half-maximal activation at 3.4 mM [3]; <1>, requirement for Triton X-series detergents [5]; <4>, stimulates [3]; <3>, strong stimulation [3]; <4>, marked decrease of activity when a small amount of detergent is added, activity gradually increases as further detergent is added, being fully restored when the

concentration reaches 2% [4]; <1>, activates [8,11]; <2>, no activity in absence [10]; <7>, 0.05-0.1%, activates [18]) [3, 4, 5, 8, 9, 10, 11, 18]
Triton X-102 <1> (<1>, stimulates [5]) [5]
Triton X-114 <1> (<1>, stimulates [5]) [5]
Triton X-45 <1> (<1>, stimulates [5]) [5]
Tween 80 <3> (<3>, strong stimulation [3]) [3]
capric acid <1> (<1>, enhancement factor: 0.04 in absence of Triton X-100, 1.0 in presence of 0.1% Triton X-100 [12]) [12]
cardiolipin <1, 4> (<1>, Lactobacillus plantarum, E. coli and bovine cardiolipin. Good activator in absence of detergent [3]; <4>, weak stimulation [3]; <4>, marked effect in activating lipid-depleted enzyme [4]; <4>, poor activator [9]; <1>, activates [11]; <1>, from bovine, enhancement factor: 0.8 in absence of Triton X-100, 0.1 in presence of 0.1% Triton X-100 [12]) [3, 4, 9, 11, 12]
cetyl sulfate <1> (<1>, activates [12]; <1>, half-maximal activation at 0.04 mM [3]) [3, 23]
deoxycholate <1, 4> (<1>, activates [12]; <1>, half-maximal activation at 0.33 mM [3]; <3,4>, weak stimulation [3]; <1>, maximal stimulation at 0.2% v/v or 4.8 mM [8]; <4>, poor activator [9]) [3, 8, 9, 12]
dipalmitoylphosphatidic acid <1> (<1>, enhancement factor: 0.2 in absence of Triton X-100, 1.2 in presence of 0.1% Triton X-100 [12]) [12]
dodecyl sulfate <1> (<1>, half-maximal activation at 0.075 mM [3]; <1>, activates [12]) [3, 12]
lauric acid <1> (<1>, enhancement factor: 0.1 in absence of Triton X-100, 2.3 in presence of 0.1% Triton X-100 [12]) [12]
lecithin <3, 4> (<4>, from egg, stimulates [3,9]; <4>, 14fold stimulation at 2 mM [9]; <3>, weak stimulation [3]) [3, 9]
myristic acid <1> (<1>, activates [11,12]; <1>, enhancement factor: 0.2 in absence of Triton X-100, 2.1 in presence of 0.1% Triton X-100 [12]) [11, 12]
oleic acid <1> (<1>, enhancement factor: 0.9 in absence of Triton X-100, 1.3 in presence of 0.1% Triton X-100 [12]; <1>, good activator in absence of detergent [3]) [3, 12]
palmitic acid <1> (<1>, enhancement factor: 0.8 in absence of Triton X-100, 1.4 in presence of 0.1% Triton X-100 [12]) [12]
phosphatidic acid <1> (<1>, from egg, good activator in absence of detergent [3,12]; <1>, enhancement factor: 0.8 in absence of Triton X-100, 1.0 in presence of 0.1% Triton X-100 [12]) [3, 12]
phosphatidylglycerol <4> (<4>, marked effect in activating lipid-depleted enzyme [4]) [4]
phosphatidylserine <1> (<1>, enhancement factor: 0.05 in absence of Triton X-100, 0.4 in presence of 0.1% Triton X-100 [12]) [12]
phospholipid <4> (<4>, phospholipid extract from Micrococcus luteus stimulates [9]) [9]
polyamines <3> (<3>, stimulate in presence of Mg^{2+} [2]) [2]
putrescine <3> (<3>, stimulates [2]) [2]
stearic acid <1> (<1>, enhancement factor: 0.07 in absence of Triton X-100, 0.8 in presence of 0.1% Triton X-100 [12]) [12]

trans,trans,trans-geranylgeranyl diphosphate <1> (<1>, stimulates activity with farnesyl diphosphate as substrate [11]) [11]
trans,trans-farnesyl diphosphate <1> (<1>, stimulates activity with farnesyl diphosphate as substrate [11]) [11]
Additional information <4> (<4>, enzyme is inactive in absence of added effectors [9]) [9]

Metals, ions

Ca^{2+} <4> (<4>, weak stimulation [19]) [19]
Co^{2+} <1, 4> (<1,3,4>, divalent cation required [3]; <1>, enzyme is equally stimulated by 0.2 mM Mg^{2+}, 0.2 mM Co^{2+} or 0.1 mM Mn^{2+} [3]; <4>, enzyme is optimally stimulated by 0.1 mM Mg^{2+}, 0.05 mM Mn^{2+} or 0.2 mM Co^{2+} [3,9]; <1>, maximal stimulation at 0.1-0.3 mM [8]; <4>, weak stimulation [19]) [3, 8, 9, 19]
K^+ <1, 3> (<1,3>, stimulates in presence of Mg^{2+} [1,2,3]) [1, 2, 3]
Mg^{2+} <1-4, 7> (<1>, required [1]; <2>, Mg^{2+} or Mn^{2+} required. Optimal concentration for Mg^{2+} is 1.0 mM [10]; <1,3,4>, divalent cation required [2,3]; <1>, maximal stimulation at 0.1-0.3 mM [8]; <1>, enzyme is equally stimulated by 0.2 mM Mg^{2+}, 0.2 mM Co^{2+} or 0.1 mM Mn^{2+} [3]; <4>, enzyme is optimally stimulated by 0.1 mM Mg^{2+}, 0.05 mM Mn^{2+} or 0.2 mM Co^{2+} [3]; <3>, best stimulation by Mg^{2+}. Mn^{2+} and Zn^{2+} may partially replace Mn^{2+} [3]; <4>, maximal stimulation at 0.1 mM [9]; <7>, required, optimal concentration is 0.5-2 mM [18]; <4>, divalent metal ion required, Mg^{2+} is most effective, optimal concentration is 240 nM [19]) [1, 2, 3, 8-10, 18, 19]
Mn^{2+} <1-4> (<3>, can partially replace Mg^{2+} [1,2,3]; <1>, enzyme is equally stimulated by 0.2 mM Mg^{2+}, 0.2 mM Co^{2+} or 0.1 mM Mn^{2+} [3]; <4>, enzyme is optimally stimulated by 0.1 mM Mg^{2+}, 0.05 mM Mn^{2+} or 0.2 mM Co^{2+} [3,9]; <3>, best stimulation by Mg^{2+}. Mn^{2+} and Zn^{2+} may partially replace Mn^{2+} [3]; <2>, maximal stimulation at 0.05-0.1 mM, Mg^{2+} or Mn^{2+} required [10]; <1>, effective in stimulation at 0.05-0.1 mM [8]; <2>, Mn^{2+} or Mg^{2+} required [10]; <4>, weak stimulation [19]) [1, 3, 8, 9, 10, 19]
NH_4^+ <1, 3> (<1,3>, stimulates in presence of Mg^{2+} [1,2,3]) [1, 2, 3]
Na^+ <1, 3> (<1,3>, stimulates in presence of Mg^{2+} [1,2,3]) [1, 2, 3]
Zn^{2+} <1, 3, 4> (<3>, can partially replace Mg^{2+} [1,2,3]; <1>, maximal stimulation at 0.1-0.3 mM [8]; <4>, optimal concentration for activation is about 240 nM [19]) [2, 3, 8, 19]

Turnover number (min^{-1})

0.012 <6> (isopentenyl diphosphate) [14]
0.075 <4> (farnesyl diphosphate, <4>, mutant enzyme N77A, pH 7.5, 37°C [15]) [15]
0.093 <4> (farnesyl diphosphate, <4>, mutant enzyme N77Q, pH 7.5, 37°C [15]) [15]
0.12 <4> ((Z,E,E)-geranylgeranyl diphosphate, <4>, mutant enzyme N77A, pH 7.5, 37°C [15]) [15]
0.1386 <4> ((Z,E,E)-geranylgeranyl diphosphate, <4>, mutant enzyme N77Q, pH 7.5, 37°C [15]) [15]

0.162 <4> (isopentenyl diphosphate, <4>, mutant enzyme N77Q, pH 7.5, 37°C [15]) [15]

0.163 <4> (farnesyl diphosphate, <4>, mutant enzyme N77D, pH 7.5, 37°C [15]) [15]

0.249 <4> ((Z,E,E)-geranylgeranyl diphosphate, <4>, mutant enzyme N77D, pH 7.5, 37°C [15]) [15]

0.2838 <4> (isopentenyl diphosphate, <4>, mutant enzyme N77D, pH 7.5, 37°C [15]) [15]

0.984 <4> (isopentenyl diphosphate, <4>, mutant enzyme N77A, pH 7.5, 37°C [15]) [15]

1.002 <4> (farnesyl diphosphate, <4>, mutant enzyme D29A, pH 7.5, 37°C [20]) [20]

3.252 <4> (farnesyl diphosphate, <4>, mutant enzyme R33A, pH 7.5, 37°C [20]) [20]

9.06 <4> (farnesyl diphosphate, <4>, mutant enzyme W78D, pH 7.5, 37°C [15]) [15]

12.96 <4> (farnesyl diphosphate, <4>, mutant enzyme R42G, pH 7.5, 37°C [20]) [20]

18 <4> ((Z,E,E)-geranylgeranyl diphosphate, <4>, mutant enzyme W78D, pH 7.5, 37°C [15]) [15]

23.82 <4> (isopentenyl diphosphate, <4>, mutant enzyme W78D, pH 7.5, 37°C [15]) [15]

26.04 <4> (farnesyl diphosphate, <4>, mutant enzyme N78R, pH 7.5, 37°C [15]) [15]

27.36 <4> (farnesyl diphosphate, <4>, mutant enzyme W78I, pH 7.5, 37°C [15]) [15]

27.72 <4> (farnesyl diphosphate, <4>, mutant enzyme G32R, pH 7.5, 37°C [20]) [20]

30.3 <4> (farnesyl diphosphate, <4>, mutant enzyme E76Q, pH 7.5, 37°C [20]) [20]

46.98 <4> (isopentenyl diphosphate, <4>, mutant enzyme W78I, pH 7.5, 37°C [15]) [15]

48.18 <4> (isopentenyl diphosphate, <4>, mutant enzyme W78R, pH 7.5, 37°C [15]) [15]

53.94 <4> ((Z,E,E)-geranylgeranyl diphosphate, <4>, mutant enzyme W78R, pH 7.5, 37°C [15]) [15]

57.6 <4> (farnesyl diphosphate, <4>, mutant enzyme G32R/R42G, pH 7.5, 37°C [20]) [20]

65.4 <4> (farnesyl diphosphate, <4>, mutant enzyme E84Q, pH 7.5, 37°C [20]) [20]

73.2 <4> (farnesyl diphosphate, <4>, wild-type enzyme, pH 7.5, 37°C [15]) [15]

82.2 <4> ((Z,E,E)-geranylgeranyl diphosphate, <4>, mutant enzyme W78I, pH 7.5, 37°C [15]) [15]

99.6 <4> (farnesyl diphosphate, <4>, wild-type enzyme, pH 7.5, 37°C [20]) [20]

106.2 <4> ((Z,E,E)-geranylgeranyl diphosphate, <4>, wild-type enzyme, pH 7.5, 37°C [15]) [15]

111 <4> (isopentenyl diphosphate, <4>, wild-type, pH 7.5, 37°C [15]) [15]

Additional information <2, 4, 6> [14, 19, 21, 22, 23]

Specific activity (U/mg)

0.000424 <3> [2]

0.004066 <4> [9]

1.534 <4> [16]

Additional information <1> [1, 3, 8]

K_m-Value (mM)

0.00013 <1> ((E,E)-farnesyl diphosphate, <1>, pH 7.5, 35°C [1,3,6]) [1, 3, 6]

0.00017 <1> ((E,E)-(2-diazo-3-trifluoropropionyloxy)geranyl diphosphate, <1>, pH 7.5, 35°C [6]) [6]

0.0004 <2> (farnesyl diphosphate, <2>, wild-type enzyme, pH 7.5, 25°C [13,21,22,23]; <2>, mutant enzyme E73A, N74A and E81A, pH 7.5, 25°C [21]; <2>, mutant enzyme W91F, pH 7.5, 25°C [22]; <2>, mutant enzymes E73A, D150A and D218A, pH 7.5, 25°C [23]) [13, 21, 2, 23]

0.0005 <2> (farnesyl diphosphate, <2>, mutant enzyme W221F, pH 7.5, 25°C [22]; <2>, mutant enzyme D26A, pH 7.5, 25°C [23]) [22, 23]

0.0007 <2> (farnesyl diphosphate, <2>, pH 7.5, 30°C [10]; <2>, mutant enzyme W149F, pH 7.5, 25°C [22]; <2>, mutant enzyme E213A, pH 7.5, 25°C [23]) [10, 22, 23]

0.001 <2> (farnesyl diphosphate, <2>, mutant enzyme S71A, pH 7.5, 25°C [21]) [21]

0.0016 <2> (farnesyl diphosphate, <2>, mutant enzyme R77A, pH 7.5, 25°C [21]) [21]

0.0018 <2> (farnesyl diphosphate, <2>, mutant enzyme W75F, pH 7.5, 25°C [22]) [22]

0.00192 <1> (isopentenyl diphosphate, <1>, pH 7.5, 35°C [1,3,6]) [1, 3, 6]

0.002 <2> (farnesyl diphosphate, <2>, mutant enzyme W31F, pH 7.5, 25°C [22]) [22]

0.0023 <4> ((Z,E,E)-geranylgeranyl diphosphate, <4>, mutant enzyme E76Q, pH 7.5, 37°C [20]) [20]

0.0023 <4> (farnesyl diphosphate, <4>, mutant enzyme E76Q, pH 7.5, 37°C [20]) [20]

0.0032 <2> (farnesyl diphosphate, <2>, mutant enzyme W75A, pH 7.5, 25°C [21]) [21]

0.0032 <1> (geranyl diphosphate, <1>, pH 7.5, 35°C [11]) [11]

0.0033 <4> (farnesyl diphosphate, <4>, mutant enzyme G32R, pH 7.5, 37°C [20]) [20]

0.0034 <2> (farnesyl diphosphate, <2>, mutant enzyme W207F, pH 7.5, 25°C [22]) [22]

0.0035 <4> (farnesyl diphosphate, <4>, mutant enzyme D221A, pH 7.3, 37°C [19]) [19]

0.0036 <4> (farnesyl diphosphate, <4>, wild-type enzyme, pH 7.5, 37°C [20]) [20]

0.0041 <2> (isopentenyl diphosphate, <2>, pH 7.5, 25°C [13,21,23]) [13, 21, 23]

0.0043 <4> ((Z,E,E)-geranylgeranyl diphosphate, <4>, mutant enzyme G32R, pH 7.5, 37°C [20]) [20]

0.0044 <2> (isopentenyl diphosphate, <2>, mutant enzyme D218A, pH 7.5, 25°C [23]) [23]

0.0045 <4> ((Z,E,E)-geranylgeranyl diphosphate, <4>, mutant enzyme R42G, pH 7.5, 37°C [20]) [20]

0.0049 <4> (farnesyl diphosphate, <4>, mutant enzyme R42G, pH 7.5, 37°C [20]) [20]

0.005 <4> ((Z,E,E)-geranylgeranyl diphosphate, <4>, mutant enzyme N77Q, pH 7.5, 37°C [15]; <4>, mutant enzyme D29A, pH 7.5, 37°C [20]) [15, 20]

0.005 <4> (farnesyl diphosphate, <4>, mutant enzyme G32R/R42G, pH 7.5, 37°C [20]) [20]

0.0053 <4> ((Z,E,E)-geranylgeranyl diphosphate, <4>, mutant enzyme S74A, pH 7.3, 37°C [19]) [19]

0.0054 <4> (farnesyl diphosphate, <4>, mutant enzyme R216Q, pH 7.3, 37°C [19]) [19]

0.0058 <4> ((Z,E,E)-geranylgeranyl diphosphate, <4>, mutant enzyme R216Q, pH 7.3, 37°C [19]) [19]

0.0059 <4> ((Z,E,E)-geranylgeranyl diphosphate, <4>, mutant enzyme G32R/R42G, pH 7.5, 37°C [20]) [20]

0.0059 <2, 4> (isopentenyl diphosphate, <4>, mutant enzyme G32R/R42G, pH 7.5, 37°C [20]; <2>, mutant enzyme W31F, pH 7.5, 25°C [22]) [20, 22]

0.006 <4> ((Z,E,E)-geranylgeranyl diphosphate, <4>, mutant enzyme R197S, pH 7.3, 37°C [19]; <4>, mutant enzyme E84Q, pH 7.5, 37°C [20]) [19, 20]

0.006 <4> (isopentenyl diphosphate, <4>, mutant enzyme G32R, pH 7.5, 37°C [20]) [20]

0.0061 <4> ((Z,E,E)-geranylgeranyl diphosphate, <4>, mutant enzyme R80A, pH 7.5, 37°C [20]) [20]

0.0066 <4> (farnesyl diphosphate, <4>, mutant enzyme E193Q, pH 7.3, 37°C [19]) [19]

0.007 <4> ((E,E)-farnesyl diphosphate, <4>, pH 7.4, 35°C [9]) [9]

0.0072 <2> (isopentenyl diphosphate, <2>, mutant enzyme W91F, pH 7.5, 25°C [22]) [22]

0.0075 <4> (geranylgeranly diphosphate, <4>, mutant enzyme D226A, pH 7.3, 37°C [19]) [19]

0.0078 <4> (isopentenyl diphosphate, <4>, wild-type enzyme, pH 7.5, 37°C [15,20]; <4>, wild-type enzyme, pH 7.3, 37°C [19]) [15, 19, 20]

0.008 <4> ((E,E)-farnesyl diphosphate) [3]

0.008 <4> ((E,E,E)-geranylgeranyl diphosphate, <4>, pH 7.5, 35°C [9]) [9]

0.008 <4> ((Z,E,E)-geranylgeranyl diphosphate, <4>, mutant enzyme N77D, pH 7.5, 37°C [15]) [15]

0.008 <4> ((Z,E,E)-geranylgeranyl diphosphate, <4>, mutant enzyme W78D, pH 7.5, 37°C [15]) [15]

0.008 <2> (isopentenyl diphosphate, <2>, mutant enzyme N74A, pH 7.5, 25°C [21]) [21]

0.0081 <4> ((Z,E,E)-geranylgeranyl diphosphate, <4>, mutant enzyme E193Q, pH 7.3, 37°C [19]) [19]

0.0081 <4> ((Z,E,E)-geranylgeranyl diphosphate, <4>, mutant enzyme R203S, pH 7.3, 37°C [19]) [19]

0.0082 <4> ((Z,E,E)-geranylgeranyl diphosphate, <4>, wild-type enzyme, pH 7.5, 37°C [15,20]; <4>, wild-type enzyme, pH 7.3, 37°C [19]) [15, 19, 20]

0.0083 <4> (farnesyl diphosphate, <4>, wild-type enzyme, pH 7.3, 37°C [19]) [19]

0.0083 <4> (farnesyl diphosphate, <4>, wild-type, pH 7.5, 37°C [15]) [15]

0.009 <4> (isopentenyl diphosphate, <4>, mutant enzyme N77Q, pH 7.5, 37°C [15]) [15]

0.0091 <3> ((E,E)-farnesyl diphosphate, <3>, pH 8.5, 37°C [2]) [2, 3]

0.0092 <4> (farnesyl diphosphate, <4>, mutant enzyme F73A, pH 7.3, 37°C [19]) [19]

0.0093 <3> ((E,E,E)-geranylgeranyl diphosphate, <3>, pH 8.5, 37°C [2]) [2]

0.0094 <4> ((Z,E,E)-geranylgeranyl diphosphate, <4>, mutant enzyme F73A, pH 7.3, 37°C [19]) [19]

0.0095 <4> ((Z,E,E)-geranylgeranyl diphosphate, <4>, mutant enzyme R33A, pH 7.5, 37°C [20]) [20]

0.0096 <4> (farnesyl diphosphate, <4>, mutant enzyme S74A, pH 7.3, 37°C [19]) [19]

0.0102 <4> ((Z,E,E)-geranylgeranyl diphosphate, <4>, mutant enzyme N77A, pH 7.5, 37°C [15]) [15]

0.0104 <4> (farnesyl diphosphate, <4>, mutant enzyme N77Q, pH 7.5, 37°C [15]) [15]

0.0112 <4> (farnesyl diphosphate, <4>, mutant enzyme D226A, pH 7.3, 37°C [19]) [19]

0.0114 <4> (farnesyl diphosphate, <4>, mutant enzyme E84Q, pH 7.5, 37°C [20]) [20]

0.0114 <4> (isopentenyl diphosphate, <4>, mutant enzyme E76Q, pH 7.5, 37°C [20]) [20]

0.0116 <4> (farnesyl diphosphate, <4>, mutant enzyme E201Q, pH 7.3, 37°C [19]) [19]

0.012 <4> (isopentenyl diphosphate, <4>, mutant enzyme N77A, pH 7.5, 37°C [15]) [15]

0.0131 <4> (isopentenyl diphosphate, <4>, mutant enzyme W78D, pH 7.5, 37°C [15]) [15]

0.0136 <4> (farnesyl diphosphate, <4>, mutant enzyme R203S, pH 7.3, 37°C [19]) [19]

0.0139 <4> (farnesyl diphosphate, <4>, mutant enzyme N77D, pH 7.5, 37°C [15]) [15]

0.014 <1, 2> (isopentenyl diphosphate, <1>, pH 7.5, 35°C [11]; <2>, mutant enzyme W221F, pH 7.5, 25°C [22]) [11, 22]

0.0141 <4> (isopentenyl diphosphate, <4>, mutant enzyme W78I, pH 7.5, 37°C [15]; <2>, mutant enzyme D26A, pH 7.5, 25°C [23]) [15, 23]

0.0147 <4> (farnesyl diphosphate, <4>, mutant enzyme N77A, pH 7.5, 37°C [15]) [15]

0.015 <4> ((Z,E,E)-geranylgeranyl diphosphate, <4>, mutant enzyme W78R, pH 7.5, 37°C [15]) [15]

0.015 <4> (isopentenyl diphosphate, <4>, mutant enzyme N77D, pH 7.5, 37°C [15]) [15]

0.0154 <3> ((Z,E,E)-geranylgeranyl diphosphate, <3>, pH 8.5, 37°C [2]) [2]

0.0157 <2> (isopentenyl diphosphate, <2>, mutant enzyme R77A, pH 7.5, 25°C [21,23]) [21, 23]

0.0162 <2> (isopentenyl diphosphate, <2>, mutant enzyme E73A, pH 7.5, 25°C [21]) [21, 23]

0.0165 <4> (isopentenyl diphosphate, <4>, mutant enzyme W78R, pH 7.5, 37°C [15]) [15]

0.017 <4> (isopentenyl diphosphate, <4>, mutant enzyme D4226A, pH 7.3, 37°C [19]) [19]

0.0177 <4> ((Z,E,E)-geranylgeranyl diphosphate, <4>, mutant enzyme E201Q, pH 7.3, 37°C [19]) [19]

0.0177 <4> (isopentenyl diphosphate, <4>, mutant enzyme D221A, pH 7.3, 37°C [19]) [19]

0.018 <4> (farnesyl diphosphate, <4>, mutant enzyme R197S, pH 7.3, 37°C [19]) [19]

0.0181 <4> ((Z,E,E)-geranylgeranyl diphosphate, <4>, mutant enzyme D221A, pH 7.3, 37°C [19]) [19]

0.0183 <4> (isopentenyl diphosphate, <4>, mutant enzyme E201Q, pH 7.3, 37°C [19]) [19]

0.0189 <3> ((Z,E)-farnesyl diphosphate, <3>, pH 8.5, 37°C [2]) [2]

0.019 <4> ((Z,E,E)-geranylgeranyl diphosphate, <4>, mutant enzyme W78I, pH 7.5, 37°C [15]) [15]

0.02 <2> (isopentenyl diphosphate, <2>, mutant enzyme W149F, pH 7.5, 25°C [22]) [22]

0.0204 <4> (isopentenyl diphosphate, <4>, mutant enzyme E193Q, pH 7.3, 37°C [19]) [19]

0.023 <2> (isopentenyl diphosphate, <2>, mutant enzyme W207F, pH 7.5, 25°C [22]) [22]

0.0231 <4> (farnesyl diphosphate, <4>, mutant enzyme R80A, pH 7.5, 37°C [20]) [20]

0.0242 <4> (isopentenyl diphosphate, <4>, mutant enzyme R42G, pH 7.5, 37°C [20]) [20]

0.026 <2> (isopentenyl diphosphate, <2>, mutant enzyme W75F, pH 7.5, 25°C [22]) [22]

0.0267 <3> (neryl diphosphate, <3>, pH 8.5, 37°C [2]) [2]

0.027 <6> (isopentenyl diphosphate) [14]

0.029 <2> (isopentenyl diphosphate, <2>, pH 7.5, 30°C [10]) [10]

0.0307 <4> (isopentenyl diphosphate, <4>, mutant enzyme E84Q, pH 7.5, 37°C [20]) [20]

0.032 <1> (neryl diphosphate, <1>, pH 7.5, 35°C [11]) [11]

0.0332 <4> (farnesyl diphosphate, <4>, mutant enzyme D29A, pH 7.5, 37°C [20]) [20]

0.037 <1> ((E,E,E)-geranylgeranyl diphosphate, <1>, pH 7.5, 35°C [11]) [11]

0.0452 <4> (isopentenyl diphosphate, <4>, mutant enzyme D29A, pH 7.5, 37°C [20]) [20]

0.046 <2> (isopentenyl diphosphate, <2>, mutant enzyme W75A, pH 7.5, 25°C [21]) [21]

0.0491 <4> (isopentenyl diphosphate, <4>, mutant enzyme R80A, pH 7.5, 37°C [20]) [20]

0.0495 <4> (farnesyl diphosphate, <4>, mutant enzyme W78I, pH 7.5, 37°C [15]) [15]

0.0526 <3> ((Z,Z,E,E)-farnesylgeranyl diphosphate, <3>, pH 8.5, 37°C [2]) [2]

0.0561 <4> (isopentenyl diphosphate, <4>, mutant enzyme R197S, pH 7.3, 37°C [19]) [19]

0.0588 <3> ((Z,E,E,E)-farnesylgeranyl diphosphate, <3>, pH 8.5, 37°C [2]) [2]

0.0621 <4> (isopentenyl diphosphate, <4>, mutant enzyme R203S, pH 7.3, 37°C [19]) [19]

0.088 <2> (isopentenyl diphosphate, <2>, mutant enzyme E81A, pH 7.5, 25°C [21]) [21]

0.0909 <4> (isopentenyl diphosphate, <4>, mutant enzyme E216Q, pH 7.3, 37°C [19]) [19]

0.103 <1> ((E,E)-farnesyl diphosphate, <1>, pH 7.5, 35°C [11]) [11]

0.1207 <4> (farnesyl diphosphate, <4>, mutant enzyme W78R, pH 7.5, 37°C [15]) [15]

0.1268 <4> (isopentenyl diphosphate, <4>, mutant enzyme S74A, pH 7.3, 37°C [19]) [19]

0.133 <2> (isopentenyl diphosphate, <2>, mutant enzyme S71A, pH 7.5, 25°C [21]) [21]

0.134 <1> ((Z,E,E)-geranylgeranyl diphosphate, <1>, pH 7.5, 35°C [11]) [11]

0.144 <4> (farnesyl diphosphate, <4>, mutant enzyme R33A, pH 7.5, 37°C [20]) [20]

0.155 <4> (farnesyl diphosphate, <4>, mutant enzyme N78D, pH 7.5, 37°C [15]) [15]

0.2525 <4> (isopentenyl diphosphate, <4>, mutant enzyme F73A, pH 7.3, 37°C [19]) [19]

0.27 <2> (isopentenyl diphosphate, <2>, mutant enzyme D150A, pH 7.5, 25°C [23]) [23]

0.28 <2> (isopentenyl diphosphate, <2>, mutant enzyme D213A, pH 7.5, 25°C [23]) [23]

0.305 <1> ((Z,E)-farnesyl diphosphate, <1>, pH 7.5, 35°C [11]) [11]

K_i-Value (mM)

0.0002 <2> ((S)-farnesyl thiopyrophosphate) [22]

17 <4> (SO_4^{2-}) [20]

pH-Optimum

6 <7> [18]

7.5 <1, 4> (<1>, and a second optimum at pH 10.2, that is 40% of the optimum at pH 7.5 [3,8]) [3, 8, 9]

7.5-9 <2> [10]
8.5 <3> [2, 3]
10.2 <1> (<1>, and a second optimum at pH 7.5 that is higher than the optimum at pH 10.2 [3,8]) [3, 8]

pH-Range
7.5-9 <4> (<4>, pH 7.5: activity maximum, pH 9.0: about 80% of maximal activity [9]) [9]

Temperature optimum (°C)
35-45 <1> [8]
60 <7> [18]

4 Enzyme Structure

Molecular weight
47000-49000 <4> (<4>, gel filtration [9]) [9]
52000-58000 <1> (<1>, sucrose density gradient centrifugation [3]) [3]
53000-60000 <1> (<1>, gel filtration [3,8]) [3, 8]
56000 <1> (<1>, gel filtration, sucrose density gradient centrifugation [1,3]) [1, 3]
57000 <3> (<3>, gel filtration [2]) [2]

Subunits
dimer <1, 2> (<1,2>, 2 * 30000, SDS-PAGE [1,3,7]) [1, 3, 7]

5 Isolation/Preparation/Mutation/Application

Localization
cytosol <2> [7]
membrane <4> [16]
soluble <1, 4> [5, 9, 12]
Additional information <6> (<6>, the recombinant enzyme expressed in E. coli mostly exists in pellet in the absence of detergents, a low quantity of soluble enzyme is purified [14]) [14]

Purification
<1> (partial [5,8]) [1, 3, 5, 8]
<2> (partial [10]) [7, 10, 21]
<3> (partial [2]) [2]
<4> [9, 15, 16]
<6> [14]

Renaturation
<6> (the recombinant protein in the pellet is solubilized with 7 M urea and purified using nickel-nitrilotriacetic acid under denaturing conditions. The protein refolding is achieved via the stepwise dialysis to remove the denatur-

ant in the presence of 6 mM β-mercaptoethanol. Alternatively, on-column refolding is carried out in a single step to obtain the active protein in large quantities. β-Mercaptoethanol and Triton are both required in this quick refolding process [14]) [14]

Crystallization

<2> (hanging-drop method [21]) [21]

<4> (the overall structure is determined at 2.2 A resolution by multiple isomorphous replacement with anomalous scattering [17]) [17]

Cloning

<2> (expression in Escherichia coli [16]; expression of mutant enzmyes W31F, W75F, W91F, W149F, W207F, and W221F as His-tag fusion proteins using Escherichia coli BL21 as host cells [22]) [16, 22]

<4> (expression in Escherichia coli, no sequence similarity between E-prenyl diphosphate and Z-prenyl diphosphate synthases [16]; expression of mutant enzymes R197S, R203S and E216Q in Escherichia coli [19]; mutated enzymes are overproduced in Escherichia coli [20]) [16, 19, 20]

<6> (expression in Escherichia coli. Thioredoxin and His tag are utilized to increase the solubility of the recombinant protein and facilitate its purification using Ni-nitrilotriacetic acid column [14]) [14]

<7> [18]

Engineering

D150A <2> (<2>, turnover-number is 44% of the activity of the wild-type enzyme, K_m value for farnesyl diphosphate is equal to that of the wild-type enzyme, K_m-value for isopentenyl diphosphate is about 50fold higher than that of the wild-type enzyme [23]) [23]

D218A <2> (<2>, turnover-number is about 9% of the activity of the wild-type enzyme, K_m value for farnesyl diphosphate is equal to that of the wild-type enzyme, K_m-value for isopentenyl diphosphate is comparable to that of the wild-type enzyme [23]) [23]

D221A <4> (<4>, comparable K_m-values for farnesyl diphosphate and geranylgeranyl diphosphate with those of the wild-type enzyme. K_m-value for isopentenyl diphosphate within moderate folds to that of the wild-type and slightly decreased enzymatic activity [19]) [19]

D226A <4> (<4>, comparable K_m-values for farnesyl diphosphate and geranylgeranyl diphosphate with those of the wild-type enzyme. K_m-value for isopentenyl diphosphate within moderate folds to that of the wild-type and slightly decreased enzymatic activity [19]) [19]

D26A <2> (<2>, turnover-number is less than 1% of that of the wild-type enzyme, K_m value for farnesyl diphosphate is comparable to that of the wild-type enzyme, K_m-value for isopentenyl diphosphate is 3.5fold higher than that of the wild-type enzyme [23]) [23]

D29A <4> (<4>, K_m-value for farnesyl diphosphate is 9.2fold higher than that for the wild-type enzyme. K_m-value for (Z,E,E)-geranylgeranyl diphosphate is 1.6fold lower than that of the wild-type enzyme. K_m-value for isopentenyl diphosphate is 5.8fold higher than that of the wild-type enzyme. The turnover-number for farnesyl diphosphate is 1% of that of the wild-type enzyme [20]) [20]

E193Q <4> (<4>, comparable K_m-values for farnesyl diphosphate and geranylgeranyl diphosphate with those of the wild-type enzyme. K_m-value for isopentenyl diphosphate within moderate folds to that of the wild-type and slightly decreased enzymatic activity [19]) [19]

E201Q <4> (<4>, comparable K_m-values for farnesyl diphosphate and geranylgeranyl diphosphate with those of the wild-type enzyme. K_m-value for isopentenyl diphosphate within moderate folds to that of the wild-type and slightly decreased enzymatic activity [19]) [19]

E213A <2> (<2>, turnover-number is less than 1% of the activity of the wild-type enzyme, K_m value for farnesyl diphosphate is 1.75fold higher than that of the wild-type enzyme, K_m-value for isopentenyl diphosphate is about 68.3fold higher than that of the wild-type enzyme [23]) [23]

E216Q <4> (<4>, comparable K_m-values for farnesyl diphosphate and geranylgeranyl diphosphate with those of the wild-type enzyme. K_m-value for isopentenyl diphosphate is about 11.6fold higher than that for the wild-type enzyme, turnover number is about 18fold lower. The major products contain C_{35} and C_{50} prenyl chain length [19]) [19]

E73A <2> (<2>, turnover-number is 12% of that of the wild-type enzyme, K_m value for farnesyl diphosphate is equal to that of the wild-type enzyme, K_m-value for isopentenyl diphosphate is about 4fold higher than that of the wild-type enzyme [23]) [23]

E73A <2> (<2>, turnover-number is 12% of that of the wild-type enzyme, K_m-value for farnesyl diphosphate is equal to that of the wild-type enzyme, the K_m-value for isopentenyl diphosphate is 3.9fold higher than that of the wild-type enzyme [21]) [21]

E76Q <4> (<4>, K_m-value for farnesyl diphosphate is 1.6fold higher than that for the wild-type enzyme. K_m-value for (Z,E,E)-geranylgeranyl diphosphate is 3.6fold lower than that of the wild-type enzyme. K_m-value for isopentenyl diphosphate is 1.5fold higher than that of the wild-type enzyme. The turnover-number for farnesyl diphosphate is 30% of that of the wild-type enzyme [20]) [20]

E81A <2> (<2>, turnover-number is 16% of that of the wild-type enzyme, K_m-value for farnesyl diphosphate is equal to that of the wild-type enzyme, the K_m-value for isopentenyl diphosphate is 21.5fold higher than that of the wild-type enzyme [21]) [21]

E84Q <4> (<4>, K_m-value for farnesyl diphosphate is 3.2fold higher than that for the wild-type enzyme. K_m-value for (Z,E,E)-geranylgeranyl diphosphate is 1.3fold lower than that of the wild-type enzyme. K_m-value for isopentenyl diphosphate is 3.98fold higher than that of the wild-type enzyme. The turnover-number for farnesyl diphosphate is 65% of that of the wild-type enzyme [20]) [20]

F73A <4> (<4>, comparable K_m-values for farnesyl diphosphate and geranylgeranyl diphosphate with those of the wild-type enzyme. 32fold increase in K_m-value for isopentenyl diphosphate and about 17fold decrease in turnover-number. A mixture of C_{30} and C_{50} products is formed [19]) [19]

G32R <4> (<4>, K_m-value for farnesyl diphosphate is comparable to that of the wild-type enzyme. K_m-value for (Z,E,E)-geranylgeranyl diphosphate is 1.9fold lower than that of the wild-type enzyme. K_m-value for isopentenyl diphosphate is comparable to that of the wild-type enzyme. The turnover-number for farnesyl diphosphate is 27% of that of the wild-type enzyme [20]) [20]

G32R/R42G <4> (<4>, K_m-value for farnesyl diphosphate is 1.4fold higher than that for the wild-type enzyme. K_m-value for (Z,E,E)-geranylgeranyl diphosphate is 1.4fold lower than that of the wild-type enzyme. K_m-value for isopentenyl diphosphate is 1.3fold higher than that of the wild-type enzyme. The turnover-number for farnesyl diphosphate is 58% of that of the wild-type enzyme [20]) [20]

L137A <2> (<2>, mutant enzyme produces C_{55}-polyprenyl diphosphate, C_{60}-polyprenyl diphosphate and C_{65}-polyprenyl diphosphate in the ratio 55:41:4 in presence of Triton, compared to the wild-type enzyme which produces C_{55}-polyprenyl diphosphate as the major product. In absence of Triton the mutant enzyme produces C_{70} and C_{75}-polyprenyl diphosphate is the major products [21]) [21]

N74A <2> (<2>, turnover-number is less than 1% of that of the wild-type enzyme, K_m-value for farnesyl diphosphate is equal to that of the wild-type enzyme, the K_m-value for isopentenyl diphosphate is 2fold higher than that of the wild-type enzyme [21]) [21]

N77A <4> (<4>, dramatic decrease in catalytic activity, but K_m-values for both allylic and homoallylic substrates are comparable to those of the wild-type [15]) [15]

N77D <4> (<4>, dramatic decrease in catalytic activity, but K_m-values for both allylic and homoallylic substrates are comparable to those of the wild-type [15]) [15]

N77Q <4> (<4>, dramatic decrease in catalytic activity, but K_m-values for both allylic and homoallylic substrates are comparable to those of the wild-type [15]) [15]

R197S <4> (<4>, comparable K_m-values for farnesyl diphosphate and geranylgeranyl diphosphate with those of the wild-type enzyme. K_m-value for isopentenyl diphosphate is about 7.2fold higher than that for the wild-type enzyme, turnover-number is 1200fold lower. Compared with the wild-type enzyme lower reaction rate in catalysis and fewer product produced after 6 h reaction [19]) [19]

R203S <4> (<4>, comparable K_m-values for farnesyl diphosphate and geranylgeranyl diphosphate with those of the wild-type enzyme. K_m-value for isopentenyl diphosphate is about 8fold higher than that for the wild-type enzyme, turnover-number is 1200fold lower. Compared with the wild-type enzyme lower reaction rate in catalysis and fewer product produced after 6 h reaction [19]) [19]

R33A <4> (<4>, K_m-value for farnesyl diphosphate is 40fold higher than that for the wild-type enzyme. K_m-value for (Z,E,E)-geranylgeranyl diphosphate is 1.2fold lower than that of the wild-type enzyme. The turnover-number for farnesyl diphosphate is 3% of that of the wild-type enzyme. Different product distribution pattern compared to the wild-type enzyme [20]) [20]

R42G <4> (<4>, K_m-value for farnesyl diphosphate is 1.4fold higher than that for the wild-type enzyme. K_m-value for (Z,E,E)-geranylgeranyl diphosphate is 1.8fold lower than that of the wild-type enzyme. K_m-value for isopentenyl diphosphate is 3.1fold higher than that of the wild-type enzyme. The turnover-number for farnesyl diphosphate is 13% of that of the wild-type enzyme [20]) [20]

R77A <2> (<2>, turnover-number is less than 1% of that of the wild-type enzyme, K_m-value for farnesyl diphosphate is 4fold higher than that of the wild-type enzyme, the K_m-value for isopentenyl diphosphate is 3.8fold higher than that of the wild-type enzyme [21]) [21]

R80A <4> (<4>, K_m-value for farnesyl diphosphate is 6.4fold higher than that for the wild-type enzyme. K_m-value for (Z,E,E)-geranylgeranyl diphosphate is 1.3fold lower than that of the wild-type enzyme. K_m-value for isopentenyl diphosphate is 6.3fold higher than that of the wild-type enzyme. The turnover-number for farnesyl diphosphate is less than 1% of that of the wild-type enzyme [20]) [20]

S71A <2> (<2>, turnover-number is 4.4% of that of the wild-type enzyme, K_m-value for farnesyl diphosphate is 2.5fold higher than that of the wild-type enzyme, the K_m-value for isopentenyl diphosphate is 32.4fold higher than that of the wild-type enzyme [21]) [21]

S74A <4> (<4>, comparable K_m-values for farnesyl diphosphate and geranylgeranyl diphosphate with those of the wild-type enzyme. 16fold increase in K_m-value for isopentenyl diphosphate and about 12fold decrease in turnover-number. The major products contain C_{35} and C_{50} prenyl chain length [19]) [19]

W149F <2> (<2>, turnover-number is 68% of that of the wild-type enzyme, K_m-value for farnesyl diphosphate is 1.75fold higher than that of the wild-type enzyme, K_m-value for isopentenyl diphosphate is 4.9fold higher than that of the wild-type enzyme [22]) [22]

W207F <2> (<2>, turnover-number is 60% of that of the wild-type enzyme, K_m-value for farnesyl diphosphate is 8.5fold higher than that of the wild-type enzyme, K_m-value for isopentenyl diphosphate is 5.6fold higher than that of the wild-type enzyme [22]) [22]

W221F <2> (<2>, turnover-number is 140% of that of the wild-type enzyme, K_m-value for farnesyl diphosphate is 1.3fold higher than that of the wild-type enzyme, K_m-value for isopentenyl diphosphate is 3.4fold higher than that of the wild-type enzyme [22]) [22]

W31F <2> (<2>, turnover-number is 44% of that of the wild-type enzyme, K_m-value for farnesyl diphosphate is 5fold higher than that of the wild-type enzyme, K_m-value for isopentenyl diphosphate is 1.4fold higher than that of the wild-type enzyme [22]) [22]

W75A <2> (<2>, turnover-number is 4.4% of that of the wild-type enzyme, K_m-value for farnesyl diphosphate is 8fold higher than that of the wild-type enzyme, the K_m-value for isopentenyl diphosphate is 11.2fold higher than that of the wild-type enzyme [21]) [21]

W75F <2> (<2>, turnover-number is 44% of that of the wild-type enzyme, K_m-value for farnesyl diphosphate is 4.5fold higher than that of the wild-type enzyme, K_m-value for isopentenyl diphosphate is 6.3fold higher than that of the wild-type enzyme [22]) [22]
W78D <4> (<4>, moderate levels of enzymatic activity and comparable K_m-values for isopentenyl diphosphate to that of the wild-type. 18.7fold increase in K_m-value for farnesyl diphosphate [15]) [15]
W78I <4> (<4>, moderate levels of enzymatic activity and comparable K_m-values for isopentenyl diphosphate to that of the wild-type. 6-fold increase in K_m-value for farnesyl diphosphate, 2.3fold increase in K_m-value for (Z,E,E)-geranylgeranyl diphosphate [15]) [15]
W78R <4> (<4>, moderate levels of enzymatic activity and comparable K_m-values for isopentenyl diphosphate to that of the wild-type. 14.5-fold increase in K_m-value for farnesyl diphosphate, 1.8fold increase in K_m-value for (Z,E,E)-geranylgeranyl diphosphate [15]) [15]
W91F <2> (<2>, turnover-number is equal to that of the wild-type enzyme, K_m-value for farnesyl diphosphate is equal to that of the wild-type enzyme, K_m-value for isopentenyl diphosphate is 1.8fold higher than that of the wild-type enzyme [22]) [22]

6 Stability

Temperature stability
70 <7> (<7>, 1 h, more than 90% of the activity is retained [18]) [18]

General stability information
<1, 2>, inactivation by irradiation with UV light in presence of (E,E)-(2-diazo-3-trifluoropropionyloxy)geranyl diphosphate, (E,E)-farnesyl diphosphate protects, isopentenyl diphosphate and Mg^{2+} are required for inactivation in presence of the photolabile substrate [6, 7]

Storage stability
<3>, 0°C, 24 h, 56% loss of activity, partially purified enzyme [2]

References

[1] Muth, J.D.; Allen, C.M.: Undecaprenyl pyrophosphate synthetase from Lactobacillus plantarum: a dimeric protein. Arch. Biochem. Biophys., **230**, 49-60 (1984)
[2] Takahashi, I.; Ogura, K.: Prenyltransferases of Bacillus subtilis: undecaprenyl pyrophosphate synthetase and geranylgeranyl pyrophosphate synthetase. J. Biochem., **92**, 1527-1537 (1982)
[3] Allen, C.M.: Purification and characterization of undecaprenylpyrophosphate synthetase. Methods Enzymol., **110**, 281-299 (1985)

[4] Koyama, T.; Yoshida, I.; Ogura, K.: Undecaprenyl diphosphate synthase from Micrococcus luteus B-P 26: essential factors for the enzymatic activity. J. Biochem., **103**, 867-871 (1988)

[5] Keenan, M.V.; Allen, C.M.: Characterization of undecaprenyl pyrophosphate synthetase from Lactobacillus plantarum. Arch. Biochem. Biophys., **161**, 375-383 (1974)

[6] Baba, T.; Allen, C.M.: Inactivation of undecaprenylpyrophosphate synthetase with a photolabile analogue of farnesyl pyrophosphate. Biochemistry, **23**, 1312-1322 (1984)

[7] Baba, T.; Muth, J.; Allen, C.M.: Photoaffinity labeling of undecaprenyl pyrophosphate synthetase with a farnesyl pyrophosphate analogue. J. Biol. Chem., **260**, 10467-10473 (1985)

[8] Allen, C.M.; Keenan, M.V.; Sack, J.: Lactobacillus plantarum undecaprenyl pyrophosphate synthetase: purification and reaction requirements. Arch. Biochem. Biophys., **175**, 236-248 (1976)

[9] Baba, T.; Allen, C.M.: Prenyl transferases from Micrococcus luteus: characterization of undecaprenyl pyrophosphate synthetase. Arch. Biochem. Biophys., **200**, 474-484 (1980)

[10] Fujisaki, S.; Nishino, T.; Katasuki, H.: Isoprenoid synthesis in Escherichia coli. Separation and partial purification of four enzymes involved in the synthesis. J. Biochem., **99**, 1327-1337 (1986)

[11] Baba, T.; Allen, C.M.: Substrate specificity of undecaprenyl pyrophosphate synthetase from Lactobacillus plantarum. Biochemistry, **17**, 5598-5604 (1978)

[12] Allen, C.M.; Muth, J.D.: Lipid activation of undecaprenyl pyrophosphate synthetase from Lactobacillus plantarum. Biochemistry, **16**, 2908-2915 (1977)

[13] Pan, J.J.; Chiou, S.T.; Liang, P.H.: Product distribution and pre-steady-state kinetic analysis of Escherichia coli undecaprenyl pyrophosphate synthase reaction. Biochemistry, **39**, 10936-10942 (2000)

[14] Chang, S.Y.; Tsai, P.C.; Tseng, C.S.; Liang, P.H.: Refolding and characterization of a yeast dehydrodolichyl diphosphate synthase overexpressed in Escherichia coli. Protein Expr. Purif., **23**, 432-439 (2001)

[15] Fujikura, K.; Zhang, Y.W.; Yoshizaki, H.; Nishino, T.; Koyama, T.: Significance of Asn-77 and Trp-78 in the catalytic function of undecaprenyl diphosphate synthase of Micrococcus luteus B-P 26. J. Biochem., **128**, 917-922 (2000)

[16] Shimizu, N.; Koyama, T.; Ogura, K.: Molecular cloning, expression, and purification of undecaprenyl diphosphate synthase. No sequence similarity between E- and Z-prenyl diphosphate synthases. J. Biol. Chem., **273**, 19476-19481 (1998)

[17] Fujihashi, M.; Zhang, Y.W.; Higuchi, Y.; Li, X.Y.; Koyama, T.; Miki, K.: Crystal structure of cis-prenyl chain elongating enzyme, undecaprenyl diphosphate synthase. Proc. Natl. Acad. Sci. USA, **98**, 4337-4342 (2001)

[18] Hemmi, H.; Yamashita, S.; Shimoyama, T.; Nakayama, T.; Nishino, T.: Cloning, expression, and characterization of cis-polyprenyl diphosphate

synthase from the thermoacidophilic archaeon Sulfolobus acidocaldarius. J. Bacteriol., **183**, 401-404 (2001)

[19] Kharel, Y.; Zhang, Y.W.; Fujihashi, M.; Miki, K.; Koyama, T.: Identification of significant residues for homoallylic substrate binding of Micrococcus luteus B-P 26 undecaprenyl diphosphate synthase. J. Biol. Chem., **276**, 28459-28464 (2001)

[20] Fujikura, K.; Zhang, Y.W.; Fujihashi, M.; Miki, K.; Koyama, T.: Mutational analysis of allylic substrate binding site of Micrococcus luteus B-P 26 undecaprenyl diphosphate synthase. Biochemistry, **42**, 4035-4041 (2003)

[21] Ko, T.P.; Chen, Y.K.; Robinson, H.; Tsai, P.C.; Gao, Y.G.; Chen, A.P.C.; Wang, A.H.J.; Liang, P.H.: Mechanism of product chain length determination and the role of a flexible loop in Escherichia coli undecaprenyl-pyrophosphate synthase catalysis. J. Biol. Chem., **276**, 47474-47482 (2001)

[22] Chen, Y.H.; Chen, A.P.C.; Chen, C.T.; Wang, A.H.J.; Liang, P.H.: Probing the conformational change of Escherichia coli undecaprenyl pyrophosphate synthase during catalysis using an inhibitor and tryptophan mutants. J. Biol. Chem., **277**, 7369-7376 (2002)

[23] Pan, J.J.; Yang, L.W.; Liang, P.H.: Effect of site-directed mutagenesis of the conserved aspartate and glutamate on E. coli undecaprenyl pyrophosphate synthase catalysis. Biochemistry, **39**, 13856-13861 (2000)

1 Nomenclature

EC number
2.5.1.32

Systematic name
geranylgeranyl-diphosphate:geranylgeranyl-diphosphate geranylgeranyltransferase

Recommended name
phytoene synthase

Synonyms
PSY
Psase
phytoene synthetase
phytoene-synthetase
prephytoene-diphosphate synthase
synthetase, phytoene

CAS registry number
50936-61-3
57219-66-6

2 Source Organism

<1> *Lycopersicon esculentum* (Mill. Cv. Ailsa Craig [18]) [1, 5, 6, 7, 12, 16, 18]
<2> *Mycobacterium sp.* [2]
<3> *Capsicum annuum* (a single polypeptide catalyzes the 2-step reaction
     from geranylgeranyl diphosphate to phytoene [3]) [3]
<4> *Capsicum sp.* (phytoene synthase complex consists of isopentenyl diphos-
     phate isomerase, geranylgeranyl diphosphate synthase (EC 2.5.1.1) and
     phytoene synthase (EC 2.5.1.32) [4]) [4]
<5> *Synechococcus sp.* (PCC7942, a single polypeptide catalyzes the 2-step re-
     action from geranylgeranyl diphosphate to phytoene [8]) [8]
<6> *Mycobacterium sp.* [9]
<7> *Erwinia uredovora* (20D3, ATCC 19321 [10]) [10, 11, 20, 21]
<8> *Narcissus pseudonarcissus* [13, 14]
<9> *Erwinia herbicola* [15]
<10> *Sinapis alba* [17]
<11> *Haematococcus pluvialis* [19]
<12> *Narcissus sp.* [22]

<13> *Ocymum basilicum* [22]
<14> *Vicia faba* [22]

3 Reaction and Specificity

Catalyzed reaction
2 geranylgeranyl diphosphate = diphosphate + prephytoene diphosphate
prephytoene diphosphate = phytoene + diphosphate

Reaction type
alkenyl group tranfer

Natural substrates and products
S geranylgeranyl diphosphate + geranylgeranyl diphosphate <1, 4, 6, 8, 9, 11> (<4>, the enzyme is involved in the pathway of phytoene synthesis [4]; <6>, first fully photoinduced step in carotenogenesis [9]; <8>, first of four specific enzymes necessary for β-carotene biosynthesis in plants [14]; <9>, first pathway-specific step in carotenoid biosynthesis [15]; <1>, phytoene synthase-2 enzyme activity in tomato does not contribute to carotenoid synthesis in ripening fruit [18]; <11>, key enzyme in astaxanthin biosynthesis. Application of environmental stress results in increased steady-state mRNA level, light-induced expression of the gene may be under photosynthetic control [19]) [4, 9, 14, 15, 18, 19]
P prephytoene diphosphate + diphosphate

Substrates and products
S geranylgeranyl diphosphate + geranylgeranyl diphosphate <1-14> (<3, 5, 7, 9>, the same enzyme catalyzes two consecutive reactions: 1. the synthesis of prephytoene diphosphate and 2. the synthesis of phytoene [3, 8, 11, 15]) (Reversibility: ? <1-14> [1-22]) [1-22]
P prephytoene diphosphate + diphosphate <1, 2, 3> [1, 2, 3, 4]
S prephytoene diphosphate <3, 5, 7, 9> (<3, 5, 7, 9>, the same enzyme catalyzes two consecutive reactions: 1. the synthesis of prephytoene diphosphate and 2. the synthesis of phytoene [3, 8, 11, 15]; <7>, the phytoene formed in vitro is present in both a 15-cis and all-trans isomeric configuration [11]) (Reversibility: ? <3, 5, 7, 9> [3, 8, 11, 15]) [3, 8, 11, 15]
P phytoene + diphosphate <3, 9> (<9>, 15,15'-Z-phytoene is the sole product [15]) [3, 15]

Inhibitors
ADP <1> [16]
AMP <1> [16]
CGA 103586 <12, 13, 14> [22]
GTP <1> [16]
Mn^{2+} <1> (<1>, above 2 mM [16]) [16]
$NADP^+$ <1> [16]
NADPH <1> [6, 16]
NEM <1> [5]

β-carotene <1> [16]
chlorophyll <1> [16]
fructose <8> [13]
iodoacetamide <1> [5]
p-hydroxymercuribenzoate <1> [5]
phosphate <1, 7> [11, 16]
squalestatin <1, 7> (<7>, IC50: 0.015 mM [11]) [11, 16]

Cofactors/prosthetic groups

ATP <1, 7> (<1>, 1.3 mM, 7fold stimulation, most probably an allosteric effector [1]; <1>, stimulates [5]; <7>, required [11]) [1, 5, 11]

Activating compounds

ATP <1, 7> (<1>, Mn^{2+} and ATP are essential for catalytic activity, the effect is stoichiometric from 0.5-2 mM. K_m for ATP is 2.0 mM [16]; <1>, 1.3 mM, 7fold stimulation, most probably an allosteric effector [1]; <1>, stimulates [5]; <7>, required [11]) [1, 5, 11, 16]
CTP <1> (<1>, 0.67 mM, 2.5fold stimulation [1]) [1]
GTP <1> (<1>, 0.67 mM, 2.5fold stimulation [1]) [1]
$NADP^+$ <1> (<1>, weak stimulation [5]) [5]
Triton X-100 <1> (<1>, 0.1 w/v, 5fold stimulation [16]) [16]
Tween 60 <1> (<1>, 0.1 w/v, 5fold stimulation [16]) [16]
Tween 80 <9> (<9>, detergent required, maximal activity at 0.08% [15]) [15]
UTP <1> (<1>, 0.67 mM, 2.5fold stimulation [1]) [1]
dithiothreitol <1> (<1>, activates [5]) [5]
galactose <8> (<8>, or galactose moieties in galactolipids are required [13]) [13]
Additional information <10> (<10>, enzyme is localized at thylakoid membranes in mature chloroplasts. Under certain light conditions, e.g. far-red light, the increase in PSY mRNA and protein levels is not accompanied by an increase in enzymatic activity. Under those conditions the enzyme is localized in the prolamellar body fraction in a mostly enzymatically inactive form. Subsequent illumination of dark-grown and/or in far-red light seedlings with white light causes the decay of these structures and a topological relocalization of PSY to developing thylakoids which results in its enzymatic activation. The light-dependent mechanism of regulation may contribute to ensuring a spatially and temporally coordinated increase in both carotenoid and chlorophyll contents [17]) [17]

Metals, ions

Mg^{2+} <1, 7, 8> (<1>, required [5]; <7>, Mn^{2+} or Mg^{2+} required [11]; <8>, stimulatory in presence of Mn^{2+} [13]) [5, 11, 13]
Mn^{2+} <1, 3, 7, 8, 9> (<1,9>, required [1,15]; <3>, absolute requirement for Mn^{2+}, half-maximal activity at 0.2 mM. No other divalent cation can substitute efficiently for Mn^{2+} [3]; <7>, Mn^{2+} or Mg^{2+} required [11]; <8>, absolute requirement, optimum at 1 mM [13]; <9>, maximal activity at 0.25 mM [15]; <1>, Mn^{2+} and ATP in combination are essential for catalytic activity, the effect is stoichiometric from 0.5-2.0 mM. K_m for Mn^{2+} is 0.4 mM [16]) [1, 3, 5, 11, 13, 15, 16]

Specific activity (U/mg)

0.067 <3> [3]

K$_m$-Value (mM)

0.001 <4> (geranyl diphosphate, <4>, pH 7.6 [4]) [4]

0.0027 <3> (prephytoene diphosphate, <3>, pH 7.6, 30°C [3]) [3]

0.003 <3> (geranylgeranyl diphosphate, <3>, pH 7.6, 30°C [3]) [3]

0.005 <1> (geranylgeranyl diphosphate, <1>, pH 8.0, 28°C, enzyme from chloroplast [16]) [16]

0.01 <1> (geranylgeranyl diphosphate, <1>, pH 8.0, 28°C, enzyme from chromoplast [16]) [16]

0.026 <1> (geranylgeranyl diphosphate) [5]

K$_i$-Value (mM)

0.000007 <1> (sqalestatin, <1>, pH 8.0, 28°C, chloroplast enzyme [16]) [16]

0.000009 <1> (sqalestatin, <1>, pH 8.0, 28°C, chromoplast enzyme [16]) [16]

0.0002 <1> (NADP$^+$, <1>, pH 8.0, 28°C, chloroplast enzyme and chromoplast enzyme [16]) [16]

0.1 <1> (NADPH, <1>, pH 8.0, 28°C, chloroplast enzyme [16]) [16]

0.2 <1> (phosphate, <1>, pH 8.0, 28°C, chloroplast enzyme and chromoplast enzyme [16]) [16]

0.4 <1> (NADPH, <1>, pH 8.0, 28°C, chromoplast enzyme [16]) [16]

0.5 <1> (phosphate, <1>, pH 8.0, 28°C, chromoplast enzyme [16]) [16]

1 <1> (AMP, <1>, pH 8.0, 28°C, chloroplast enzyme [16]) [16]

3 <1> (ADP, <1>, pH 8.0, 28°C, chloroplast enzyme [16]) [16]

4.3 <1> (ADP, <1>, pH 8.0, 28°C, chromoplast enzyme [16]) [16]

4.3 <1> (GTP, <1>, pH 8.0, 28°C, chloroplast enzyme [16]) [16]

pH-Optimum

6.5 <1> (<1>, enzyme from chloroplast [16]) [16]

6.8 <1> [6]

7-7.6 <4> [4]

7.5 <1> (<1>, enzyme from chromoplast [16]) [16]

7.6 <3> [3]

8.2 <9> [15]

pH-Range

6.4-7.3 <1> (<1>, pH 6.4: about 45% of maximal activity, pH 7.3: about 70% of maximal activity [6]) [6]

6.8-8.8 <9> (<9>, about 70% of maximal activity at pH 6.8 and pH 8.8 [15]) [15]

Temperature optimum (°C)

23-32 <1> [6]

Temperature range (°C)

12-37 <1> (<1>, 12°C: about 60% of maximal activity, 37°C: about 20% of maximal activity [6]) [6]

4 Enzyme Structure

Molecular weight

43000 <1> (<1>, native PAGE [16]) [16]
200000 <1> (<1>, gel filtration [1,5]; <1>, phytoene synthesizing enzyme system, gel filtration [6]) [1, 5, 6]

Subunits

? <5, 9> (<9>, x * 34500, recombinant enzyme, SDS-PAGE [15]; <5>, x * 35800, calculation from nucleotide sequence [8]) [8, 15]
dimer <4> (<4>, 2 * 36000-37000 [4]) [4]
monomer <3> (<3>, 1 * 47500, SDS-PAGE [3]) [3]
Additional information <1> (<1>, the enzyme is functional in a monomeric state, under optimal native conditions it is associated with a large protein complex, at least 200000 Da, which contains other terpenoid enzymes [16]) [16]

Posttranslational modification

proteolytic modification <1> (<1>, expression during fruit ripening as a 47000 Da protein which, upon import into isolated chloroplasts, is processed to a mature 42000 Da form [7]; <1>, the size of the transit peptide of phytoene synthase from ripe fruit is approximatyle 9000 Da, corresponding to about 80 amino acid residues [12]) [7, 12]

5 Isolation/Preparation/Mutation/Application

Source/tissue

cotyledon <10> [17]
endosperm <8> [14]
flower <8> (<8>, inner corona of, in all stages the steady-state phytoene synthase transcript levels are low [13]) [13]
fruit <1, 3, 4> [1, 3, 4, 5, 6, 12, 16, 18]
leaf <1> (<1>, low activity in young leaf [16]) [16]
Additional information <8> (<8>, more activity in green leaves [13]) [13]

Localization

chloroplast <4, 10> (<4>, more highly expressed in chromoplasts compared to chloroplasts [4]; <10>, localized at thylakoid membranes in mature chloroplasts [17]) [4, 16, 17]
chromoplast <1, 3, 4, 8> (<3>, stroma [3]; <4>, more highly expressed in chromoplasts compared to chloroplasts [4]; <1>, peripheral plastid membrane [7]) [3, 4, 7, 13]
etioplast prolamellar body <10> (<10>, under certain light conditions, e.g. far-red light, the increase in PSY mRNA and protein levels are not accompanied by an increase in enzymatic activity. Under those conditions the enzyme is localized in the prolamellar body fraction in a mostly enzymatically inactive form [17]) [17]

membrane <1, 8> (<1>, peripheral plastid membrane [7]; <8>, the soluble enzyme is complexed and inactive, the membrane-bound enzyme is active [13]; <1>, membrane-associated. Treatment with high ionic strength buffer or mild non-ionic detergent is required for solubilization [16]) [7, 13, 16] plastid <1> [1, 5, 6, 17]
soluble <8, 9> (<8>, the soluble enzyme is complexed and inactive, the membrane-bound enzyme is active [13]; <9>, upon induction the recombinant enzyme constitutes 5-10% of the total soluble protein [15]) [13, 15]

Purification

<1> (partial [16]) [5, 6, 16]
<3> [3]
<4> [4]
<7> [11]
<8> [13]
<9> (to facilitate purification of the recombinant enzyme, the structural gene for Psase is modified by site-directed mutagenesis to incorporate a C-terminal Glu-Glu-Phe tripeptide to allow purification by immunoaffinity chromatography on an immobilized monoclonal anti-α-tubulin antibody YL1/2 column [15]) [15]

Cloning

<1> (full length and truncated cDNA expression constructs of the phytoene synthase gene from tomato ligated into a pUC8 cloning vector, expression in Escherichia coli carrying Erwinia uredovora geranylgeranyl diphosphate synthase gene [12]) [7, 12]
<5> (expression in Escherichia coli [8]) [8]
<7> (expression in Escherichia coli. Two genes for the two reaction steps: 1. formation of phytoene diphosphate from geranylgeranyl diphosphate and 2. formation of phytoene from prephytoene diphosphate [10]; overexpressed to about 20% of the total cellular protein in Escherichia coli. Formation of the active phytoene synthase has the effect of suppressing the growth of the expressing strain. Inhibition of growth arises from the depletion of the substrate geranylgeranyl diphosphate which, in Escherichia coli, is necesssary for the synthesis of essential propyldiphosphate derivatives. In order to overcome the poor growth characteristics of the phytoene expressing strain, the enzyme level is increased by co-expressing the isoprenoid biosynthetic genes crtE and idi, encoding Erwinia geranylgeranyl diphosphate synthase and Rhodobacter isopentenyl diphosphate isomerase, respectively [11]; Agrobacterium-mediated transformation of hypocotyl explants of Brassica napus, 50fold increase in carotenoid levels [20]; overexpression in Lycopersicon esculentum, fruit-specific expression is achieved by using the tomato polygalacturonase promoter, and the enzyme is targeted to the chromoplast by the tomato phytoene synthase-1 transit sequence, total fruit carotenoids of primary transformants are 2-4fold higher than the control [21]) [10, 11, 20, 21]
<8> (overexpressed in insect Sf9 cells using the baculovirus lipofection system. The full-length overexpressed enzyme exhibits very reduced catalytic activity compared with an overexpressed N-truncated form, with its transit

sequence of 214 base pairs removed from the psy cDNA at the 5'-end by site-directed mutagenesis. The shortened form readily binds to lipid bilayers [13]; the Japonica rice model variety Taipei 309 is transformed by microprojectile bombardment with a cDNA coding for phytoene synthase under the control of either a constitutive or an endosperm-specific promoter. In transgenic rice plants the enzyme is active [14]) [13, 14]

<9> (expression in Escherichia coli [15]) [15]

<11> (isolation of the gene [19]) [19]

Engineering
Additional information <1, 8, 9> (<1>, the size of the transit peptide of phytoene synthase from ripe fruit is approximatyle 9000 Da, corresponding to about 80 amino acid residues. Removal of further N-terminal amino acids up to 114 from the enzyme, to yield a protein of apparent molecular mass 3400 Da, increases its catalytic activity in E. coli [12]; <8>, the full-length overexpressed enzyme exhibits very reduced catalytic activity compared with an overexpressed N-truncated form, with its transit sequence of 214 base pairs removed from the psy cDNA at the 5'-end by site-directed mutagenesis. The shortened form readily binds to lipid bilayers [13]; <9>, to facilitate purification of the recombinant enzyme, the structural gene for Psase is modified by site-directed mutagenesis to incorporate a C-terminal Glu-Glu-Phe tripeptide to allow purification by immunoaffinity chromatography on an immobilized monoclonal anti-α-tubulin antibody YL1/2 column [15]) [12, 13, 15]

Application
nutrition <7, 8> (<8>, engineering of a critical step in provitamin A biosynthesis in a non-photosynthetic, carotenoid-lacking plant tissue, important implications for long-term prospects of overcoming worldwide vitamin A deficiency [14]; <7>, 50fold increase in carotenoid levels in green embryos of Brassica napus after overexpression of bacterial phytoene synthase. Brassica and perhaps other oil seed crops may be used as commercial sources of carotenoids [20]; <7>, elevation of lycopene in tomato fruit by genetic manipulation of carotenoid biosynthesis using the fruit-specific expression of a bacterial phytoene synthase [21]) [14, 20, 21]

6 Stability

Temperature stability
4 <1> (<1>, 10 h, stable [16]) [16]

Storage stability
<1>, -70°C, 50 mM Tris, pH 7.5, 10% v/v glycerol, 80% of the activity retained [16]

<1>, activity can be maintained if 30% glycerol and 10 mM dithiothreitol are present during storage of the enzyme system [6]

<3>, -20°C, 50 mM Tris-HCl buffer, pH 7.6, 2 mM dithiothreitol, 0.3 M NaCl, 30% glycerol, stable for 1 year [3]
<9>, flash frozen in liquid N_2 and stored at -80°C, stable for at least 6 months [15]

References

[1] Maudinas, B.; Bucholtz, M.L.; Papastephanou, C.; Katiyar, S.S.; Briedis, A.V.; Porter, J.W.: Adenosine 5-triphosphate stimulation of the activity of a partially purified phytoene synthetase complex. Biochem. Biophys. Res. Commun., **66**, 430-436 (1975)

[2] Gregonis, D.E.; Rilling, H.C.: The stereochemistry of trans-phytoene synthesis. Some observations on lycopersene as a carotene precursor and a mechanism for the synthesis of cis- and trans-phytoene. Biochemistry, **13**, 1538-1542 (1974)

[3] Dogbo, O.; Laferriere, A.; D'Harlingue, A.; Camara, B.: Carotenoid biosynthesis: isolation and characterization of a bifunctional enzyme catalyzing the synthesis of phytoene. Proc. Natl. Acad. Sci. USA, **85**, 7054-7058 (1988)

[4] Camara, B.: Plant phytoene synthase complex: component enzymes, immunology, and biogenesis. Methods Enzymol., **214**, 352-365 (1993)

[5] Jones, B.L.; Porter, J.W.: Enzymatic synthesis of phytoene. Methods Enzymol., **110**, 209-220 (1985)

[6] Maudinas, B.; Bucholtz, M.L.; Papastephanou, C.; Katiyar, S.S.; Briedis, A.V.; Porter, J.W.: The partial purification and properties of a phytoene synthesizing enzyme system. Arch. Biochem. Biophys., **180**, 354-362 (1977)

[7] Bartley, G.E.; Viitanen, P.V.; Bacot, K.O.; Scolnik, P.A.: A tomato gene expressed during fruit ripening encodes an enzyme of the carotenoid biosynthesis pathway. J. Biol. Chem., **267**, 5036-5039 (1992)

[8] Chamovitz, D.; Misawa, N.; Sandmann, G.; Hirschberg, J.: Molecular cloning and expression in Escherichia coli of a cyanobacterial gene coding for phytoene synthase, a carotenoid biosynthesis enzyme. FEBS Lett., **296**, 305-310 (1992)

[9] Gregonis, D.E.; Rilling, H.C.: Photoinduced prephytoene pyrophosphate synthesis in a mycobacterium sp. Biochem. Biophys. Res. Commun., **54**, 449-454 (1973)

[10] Misawa, N.; Nakagawa, M.; Kobayashi, K.; Yamano, S.; Izawa, Y.; Nakamura, K.; Harashima, K.: Elucidation of the Erwinia uredovora carotenoid biosynthetic pathway by functional analysis of gene products expressed in Escherichia coli. J. Bacteriol., **172**, 6704-6712 (1990)

[11] Neudert, U.; Martinez-Ferez, I.M.; Fraser, P.D.; Sandmann, G.: Expression of an active phytoene synthase from Erwinia uredovora and biochemical properties of the enzyme. Biochim. Biophys. Acta, **1392**, 51-58 (1998)

[12] Misawa, N.; Truesdale, M.R.; Sandmann, G.; Fraser, P.D.; Bird, C.; Schuch, W.; Bramley, P.M.: Expression of a tomato cDNA coding for phytoene synthase in Escherichia coli, phytoene formation in vivo and in vitro, and

functional analysis of the various truncated gene products. J. Biochem., **116**, 980-985 (1994)

[13] Schledz, M.; Al-Babili, S.; Lintig, J.v.; Haubruck, H.; Rabbani, S.; Kleinig, H.; Beyer, P.: Phytoene synthase from Narcissus pseudonarcissus: functional expression, galactolipid requirement, topological distribution in chromoplasts and induction during flowering. Plant J., **10**, 781-792 (1996)

[14] Burkhardt, P.K.; Beyer, P.; Wunn, J.; Kloti, A.; Armstrong, G.A.; Schledz, M.; Von Lintig, J.; Potrykus, I.: Transgenic rice (Oryza sativa) endosperm expressing daffodil (Narcissus pseudo-narcissus) phytoene synthase accumulates phytoene, a key intermediate of provitamin A biosynthesis. Plant J., **11**, 1071-1078 (1997)

[15] Iwata-Reuyl, D.; Math, S.K.; Desai, S.B.; Poulter, C.D.: Bacterial phytoene synthase: molecular cloning, expression, and characterization of Erwinia herbicola phytoene synthase. Biochemistry, **42**, 3359-3365 (2003)

[16] Fraser, P.D.; Schuch, W.; Bramley, P.M.: Phytoene synthase from tomato (Lycopersicon esculentum) chloroplasts - partial purification and biochemical properties. Planta, **211**, 361-369 (2000)

[17] Welsch, R.; Beyer, P.; Hugueney, P.; Kleinig, H.; von Lintig, J.: Regulation and activation of phytoene synthase, a key enzyme in carotenoid biosynthesis, during photomorphogenesis. Planta, **211**, 846-854 (2000)

[18] Fraser, P.D.; Kiano, J.W.; Truesdale, M.R.; Schuch, W.; Bramley, P.M.: Phytoene synthase-2 enzyme activity in tomato does not contribute to carotenoid synthesis in ripening fruit. Plant Mol. Biol., **40**, 687-698 (1999)

[19] Fraser, P.D.; Kiano, J.W.; Truesdale, M.R.; Schuch, W.; Bramley, P.M.: Phytoene synthase-2 enzyme activity in tomato does not contribute to carotenoid synthesis in ripening fruit. Plant Mol. Biol., **40**, 687-698 (1999)

[20] Shewmaker, C.K.; Sheehy, J.A.; Daley, M.; Colburn, S.; Ke, D.Y.: Seed-specific overexpression of phytoene synthase: increase in carotenoids and other metabolic effects. Plant J., **20**, 401-412X (1999)

[21] Fraser, P.D.; Romer, S.; Shipton, C.A.; Mills, P.B.; Kiano, J.W.; Misawa, N.; Drake, R.G.; Schuch, W.; Bramley, P.M.: Evaluation of transgenic tomato plants expressing an additional phytoene synthase in a fruit-specific manner. Proc. Natl. Acad. Sci. USA, **99**, 1092-1097 (2002)

[22] Oberhauser, V.; Gaudin, J.; Fonne-Pfister, R.; Schar, H.-P.: New target enzyme(s) for bisphosphonates: inhibition of geranylgeranyl diphosphate synthase. Pestic. Biochem. Physiol., **60**, 111-117 (1998)

trans-Pentaprenyltranstransferase 2.5.1.33

1 Nomenclature

EC number
2.5.1.33

Systematic name
all-trans-pentaprenyl-diphosphate:isopentenyl-diphosphate pentaprenyltrans-transferase

Recommended name
trans-pentaprenyltranstransferase

Synonyms
HPS
HexPP synthase
HexPS
all-trans-hexaprenyl-diphosphate synthase
hexaprenyl diphosphate synthase
hexaprenyl pyrophosphate synthetase
hexaprenyl-diphosphate synthase
hexaprenylpyrophosphate synthetase
synthetase, hexaprenyl pyrophosphate

CAS registry number
83745-07-7

2 Source Organism

<1> *Micrococcus luteus* (B-P 26 [1-5]) [1-5, 7, 8]
<2> *Sulfolobus solfataricus* [6, 8]

3 Reaction and Specificity

Catalyzed reaction
all-trans-pentaprenyl diphosphate + isopentenyl diphosphate = diphosphate + all-trans-hexaprenyl diphosphate

Reaction type
alkenyl group transfer

30

Natural substrates and products

S farnesyl diphosphate + isopentenyl diphosphate <1> (<1>, probably produces the isoprenoid precursor for the biosynthesis of menaquinone-6 [2]; <1>, the enzyme produces (all-E)-hexaprenyl diphosphate, the precursor of the prenyl side chain of menaquinone-6 [7]) [2, 7]

P diphosphate + all-trans-hexaprenyl diphosphate

Substrates and products

S (E,E,E)-geranylgeranyl diphosphate + isopentenyl diphosphate <1> (<1>, about 35% of the activity with (E,E)-farnesyl diphosphate [1]) (Reversibility: ? <1> [1]) [1]

P diphosphate + all-trans-hexaprenyl diphosphate <1> [1]

S dimethylallyl diphosphate + isopentenyl diphosphate <1> (<1>, 8% of the activity with (E,E)-farnesyl diphosphate [1]) (Reversibility: ? <1> [1]) [1]

P diphosphate + ?

S farnesyl diphosphate + 3-ethylbut-3-enyl diphosphate <1> (Reversibility: ? <1> [5,8]) [5, 8]

P diphosphate + (all-E)-3-ethyl-7,11,15-trimethylhexadeca-2,6,10,14-tetraenyl diphosphate + (all-E)-3,7-diethyl-11,15,19-trimethyleicosa-2,6,10,14,18-pentaenyl diphosphate <1> [5, 8]

S farnesyl diphosphate + 3-propylbut-3-enyl diphosphate <1> (Reversibility: ? <1> [5]) [5]

P ? <1> (<1>, mixture of corresponding single and double condensation products, e.g. 3-propylgeranylgeranyl diphosphate and 3,7-dipropylgeranylfarnesyl diphosphate [5]) [5]

S farnesyl diphosphate + isopentenyl diphosphate <1, 2> (Reversibility: ? <1,2> [1-7]) [1-7]

P diphosphate + all-trans-hexaprenyl diphosphate <1> (<1>, amount of C_{30} product to C_{25} product increases as Mg^{2+} concentration is elevated, chain length never exceeds C_{30}, even at 20 mM Mg^{2+} [1]; <1>, the distribution of the polyprenyl diphosphate synthesized varies, depending on concentration of Mg^{2+}. At 30 mM Mg^{2+} more than 90% of product is the C_{30} compound [2]) [1, 2, 5]

S farnesyldiphosphate + but-3-enyl diphosphate <1> (Reversibility: ? <1> [5,8]) [5, 8]

P diphosphate + E-norgeranylgeranyl diphosphate <1> [5, 8]

S geranyl diphosphate + isopentenyl diphosphate <1> (<1>, 18% of the activity with (E,E)-farnesyl diphosphate [1]) (Reversibility: ? <1> [1]) [1]

P diphosphate + ?

S geranylgeranyl diphosphate + isopentenyl diphosphate <2> (Reversibility: ? <2> [8]) [8]

P diphosphate + hexaprenyl diphosphate <2> (<2>, at increasing concentration of isopentenyl diphosphate a significant amount of heptaprenyl diphosphate is produced, although hexaprenyl diphosphate remains the main product [8]) [8]

Inhibitors

1,2-cyclohexanedione <1> (<1>, rapid loss of component B activity, component A is resistant [4]) [4]

2,3-butanedione <1> (<1>, rapid loss of component B activity, component A is resistant [4]) [4]

NEM <1> (<1>, rapid loss of component B activity, component A is resistant, farnesyl diphosphate, isopentenyl diphosphate, farnesyl monophosphate and inorganic diphosphate protect against inactivation, Mg^{2+} is essential for protection by isopentenyl diphosphate and inorganic diphosphate, in absence of component A, component B activity is not protected by any substrate or its analogues [4]) [4]

iodoacetamide <1> (<1>, rapid loss of component B activity, component A is resistant [4]) [4]

p-chloromercuribenzoate <1> (<1>, rapid loss of component B activity, component A is resistant) [4]

phenylglyoxal <1> (<1>, rapid loss of component B activity, component A is resistant [4]) [4]

trypsin <1> [1]

Activating compounds

Triton X-100 <1> (<1>, required for full activity [8]) [8]

Metals, ions

Mg^{2+} <1, 2> (<2>, divalent cation required, Mg^{2+} is preferred to Mn^{2+} or Ca^{2+} [8]; <1>, amount of C_{30} product to C_{25} product increases as Mg^{2+} concentration is elevated, chain length never exceeds C_{30} [1,2]; <1>, at 30 mM Mg^{2+} more than 90% of product is the C_{30} compound [1,2]; <1>, Mg^{2+} is essential for protection by isopentenyl diphosphate and inorganic diphosphate against N-ethylmaleimide inactivation [4]) [1, 2, 4, 8]

Specific activity (U/mg)

0.00184 <2> [8]

K_m-Value (mM)

0.0014 <1> ((E,E)-farnesyl diphosphate) [2]

0.004 <1> ((E,E,E)-geranylgeranyl diphosphate) [2]

0.025 <1> (isopentenyl diphosphate) [2]

pH-Optimum

6 <2> [8]

6-7 <1> [2]

4 Enzyme Structure

Molecular weight

50000 <1> (<1>, gel filtration, complex of components I and II [3]) [3]

Subunits

? <2> (<2>, x * 32000, SDS-PAGE [8]; <2>, x * 32274, calculation from nucleotide sequence [8]) [8]
dimer <1> (<1>, x * 20000 + x * 60000, gel filtration [1]; <1>, the enzyme is composed of two essential components: component A and B [4]) [1, 4]

5 Isolation/Preparation/Mutation/Application

Purification

<1> (B-P 26) [1, 2]
<2> [8]

Cloning

<1> (cloning, expression, and characterization of the genes encoding the two essential protein components A and B [7]) [7]
<2> (expression in Escherichia coli [8]) [8]

Engineering

Additional information <2> (<2>, introduction of several sets of amino acid substitution into the chain-length determination region of the enzyme to mimic the product determination mechanism of various types of short chain enzymes [6]) [6]

6 Stability

Temperature stability

50 <1> (<1>, 5 min, 25% loss of component A activity, 75% loss of component B activity [1]; <1>, 5 min, about 90% loss of component A activity, about 10% loss of component B activity [2]) [1, 2]
80 <2> (<2>, 2 h, stable [8]) [8]

Storage stability

<1>, -20°C, stable for several months, components A and B [2]

References

[1] Fujii, H.; Koyama, T.; Ogura, K.: Hexaprenyl pyrophosphate synthetase from Micrococcus luteus B-P 26. Separation of two essential components. J. Biol. Chem., 257, 14610-14612 (1982)
[2] Fujii, H.; Koyama, T.; Ogura, K.: Hexaprenylpyrophosphate synthetase of Micrococcus luteus B-P 26. Methods Enzymol., 110, 192-198 (1985)
[3] Yoshida, I.; Koyama, T.; Ogura, K.: Formation of a stable and catalytically active complex of the two essential components of hexaprenyl diphosphate synthase from Micrococcus luteus B-P 26. Biochem. Biophys. Res. Commun., 160, 448-452 (1989)

[4] Yoshida, I.; Koyama, T.; Ogura, K.: Protection of hexaprenyl-diphosphate synthase of Micrococcus luteus B-P 26 against inactivation by sulphydryl reagents and arginine-specific reagents. Biochim. Biophys. Acta, **995**, 138-143 (1989)

[5] Nagaki, M.; Kuwahara, K.; Kimura, K.; Kawakami, J.; Maki, Y.; Ito, S.; Morita, N.; Nishino, T.; Koyama, T.: Substrate specificities of medium-prenylchain elongating enzymes, hexaprenyl- and heptaprenyl diphosphate synthases. J. Mol. Catal. B, **22**, 97-103 (2003)

[6] Hemmi, H.; Noike, M.; Nakayama, T.; Nishino, T.: Change of product specificity of hexaprenyl diphosphate synthase from Sulfolobus solfataricus by introducing mimetic mutations. Biochem. Biophys. Res. Commun., **297**, 1096-1101 (2002)

[7] Shimizu, N.; Koyama, T.; Ogura, K.: Molecular cloning, expression, and characterization of the genes encoding the two essential protein components of Micrococcus luteus B-P 26 hexaprenyl diphosphate synthase. J. Bacteriol., **180**, 1578-1581 (1998)

[8] Hemmi, H.; Ikejiri, S.; Yamashita, S.; Nishino, T.: Novel medium-chain prenyl diphosphate synthase from the thermoacidophilic archaeon Sulfolobus solfataricus. J. Bacteriol., **184**, 615-620 (2002)

Tryptophan dimethylallyltransferase

1 Nomenclature

EC number

2.5.1.34

Systematic name

dimethylallyl-diphosphate:L-tryptophan dimethylallyltransferase

Recommended name

tryptophan dimethylallyltransferase

Synonyms

4-(γ,γ-dimethylallyl)tryptophan synthase

DMAT synthase

DMAT synthetase

dimethylallylpyrophosphate:L-tryptophan dimethylallyltransferase

dimethylallylpyrophosphate:tryptophan dimethylallyl transferase

dimethylallylpyrophosphate:tryptophan dimethylallyltransferase

dimethylallyltransferase, tryptophan

dimethylallyltryptophan synthetase

CAS registry number

55127-01-0

2 Source Organism

<1> *Claviceps sp.* (SD58 [1,3,5]) [1-3, 5]

<2> *Claviceps purpurea* [4, 6]

3 Reaction and Specificity

Catalyzed reaction

dimethylallyl diphosphate + L-tryptophan = diphosphate + 4-(3-methylbut-2-enyl)-L-tryptophan (<1>, random sequential mechanism [5]; <1>, sequential mechanism [1]; <1>, stereochemistry, direct attack of dimethylallyl di-phosphate on C-4 of the indole [2])

Reaction type

alkenyl group transfer

Natural substrates and products

S dimethylallyl diphosphate + L-tryptophan <1, 2> (<1>, first pathway-specific enzyme of ergot alkaloid biosynthesis [1-3,6]) [1-3, 6]

P diphosphate + L-4-(γ,γ-dimethylallyl)tryptophan

Substrates and products

S dimethylallyl diphosphate + 5-methyltryptophan <1> (<1>, 7.4% of the activity with L-tryptophan [1]) (Reversibility: ? <1> [1]) [1]

P diphosphate + 4-(3-methylbut-2-enyl)-5-methyl-L-tryptophan

S dimethylallyl diphosphate + 6-methyltryptophan <1> (<1>, 9.8% of the activity with tryptophan [1]) (Reversibility: ? <1> [1]) [1]

P diphosphate + 4-(3-methylbut-2-enyl)-6-methyl-L-tryptophan

S dimethylallyl diphosphate + 7-methyltryptophan <1> (<1>, 40.5% of the activity with tryptophan [1]) (Reversibility: ? <1> [1]) [1]

P diphosphate + 4-(3-methylbut-2-enyl)-7-methyl-L-tryptophan

S dimethylallyl diphosphate + L-tryptophan <1, 2> (Reversibility: ? <1,2> [1-6]) [1-6]

P diphosphate + L-4-(γ,γ-dimethylallyl)tryptophan <1, 2> [1, 4]

S dimethylallyl diphosphate + bishomotryptophan <1> (<1>, 2% of the activity with tryptophan [1]) (Reversibility: ? <1> [1]) [1]

P diphosphate + bis[4-(3-methylbut-2-enyl)]-L-homotryptophan

Inhibitors

1,10-phenanthroline <1> [3, 5]

2,2'-bipyridyl <1> [3, 5]

8-hydroxyquinoline <1> [3, 5]

8-mercaptoquinoline <1> [3, 5]

Ca^{2+} <1> (<1>, no inhibition, even at 5 mM [1]; <1>, 30 mM inhibits [3]) [3]

Co^{3+} <1> [1]

EDTA <1> [3, 5]

Hg^{2+} <1> [1]

L-4-(γ,γ-dimethylallyl)tryptophan <1> (<1>, product inhibition, competitive to dimethylallyl diphosphate, noncompetitive to L-tryptophan [1]) [1]

Mg^{2+} <1> (<1>, inhibition above 5 mM [1]) [1]

Mn^{2+} <1> [1]

Zn^{2+} <1> [1]

acetylacetone <1> [3, 5]

chelating agents <1> (<1>, in presence of excess of Ca^{2+} [3,5]) [3, 5]

diphosphate <1> (<1>, noncompetitive to dimethylallyl diphosphate [1]) [1]

dithiothreitol <1> [3, 5]

nitrilotriacetic acid <1> [3, 5]

Metals, ions

Ca^{2+} <1, 2> (<2>, maximal stimulation at 4-10 mM [4]; <1>, partial activation [1]) [1, 4]

Cu^{2+} <1> (<1>, slight stimulation [1]) [1]

Fe^{2+} <1> (<1>, activates [1]) [1]

Li^+ <1> (<1>, slight stimulation [1]) [1]

Mg^{2+} <1, 2> (<1>, activates [1]; <2>, maximal stimulation at 4-10 mM [4]; <1>, optimal concentration: 3-5 mM [1]) [1, 4]

Na^+ <1> (<1>, slight stimulation [1]) [1]

Additional information <1> (<1>, metal ion, free in solution not required for activity [5]; <1>, enzyme contains a divalent cation which is most likely involved in catalysis [3,5]) [3, 5]

Specific activity (U/mg)

0.309 <1> [1]

0.35 <1> [3, 5]

0.5 <2> [4]

K_m-Value (mM)

0.007 <1> (dimethylallyl diphosphate, <1>, in presence of Ca^{2+} [3]) [3]

0.008 <2> (dimethylallyl diphosphate, <2>, pH 7.8, 30°C, in presence of 4 mM $CaCl_2$ [4]) [4]

0.009 <1> (tryptophan, <1>, in presence of Ca^{2+} [3]) [3]

0.012 <2> (L-tryptophan, <2>, pH 7.8, 30°C, in presence of 4 mM $MgCl_2$ [4]) [4]

0.014 <2> (dimethylallyl diphosphate) [4]

0.017 <2> (L-tryptophan, <2>, pH 7.8, 30°C, in presence of 4 mM $CaCl_2$ [4]) [4]

0.04 <2> (L-tryptophan, <2>, pH 7.8, 30°C [4]) [4]

0.067 <1> (L-tryptophan, <1>, 30°C [1]) [1]

0.2 <1> (dimethylallyl diphosphate, <1>, 30°C [1]) [1]

Additional information <1> [3]

pH-Optimum

7-7.5 <1> [1]

7.5 <2> [4]

4 Enzyme Structure

Molecular weight

70000 <1> (<1>, gel filtration [5]) [5]

105000 <2> (<2>, gel filtration [4]) [4]

210000 <1> (<1>, gel filtration [1]) [1]

Subunits

? <2> (<2>, x * 51824, calculation from nucleotide sequence [6]) [6]

dimer <1, 2> (<2>, α_2, 2 * 53000, SDS-PAGE [4]; <1>, 2 * 34000, SDS-PAGE [5]) [4, 5]

monomer <1> (<1>, 1 * 70000, SDS-PAGE [1]) [1]

5 Isolation/Preparation/Mutation/Application

Source/tissue
    mycelium <1, 2> [1, 3, 4]

Purification
    <1> [1, 3, 5]
    <2> [4]

Crystallization
    <1> [3, 5]

Cloning
    <2> (expression in Saccharomyces cerevisiae [6]) [6]

6 Stability

Temperature stability
    50 <1> (<1>, 1 h, complete loss of activity [1]) [1]

General stability information
    <1>, repeated thawing and freezing greatly accelerates denaturation [1]

Storage stability
    <1>, -20°C, 10 mM Tris-HCl, pH 8.0, 20 mM $CaCl_2$, 20 mM 2-mercaptoethanol, 10% glycerol, 2 mg/ml enzyme concentration, stable for 1 week [1]
    <1>, -20°C, 50% loss of activity after several weeks [5]
    <2>, -20°C, glycerol-containing buffer, stable for more than 6 months [4]

References

[1] Lee, S.L.; Floss, H.G.; Heinstein, P.: Purification and properties of dimethylallylpyrophosphate:tryptopharm dimethylallyl transferase, the first enzyme of ergot alkaloid biosynthesis in Claviceps. sp. SD 58. Arch. Biochem. Biophys., **177**, 84-94 (1976)

[2] Shibuya, M.; Chou, H.M.; Fountoulakis, M.; Hassam, S.; Kim, S.-U.; Kobayashi, K.; Otsuka, H.; Rogalska, E.; Cassady, J.M.; Floss, H.G.: Stereochemistry of the isoprenylation of tryptophan catalyzed by 4-(γ,γ-dimethylallyl)tryptophan synthase from Claviceps, the first pathway-specific enzyme in ergot alkaloid biosynthesis. J. Am. Chem. Soc., **112**, 297-304 (1990)

[3] Rilling, H.C.: Dimethylallylpyrophosphate: L-tryptophan dimethylallyltransferase. Methods Enzymol., **110**, 335-340 (1985)

[4] Gebler, J.C.; Poulter, C.D.: Purification and characterization of dimethylallyl tryptophan synthase from Claviceps purpurea. Arch. Biochem. Biophys., **296**, 308-313 (1992)

[5] Cress, W.A.; Chayet, L.T.; Rilling, H.C.: Crystallization and partial characterization of dimethylallyl pyrophosphate: L-tryptophan dimethylallyltransferase from Claviceps sp. SD58. J. Biol. Chem., **256**, 10917-10923 (1981)

[6] Tsai, H.-F.; Wang, H.; Gebler, J.C.; Poulter, C.D.; Schardl, C.L.: The Claviceps purpurea gene encoding dimethylallyltryptophan synthase, the committed step for ergot alkaloid biosynthesis. Biochem. Biophys. Res. Commun., **216**, 119-125 (1995)

Aspulvinone dimethylallyltransferase 2.5.1.35

1 Nomenclature

EC number
2.5.1.35

Systematic name
dimethylallyl-diphosphate:aspulvinone-E dimethylallyltransferase

Recommended name
aspulvinone dimethylallyltransferase

Synonyms
dimethylallyl pyrophosphate:aspulvinone dimethylallyltransferase
dimethylallyltransferase, aspulvinone

CAS registry number
67584-68-3

2 Source Organism

<1> *Aspergillus terreus* (IAM 2054 [1,2]) [1, 2]

3 Reaction and Specificity

Catalyzed reaction
2 dimethylallyl diphosphate + aspulvinone E = 2 diphosphate + aspulvinone
H

Reaction type
alkenyl group transfer

Natural substrates and products
  S Additional information <1> (<1>, enzyme is involved in biosynthesis of
  aspulvinone pigments [1]) [1]
  P ?

Substrates and products
  S (E)-3-methylpent-2-enyl diphosphate + aspulvinone E <1> (<1>, 35% of
  the activity with dimethylallyl diphosphate [1]) (Reversibility: ? <1>
  [1,2]) [1, 2]
  P ?
  S (E)-butenyl diphosphate + aspulvinone E <1> (<1>, 10% of the activity
  with dimethylallyl diphosphate [1]) (Reversibility: ? <1> [1,2]) [1, 2]

P ?

S cycloheptylideneethyl diphosphate + aspulvinone E <1> (<1>, 10% of the activity with dimethylallyl diphosphate [1]) (Reversibility: ? <1> [1,2]) [1, 2]

P ?

S cyclohexylideneethyl diphosphate + aspulvinone E <1> (<1>, 35% of the activity with dimethylallyl diphosphate [1]) (Reversibility: ? <1> [1,2]) [1, 2]

P ?

S cyclopentylideneethyl diphosphate + aspulvinone E <1> (<1>, 35% of the activity with dimethylallyl diphosphate [1]) (Reversibility: ? <1> [1,2]) [1, 2]

P ?

S dimethylallyl diphosphate + aspulvinone E <1> (<1>, i.e. dihydroxypulvinone [2]) (Reversibility: ? <1> [1,2]) [1, 2]

P diphosphate + aspulvinone H <1> [1]

S dimethylallyl diphosphate + aspulvinone G <1> (<1>, i.e. trihydroxypulvinone, 71% of the activity with aspulvinone E [2]) (Reversibility: ? <1> [1,2]) [1, 2]

P ?

S dimethylallyl diphosphate + aspulvinone I <1> (<1>, 75% of the activity with aspulvinone E [1]) (Reversibility: ? <1> [1]) [1]

P ?

S dimethylallyl diphosphate + aspulvinone I-c <1> (<1>, 13% of the activity with aspulvinone E [1]) (Reversibility: ? <1> [1]) [1]

P ?

S dimethylallyl diphosphate + atromentic acid <1> (<1>, 7% of the activity with aspulvinone E [1]) (Reversibility: ? <1> [1]) [1]

P ?

S dimethylallyl diphosphate + pulvinone <1> (<1>, 6% of the activity with aspulvinone E [1]) (Reversibility: ? <1> [1]) [1]

P ?

Inhibitors

$(NH_4)_2SO_4$ <1> [2]

aspulvinone E dimethyl ether <1> [1]

bromophenol blue <1> [1, 2]

diphosphate <1> (<1>, noncompetitive against aspulvinone, mixed type against dimethylallyl diphosphate [1]) [1, 2]

isopentenyl diphosphate <1> [1]

lecithin <1> (<1>, 0.025%, 69% inhibition [2]) [2]

Activating compounds

Triton X-100 <1> (<1>, 0.025%, activates 1.5times [2]) [2]

Tween 80 <1> (<1>, 0.025%, activates 2.3times [2]) [2]

Specific activity (U/mg)

0.0294 <1> [1, 2]

K$_m$-Value (mM)
  0.0077 <1> (aspulvinone G) [1, 2]
  0.0137 <1> (aspulvinone E) [1, 2]
  0.04 <1> (dimethylallyl diphosphate) [1, 2]

K$_i$-Value (mM)
  0.0178 <1> (diphosphate, <1>, inhibition against dimethylallyl diphosphate
  [1]) [1, 2]
  0.05 <1> (diphosphate, <1>, inhibition against aspulvinone [1]) [1, 2]

pH-Optimum
  7 <1> [1]

4 Enzyme Structure

Molecular weight
  240000-270000 <1> (<1>, gel filtration [1,2]) [1, 2]

Subunits
  hexamer <1> (<1>, 6 * 45000, SDS-PAGE [1,2]) [1, 2]

5 Isolation/Preparation/Mutation/Application

Source/tissue
  mycelium <1> [1, 2]

Purification
  <1> [1, 2]

6 Stability

General stability information
  <1>, addition of 2-mercaptoethanol does not affect stability [1]

Storage stability
  <1>, -20°C, stable for 1 week [1, 2]
  <1>, 4°C, 50% loss of activity after 3 weeks [1, 2]

References

[1] Takahashi, I.; Ojima, N.; Ogura, K.; Seto, S.: Purification and characterization
    of dimethylallyl pyrophosphate:aspulvinone dimethylallyltransferase from
    Aspergillus terreus. Biochemistry, **17**, 2696-2702 (1978)
[2] Sagami, I.; Ojima, N.; Ogura, K.; Seto, S.: Aspulvinone dimethylallyltransfer-
    ase. Methods Enzymol., **110**, 320-326 (1985)

Trihydroxypterocarpan dimethylallyltransferase

2.5.1.36

1 Nomenclature

EC number
2.5.1.36

Systematic name
dimethylallyl-diphosphate:(6aS,11aS)-3,6a,9-trihydroxypterocarpan dimethyl-allyltransferase

Recommended name
trihydroxypterocarpan dimethylallyltransferase

Synonyms
dimethylallylpyrophosphate:3,6a,9-trihydroxypterocarpan dimethylallyltrans-ferase
dimethylallylpyrophosphate:trihydroxypterocarpan dimethylallyl transferase
dimethylallytransferase, trihydroxypterocarpan
glyceollin synthase

CAS registry number
70851-94-4

2 Source Organism

<1> *Glycine max* (elicitor-treated [1,2,3]) [1-3]

3 Reaction and Specificity

Catalyzed reaction
dimethylallyl diphosphate + (6aS,11aS)-3,6a,9-trihydroxypterocarpan = di-phosphate + 2-dimethylallyl-(6aS,11aS)-3,6a,9-trihydroxypterocarpan
dimethylallyl diphosphate + (6aS,11aS)-3,6a,9-trihydroxypterocarpan = di-phosphate + 4-dimethylallyl-(6aS,11aS)-3,6a,9-trihydroxypterocarpan

Reaction type
alkenyl group transfer

Natural substrates and products
S dimethylallyl diphosphate + (6aS,11aS)-3,6a,9-trihydroxypterocarpan <1> (<1>, enzyme is involved in synthesis of 2-dimethylallyl-trihydroxy-pterocarpan and 4-dimethylallyl-trihydroxypterocarpan, putative pre-

cursor of the glyceollins [1]; <1>, introduction of the dimethylallyl substituent into trihydroxypterocarpan increases fungitoxicity against Cladosporium cucumerinum [2]; <1>, involved in biosynthesis of glyceollins [3]) (Reversibility: ? <1> [1-3]) [1-3]

P   diphosphate + 2-dimethylallyl-trihydroxypterocarpan + 4-dimethylallyl-trihydroxypterocarpan

Substrates and products

S   dimethylallyl   diphosphate   +   (6aS,11aS)-3,6a,9-trihydroxypterocarpan <1> (Reversibility: ? <1> [1,2,3]) [1, 2, 3]

P   diphosphate + 2-dimethylallyl-trihydroxypterocarpan + 4-dimethylallyl-trihydroxypterocarpan <1> [1-3]

5   Isolation/Preparation/Mutation/Application

Source/tissue

cell culture <1> [1, 2]
cotyledon <1> [1, 2, 3]
hypocotyl <1> [1]

Localization

particle-bound <1> [3]

References

[1] Leube, J.; Grisebach, H.: Further studies on induction of enzymes of phytoalexin synthesis in soybean and cultured soybean cells. Z. Naturforsch. C, **38**, 730-735 (1983)

[2] Zaehringer, U.; Schaller, E.; Grisebach, H.: Induction of phytoalexin synthesis in soybean. Structure and reactions of naturally occuring and enzymatically prepared prenylated pterocarpans from elicitor-treated cotyledons and cell cultures of soybean. Z. Naturforsch. C, **36**, 234-241 (1981)

[3] Zaehringer, U.; Ebel, J.; Mulheirn, L.J.; Lyne, R.L.; Grisebach, H.: Induction of phytoalexin synthesis in soybean. Dimethylallylpyrophosphate:trihydroxypterocarpan dimethylallyl transferase from elicitor-induced cotyledons. FEBS Lett., **101**, 90-92 (1979)

Leukotriene-C$_4$ synthase

1 Nomenclature

EC number

2.5.1.37 (transferred to EC 4.4.1.20)

Recommended name

leukotriene-C$_4$ synthase

Isonocardicin synthase 2.5.1.38

1 Nomenclature

EC number
2.5.1.38

Systematic name
S-adenosyl-L-methionine:nocardicin-E 3-amino-3-carboxypropyltransferase

Recommended name
isonocardicin synthase

Synonyms
aminocarboxypropyltransferase, nocardicin
nocardicin aminocarboxypropyltransferase

CAS registry number
118246-74-5

2 Source Organism

<1> *Nocardia uniformis* (ssp. tsuyamanensis, ATCC 21806) [1]

3 Reaction and Specificity

Catalyzed reaction
S-adenosyl-L-methionine + nocardicin E = 5'-methylthioadenosine + isono-cardicin A

Reaction type
3-amino-3-carboxypropyl group transfer

Natural substrates and products
S   S-adenosyl-L-methionine + nocardicin E <1> (<1>, enzyme is involved in the biosynthesis of the β-lactam antibiotic nocardicin A [1]) [1]
P   ?

Substrates and products
S   S-adenosyl-L-methionine + nocardicin E <1> (Reversibility: ? <1> [1]) [1]
P   5'-methylthioadenosine + isonocardicin A

5 Isolation/Preparation/Mutation/Application

Purification

<1> (ssp. tsuyamanensis, ATCC 21806, partial) [1]

6 Stability

General stability information

<1>, partially purified enzyme is stable after dialysis against 50 mM phosphate buffer, pH 7.5, containing 20% glycerol and 10 mM 2-mercaptoethanol [1]

References

[1] Wilson, B.A.; Bantia, S.; Salituro, G.M.; Reeve, A.M.; Townsend, C.A.: Cell-free biosynthesis of nocardicin A from nocardicin E and S-adenosylmethionine. J. Am. Chem. Soc., **110**, 8238-8239 (1988)

4-Hydroxybenzoate nonaprenyltransferase 2.5.1.39

1 Nomenclature

EC number
2.5.1.39

Systematic name
solanesyl-diphosphate:4-hydroxybenzoate nonaprenyltransferase

Recommended name
4-hydroxybenzoate nonaprenyltransferase

Synonyms
4-hydroxybenzoate transferase
dimethylallyltransferase, p-hydroxybenzoate poly-
nonaprenyl-4-hydroxybenzoate transferase
p-hydroxybenzoate dimethylallyltransferase
p-hydroxybenzoate polyprenyltransferase
p-hydroxybenzoic acid-polyprenyl transferase
p-hydroxybenzoic-polyprenyl transferase

CAS registry number
9030-77-7

2 Source Organism

<1> *Rattus norvegicus* (male Sprague-Dawley [2,8]) [1-3, 8]
<2> *Cavia porcellus* (guinea pig [3]) [3]
<3> *Pseudomonas putida* (IAM 1219 [5]) [4, 5]
<4> *Escherichia coli* [6, 7, 10]
<5> *Schizosaccharomyces pombe* (wild-type strain SP870, diploid strains SP826 and TP4-1D/TP4-5A [9]) [9]

3 Reaction and Specificity

Catalyzed reaction
solanesyl diphosphate + 4-hydroxybenzoate = diphosphate + nonaprenyl-4-hydroxybenzoate

Reaction type
alkenyl group transfer

48

Natural substrates and products

S solanesyl diphosphate + 4-hydroxybenzoate <1, 4, 5> (<1, 4, 5> involved in the biosynthesis of ubiquinone [1, 7, 9]; <1> key enzyme in ubiquinone synthesis, it catalyzes the step which brings together the precursor of the benzoquinone ring and the polyisoprenoid side chain, in rat liver the major product is 3-nonaprenyl-4-hydroxybenzoate, in human and guinea pig it is 3-decaprenyl-4-hydroxybenzoate [2]; <4> involved in biosynthesis of ubiquinone and vitamin E [10]) [1, 2, 7, 9, 10]

P diphosphate + 3-solanesyl-4-hydroxybenzoate

Substrates and products

S farnesyl diphosphate + 4-hydroxybenzoate <3, 4> (<3> at 1% of the activity with nonaprenyl diphosphate [4]) (Reversibility: ? <3, 4> [4, 7]) [4, 7]

P 3-farnesyl-4-hydroxybenzoate + diphosphate <4> [7]

S geranyl diphosphate + 4-hydroxybenzoate <4> (Reversibility: ? <4> [6, 7, 10]) [6, 7, 10]

P 3-geranyl-4-hydroxybenzoate + diphosphate <4> [7]

S geranylgeranyl diphosphate + 4-hydroxybenzoate <4> (<4> various derivatives of 4-hydroxybenzoate can be used as substrate [10]) (Reversibility: ? <4> [10]) [10]

P geranylgeranyl-4-hydroxybenzoate + diphosphate

S hexaprenyl diphosphate + 4-hydroxybenzoate <3> (<3> at 68% of the activity with nonaprenyl diphosphate [4]) (Reversibility: ? <3> [4]) [4]

P hexaprenyl-4-hydroxybenzoate + diphosphate

S pentaprenyl diphosphate + 4-hydroxybenzoate <3> (<3> 86% of the activity with nonaprenyl diphosphate [4]) (Reversibility: ? <3> [4]) [4]

P pentaprenyl-4-hydroxybenzoate + diphosphate

S phytyl diphosphate + 4-hydroxybenzoate <3> (<3> at 1% of the activity with nonaprenyl diphosphate [4]) (Reversibility: ? <3> [4]) [4]

P phytyl-4-hydroxybenzoate + diphosphate

S solanesyl diphosphate + 4-hydroxybenzoate <1-4> (<1-3> i.e. nonaprenyl diphosphate [1-5]) (Reversibility: ? <1-4> [1-5,7]) [1-5, 7]

P diphosphate + 3-solanesyl-4-hydroxybenzoate <1-4> (<1,2> 3-nonaprenyl-4-hydroxybenzoate [3]; <1> small amounts of decaprenyl-4-hydroxybenzoate and octaprenylhydroxybenzoate can be found, depending on source of enzyme and substrate used to provide polyprenyl side chains [2]) [1-5, 7]

S tetraprenyl diphosphate + 4-hydroxybenzoate <3> (<3> 6% of the activity with nonaprenyl diphosphate [4]) (Reversibility: ? <3> [4]) [4]

P tetraprenyl-4-hydroxybenzoate + diphosphate

S Additional information <1, 2, 4> (<1,2> no substrate: nonaprenyl monophosphate [1,3]; <1,2> detergents can affect the specificity of the mitochondrial system synthesizing polyprenyl diphosphates [3]; <4> no substrate: geranyl monophosphate [10]) [1, 3, 10]

P ?

Inhibitors

2,4-dinitrophenol <3> (<3> 16% inhibition at 1 mM [5]) [5]
4-aminobenzoic acid <1> [2]
4-chlorobenzoic acid <1> [2]
Ag^+ <3> (<3> complete inhibition at 0.1 mM [5]) [5]
Brij W1 <1> (<1> inhibition at 0.5% or more, activation at 0.05% [2]) [2]
Ca^{2+} <1, 3> (<1> Mg^{2+} can overcome the inhibition [2]; <3> 46% inhibition at 1 mM [5]) [2, 5]
Cd^{2+} <3> (<3> complete inhibition at 0.1 mM [5]) [5]
Co^{2+} <3> (<3> 28% inhibition at 0.1 mM [5]) [5]
Cu^{2+} <3> (<3> complete inhibition at 0.1 mM [5]) [5]
Fe^{2+} <3> (<3> 68% inhibition at 0.1 mM [5]) [5]
Hg^{2+} <3> (<3> complete inhibition at 0.1 mM [5]) [5]
KCN <1> [2]
Mn^{2+} <3> (<3> 68% inhibition at 0.1 mM [5]) [5]
N-ethylmaleimide <3> (<3> 67% inhibition at 1 mM [5]) [5]
NaCl <4> (<4> increasing concentrations in the presence of 5 mM Mg^{2+} result in inhibition [7]) [7]
NaF <1> [2]
Nonidet P-40 <1> (<1> inhibition at 0.5% or more, activation at 0.05% [2]) [2]
Triton X-100 <1, 4> (<1> inhibition at 0.5% or more, activation at 0.01% [2]; <1> complete inhibition at 0.5% [1]) [1, 2, 7]
Tween 80 <1, 4> (<1> inhibition at 0.5% or more, activation at 0.01% [2]) [1, 2, 7]
Zn^{2+} <1, 3> (<3> complete inhibition at 0.1 mM [5]) [1, 5]
acetyl-CoA <3> (<3> weak [4]) [4]
bacitracin <1> (<1> strong inhibitor [2]) [2]
β-octyl glucoside <1> [1]
cis-vaccenic acid <3> (<3> synonym: cis-11-octadecenoic acid [4]) [4]
deoxycholate <1> (<1> complete inhibition at 0.5% [1]) [1]
diphenylamine <1> [2]
iodoacetamide <1> [2]
isopentenyl diphosphate <3> [4]
p-chloromercuribenzoate <1, 3> (<3> 95% inhibition at 0.01 mM [5]) [2, 4, 5]
palmitoyl-CoA <3> [4]
sodium deoxycholate <1, 4> (<1> inhibition at 0.5% or more, activation at 0.05% [2]) [1, 2, 7]
Additional information <1> (<1> no inhibitor: taurodeoxycholate [1]) [1]

Activating compounds

Brij W1 <1, 2> (<1,2> activates the enzyme in mitochondria having previously been aged by freezing at -20°C for periods ranging from 1 h to several days [3]; <1> inhibition at 0.5% or more, activation at 0.05% [2]) [2, 3]
CHAPS <4> (<4> 46% increase at 0.01% [7]) [7]

Nonidet P-40 <1, 2> (<1,2> activates the enzyme in mitochondria which is aged by freezing at -20°C for periods ranging from 1 h to several days [3]; <1> inhibition at 0.5% or more, activation at 0.05% [2]) [2, 3]

Triton X-100 <1, 2> (<1,2> activates the enzyme in mitochondria which is aged by freezing at -20°C for periods ranging from 1 h to several days [3]; <1> inhibition at 0.5% or more, activation at 0.01% [2]) [2, 3]

Tween 80 <1, 2> (<1,2> activates the enzyme in mitochondria which is aged by freezing at -20°C for periods ranging from 1 h to several days [3]; <1> inhibition at 0.5% or more, activation at 0.01% [2]) [2, 3]

cardiolipin <3> [5]

dipalmitoylphosphatidylcholine <3> [5]

phosphatidylethanolamine <3> [5]

phosphatidylglycerol <3> [5]

phospholipid <3> (<3> required, 3fold activation [4]; <3> strongly activated by phospholipid extracted from the bacterial cell, phosphatidylethanolamine is most effective [5]) [4, 5]

sodium deoxycholate <1, 2> (<1,2> activates the enzyme in mitochondria which is aged by freezing at -20°C for periods ranging from 1 h to several days [3]; <1> inhibition at 0.5% or more, activation at 0.05% [2]) [2, 3]

Metals, ions

Co^{2+} <4> (<4> can replace Mg^{2+} to some extent [7]) [7]

Mg^{2+} <1-4> (<3> required [5]; <1,2> activation [2,3]; <1> maximal activation: 10 mM [2]; <4> optimum between 20 and 100 mM [7]) [2, 3, 5, 6, 7]

Mn^{2+} <4> (<4> can replace Mg^{2+} to some extent [7]) [7]

Specific activity (U/mg)

0.0006 <4> (<4> in the presence of 5 mM Co^{2+} [7]) [7]

0.0013 <4> (<4> in the presence of 5 mM Mn^{2+} [7]) [7]

0.014 <4> (<4> in the presence of 5 mM Mg^{2+} [7]) [7]

Additional information <3> [4, 5]

K_m-Value (mM)

0.000833 <1> (nonaprenyl diphosphate, <1> pH 7.4, 25°C or 37°C [2]) [2]

0.00114 <1> (4-hydroxybenzoate, <1> pH 7.4, 25°C or 37°C [2]) [2]

0.022 <4> (farnesyl diphosphate, <4> pH 7.8, 37°C [7]) [7]

0.031 <4> (solanesyl diphosphate, <4> pH 7.8, 37°C [7]) [7]

0.054 <3> (4-hydroxybenzoate, <3> pH 7, 37°C [5]) [5]

0.14 <3> (nonaprenyl diphosphate, <3> pH 7, 37°C [5]) [5]

0.254 <4> (geranylgeranyl diphosphate, <4> pH 7.8, 37°C [7]) [7]

pH-Optimum

7-7.2 <3> [5]

7.4 <1> [2]

7.5 <1> (<1> microsomes [1]) [1]

7.8 <1, 4> (<1,4> assay at [6,8]) [6, 7, 8]

8-9 <1> (<1> mitochondria [1]) [1]

pH-Range

6.7-9.4 <4> (<4> 50% activity at pH 6.7 and pH 9.4 [7]) [7]

Temperature optimum (°C)

37 <1, 3, 4> (<1,3,4> assay at [1,5,6]) [1, 5, 6, 8]

5 Isolation/Preparation/Mutation/Application

Source/tissue

heart <1> [2]
kidney <1> [1, 2]
liver <1, 2> [1-3, 8]
spleen <1> [1]

Localization

Golgi vesicle <1> (<1> Golgi III vesicles, high specific activity, associated
with inner luminal surface of microsomal vesicles [1]) [1]
cytoplasmic membrane <1> [1]
lysosome <1> [1]
membrane <3, 4> [5, 6, 7]
microsome <1> (<1> smooth II microsome, high specific activity, associated
with inner luminal surface of microsomal vesicles [1]) [1, 8]
mitochondrial inner membrane <1> [2]
mitochondrion <1, 2, 5> (<1> low activity [1]) [1, 2, 3, 9]
peroxisome <1> [8]
Additional information <1> (<1> at low levels: rough and smooth I microsomes, mitochondria, not: peroxisomes, cytosol [1]) [1]

Purification

<3> (partial [4]) [4]

Cloning

<4> (cloned under control of different promoters, expressed in Nicotiana ta-
bacum [6]; wild-type and disrupted gene, fusion with green fluorescent protein [9]) [6, 7, 9]

6 Stability

General stability information

<1>, isolated mitochondria can be stored frozen for 2-3 weeks prior to pre-
paration of membrane fragments without any loss of activity, preparation of
mitochondrial fragments loses activity after 2 cycles of freezing and thawing
[2]

Storage stability

<4>, -25°C, stable for several months, loss of 20% activity after one year [10]
<1, 2>, -20°C, activity is maintained for several months [3]

References

[1] Kalen, A.; Applekvist, E.L.; Chojnacki, T.; Dallner, G.: Nonaprenyl-4-hy-droxybenzoate transferase, an enzyme involved in ubiquinone biosynthesis, in the endoplasmic reticulum-Golgi system of rat liver. J. Biol. Chem., 265, 1158-1164 (1990)

[2] Gupta, A.; Rudney, H.: 4-Hydroxybenzoate polyprenyltransferase from rat liver. Methods Enzymol., 110, 327-334 (1985)

[3] Nishino, T.; Rudney, H.: Effects of detergents on the properties of 4-hy-droxybenzoate. Polyprenyl transferase and the specificity of the polyprenyl pyrophosphate synthetic system in mitochondria. Biochemistry, 16, 605-609 (1977)

[4] Kawahara, K.; Koizumi, N.; Kawaji, H.; Oishi, K.; Aida, K.; Uchida, K.: Partial purification and characterization of 4-hydroxybenzoate-polyprenyl-transferase in ubiquinone biosynthesis of Pseudomonas putida. Agric. Biol. Chem., 55, 2307-2311 (1991)

[5] Uchida, K.; Koizumi, N.; Kawaji, H.; Kawahara, K.; Aida, K.: Solubilization of 4-hydroxybenzoate-polyprenyltransferase from cell membrane of Pseudomonas putida and its properties. Agric. Biol. Chem., 55, 2299-2305 (1991)

[6] Boehm, R.; Sommer, S.; Severin, K.; Li, S.M.; Heide, L.: Active expression of the ubiA gene from E. coli in tobacco: influence of plant ER-specific signal peptides on the expression of a membrane-bound prenyltransferase in plant cells. Transgenic Res., 9, 477-486 (2000)

[7] Melzer, M.; Heide, L.: Characterization of polyprenyldiphosphate: 4-hy-droxybenzoate polyprenyltransferase from Escherichia coli. Biochim. Biophys. Acta, 1212, 93-102 (1994)

[8] Tekle, M.; Bentinger, M.; Nordman, T.; Appelkvist, E.L.; Chojnacki, T.; Olsson, J.M.: Ubiquinone biosynthesis in rat liver peroxisomes. Biochem. Biophys. Res. Commun., 291, 1128-1133 (2002)

[9] Uchida, N.; Suzuki, K.; Saiki, R.; Kainou, T.; Tanaka, K.; Matsuda, H.; Kawamukai, M.: Phenotypes of fission yeast defective in ubiquinone production due to disruption of the gene for p-hydroxybenzoate polyprenyl diphosphate transferase. J. Bacteriol., 182, 6933-6939 (2000)

[10] Wessjohann, L.; Sontag, B.: Prenylation of benzoic acid derivatives catalyzed by a transferase from Escherichia coli overproduction: method development and substrate specificity. Angew. Chem., 35, 1697-1699 (1996)

Aristolochene synthase

1 Nomenclature

EC number
2.5.1.40 (transferred to EC 4.2.3.9)

Recommended name
aristolochene synthase

Phosphoglycerol geranylgeranyltransferase

1 Nomenclature

EC number
2.5.1.41

Systematic name
geranylgeranyl diphosphate:sn-glyceryl phosphate geranylgeranyltransferase

Recommended name
phosphoglycerol geranylgeranyltransferase

Synonyms
geranylgeranyl-transferase
geranylgeranyltransferase II
geranylgeranyltransferase, geranylgeranyloxyglycerol phosphate
glycerol phosphate feranylgeranyltransferase
prenyltransferase

CAS registry number
124650-68-6

2 Source Organism

<1> *Methanobacterium thermoautotrophicum* (strain Marburg [1]) [1]

3 Reaction and Specificity

Catalyzed reaction
geranylgeranyl diphosphate + sn-glyceryl phosphate = diphosphate + sn-3-O-
(geranylgeranyl)glyceryl 1-phosphate

Reaction type
alkenyl group transfer

Natural substrates and products
S   Additional information <1> (<1>, probably involved in the biosynthesis
of membrane lipids in Archaebacteriae [1]) [1]
P   ?

Substrates and products

S   geranylgeranyl diphosphate + sn-glyceryl phosphate <1> (Reversibility: ?
<1> [1]) [1]

P   diphosphate + sn-3-O-(geranylgeranyl)glyceryl 1-phosphate

5   Isolation/Preparation/Mutation/Application

Source/tissue

cell extract <1> (<1>, from an early stationary phase culture [1]) [1]

Localization

membrane <1> [1]

References

[1] Zhang, D.-L.; Daniels, L.; Poulter, C.D.: Biosynthesis of archaebacterial membranes. Formation of isoprene ethers by a prenyl transfer reaction. J. Am.
Chem. Soc., **112**, 1264-1265 (1990)

Geranylgeranylglycerol-phosphate geranylgeranyltransferase

2.5.1.42

1 Nomenclature

EC number

2.5.1.42

Systematic name

geranylgeranyl diphosphate:sn-3-O-(geranylgeranyl)glycerol 1-phosphate geranylgeranyltransferase

Recommended name

geranylgeranylglycerol-phosphate geranylgeranyltransferase

Synonyms

geranylgeranyltransferase II

geranylgeranyltransferase, geranylgeranyloxyglycerol phosphate

CAS registry number

124650-68-6

2 Source Organism

<1> *Methanobacterium thermoautotrophicum* (strain Marburg [1]) [1]

3 Reaction and Specificity

Catalyzed reaction

geranylgeranyl diphosphate + sn-3-O-(geranylgeranyl)glycerol 1-phosphate = diphosphate + 2,3-bis-O-(geranylgeranyl)glycerol 1-phosphate

Reaction type

alkenyl group transfer

Natural substrates and products

S Additional information <1> (<1>, probably involved in the biosynthesis of membrane lipids in archaebacteriae [1]) [1]

P ?

Substrates and products

S geranylgeranyl diphosphate + (S)-3-O-(geranylgeranyl)glycerol 1-phosphate <1> (Reversibility: ? <1> [1]) [1]

P diphosphate + 2,3-bis-O-(geranylgeranyl)glycerol 1-phosphate

5 Isolation/Preparation/Mutation/Application

Source/tissue
   cell extract <1> (<1>, from an early stationary stage cell culture [1]) [1]

Localization
   membrane <1> [1]

References

[1] Zhang, D.-L.; Daniels, L.; Poulter, C.D.: Biosynthesis of archaebacterial membranes. Formation of isoprene ethers by a prenyl transfer reaction. J. Am. Chem. Soc., **112**, 1264-1265 (1990)

Nicotianamine synthase

1 Nomenclature

EC number

2.5.1.43

Systematic name

S-adenosyl-L-methionine:S-adenosyl-L-methionine:S-adenosyl-methionine 3-amino-3-carboxypropyltransferase

Recommended name

nicotianamine synthase

Synonyms

AtNAS1

AtNAS2

AtNAS3

HvNAS1

HvNAS2

HvNAS3

HvNAS4

HvNAS6

HvNAS7

NA synthase

NAS

OsNAS1

OsNAS2

OsNAS3

S-adenosyl-L-methionine:S-adenosyl-L-methionine:S-adenosyl-methionine 3-amino-3-carboxypropyltransferase

chloronerva

nicotianamine synthase

synthase, nicotianamine

CAS registry number

161515-44-2

2 Source Organism

<1> *Hordeum vulgare* [1, 2, 4, 5, 7]

<2> *Oryza sativa* (cv Alice [5]) [1, 5]

<3> *Zea mays* [1, 6]

<4> *Nicotiana plumbaginifolia* [3]
<5> *Nicotiana tabacum* [3]
<6> *Nicotiana rustica* [3]
<7> *Nicotiana megalosiphon* [3]
<8> *Nicotiana debneyi* [3]
<9> *Arabidopsis thaliana* [8]

3 Reaction and Specificity

Catalyzed reaction
3 S-adenosyl-L-methionine = 3 S-methyl-5'-thioadenosine + nicotianamine

Reaction type
alkenyl group transfer

Natural substrates and products
S S-adenosyl-L-methionine <1, 4-8> (<1>, enzyme activity is induced at the 3rd day after withholding Fe supply and declines within one day following the supply of Fe^{3+}-epihydroxymugineic acid [1]; <1>, enzyme in the biosynthetic pathway of mugineic acids [1]; <1>, synthesis of nicotianamide, the key precursor of the mugineic acid family phytosiderophores [2]; <1>, strongly induced by Fe-deficiency treatment [2]; <4-8>, enzyme activity is not induced by Fe-deficiency [3]; <1>, key enzyme in the biosynthetic pathway for the mugineic acid family of phytosiderophores [4]; <3>, in agreement with the increased secretion of phytosiderophores with Fe deficiency, ZmNAS1 and ZmNAS2 are positively expressed only in Fe-deficient roots. In contrast ZmNAS3 is expressed under Fe-sufficient conditions, and is negatively regulated by Fe deficiency [6]) (Reversibility: ? <1, 4-8> [1-4]) [1-4, 6]
P ?

Substrates and products
S S-adenosyl-L-methionine <1-8> (Reversibility: ? <1-8> [1-4]) [1-4]
P 5'-S-methyl-5'-thioadenosine + nicotianamine <1-8> [1-4]

Specific activity (U/mg)
Additional information <1> [2]

pH-Optimum
9 <1> [2]

pH-Range
7.5-10 <1> (<1>, pH 7.5: about 20% of maximal activity, pH 10.0: about 50% of maximal activity [1]) [1]

4 Enzyme Structure

Molecular weight
35000 <1> (<1>, gel filtration [4]) [4]
40000-50000 <1> (<1>, gel filtration [2]) [2]

Subunits
? <1> (<1>, x * 30000, SDS-PAGE [2]) [2]

5 Isolation/Preparation/Mutation/Application

Source/tissue
leaf <4-8> (<4-8>, enzyme activity is not induced by Fe-deficiency [3]) [3]
root <1-8> (<1>, primary roots [2]; <1>, Fe-deficient [4]; <1>, enzyme activity is induced at the 3rd day after withholding Fe supply and decline within one day following the supply of Fe^{3+}-epihydroxymugineic acid [1]; <4-8>, enzyme activity is not induced by Fe-deficiency [3]) [1-5]

Purification
<1> [2]

Cloning
<1> (expression in Escherichia coli [4]; expression of a barley HvNAS1 nicotianamine synthase gene promoter-gus fusion gene in transgenic tobacco is induced by Fe-deficiency in root [7]) [4, 5, 7]
<2> (three cDNA clones osnas1, asnas2 and osnas3 from Fe-deficient rotts and a genomic fragment containing both OsNAS1 and OsNAS2 [5]) [5]
<3> (three nicotianamine synthase genes: ZmNAS1, ZmNAS2 and ZmNAS3, fusion to the maltose-binding protein and production of the resulting fusion protein in Escherichia coli, ZmNAS1 and ZmNAS3 show nicotianamine synthase activity, ZmNAs2 does not [6]) [6]
<9> (analysis of upstream region of nicotianamide synthase gene from Arabidopsis thaliana: presence of putative ERE-like sequence [8]) [8]

Application
agriculture <1> (<1>, the nicotianamine synthase gene may be a suitable candidate for making a transgenic plant tolerant to Fe-deficiency [1]) [1]

6 Stability

pH-Stability
4 <1> (<1>, overnight in 50 mM Tris, 1 mM EDTA, 3 mM dithiothreitol, pH 8.7 with HCl, 90% loss of activity [2]) [2]
5.5 <1> (<1>, overnight in 50 mM Tris, 1 mM EDTA, 3 mM dithiothreitol, pH 8.7 with HCl, 80% loss of activity [2]) [2]
7 <1> (<1>, activity is almost completely recovered after overnight exposure in 50 mM Tris, 1 mM EDTA, 3 mM dithiothreitol, pH 8.7 with HCl [2]) [2]

General stability information
<1>, thiol proteases such as papain might digest nicotianamide synthase in crude extracts [1]

Storage stability
<1>, 4°C, in 50 mM Tris, 1 mM EDTA, 3 mM dithiothreitol, pH 8.7 with HCl, stable for about 1 week [2]

References

[1] Higuchi, K.; Kanazawa, K.; Nishizawa, N.-K.; Mori, S.: The role of nicotiana-mine synthase in response to Fe nutrition status in Gramineae. Plant Soil, **178**, 171-177 (1996)

[2] Higuchi, K.; Kanazawa, K.; Nishizawa, N.-K.; Chino, M.; Mori, S.: Purifica-tion and characterization of nicotianamine synthase from Fe-deficient barley roots. Plant Soil, **165**, 173-179 (1994)

[3] Higuchi, K.; Nishizawa, N.-K.; Yamaguchi, H.; Roemheld, V.; Marschner, H.; Mori, S.: Response of nicotianamine synthase activity to Fe-deficiency in tobacco plants as compared with barley. J. Exp. Bot., **46**, 1061-1063 (1995)

[4] Higuchi, K.; Suzuki, K.; Nakanishi, H.; Yamaguchi, H.; Nishizawa, N.-K.; Mori, S.: Cloning of nicotianamine synthase genes, novel genes involved in the biosynthesis of phytosiderophores. Plant Physiol., **119**, 471-479 (1999)

[5] Higuchi, K.; Watanabe, S.; Takahashi, M.; Kawasaki, S.; Nakanishi, H.; Nishi-zawa, N.K.; Mori, S.: Nicotianamine synthase gene expression differs in bar-ley and rice under Fe-deficient conditions. Plant J., **25**, 159-167 (2001)

[6] Mizuno, D.; Higuchi, K.; Sakamoto, T.; Nakanishi, H.; Mori, S.; Nishizawa, N.K.: Three nicotianamine synthase genes isolated from maize are differen-tially regulated by iron nutritional status. Plant Physiol., **132**, 1989-1997 (2003)

[7] Higuchi, K.; Tani, M.; Nakanishi, H.; Yoshiwara, T.; Goto, F.; Nishizawa, N.K.; Mori, S.: The expression of a barley HvNAS$_1$ nicotinamine synthase gene promoter-gus fusion gene in transgenic tobacco is induced by Fe-deficiency in root. Biosci. Biotechnol. Biochem., **65**, 1692-1696 (2001)

[8] Suzuki, K.; Nakanishi, H.; Nishizawa, N.K.; Mori, S.: Analysis of upstream region of nicotinamine synthase gene from Arabidopsis thaliana: presence of putative ERE-like sequence. Biosci. Biotechnol. Biochem., **65**, 2794-2797 (2001)

Homospermidine synthase

1 Nomenclature

EC number
2.5.1.44

Systematic name
putrescine:putrescine 4-aminobutyltransferase (ammonia-forming)

Recommended name
homospermidine synthase

Synonyms
HSPD synthase
HSS
synthase, homospermidine

CAS registry number
76106-84-8

2 Source Organism

<1> *Rhodopseudomonas viridis* (N.C.I.B. 10028 [1]; DSM 134 [3, 5, 6]; expressed in Escherichia coli strain BL21/pHsRvT72.2 [5,6]; accession no. L77975) [1, 3, 5, 6]
<2> *Lathyrus sativus* (grass pea) [2]
<3> *Santalum album* [2]
<4> *Acinetobacter tartarogenes* (ATCC 31105 [4]) [4]

3 Reaction and Specificity

Catalyzed reaction
2 putrescine = sym-homospermidine + NH_3 + H^+ (the reaction occurs in three steps: i. NAD-dependent dehydrogenation of putrescine, ii. transfer of the 4-aminobutylidene group from dehydroputrescine to a second molecule of putrescine, iii. reduction of the imine intermediate to form homospermidine. Hence the overall reaction is transfer of a 4-aminobutyl group. In the presence of putrescine, spermidine can function as a donor of the aminobutyl group, in which case, propane-1,3-diamine is released instead of ammonia. Differs from EC 2.5.1.45, homospermidine synthase, spermidine-specific, which cannot use putrescine as donor of the aminobutyl group)
2. Reaction: putrescine + spermidine = sym-homospermidine + propane 1,3-diamine

Reaction type

aminobutyl group transfer

Natural substrates and products

S putrescine + putrescine <1-4> (<1> one molecule of putrescine is oxidized by NAD$^+$ to form enzyme-bound 4-aminobutyraldehyde. This intermediate reacts with a second molecule of putrescine to form a Schiff base which is reduced by NADH (formed from NAD$^+$ in the first part of the reaction) to give homospermidine [1]; <2,3> as partially purified enzymes have been used in the assay the dependence on spermidine for the homospermidine synthesis may have been overlooked. If these enzymes should proof to be spermidine-dependent like other enzymes from plant sources, they must be classified as EC 2.5.1.45 [2]) (Reversibility: ? <1-4> [1-6]) [1-6]

P sym-homospermidine + NH$_3$ <1-4> [1-6]

Substrates and products

S putrescine + 1,3-diaminopropane <1> (Reversibility: ? <1> [3]) [3, 6]

P homospermidine + spermidine <1> (<1>, low but significant amounts of homologous polyamines in addition to homospermidine [3]) [3, 6]

S putrescine + 1,6 diaminohexane <1> (Reversibility: ? <1> [3, 6]) [3, 6]

P homospermidine + N-(4-aminobutyl)-1,6-diaminohexane <1> (<1>, low but significant amounts of homologous polyamines in addition to homospermidine [3]) [3, 6]

S putrescine + 1,7-diaminoheptane <1> [3, 6]

P homospermidine + N-(4-aminobutyl)-1,7-diaminoheptane <1> (<1>, low but significant amounts of homologous polyamines in addition to homospermidine [3]) [3, 6]

S putrescine + cadaverine <1> (Reversibility: ? <1> [3, 6]) [3, 6]

P homospermidine + N-(4-aminobutyl)-1,5-diaminopentane <1> (<1>, low but significant amounts of homologous polyamines in addition to homospermidine [3]) [3, 6]

S putrescine + putrescine <1-4> (<1> one molecule of putrescine is oxidized by NAD$^+$ to form enzyme-bound 4-aminobutyraldehyde. This intermediate reacts with a second molecule of putrescine to form a Schiff base which is reduced by NADH (formed from NAD$^+$ in the first part of the reaction) to give homospermidine [1]; <2,3> as partially purified enzymes have been used in the assay the dependence on spermidine for the homospermidine synthesis may have been overlooked. If these enzymes should proof to be spermidine-dependent like other enzymes from plant sources, they must be classified as EC 2.5.1.45 [2]) (Reversibility: ? <1-4> [1-6]) [1-6]

P sym-homospermidine + NH$_3$ <1-4> [1-6]

S spermidine <1> (Reversibility: ? <1> [3, 6]) [3, 6]

P homospermidine + putrescine + 1,3-diaminopropane <1> [3, 6]

S spermidine + 1,6-diaminohexane <1> (Reversibility: ? <1> [3, 6]) [3, 6]

P homospermidine + N-(4-aminobutyl)-1,6-diaminohexane <1> [3, 6]

S spermidine + cadaverine <1> (Reversibility: ? <1> [3, 6]) [3, 6]

P   homospermidine + N-(4-aminobutyl)-1,5-diaminopentane <1> [3, 6]
S   spermidine + putrescine <1> (Reversibility: ? <1> [3, 6]) [3, 6]
P   homospermidine + 1,3-diaminopropane <1> [3, 6]

Inhibitors

1,3-diaminopropane <1, 2> (<1>, strong competitive inhibitor, K_i: 0.002 mM [1]) [1, 2]
1,5-diaminopentane <1> (<1>, weak inhibition [1]) [1]
N-ethylmaleimide <2, 4> (<2>, inhibition suggests, that the enzyme is of a thiol nature [2]; <4> preincubation, 1 mM, 30 min prior to assay causes 91% inhibition [4]) [2, 4]
NADH <1> (<1>, competitive inhibitor, K_i: 0.0015 mM [1]; <4>, K_i: 0.006 mM [4]) [1, 4]
cadaverine <2> [2]
iodoacetamide <4> (<4> preincubation, 1 mM, 30 min prior to assay causes 50% inhibition [4]) [4]
spermidine <2> [2]
Additional information <1> (<1> 4-aminobutyraldehyde, postulated intermediate, no inhibition [1]) [1]

Cofactors/prosthetic groups

NAD^+ <1-4> (<1>, cannot be replaced by $NADP^+$, NADPH, NADH [1]; <1>, required in catalytic amounts with a K_m-value of 0.0025 mM [1]; <2>, absolute requirement [2]; <1>, NAD^+ seems to function as hydride acceptor in the first part of the reaction and subsequently as hydride donor in the second part [3]) [1-6]
$NADP^+$ <1> (<1>, 68% of the activity with NAD^+ [6]) [6]

Metals, ions

K^+ <1, 4> (<1>, optimal activity with 40 mM, Na^+ and Rb^+ are less effective [1]; <4> required for full activity, optimal activity at 50 mM [4]; <1> activates, optimal concentration at 50 mM [6]) [1, 4, 6]
Na^+ <1> (<1>, less effective than K^+ in activation [1]) [1]
Rb^+ <1> (<1>, less effective than K^+ in activation [1]) [1]

Specific activity (U/mg)

0.00025 <2> [2]
0.0038 <3> [2]
0.6-1.2 <1> [3]
4.97 <1> [5]
8.41 <4> [4]

K_m-Value (mM)

0.001 <1> (NAD^+) [6]
0.0025 <1> (NAD^+) [1]
0.018 <1> (NAD^+) [5]
0.2 <1> (putrescine) [1]
0.26 <1> (putrescine) [3]
0.27 <1> (putrescine) [6]

0.28 <4> (putrescine) [4]
1.5 <1> (spermidine, in presence of 1 mM putrescine, <1>) [3, 6]
1.7 <1> (putrescine, <1> in presence of 1 mM spermidine [3]) [3, 6]
3 <2> (putrescine) [2]

K_i-Value (mM)
0.95 <3> (homospermidine) [4]
2.2 <3> (spermine) [4]

pH-Optimum
8.4 <2> (<2> in Tris/HCl buffer [2]) [2]
8.8-9 <1> (<1> in 50 mM potassium phosphate or Bis-Tris-propane buffers [6]) [6]

Temperature optimum (°C)
37 <2> (<2>, reaction linear up to 60 min and is proportional to the amount of enzyme protein (0.1-0.5 mg) [2]) [2]
45 <1> (<1> increasing activity up to 45°C, sharp decrease at higher temperatures [6]) [6]

4 Enzyme Structure

Molecular weight
73000 <1> (<1> gel filtration) [1]
75000 <2> (<2> gel filtration) [2]
100000 <1> (<1> gel filtration) [5]
102000 <4> (<4> gel filtration [4]) [4]

Subunits
dimer <1, 4> (<1> 2 * 52000, SDS-PAGE [5]; <1> 2 * 52600, calculated from the amino acid sequence [5]; <4> 2 * 52000, SDS-PAGE [4]) [4, 5]

5 Isolation/Preparation/Mutation/Application

Source/tissue
culture medium <1> (<1>, crude extract after ultrasonication and centrifugation [1]) [1, 3-6]
leaf <3> [2]
seedling <2> [2]

Purification
<1> (200fold [1]; partial [3]; 224fold [5]; overexpressed Escherichia coli strain BL21 [5]) [1, 3, 5]
<2> (partial, 100fold) [2]
<4> [4]

Cloning

<1> (overexpressed in Escherichia coli BL21, which originally does not possess HSS activity. 40-50% of the soluble protein in crude extracts are detected as homospermidine synthase) [5]

6 Stability

Temperature stability

-18 <1> (<1>, storage as acetone dry-powder, stable for several months [6]) [6]

4 <1, 2, 4> (<2>, stable for at least 2 weeks in the presence of 1 mM NAD$^+$ [2]; <1>, stable for at least several weeks [3]) [2-4]

Storage stability

<1>, -18°C, storage as acetone dry-powder, without significant loss of activity [3]

<4>, 1°C, 20 mM, Tris/HCl, pH 7.5, 10 mM 2-mercaptoethanol, 0.5 mM putrescine, 0.02% NaN$_3$ [4]

References

[1] Tait, G.H.: The formation of homospermidine by an enzyme from Rhodopseudomonas viridis. Biochem. Soc. Trans., 7, 199-200 (1979)

[2] Srivenugopal, K.S.; Adiga, P.R.: Enzymic synthesis of sym-homospermidine in Lathyrus sativus (grass pea) seedlings. Biochem. J., 190, 461-464 (1980)

[3] Böttcher, F.; Ober, D.; Hartmann, T.: Biosynthesis of pyrrolizidine alkaloids: putrescine and spermidine are essential substrates of enzymic homospermidine formation. Can. J. Chem., 72, 80-84 (1994)

[4] Yamamoto, S.; Nagata, S.; Kusaba, K.: Purification and characterization of homospermidine synthase in Acinetobacter tartarogenes ATCC 31105. J. Biochem., 114, 45-49 (1993)

[5] Tholl, D.; Ober, D.; Martin, W.; Kellermann, J.; Hartmann, T.: Purification, molecular cloning and expression in Escherichia coli of homospermidine synthase from Rhodopseudomonas viridis. Eur. J. Biochem., 240, 373-379 (1996)

[6] Ober, D.; Tholl, D.; Martin, W.; Hartmann, T.: Homospermidine synthase of Rhodopseudomonas viridis: substrate specificity and effects of the heterologously expressed enzyme of polyamine metabolism of Escherichia coli. J. Gen. Appl. Microbiol., 42, 411-420 (1996)

Homospermidine synthase (spermidine-specific)

2.5.1.45

1 Nomenclature

EC number

2.5.1.45

Systematic name

spermidine:putrescine 4-aminobutyltransferase (propane-1,3-diamine-forming)

Recommended name

homospermidine synthase (spermidine-specific)

Synonyms

HSS

synthase, homospermidine

synthase, homospermidine (Senecio vernalis root gene HSS1)

CAS registry number

259168-77-9

76106-84-8

2 Source Organism

<1> *Eupatorium cannabinum* [1, 4]

<2> *Senecio vernalis* (SwissProt-ID: Q9SC13) [2]

<3> *Senecio vulgaris* (SwissProt-ID: Q9M4B0) [3]

<4> *Eupatorium pauciflorum* [4]

<5> *Senecio jacobaea* [4]

<6> *Senecio vulgaris* [5, 6]

<7> *Senecio vernalis* [7, 8]

3 Reaction and Specificity

Catalyzed reaction

spermidine + putrescine = sym-homospermidine + propane-1,3-diamine (the reaction of this enzyme occurs in three steps: i. NAD-dependent dehydrogenation of spermidine, ii. transfer of the 4-aminobutylidene group from dehydrospermidine to putrescine, iii. reduction of the imine intermediate to form homospermidine. Hence the overall reaction is transfer of a 4-aminobutyl group. This enzyme is more specific than EC 2.5.1.44, homospermidine synthase, which is found in bacteria, as it cannot use putrescine as donor of the 4-aminobutyl group. Forms part of the biosynthetic pathway of the poisonous pyrrolizidine alkaloids of the ragworts, Senecio)

Reaction type
aminobutyl group transfer

Natural substrates and products
S putescine + spermine <1-7> (<3> the plant homospermidine synthase is essentially dependent on spermidine as aminobutyl donor and is unable to synthesize homospermidine from two molecules of putrescine [3]) [1-8]
P sym-homospermine + propane-1,3-diamine

Substrates and products
S putrescine + spermidine <1-7> (<3> the plant homospermidine synthase is essentially dependent on spermidine as aminobutyl donor and is unable to synthesize homospermidine from two molecules of putrescine [3]) (Reversibility: ? <1-7> [1-8]) [1-8]
P N-(4-aminobutyl)butane-1,4-diamine + propane-1,3-diamine <1-7> [1-8]

Inhibitors
1,3-diaminopropane <1> (<1> competitive [4]) [4]
cadaverine <1> (<1> mixed type inhibition [4]) [4]
homospermidine <1> (<1> competitive [4]) [4]
spermidine <1> (<1> competitive [4]) [4]
spermine <1> (<1> mixed type inhibition [4]) [4]

Cofactors/prosthetic groups
NAD^+ <1-3> (<1> the NADH formed in the first part of the reaction, i.e. oxidative deamination of putrescine, is cosubstrate of the second part of the reaction, i.e. hydrogenation of an intermolecular Schiff's base between the semialdehyde in the reaction and the second putrescine unit [4]) [1-4]
NADH <1-3> [1-3]

Turnover number (min^{-1})
0.192 <7> (putrescine, <7>, + spermidine [7]) [7]

Specific activity (U/mg)
0.0006 <1> [1]
0.047 <1> [4]

K$_m$-Value (mM)
0.003 <1> (NAD^+, <1> NAD^+ functions as an enzyme-bound redoxsystem that is reduced in the first and reoxidized in the second part of the overall reaction. It functions as a coenzyme and not a cosubstrate [4]) [4]
0.012 <1> (spermidine, <1> at 0.1 mM putrescine [1]) [1]
0.0135 <1> (putrescine, <1> in the absence of spermidine [4]) [4]
0.015 <1> (putrescine, <1> at 0.1 mM spermidine [1]) [1]
Additional information <7> [7]

K$_i$-Value (mM)
  0.0022 <1> (spermine) [4]
  0.0063 <1> (1,3-diaminopropane) [4]
  0.058 <1> (cadaverine) [4]
  0.094 <1> (spermidine) [4]
  0.95 <1> (homospermidine) [4]

pH-Optimum
  9.3 <7> [7]

pH-Range
  8.7-9.9 <7> (<7>, about 55% of maximal activity at pH 8.7 and at pH 9.9 [7])
  [7]

4 Enzyme Structure

Molecular weight
  174000 <2> (<2> gel filtration [2]) [2]

Subunits
  ? <7> (<7>, x * 41000, SDS-PAGE [7]) [7]
  tetramer <2> (<2> 4 * 44500, SDS-PAGE [2]; <2> 4 * 40700, calculation from
  sequence of cDNA [2]; <3> 4 * 45000, SDS-PAGE [3]; <3> 4 * 40740, calculation from sequence of cDNA [3]) [2, 3]

5 Isolation/Preparation/Mutation/Application

Source/tissue
  root <7> (<7>, expressed at high levels independently of the root age [8]) [8]
  root culture <1-6> [1-6]
  Additional information <7> (<7>, not expressed in the aerial parts of the
  plant like buds, flower heads, leaves, or stems [8]) [8]

Localization
  cytosol <7> [8]
  soluble <7> [8]

Purification
  <1> [1]
  <2> [2]

Cloning
  <2> [2]
  <3> (expressed in Escherichia coli [3]) [3]

6 Stability

General stability information
<1>, increased instablity with higher purification grade [4]

Storage stability
<1>, 4°C, crude extract, 0.5-1 mM spermidine, 0.5-1 mM NAD$^+$, 1 week, 30% loss of activity [4]

References

[1] Böttcher, F.; Ober, D.; Hartmann, T.: Biosynthesis of pyrrolizidine alkaloids: putrescine and spermidine are essential substrates of enzymatic homospermidine formation. Can. J. Chem., 72, 80-85 (1994)

[2] Ober, D.; Hartmann, T.: Homospermidine synthase, the first pathway-specific enzyme of pyrrolizidine alkaloid biosynthesis, evolved from deoxyhypusine synthase. Proc. Natl. Acad. Sci. USA, 96, 14777-14782 (1999)

[3] Ober, D.; Harms, R.; Hartmann, T.: Cloning and expression of homospermidine synthase from Senecio vulgaris: a revision. Phytochemistry, 55, 311-316 (2000)

[4] Böttcher, F.; Adolph, R.-D.; Hartmann, T.: Homospermidine synthase, the first pathway-specific enzyme in pyrrolizidine alkaloid biosynthesis. Phytochemistry, 32, 679-689 (1993)

[5] Graser, G.; Hartmann, T.: Biosynthetic incorporation of the aminobutyl group of spermidine into pyrrolizidine alkaloids. Phytochemistry, 45, 1591-1595 (1997)

[6] Graser, G.; Witte, L.; Robins, D.J.; Hartmann, T.: Incorporation of chirally deuterated putrescines into pyrrolizidine alkaloids: a reinvestigation. Phytochemistry, 47, 1017-1024 (1998)

[7] Ober, D.; Harms, R.; Witte, L.; Hartmann, T.: Molecular evolution by change of function. Alkaloid-specific homospermidine synthase retains all properties of deoxyhypusine synthase except binding the eIF5A precursor protein. J. Biol. Chem., 278, 12805-12812 (2003)

[8] Moll, S.; Anke, S.; Kahmann, U.; Haensch, R.; Hartmann, T.; Ober, D.: Cell-specific expression of homospermidine synthase, the entry enzyme of the pyrrolizidine alkaloid pathway in Senecio vernalis, in comparison with its ancestor, deoxyhypusine synthase. Plant Physiol., 130, 47-57 (2002)

Deoxyhypusine synthase

1 Nomenclature

EC number

2.5.1.46

Systematic name

spermidine:eIF5A-lysine 4-aminobutyltransferase (propane-1,3-diamine-forming)

Recommended name

deoxyhypusine synthase

Synonyms

DHS

deoxyhypusine synthase (Caulobacter crescentus gene CC0359)

deoxyhypusine synthase (Halobacterium strain NRC-1 gene dhs)

deoxyhypusine synthase (Nicotiana tabacum gene DHS1)

deoxyhypusine synthase (Senecio vernalis gene DHS1)

deoxyhypusine synthase (human clone 30649 gene DHPS subunit reduced)

synthase, deoxyhypusine

CAS registry number

127069-31-2 (deoxyhypusine synthase)

171041-87-5 (human clone 30649 gene DHPS subunit reduced)

253180-41-5 (Nicotiana tabacum gene DHS1)

259168-76-8 (Senecio vernalis gene DHS1)

302492-11-1 (Halobacterium strain NRC-1 gene dhs)

332963-67-4 (Caulobacter crescentus gene CC0359)

2 Source Organism

<1> *Rattus norvegicus* [1, 18, 19]

<2> *Homo sapiens* [2, 3, 7, 8, 11, 12, 13, 14, 16, 22, 24, 25, 26, 28]

<3> *Nicotiana tabacum* [4]

<4> *Senecio vernalis* [5, 23]

<5> *Saccharomyces cerevisiae* [6, 9, 15, 17, 20, 21]

<6> *Neurospora crassa* [3, 9]

<7> *Lycopersicon esculentum* [10]

<8> *yeast* [24]

<9> *Cricetulus griseus* [24]

<10> *Gallus gallus* [24]

<11> *Senecio vernalis* (SwissProt-ID: Q9SC14) [27]
<12> *Nicotiana tabacum* (SwissProt-ID: Q9SC80) [27]

3 Reaction and Specificity

Catalyzed reaction
[eIF5A-precursor]-lysine + spermidine = [eIF5A-precursor]-deoxyhypusine
+ propane-1,3-diamine (the eukaryotic initiation factor eIF5A contains a hypusine residue that is essential for activity. This enzyme catalyses the first
reaction of hypusine formation from one specific lysine residue of the eIF5A
precursor, the second reaction being catalysed by EC 1.14.99.29, deoxyhypusine monooxygenase. The reaction of this enzyme occurs in four steps: i.
NAD-dependent dehydrogenation of spermidine, ii. formation of an en-
zyme-imine intermediate by transfer of the 4-aminobutylidene group from
dehydrospermidine to the active site lysine residue, Lys329 for the human
enzyme, iii. transfer of the same 4-aminobutylidene group from the enzyme
intermediate to the eIF5A precursor, iv. reduction of the eIF5A-imine intermediate to form a deoxyhypusine residue. Hence the overall reaction is transfer of a 4-aminobutyl group. For the plant enzyme, homospermidine can substitute for spermidine and putrescine can substitute for the lysine residue of
the eIF5A precursor)

Reaction type
aminobutyl group transfer

Natural substrates and products
S [eIF5A-precursor]-lysine + spermidine <1-6, 8, 9, 10> (<2>, first step in
the biosynthesis of hypusine [2, 28]; <4>, posttranslational activation of
the initiation factor 5A [23]; <8, 9, 10>, the enzyme catalyzes the first
reaction in a two-step conversion of inactive eIF-5A to its active form
[24]; <2>, posttranslational activation of eIF-5A which has a specialized
role in proliferation and in serum starvation-induced apoptosis in endothelial cells [25]) [1, 2, 4-9, 23, 24, 25, 28]
P [eIF5A-precursor]-deoxyhypusine + propane-1,3-diamine

Substrates and products
S 1-ethylspermidine + [eIF5A-precursor]-lysine <2> (Reversibility: ? <2>
[26]) [26]
P [eIF5A-precursor]-deoxyhypusine + ? <2> [26]
S 1-ethylspermidine + putrescine <2> (Reversibility: ? <2> [26]) [26]
P homospermidine + ? <2> [26]
S 1-methylspermidine + [eIF5A-precursor]-lysine <2> (Reversibility: ? <2>
[26]) [26]
P [eIF5A-precursor]-deoxyhypusine + ? <2> [26]
S 1-methylspermidine + putrescine <2> (Reversibility: ? <2> [26]) [26]
P homospermidine + ? <2> [26]

S 8-ethylspermidine + [eIF5A-precursor]-lysine <2> (Reversibility: ? <2> [26]) [26]

P [eIF5A-precursor]-ethyldeoxyhypusine + ? <2> [26]

S 8-ethylspermidine + putrescine <2> (Reversibility: ? <2> [26]) [26]

P ethylhomospermidine + ? <2> [26]

S 8-methylspermidine + [eIF5A-precursor]-lysine <2> (Reversibility: ? <2> [26]) [26]

P [eIF5A-precursor]-methyldeoxyhypusine + ? <2> [26]

S 8-methylspermidine + putrescine <2> (Reversibility: ? <2> [26]) [26]

P methylhomospermidine + ? <2> [26]

S N-(3-aminopropyl)-1,4-diamino-cis-but-2-ene + putrescine <2> (Reversibility: ? <2> [26]) [26]

P cis-deoxyhypusine + ? <2> [26]

S N-(3-aminopropyl)-1,4-diamino-cis-but-2-ene +[eIF5A-precursor]-lysine <2> (Reversibility: ? <2> [26]) [26]

P [eIF5A-precursor]-cis-deoxyhypusine + ? <2> [26]

S N-(3-aminopropyl)-1,4-diamino-trans-but-2-ene + [eIF5A-precursor]-lysine <2> (Reversibility: ? <2> [26]) [26]

P [eIF5A-precursor]-cis-deoxyhypusine + ? <2> [26]

S N-(3-aminopropyl)-1,4-diamino-trans-but-2-ene + putrescine <2> (Reversibility: ? <2> [26]) [26]

P trans-deoxyhypusine + ? <2> [26]

S [eIF5A-precursor]-lysine + homospermidine <3> (Reversibility: ? <3> [4]) [4]

P [eIF5A-precursor]-deoxyhypusine + putrescine <3> [4]

S [eIF5A-precursor]-lysine + spermidine <1-10> (<2>, reversal of reaction in presence of NAD$^+$ [26]) (Reversibility: r <2> [26]; ? <1-10> [1, 2, 4-25, 27, 28]) [1, 2, 4-26]

P [eIF5A-precursor]-deoxyhypusine + propane-1,3-diamine <1-7> (i.e. [eIF5A-precursor]-N^6-(4-aminobutyl)lysine) [1, 2, 4-28]

S aminopropylcadaverine + [eIF5A-precursor]-lysine <2> (Reversibility: ? <2> [26]) [26]

P [eIF5A-precursor]-homodeoxyhypusine + ? <2> [26]

S caldine + putrescine <2> (Reversibility: ? <2> [26]) [26]

P spermidine + ? <2> [26]

S homospermidine + [eIF5A-precursor]-lysine <2> (Reversibility: ? <2> [26]) [26]

P [eIF5A-precursor]-deoxyhypusine + ? <2> [26]

S homospermidine + putrescine <2, 4> (Reversibility: ? <2,4> [23,26]) [23, 26]

P homospermidine + ? <2> [26]

S spermidine <1, 2, 5> (<1, 5>, reaction in absence of eIF5A-precursor [1, 6, 17, 18]) (Reversibility: ? <1, 2, 5> [1, 6, 17, 18, 28]) [1, 6, 17, 18, 28]

P 1,3-diaminopropane + Δ^1-pyrroline <1, 2, 5> [1, 6, 17, 18, 28]

S spermidine + putrescine <2, 3, 4> (Reversibility: ? <2, 3, 4> [4, 5, 23, 26]) [4, 5, 23, 26]

P homospermidine + ? <2, 3, 4> [4, 5, 23, 26]

Inhibitors

1,3-diaminopropane <1, 5> [19, 20]

1,7-diaminoheptane <1, 2, 5> (<2>, inhibits generation of spermidine from [eIF5A-precursor]-deoxyhypusine [26]) [19, 20, 26]

1,8-diaminooctane <1, 5> [19, 20]

1,9-diaminononane <5> [20]

1-(3-aminopropyl)-4-aminomethylpiperidine <1> (<1> 1 mM, less than 50% inhibition [19]) [19]

1-amino-7-guanidinoheptane <2> (<2> competitive, inhibits binding of spermidine to the enzyme [2]) [2]

1-aminooctane <1> (<1> 1 mM, less than 50% inhibition [19]) [19]

1-methylspermidine <1> (<1> 0.184 mM, 50% inhibition [19]) [19]

5,5'-dimethylspermidine <1> (<1>, 1 mM, less than 50% inhibition [19]) [19]

6,6'-difluorospermidine <1, 2> (<1> 0.014 mM, 50% inhibition [19]; <2>, inhibition of deoxyhypusine synthesis [26]) [19, 26]

6-fluorospermidine <1> (<1> 0.048 mM, 50% inhibition [19]) [19]

7,7-difluorospermidine <1> (<1> 1 mM, less than 50% inhibition [19]) [19]

Cu^{2+} <5> [20]

FAD <5> [20]

FMN <5> [20]

Fe^{2+} <5> [20]

Fe^{3+} <5> [20]

Mn^{2+} <5> (<5> weak [20]) [20]

N,N'-bis-benzyldiaminooctane <1> (<1> 1 mM, less than 50% inhibition [19]) [19]

N-(2-cyanoethyl)-1,3-diaminopropane <1> (<1> 1 mM, less than 50% inhibition [19]) [19]

N-(3-aminopropyl)-1,3-diaminopropane <1> (<1> 1 mM, less than 50% inhibition [19]) [19]

N-(3-aminopropyl)-1,4-diamino-cis-but-2-ene <2> [26]

N-(3-aminopropyl)-1,4-diamino-trans-but-2-ene <2> [26]

N-(3-aminopropyl)-cadaverine <1> [19]

N-(3-aminopropyl)-cis-1,4-diaminocyclohexane <1> (<1> 1 mM, less than 50% inhibition [19]) [19]

N-(3-aminopropyl)-trans-1,4-diaminocyclohexane <1> (<1> 1 mM, less than 50% inhibition [19]) [19]

N-(3-cyanopropyl)-1,3-diaminopropane <1> (<1> 1 mM, less than 50% inhibition [19]) [19]

N-butyl-1,3-diaminopropane <1> (<1> 1 mM, less than 50% inhibition [19]) [19]

N^1,N^3-bis-guanyl-1,3-diaminopropane <1> (<1> 1 mM, less than 50% inhibition [19]) [19]

N^1,N^3-bis-tert-butyloxycarbonylspermidine <1> (<1> 1 mM, less than 50% inhibition [19]) [19]

N^1,N^6-bis-guanyl-1,6-diaminohexane <1> [19]

N^1,N^7-bis-allylcaldine <1> (<1> 1 mM, less than 50% inhibition [19]) [19]

N^1,N^7-bis-benzylcaldine <1> (<1> 1 mM, less than 50% inhibition [19]) [19]

N^1,N^7-bis-dimethylcaldine <1> (<1> 1 mM, less than 50% inhibition [19])
[19]
N^1,N^7-bis-guanyl-1,7-diaminoheptane <1> [19]
N^1,N^7-bis-guanylcaldine <1> [19]
N^1,N^8-bis-guanyl-1,8-diaminooctane <1> [19]
N^1-acetylspermidine <1> (<1> 1 mM, less than 50% inhibition [19]) [19]
N^1-ethylspermidine <1> (<1> 1 mM, less than 50% inhibition [19]) [19]
N^1-ethylspermidine <2> (<2>, inhibits generation of spermidine from
[eIF5A-precursor]-deoxyhypusine [26]) [26]
N^1-guanyl-1,3-diaminopropane <1> (<1> 1 mM, less than 50% inhibition
[19]) [19]
N^1-guanyl-1,7-diaminoheptane <1, 2> (<2>, inhibits generation of spermidine from [eIF5A-precursor]-deoxyhypusine [26]) [19, 25, 26]
N^1-guanyl-1,8-diaminooctane <1> [19]
N^1-guanylcaldine <1> [19]
N^1-guanylspermine <1> [19]
N^1-methylspermidine <2> (<2>, inhibits generation of spermidine from
[eIF5A-precursor]-deoxyhypusine [26]) [26]
N^3-ethylspermidine <1> (<1> 1 mM, less than 50% inhibition [19]) [19]
N^4-acetylspermidine <1> (<1> 1 mM, less than 50% inhibition [19]) [19]
N^4-benzoylspermidine <1> (<1> 1 mM, less than 50% inhibition [19]) [19]
N^4-benzylspermidine <1> (<1> 1 mM, less than 50% inhibition [19]) [19]
N^4-bromoacetylspermidine <1> (<1> 1 mM, less than 50% inhibition [19])
[19]
N^4-methylcaldine <1> (<1> 1 mM, less than 50% inhibition [19]) [19]
NADH <5> [20]
NEM <6> [9]
Zn^{2+} <5> (<5> weak [20]) [20]
aminopropylcadaverine <2> (<2>, inhibition of deoxyhypusine synthesis
[26]) [26]
caldine <1, 2> (<2>, inhibition of deoxyhypusine synthesis [26]) [19, 26]
guazatine <1> [19]
hirudonine <1> [19]
homospermidine <2> (<2>, inhibition of deoxyhypusine synthesis [26]) [26]
iodoacetate <6> [9]
putrescine <1> (<1> 1 mM, less than 50% inhibition [19]) [19]
spermidine <1, 2, 5> (<2>, inhibits generation of spermidine from [eIF5A-precursor]-deoxyhypusine [26]) [19, 20, 26]

Cofactors/prosthetic groups

NAD^+ <1-7> (<2> the NADH generated in the first step of the reaction remains enzyme-associated during the reaction and the hydride ion generated by the oxidation of spermidine is preserved for the reduction of the eIF5A-imine intermediate [7]; <2> one enzyme tetramer can bind up to four each of NAD^+ and spermidine, the binding of spermidine being dependent on NAD^+
[11]; <2> induces spermidine binding [12]; <2> not dependent on NAD^+ for

binding of eIF5A [13]; <1> specific for NAD$^+$ [18]; <2>, strict requirement [28]) [1, 2, 4-22, 26, 28]

NADH <1-7> (<2> induces spermidine binding [12]) [1, 2, 4-12]

Turnover number (min^{-1})

0.132 <4> (eIF5A-precursor, <4>, + spermidine [23]) [23]

Specific activity (U/mg)

0.000137 <1> [1]

0.00325 <2> [22]

0.0034 <3> (<3> with donor homospermidine [4]) [4]

0.0091 <3> (<3> with donor homospermidine [4]) [4]

0.0101 <4> (<4> with acceptor eIF5A-precursor [5]) [5]

0.0131 <5> [17]

0.015 <2> [28]

0.019 <2> [8]

0.0211 <3> (<3> with donor spermidine [4]) [4]

0.0442 <4> (<4> with acceptor putrescine [5]) [5]

K$_m$-Value (mM)

0.00008 <1> (eIF5A-precursor) [1]

0.0006 <2> (eIF5A-precursor) [28]

0.0006 <2> (eIF5A-precursor) [8]

0.001 <1> (spermidine) [1]

0.0013 <4> (eIF5A-precursor) [23]

0.0015 <2> (putrescine) [26]

0.0048 <2> (NAD$^+$) [28]

0.0048 <2> (NAD$^+$) [8]

0.0066 <2> ([eIF5A-precursor]-deoxyhypusine, <2>, pH 7.5 [26]) [26]

0.0072 <2> (spermidine) [28]

0.0072 <2> (spermidine) [8]

0.00726 <2> (spermidine, <2>, reaction with eIF5A-precursor [26]) [26]

0.0083 <5> (spermidine, <5> yeast recombinant eIF-5Aa [17]) [17]

0.0086 <2> ([eIF5A-precursor]-deoxyhypusine, <2>, pH 9.5 [26]) [26]

0.0125 <5> (spermidine, <5> yeast recombinant eIF-5Ab [17]) [17]

0.0193 <2> (homospermine) [26]

0.028 <2> (spermidine, <2>, reaction with putrescine [26]) [26]

0.03 <1> (NAD$^+$) [1]

0.0954 <2> (1,3-diaminopropane, <2>, pH 7.5 [26]) [26]

0.1058 <2> (1,3-diaminopropane, <2>, pH 9.5 [26]) [26]

0.203 <2> (putrescine) [26]

0.56 <5> (NAD$^+$, <5> yeast recombinant eIF-5Aa [17]) [17]

0.81 <5> (NAD$^+$, <5> yeast recombinant eIF-5Ab [17]) [17]

1.12 <2> (eIF5A-precursor) [26]

Additional information <4> [23]

K$_i$-Value (mM)

0.0000097 <1> (N^1-guanyl-1,7-diaminoheptane) [19]

0.00024 <1> (N^1-guanyl-1,8-diaminooctane) [19]

0.00033 <1> (N^1-guanylspermine) [19]
0.00074 <1> (N^1-guanylcaldine) [19]
0.0009 <5> (1,7-diaminoheptane) [20]
0.0017 <1> (N^1,N^7-bis-guanyl-1,7-diaminoheptane) [19]
0.00191 <5> (1,8-diaminooctane) [20]
0.00489 <1> (hirudonine) [19]
0.00565 <1> (N^1,N^8-bis-guanyl-1,8-diaminooctane) [19]
0.00649 <5> (spermidine) [20]
0.00752 <5> (1,9-diaminononane) [20]
0.012 <1> (guazatine) [19]
0.0129 <5> (1,3-diaminopropane) [20]
0.035 <1> (N^1,N^6-bis-guanyl-1,6-diaminohexane) [19]
0.1545 <1> (N^1,N^7-bis-guanylcaldine) [19]
0.85 <5> (NADH) [20]
1.28 <5> (FMN) [20]
2.27 <5> (FAD) [20]

pH-Optimum

9 <2> [22]
9-9.5 <2> (<2> optimum for spermidine-binding [12]) [12]
9.3 <4> (<4>, glycine-NaOH buffer [23]) [23]
9.5 <5> [20]
9.5-9.6 <1> [18]

pH-Range

8.5-11 <5> (<5>, pH 8.5: about 45% of maximal activity, pH 11.0: about 50% of maximal activity [20]) [20]
8.8-9.9 <4> (<4>, pH 8.8: about 35% of maximal activity, pH 9.9: about 60% of maximal activity [23]) [23]

Temperature optimum (°C)

30 <5> [20]

Temperature range (°C)

20-37 <5> (<5>, 20°C: about 75% of maximal activity, 37°C: about 25% of maximal activity [20]) [20]

4 Enzyme Structure

Molecular weight

144000-180000 <1> (<1> gel filtration, sedimentation equilibrium centrifugation [18]) [18]
150000 <2> (<2> gel filtration [22]) [22]
165000-172000 <5> (<5> gel filtration sedimentation equilibrium centrifugation [17,20]) [17, 20]
180000 <6> (<6> gel filtration [9]) [9]
180000-190000 <1> (<1> gel filtration [1]) [1]
190000 <3> (<3> gel filtration [4]) [4]

Subunits

? <2, 4> (<2,4>, x * 41000, SDS-PAGE [23,28]) [23, 28]
tetramer <1, 2, 3, 5, 6> (<3> 4 * 42074, calculation from amino acid sequence [4]; <6> 4 * 40000, SDS-PAGE [9]; <5> 4 * 43000, SDS-PAGE [17,20]; <5> 4* 42998, MALDI mass spectrometry [17]; <1> 4 * 40800, MALDI mass spectrometry [18]; <2> 4 * 41000, SDS-PAGE [22]; <2>, four identical subunits arranged in 222 symmetry [14]) [4, 9, 14, 17, 18, 20, 22]

Posttranslational modification

phosphoprotein <8, 9, 10> (<8,9,10>, phosphorylated by protein kinase C in vitro [24]) [24]

5 Isolation/Preparation/Mutation/Application

Source/tissue

3T3 cell <2> [24]
CHO-K1 cell <9> [24]
HeLa cell <2> [22]
embryo <10> [24]
flower <11> (<11,12>, low level [27]) [27]
flower bud <11> [27]
fruit <12> [27]
leaf <11, 12> (<11>, young, low level [27]) [27]
ovary <12> [27]
petal <12> [27]
root <11, 12> (<11>, high activity [27]) [27]
shoot <11, 12> [27]
shoot tip <4> [5]
stamen <12> [27]
stem <11> (<11>, young or old, low level [27]) [27]
testis <1> [1, 18, 19]
umbilical vein endothelium <2> [25]

Localization

cytoplasm <1, 5, 6> [1, 9, 17]

Purification

<1> [1]
<2> [8, 11]
<2> (recombinant enzyme [28]) [28]
<4> (recombinant enzyme [23]) [23]
<5> [17]

Crystallization

<2> (vapor diffusion in hanging drops, crystal structure of the enzyme-NAD complex at 2.2 A resolution [14]) [14]

Cloning

<2> (expression in Escherichia coli, the deduced amino acid sequence shows a high degree of identity to that of yeast enzyme and to the known sequences of tryptic peptides from the rat and Neurospora enzymes [28]) [28]
<3> [4]
<4> [5]
<5> [15]
<7> [10]

Engineering

D238A <2> (<2> NAD-site mutant, less than 5% activity of wild-type [11]) [11]
D243A <2> (<2> spermidine-site mutant, no spermidine-binding [11]) [11]
D313A <2> (<2> unable to bind eIF5A-precursor, no synthesis of deoxyhypusine [11]) [11]
D316A <2> (<2> spermidine-site mutant, no spermidine-binding [11]) [11]
D342A <2> (<2> NAD-site mutant, less than 5% activity of wild-type [11]) [11]
E137A <2> (<2> NAD-site mutant, less than 5% activity of wild-type [11]) [11]
E323A <2> (<2> spermidine-site mutant, no spermidine-binding [11]) [11]
G283A <2> (<2> NAD-site mutant, activity similar to wild-type [11]) [11]
H288A <2> (<2> spermidine-site mutant, no spermidine-binding [11]; <2> 6% of wild-type activity [16]) [11, 16]
K305A <2> (<2> spermidine-site mutant with retained avtivity [11]) [11]
K308A <5> (<5> very low reaction rate [6]) [6]
K308R <5> (<5> low reaction rate [6]) [6]
K308R/K350E <5> (<5> no activity [15]) [15]
K308R/K350P <5> (<5> no activity [15]) [15]
K308R/K350R <5> (<5> no activity [15]) [15]
K329A <2> (<2> spermidine-site mutant, 18-24% spermidine binding capacity [11]) [11]
K329R <2> (<2> no activity [7,16]) [7, 16]
K350A <5> (<5> no activity [6]) [6]
K350R <5> (<5> no activity [6,15]) [6, 15]
S317A <2> (<2> NAD-site mutant, activity similar to wild-type [11]) [11]
T308A <2> (<2> NAD-site mutant, activity similar to wild-type [11]) [11]
W327A <2> (<2> spermidine-site mutant, no spermidine-binding [11]) [11]
Additional information <2> (<2>, variant recombinant proteins with i. a truncation of 48 or 97 NH_2-terminal amino acids, ii. A truncation of 39 COOH-terminal amino acids, or iii. An internal deletion (ASp262-Ser317) are inactive. A chimeric protein consisting of the complete human sequence and 16 amino acids of the yeast sequence, Gln197-Asn212, not present in the human enzyme, inserted between Glu193 and Gln194 exhibit moderate activity [28]) [28]

6 Stability

Temperature stability

4 <5> (<5> 1 h, 35% loss of activity, NAD$^+$ protects [20]) [20]
65 <5> (<5> 15 min, almost complete loss of activity, NAD$^+$ protects [20]) [20]

General stability information

<6>, freeze-thawing of the partially purified enzyme inactivates, no stabilization by glycerol [9]

References

[1] Wolff, E.C.; Park, M.H.; Folk, J.E.: Cleavage of spermidine as the first step in deoxyhypusine synthesis. The role of NAD$^+$. J. Biol. Chem., **265**, 4793-4799 (1990)

[2] Wolff, E.C.; Folk, J.E.; Park, M.H.: Enzyme-substrate intermediate formation at lysine 329 of human deoxyhypusine synthase. J. Biol. Chem., **272**, 15865-15871 (1997)

[3] Chen, K.Y.; Liu, A.Y.C.: Biochemistry and function of hypusine formation on eukaryotic initiation factor 5A. Biol. Signals, **6**, 105-109 (1997)

[4] Ober, D.; Hartmann, T.: Deoxyhypusine synthase from tobacco. cDNA isolation, characterization, and bacterial expression of an enzyme with extended substrate specificity. J. Biol. Chem., **274**, 32040-32047 (1999)

[5] Ober, D.; Hartmann, T.: Homospermidine synthase, the first pathway-specific enzyme of pyrrolizidine alkaloid biosynthesis, evolved form deoxyhypusine synthase. Proc. Natl. Acad. Sci. USA, **96**, 14777-14782 (1999)

[6] Wolff, E.C.; Park, M.H.: Identification of lysine350 of yeast deoxyhypusine synthase as the site of enzyme intermediate formation. Yeast, **15**, 43-50 (1999)

[7] Wolff, E.C.; Wolff. J.; Park, M.H.: Deoxyhypusine synthase generates and uses bound NADH in a transient hydride transfer mechanism. J. Biol. Chem., **275**, 9170-9177 (2000)

[8] Joe, Y.A.; Wolff, E.C.; Park, M.H.: Cloning and expression of human deoxyhypusine synthase cDNA: structure-function studies with the recombinant enzyme and mutant proteins. J. Biol. Chem., **270**, 22386-22392 (1995)

[9] Tao, Y.; Chen, K.Y.: Molecular cloning and functional expression of Neurospora deoxyhypusine synthase cDNA and identification of yeast deoxyhypusine synthase cDNA. J. Biol. Chem., **270**, 23984-23987 (1995)

[10] Wang, T.W.; Lu, L.; Wang, D.; Thompson, J.E.: Isolation and characterization of senescence-induced cDNAs encoding deoxyhypusine synthase and eucaryotic translation initiation factor 5A from tomato. J. Biol. Chem., **276**, 17541-17549 (2001)

[11] Lee, C.H.; Um, P.Y.; Park, M.H.: Structure-function studies of human deoxyhypusine synthase: identification of amino acid residues critical for the binding of spermidine and NAD. Biochem. J., **355**, 841-849. (2001)

[12] Lee, C.H.; Park, M.H.: Human deoxyhypusine synthase: interrelationship between binding of NAD and substrates. Biochem. J., **352**, 851-857. (2000)

[13] Lee, Y.B.; Joe, Y.A.; Wolff, E.C.; Dimitriadis, E.K.; Park, M.H.: Complex formation between deoxyhypusine synthase and its protein substrate, the eukaryotic translation initiation factor 5A (eIF5A) precursor. Biochem. J., **340**, 273-281. (1999)

[14] Liao, D.-I.; Wolff, E.C.; Park, M.H.; Davies, D.R.: Crystal structure of the NAD complex of human deoxyhypusine synthase: an enzyme with a ball-and-chain mechanism for blocking the active site. Structure, **6(1)**, 23-32 (1998)

[15] Park, M.H.; Joe, Y.A.; Kang, K.R.: Deoxyhypusine synthase activity is essential for cell viability in the yeast Saccharomyces cerevisiae. J. Biol. Chem., **273**, 1677-1683. (1998)

[16] Joe, Y.A.; Wolff, E.C.; Lee, Y.B.; Park, M.H.: Enzyme-substrate intermediate at a specific lysine residue is required for deoxyhypusine synthesis. The role of Lys329 in human deoxyhypusine synthase. J. Biol. Chem., **272**, 32679-32685 (1997)

[17] Kang, K.R.; Wolff, E.C.; Park, M.H.; Folk, J.E.; Chung, S.I.: Identification of YHR068w in Saccharomyces cerevisiae chromosome VIII as a gene for deoxyhypusine synthase. Expression and characterization of the enzyme. J. Biol. Chem., **270**, 18408-18412 (1995)

[18] Wolff, E.C.; Lee, Y.B.; Chung, S.I.; Folk, J.E.; Park, M.H.: Deoxyhypusine synthase from rat testis: purification and characterization. J. Biol. Chem., **270**, 8660-8666. (1995)

[19] Jakus, J.; Wolff, E.C.; Park, M.H.; Folk, J.E.: Features of the spermidine-binding site of deoxyhypusine synthase as derived from inhibition studies. Effective inhibition by bis- and mono-guanylated diamines and polyamines. J. Biol. Chem., **268**, 13151-13159 (1993)

[20] Abid, M.R.; Sasaki, K.; Titani, K.; Miyazaki, M.: Biochemical and immunological characterization of deoxyhypusine synthase purified from the yeast Saccharomyces carlsbergensis. J. Biochem., **121**, 769-778. (1997)

[21] Abid, R.; Ueda, K.; Miyazaki, M.: Novel features of the functional site and expression of the yeast deoxyhypusine synthase. Biol. Signals, **6**, 157-165. (1997)

[22] Klier, H.; Csonga, R.; Steinkasserer, A.; Woehl, T.; Lottspeich, F.; Eder, J.: Purification and characterization of human deoxyhypusine synthase from HeLa cells. FEBS Lett., **364**, 207-210 (1995)

[23] Ober, D.; Harms, R.; Witte, L.; Hartmann, T.: Molecular evolution by change of function. Alkaloid-specific homospermidine synthase retains all properties of deoxyhypusine synthase except binding the eIF5A precursor protein. J. Biol. Chem., **278**, 12805-12812 (2003)

[24] Kang, K.R.; Kim, J.-S.; Chang, S.II; Park, M.H.; Kim, Y.W.; Lim, D.; Lee, S.-Y.: Deoxyhypusine synthase is phosphorylated by protein kinase C in vivo as well as in vitro. Exp. Mol. Med., **34**, 489-495 (2002)

[25] Lee, Y.; Kim, H.-K.; Park, H.-E.; Park, M.H.; Joe, Y.A.: Effect of N^1-guanyl-1,7-diaminoheptane, an inhibitor of deoxyhypusine synthase, on endothe-

lial cell growth, differentiation and apoptosis. Mol. Cell. Biochem., **237**, 69-76 (2002)

[26] Park, J.-H.; Wolff, E.C.; Folk, E.C.; Park, M.H.: Reversal of the deoxyhypusine synthesis reaction. Generation of spermidine or homospermidine from deoxyhypusine by deoxyhypusine synthase. J. Biol. Chem., **278**, 32683-32691 (2003)

[27] Moll, S.; Anke, S.; Kahmann, U.; Hänsch, R.; Hartmann, T.; Ober, D.: Cell-specific expression of homospermidine synthase, the entry enzyme of the pyrrolizidine alkaloid pathway in Senecio vernalis, in comparison with its ancestor, deoxyhypusine synthase. Plant Physiol., **130**, 47-57 (2002)

[28] Joe, Y.A.; Wolff, E.C.; Park, M.H.: Cloning and expression of human deoxyhypusine synthgase cDNA. Structure-function studies with the recombinant enzyme and mutant proteins. J. Biol. Chem., **270**, 22386-22392 (1995)

Cysteine synthase 2.5.1.47

1 Nomenclature

EC number
2.5.1.47

Systematic name
O^3-acetyl-L-serine:hydrogen-sulfide 2-amino-2-carboxyethyltransferase

Recommended name
cysteine synthase

Synonyms
Csase
EC 4.2.99.8 (formerly)
O-acetyl-L-serine acetate-lyase (adding hydrogen sulfide)
O-acetyl-L-serine sulfhydrylase
O-acetyl-L-serine sulfohydrolase
O-acetyl-L-serine(thiol)lyase
O-acetylserine (thiol)-lyase
O-acetylserine (thiol)lyase
O-acetylserine sulfhydrylase
O-acetylserine sulfhydrylase A
O-acetylserine(thiol)lyase
O-acetylserine-O-acetylhomoserine sulfhydro-lyase
OAS Shase
OAS-TL
OASS
OASTL
S-sulfocysteine synthase
acetylserine sulfhydrylase
cysteine synthetase
synthase, cysteine

CAS registry number
37290-89-4

2 Source Organism

<1> *Salmonella typhimurium* (one enzyme free, one associated with serine transacetylase in cysteine synthase [1]; S-sulfocysteine synthase is identical with cysteine synthase B [22]) [1, 3, 11, 22, 26, 44, 49, 50, 56, 65, 66, 67]

<2> *Escherichia coli* [1, 26, 43, 51, 57]
<3> *Saccharomyces cerevisiae* (low molecular weight enzyme, wild type and Cys auxotroph strain [13]; low molecular weight enzyme is the same protein as serine sulfhydrylase [15]) [2, 4, 6, 7, 13, 15, 19]
<4> *Brassica rapa* [2]
<5> *Neurospora crassa* [2]
<6> *Raphanus sativus* [5]
<7> *Paracoccus denitrificans* [8]
<8> *Triticum aestivum* [9, 55]
<9> *Trifolium repens* [10]
<10> *Pisum sativum* (three isoforms [29]) [10, 29]
<11> *Spinacia oleracea* (L. cv. Medina [53]; three isoenzymes [33]; isoenzyme 1' [45]) [12, 25, 33, 42, 45-48, 53, 53, 55]
<12> *Synechococcus sp.* (strain 6301, two isoenzymes [14]) [14]
<13> *Chromatium vinosum* (two isoenzymes, one with additional S-sulfocysteine synthase activity [16]) [16]
<14> *Enterobacter sp.* (strain 10-1 [17,18]) [17, 18]
<15> *Hordeum vulgare* [21]
<16> *Zea mays* [23, 55]
<17> *Bacillus sphaericus* [24]
<18> *Brassica juncea* [27]
<19> *Chlamydomonas reinhardtii* (two isoenzymes [35]) [30, 35, 37]
<20> *Chlorella fusca* [31]
<21> *Citrullus vulgaris* (two isoforms [28]) [28]
<22> *Rhodomicrobium vannielii* [20]
<23> *Rhodopseudomonas acidophila* [20]
<24> *Rhodocyclus purpureus* [20]
<25> *Thiocystis violacea* [20]
<26> *Rhodospirillum tenue* (two isoenzymes, one with additional S-sulfocysteine synthase activity [20]) [20]
<27> *Penicillium chrysogenum* [32]
<28> *Allium tuberosum* (two isoforms, immunologically distinct [34]) [34]
<29> *Xanthium pennsylvanicum* (three isoforms [36]) [36]
<30> *Capsicum annuum* [38]
<31> *Arabidopsis thaliana* (ecotype Columbia [54]) [39, 41, 54, 55, 60]
<32> *Datura innoxia* (three isoenzymes [40]) [40]
<33> *Lathyrus sativus* [52]
<34> *Lathyrus odoratus* [52]
<35> *Thermus thermophilus* (HB8 [58]) [58]
<36> *Methanosarcina thermophila* [59, 61]
<37> *Arabidopsis sp.* [62]
<38> *Solanum tuberosum* (cv. Danshaku [63]) [63]
<39> *Aeropyrum pernix* (K1 [64]) [64]

3 Reaction and Specificity

Catalyzed reaction

O^3-acetyl-L-serine + hydrogen sulfide = L-cysteine + acetate (<3>, mechanism similar to those of other pyridoxal enzymes [6]; <12>, model of reaction mechanism [14]; <1,11> mechanism [42, 49]; <2>, mechanism of reverse reaction [43]; <1>, mechanism interpreted in structural context, conformational changes during catalysis [44]; <1>, identification and spectral characterization of the external aldimine of the reaction, mechanism [50])

Reaction type

C-O bond cleavage
elimination

Natural substrates and products

S   L-Cys + acetate <2> (<2>, involved in mobilization of sulfide from cysteine for Fe-S cluster formation, significance in vivo unclear [43]) [43]
P   ?
S   O-acetyl-L-Ser + isoxazylin-5-one <33, 34> (<33, 34>, synthesis of precursor of neurotoxin β-N-oxalyl-L-α,β-diaminopropionic acid [52]) [52]
P   ?
S   O-acetyl-L-Ser + sulfide <1-7, 11, 12, 20, 26, 30-32, 36> (<1, 2, 6, 11, 36>, final step in Cys synthesis [1, 3, 5, 55, 59, 61]; <1>, involved in thiosulfate assimilation [22]; <3>, key role in metabolism of S-containing amino acids [4]; <3>, functions as a Cys synthase rather than as a homocysteine synthase in vivo [6]; <7>, controlled by feedback inhibition, adaptively significant as sulfide removal mechanism [8]; <11, 12, 20, 31>, last step of assimilatory sulfate reduction [12, 14, 31, 39]; <26>, repressed during growth with sulfide or thiosulfide as sulfur source [20]; <1>, enzyme transcription repressed by L-cystine, derepressed by limiting sulfide concentrations [26]; <20>, activity varies between sulfur sources, enzyme formation regulated by L-Cys concentration [31]; <30>, involved in synthesis of antioxidants such as glutathione during fruit development [38]; <32>, involved in glutathione formation [40]; <31>, the cysteine synthase complex functions as a molecular sensor system that monitors the sulfur status of the cell and controls sulfate assimilation and cysteine synthesis according to the availability of sulfate [60]) [1-6, 8, 12, 14, 20, 22, 26, 31, 38-40, 55, 59, 60, 61]
P   L-Cys + acetate
S   cysteine + CN⁻ <29> (<29>, involved in cyanide metabolism during seed germination [36]) [36]
P   ?
S   cysteine + dithiothreitol <37> (<37>, the side reaction of the enzyme seems to contribute massively to the total H_2S release of higher plants at least at higher pH values [62]) [62]
P β-cyanoalanine + H_2S

S   Additional information <31> (<31>, enzyme is induced in leaves exposed to salt stress. The results suggest that the plant enzyme is responding to the salt stress by inducing cysteine biosynthesis as a protection against high ion concentrations [54]) [54]

P   ?

Substrates and products

S   3-chloro-L-alanine + NaHS <17> (<17>, β-replacement reaction, enzyme can be induced by 3-chloro-L-alanine [24]) (Reversibility: ? <17> [24]) [24]

P   L-cysteine + ? <17> [24]

S   L-Cys + acetate <1, 2> (<1>, equilibrium constant [3]) (Reversibility: r <1, 2> [1, 3, 43]) [1, 3, 43]

P   O-acetyl-L-Ser + H_2S <1, 2> [1, 3, 43]

S   L-azaserine + sulfide <35> (<35>, 44% of the activity with O-acetyl-Ser [58]) (Reversibility: ? <35> [58]) [58]

P   ?

S   L-homocysteine + L-serine <39> (Reversibility: ? <39> [64]) [64]

P   L-cystathionine + H_2O <39> [64]

S   L-homoserine + sulfide <35> (<35>, 1.6% of the activity with O-acetyl-L-serine [58]) (Reversibility: ? <35> [58]) [58]

P   ?

S NaN_3 + O-acetyl-Ser <2, 15> (Reversibility: ? <2,15> [21,51]) [21, 51]

P   β-azidoalanine + sodium acetate <2, 15> (<2> mutagenic [51]; <15>, mutagenic in Salmonella typhimurium [21]) [21, 51]

S   O-acetyl-L-Ser + 2-propene-1-thiol <10, 11, 18, 21> (<11, 18>, 18% of activity with sulfide [25,2 7]; <21>, 6.2% of activity with sulfide, isoenzyme 2 [28]; <10>, 2.6% and 6.8% of activity with sulfide, isoenzyme 1 and 2, respectively [29]) (Reversibility: ? <10, 11, 18, 21> [25, 27-29]) [25, 27-29]

P   S-allyl-L-cysteine + ? <11, 18, 21> [25, 27-29]

S   O-acetyl-L-Ser + 5-mercapto-2-nitrobenzoate <1> (Reversibility: ? <1> [49]) [49]

P   S-(3-carboxy-4-nitrophenyl)-L-cysteine + ? <1> [49]

S   O-acetyl-L-Ser + H_2S <1, 2, 10, 11, 14, 18, 21, 31, 35, 36, 37, 38, 39> (<1>, equilibrium constant [3]) (Reversibility: r <1, 2, 11> [1, 3, 43, 53]; ? <1, 2, 10, 11, 14, 18, 21, 31, 35, 36, 37, 38, 39> [17, 25, 27-29, 54, 55, 56, 57, 58, 59, 60, 61, 62, 63, 64, 65, 66]) [1, 3, 17, 22, 25, 27-29, 43, 53, 54, 55, 56, 57, 58, 59, 60, 61, 62, 63, 64, 65, 66]

P   L-Cys + acetate <1, 2, 10, 11, 14, 18, 21> [1, 3, 17, 22, 25, 27-29, 43, 53]

S   O-acetyl-L-Ser + NaCN <14, 18, 28> (<18>, 12.3% of activity with sulfide [27]; <28>, 6.84% and 7.64% of activity with sulfide, isoenzymes 1 and 2, respectively [34]) (Reversibility: ? <14, 18, 28> [17, 27, 34]) [17, 27, 34]

P   β-cyanoalanine + sodium acetate <14, 18, 28> [17, 27, 34]

S   O-acetyl-L-Ser + mercaptoacetic acid <10, 11, 18, 21> (<11>, 2.5% of activity with sulfide [25]; <18>, 6.1% of activity with sulfide [27]; <21>, 2.2% of activity with sulfide, isoenzyme 1 and 2 [28]; <10>, 2.3% and

1.6% of activity with sulfide, isoenzymes 1 and 2, respectively [29]) (Reversibility: ? <11, 18, 21> [25, 27, 28]) [25, 27-29]

P S-carboxymethyl-L-cysteine <10, 11, 18, 21> [25, 27-29]

S O-acetyl-L-Ser + methyl mercaptan <1, 3-5, 10, 11, 21> (<11>, 32% of activity with sulfide [25]; <21>, 21.5% and 77% of activity with sulfide, isoenzymes 1 and 2, respectively [28]; <10>, 4% and 1% of activity with sulfide, isoenzymes 1 and 2, respectively [29]) (Reversibility: ir <1, 3-5> [1, 2]; ? <1, 10, 11, 21> [22, 25, 28, 29]) [1, 2, 22, 25, 28, 29]

P S-methylcysteine + acetate <1, 3-5, 10, 11, 21> (<1>, product identification uncertain [1]) [1, 2, 22, 25, 28, 29]

S O-acetyl-L-Ser + thiosulfate <1, 39> (Reversibility: ? <1,39> [22,64]) [22, 64]

P S-sulfocysteine + sodium acetate <1> [22]

S O-acetyl-Ser + selenide <7> (<7>, maximal 40% rate of cysteine synthesis [8]) (Reversibility: ? <7> [8]) [8]

P selenocysteine + acetate <7> [8]

S O-acetylhomoserine + H_2S <3, 35> (<3>, not, low molecular weight enzyme [13]; <35>, 2.4% of the activity with O-acetyl-L-Ser [58]) (Reversibility: ? <3,35> [4,58]) [4, 58]

P homocysteine + ? <3> [4]

S O-succinyl-L-homoserine + sulfide <35> (<35>, 3.6% of the activity with O-acetyl-L-serine [58]) (Reversibility: ? <35> [58]) [58]

P ?

S Ser + sulfide <35> (<35>, 1.8% of the activity with O-acetyl-L-serine [58]) (Reversibility: ? <35> [58]) [58]

P ?

S chloroalanine + sulfide <11, 35> (<11>, 3-6% of the activity of the cysteine synthase reaction [53]; <35>, β-chloroalanine, 6% of the activity with O-acetyl-L-Ser [58]) (Reversibility: r <11,35> [53,58]) [53, 58]

P cysteine + chloride <11> [53]

S cyanide + cysteine <38> (Reversibility: ? <38> [63]) [63]

P β-cyanoalanine + sulfide <38> [63]

S cysteine + CN^- <11, 29> (Reversibility: ? <11,29> [36,53]) [36, 53]

P cyanoalanine + H_2S <11, 29> (<11>, H_2S i.e. bisulfide [53]) [36, 53]

S cysteine + dithiothreitol <37, 39> (Reversibility: ? <37,39> [62,64]) [62, 64]

P β-cyanoalanine + H_2S

S Additional information <2, 11, 18, 21, 28> (<11, 18, 21>, β-substituted alanines, low activity [25, 27, 28]; <28>, isoenzyme 1 and 2 have different substrate specificities towards various β-substituted L-Cys [34]; <2>, several nucleophiles may stimulate sulfide formation [43]; <2>, activity of the enzyme bound to serine acetyltransferase is lower than that of the free enzyme [57]) [25, 27, 28, 34, 43, 57]

P ?

Inhibitors

$(NH_4)_6Mo_7O_{24}$ <2> (<2>, 1 mM, 97% inhibition [57]) [57]

1,10-phenanthroline <14> (<14>, 14% inhibition at 1 mM [18]) [18]

5,5'-dithiobis(2-nitrobenzoic acid) <8, 35> (<8>, non-competitive [9]; <35>, 1 mM, 92% inactivation [58]) [9, 58]

$AgNO_3$ <2> (<2>, 1 mM, 22% inhibition [57]) [57]

Cd^{2+} <14> (<14>, 55% inhibition at 1 mM [18]) [18]

Cl^- <1, 14> (<14>, $HgCl_2$ [18]) [18, 56]

Co^{2+} <14> (<14>, complete inhibition at 1 mM [18]) [18]

Cys <2> [57]

D-cycloserine <1> (<1>, 82% loss of activity at 5 mM [22]) [22]

EDTA <2> (<2>, 1 mM, 16% inhibition [57]) [57]

$FeSO_4$ <2> (<2>, 1 mM, 96% inhibition [57]) [57]

KCN <35> (<35>, 15.6% inhibition by 1 mM [58]) [58]

L-cysteine <1, 8-10, 11, 12, 13, 17, 19, 26> (<1>, not inhibitory up to 3.7 mM [3]; <8>, non-competitive [9]; <9>, 66% inhibition at 10 mM [10]; <13>, 35% inhibition at 4.5 mM [16]; <26>, 28-41% inhibition at 4.5 mM, isoenzyme-dependent [20]; <19>, 50% inhibition at 5 mM, only isoenzyme 1 [35]; <11>, substrate inhibition [53]) [3, 9-11, 14, 16, 20, 24, 30, 35, 53]

L-homocysteine <7> (<7>, competitive to sulfide [8]) [8]

L-homoserine <6, 8> (<8>, non-competitive [9]) [5, 9]

NEM <8> (<8>, non-competitive [9]) [9]

NH_2OH <1> (<1>, 97% loss of activity at 10 mM [22]) [22]

NaN_3 <15> [21]

Ni^{2+} <14> (<14>, complete inhibition at 1 mM [18]) [18]

O-acetylserine <2, 8, 11, 15, 19, 36> (<15>, at 150 mM [21]; <8>, above 72 mM [9]; <11,36>, substrate inhibition [53,61]) [9, 21, 35, 53, 57, 61]

S-methylcysteine <6> (<6>, slight inhibition [5]) [5]

S-sulfocysteine <13, 26> (<1>, 26% loss of activity at 5 mM [22]; <13>, 52% inhibition at 4.5 mM [16]; <26>, 24% inhibition at 4.5 mM, isoenzyme 1 [20]) [16, 20]

SO_3^{2-} <7> (<7>, competitive to sulfide [8]) [8]

SO_4^{2-} <1> [56]

Zn^{2+} <2, 14> (<2>, 1 mM, 94% inhibition [57]; <14>, 95% inhibition at 1 mM [18]) [18, 57]

acetate <2> [57]

acetyl-CoA <1> [11]

aminooxyacetate <19> (<19>, 57% and 64% inhibition at 1 mM, isoenzymes 1 and 2, respectively [35]) [35]

chloroalanine <11> (<11>, substrate inhibition [53]) [53]

cystathionine <7, 9, 10> (<7>, competitive to sulfide [8]; <9>, 91% inhibition at 10 mM [10]) [8, 10]

hydroxylamine <6, 11, 14, 19, 35> (<14>, 31% inhibition at 1 mM [18]; <11>, complete inhibition at 10 mM, isoenzymes 1 and 2, 90% inhibition at 10 mM, isoenzyme 3 [33]; <11>, 57% loss of activity at 10 mM, isoenzyme 1' [45]; <19>, 35% and 48% inhibition at 5 mM, isoenzymes 1 and 2, respectively [35]; <35>, 10 mM, 71.2% inhibition [58]) [5, 18, 30, 33, 35, 45, 58]

methionine <6, 7, 8, 19> (<7,8>, competitive to sulfide [8,9]; <6>, slight inhibition [5]; <19>, 46% and 37% inhibition at 1 mM, isoenzymes 1 and 2, respectively [35]) [5, 8, 9, 30, 35]
monoiodoacetic acid <35> (<35>, 1 mM, complete inactivation [58]) [58]
p-chloromercuribenzoate <2, 8, 14, 19> (<8>, non-competitive [9]; <14>, 40% inhibition at 1 mM [18]; <19>, 14% and 4% inhibition at 1 mM, isoenzymes 1 and 2, respectively [35]; <2>, 1 mM, 59% inhibition [57]) [9, 18, 35, 57]
p-chloromercuriphenylsulfonic acid <14> (<14>, 46% inhibition at 1 mM [18]) [18]
p-hydroxymercuribenzoate <19> [30]
phenylhydrazine <35> (<35>, 73% inhibition by 1 mM, 97.4% inhibition by 10 mM [58]) [58]
pyridoxal <3> [6]
pyridoxal hydrochloride <3> (<3>, 54% inhibition at 1 mM [6]) [6]
semicarbazide <1, 35> (<1>, 60% loss of activity at 1 mM [22]; <35>, 13% inhibition by 1 mM, 38.6% inhibition by 10 mM [58]) [22, 58]
serine <7> (<7>, competitive to O-acetylserine [8]) [8]
sodium borohydride <14> (<14>, 59% inhibition at 1 mM [18]) [18]
sulfide <2, 8, 11, 19, 26, 32> (<8>, above 107 mM [9]; <32>, above 200 mM [40]; <11>, substrate inhibition [53]) [9, 20, 35, 40, 53, 57]
thiourea <14> (<14>, 34% inhibition at 1 mM [18]) [18]

Cofactors/prosthetic groups
pyridoxal 5'-phosphate <1, 3, 6, 7, 9-11, 14, 18, 19, 28, 31, 32, 35, 36, 39> (<1>, 1 per subunit of free enzyme, 4 per of cysteine synthase complex [1]; <1>, 2.1 mol per mol of dimeric enzyme [3]; <3>, stoichiometry, binds to Lys, protects against inactivation by heat, urea, and trypsin, four binding sites per mol apoenzyme, association constant [6]; <9,10>, restores activity after dialysis against cysteine [10]; <14>, 0.75 mol per mol subunit [18]; <11,18>, one per subunit [25,27]; <19>, protects against inactivation [30]; <28>, one per subunit [34]; <32>, responsible for O-acetylserine binding [40]; <31>, 4 per cysteine synthase complex [41]; <11>, pyridoxal 5'-phosphate containing catalytic sites are not equivalent [42]; <1>, pyridoxal 5'-phosphate-binding site covalently bound, buried deeply within protein [44]; <11>, 1.1 mol per mol subunit [47]; <11>, Lys involved in cofactor binding [48]; <35>, K_m of the cofactor at pH 8.0: 0.029 mM [58]; <36>, enzyme contains pyridoxal phosphate [61]; <39>, enzyme contains 1.2 pyridoxal phosphate per subunit [64]; <1>, enzyme contains pyridoxal 5'-phosphate [67]) [1, 3, 5, 6, 8, 10, 18, 22, 25, 27, 29, 30, 34, 40-42, 44, 47, 48, 58, 61, 64, 67]

Activating compounds
dithiothreitol <2> (<2>, slight activation [57]) [57]
iodoacetamide <2> (<2>, slight activation [57]) [57]

Metals, ions
Mg^{2+} <2> (<2>, slight activation [57]) [57]

Turnover number (min^{-1})

1440 <39> (O-acetyl-L-Ser, <39>, synthesis of S-sulfo-L-cysteine [64]) [64]
1440 <39> (thiosulfate, <39>, synthesis of S-sulfo-L-cysteine [64]) [64]
12120 <39> (O-acetyl-L-Ser, <39>, sulfhydrylation of O-acetyl-L-serine [64])
[64]
12120 <39> (S^{2-}, <39>, sulfhydrylation of O-acetyl-L-serine [64]) [64]
44000 <1, 2> (O-acetyl-L-Ser) [1]

Specific activity (U/mg)

0.0283 <3> (<3>, Cys-auxotroph strain, low molecular weight enzyme [13])
[13]
0.129 <36> [61]
0.589 <3> (<3>, low molecular weight enzyme [15]) [15]
1.01 <28> (<28>, isoenzyme 2 [34]) [34]
1.02 <3> [4]
1.36 <28> (<28>, isoenzyme 1 [34]) [34]
1.91 <9> [10]
2.43 <13> (<13>, isoenzyme 1 [16]) [16]
5.7 <8> [9]
6.3 <2> (<2>, strain K 12 [51]) [51]
8.1 <2> (<2>, transformed strain cysk [51]) [51]
15.4 <19> (<19>, isoenzyme 2 [35]) [35]
25.2 <3> [6]
41.15 <19> [30]
50.9 <39> [64]
96 <11> (<11>, chloroplast enzyme [47]) [47]
130 <26> (<26>, isoenzyme 1 [20]) [20]
143 <26> (<26>, isoenzyme 2 [20]) [20]
245.6 <13> (<13>, isoenzyme 2 [16]) [16]
292 <1> (<1>, complex-bound enzyme [1]) [1]
303 <1> [22]
336 <19> (<19>, isoenzyme 1 [35]) [35]
337.4 <35> [58]
400 <11, 30> (<11>, isoenzyme 1 [33]) [33, 38]
410 <18> [27]
543 <11> [25]
649 <21> (<21>, isoenzyme 1 [28]) [28]
691 <21> (<21>, isoenzyme 2 [28]) [28]
834 <11> [45]
852 <10> (<10>, isoenzyme 2 [29]) [29]
880 <32> (<32>, isoenzyme 1 and 2 [40]) [40]
933 <10> (<10>, isoenzyme 1 [29]) [29]
1024 <6> [5]
1100 <1> (<1>, free enzyme [1,3]) [1, 3]
1400 <11> (<11>, isoenzyme 2 [33]) [33]
1431 <11> (<11>, isoenzyme 3 [33]) [33]
5500 <15> [21]

Additional information <2-4, 7, 11, 30, 31> (<3>, activity buffer-dependent, substrate unstable in Tris-HCl buffer [4]; <4>, cysteine and methylcysteine formation of crude leaf extracts [2]; <7>, activity increased during growth on cystine, as compared to growth on sulfate [8]; <30>, maximal activity in red fruit chromoplasts [38]; <31>, activity modified by protein-protein interactions within cystein synthase complex [41]; <2>, activity depends on presence of nucleophiles [43]; <11>, activities of various recombinant enzymes [48]) [2, 4, 8, 38, 41, 43, 48]

K_m-Value (mM)

0.006 <2> (S^{2-}, <2>, pH 7.5, 25°C, free enzyme [57]) [57]

0.013 <2> (S^{2-}, <2>, pH 7.5, 25°C, enzyme bound to serine acetyltransferase [57]) [57]

0.02 <30> (S^{2-}) [38]

0.022 <11> (S^{2-}, <11>, pH 8.0, 30°C [25]) [25]

0.033 <21> (S^{2-}, <21>, pH 8.0, 30°C, isoenzyme 1 [28]; <10>, pH 7.8, 35°C [29]) [28, 29]

0.037 <8> (O-acetyl-L-Ser, <8>, pH 6.8, 37°C [9]) [9]

0.038 <10> (S^{2-}, <10>, pH 7.8, 35°C, isoenzyme 2 [29]) [29]

0.0384 <38> (cysteine, <38>, pH 8.5, enzyme form PCS-2 [63]) [63]

0.043 <18> (S^{2-}, <18>, pH 8.0, 30°C [27]) [27]

0.045 <29> (O-acetyl-L-Ser, <29>,isoenzyme 3 [36]) [36]

0.05 <29> (O-acetyl-L-Ser, <29>, isoenzymes 1 and 2 [36]) [36]

0.05 <35> (S^{2-}, <35>, pH 7.8, 50°C [58]) [58]

0.0512 <38> (cysteine, <38>, pH 8.5, enzyme form PCS-1 [63]) [63]

0.0526 <37> (cysteine, <37>, pH 7.5, 37°C, isoenzyme C [62]) [62]

0.0986 <37> (cysteine, <37>, isoenzyme A [62]) [62]

0.1 <1> (S^{2-}, <1>, less than [1]) [1]

0.108 <11> (cysteine, <11>, pH 8.0, 26°C, cysteine synthase A [53]) [53]

0.113 <37> (cysteine, <37>, pH 7.5, 37°C, isoenzyme B [62]) [62]

0.12 <1, 2> (acetyl-CoA) [1]

0.201 <11> (cysteine, <11>, pH 8.0, 26°C, cysteine synthase B [53]) [53]

0.24 <10> (S^{2-}, <10>, pH 7.8, 35°C [10]) [10]

0.25 <11, 27> (S^{2-}, <11>, pH 7.5, 25°C, chloroplast enzyme [47]; <27>, pH 7.5, 25°C [32]) [32, 47]

0.37 <31> (S^{2-}, <31>, pH 7.0, 25°C, recombinant free enzyme [41]) [41]

0.373 <29> (O-acetyl-L-Ser, <29>,isoenzyme 2 [36]) [36]

0.4 <13> (S^{2-}, <13>, pH 7.4, 37°C, enzyme 1 [16]) [16]

0.5 <14, 36> (S^{2-}, <14>, pH 7.4, 30°C [18]; <36>, 40°C [61]) [18, 61]

0.51 <9> (S^{2-}, <9>, pH 7.8, 35°C [10]) [10]

0.547 <29> (O-acetyl-L-Ser, <29>,isoenzyme 3 [36]) [36]

0.55 <11, 31> (S^{2-}, <11>, pH 8.0, 25°C, isoenzyme 1 [33]; <31>, pH 7.5, 25°C, recombinant complex-bound enzyme [41]) [33, 41]

0.59 <8> (S^{2-}, <8>, pH 6.8, 37°C [9]) [9]

0.6 <26> (S^{2-}, <26>, isoenzyme 1 and 2 [20]) [20]

0.64 <31> (O-acetyl-L-Ser, <31>, pH 7.5, 25°C, recombinant free enzyme [41]) [41]

0.66 <11> (S^{2-}, <11>, pH 8.0, 25°C, isoenzyme 1' [45]) [45]

0.7 <1, 2> (L-Ser) [1]

0.7 <12, 37> (O-acetyl-L-Ser, <12>, pH 7.5, 37°C, isoenzyme 1 and 2 [14]; <37>, pH 7.5, 37° C, isoenzyme B [62]) [14, 62]

0.75 <19> (S^{2-}, <19>, pH 7.5, 50°C [30]) [30]

0.785 <11> (chloroalanine, <11>, pH 8.0, 26°C, cysteine synthase B [53]) [53]

0.8 <12> (S^{2-}, <12>, pH 7.5, 37°C, isoenzyme 1 and 2 [14]) [14]

0.998 <38> (S^{2-}, <38>, pH 7.5, enzyme form PCS-2 [63]) [63]

1 <13> (O-acetyl-L-Ser, <13>, pH 7.4, 37°C, isoenzyme 1 [16]) [16]

1.03 <11> (O-acetylserine, <11>, pH 8.0, 26°C, cysteine synthase B [53]) [53]

1.053 <11> (chloroalanine, <11>, pH 8.0, 26°C, cysteine synthase A [53]) [53]

1.25 <7, 11> (S^{2-}, <11>, pH 8.0, 25°C, isoenzyme 2 [33]) [8, 33]

1.29 <29> (O-acetyl-L-Ser, <29>, isoenzyme 1 [36]) [36]

1.3 <11, 27> (O-acetyl-L-Ser, <11>, pH 7.5, 25°C, chloroplast enzyme [47]; <27>, pH 6.8, 36°C [32]) [32, 47]

1.355 <11> (O-acetylserine, <11>, pH 8.0, 26°C, cysteine synthase A [53]) [53]

1.5 <21> (O-acetyl-L-Ser, <21>, pH 8.0, 30°C, isoenzyme 2 [28]) [28]

1.57 <38> (S^{2-}, <38>, pH 7.5, enzyme form PCS-1 [63]) [63]

1.6 <13> (S^{2-}, <13>, pH 7.4, 37°C, isoenzyme 2 [16]) [16]

1.83 <38> (O-acetyl-L-serine, <38>, pH 7.5, enzyme form PCS-2 [63]) [63]

2 <19, 30> (O-acetyl-L-Ser, <19>, pH 7.5, 50°C [30]) [30, 38]

2.1 <10, 33, 37> (O-acetyl-L-Ser, <10>, pH 8.0, 30°C, isoenzyme 1 [29]; <33>, formation of β-(isoxazylin5-on-4-yl)-L-alanine [52]; <37>, pH 7.5, 37°C, isoenzyme A [62]) [29, 52, 62]

2.17 <6> (O-acetyl-L-Ser) [5]

2.3 <15> (NaN$_3$, <15>, pH 7.6, 25°C [21]) [21]

2.3 <10> (O-acetyl-L-Ser, <10>, pH 8.0, 30°C, isoenzyme 2 [29]) [29]

2.5 <18> (O-acetyl-L-Ser, <18>, pH 8.0, 30°C [27]) [27]

2.5 <11> (S^{2-}, <11>, pH 8.0, 25°C, isoenzyme 3 [33]) [33]

2.6 <21> (O-acetyl-L-Ser, <21>, pH 8.0, 30°C, isoenzyme 1 [28]) [28]

2.7 <7> (O-acetyl-L-Ser, <7>, pH 7.5, 30°C [8]) [8]

2.9 <11, 26> (O-acetyl-L-Ser, <26>, isoenzyme 1 [20]; <11>, pH 8.0, 30°C [25]) [20, 25]

2.99 <38> (O-acetyl-L-serine, <38>, pH 7.5, enzyme form PCS-1 [63]) [63]

3.1 <10> (O-acetyl-L-Ser, <10>, pH 7.8, 35°C [10]) [10]

3.5 <9> (O-acetyl-L-Ser, <9>, pH 7.8, 35°C [10]) [10]

3.57 <11> (O-acetyl-L-Ser, <11>, pH 8.0, 25°C, isoenzyme 1 [33]) [33]

3.6 <1> (S^{2-}) [22]

3.8 <33> (O-acetyl-L-Ser, <33>, formation of β-(isoxazylin5-on-2-yl)-L-alanine [52]) [52]

3.9 <37> (O-acetyl-L-Ser, <37>, pH 7.5, 37°C, isoenzyme C [62]) [62]

4.28 <31> (O-acetyl-L-Ser, <31>, pH 7.5, 25°C, recombinant complex-bound enzyme [41]) [41]

4.8 <2, 35> (O-acetyl-L-Ser, <2>, pH 7.5, 25°C, free enzyme [57]; <35>, pH 7.8, 50°C [58]) [57, 58]

5 <1> (O-acetyl-L-Ser, <1>, pH 7.2-7.4, 25°C [3]; <1>, free enzyme, [1,3]; <1>, independent of sulfide concentration [3]) [1, 3]

5.12 <3> (O-acetyl-L-Ser, <3>, pH 7.8 [4]) [4]

5.2 <14> (NaCN, <14>, pH 7.4, 30°C [18]) [18]

5.2 <19> (S^{2-}, <19>, pH 7.5, 50°C, isoenzyme 1 and 2 [35]) [35]

5.26 <11> (O-acetyl-L-Ser, <11>, pH 8.0, 25°C, isoenzyme 3 [33]) [33]

5.56 <11> (O-acetyl-L-Ser, <11>, pH 8.0, 25°C, isoenzyme 2 [33]) [33]

6.67 <3> (O-acetylhomoserine) [4]

6.7-8.3 <14> (O-acetyl-L-Ser, <14>, pH 7.4, 30°C, cosubstrate-dependent [18]) [18]

7.1 <1> (O-acetyl-L-Ser) [22]

8.2 <15> (O-acetyl-L-Ser, <15>, pH 7.6, 25°C [21]) [21]

8.3 <11> (O-acetyl-L-Ser, <11>, pH 8.0, 25°C, isoenzyme 1' [45]) [45]

9 <28> (O-acetyl-L-Ser, <28>, isoenzyme 2 [34]) [34]

9.6 <19> (O-acetyl-L-Ser, <19>, pH 7.5, 50°C, isoenzyme 1 and 2 [35]) [35]

14.2 <38> (CN^-, <38>, pH 8.5, enzyme form PCS-1 [63]) [63]

15 <13> (O-acetyl-L-Ser, <13>, pH 7.4, 37°C, isoenzyme 2 [16]) [16]

15.1 <38> (CN^-, <38>, pH 8.5, enzyme form PCS-2 [63]) [63]

20 <1> (O-acetyl-L-Ser, <1>, complex-bound enzyme [1]) [1]

21 <39> (thiosulfate, <39>, synthesis of S-sulfo-L-cysteine [64]) [64]

24 <39> (O-acetyl-L-Ser, <39>, synthesis of S-sulfo-L-cysteine [64]) [64]

27 <2> (O-acetyl-L-Ser, <2>, pH 7.5, 25°C, enzyme bound to serine acetyltransferase [57]) [57]

28 <39> (O-acetyl-L-Ser, <39>, sulfhydrylation of O-acetyl-L-serine [64]) [64]

50 <26> (O-acetyl-L-Ser, <26>, isoenzyme 2 [20]) [20]

Additional information <1, 9-11, 14, 16, 19, 20, 32> (<1>, Hill numbers [11]; <9,10>, various methods compared [10]; <14>, cysteine-forming activity 245 times greater than β-cyanoalanine-forming activity [17]; <20>, activity varies between sulfur sources [31]; <19>, not significantly altered by immobilization [37]; <32>, no Michaelis-Menten-kinetics [40]; <11>, positive kinetic cooperativity with respect to O-acetylserine in the presence of sulfide [42]) [10, 11, 17, 23, 31, 37, 40, 42]

K_i-Value (mM)

0.011 <2> (S^{2-}, <2>, pH 7.5, 25°C, free enzyme [57]) [57]

0.025 <1> (sulfide, <1>, pH 8 [56]) [56]

0.11 <2> (S^{2-}, <2>, pH 7.5, 25°C, enzyme bound to serine acetyltransferase [57]) [57]

0.63 <8> (5,5'-dithiobis(2-nitrobenzoic) acid, <8>, pH 6.8, 37°C [9]) [9]

0.8 <11> (chloroalanine, <11>, pH 9.8, 26°C, cysteine synthase B [53]) [53]

1 <1> (acetyl-CoA) [11]

1.135 <11> (chloroalanine, <11>, pH 9.8, 26°C, cysteine synthase A [53]) [53]

1.32 <8> (L-homoserine, <8>, pH 6.8, 37°C [9]) [9]

1.43 <8> (N-ethylmaleimide, <8>, pH 6.8, 37°C [9]) [9]

1.75 <8> (methionine, <8>, pH 6.8, 37°C [9]) [9]

2.074 <11> (L-cysteine, <11>, pH 8.0, 26°C, cysteine synthase A [53]) [53]

2.113 <11> (cysteine, <11>, pH 8.0, 26°C, cysteine synthase B [53]) [53]
2.27 <8> (L-cysteine, <8>, pH 6.8, 37°C [9]) [9]
2.3 <15> (NaN$_3$, <15>, pH 7.4, 25°C [21]) [21]
2.86 <8> (p-chloromercuribenzoate, <8>, pH 6.8, 37°C [9]) [9]
5.546 <11> (chloroalanine, <11>, pH 8.0, 26°C, cysteine synthase B [53]) [53]
6.5 <19> (L-methionine, <19>, pH 7.5, 50°C [35]) [35]
7.1 <2> (O-acetyl-L-Ser, <2>, pH 7.5, 25°C, free enzyme [57]) [57]
8.6 <2> (cysteine, <2>, pH 7.5, 25°C, free enzyme [57]) [57]
18 <2> (O-acetyl-L-Ser, <2>, pH 7.5, 25°C, enzyme bound to serine acetyl-transferase [57]) [57]
19.56 <11> (O-acetylserine, <11>, pH 8.0, 26°C, cysteine synthase B [53]) [53]
23 <1> (SO$_4^{2-}$, <1>, pH 6.5, 25°C [56]) [56]
32.25 <11> (O-acetylserine, <11>, pH 8.0, 26°C, cysteine synthase A [53]) [53]
38 <1> (Cl, <1>, pH 6.5, 25°C [56]) [56]
48 <2> (cysteine, <2>, pH 7.5, 25°C, enzyme bound to serine acetyltransferase [57]) [57]
70.92 <11> (chloroalanine, <11>, pH 8.0, 26°C, cysteine synthase A [53]) [53]
160 <2> (acetate, <2>, pH 7.5, 25°C, free enzyme [57]) [57]
340 <2> (acetate, <2>, pH 7.5, 25°C, enzyme bound to serine acetyltransferase [57]) [57]
Additional information <1> [56]

pH-Optimum
6.7 <39> (<39>, sulfhydrylation of O-acetyl-L-serine [64]) [64]
6.8 <8> (<8>, phosphate buffer [9]) [9]
7 <35, 37> (<37>, recombinant isoenzyme C [62]) [58, 62]
7.2 <1, 13, 27> (<13>, isoenzyme 1 [16]) [16, 22, 32]
7.3-8.2 <26> (<26>, isoenzyme 1 [20]) [20]
7.5 <7, 11, 12, 19> (<11>, isoenzyme 1', lower than isoenzymes one, two, and three [45]) [8, 14, 21, 30, 45]
7.6 <32> (<32>, all isoenzymes [40]) [40]
7.7 <26> [20]
7.8 <9, 20> (<9,10>, potassium phosphate buffer [10]) [10, 31]
7.9 <33, 34> [52]
8 <8, 11, 15, 18, 21> (<11>, broad optimum around [33]; <8>, Tris-HCl buffer [9]; <37>, recombinant isoenzyme A and B [62]) [9, 21, 27, 28, 33, 47, 62]
8-9 <2> (<2>, free enzyme [57]) [57]
8.1-8.8 <39> (<39>, synthesis of L-cystathionine [64]) [64]
8.3 <11> [25]
8.4 <3> [4]
8.5 <19, 29> [35, 36]
9-11.5 <11> [53]
9.5 <2> (<2>, enzyme bound to serine acetyltransferase [57]) [57]
9.75 <14> [18]
10.5 <17> [24]

pH-Range

5.5-8 <35> (<35>, about 50% of maximal activity at pH 5.5 and pH 8.0 [58])
[58]

5.8-9.4 <1> [3]

7.3-8.2 <13> (<13>, isoenzyme 2 [16]) [16]

8-10 <14> [18]

Additional information <19> (<19>, altered by immobilization [37]) [37]

Temperature optimum (°C)

37 <8> [9]

40-60 <36> [61]

50 <11, 12, 19> (<11>, chloroplast enzyme [47]; <19>, soluble enzyme,
DEAE-immobilized enzyme and silica-immobilized enzyme [37]) [14, 30, 35,
47]

55 <35> [58]

60 <19> (<19>, vinylacetate-epoxy-immobilized enzyme and vinylacetate-hy-
droxy-immobilized enzyme [37]) [37]

70 <39> (<39>, maximal for O-acetyl-L-serine sulfhydrylation at both 70°C
and at 80°C [64]) [64]

80 <39> (<39>, synthesis of L-cystathionine and sulfhydrylation of L-Ser.
Maximal for O-acetyl-L-serine sulfhydrylation at both 70°C and at 80°C
[64]) [64]

90 <39> (<39>, synthesis of S-sulfo-L-cysteine [64]) [64]

Temperature range (°C)

35-67 <35> (<35>, about 50% of maximal activity at 35°C and at 67°C [58])
[58]

42-58 <32> (<32>, all isoenzymes [40]) [40]

45-65 <12> [14]

4 Enzyme Structure

Molecular weight

46000 <26> (<26>, isoenzyme 2, gel filtration [20]) [20]

50000 <13> (<13>, isoenzyme 2, gel filtration [16]) [16]

52000 <10, 18, 29> (<10,18,29>, gel filtration [27,29,36]) [27, 29, 36]

55000 <1> (<1>, cysteine synthase B, gel filtration [22]) [22]

56000 <12, 13> (<12>, isoenzyme 1 and 2, gel filtration [14]; <13>, isoenzyme 1, gel filtration [16]) [14, 16]

57000 <26> (<26>, isoenzyme 1, gel filtration [20]) [20]

58000 <21> (<21>, gel filtration [28]) [28]

59000 <27> (<27>, one of two bands gel filtration [32]) [32]

60000 <11> (<11>, gel filtration [25]) [25]

63000 <11, 32> (<11>, all isoenzymes, gel filtration [33]; <32>, isoenzymes
one and three, gel filtration [40]) [33, 40]

64000 <11> (<11>, chloroplast enzyme, gel filtration [48]) [48]

65000 <19> (<19>, sedimentation equilibrium [30]) [30]

66000 <6> (<6>, gel filtration [5]) [5]
68000 <1, 11, 14, 27, 35> (<1>, free enzyme, sedimentation equilibrium [1,3];
<14>, gel filtration [17]; <27>, one of two bands, gel filtration [32]; <11>,
chloroplast enzyme, gel filtration [47]; <11>, isoenzyme 1', gel filtration
[45]; <35>, gel filtration [58]) [1, 3, 17, 32, 45, 47, 58]
70580 <39> (<39>, sedimentation equilibrium [64]) [64]
73000 <28> (<28>, isoenzyme 2, gel filtration [34]) [34]
81000 <28> (<28>, isoenzyme 1, gel filtration [34]) [34]
86000 <32> (<32>, isoenzyme 2, gel filtration [40]) [40]
93000 <36> (<36>, gel filtration [61]) [61]
96000 <3> (<3>, wild type strain, low molecular weight enzyme, sedimentation equilibrium [13]) [13]
99000 <3> (<3>, Cys auxotroph strain, sedimentation equilibrium [13]) [13]
200000 <3> (<3>, gel filtration [7]) [7]
300000 <31> (<31>, bienzyme complex with serine acetyltransferase, gel filtration [41]) [41]
309000 <1> (<1>, cysteine synthetase complex, sedimentation equilibrium
[1]) [1]
310000 <11> (<11>, cysteine synthetase complex, gel filtration [47]) [47]
1390000 <1> (<1>, cysteine synthetase complex, sedimentation equilibrium
[11]) [11]

Subunits
dimer <1, 6, 10, 11, 14, 18, 21, 28, 30-32, 35, 36, 39> (<10,18>, 2 * 26000, SDS-
PAGE [27,29]; <1,21>, 2 * 29000, SDS-PAGE [22,28]; <11>, 2 * 32000, SDS-
PAGE [25]; <32>, 2 * 32000, isoenzymes 1 and 3, SDS-PAGE [40]; <32>, 1 *
32000 + 1 * 31200, isoenzyme 2, SDS-PAGE [40]; <6>, 2 * 33000, SDS-PAGE [5];
<1,11,14,35>, 2 * 34000, SDS-PAGE [3,17,33,45,58]; <11,30>, 2 * 35000, SDS-
PAGE [38,42,47,48]; <36>, 2 * or 3 * 36000, SDS-PAGE [61]; <31>, 2 * 36000
[41]; <28>, 2 * 36000, isoenzyme 2, SDS-PAGE [34]; <28>, 2 * 40000, isoenzyme 1, SDS-PAGE [34]; <39>, 2 * 42000, SDS-PAGE [64]) [1, 3, 5, 17, 22, 25,
27, 28, 29, 33, 34, 38, 40-42, 45, 47, 48, 58, 61, 64]
tetramer <3> (<3>, 4 * 51000, SDS-PAGE, association does not require disulfide linkage [7]) [7]
trimer <36> (<36>, 3 or 2 * 36000, SDS-PAGE [61]) [61]
Additional information <1, 2, 3, 10, 11, 14, 17, 21, 28, 29, 31, 32> (<1>, amino
acid analysis, N-terminal amino acid is serine [3,26]; <1>, at least tetramer,
equilibrium between aggregated and disintegrated cysteine synthetase complex can be shifted by various effectors, has consequences for enzyme function, model of structural-functional relationships [11]; <3>, amino acid analysis, one sulfhydryl group per subunit, essential for restoration of native subunit structure [7]; <2, 10, 11, 14, 18, 21, 28, 29, 31, 32>, amino acid composition [17, 25-29, 33, 34, 36, 39, 40, 47, 48]; <17>, cysteine synthase complex
contains 2 mol O-acetylserine sulfhydrylase per mol serine-O-transacetylase
[24]; <21>, isoenzymes have different cysteine and methionine content [28];
<31>, two dimeric enzyme molecules form complex with tetrameric serine
acetyltransferase, complex does not dissociate during catalysis [41]; <1>,

three-dimensional model of enzyme structure, catalytic sites [44]; <11>, isoenzyme 1', N-terminal amino acid sequence differs from those of enzymes
one, two, and three [45]; <11>, all isoforms, but particularly isoenzyme 2 can
form cysteine synthetase complex with serine acetyltransferase, as revealed by
polyclonal antibody studies [46]) [3, 7, 17, 24-29, 33, 34, 36, 39-41, 44-46]

5 Isolation/Preparation/Mutation/Application

Source/tissue
cell suspension culture <32> [40]
fruit <30> [38]
hypocotyl <11> [12]
leaf <4, 8, 11, 16, 18, 28, 31> (<11>, two isoforms [12]; <31>, enzyme is
induced in leaves exposed to salt stress [54]) [2, 9, 12, 23, 25, 27, 33, 34, 46-
48, 53, 54]
resting cell <17> [24]
root <6, 11> (<11>, 30-35% of activity [12]) [5, 12]
seed <11, 29> (<11>, hydrated seeds [45]) [36, 45]
seedling <10, 21> (<10>, maximal activity [29]) [28, 29]
tuber <38> [63]

Localization
chloroplast <1, 9-11, 16, 31> (<9,10>, 68-86% of activity, stroma [10]; <11>,
different from cytosolic enzyme [12]; <11>, different from cytosolic and mitochondrial enzymes, as revealed by polyclonal antibody studies, 3-5% of
total activity [47]; <16> both mesophyll and bundle sheath chloroplasts
[23]; <31>, bienzyme complex [41]) [10-12, 23, 33, 39, 41, 47, 48]
chromoplast <30> [38]
cytoplasm <37> (<37>, isoenzyme A [62]) [62]
cytosol <11, 14, 29, 31, 38> (<11>, different from chloroplast enzyme [12];
<38>, enzyme form PCS-1 [63]) [12, 17, 36, 39, 54, 55, 63]
mitochondrion <11, 37> (<11>, all isoenzymes [45]; <37>, isoenzyme C
[62]) [45, 62]
plastid <37, 38> (<37>, isoenzyme B [62]; <38>, enzyme form PCS-2 [63])
[62, 63]
proplastid <11> [12]
soluble <7> [8]

Purification
<1> [1, 3]
<3> [4, 6, 13, 19]
<6> [5]
<7> [8]
<8> [9]
<9> [10, 29]
<10> [10, 29]
<11> (isoenzyme A and B [53]) [12, 25, 33, 42, 45, 53]

<12> [14]
<13> [16]
<14> [17]
<15> [21]
<18> [27]
<19> [30]
<21> [28]
<26> [20]
<27> [32]
<28> [34]
<29> [36]
<30> [38]
<31> [41]
<32> [40]
<33> [52]
<34> [52]
<35> [58]
<36> [61]
<37> (recombinant enzyme [62]) [62]
<39> [64]

Renaturation
<9, 10> [10]

Crystallization
<1> (structural model based on crystallographic data [44]; crystal structure of the enzyme with chloride bound at an allosteric site and sulfate bound at the active site) [44, 67]
<14> [17]

Cloning
<1> [26]
<2> [26, 51]
<11> (expressed in Escherichia coli [48]; construction of transgenic Nicotiana tabacum carrying either spinach cytosolic cDNA, designated 3F plants, or chimeric CSAse A cDNA fused with the sequence for chloroplast-targeting transit peptide of pea Rubisco small subunit, designated 4F plants. Generation of F1 transgenic tobacco, highly tolerant to sulfur-containing pollutants, in which Csase activities are enhanced both in cytosol and in the chloroplasts by crossing 3F plants with 4F plants [55]) [42, 48, 55]
<30> [38]
<31> (expressed in Escherichia coli [41]; overexpression of the Atcys-3A gene of the cytosolic isoform in Saccharomyces cerevisiae can support the growth of the yeast cells at high concentrations of sodium chloride, suggesting that the plant protein is able to confer salt tolerance in yeast [54]) [39, 41, 54]
<37> (expression of isoenzyme A, B and C in Escherichia coli [62]) [62]
<38> [63]
<39> (expression in Escherichia coli [64]) [64]

Engineering

S272A <1> (<1>, mutant enzyme catalyzes the overall reaction, first half-reaction is decreased by factor 3, the decrease in rate of elimination is compensated by an increase in affinity for O-acetyl-L-Ser [66]) [66]

S272D <1> (<1>, mutant enzyme catalyzes the overall reaction [66]) [66]

Application

pharmacology <27> (<27>, involved in β-lactam synthesis [32]) [32]

synthesis <19> (<19>, synthesis of L-Cys, therefore immobilization [37]) [37]

6 Stability

pH-Stability

6-10 <14> (<14> 37°C, 60 min, stable [18]) [18]

8-12 <35> (<35>, 50°C, 30 min, stable [58]) [58]

Temperature stability

0 <2> (<2>, 12 h, about 20% loss of activity [57]) [57]

2 <8> (<8>, inactivated above, after a few days [9]) [9]

20-40 <2> (<2>, 12 h, stable [57]) [57]

21 <36> (<36>, pH 8.0, 50 mM dithiothreitol, 0.08 mM pyridoxal phosphate, 30 min, stable [61]) [61]

40 <3> (<3>, stable below [4]) [4]

50 <11, 14, 15> (<14>, 10 min, unstable above [18]; <15>, 10 min, no loss of activity [21]; <11>, 2 min, all isoenzymes stable [33]) [18, 21, 33]

55 <3> (<3>, 5 min, 45% loss of activity, without pyridoxal 5'-phosphate [6]) [6]

60 <2, 11, 14, 15> (<11>, 1 min, no loss of activity [45]; <14>, 10 min, no loss of activity [18]; <15>, 10 min, 85% loss of activity [21]; <2>, 12 h, about 15% loss of activity [57]) [18, 21, 45, 57]

62 <1> (<1>, 10 min, 5% loss of activity [1]) [1]

65 <1, 3, 11> (<3>, 5 min, complete loss of activity [4]; <3>, 5 min, complete loss of activity, without pyridoxal 5'-phosphate [6]; <1>, 5 min, unstable above [22]; <11>, 2 min, isoenzyme 1 loses 50% loss of activity, isoenzymes 2 loses 25% of its activity, no loss of activity, isoenzyme 3 [33]) [4, 6, 22, 33]

67 <1> (<1>, 10 min, 50% loss of activity [1]) [1]

70 <1, 6, 19, 35> (<1>, 10 min, 70% loss of activity [1]; <6>, 3 min, stable [5]; <19>, 10 min, 10% loss of activity of immobilized enzyme, 65-80% loss of activity of native enzyme [37]; <35>, pH 7.8, 60 min, stable [58]) [1, 5, 37, 58]

77 <1> (<1>, 5 min, complete loss of activity [22]) [22]

80 <6, 7, 35> (<6,7>, 3 min, unstable [5,8]; <35>, 60 min, 10 mM dithiothreitol, less than 10% of activity [58]) [5, 8, 58]

100 <39> (<39>, pH 6.1 and 6.7, 6 h, 10% loss of activity [64]) [64]

Additional information <3> (<3>, pyridoxal 5'-phosphate stabilizes against heat inactivation [6]) [6]

General stability information

<1>, holo-O-acetylserine sulfhydrylase exhibits greater conformational stability than the apoenzyme form. Role of pyridoxal 5'-phosphate in the structural stabilization of O-acetylserine sulfhydrylase [65]

<6>, relatively stable [5]

<11>, chloroplast enzyme very unstable [47]

<19>, very unstable [30]

<36>, stable to several freeze-thaw cycles [61]

Storage stability

<1>, -20°C, Tris-HCl buffer, pH 7.6, bovine serum albumine, 8 months, 10% loss of activity [1]

<1>, 20°C, room temperature, Tris-HCl buffer, pH 7.6, bovine serum albumine, several days, no loss of activity [1]

<3>, -20°C, potassium phosphate buffer, pH 6.5, 1 mM EDTA, at least several days, no loss of activity [6]

<8>, -15°C, phosphate buffer, pH 7.2, several months, no loss of activity [9]

<8>, 2°C, inactivated above, after a few days [9]

<9>, -15°C, potassium phosphate buffer, dithiothreitol, one week, 50% loss of activity [10]

<11>, 0°C, potassium phosphate buffer, pH 8.0, 2-mercaptoethanol, EDTA, several months, no loss of activity [25]

<18>, 0°C, potassium phosphate buffer, pH 8, mercaptoethanol, EDTA, three months, no loss of activity [27]

<19>, -20°C, phosphate buffer, pH 7.5, glycerol, pyridoxal-5'-phosphate, 8 days, 50% loss of activity [30]

<30>, -20°C, polybuffer 74, pH 4.0, glycerol, 4 months [38]

<32>, -70°C, Tris-HCl, pH 8.1, glycerol, 6 months, no loss of activity [40]

<36>, -20°C, complete loss of activity after 2 months [61]

References

[1] Kredich, N.M.; Becker, M.A.: Cysteine biosynthesis: serine transacetylase and O-acetylserine sulfhydrylase. Methods Enzymol., **17 B**, 459-470 (1971)

[2] Thompson, J.F.; Moore, D.P.: Enzymatic synthesis of cysteine and S-methylcysteine in plant extracts. Biochem. Biophys. Res. Commun., **31**, 281-287 (1968)

[3] Becker, M.A.; Kredich, N.M.; Tomkins, G.M.: The purification and characterization of O-acetylserine sulfhydrylase-A from Salmonella typhimurium. J. Biol. Chem., **244**, 2418-2427 (1969)

[4] Yamagata, S.; Takeshima, K.; Naiki, N.: Evidence for the identity of O-acetylserine sulfhydrylase with O-acetylhomoserine sulfhydrylase in yeast. J. Biochem., **75**, 1221-1229 (1974)

[5] Tamura, G.; Iwasawa, T.; Masada, M.; Fukushima, K.: Some properties of cysteine synthase from radish roots. Agric. Biol. Chem., **40**, 637-638 (1976)

[6] Yamagata, S.; Takeshima, K.: O-Acetylserine and O-acetylhomoserine sulf-hydrylase of yeast. Further purification and characterization as a pyridoxal enzyme. J. Biochem., **80**, 777-785 (1976)

[7] Yamagata, S.: O-Acetylserine and O-acetylhomoserine sulfhydrylase of yeast. Subunit structure. J. Biochem., **80**, 787-797 (1976)

[8] Burnell, J.N.; Whatley, F.R.: Sulphur metabolism in Paracoccus denitrificans. Purification, properties and regulation of serine transacetylase, O-acetylserine sulphydrylase and β-cystathionase. Biochim. Biophys. Acta, **481**, 246-265 (1977)

[9] Ascano, A.; Nicholas, D.J.D.: Purification and properties of O-acetyl-L-serine sulphhydrolase from wheat leaves. Phytochemistry, **16**, 889-893 (1977)

[10] Hock Ng, B.; Anderson, J.W.: Chloroplast cysteine synthases of Trifolium repens and Pisum sativum. Biochemistry, **17**, 879-885 (1978)

[11] Cook, P.F.; Wedding, R.T.: Cysteine synthetase from Salmonella typhimurium LT-2. Aggregation, kinetic behavior, and effect of modifiers. J. Biol. Chem., **253**, 7874-7879 (1978)

[12] Fankhauser, H.; Brunold, C.: Localization of O-acetyl-L-serine sulphydrylase in Spinacia oleracea L. Plant Sci. Lett., **14**, 185-192 (1979)

[13] Yamagata, S.: Occurrence of low molecular weight O-acetylserine sulfhydrylase in the yeast Saccharomyces cerevisiae. J. Biochem., **88**, 1419-1423 (1980)

[14] Diessner, W.; Schmidt, A.: Isoenzymes of cysteine synthase in the cyanobacterium Synechococcus 6301. Z. Pflanzenphysiol., **102**, 57-68 (1981)

[15] Yamagata, S.: Low-molecular-weight O-acetylserine sulfhydrylase and serine sulfhydrylase of Saccharomyces cerevisiae are the same protein. J. Bacteriol., **147**, 688-690 (1981)

[16] Hensel, G.; Trueper, H.G.: O-acetylserine sulfhydrylase and S-sulfocysteine synthase activities of Chromatium vinosum. Arch. Microbiol., **130**, 228-233 (1981)

[17] Yanase, H.; Sakai, T.; Tonumura, K.: Purification and crystallization of a β-cyanoalanine-forming enzyme from Enterobacter sp. 10-1. Agric. Biol. Chem., **46**, 355-361 (1982)

[18] Yanase, H.; Sakai, T.; Tonomura, K.: Some properties of a β-cyanoalanine-forming enzyme of Enterobacter sp. 10-1. Agric. Biol. Chem., **46**, 363-369 (1982)

[19] Yamagata, S.; Kawai, T.; Takeshima, K.: Partial purification and comparison of some properties of L-serine sulfhydro-lyase of Saccharomyces cerevisiae. Biochim. Biophys. Acta, **701**, 334-338 (1982)

[20] Hensel, G.; Trueper, H.G.: O-Acetylserine sulfhydrylase and S-sulfocysteine synthase activities of Rhodospirillum tenue. Arch. Microbiol., **134**, 227-232 (1983)

[21] Rosichan, J.L.; Blake, N.; Stallard, R.; Owais, W.M.; Kleinhofs, A.; Nilan, R.A.: O-acetylserine (thiol)-lyase from barley converts sodium azide to a mutagenic metabolite. Biochim. Biophys. Acta, **748**, 367-373 (1983)

[22] Nakamura, T.; Iwahashi, A.; Eguchi, Y.: Enzymatic proof for the identity of the S-sulfocysteine synthase and cysteine synthase B of Salmonella typhimurium. J. Bacteriol., **158**, 1122-1127 (1984)

[23] Burnell, J.N.: Sulfate assimilation in C_4 plants. Plant Physiol., **75**, 873-875 (1984)

[24] Nagasawa, T.; Dhillon, G.S.; Ishii, T.; Yamada, H.: Enzymatic synthesis of L-cysteine by O-acetylserine sulfhydrylase of 3-chloro-L-alanine resistant Bacillus sphaericus L-118. J. Bacteriol., **2**, 365-367 (1985)

[25] Murakoshi, I.; Ikegami, F.; Kaneko, M.: Purification and properties of cysteine synthase from Spinacia oleracea. Phytochemistry, **24**, 1907-1911 (1985)

[26] Byrne, C.R.; Monroe, R.S.; Ward, K.A.; Kredich, N.M.: DNA sequences of the cysK regions of Salmonella typhimurium and Escherichia coli and linkage of the cysK regions to ptsH. J. Bacteriol., **170**, 3150-3157 (1988)

[27] Ikegami, F.; Kaneko, M.; Kobori, M.; Murakoshi, I.: Purification and characterization of cysteine synthase from Brassica juncea. Phytochemistry, **27**, 3379-3383 (1988)

[28] Ikegami, F.; Kaneko, M.; Kamiyama, H.; Murakoshi, I.: Purification and characterization of cysteine synthases from Citrullus vulgaris. Phytochemistry, **27**, 697-701 (1988)

[29] Ikegami, F.; Kaneko, M.; Lambein, F.; Kuo, Y.-H.; Murakoshi, I.: Difference between uracilylalanine synthases and cysteine synthases in Pisum sativum. Phytochemistry, **26**, 2699-2704 (1987)

[30] Leon, J.; Romero, L.C.; Galvan, F.; Vega, J.M.: Purification and physicochemical characterization of O-acetyl-L-serine sulfhydrylase from Chlamydomonas reinhardtii. Plant Sci., **53**, 93-99 (1953)

[31] Krauss, F.; Schmidt, A.: Sulfur sources for growth of Chlorella fusca and their influence on key enzymes of sulfur metabolism. J. Gen. Microbiol., **133**, 1209-1319 (1987)

[32] Ostergaard, S.; Theilgaard, H.B.A.; Nielsen, J.: Identification and purification of O-acetyl-L-serine sulphhydrylase in Penicillium chrysogenum. Appl. Microbiol. Biotechnol., **50**, 663-668 (1998)

[33] Yamaguchi, T.; Masada, M.: Comparative studies on cysteine synthase isozymes from spinach leaves. Biochim. Biophys. Acta, **1251**, 91-98 (1995)

[34] Ikegami, F.; Itagaki, S.; Murakoshi, I.: Purification and characterization of two forms of cysteine synthase from Allium tuberosum. Phytochemistry, **32**, 31-34 (1993)

[35] Leon, J.; Vega, J.M.: Separation and regulatory properties of O-acetyl-L-serine sulfhydrylase isoenzymes from Chlamydomonas reinhardtii. Plant Physiol. Biochem., **29**, 595-599 (1991)

[36] Maruyama, A.; Ishizawa, K.; Takagi, T.; Esashi, Y.: Cytosolic β-cyanoalanine synthase activity attributed to cysteine synthases in cocklebur seeds. Purification and characterization of cytosolic cysteine synthases. Plant Cell Physiol., **39**, 671-680 (1998)

[37] Leon, J.; Vega, J.M.: Effect of immobilization on the kinetic ans stability properties of o-acetyl-L-serine sulfhydrylase from Chlamydomonas reinhardtii. Biocatalysis, **7**, 29-35 (1992)

[38] Roemer, S.; d'Harlingue, A.; Camara, B.; Schantz, R.; Kuntz, M.: Cysteine synthase from Capsicum annuum chromoplasts. Characterization and

cDNA cloning of an up-regulated enzyme during fruit development. J. Biol. Chem., **267**, 17966-17970 (1992)

[39] Hell, R.; Bork, C.; Bogdanova, N.; Frolov, I.; Hauschild, R.: Isolation and characterization of two cDNAs encoding for compartment specific isoforms of O-acetylserine (thiol) lyase from Arabidopsis thaliana. FEBS Lett., **351**, 257-262 (1994)

[40] Kuske, C.R.; Ticknor, L.O.; Guzman, E.; Gurley, L.R.; Valdez, J.G.; Thompson, M.E.; Jackson, P.J.: Purification and characterization of O-acetylserine sulfhydrylase isoenzymes from Datura innoxia. J. Biol. Chem., **269**, 6223-6232 (1994)

[41] Droux, M.; Ruffet, M.-L.; Douce, R.; Job, D.: Interactions between serine acetyltransferase and O-acetylserine (thiol) lyase in higher plants–structural and kinetic properties of the free and bound enzymes. Eur. J. Biochem., **255**, 235-245 (1998)

[42] Rolland, N.; Ruffet, M.-L.; Job, D.; Douce, R.; Droux, M.: Spinach chloroplast 0-acetylserine (thiol)-lyase exhibits two catalytically non-equivalent pyridoxal-5 -phosphate-containing active sites. Eur. J. Biochem., **236**, 272-282 (1996)

[43] Flint, D.H.; Tuminello, J.F.; Miller, T.J.: Studies on the synthesis of the Fe-S cluster of dihydroxy-acid dehydratase in Escherichia coli crude extract. Isolation of O-acetylserine sulfhydrylases A and B and β-cystathionase based on their ability to mobilize sulfur from cysteine and to participate. J. Biol. Chem., **271**, 16053-16067 (1996)

[44] Burkhard, P.; Rao, G.S.J.; Hohenester, E.; Schnackerz, K.D.; Cook, P.F.; Jansonius, J.N.: Three-dimensional structure of O-acetylserine sulfhydrylase from Salmonella typhimurium. J. Mol. Biol., **283**, 121-133 (1998)

[45] Yamaguchi, T.; Zhu, X.; Masada, M.: Purification and characterization of a novel cysteine synthase isozyme from spinach hydrated seeds. Biosci. Biotechnol. Biochem., **62**, 501-507 (1998)

[46] Zhu, X.; Yamaguchi, T.; Masada, M.: Complexes of serine acetyltransferase and isoenzymes of cysteine synthase in spinach leaves. Biosci. Biotechnol. Biochem., **62**, 947-952 (1997)

[47] Droux, M.; Martin, J.; Sajus, P.; Douce, R.: Purification and characterization of O-acetylserine (thiol) lyase from spinach chloroplasts. Arch. Biochem. Biophys., **295**, 379-390 (1992)

[48] Rolland, N.; Droux, M.; Lebrun, M.; Douce, R.: O-acetylserine(thiol)lyase from spinach (Spinacia oleracea L.) leaf: cDNA cloning, characterization, and overexpression in Escherichia coli of the chloroplast isoform. Arch. Biochem. Biophys., **300**, 213-222 (1993)

[49] Tai, C.-H.; Nalabolu, S.R.; Jacobson, T.M.; Minter, D.E.; Cook, P.F.: Kinetic mechanisms of the A and B isozymes of O-acetylserine sulfhydrylase from Salmonella typhimurium LT-2 using the natural and alternative reactants. Biochemistry, **32**, 6433-6442 (1993)

[50] Schnackerz, K.D.; Tai, C.-H.; Simmons, J.W.; Jacobson, T.M.; Rao, G.S.J.; Cook, P.F.: Identification and spectral characterization of the external aldimine of the O-acetylserine sulfhydrylase reaction. Biochemistry, **34**, 12152-12160 (1995)

[51] Owais, W.M.; Gharaibeh, R.: Cloning of the E. coli O-acetylserine sulfhy-drylase gene: ability of the clone to produce a mutagenic product from azide and O-acetylserine. Mut. Res., **245**, 151-155 (1990)

[52] Ikegami, F.; Kamiya, M.; Kuo, Y.-H.; Lambein, F.; Murakoshi, I.: Enzymatic synthesis of two isoxazolylalanine isomers by cysteine synthases in Lathyrus species. Biol. Pharm. Bull., **19**, 1214-1215 (1996)

[53] Warrilow, A.G.S.; Hawkesford, M.J.: Cysteine synthase (O-acetylserine (thiol) lyase) substrate specificities classify the mitochondrial isoform as a cyanoalanine synthase. J. Exp. Bot., **51**, 985-993 (2000)

[54] Romero, L.C.; Dominguez-Solis, J.R.; Gutierrez-Alcala, G.; Gotor, C.: Salt regulation of O-acetylserine(thiol)lyase in Arabidopsis thaliana and in-creased tolerance in yeast. Plant Physiol. Biochem., **39**, 643-647 (2001)

[55] Noji, M.; Saito, K.: Molecular and biochemical analysis of serine acetyl-transferase and cysteine synthase towards sulfur metabolic engineering in plants. Amino Acids, **22**, 231-243 (2002)

[56] Tai, C.-H.; Burkhard, P.; Gani, D.; Jenn, T.; Johnson, C.; Cook, P.F.: Charac-terization of the allosteric anion-binding site of O-acetylserine sulfhydry-lase. Biochemistry, **40**, 7446-7452 (2001)

[57] Mino, K.; Yamanoue, T.; Sakiyama, T.; Eisaki, N.; Matsuyama, A.; Nakanishi, K.: Effects of bienzyme complex formation of cysteine synthetase from Escherichia coli on some properties and kinetics. Biosci. Biotechnol. Bio-chem., **64**, 1628-1640 (2000)

[58] Mizuno, Y.; Miyashita, Y.; Yamagata, S.; Iwama, T.; Akamatsu, T.: Cysteine synthase of an extremely thermophilic bacterium, Thermus thermophilus HB8. Biosci. Biotechnol. Biochem., **66**, 549-557 (2002)

[59] Borup, B.; Ferry, J.G.: Cysteine biosynthesis in the Archaea: Methanosarcina thermophila utilizes O-acetylserine sulfhydrylase. FEMS Microbiol. Lett., **189**, 205-210 (2000)

[60] Mino, K.; Ishikawa, K.: Characterization of a novel thermostable O-acetyl-serine sulfhydrylase from Aeropyrum pernix K1. J. Bacteriol., **185**, 2277-2284 (2003)

[61] Borup, B.; Ferry, J.G.: O-Acetylserine sulfhydrylase from Methanosarcina thermophila. J. Bacteriol., **182**, 45-50 (2000)

[62] Burandt, P.; Schmidt, A.; Papenbrock, J.: Three O-acetyl-L-serine(thiol)lyase isoenzymes from Arabidopsis catalyse cysteine synthesis and cysteine de-sulfuration at different pH values. J. Plant Physiol., **159**, 111-119 (2002)

[63] Maruyama, A.; Saito, K.; Ishizawa, K.: β-Cyanoalanine synthase and cy-steine synthase from potato: molecular cloning, biochemical characteriza-tion, and spatial and hormonal regulation. Plant Mol. Biol., **46**, 749-760 (2001)

[64] Mino, K.; Ishikawa, K.: Characterization of a novel thermostable O-acetyl-serine sulfhydrylase from Aeropyrum pernix K1. J. Bacteriol., **185**, 2277-2284 (2003)

[65] Bettati, S.; Benci, S.; Campanini, B.; Raboni, S.; Chirico, G.; Beretta, S.; Schnackerz, K.D.; Hazlett, T.L.; Gratton, E.; Mozzarelli, A.: Role of pyridoxal 5'-phosphate in the structural stabilization of O-acetylserine sulfhydrylase. J. Biol. Chem., **275**, 40244-40251 (2000)

[66] Daum, S.; Tai, C.-H.; Cook, P.F.: Characterization of the S272A,D site-directed mutations of O-acetylserine sulfhydrylase: involvement of the pyridine ring in the α,β-elimination reaction. Biochemistry, 42, 106-113 (2003)

[67] Burkhard, P.; Tai, C.H.; Jansonius, J.N.; Cook, P.F.: Identification of an allosteric anion-binding site on O-acetylserine sulfhydrylase: structure of the enzyme with chloride bound. J. Mol. Biol., 303, 279-286 (2000)

Cystathionine γ-synthase 2.5.1.48

1 Nomenclature

EC number
2.5.1.48

Systematic name
O^4-succinyl-L-homoserine:L-cysteine S-(3-amino-3-carboxypropyl)transferase

Recommended name
cystathionine γ-synthase

Synonyms
CS,26
CgS
EC 4.2.99.9 (formerly)
O-succinyl-L-homoserine succinate-lyase (adding cysteine)
O-succinylhomoserine (thiol)-lyase
O-succinylhomoserine synthase
O-succinylhomoserine synthetase
cystathionine synthase
cystathionine synthetase
cystathionine-γ-synthase
homoserine O-transsuccinylase
homoserine transsuccinylase
synthase, cystathionine γ-

CAS registry number
9030-70-0

2 Source Organism

<1> *Escherichia coli* (enzyme type I, preference for O-succinyl-L-homoserine + L-cysteine as substrates [15]) [1, 13, 15, 19, 22, 23, 28]
<2> *Salmonella typhimurium* (enzyme type I, preference for O-succinyl-L-homoserine + L-cysteine as substrates [15]) [2, 3, 5, 6, 7, 8, 9, 10, 11, 15, 19, 21, 22, 23]
<3> *Lemna paucicostata* [12, 16]
<4> *Erwinia carotovora* (enzyme type I, preference for O-succinyl-L-homoserine + L-cysteine as substrates [15]) [15, 18]
<5> *Pseudomonas putida* (enzyme type I, preference for O-succinyl-L-homoserine + L-cysteine as substrates [15]) [15]

<6> *Acinetobacter calcoaceticus* (enzyme type I, preference for O-succinyl-L-homoserine + L-cysteine as substrates [15]) [15]

<7> *Pseudomonas dacunhae* (enzyme type I, preference for O-succinyl-L-homoserine + L-cysteine as substrates [15]) [15]

<8> *Agrobacterium agrobacter* (enzyme type II, preference for O-acetyl-L-homoserine + L-cysteine as substrates [15]) [15]

<9> *Agrobacterium tumefaciens* (enzyme type II, preference for O-acetyl-L-homoserine + L-cysteine as substrates [15]) [15]

<10> *Bacillus subtilis* (enzyme type II, preference for O-acetyl-L-homoserine + L-cysteine as substrates [15]) [15]

<11> *Bacillus sphaericus* (enzyme type II, preference for O-acetyl-L-homoserine + L-cysteine as substrates [15]) [15, 17, 18]

<12> *Bacillus pumilus* (enzyme type II, preference for O-acetyl-L-homoserine + L-cysteine as substrates [15]) [15]

<13> *Alcaligenes faecalis* (two enzymes: enzyme type I, preference for O-succinyl-L-homoserine + L-cysteine as substrates and enzyme type II, preference for O-acetyl-L-homoserine + L-cysteine as substrates [15]) [15]

<14> *Bacillus aneurinolyticus* [15]

<15> *Streptomyces phaerochromogenes* [14]

<16> *Neurospora crassa* [4]

<17> *Astragalus racemosus* [20]

<18> *Astragalus bisulcatus* [20]

<19> *Neptunia amplexicaulis* [20]

<20> *Astragalus sinicus* [20]

<21> *Astragalus hamosus* [20]

<22> *Saccharomyces cerevisiae* [24, 25]

<23> *Triticum aestivum* (wheat) [26]

<24> *Spinacia oleracea* (spinach) [27]

<25> *Arabidopsis thaliana* [29, 33, 34]

<26> *Solanum tuberosum* [30, 36]

<27> *Corynebacterium glutamicum* [31]

<28> *Nicotiana tabacum* [32, 35]

3 Reaction and Specificity

Catalyzed reaction

O^4-succinyl-L-homoserine + L-cysteine = L-cystathionine + succinate (<2>, can also use hydrogen sulfide and methanethiol as substrates producing homocysteine and methionine respectively. In the absence of thiol, the enzyme can also catalyse β,γ-elimination to form 2-oxobutanoate, succinate and ammonia [2]; <23>, ping-pong mechanism [26]; <28>, ping-pong mechanism [32])

Reaction type
elimination
γ-replacement

108

Natural substrates and products

S   O-succinyl-L-homoserine + L-cysteine <1, 2, 16, 25, 26, 27, 28> (<2> essential step in bacterial methionine biosynthesis [2,6]; <25>, key enzyme of Met biosynthesis [34]; <16>, methionine biosynthesis [4]; <1>, key intermediate in transsulfuration pathways [19]; <26>, the expression of the enzyme is light-inducible [30]; <27>, the second enzyme of the methionine biosynthetic pathway [31]; <28>, the enzyme catalyzes the first reaction specific for methionine biosynthesis [32]; <25>, constitutive overexpression of the enzyme in Arabidopsis leads to accumulation of soluble methionine and S-methylmethionine [33]; <26>, the enzyme catalyzes a near-equilibrium reaction and does not display features of a pathway-regulating enzyme [36]) [2, 4, 6, 19, 30, 31, 32, 33, 34, 36]

P   O-acetyl-L-homocysteine + L-cysteine

Substrates and products

S   L-cystathionine + H_2O <2> (<2>, β-elimination [6]) (Reversibility: ? <2> [6]) [6]

P   L-cysteine + NH_3 + 2-oxobutyrate <2> [6]

S   L-cystathionine + H_2O <2, 24> (<2>, β-elimination [6]) (Reversibility: ? <2,24> [6,27]) [6, 27]

P   L-homocysteine + NH_3 + pyruvate <2, 24> [6, 27]

S   L-cystathionine + L-cysteine <2> [6]

P   L-cystathionine + L-cysteine <2> (<2>, observed via 35S-exchange [6]) [6]

S   O-acetyl-L-homoserine + 2-mercaptoethanol <11> (<11>, 150% relative activity to O-acetyl-L-homoserine + L-cysteine [17]) (Reversibility: ? <11> [17]) [17]

P   S-hydroxyethyl-L-homocysteine + acetate <11> [17]

S   O-acetyl-L-homoserine + 2-mercaptopropionate <11> (<11>, 50% relative activity to O-acetyl-L-homoserine + L-cysteine [17]) (Reversibility: ? <11> [17]) [17]

P   S-methylcarboxymethyl-L-homoserine + acetate <11> [17]

S   O-acetyl-L-homoserine + 3-mercaptopropionate <11> (<11>, 65% relative activity to O-acetyl-L-homoserine + L-cysteine [17]) (Reversibility: ? <11> [17]) [17]

P   S-carboxyethyl-L-homocysteine + acetate <11> [17]

S   O-acetyl-L-homoserine + CH_3SH <11> (<11>, 11% relative activity to O-acetyl-L-homoserine + L-cysteine [17]) (Reversibility: ? <11> [17]) [17]

P   L-methionine + acetate <11> [17]

S   O-acetyl-L-homoserine + D-Cys <11> (<11>, 101% relative activity to O-acetyl-L-homoserine + L-cysteine [17]) (Reversibility: ? <11> [17]) [17]

P   D-allocystathionine + acetate <11> [17]

S   O-acetyl-L-homoserine + D-homocysteine <11> (<11>, 125% relative activity to O-acetyl-L-homoserine + L-cysteine [17]) (Reversibility: ? <11> [17]) [17]

P   meso-homolanthionine + acetate <11> [17]

S O-acetyl-L-homoserine + L-Cys ethyl ester <11> (<11>, 15% relative activity to O-acetyl-L-homoserine + L-cysteine [17]) (Reversibility: ? <11> [17]) [17]

P S-(L-2-amino-2-ethoxycarbonyl)-L-homocysteine + acetate <11> (<11>, i.e. L-cystathionine monoethyl ester + acetate [17]) [17]

S O-acetyl-L-homoserine + L-Cys methyl ester <11, 28> (<11>, 20% relative activity to O-acetyl-L-homoserine + L-cysteine [17]; <28>, 26% of the activity with L-Cys [32]) (Reversibility: ? <11,28> [17,32]) [17, 32]

P S-(L-2-amino-2-methoxycarbonyl)-L-homocysteine + acetate <11> (<11>, i.e. L-cystathionine monomethyl ester + acetate [17]) [17]

S O-acetyl-L-homoserine + L-cysteine <2, 8-15> (<2> γ-replacement [7]) (Reversibility: ? <2,8-15> [7,9,14,15,17]) [7, 9, 14, 15, 17]

P L-cystathionine + acetate <2, 8-14> [7, 9, 15, 17]

S O-acetyl-L-homoserine + L-homocysteine <11> (<11>, 92% relative activity to O-acetyl-L-homoserine + L-cysteine [17]) (Reversibility: ? <11> [17]) [17]

P L-homolanthionine + acetate <11> [17]

S O-acetyl-L-homoserine + N-acetyl-L-Cys <11> (<11>, 12% relative activity to O-acetyl-L-homoserine + L-cysteine [17]) (Reversibility: ? <11> [17]) [17]

P S-(L-acetylamino-2-carboxyethyl)-L-homocysteine + acetate <11> (<11>, i.e. mono-N-acetyl-L-cystathionine + acetate [17]) [17]

S O-acetyl-L-homoserine + S^{2-} <11> (<11>, 163% relative activity to O-acetyl-L-homoserine + L-cysteine [17]) (Reversibility: ? <11> [17]) [17]

P L-homocysteine + acetate <11> [17]

S O-acetyl-L-homoserine + allyl mercaptan <11> (<11>, 4% relative activity to O-acetyl-L-homoserine + L-cysteine [17]) (Reversibility: ? <11> [17]) [17]

P S-allyl-L-homocysteine + acetate <11> [17]

S O-acetyl-L-homoserine + benzyl mercaptan <11> (<11>, 41% relative activity to O-acetyl-L-homoserine + L-cysteine [17]) (Reversibility: ? <11> [17]) [17]

P S-benzyl-L-homocysteine + acetate <11> [17]

S O-acetyl-L-homoserine + ethyl mercaptan <11> (<11>, 24% relative activity to O-acetyl-L-homoserine + L-cysteine [17]) (Reversibility: ? <11> [17]) [17]

P L-thionine + acetate <11> [17]

S O-acetyl-L-homoserine + isobutyl mercaptan <11> (<11>, 13% relative activity to O-acetyl-L-homoserine + L-cysteine [17]) (Reversibility: ? <11> [17]) [17]

P S-isobutyl-L-homocysteine + acetate <11> [17]

S O-acetyl-L-homoserine + isopropyl mercaptan <11> (<11>, 1% relative activity to O-acetyl-L-homoserine + L-cysteine [17]) (Reversibility: ? <11> [17]) [17]

P S-isopropyl-L-homocysteine + acetate <11> [17]

S O-acetyl-L-homoserine + *m*-thiocresol <11> (<11>, 150% relative activity to O-acetyl-L-homoserine + L-cysteine [17]) (Reversibility: ? <11> [17]) [17]

P S-*m*-tolyl-L-homocysteine + acetate <11> [17]

S O-acetyl-L-homoserine + n-butyl mercaptan <11> (<11>, 5% relative activity to O-acetyl-L-homoserine + L-cysteine [17]) (Reversibility: ? <11> [17]) [17]

P S-n-butyl-L-homocysteine + acetate <11> [17]

S O-acetyl-L-homoserine + n-propyl mercaptan <11> (<11>, 13% relative activity to O-acetyl-L-homoserine + L-cysteine [17]) (Reversibility: ? <11> [17]) [17]

P S-n-propyl-L-homocysteine + acetate <11> [17]

S O-acetyl-L-homoserine + o-thiocresol <11> (<11>, 3% relative activity to O-acetyl-L-homoserine + L-cysteine [17]) (Reversibility: ? <11> [17]) [17]

P S-*o*-tolyl-L-homocysteine + acetate <11> [17]

S O-acetyl-L-homoserine + p-nitrothiophenol <11> (<11>, 41% relative activity to O-acetyl-L-homoserine + L-cysteine [17]) (Reversibility: ? <11> [17]) [17]

P S-p-nitrophenyl-L-homocysteine + acetate <11> [17]

S O-acetyl-L-homoserine + p-thiocresol <11> (<11>, 25% relative activity to O-acetyl-L-homoserine + L-cysteine [17]) (Reversibility: ? <11> [17]) [17]

P S-p-tolyl-L-homocysteine + acetate <11> [17]

S O-acetyl-L-homoserine + phenyl mercaptan <11> (<11>, 98% relative activity to O-acetyl-L-homoserine + L-cysteine [17]) (Reversibility: ? <11> [17]) [17]

P S-phenyl-L-homocysteine + acetate <11> [17]

S O-acetyl-L-homoserine + sec-butyl mercaptan <11> (<11>, 5% relative activity to O-acetyl-L-homoserine + L-cysteine [17]) (Reversibility: ? <11> [17]) [17]

P S-sec-butyl-L-homocysteine + acetate <11> [17]

S O-acetyl-L-homoserine + tert-butyl mercaptan <11> (<11>, 2% relative activity to O-acetyl-L-homoserine + L-cysteine [17]) (Reversibility: ? <11> [17]) [17]

P S-tert-butyl-L-homosysteine + acetate <11> [17]

S O-acetyl-L-homoserine + thioacetic acid <11> (<11>, 3% relative activity to O-acetyl-L-homoserine + L-cysteine [17]) (Reversibility: ? <11> [17]) [17]

P S-acetyl-L-homocysteine + acetate <11> [17]

S O-acetyl-L-homoserine + thioglycolate ethyl ester <11> (<11>, 348% relative activity to O-acetyl-L-homoserine + L-cysteine [17]) (Reversibility: ? <11> [17]) [17]

P S-ethoxycarbonyl-L-homocysteine + acetate <11> [17]

S O-acetyl-L-homoserine + thioglycolate n-butyl ester <11> (<11>, 274% relative activity to O-acetyl-L-homoserine + L-cysteine [17]) (Reversibility: ? <11> [17]) [17]

P S-n-butoxycarbonyl-L-homocysteine + acetate <11> [17]

S O-acetyl-L-homoserine + thiosalicylic acid <11> (<11>, 63% relative activity to O-acetyl-L-homoserine + L-cysteine [17]) (Reversibility: ? <11> [17]) [17]

P S-*o*-carboxyphenyl-L-homocysteine + acetate <11> [17]

S O-malonyl-L-homoserine + L-cysteine <11> (<11>, 14% relative activity to O-acetyl-L-homoserine + L-cysteine [17]) (Reversibility: ? <11> [17]) [17]

P L-cystathionine + malonate <11> [17]

S O-phenyl-L-homoserine + L-cysteine <11> (<11>, 1% relative activity to O-acetyl-L-homoserine + L-cysteine [17]) (Reversibility: ? <11> [17]) [17]

P L-cystathionine + phenol <11> [17]

S O-phospho-L-homoserine + L-cysteine <3, 17-21, 23, 24> (Reversibility: ? <3, 17-21, 23, 24> [12, 16, 20, 26, 27]) [12, 16, 20, 26, 27]

P L-cystathionine + phosphate <17-21, 23, 24> [20, 26, 27]

S O-phospho-L-homoserine + L-cysteine ethyl ester <23> (<23>, 1% relative activity to O-phospho-L-homoserine + L-cysteine [26]) (Reversibility: ? <23> [26]) [26]

P S-(L-2-amino-2-ethoxycarbonyl)-L-homocysteine + phosphate

S O-phospho-L-homoserine + L-cysteine methyl ester <23> (<23>, 15% relative activity to O-phospho-L-homoserine + L-cysteine [26]) (Reversibility: ? <23> [26]) [26]

P S-(L-2-amino-2-methoxycarbonyl)-L-homocysteine + phosphate

S O-phospho-L-homoserine + L-selenocysteine <17> (Reversibility: ? <17> [20]) [20]

P L-selenocystathionine + phosphate <17> [20]

S O-phospho-L-homoserine + S^{2-} <23, 24> (<23>, 13% relative activity to O-phospho-L-homoserine + L-cysteine [26]) (Reversibility: ? <23, 24> [26, 27]) [26, 27]

P L-homocysteine + phosphate <23> [26]

S O-succinyl-L-homoserine + 2-mercaptoethanol <28> (<28>, 32% of the activity with L-cysteine [32]) (Reversibility: ? <28> [32]) [32]

P S-ethyl-L-homocysteine + succinate

S O-succinyl-L-homoserine + CH₃SH <2> (Reversibility: ? <2> [2,7]) [2, 7]

P L-methionine + succinate <2> [2, 7]

S O-succinyl-L-homoserine + D-cysteine <28> (<28>, 20% of the activity with L-cysteine [32]) (Reversibility: ? <28> [32]) [32]

P ?

S O-succinyl-L-homoserine + H₂O <1, 2> (<2>, γ-elimination, in absence of L-cysteine [2,6]) (Reversibility: ? <1,2> [2, 3, 5-11, 13, 21, 22]) [2, 3, 5-11, 13, 21, 22]

P succinate + 2-oxobutyrate + NH₃ <1, 2> [2, 5, 8, 9, 10, 11, 13, 21, 22]

S O-succinyl-L-homoserine + H₂S <2> (Reversibility: ? <2> [2,7]) [2, 7]

P L-homocysteine + succinate <2> [2, 7]

S O-succinyl-L-homoserine + L-cysteine <1-7, 11, 13, 16, 22, 26, 27, 28> (Reversibility: ? <1-7, 11, 13, 16, 22, 26, 27, 28> [1-12, 15, 17, 18, 21, 22, 24, 30, 31, 35]) [1-12, 15, 17, 18, 21, 22, 24, 30, 31, 35]

P cystathionine + succinate <1, 2, 3, 4, 5, 6, 7, 11, 13, 16, 22> [1-12, 15, 17, 18, 21, 22, 24]

S O-succinyl-L-homoserine + β-mercaptopropionate <2> (Reversibility: ? <2> [8]) [8]

P S-carboxyethyl-L-homocysteine + succinate <2> [8]

S O-succinyl-L-homoserine + sodium sulfide <28> (<28>, 62% of the activity with L-cysteine [32]) (Reversibility: ? <28> [32]) [32]

P L-homocysteine + sodium succinate

S O-succinyl-L-serine + H_2O <2> (<2>, β-elimination [5,9]) (Reversibility: ? <2> [5,9]) [5, 9]

P succinate + pyruvate + NH_3 <2> [5, 9]

S O-succinyl-L-serine + L-cysteine <2> (Reversibility: ? <2> [5]) [5]

P lanthionine + succinate <2> [5]

S O-succinyl-L-serine + L-homocysteine <2> (<2>, β-replacement, very low reaction rate [5,9]) (Reversibility: ? <2> [5,9]) [5, 9]

P succinate + L-cystathionine <2> [5, 9]

S Additional information <1, 2, 11> (<2>, exchange of tritium from water into both α and β positions of a wide variety of amino acids which are otherwise no substrates [9]; <11>, β-elimination and γ-elimination barely catalyzed [17]; <1> catalyzes α or β proton exchange with a number of amino acids and L-allylglycine [21,23]) [9, 17, 21, 23]

P ?

Inhibitors

(Z)-3-(2-phosphonethen-1-yl)pyridine-2-carboxylic acid <23> (<23>, K_i: 0.04 mM [26]) [26]

3-(phosphonomethyl)pyridine-2-carboxylic acid <28> [32]

3-methyl-2-benzothiazolinone hydrazone <11> [17]

4-(phosphonomethyl)pyridine-2-carboxylic acid <23, 28> (<23>, K_i: 0.045 mM [26]) [26, 32]

D-cycloserine <11> [17]

D-cysteine <2> (<2>, inhibition of γ-elimination [7]) [7]

DL-(E)-2-amino-5-phosphono-3-pentenoic acid <23, 28> (<23>, K_i: 0.0011 mM [26]) [26, 32]

DL-propargylglycine <28> (<28>, irreversible [32]) [32]

L-cysteine <2> (<2>, complete inhibition of γ-elimination and β-elimination [5]; <2> inhibition of β-elimination [6]) [5, 6]

L-homocysteine <2> (<2>, partial inhibition of β-elimination and γ-elimination [5]) [5, 7]

N-ethylmaleimide <2> (<2>, inhibition of γ-elimination [7]) [7]

S-adenosyl-L-homocysteine <16> [4]

S-adenosyl-L-methionine <16, 23> (<16>, 0.003 mM, 34% inhibition, 0.1 mM, 96% inhibition, physiological inhibitor [4]; <23>, 9.5 mM, 50% inhibition [26]) [4, 26]

aminooxyacetic acid <23> [26]

β-mercaptopropionate <2> (<2>, competitive inhibitor of γ-replacement reaction [7]; <2>, also a substrate for γ-replacement reaction [8]) [7, 8]
chloromercuriphenylsulfonate <2> (<2>, inhibition of γ-elimination [7]) [7]
cystathionine <2> (<2>, competitive, γ-elimination [7]) [7]
dithiothreitol <2> (<2>, 10% inhibition of β-elimination from O-succinyl-L-serine [5]) [5]
hydroxylamine <11, 23> [17, 26]
phenylhydrazine <11> [17]
propargylglycine <2, 3, 23, 24> (<2> i.e. 2-amino-4-pentynoate [11]; <3>, pseudo-first order inactivation kinetics [12]) [11, 12, 23, 26, 27]
semicarbazide <11> [17]
Additional information <11, 23> (<11> not inhibited by p-chloromercuribenzoate, iodoacetate, N-ethylmaleimide, 5,5'-dithiobis(2-nitrobenzoate), NaCN [17]; <23>, no inhibition by 2-amino-1-hydroxycyclobutane-1-acetic acid [26]) [17, 26]

Cofactors/prosthetic groups
pyridoxal 5'-phosphate <1, 2, 11, 23, 28> (<2> K_m: 0.00004 mM [6]; <2>, 4 mol tightly bound per mol of enzyme [7]; <11>, 4 mol per mol of enzyme, K_m: 0.021 mM [17]; <1>, binding site [19]; <1> 4 mol per tetrameric holoenzyme [22]; <28>, enzyme is dependent on [32]) [3, 6, 7, 11, 17, 19, 22, 26, 32]

Turnover number (min^{-1})
8 <2> (O-succinyl-L-serine, <2> γ-replacement reaction, pH 8.2, 37°C [5]) [5]
90 <2> (O-succinyl-L-serine, <2> γ-elimination reaction, final, inhibited reaction phase, pH 7.3, 30°C [5]) [5]
230 <2> (O-succinyl-L-serine, <2> γ-elimination reaction, initial reaction phase, pH 7.3, 30°C [5]) [5]
240 <2> (O-succinyl-L-homoserine, <2> γ-elimination reaction, pH 7.3, 30°C [5]) [5]
460 <1> (cystathionine, <1>, pH 8.2, 25°C, 460 mol product formed per mol of subunit [22]) [22]
760 <2> (O-succinyl-L-homoserine, <2> γ-elimination reaction, pH 8.2, 37°C [5]) [5]
3800 <2> (O-succinyl-L-homoserine, <2> γ-replacement reaction, pH 8.2, 37°C [5]) [5]

Specific activity (U/mg)
3.7 <27> (<27>, plasmid pSL109 [31]) [31]
9 <1> [22]
10 <1> [13]
10.5 <2> (<2>, O-succinyl-L-homoserine + L-cysteine, γ-replacement reaction [6]) [6]
13.11 <24> [27]
18.2 <2> [7]

18.7 <2> (<2>, O-succinyl-L-homoserine + L-cysteine, γ-replacement reaction [5]) [5]

37 <27> (<27>, plasmid pSL123 [31]) [31]

118 <11> [17]

K$_m$-Value (mM)

0.05 <2> (O-succinyl-L-homoserine, <2>, pH 7.3, 30°C, reaction with H$_2$S [2]) [2]

0.07 <17> (L-selenocysteine) [20]

0.07 <2> (O-succinyl-L-homoserine, <2>, pH 7.3, 30°C, γ-replacement reaction [2]) [2]

0.14 <17> (L-cysteine) [20]

0.18 <24> (L-cysteine, <24>, pH 7.5, 30°C [27]) [27]

0.23 <28> (L-cysteine) [32]

0.3 <2> (O-succinyl-L-homoserine, <2>, pH 7.3, 30°C, γ-elimination reaction [2]) [2]

0.33 <1> (O-succinyl-L-homoserine, <1>, pH 8.2, 25°C, γ-elimination reaction [22]) [22]

0.5 <23> (L-cysteine, <23>, pH 7.5, 37°C [26]) [26]

0.55 <28> (sodium sulfide) [32]

0.6 <24> (S^{2-}, <24>, pH 7.5, 30°C [27]) [27]

0.7 <2> (O-succinyl-L-serine, <2>, final, inhibited reaction phase [5]) [5]

0.89 <28> (L-cysteine methyl ester) [32]

1.2 <2> (O-succinyl-L-serine, <2>, initial reaction phase [5]) [5]

1.4 <24> (O-phospho-L-homoserine, <24>, pH 7.5, 30°C [27]) [27]

2.6 <28> (2-mercaptoethanol) [32]

2.85 <17> (O-phospho-L-serine) [20]

3 <2> (H$_2$S, <2>, pH 7.3, 30°C, γ-replacement reaction [2]) [2]

3.6 <23> (O-phospho-L-homoserine, <23>, pH 7.5, 37°C [26]) [26]

4 <2> (O-succinyl-L-homoserine, <2>, pH 7.3, 30°C, γ-replacement reaction [2]) [2]

4.07 <4> (L-cysteine) [18]

7.1 <28> (O-phospho-L-homoserine) [32]

9 <2> (O-succinyl-L-homoserine, <2>, pH 7.3, 30°C, reaction wirh CH$_3$SH [2]) [2]

13.34 <24> (O-succinyl-L-homoserine, <24>, pH 7.5, 30°C [27]) [27]

45.5 <4, 11> (O-succinyl-L-homoserine) [18]

64.5 <4, 11> (O-acetyl-L-homoserine) [18]

100 <2> (CH$_3$SH, <2>, pH 7.3, 30°C, γ-replacement reaction [2]) [2]

Additional information <2> (<2> kinetic analysis of the reaction O-succinyl-L-homoserine + β-mercaptopropionate [8]) [8]

K$_i$-Value (mM)

0.0011 <23> (DL-(E)-2-amino-5-phosphono-3-pentenoic acid, <23>, pH 7.5, 37°C [26]) [26]

0.018 <28> (DL-propargylglycine) [32]

0.027 <28> (DL-(E)-2-amino-5-phosphono-3-pentenoic acid) [32]

0.04 <23> ((Z)-3-(2-phosphonethen-1-yl)pyridine-2-carboxylic acid, <23>, pH 7.5, 37°C [26]) [26]

0.045 <23> (4-(phosphonomethyl)pyridine-2-carboxylic acid, <23>, pH 7.5, 37°C [26]) [26]

0.2 <28> (3-(phosphonomethyl)pyridine-2-carboxylic acid) [32]

0.3 <28> (4-(phosphonomethyl)pyridine-2-carboxylic acid) [32]

0.45 <28> ((Z)-3-(2-phosphonoethen-1-yl)pyridine-2-carboxylic acid) [32]

9.5 <23> (S-adenosyl-L-methionine, <23>, pH 7.5, 37°C [26]) [26]

pH-Optimum

7.4 <24> [27]

7.5 <11, 23> [17, 26]

7.5-8 <4, 11> [18]

8.2 <2> (<2>, O-succinyl-L-homoserine + H_2O [5]) [5]

pH-Range

6.5 <2> (<2>, O-succinyl-L-homoserine + H_2O, 12% of the reaction rate at pH 8.2 [5]; <2>, O-succinyl-L-serine + H_2O, 28% of the reaction rate at pH 8.2 [5]) [5]

Temperature optimum (°C)

40 <11> [17, 18]

50 <4> [18]

4 Enzyme Structure

Molecular weight

160000 <1, 2> (<2>, sedimentation equilibrium centrifugation [7]; <1>, gel filtration [13]; <1>, sedimentation equilibrium centrifugation [22]) [7, 13, 22]

165000 <11> (<11>, gel filtration [17]) [17]

194000 <22> (<22>, gel filtration [25]) [25]

215000 <24> (<24>, gel filtration [27]) [27]

Subunits

? <27> (<27>, x * 41655, SDS-PAGE [31]) [31]

tetramer <1, 11, 22, 24> (<1> , 4 * 40000, SDS-PAGE [13]; <11>, 4 * 43000, SDS-PAGE [17]; <1>, 4 * 39000, SDS-PAGE [22]; <1>, 4 * 41503, calculation from sequence of DNA [22]; <22> 4 * 41200, SDS-PAGE [24]; <22>, 4 * 48000, SDS-PAGE [25]; <24>, 4 * 50000 or 2 * 50000 + 2 * 53000, SDS-PAGE [27]) [13, 17, 22, 24, 25, 27]

Posttranslational modification

proteolytic modification <24> (<24>, the 50000 Da polypeptide may be derived from the 53000 Da polypeptide by proteolysis [27]) [27]

5 Isolation/Preparation/Mutation/Application

Source/tissue
flower <26> [30]
leaf <17-21, 24, 26> (<26>, mature [30]) [20, 27, 30, 36]
root <26> (<26>, low activity [30]) [30]
seed <23> [26]
stem <26> (<26>, low activity [30]) [30]
stolon <26> (<26>, low activity [30]) [30]
tuber <26> (<26>, low activity [30]) [30]

Localization
chloroplast <23, 24> [26, 27]

Purification
<1> [1, 13, 22]
<2> (80% homogenous [7]) [5, 7]
<4> [18]
<11> [17, 18]
<22> (simultaneous purification of EC 4.4.1.1 and EC 4.2.99.2 [24]) [24]
<24> [27]
<28> (recombinant enzyme [32]) [32]

Crystallization
<1> (crystal structure at 1.5 A resulution [28]) [28]
<28> (crystals grown by sitting drop vapour diffusion against a reservoir containing 100 mM MES-NaOH [35]) [32, 35]

Cloning
<22> (expression of the CYS3 gene in Escherichia coli, the protein shows activities of EC 4.4.1.1, EC 4.4.1.8, and EC 2.5.1.8 [25]) [25]
<25> (expression in Escherichia coli [29]; transgenic plants overexpressing the enzyme under the control of the cauliflower mosaic virus 35S promoter show increased soluble Met and its metabolite S-methyl-Met [33]; overexpression of full-length enzyme and its truncated version that lacks the N-terminal region in transgenic Nicotiana tabacum plants. Transgenic plants expressing both types of enzyme have a significant higher level of S-methyl-Met, and Met content in their proteins. Plants expressing full-length enzyme show the same phenotype and developmental pattern as wild-type plants, those expressing the truncated length enzyme show a severely abnormal phenotype. The N-terminal region plays a role in protecting plants from a high level of Met catabolic products such as ethylene [34]) [29, 33, 34]
<26> (the truncated cDNA without putative leader peptide, when cloned into a bacterial expression vector, complements the Escherichia coli metB1 mutant strain LE392 [30]) [30]
<27> (expression in Escherichia coli [31]) [31]
<28> (overexpression in Escherichia coli [32]) [32]

Engineering

Additional information <25> (<25>, overexpression of full-length enzyme and its truncated version that lacks the N-terminal region in transgenic Nicotiana tobacum plants. Transgenic plants expressing both types of enzyme have a significant higher level of S-methyl-Met, and Met content in their proteins. Plants expressing full-length enzyme show the same phenotype and developmental pattern as wild-type plants, those expressing the truncated length enzyme show a severely abnormal phenotype [34]) [34]

Application

synthesis <1, 4, 11> (<4,11>, production of L-cystathionine [18]; <1>, production of α or β deuterated amino acids [23]) [18, 23]

6 Stability

General stability information

<16>, EDTA and dithiothreitol stabilize [4]

Storage stability

<2>, -15°C, protein concentration 10 mg/ml, 6-12 months stable [7]
<11>, -20°C, 10 mM potassium phosphate buffer, pH 7.5, 50% glycerol [17]
<24>, -20°C, 20 mM MOPS-NaOH, pH 7.5, 1 mM EDTA, 1 mM DTT, 10% v/v glycerol, 5 days, 50% loss of activity [27]

References

[1] Wiebers, J.L.; Garner, H.R.: Acyl derivatives of homoserine as substrates for homocysteine synthesis in Neurospora crassa, yeast and Escherichia coli. J. Biol. Chem., 242, 5644-5649 (1967)

[2] Flavin, M.; Slaughter, C.: Enzymatic synthesis of homocysteine or methionine directly from O-succinyl-homoserine. Biochim. Biophys. Acta, 132, 400-405 (1967)

[3] Guggenheim, S.; Flavin, M.: Proton retention in the γ-elimination reaction catalysed by cystathionine γ-synthase. Biochim. Biophys. Acta, 151, 664-669 (1968)

[4] Kerr, D.; Flavin, M.: Inhibition of cystathionine γ-synthase by S-adenosylmethionine: A control mechanism for methionine synthesis in Neurospora. Biochim. Biophys. Acta, 177, 177-179 (1969)

[5] Guggenheim, S.; Flavin, M.: Cystathionine γ-synthase from Salmonella. β elimination and replacement reactions and inhibition by O-succinylserine. J. Biol. Chem., 244, 3722-3727 (1969)

[6] Guggenheim, S.; Flavin, M.: Cystathionine γ-synthase. A pyridoxal phosphate enzyme catalyzing rapid exchanges of β and α hydrogen atoms in amino acids. J. Biol. Chem., 244, 6217-6227 (1969)

[7] Kaplan, M.; Guggenheim, S.: Cystathionine γ-synthase (Salmonella). Methods Enzymol., 17B, 425-433 (1971)

[8] Posner B.I.: Cystathionine-synthase: studies on the replacement reaction. Biochim. Biophys. Acta, 276, 277-283 (1972)

[9] Posner, B.I.; Flavin, M.: Cystathionine-synthase. Studies of hydrogen exchange reactions. J. Biol. Chem., 247, 6402-6411 (1972)

[10] Posner, B.I.; Flavin, M.: Cystathionine-synthase. The nature of intramolecular proton shifts in the elimination reaction. J. Biol. Chem., 247, 6412-6419 (1972)

[11] Marcotte,P.; Walsh, C.: Active site-directed inactivation of cystathionine γ-synthetase and glutamic pyruvic transaminase by propargylglycine. Biochem. Biophys. Res. Commun., 62, 677-682 (1975)

[12] Thompson, G.A.; Datko, A.H.; Mudd, S.H.: Methionine synthesis in Lemna. Inhibition of cystathionine γ-synthase by propargylglycine. Plant Physiol., 70, 1347-1352 (1982)

[13] Tran, S.V.; Schaeffer, E.; Bertrand, O.; Mariuzza, R.; Ferrara, P.: Appendix. Purification, molecular weight, and NH$_2$-terminal sequence of cystathionine γ-synthase of Escherichia coli. J. Biol. Chem., 258, 14872-14873 (1983)

[14] Nagasawa, T.; Kanzaki, H.; Yamada, H.: Cystathionine γ-lyase of Streptomyces phaeochromogenes. The occurrence of cystathionine γ-lyase in filamentous bacteria and its purification and characterization. J. Biol. Chem., 259, 10393-10403 (1984)

[15] Kanzaki, H.; Kobayashi, M.; Nagasawa, T.; Yamada, H.: Distribution to two kinds of cystathionine γ-synthase in various bacteria. FEMS Microbiol. Lett., 33, 65-68 (1987)

[16] Giovanelli, J.; Mudd, S.H.; Datko, A.H.; Thompson, G. A.: Effects of orthophosphate and adenosine 5'-phosphate on threonine synthase and cystathionine γ-synthase of Lemna paucicostata Hegelm. 6746. Plant Physiol., 81, 577-583 (1986)

[17] Kanzaki, H.; Kobayashi, M.; Nagasawa, T.; Yamada, H.: Purification and characterization of cystathionine γ-synthase type II from Bacillus sphaericus. Eur. J. Biochem., 163, 105-112 (1987)

[18] Kanzaki, H.; Kobayashi, M.; Nagasawa, T.; Yamada, H.: Production of L-cystathionine using bacterial cystathionine γ-synthase. Appl. Microbiol. Biotechnol., 25, 322-326 (1987)

[19] Martel, A.; Bouthier de la Tour, C.; le Goffic, F: Pyridoxal 5 phosphate binding site of Escherichia coli β cystathionase and cystathionine γ synthase comparison of their sequences. Biochem. Biophys. Res. Commun., 147, 565-571 (1987)

[20] Dawson, J.C.; Anderson, J.W.: Comparative enzymology of cystathionine and selenocystathionine synthesis of selenium-accumulator and non-accumulator plants. Phytochemistry, 28, 51-55 (1989)

[21] Brzovic, P.; Litzenberger Holbrook, E; Greene, R.C.; Dunn, M.F.: Reaction mechanism of Escherichia coli cystathionine γ-synthase: direct evidence for a pyridoxamine derivative of vinylglyoxylate as a key intermediate in pyridoxal phosphate dependent γ-elimination and γ-replacement reactions. Biochemistry, 29, 442-451 (1990)

[22] Litzenberger Holbrook, E.; Greene, R.C.; Heilig Krueger, J.: Purification and properties of cystathionine γ-synthase from overproducing strains of Escherichia coli. Biochemistry, 29, 435-442 (1990)

[23] Homer, R.J.; Kim, M.S.; LeMaster, D.M.: The use of cystathionine γ-synthase in the production of α and chiral β deuterated amino acids. Anal. Biochem., 215, 211-215 (1993)

[24] Ono, B.-I.; Ishi, N.; Naito, K.; Miyoshi, S.-I.; Shinoda, S.; Yamamoto, S.; Ohmori, S.: Cystathionine γ-lyase of Saccharomyces cerevisiae: structural gene and cystathionine γ-synthase activity. Yeast, 9, 389-397 (1993)

[25] Yamagata, S.; DÀndrea, R.J.; Fujisaki, S.; Isaji, M.; Nakamura, K.: Cloning and bacterial expression of the CYS3 gene encoding cystathionine γ-lyase of Saccharomyces cerevisiae and the physicochemical and enzymatic properties of the protein. J. Bacteriol., 175, 4800-4808 (1993)

[26] Kreft, B.D.; Townsend, A.; Pohlenz, H.-D.; Laber, B.: Purification and properties of cystathionine γ-synthase from wheat (Triticum aestivum L.). Plant Physiol., 104, 1215-1220 (1994)

[27] Ravanel, S.; Droux, M.; Douce, R.: Methionine biosynthesis in higher plants. I. Purification and characterization of cystathionine γ-synthase from spinach chloroplasts. Arch. Biochem. Biophys., 316, 572-584 (1995)

[28] Clausen, T.; Huber, R.; Prade, L.; Wahl, M.C.; Messerschmidt, A.: Crystal structure of Escherichia coli cystathionine γ-synthase at 1.5 A resolution. EMBO J., 17, 6827-6838 (1998)

[29] Ravanel, S.; Gakiere, B.; Job, D.; Douce, R.: Cystathionine γ-synthase from Arabidopsis thaliana: purification and biochemical characterization of the recombinant enzyme overexpressed in Escherichia coli. Biochem. J., 331, 639-648 (1998)

[30] Ominato, K.; Akita, H.; Suzuki, A.; Kijima, F.; Yoshino, T.; Yoshino, M.; Chiba, Y.; Onouchi, H.; Naito, S.: Identification of a short highly conserved amino acid sequence as the functional region required for posttranscriptional autoregulation of the cystathionine γ-synthase gene in Arabidopsis. J. Biol. Chem., 277, 36380-36386 (2002)

[31] Hwang, B.-J.; Kim, Y.; Kim, H.-B.; Hwang, H.-J.; Kim, J.-H.; Lee, H.-S.: Analysis of Corynebacterium glutamicum methionine biosynthetic pathway: isolation and analysis of metB encoding cystathionine γ-synthase: Isolation and analysis of metB encoding cystathionine γ synthase. Mol. Cells, 9, 300-308 (1999)

[32] Clausen, T.; Wahl, M.C.; Messerschmidt, A.; Huber, R.; Fuhrmann, J.C.; Laber, B.; Streber, W.; Steegborn, C.: Cloning, purification, and characterization of cystathionine γ-synthase from Nicotiana tabacum. Biol. Chem., 380, 1237-1242 (1999)

[33] Kim, J.; Lee, M.; Chalam, R.; Martin, M.N.; Leustek, T.; Boerjan, W.: Constitutive overexpression of cystathionine γ-synthase in Arabidopsis leads to accumulation of soluble methionine and S-methylmethionine. Plant Physiol., 128, 95-107 (2002)

[34] Hacham, Y.; Avraham, T.; Amir, R.: The N-terminal region of Arabidopsis cystathionine γ-synthase plays an important regulatory role in methionine metabolism. Plant Physiol., 128, 454-462 (2002)

[35] Steegborn, C.; Messerschmidt, A.; Laber, B.; Streber, W.; Huber, R.; Clausen, T.: The crystal structure of cystathionine γ-synthase from Nicotiana tabacum reveals its substrate and reaction specificity. J. Mol. Biol., **290**, 983-996 (1999)

[36] Kreft, O.; Hoefgen, R.; Hesse, H.: Functional analysis of cystathionine γ-synthase in genetically engineered potato plants. Plant Physiol., **131**, 1843-1854 (2003)

O-Acetylhomoserine aminocarboxypropyltransferase

2.5.1.49

1 Nomenclature

EC number
2.5.1.49

Systematic name
O-acetyl-L-homoserine:methanethiol 3-amino-3-carboxypropyltransferase

Recommended name
O-acetylhomoserine aminocarboxypropyltransferase

Synonyms
AHS <3> [11]
EC 4.2.99.10 (formerly)
MS <2> [19]
O-acetyl-L-homoserine (thiol)-lyase
O-acetyl-L-homoserine acetate-lyase (adding methanethiol)
O-acetyl-L-homoserine sulfhydrolase
O-acetylhomoserine sulfhydrolase
OAH <8, 10> [5, 8, 12]
OAH SHLase <1> [16]
OAH sulfhydrylase <11> [13]
OAHS <5> [20]
OAS-OAH sulfhydrylase <10> [6, 8, 10]
homocysteine synthase <11> [13]
methionine synthase

CAS registry number
37290-90-7

2 Source Organism

<1> *Aspergillus nidulans* (strain sG8, pabaA2, YA1 [15,16]) [15, 16]
<2> *Bos taurus* (cattle, Holstein steer [19]) [19]
<3> *Brevibacterium flavum* (No. 2247, ATCC 14067 [11]) [11]
<4> *Corynebacterium acetophilum* (strain A51 [9]) [9]
<5> *Leptospira meyeri* (MetY protein sequence [20]) [20]
<6> *Neurospora crassa* (wild-type, strain EM 5297 [3]) [2, 3]
<7> *Neurospora sp.* (wild-type, strain me-2 and me-7, methionineless mutants [1]) [1]
<8> *Pseudomonas sp.* (FM518 [12]) [12]

<9> *Saccharomyces cerevisiae* (Met17 protein sequence [20] (SwissProt-ID: P06106)) [20]
<10> *Saccharomyces cerevisiae* (baker's yeast [4, 5, 7, 8, 14]; mutant strain No. 17 [10]) [4-8, 10, 14, 17]
<11> *Schizosaccharomyces pombe* (fission yeast, IFO-0363 [13]) [13]
<12> *Thermus thermophilus* (HB8, HB27 [18]; nucleotide sequence accession no. [18]; DDBJ: AB049221, EMBL:AB049221 [18]) [18]

3 Reaction and Specificity

Catalyzed reaction

O-acetyl-L-homoserine + methanethiol = L-methionine + acetate (Also reacts with other thiols and H_2S, producing homocysteine or thioethers. The name methionine synthase is more commonly applied to EC 2.1.1.13 5-methyltetra-hydrofolate-homocysteine S-methyltransferase. The enzyme from baker's yeast also catalyses the reaction of EC 2.5.1.47 cysteine synthase, but more slowly)

Reaction type

elimination (C-O bond cleavage)

Natural substrates and products

S   O-acetyl-L-homoserine + H_2S <1, 3-6, 8-11> (<8> direct sulfhydrylation pathway, biosynthesis of homocysteine [12]; <5> direct sulfhydrylation pathway, methionine biosynthesis [20]) [2-5, 8, 9, 11-13, 15, 20]
P   L-homocysteine + acetic acid
S   O-acetyl-L-homoserine + S^{2-} <10> (<10> physiological substrate NaHS [14]) [14]
P   L-homocysteine + acetic acid
S   O-acetyl-L-homoserine + methanethiol <3, 6, 7, 10> (<6> pathway for formation of methionine, incorporation of sulfur, possible temporary function for this enzyme may be formation of methionine from methyl-mercaptan at times of exhaustion of an exogenous source of sulfur [3]; <3> methionine biosynthetic pathway [11]) [1, 3, 5, 11]
P   L-methionine + acetate
S   O-acetyl-L-serine + S^{2-} <10> (<10> physiological substrate NaHS [14]) [14]
P   L-cysteine + acetic acid

Substrates and products

S   L-homoserine + H_2S <11, 12> (Reversibility: ? <11,12> [13,18]) [13, 18]
P   homocysteine + ?
S   L-serine-O-sulfate + Se^{2-} <10> (Reversibility: ? <10> [14]) [14]
P   L-selenocystine + ?
S   O-acetyl-L-homoserine + 2-mercaptoethanol <1> (Reversibility: ? <1> [16]) [16]
P   ?

S O-acetyl-L-homoserine + H$_2$S <1, 3-6, 8-11> (Reversibility: ? <1, 3-6, 8-11> [2-5, 8, 9, 11-15, 20]) [2-5, 8, 9, 11-15, 20]
P L-homocysteine + acetic acid <1, 3-6, 8-11> [3, 5, 8, 9, 11-15, 20]
S O-acetyl-L-homoserine + Se^{2-} <10> (Reversibility: ? <10> [14]) [14]
P L-selenohomocystine + ?
S O-acetyl-L-homoserine + ethanol <4> (Reversibility: ? <4> [9]) [9]
P O-ethyl-homoserine + acetic acid <4> [9]
S O-acetyl-L-homoserine + ethyl mercaptan <10> (Reversibility: ? <10> [4,5]) [4, 5]
P L-ethionine + acetic acid <10> [4, 5]
S O-acetyl-L-homoserine + methanethiol <6, 7, 10> (<10> relative activity 100% [5]) (Reversibility: ? <6,7,10> [1,3,5]) [1, 3, 5]
P L-methionine + acetic acid <6, 7, 10> [1, 3, 5]
S O-acetyl-L-homoserine + methanol <4> (Reversibility: ? <4> [9]) [9]
P O-methyl-homoserine + acetic acid <4> [9]
S O-acetyl-L-homoserine + methyl mercaptan <4, 6, 7, 10> (Reversibility: ? <4, 6, 7, 10> [1-5, 8, 9]) [1-5, 8, 9]
P L-methionine + acetic acid <4, 6, 7, 10> [1-5, 8, 9]
S O-acetyl-L-homoserine + methylselenide <1> (Reversibility: ? <1> [16]) [16]
P seleno-methionine + acetic acid
S O-acetyl-L-homoserine + n-butanol <4> (Reversibility: ? <4> [9]) [9]
P O-butyl-homoserine + acetic acid <4> [9]
S O-acetyl-L-homoserine + n-pentanol <4> (Reversibility: ? <4> [9]) [9]
P O-pentyl-homoserine + acetic acid <4> [9]
S O-acetyl-L-homoserine + n-propanol <4> (Reversibility: ? <4> [9]) [9]
P O-propyl-homoserine + acetic acid <4> [9]
S O-acetyl-L-homoserine + thioglycerol <1> (Reversibility: ? <1> [16]) [16]
P ?
S O-acetyl-L-homoserine + thioglycolic acid <1> (Reversibility: ? <1> [16]) [16]
P ?
S O-acetyl-L-serine + H$_2$S <10> (Reversibility: ? <10> [5]) [5]
P S-methylcysteine + ? <10> [5]
S O-acetyl-L-serine + H$_2$S <1, 4, 5, 10-12> (Reversibility: ? <1, 4, 5, 10-12> [8, 9, 13-16, 18, 20]) [8, 9, 13-16, 18, 20]
P L-cysteine + acetic acid <1, 4, 5, 10-12> [8, 9, 13-16, 18, 20]
S O-acetyl-L-serine + NaHSe$_2$ <10> (Reversibility: ? <10> [14]) [14]
P L-selenocystine + ?
S O-acetyl-L-serine + Se^{2-} <10> (Reversibility: ? <10> [14]) [14]
P L-selenocystine + ?
S O-acetyl-L-serine + methanethiol <10> (<10> reaction only in phosphate buffer, relative activity compared to O-acetyl-L-homoserine 14.8% [5,8]) (Reversibility: ? <10> [5,8]) [5, 8]
P ?
S O-acetyl-L-serine + methyl mercaptan <10> (Reversibility: ? <10> [5]) [5]
P S-methylcysteine + ? <10> [5]

S O-phospho-L-serine + H_2S <12> (Reversibility: ? <12> [18]) [18]

P ?

S O-succinyl-L-homoserine + H_2S <11, 12> (Reversibility: ? <11, 12> [13, 18]) [13, 18]

P homocysteine + ?

S Additional information <3, 4, 6, 8, 10-12> (<3> shows with O-acetylserine 1% of the activity with O-acetylhomoserine, no activity with O-succinyl-L-homoserine, homoserine or serine [11]; <4> O-acetyl-L-serine, L-homoserine, O-succinyl-L-homoserine and O-acetyl-L-threonine are no substrates [9]; <6> L-homoserine, O-succinyl-DL-homoserine, O-phosphoryl-L-homoserine, L-serine, O-acetyl-L-serine, O-phosphoryl-L-serine react at less than 1% the rate with O-acetyl-L-homoserine [2]; <6> no cystathione formed from cysteine, no α-ketobutyrate formed from O-acetylhomoserine, negligible activity with homoserine and serine, O-succinylhomoserine, O-phosporylhomoserine, O-acetylserine and O-phosphorylserine are no substrates [3]; <8> L-homoserine and O-succinyl-L-homoserine are no substrates [12]; <10> cystathionine is not synthesized from O-acetylhomoserine and cysteine, O-acetyl-L-serine is no substrate [4]; <10> O-acetyl-L-threonine, O-acetyl-L-tyrosine, O-acetyl-L-hydroxyproline and O-succinyl-DL-homoserine are no substrates [4,5]; <11> does not synthesize cystathionine from O-acetyl-L-homoserine and L-cysteine, cannot utilize cysteine as co-substrate in place of H_2S, O-acetyl-L-serine, O-acetyl-L-tyrosine, O-acetyl-L-threonine, O-phospho-DL-threonine, O-phospho-L-serine, L-serine-O-sulfate and O-acetyl-L-hydroxyproline are no substrates [13]; <12> low cystathione β-lyase reaction, relative activity 0.12%, no cystathione γ-lyase reaction [18]) [2-5, 9, 11-13, 18]

P ?

Inhibitors

DL-C-propagylglycine <1, 12> (<1> only 25% inhibition [15]) [15, 18]

DL-ethionine <11> [13]

KCN <4, 11> [9, 13]

L-homoserine <3, 4, 10> (<10> competitive inhibition [4]) [4, 9, 11]

L-methionine <1, 3, 6, 8, 10-12> (<8, 10, 11> competitive inhibition [4, 6, 12, 13]; <1> only 20% inhibition [15]) [3, 4, 6, 11-13, 15, 18]

L-penicillamine <10, 11> [13, 17]

Na_2S <4> [9]

Na_2Se <10> [14]

O-acetyl-L-serine <3, 4, 10, 11> (<10> competitive inhibition [4]) [4, 9, 11, 13]

O-acetyl-L-threonine <4> [9]

O-ethylhomoserine <4> [9]

O-phospho-L-serine <11> [13]

O-succinyl-DL-homoserine <4, 10> (<10> competitive inhibition [4]) [4, 9]

S-adenosyl-L-homocysteine <6, 11> [3, 13]

S-adenosyl-L-methionine <6, 8, 11> [3, 12, 13]

Se^{2-} <10> [14]

butanol <4> [9]
hydrazine <4> [9]
hydroxylamine hydrochloride <1, 4, 10-12> [9, 13, 16-18]
isopropanol <4> [9]
methanol <4> [9]
methylmercaptan <4> [9]
n-propanol <4> [9]
p-chloromercuribenzoate <4> [9]
phenylhydrazine <1, 4, 10-12> [9, 13, 16-18]
pyridoxal <10> (<10> inhibits the activity competitively with respect to pyridoxal 5'-phosphate [8]) [8]
semicarbazide <10-12> [13, 17, 18]
Additional information <1, 3, 6, 10, 11> (<10> L-serine, glycine and L-alanine are not inhibitors, dithiothreitol and p-chloromercuribenzoate shows no
inhibitory effect [4]; <6> no end product inhibition by methionine or S-adenosylmethionine observed [2]; <3> L-serine, glycine, cystathione and S-adenosyl-L-methionine are no inhibitors [11]; <11> p-chloromercuribenzoic
acid, 5,5'-dithio-bis(2-nitrobenzoic acid) and monoiodoacetic acid have no
inhibitory effect [13]; <1> no sulfhydryl reagents inhibits the enzyme, S-adenosylmethionine is no inhibitor [15]) [2, 4, 11, 13, 15]

Cofactors/prosthetic groups

pyridoxal 5'-phosphate <1, 3, 4, 6, 10-12> (<10> absolutely required for activity, K_m 0.00294 mM [4]; <6> only required after enzyme is resolved by
dialysis against cysteine [2]; <4> tightly bound [9]) [2-4, 7-9, 11, 13-15, 18]

Activating compounds

DL-cystathionine <3> (<3> 10 mM addition, AHS activity 105% [11]) [11]
dithiothreitol <10> (<10> 2 mM, catalytic activity 124% [7]) [7]
ferricyanide <10> (<10> 1 mM, catalytic activity 102-119% [7]) [7]
pyridoxal 5'-phosphate <3, 11> [11, 13]

Specific activity (U/mg)

0.00472 <3> [11]
0.06 <12> (<12> cysthathione β-lyase reaction [18]) [18]
0.108 <4> (<4> O-alkylhomoserine synthesizing activity [9]) [9]
0.58 <12> (<12> L-homoserine as substrate [18]) [18]
0.72 <12> (<12> O-succinyl-L-homoserine as substrate [18]) [18]
1.02 <10> (<10> substrate O-acetyl-L-serine [5]) [5]
1.33 <12> (<12> L-serine as substrate [18]) [18]
1.57 <12> (<12> O-phospho-L-serine as substrate [18]) [18]
2.12 <12> (<12> O-acetyl-L-serine as substrate [18]) [18]
4 <6> [2, 3]
4.2 <6> (<6> O-acetylhomoserine sulfhydrylation [3]) [3]
4.4 <1> [16]
6.88 <10> (<10> substrate O-acetyl-L-homoserine [5]) [5]
7.45 <10> (<10> mutant strain No. 17, O-acetylhomoserine sulfhydrylase activity [10]) [10]

7.7 <6> (<6> O-acetylhomoserine methylsulfhydrylation [3]) [3]
8.96 <1> [15]
16.7 <4> (<4> O-acetylhomoserine sulfhydrylase activity [9]) [9]
19.2 <10> [4]
25.2 <10> [8]
28.3 <10> (<10> mutant strain No. 17, O-acetylserine sulfhydrylase activity [10]) [10]
30 <10> [7]
40.7 <12> (<12> recombinant enzyme [18]) [18]
49.82 <12> (<12> O-acetyl-L-homoserine as substrate [18]) [18]

K_m-Value (mM)

0.053 <11> (H_2S) [13]
0.083 <3> (H_2S) [11]
0.52 <10> (S^{2-}) [14]
0.7 <6> (H_2S) [2, 3]
0.7 <10> (S^{2-}) [14]
0.8 <6> (methyl-mercaptan) [2, 3]
1.1 <4> (O-acetyl-L-homoserine, <4> O-alkylhomoserine synthesis [9]) [9]
1.2 <10> (NaHSe, <10> L-serine-O-sulfate as acceptor [14]) [14]
1.3 <12> (H_2S) [18]
2 <3> (O-acetyl-L-homoserine) [11]
2.4 <4> (H_2S) [9]
2.5 <10> (O-acetyl-L-serine, <10> NaHS as donor [14]) [14]
3.45 <10> (O-acetyl-L-serine) [6]
3.6 <8> (O-acetyl-L-serine) [12]
4 <10> (L-serine-O-sulfate, <10> NaHS or Na_2Se as donor [14]) [14]
4.1 <10> (O-acetyl-L-homoserine, <10> NaHS as donor [14]) [14]
4.55 <10> (O-acetyl-L-homoserine) [4]
5 <10> (O-acetyl-L-serine, <10> Na_2Se as donor [14]) [14]
5.12 <10> (O-acetyl-L-serine) [5]
5.25 <10> (O-acetyl-L-homoserine, <10> Na_2Se as donor [14]) [14]
6.06 <10> (O-acetyl-L-homoserine) [6]
6.67 <10> (O-acetyl-L-homoserine) [5]
7 <6> (O-acetyl-L-homoserine) [2, 3]
8.7 <10> (O-acetyl-L-homoserine, <10> wild-type enzyme [17]) [17]
8.9 <10> (Se2-) [14]
9 <10> (O-acetyl-L-homoserine, <10> recombinant enzyme [17]) [17]
10 <10> (Na_2Se, <10> L-serine-O-sulfate as acceptor [14]) [14]
10.4 <11> (L-homoserine) [13]
11.1 <11> (O-succinyl-L-homoserine) [13]
12.5 <11> (O-acetyl-L-homoserine) [13]
13 <4> (O-acetyl-L-homoserine, <4> O-acetylhomoserine sulfhydrylation [9]) [9]
40 <4> (ethanol) [9]

K_i-Value (mM)

0.01 <6> (S-adenosylmethionine) [3]

0.9 <10> (L-methionine) [4]
1.6 <8> (L-methionine, <8> O-acetylserine sulfhydrylase activity [12]) [12]
2.6 <11> (L-methionine) [13]
3.6 <10> (L-methionine, <10> O-acetyl-L-serine sulfhydrylase activity, wild-type and mutant enzymes [6]) [6]
3.7 <10> (L-methionine, <10> mutant enzyme [6]) [6]
3.8 <10> (L-methionine, <10> wild-type [6]) [6]
15 <8> (L-methionine, <8> O-acetylhomoserine sulfhydrylase activity [12]) [12]
16 <10> (L-homoserine) [4]
19 <10> (O-succinyl-DL-homoserine) [4]
21.8 <10> (O-acetyl-L-serine) [4]
22 <10> (Se^{2-}) [14]
23 <10> (Na_2Se) [14]
57 <10> (NaHS, <10> O-acetlyhomoserine as substrate [14]) [14]
70 <10> (NaHS, <10> L-serine-O-sulfate as substrate [14]) [14]

pH-Optimum

7.5 <1> (<1> O-acetyl-L-serine as substrate [15]) [15]
7.8 <10, 12> [4, 18]
8 <1, 6, 10, 11> (<10> optimum for γ-replacement reaction of O-acetylhomoserine with Na_2Se [14]; <1> O-acetyl-L-homoserine as substrate [15]) [3, 13-15]
8.4 <10> [5]
8.5 <4> [9]
8.7 <3> [11]

pH-Range

4-12 <12> (<12> 50% maximum activity at pH 6.3 and pH 9.3 [18]) [18]
5.5-8.5 <10> [4]
6.6-8.5 <6> (<6> 50% activity at pH 6.6, 90% activity at pH 7.5 and 8.5 [3]) [3]

Temperature optimum (°C)

42-45 <4> [9]
70 <12> [18]

Temperature range (°C)

60-85 <12> [18]

4 Enzyme Structure

Molecular weight

96000 <10> (<10> sucrose density gradient centrifugation , wild-type, O-acetylserine sulfhydrylase activity [10]) [10]
99000 <10> (<10> gel filtration, sucrose density gradient centrifugation , mutant strain No. 17, O-acetylserine sulfhydrylase activity [10]) [10]
163000 <12> (<12> gel filtration [18]) [18]

170000 <11> (<11> sucrose density gradient ultracentrifugation [13]) [13]
182000 <10> (<10> gel filtration, mutant strain No. 17, sucrose density gradient centrifugation, O-acetylhomoserine sulfhydrylase activity [10]) [10]
186000 <11> (<11> gel filtration [13]) [13]
194000 <10> (<10> sucrose density gradient centrifugation, wild-type, OAS-OAH sulfhydrylase with O-acetylserine as substrate [10]) [10]
196000 <10> (<10> sucrose density gradient centrifugation, wild-type, OAS-OAH sulfhydrylase with O-acetylhomoserine as substrate [10]) [10]
200000 <10> (<10> gel filtration, density sucrose gradient centrifugation [7]) [7]
220000 <4> (<4> gel filtration [9]) [9]
250000-260000 <1> (<1> gel filtration [15,16]) [15, 16]
267000 <1> (<1> gel filtration [16]) [16]
360000 <3> (<3> gel filtration [11]) [11]
480000 <10> (<10> gel filtration, larger than, no correct value could be estimated [4]) [4]

Subunits

dimer <4> (<4> 2 * 110000, SDS-PAGE [9]) [9]
hexamer <1> (<1> 6 * 43000, SDS-PAGE [15,16]) [15, 16]
tetramer <10, 12> (<10> 4 * 51000, SDS-PAGE [7]; <10> 4 * 57000, gel filtration in presence of urea and β-mercaptoethanol [7]; <12> 4 * 47000, SDS-PAGE [18]; <12> 4 * 46000, calculated from OAH1 sequence [18]; <12> 4 * 46059, calculated from amino acid sequence [18]) [7, 18]

5 Isolation/Preparation/Mutation/Application

Source/tissue

cardiac muscle <2> [19]
kidney <2> [19]
large intestine <2> [19]
liver <2> [19]
lung <2> [19]
mycelium <1, 7> [1, 15, 16]
pancreas <2> [19]
rumen <2> [19]
skeletal muscle <2> [19]
skin <2> [19]
small intestine <2> [19]
spleen <2> [19]

Purification

<1> [15, 16]
<3> [11]
<4> [9]
<6> [2, 3]

<10> (partial [4,6,14]; mutant strain No. 17 [10]; wild-type and recombinant enzyme [17]) [4-8, 10, 14, 17]
<11> [13]
<12> (recombinant enzyme [18]) [18]

Cloning

<5> (MetY gene can complement the metB mutant of Escherichia coli [20]) [20]
<10> (MET17/MET25 gene cloned and overexpressed in Escherichia coli [17]) [17]
<12> (oah1 gene cloned, sequenced and overexpressed in Escherichia coli BL21(DE3) [18]) [18]

6 Stability

pH-Stability

7 <11> (<11> optimal pH für stability [13]) [13]
7.8-8 <10> (<10> most stable at [4]) [4]

Temperature stability

30 <11> (<11> more stable at 30°C than at 0°C [13]) [13]
40-65 <10> (<10> activity decreases at about 40°C and is completely lost at 65°C [5]) [5]
60 <1> (<1> stable at [15]) [15]
90 <12> (<12> very stable at high temperatures, 90% activity at 90°C for 60 min at pH 7.8 [18]) [18]

Organic solvent stability

urea <10> (<10> denatured by 2.0 M urea [8]) [8]

General stability information

<1>, existence of glycerol in the purification buffer keeps the enzyme very stable, sensitive to salts such as $(NH_4)_2SO_4$ and NaCl in absence of glycerol [15]
<4>, not stable even in presence of a cofactor [9]
<6>, enzyme appears to be quite stable [2]
<6>, enzyme remains stable after purification, no dilution inactivation is observed [3]
<10>, sensitive to trypsin digestion [17]
<11>, very unstable, can be stabilized by addition of 25% sucrose or glycerol [13]

Storage stability

<1>, -20°C, stable for at least 1 week, stable for at least 6 weeks if stabilized by addition of 20% glycerol [16]
<1>, -20°C, stable for at least 6 months [15]
<6>, -15°C, 80% of the activity remains after storage for 6 months [2, 3]
<10>, -20°C, can be stored without any loss of activity [8]
<11>, -20°C, can be stored without loss of catalytic activity [13]

References

[1] Smith, I.K.; Thompson, J.F.: Utilization of S-methylcysteine and methylmercaptan by methionineless mutants of Neurospora and the pathway of their conversion to methionine. II. Enzyme studies. Biochim. Biophys. Acta, **184**, 130-138 (1969)

[2] Kerr, D.S.: O-Acetylhomoserine sulfhydrylase (Neurospora). Methods Enzymol., **17**, 446-450 (1971)

[3] Kerr, D.S.: O-Acetylhomoserine sulfhydrylase from Neurospora. Purification and consideration of its function in homocysteine and methionine synthesis. J. Biol. Chem., **246**, 95-102 (1971)

[4] Yamagata, S.: Homocysteine in yeast. Partial purification and properties of O-acetylhomoserine sulfhydrylase. J. Biochem., **70**, 1035-1045 (1971)

[5] Yamagata, S.; Takeshima, K.; Naiki, N.: Evidence for the identity of O-acetylserine sulfhydrylase with O-acetylhomoserine sulfhydrylase in yeast. J. Biochem., **75**, 1221-1229 (1974)

[6] Yamagata, S.; Takeshima, K.; Naiki, N.: O-acetylserine and O-acetylhomoserine sulfhydrylase of yeast; studies with methionine auxotrophs. J. Biochem., **77**, 1029-1036 (1975)

[7] Yamagata, S.: O-acetylserine and O-acetylhomoserine sulfhydrylase of yeast. Subunit structure. J. Biochem., **80**, 787-797 (1976)

[8] Yamagata, S.; Takeshima, K.: O-acetylserine and O-acetylhomoserine sulfhydrylase of yeast. Further purification and characterization as a pyridoxal enzyme. J. Biochem., **80**, 777-785 (1976)

[9] Murooka,Y.; Kakihara, K.; Miwa, T.; Seto, K.; Harada, T.: O-Alkylhomoserine synthesis catalyzed by O-acetylhomoserine sulfhydrylase in microorganisms. J.Bacteriol., **130**, 62-73 (1977)

[10] Yamagata, S.: Occurence of low molecular weight O-acetylserine sulfhydrylase in the yeast Saccharomyces cerevisiae. J. Biochem., **88**, 1419-1423 (1980)

[11] Ozaki, H.; Shiio, I.: Methionine biosynthesis in Brevibacterium flavum: Properties and essential role of O-acetylhomoserine sulfhydrylase. J. Biochem., **91**, 1163-1171 (1982)

[12] Morinaga, Y.; Tani, Y.; Yamada, H.: Biosynthesis of homocysteine in a facultative methylotroph, Pseudomonas FM518. Agric. Biol. Chem., **47**, 2855-2860 (1983)

[13] Yamagata, S.: O-Acetylhomoserine sulfhydrylase of the fission yeast Schizosaccharomyces pombe: Partial purification, characterization, and its probable role in homocysteine biosynthesis. J. Biochem., **96**, 1511-1523 (1984)

[14] Chocat, P.; Esaki, N.; Tanaka, H.; Soda, K.: Synthesis of selenocystine and selenohomocystine with O-acetylhomoserine sulfhydrylase. Agric. Biol. Chem., **49**, 1143-1150 (1985)

[15] Yamagata, S.; Paszewski, A.; Lewandowska, I.: Purification and properties of O-acetyl-L-homoserine O-sulfhydrylase from Aspergillus nidulans. J. Gen. Appl. Microbiol., **36**, 137-141 (1990)

[16] Brzywczy, J.; Yamagata, S.; Paszewski, A.: Comparative studies on O-acetyl-homoserine sulfhydrylase: physiological role and characterization of the Aspergillus nidulans enzyme. Acta Biochim. Pol., **40**, 421-427 (1993)

[17] Yamagata, S.; Isaji, M.; Nakamura, K.; Fujisaki, S.; Doi, K.; Bawden, S.; D'Andrea, R.: Overexpression of the Saccharomyces cerevisiae MET17/MET25 gene in Escherichia coli and comparative characterization of the product with O-acetylserine-O-acetylhomoserine sulfhydrylase of the yeast. Appl. Microbiol. Biotechnol., **42**, 92-99 (1994)

[18] Shimizu, H.; Yamagata, S.; Masui, R.; Inoue,Y.; Shibata, T.; Yokoyama, S.; Kuramitsu, S.; Iwama, T.: Cloning and overexpression of the oah1 gene encoding O-acetyl-L-homoserine sulfhydrylase of Thermus thermophilus HB8 and characterization of the gene product. Biochim. Biophys. Acta, **1549**, 61-72 (2001)

[19] Lambert, B.D.; Titgemeyer, E.C.; Stokka, G.L.; DeBey, B.M.; Loest, C.A.: Methionine supply to growing steers affects hepatic activities of methionine synthase and betaine-homocysteine methyltransferase, but not cystathionine synthase. J. Nutr., **132**, 2004-2009 (2002)

[20] Hacham, Y.; Gophna, U.; Amir, R.: In vivo analysis of various substrates utilized by cystathionine γ-synthase and O-acetylhomoserine sulfhydrylase in methionine biosynthesis. Mol. Biol. Evol., **20**, 1513-1520 (2003)

Zeatin 9-aminocarboxyethyltransferase

1 Nomenclature

EC number
2.5.1.50

Systematic name
O^3-acetyl-L-serine:zeatin 2-amino-2-carboxyethyltransferase

Recommended name
zeatin 9-aminocarboxyethyltransferase

Synonyms
4.2.99.13 (formerly)
O-acetyl-L-serine acetate-lyase (adding N^6-substituted adenine)
β-(9-cytokinin)-alanine synthase
β-(9-cytokinin)alanine synthase
lupinic acid synthase
lupinic acid synthetase
synthetase, lupinate

CAS registry number
62683-23-2

2 Source Organism

<1> *Lupinus luteus* [1, 2]

3 Reaction and Specificity

Catalyzed reaction
O-acetyl-L-serine + zeatin = lupinate + acetate (<1>, ping pong bi bi mechanism [2])

Reaction type
aminopropanoyl-group transfer

Substrates and products
S O-acetyl-L-serine + 6-(2-hydroxyethylamino)purine <1> (<1>, 20% of the activity with zeatin [2]) (Reversibility: ? <1> [2]) [2]
P 2-amino-3-(6-(2-hydroxyethyl)aminopurin-9yl)-propanoate + acetate

S O-acetyl-L-serine + 6-(3,4-dimethoxybenzylamino)purine <1> (<1>, 7.5% of the activity with zeatin [2]) (Reversibility: ? <1> [2]) [2]
P 2-amino-3-(6-(3,4-dimethoxybenzylamino)purin-9yl)-propanoate + acetate
S O-acetyl-L-serine + 6-(6-hydroxyhexylamino)purine <1> (<1>, 64% of the activity with zeatin [2]) (Reversibility: ? <1> [2]) [2]
P 2-amino-3-(6-aminohexylpurin-9yl)-propanoate + acetate
S O-acetyl-L-serine + 6-(Δ^2-isopentenyl)aminopurine <1> (<1>, 290% of the activity with zeatin [2]) (Reversibility: ? <1> [2]) [2]
P 2-amino-3-(6-(Δ^2-isopentenyl)aminopurin-9yl)-propanoate + acetate
S O-acetyl-L-serine + 6-benzylaminopurine <1> (<1>, as active as zeatin [2]) (Reversibility: ? <1> [2]) [2]
P 2-amino-3-(6-benzylaminopurin-9yl)-propanoate + acetate
S O-acetyl-L-serine + 6-furfurylaminopurine <1> (<1>, 138% of the activity with zeatin [2]) (Reversibility: ? <1> [2]) [2]
P 2-amino-3-(6-furfurylaminopurin-9yl)-propanoate + acetate
S O-acetyl-L-serine + 6-isopentylaminopurine <1> (<1>, 99% of the activity with zeatin [2]) (Reversibility: ? <1> [2]) [2]
P 2-amino-3-(6-isopentylaminopurin-9yl)-propanoate + acetate
S O-acetyl-L-serine + 6-propylaminopurine <1> (<1>, 57% of the activity with zeatin [2]) (Reversibility: ? <1> [2]) [2]
P 2-amino-3-(6-propylaminopurin-9yl)-propanoate + acetate
S O-acetyl-L-serine + O-β-D-glucopyranosylzeatin <1> (<1>, 34% of the activity with zeatin [2]) (Reversibility: ? <1> [2]) [2]
P ?
S O-acetyl-L-serine + adenine <1> (<1>, 7.5% of the activity with zeatin [2]) (Reversibility: ? <1> [2]) [2]
P 2-amino-3-(6-aminopurin-9-yl)-propanoate + acetate
S O-acetyl-L-serine + cis-zeatin <1> (<1>, 91% of the activity with zeatin [2]) (Reversibility: ? <1> [2]) [2]
P lupinate + acetate
S O-acetyl-L-serine + dihydrozeatin <1> (<1>, 30% of the activity with zeatin [2]) (Reversibility: ? <1> [2]) [2]
P ?
S O-acetyl-L-serine + methylaminopurine <1> (<1>, 25.0% of the activity with zeatin [2]) (Reversibility: ? <1> [2]) [2]
P 2-amino-3-(6-methylaminopurin-9-yl)propanoate + acetate
S O-acetyl-L-serine + zeatin <1> (Reversibility: ? <1> [1,2]) [1, 2]
P lupinate + acetate <1> [2]

Inhibitors

2,4-dichlorophenoxy acetic acid <1> [1]
2,6-dichlorophenoxy acetic acid <1> [1]
2-(indol-3-yl)propionic acid <1> [1]
3-(indol-3-yl)propionic acid <1> [1]
4,7-dichloroindol-3-yl-acetic acid <1> [1]
4-chloroindol-3-yl acetic acid <1> [1]

5,7-dichloroindol-3-yl-acetic acid <1> [1]
N-(3-chlorophenyl)-N'-phenylurea <1> [1]
N-(3-nitrophenyl)-N'-phenylurea <1> [1]
N-benzyl-N'-(3,4-dichlorophenyl)urea <1> [1]
N-benzyl-N'-(3-chlorophenyl)urea <1> [1]
N-benzyl-N'-phenylurea <1> [1]
indol-3-yl acetic acid <1> [1]
indole-2-carboxylic acid <1> [1]
indole-3-carboxylic acid <1> [1]
phenylacetic acid <1> [1]

Specific activity (U/mg)
0.009 <1> [2]

K_m-Value (mM)
0.047 <1> (O-acetyl-serine) [2]
0.88 <1> (zeatin) [2]
2.5 <1> (O-glycosylzeatin) [2]
26 <1> (adenine) [2]

K_i-Value (mM)
0.0004 <1> (5,7-dichloroindol-3-yl-acetic acid) [1]
0.002 <1> (2,4-dichlorophenoxy acetic acid) [1]
0.002 <1> (4,7-dichloroindol-3-yl-acetic acid) [1]
0.002 <1> (4-chloroindol-3-yl acetic acid) [1]
0.023 <1> (N-benzyl-N'-(3,4-dichlorophenyl)urea) [1]
0.026 <1> (N-benzyl-N'-phenylurea) [1]
0.034 <1> (2,6-dichlorophenoxy acetic acid) [1]
0.051 <1> (N-benzyl-N'-(3-chlorophenyl)urea) [1]
0.065 <1> (phenylacetic acid) [1]
0.07 <1> (indol-3-yl acetic acid) [1]
0.091 <1> (N-(3-nitrophenyl)-N'-phenylurea) [1]
0.14 <1> (indole-2-carboxylic acid) [1]
0.22 <1> (N-(3-chlorophenyl)-N'-phenylurea) [1]
1.1 <1> (indole-3-carboxylic acid) [1]
2.8 <1> (3-(indol-3-yl)propionic acid) [1]
7.2 <1> (2-(indol-3-yl)propionic acid) [1]

pH-Optimum
8 <1> [2]

4 Enzyme Structure

Molecular weight
64500 <1> (<1>, gel filtration [2]) [2]

5 Isolation/Preparation/Mutation/Application

Source/tissue
  seed <1> [1, 2]

Purification
  <1> (partial [2]) [2]

6 Stability

Storage stability
  <1>, 4°C, 2-3 months, stable [2]

References

[1] Parker, C.W.; Entsch, B.; Letham, D.S.: Inhibitors of two enzymes which me-
    tabolize cytokinins. Phytochemistry, **25**, 303-310 (1986)
[2] Entsch, B.; Parker, C.W.; Letham, D.S.: An enzyme from lupin seeds forming
    alanine derivatives of cytokinins. Phytochemistry, **22**, 375-381 (1983)

β-Pyrazolylalanine synthase

1 Nomenclature

EC number
2.5.1.51

Systematic name
O^3-acetyl-L-serine:pyrazole 1-(2-amino-2-carboxyethyl)transferase

Recommended name
β-pyrazolylalanine synthase

Synonyms
BPA-synthase
EC 4.2.99.14 (formerly)
β-(1-pyrazolyl)alanine synthase
β-pyrazolealanine synthase
β-pyrazolylalanine synthase (acetylserine)
pyrazolealanine synthase
pyrazolylalaninase
synthase, pyrazolealanine
Additional information (<1>, may be identical with EC 4.2.1.50 [1]; <1>, not identical with EC 2.5.1.52, but similar in its reaction pattern [2])

CAS registry number
37290-81-6 (not distinguished from EC 4.2.1.50 in Chemical Abstracts)

2 Source Organism

<1> *Citrullus vulgaris* [1, 2]

3 Reaction and Specificity

Catalyzed reaction
O^3-acetyl-L-serine + pyrazole = 3-(pyrazol-1-yl)-L-alanine + acetate

Reaction type
aminopropanoyl-group transfer

Natural substrates and products
S O^3-acetyl-L-serine + pyrazole <1> (<1>, involved in biosynthesis of heterocyclic β-substituted alanines [2]) [2]
P 3-(pyrazol-1-yl)-L-alanine + acetate

Substrates and products

S O-succinyl-L-serine + pyrazole <1> (<1>, at 16% the rate of O-acetyl-L-serine [2]) (Reversibility: ? <1> [2]) [2]

P 3-(pyrazol-1-yl)-L-alanine + succinate

S O-sulfo-L-serine + pyrazole <1> (<1>, at 2.5% the rate of O-acetyl-L-serine [2]) (Reversibility: ? <1> [2]) [2]

P 3-(pyrazol-1-yl)-L-alanine + $SO_4{}^{2-}$

S O^3-acetyl-L-serine + 3-amino-1,2,4-triazole <1> (<1>, at 11% the rate of pyrazole [2]) (Reversibility: ? <1> [1]) [2]

P 3-(3-amino-1,2,4-triazol-1-yl)-L-alanine + acetate <1> [2]

S O^3-acetyl-L-serine + N-hydroxyurea <1> (<1>, at 9% the rate of pyrazole [2]) (Reversibility: ? <1> [2]) [2]

P O-ureidoserine + acetate <1> [2]

S O^3-acetyl-L-serine + pyrazole <1> (<1>, highly specific [1,2]) (Reversibility: ? <1> [1,2]) [1, 2]

P 3-(pyrazol-1-yl)-L-alanine + acetate <1> [1, 2]

Inhibitors

KCN <1> (<1>, 10 mM, almost complete inhibition [2]) [2]

hydroxylamine <1> (<1>, 50 mM, almost complete inhibition [2]) [2]

Additional information <1> (<1>, no inhibition by pyridoxal phosphate [1]) [1]

Cofactors/prosthetic groups

pyridoxal 5'-phosphate <1> (<1>, requirement, enzyme bound, 1 mol pyridoxal 5'-phosphate per mol subunit, further increase of activity by 40% by adding pyridoxal phosphate [2]) [2]

Specific activity (U/mg)

0.355 <1> [2]

K_m-Value (mM)

2.5 <1> (O-acetylserine) [2]

74 <1> (pyrazole) [2]

pH-Optimum

7.3 <1> [1]

7.3-7.4 <1> [2]

pH-Range

6.5-8.5 <1> (<1>, about half-maximal activity at pH 6.5 and 8.5 [2]) [2]

4 Enzyme Structure

Molecular weight

58000 <1> (<1>, gel filtration [2]) [2]

Subunits

dimer <1> (<1>, 2 * 32000, SDS-PAGE [2]) [2]

5 Isolation/Preparation/Mutation/Application

Source/tissue
seedling <1> (<1>, cotyledons removed, 1 week old [2]) [1, 2]
Additional information <1> (<1>, barely detectable in dry seeds [2]) [2]

Purification
<1> (partial) [1, 2]

6 Stability

General stability information
<1>, freeze-thawing or lyophilization inactivates [2]
<1>, sensitive to freezing even in the presence of 20% glycerol [2]

Storage stability
<1>, 0-2°C, stable for at least 1 month [2]
<1>, 0°C, crude enzyme preparation, 25% loss of activity within 22 h [1]

References

[1] Murakoshi, I.; Kuramoto, H.; Haginiwa, J.: The enzymic synthesis of β-substituted alanines. Phytochemistry, **11**, 177-182 (1972)
[2] Murakoshi, I.; Ikegami, F.; Hinuma, Y.; Hanma, Y.: Purification and characterization of β-(pyrazol-1-yl)L-alanine synthase from Citrullus vulgaris. Phytochemistry, **23**, 973-977 (1984)

L-Mimosine synthase   2.5.1.52

1 Nomenclature

EC number
2.5.1.52

Systematic name
O^3-acetyl-L-serine:3,4-dihydroxypyridine 1-(2-amino-2-carboxyethyl)trans-ferase

Recommended name
L-mimosine synthase

Synonyms
EC 4.2.99.15 (formerly)
O^3-acetyl-L-serine acetate-lyase (adding 3,4-dihydroxypyridin-1-yl)
synthase, L-mimosine
Additional information (not identical with EC 4.2.99.14)

CAS registry number
93229-75-5

2 Source Organism

<1> *Leucaena leucocephala* [1, 2]

3 Reaction and Specificity

Catalyzed reaction
O^3-acetyl-L-serine + 3,4-dihydroxypyridine = 3-(3,4-dihydroxypyridin-1-yl)-
L-alanine + acetate

Reaction type
aminopropanoyl-group transfer

Natural substrates and products
   S O^3-acetyl-L-serine + 3,4-dihydroxypyridine <1> (<1>, involved in bio-synthesis of heterocyclic β-substituted alanines, brings about the biosyn-thesis of L-mimosine [1,2]) [1, 2]
   P 3-(3,4-dihydroxypyridin-1-yl)-L-alanine + acetate

Substrates and products

S O-succinyl-L-serine + 3,4-dihydroxypyridine <1> (<1>, 60% of the activity with O-acetylserine [1]) (Reversibility: ? <1> [1]) [1]

P 3-(3,4-dihydroxypyridin-1-yl)-L-alanine + succinate

S O-sulfo-L-serine + 3,4-dihydroxypyridine <1> (<1>, 32% of the activity with O-acetyl-L-serine [1]) (Reversibility: ? <1> [1]) [1]

P 3-(3,4-dihydroxypyridin-1-yl)-L-alanine + sulfate

S O^3-acetyl-L-serine + 3,4-dihydroxypyridine <1> (<1>, best substrates [1]) (Reversibility: ? <1> [1,2]) [1, 2]

P 3-(3,4-dihydroxypyridin-1-yl)-L-alanine + acetate <1> (<1>, i.e. L-mimosine [1,2]) [1, 2]

S O^3-acetyl-L-serine + 3-amino-1,2,4-triazole <1> (<1>, 30% of the activity with 3,4-dihydroxypyridine [1]) (Reversibility: ? <1> [1]) [1]

P 3-(3-amino-1,2,4-triazol-1-yl)-L-alanine + acetate <1> [1]

S O^3-acetyl-L-serine + N-hydroxyurea <1> (<1>, 13% of the activity with 3,4-dihydroxypyridine [1]) (Reversibility: ? <1> [1]) [1]

P ureidoserine + acetate <1> [1]

S O^3-acetyl-L-serine + pyrazole <1> (<1>, 82% of the activity with 3,4-dihydroxypyridine [1]) (Reversibility: ? <1> [1]) [1]

P 3-(pyrazol-1-yl)-L-alanine + acetate <1> [1]

S Additional information <1> (<1>, no activity with O-phospho-L-serine and L-Ser [1,2]; <1>, no activity with α,β-diaminopropionic acid [2]) [1, 2]

P ?

Inhibitors

Additional information <1> (<1>, no substrate inhibition [2]) [2]

Cofactors/prosthetic groups

pyridoxal 5'-phosphate <1> (<1>, requirement, enzyme bound, 1 mol pyridoxal 5'-phosphate per mol subunit [1]) [1]

Specific activity (U/mg)

0.189 <1> [1]

K_m-Value (mM)

5 <1> (3,4-dihydroxypyridine) [1]

6.25 <1> (O-acetylserine) [1]

pH-Optimum

7.8 <1> (<1>, reaction with 3,4-dihydroxypyridine [1]) [1]

4 Enzyme Structure

Molecular weight

64000 <1> (<1>, gel filtration [1]) [1]

Subunits

dimer <1> (<1>, 2 * 32000, SDS-PAGE [1]) [1]

5 Isolation/Preparation/Mutation/Application

Source/tissue
  seedling <1> (<1>, cotyledons removed [1]) [1, 2]

Purification
  <1> (partial, to near homogeneity [1]) [1, 2]

References

[1] Murakoshi, I.; Ikegami, F.; Hinuma, Y.; Hanma, Y.: Purification and characterization of L-mimosine synthase from Leucaema leucocephala. Phytochemistry, **23**, 1905-1908 (1984)
[2] Murakoshi, I.; Kuramoto, H.; Haginiwa, J.: The enzymic synthesis of β-substituted alanines. Phytochemistry, **11**, 177-182 (1972)

Uracilylalanine synthase

1 Nomenclature

EC number
2.5.1.53

Systematic name
O^3-acetyl-L-serine:uracil 1-(2-amino-2-carboxyethyl)transferase

Recommended name
uracilylalanine synthase

Synonyms
O^3-acetyl-L-serine acetate-lyase
isowillardiine synthase
synthase, uracilylalanine
willardiine synthase
Additional information (not identical with EC 4.2.99.8)

CAS registry number
113573-73-2

2 Source Organism

<-3> no activity in *Citrullus vulgaris* [1]
<-2> no activity in *Lupinus luteus* [1]
<-1> no activity in *Fagus japonica* [1]
<1> *Pisum sativum* (pea, cv. Meteor [2]) [1, 2]
<2> *Lathyrus odoratus* [1]
<3> *Albizzia julibrissin* [1]
<4> *Leucaena leucocephala* [1]
<5> *Fagus crenata* [1]

3 Reaction and Specificity

Catalyzed reaction
O^3-acetyl-L-serine + uracil = 3-(uracil-1-yl)-L-alanine + acetate

Reaction type
condensing reaction

Natural substrates and products

S O^3-acetyl-L-serine + uracil <1> (<1>, involved in biosynthesis of hetero-
    cyclic β-substituted alanines [2]) [2]

P   3-(uracil-1-yl)-L-alanine + 3-(uracil-3-yl)-L-alanine + acetate

Substrates and products

S O^3-acetyl-L-serine + uracil <1-5> (<1>, concentration of uracil at around
    13 mM and 35 mM is sufficient to give maximal rates of willardiine and
    isowillardiine synthesis [1]; <1>, highly specific [2]) (Reversibility: ? <1-
    5> [1,2]) [1, 2]

P   3-(uracil-1-yl)-L-alanine + 3-(uracil-3-yl)-L-alanine + acetate <1-5> (<1-
    5>, i.e. willardiine and isowillardiine [1]) [1, 2]

Inhibitors

(2,4-dihydroxypyrimidin-1-yl)-L-alanine <1> (<1>, i.e. willardiine or 3-(ur-
acil-1-yl)-L-alanine, 30 mM, 84% inhibition of willardiine synthesis and 96%
inhibition of isowillardiine synthesis [2]) [2]

(2,4-dihydroxypyrimidin-3-yl)-L-alanine <1> (<1>, i.e. isowillardiine or 3-
(uracil-3-yl)-L-alanine, 30 mM produces 75% inhibition of willardiine synthesis and 93% inhibition of synthesis of isowillardiine [2]) [2]

KCN <1-5> (<1-5>, 100 mM, complete [1]) [1]

PCMB <1> (<1>, 0.001 mM, about 90% inhibition [2]) [2]

Tris-HCl buffer <1> [2]

hydroxylamine <1-5> (<1-5>, 100 mM, complete [1]) [1]

pyridoxal 5'-phosphate <1-5> (<1>, 0.25 mg/ml, 12% inhibition of willardine
synthase and 4% inhibition of isowillardine synthase [1]; <1>, above 0.1 mM
[2]; <1-5>, activation at 0.06 mg/ml [1]) [1, 2]

uracil <1> (<1>, substrate inhibition above 35 mM [2]) [1, 2]

Cofactors/prosthetic groups

pyridoxal 5'-phosphate <1-5> (<1>, increases activity [1]) [1, 2]

Specific activity (U/mg)

Additional information <1> [2]

K_m-Value (mM)

3.3 <1> (uracil, <1>, synthesis of isowillardiine [1,2]) [1, 2]

10 <1> (uracil, <1>, willardiine synthesis [1,2]) [1, 2]

pH-Optimum

7.8-7.9 <1-5> (<1>, O-acetylserine undergoes rapid O-acetyl-/N-acetyl shift
above pH 8 [2]) [1, 2]

pH-Range

7.5-8.2 <1> (<1>, about half-maximal activity at pH 7.5 and 8.2 [1]) [1]

Temperature optimum (°C)

30 <1> [2]

4 Enzyme Structure

Molecular weight

  50000 <1> (<1>, gel filtration [2]) [2]

5 Isolation/Preparation/Mutation/Application

Source/tissue

  seedling <1-5> (<1>, activity increases during 3-8 days growth. No activity in extracts of 1-day-old seedlings [1]) [1, 2]

Purification

  <1> (partial [2]) [2]

6 Stability

Temperature stability

  40-50 <1> (<1>, 2 min, slow decrease of activity [2]) [2]
  50-60 <1> (<1>, 2 min, rapid decrease of activity [2]) [2]
  70 <1> (<1>, 2 min, all activity ceased [2]) [2]

Storage stability

  <1>, 4°C, partially purified enzyme preparation, 30% loss of activity within 24 h, complete inactivation after 48 h [2]
  <1>, liquid N_2, crude enzyme preparation, in 3 M ammonium sulfate, 30% loss of activity within 8 weeks, beyond 8 weeks rapid decline of activity [2]

References

[1] Murakoshi, I.; Ikegami, F.; Ookawa, N.; Ariki, T.; Haginawa, J.; Kuo, Y.-H.; Lambein, F.: Biosynthesis of the uraclylalanine willdardiine and isowillardiine in higher plants. Phytochemistry, **17**, 1571-1576 (1978)
[2] Ahmmad, M.A.S.; Maskall, C.S.; Brown, E.G.: Partial purification and properties of willardiine and isowillardiine synthase activity from Pisum sativum. Phytochemistry, **23**, 265-270 (1984)

3-Deoxy-7-phosphoheptulonate synthase

2.5.1.54

1 Nomenclature

EC number

2.5.1.54

Systematic name

phosphoenolpyruvate:D-erythrose-4-phosphate C-(1-carboxyvinyl)transferase (phosphate-hydrolysing, 2-carboxy-2-oxoethyl-forming)

Recommended name

3-deoxy-7-phosphoheptulonate synthase

Synonyms

2-dehydro-3-deoxy-D-arabino-heptonate-7-phosphate D-erythrose-4-phosphate-lyase (pyruvate-phosphorylating)

2-dehydro-3-deoxy-phosphoheptanoate aldolase

2-dehydro-3-deoxy-phosphoheptonate aldolase

2-keto-3-deoxy-D-arabino-heptonic acid 7-phosphate synthetase

3-deoxy-D-arabino-2-heptulosonic acid 7-phosphate synthetase

3-deoxy-D-arabino-heptolosonate-7-phosphate synthetase

3-deoxy-D-arabino-heptulosonate 7-phosphate synthetase

3-deoxy-D-arabino-heptulosonate-7-phosphate synthase

7-phospho-2-dehydro-3-deoxy-D-arabino-heptonate D-erythrose-4-phosphate-lyase (pyruvate-phosphorylating)

7-phospho-2-keto-3-deoxy-D-arabino-heptonate D-erythrose-4-phosphate lyase (pyruvate-phosphorylating)

D-erythrose-4-phosphate-lyase

D-erythrose-4-phosphate-lyase (pyruvate-phosphorylating)

DAH7-P synthase

DAH7-P synthase (phe)

DAH7-P synthase (phe) <1> [34]

DAHP synthase

DAHP synthase-phe

DAHP synthase-trp

DAHP synthase-tyr

DAHP(Phe)

DS-Co <19> (<19> plastidic enzyme form [33]) [33]

DS-Mn <19> (<19> cytosolic enzyme form [33]) [33]

EC 4.1.2.15 (formerly)

KDPH synthase

KDPH synthetase

aldolase, phospho-2-keto-3-deoxyheptanoate

deoxy-D-arabino-heptulosonate-7-phosphate synthetase
phospho-2-dehydro-3-deoxyheptonate aldolase
phospho-2-keto-3-deoxyheptanoate aldolase
phospho-2-keto-3-deoxyheptonate aldolase
phospho-2-keto-3-deoxyheptonic aldolase
phospho-2-oxo-3-deoxyheptonate aldolase
Additional information <32> (<32> 2 distinct homology classes: AroAI and
AroAII, both being found in bacteria and higher plant chloroplasts, AroAII
can be devided into 4 subgroups, phylogenetic evolution [55]) [55]

CAS registry number
9026-94-2

2 Source Organism

<1> *Escherichia coli* (K12 [8-10,13,52]; strain W3110 [52]; strain HE 401 [13];
      mutant 83-24 [1]; deletion mutant strain H80c from strain HfrH [6]; 3
      isoenzymes [35]; tryptophan-sensitive isozyme [27,28]; phenylalanine-
      sensitive isozyme [13, 31, 34, 36, 44, 49, 56, 57]; tyrosine-sensitive iso-
      zyme [6, 7, 37, 52]) [1, 6-10, 13, 27, 28, 31, 34-37, 43, 44, 47, 49, 52, 56, 57]
<2> *Streptomyces aureofaciens* (strain Tü 24 [2, 3];) [2, 3]
<3> *Salmonella typhimurium* (strain SG12 [11]; tyrosine-sensitive isozyme
      [4,11]; operator-constitutive strain tyrO [4, 11]) [4, 11]
<4> *Ralstonia eutropha* (basonym Alcaligenes eutrophus [5]; strain H16 [5])
      [5]
<5> *bacteria* (strain H1 [5]; strain H20 [5]; strain3/2 [5]; strain 12/60/X, strain
      33/X [5]) [5]
<6> *Schizosaccharomyces pombe* (tyrosine-sensitive isozyme [12]) [12]
<7> *Nicotiana silvestris* (cytosolic isozyme [14,40]) [14, 40]
<8> *Solanum tuberosum* (cytosolic isozyme Ds-Co [40]; cv. Superior [15]) [15,
      16, 32, 40]
<9> *Saccharomyces cerevisiae* (strain RH1326 [45]; ARO3-encoded isozyme is
      a phenylalanine-sensitive isozyme [17,26]; ARO4-encoded isozyme is a
      tyrosine-sensitive isozyme [17,26,45]; Phe-sensitive isozyme [30]) [17,
      26, 30, 45]
<10> *Daucus carota* (cv. Kurodagosun [25]; cytosolic isozyme [25]; 3 enzymes:
      I, II, and II [18]) [18, 25]
<11> *Pseudomonas acidovorans* (i.e. Comamonas terrigena [19]; strain ATCC
      11299a, two distinct regulatory isoenzymes: DAHP synthase-phe and
      DAHP synthase-tyr [19]) [19]
<12> *Pseudomonas testosteroni* (strain ATCC 17409 [19]) [19]
<13> *Pseudomonas palleronii* (strain ATCC 17724 [19]) [19]
<14> *Pseudomonas aeruginosa* (2 isoenzymes: tyrosine-sensitive DAHP
      synthase-tyr and tryptophane-sensitive DAHP synthase-trp [20]) [20]
<15> *Pisum sativum* (cv. Onyx [21]; cytosolic isozyme DS-Co [40]; tyrosine-
      sensitive isozyme [21]) [21, 40]

<16> *Neurospora crassa* (strain 74A [24]; strain 74-OR23-1A [23]; tryptophan-sensitive isozyme [22,24,39]; Trp-, Phe- and Tyr-sensitive isoenzymes [23]) [22-24, 39]

<17> *Nocardia mediterranei* (strain U-32 [29]; Trp-sensitive isozyme [29]) [29]

<18> *Arabidopsis thaliana* (enzyme is encoded by duplicated genes : DHS1 and DHS2 [32]) [32, 58]

<19> *Petroselinum crispum* (plastidic isozyme DS-Mn and cytosolic isozyme DS-Co [33]) [33]

<20> *Amycolatopsis methanolica* (wild type strain WV2 contains DS I, a leaky L-Phe-requiring auxotroph mutant strain GH141 grown under L-Phe limi-tatin possesses additional DS II activity [38]) [38]

<21> *Streptomyces coelicolor* (Trp-sensitive isozyme [39]; strain A3(2) [39]) [39]

<22> *Streptomyces rimosus* (Trp-sensitive isozyme [39]) [39, 41]

<23> *Vigna radiata* (2 differentially regulated isozymes: DS-Mn and DS-Co [42]) [40, 42]

<24> *Spinacia oleracea* (cytosolic isozyme Ds-Co [40]) [40]

<25> *Corynebacterium autotrophicum* (strain 7C [5]) [5]

<26> *Pseudomonas faecalis* [5]

<27> *Brassica juncea* [46]

<28> *Escherichia coli* (Phe-sensitive isozyme [48]) [48]

<29> *Arachis hypogaea* (cv. JL-24 [50]; 2 isozymes: DS-Co and DS-Mn [50]) [50]

<30> *Bacillus sp.* (strain B-6 [51]) [51]

<31> *Corynebacterium glutamicum* (Tyr-sensitive isozyme [54]) [53, 54]

<32> *Xanthomonas campestris* (2 classes of enzymes: class AroAI and AroAII [55]; strain ATCC 33436 [55]) [55]

3 Reaction and Specificity

Catalyzed reaction

phosphoenolpyruvate + D-erythrose 4-phosphate + H_2O = 3-deoxy-D-arabi-no-hept-2-ulosonate 7-phosphate + phosphate (<1> Lys186 is involved in hy-drogen bonding with phosphoenolpyruvate [56]; <1> His268 is located in the active site [49]; <31> Cys67 and Cys145 are important for the catalytic reac-tion [53]; <1> active site cysteines: Cys61 and Cys328 [44]; <1> mechanism [1, 6, 7, 47, 57]; <1,3,9> sequential mechanism [4, 9, 26, 56]; <1, 2, 16> first substrate is phosphoenolpyruvate [2, 24, 56]; <16> rapid equilibrium ordered mechanism, in which phosphoenolpyruvate is the first substrate to bind and 3-deoxy-D-arabinoheptulosonate 7-phosphate is the second product to be re-leased [22]; <1, 2, 3, 16> ordered ping-pong bi-bi mechanism [2, 4, 7, 24])

Reaction type

condensation

Natural substrates and products

S phosphoenolpyruvate + D-erythrose 4-phosphate <1-4, 8, 10, 17-20, 27, 30, 32> (<18> regulatory role of thioredoxin of photosystem I on the chloroplastic enzyme [58]; <1,18> first enzyme in the shikimic pathway leading to biosynthesis of aromatic amino acids [56, 58]; <27> effect of external factors on regulation [46]; <27> one of the key regulator enzymes of the shikimic acid pathway [46]; <20> key regulatory enzyme in L-Phe and L-Tyr biosynthesis [38]; <30> first enzyme of aromatic amino acids, folic acid and phenazine 1-carboxylic acid biosynthesis pathways [51]; <1-3, 8, 10, 17> first enzyme of the aromatic amino acid biosynthesis [2, 4, 6, 15, 18, 25, 27, 29, 31, 34, 43]; <8> enzyme is induced by N-[phosphomonomethyl]glycine [15]; <18> DSH1 RNA levels increase in Arabidopsis leaves subjected either to physical wounding or to infiltration with pathogenic Pseudomonas syringae strains, DSH2 RNA levels are not increased by these treatments [32]; <19> the plastidic enzyme form is induced by fungal elicitor of Phytophthora megasperma, the cytosolic enzyme form is not induced [33]; <4> no repression by the aromatic amino acids [5]) (Reversibility: ? <1-4, 8, 10, 17-20, 27, 30, 32> [2, 5, 6, 15, 18, 27, 29, 31, 33, 32, 34, 38, 43, 46, 51, 55, 56, 58]) [2, 4-6, 15, 18, 27, 29, 31, 33, 32, 34, 38, 43, 46, 51, 55, 56, 58]

P 2-dehydro-3-deoxy-D-arabino-heptonate 7-phosphate + phosphate <1, 20, 30> [38, 43, 51, 56]

S Additional information <1> (<1> structure comparison between 3-deoxy-7-phosphoheptulonate synthase and 3-deoxy-D-manno-octulosonate 8-phosphate synthase, EC 4.1.2.16, reveal that they share a common ancestor and adopt the same catalytic strategy [47]) [47]

P ?

Substrates and products

S phosphoenolpyruvate + 2-deoxy-D-ribose 5-phosphate <1> (Reversibility: ? <1> [34]) [34]

P 3,5-dideoxy-D-gluco-octulosonate 8-phosphate + 3,5-dideoxy-D-manno-octulosonate 8-phosphate + phosphate <1> [34]

S phosphoenolpyruvate + D,L-lyxose <24> (<24> D-lyxose or L-lyxose, at 1.8% of the activity with D-erythrose 4-phosphate [40]) (Reversibility: ? <24> [40]) [40]

P 2-dehydro-3-deoxy-D,L-galacto-octonate + phosphate

S phosphoenolpyruvate + D-arabinose <24> (<24>, at 0.7% of the activity with D-erythrose 4-phosphate [40]) (Reversibility: ? <24> [40]) [40]

P 2-dehydro-3-deoxy-D-gluco-octonate + phosphate

S phosphoenolpyruvate + D-arabinose 5-phosphate <1, 24> (<24> at 8% of the activity with D-erythrose 4-phosphate [40]) (Reversibility: ? <1,24> [34,40]) [34, 40]

P 3-deoxy-D-manno-octulosonate 8-phosphate + phosphate <1> [34]

S phosphoenolpyruvate + D-erythrose <24> (<24> at 93% of the activity with D-erythrose 4-phosphate [40]) (Reversibility: ? <24> [40]) [40]

P 2-dehydro-3-deoxy-D-arabino-heptonate + phosphate

S phosphoenolpyruvate + D-erythrose 4-phosphate + H$_2$O <1-32> (<1> binds 1 molecule D-erythrose 4-phosphate per subunit [56]; <16> in solution D-erythrose 4-phosphate forms dimers that are in slow equilibrium with monomers [22]) (Reversibility: r <3> [4]; ir <1, 24> [1, 34, 40, 49, 56, 57]; ? <1, 2, 4-32> [2, 4-33, 35-39, 41-48, 50-55, 58]) [1-58]

P 2-dehydro-3-deoxy-D-arabino-heptonate 7-phosphate + phosphate <1-32> [1-58]

S phosphoenolpyruvate + D-fructose 1,6-diphosphate <17> (<17> same activity as with D-erythrose 4-phosphate [29]) (Reversibility: ? <17> [29]) [29]

P ?

S phosphoenolpyruvate + D-fructose 6-phosphate <17> (<17> same activity as with D-erythrose 4-phosphate [29]) (Reversibility: ? <17> [29]) [29]

P ?

S phosphoenolpyruvate + D-glucose 6-phosphate <17, 24> (<17> as effective as D-erythrose 4-phosphate [29]; <24> at 0.5% of the activity with D-erythrose 4-phosphate [40]) (Reversibility: ? <17,24> [29,40]) [29, 40]

P ?

S phosphoenolpyruvate + D-glyceraldehyde <24> (<24> at 176% of the activity with D-erythrose 4-phosphate [40]) (Reversibility: ? <24> [40]) [40]

P 2-dehydro-3-deoxy-D-threo-hexonate + phosphate

S phosphoenolpyruvate + D-ribose 5-phosphate <1, 17> (<17> as effective as D-erythrose 4-phosphate [29]) (Reversibility: ? <1,17> [27,29,34]) [27, 29, 34]

P 3-deoxy-D-altro-octulosonate 8-phosphate + phosphate <1> [34]

S phosphoenolpyruvate + D-ribulose 5-phosphate <24> (<24> at 12% of the activity with D-erythrose 4-phosphate [40]) (Reversibility: ? <24> [40]) [40]

P ?

S phosphoenolpyruvate + D-threose <24> (<24> at 92% of the activity with D-erythrose 4-phosphate [40]) (Reversibility: ? <24> [40]) [40]

P 2-dehydro-3-deoxy-D-xylo-heptonate + phosphate

S phosphoenolpyruvate + D-xylose <24> (<24> at 1% of the activity with D-erythrose 4-phosphate [40]) (Reversibility: ? <24> [40]) [40]

P 2-dehydro-3-deoxy-D-ido-octonate + phosphate

S phosphoenolpyruvate + D-xylulose 5-phosphate <17> (<17> at 67% of the activity with D-erythrose 4-phosphate [29]) (Reversibility: ? <17> [29]) [29]

P ?

S phosphoenolpyruvate + DL-glyceraldehyde 3-phosphate <1, 24> (<24> at 142% of the activity with D-erythrose 4-phosphate [40]) (Reversibility: ? <1,24> [40,57]) [40, 57]

P pyruvate + phosphate <1> [57]

S phosphoenolpyruvate + L-erythrose <24> (<24> at 70% of the activity with D-erythrose 4-phosphate [40]) (Reversibility: ? <24> [40]) [40]

P 2-dehydro-3-deoxy-L-arabino-heptonate + phosphate

S phosphoenolpyruvate + L-glyceraldehyde <24> (<24> at 212% of the activity with D-erythrose 4-phosphate [40]) (Reversibility: ? <24> [40]) [40]
P 2-dehydro-3-deoxy-L-threo-hexonate + phosphate
S phosphoenolpyruvate + L-threose <24> (<24> at 52% of the activity with D-erythrose 4-phosphate [40]) (Reversibility: ? <24> [40]) [40]
P 2-dehydro-3-deoxy-L-xylo-heptonate + phosphate
S phosphoenolpyruvate + L-xylose <24> (<24> at 1% of the activity with D-erythrose 4-phosphate [40]) (Reversibility: ? <24> [40]) [40]
P 2-dehydro-3-deoxy-L-ido-octonate + phosphate
S phosphoenolpyruvate + glycolaldehyde <24> (<24> at 245% of the activity with D-erythrose 4-phosphate [40]) (Reversibility: ? <24> [40]) [40]
P (S)-2-oxo-4,5-dihydroxypentanoate + phosphate
S phosphoenolpyruvate + glyoxylate <24> (<24> at 205% of the activity with D-erythrose 4-phosphate [40]) (Reversibility: ? <24> [40]) [40]
P (S)-4-hydroxy-2-oxo-1,5-pentanedioate + R-4-hydroxy-2-oxo-1,5-pentanedioate + phosphate
S Additional information <1> (<1> no substrates are: D-glucose 6-phosphate, D-arabinose, DL-glyceraldehyde, D-erythrose, glycoaldehyde, DL-glyceraldehyde 3-phosphate [27]; <1> no activity with D-erythrose, D-glyceraldehyde 3-phosphate, ribose 5-phosphate, glucose 6-phosphate, glucosamine 6-phosphate, N-acetylglucosamine 6-phosphate, and pyruvate [1]) [1, 27]
P ?

Inhibitors

(1R,2S)-1,2-epoxypropylphosphonic acid <3> [11]
(Z)-phosphoenol 3-fluoropyruvate <3> [11]
1,10-phenanthroline <1> (<1> activity is restored by Fe^{2+} or Zn^{2+} [28]) [28]
2,3-bisphosphoglycerate <1> [9]
2-methyl-DL-Trp <2> (<2> 0.02 mM, 30% inhibition [3]) [3]
2-phosphoglycerate <1> (<1> competitive with respect to phosphoenolpyruvate [9]) [9]
3,4-dihydroxycinnamate <23> (<23> isoenzyme DS-Co [42]) [42]
3-deoxy-D-arabino-heptonic acid 7-phosphate <1> [1]
3-methylphosphoenolpyruvate <1> [9]
3-propylphosphoenolpyruvate <1> [9]
4-methyl-DL-Trp <2> (<2> 0.02 mM, 30% inhibition [3]) [3]
5,5'-dithiobis(2-nitrobenzoate) <1> [10]
5-fluoro-DL-Trp <2> (<2> 0.02 mM, 56% inhibition [3]) [3]
5-hydroxy-DL-Trp <2> (<2> 0.02 mM, 10% inhibition [3]) [3]
6-methyl-DL-Trp <2> (<2> 0.02 mM, 6% inhibition [3]) [3]
7-aza-DL-Trp <2> (<2> 0.02 mM, 9% inhibition [3]) [3]
7-methyl-DL-Trp <2> (<2> 0.02 mM, 14% inhibition [3]) [3]
7-phospho-2-dehydro-3-deoxy-D-arabino-heptonate <10, 16> (<16> competitive against D-erythrose 4-phosphate and phosphoenolpyruvate [22]; <10> product inhibition [18]) [18, 22]

CN⁻ <1> (<1> Tyr-sensitive isozyme, strong inhibition, reactivation by divalent cations only to a small extent [37]) [37]

Co^{2+} <3, 4> (<4> weak [5]; <3> concentrations above 0.1 mM [4]) [4, 5]

Cu^{2+} <1, 4, 8> (<1> phosphoenolpyruvate protects [44]; <1> Phe-sensitive isozyme, complete inactivation at 0.02 mM, destabilization of the enzymes quarternary structure [44]; <4> weak [5]; <1> 0.02 mM, complete inactivation [31]) [5, 16, 31, 44]

D-erythrose 4-phosphate <1, 16, 17, 23> (<1> Phe-sensitive isozyme: in absence of phosphoenolpyruvate the enzyme is inhibited via formation of a covalent binding to Lys186 via a slow Schiff base reaction, mechanism [56]; <16> substrate-inhibition [24]; <23> isoenzyme DS-Mn: substrate inhibition above 0.5 mM [42]) [24, 29, 42, 56]

D-fructose 1,6-diphosphate <1> [1]

D-sedoheptulose 1,7-diphosphate <1> [1]

D-sedoheptulose 7-phosphate <1> [1]

DL-dibromotryptophan <2> (<2> 0.02 mM, 52% inhibition [3]) [3]

DL-erythro-β-methyltryptophan <2> (<2> 0.02 mM, 40% inhibition [3]) [3]

DL-homotryptophan <2> (<2> 0.02 mM, 46% inhibition [3]) [3]

EDTA <1, 2, 9, 10, 16, 21, 29, 32> (<32> strong, reversible by dialysis [55]; <1> reversible by diverse divalent metal ions with varying efficiency [35]; <1,2> reversible by Co^{2+} [2,6]; <16> phosphoenolpyruvate or 3-deoxy-D-arabino-heptulosonate 7-phosphate and Mn^{2+}, Co^{2+}, Ca^{2+} and Mg^{2+} protect [22]; <9> Co^{2+}, Zn^{2+}, Cu^{2+}, and Fe^{3+} restore activity [26]; <1> activity is restored by Fe^{2+} or Zn^{2+} [28]; <9> Co^{2+}, Mn^{2+}, Zn^{2+} and Fe^{2+} restore activity [30]; <9> phosphoenolpyruvate, 2 mM, partially protects from inactivation, does not restore activity [30]; <1> Mn^{2+}, Cd^{2+}, Co^{2+}, Fe^{2+}, Cu^{2+}, Mg^{2+} or Zn^{2+} reactivate, Fe^{2+} and Cu^{2+} only partially reactivate [37]; <3,4> no inhibition [4,5]) [2, 6, 18, 22, 26, 28, 30, 35, 37, 39, 50, 55]

Fe^{2+} <1, 8> (<1> phosphoenolpyruvate protects [44]; <1> Phe-sensitive isozyme, 60% inactivation at 0.02 mM, 90% inactivation at 0.2 mM [44]; <1> 0.02 mM: 60% inactivation. 0.2 mM: 90% inactivation [31]) [16, 31, 44]

Hg^{2+} <4> (<4> complete inhibition at 1 mM [5]) [5]

L-arogenate <23> (<23> isozyme DS-Mn, competitive against D-erythrose 4-phosphate, non-competitive against phosphoenolpyruvate [42]) [42]

Mg^{2+} <31> (<31> wild-type and mutants [53]; <31> no inhibition of mutant C67L [53]) [53]

N-[phosphomonomethyl]glycine <7, 29> (<7> cytosolic isozyme [14]; <29> inhibition of isozyme DS-Co in vitro and in vivo, isozyme DS-Mn is only slightly inhibited in vitro [50]; <7,29> i.e. glyphosate [14,50]; <7> concentration of Co^{2+} markedly increases the concentration of N-[phosphomonomethyl]glycine required for inhibition [14]) [14, 50]

N-α-methyl-DL-Trp <2> (<2> 0.02 mM, 60% inhibition [3]) [3]

N-bromosuccinimide <1> [10]

N-ethylmaleimide <1> [10]

Na$_2$SO$_4$ <4> (<4> slightly [5]) [5]

Ni^{2+} <4> (<4> weak [5]) [5]

Phe <1, 4, 5, 9, 11, 16, 20> (<1, 4, 5, 9, 11, 16, 20> Phe-sensitive isozyme [5, 10, 13, 19, 23, 28, 30, 31, 35, 36, 38, 44, 56]; <31> slight inhibition of the wild-type, activation of mutants [54]; <9> Tyr-sensitive isozyme is less inhibited than the Phe-sensitive isozyme [26]; <1> strong [13]; <4> 0.5 mM, 60% inhibition [5]; <5> strain H1, strain H20 and strain 3/2: cumulative inhibition of Tyr and Phe [5]; <5> strain 12/60/X, no inhibition [5]; <1,11> L-Phe [10,13,19]; <11> 0.04 mM, 50% inhibition of DAHP synthase-phe [19]; <9> competitive with respect to D-erythrose 4-phosphate and non-competitive to phosphoenolpyruvate [26,30]; <20> DSI [38]; <1> feed-back inhibition, Phe-sensitive isozyme [35,44]; <1, 2, 25, 29> no inhibition [2, 5, 6, 50]) [5, 10, 13, 19, 23, 26, 28, 30, 31, 35, 36, 38, 54, 56]

Trp <1, 2, 14, 16, 17, 20-22, 32> (<1, 2, 14, 16, 17, 20-22> Trp-sensitive isozyme [3, 20, 22-24, 27, 29, 38, 39, 41, 42]; <32> minor feedback inhibitor [55]; <29> isozyme DS-Mn [50]; <16> allosteric, best at pH 6.0, no inhibition at pH 7.3 [24]; <2> strongly dependent on pH [3]; <17> noncompetitive to both substrates [29]; <2,14,16> non-competitive against phosphoenolpyruvate and competitive for D-erythrose 4-phosphate [3, 20, 22]; <2> 0.02 mM, 65% inhibition [3]; <14> DAHP synthase-trp [20]; <20> DSI [38]; <23> L-Trp, weak [42]) [3, 20, 22-24, 27, 29, 38, 39, 41, 42, 50, 55]

Tyr <1, 3-6, 9, 11, 14-16, 20, 31> (<1, 3-6, 9, 11, 14-16, 20, 31> Tyr-sensitive isozyme [5-7, 10-13, 19, 20, 21, 23, 26, 37, 38, 54]; <31> wild-type and mutant S187A [54]; <1> wild-type isozyme, mutant N8K and N-terminal deletion mutant are not inhibited, N-terminus is structurally involved in the sensitivity for feedback inhibition [52]; <1, 11, 14, 15> L-Tyr [10, 19, 20, 21]; <4> 0.5 mM, 20% inhibition [5]; <5> strain H1, strain H20 and strain 3/2: cumulative inhibition of Tyr and Phe [5]; <5> strain 33/X strongly but not totally inhibited [5]; <1> 0.02 mM, 50% inhibition [6]; <1, 31> feedback inhibiton [7, 54]; <1, 3, 9, 31> noncompetitive with respect to D-erythrose 4-phosphate, competitive with respect to phosphoenolpyruvate [7, 11, 26, 54]; <11> 0.04 mM, 50% inhibition of DAHP synthase-tyr [19]; <14> DAHP synthase-tyr [20]; <20> DSI and DSII [38]; <6> at pH 6.2-7.8 [12]; <29> no inhibition [50]) [5-7, 10-13, 19, 20, 21, 23, 26, 37, 38, 52, 54]

Zn^{2+} <4, 8, 31> (<31> wild-type and mutants [53]; <31> no inhibition of mutant C67L [53]) [5, 16, 53]

α-methylphenylalanine <1> [10]

β-2-thienyl-D,L-Ala <1> [13]

β-phenylserine <1> [10]

β-thienylalanine <1> [10]

chorismate <14, 30, 32> (<32> main allosteric feedback inhibitor and regulator for enzyme of class AroAII [55]; <30> 50% at 0.12 mM [51]; <30> feedback inhibition, noncompetitive against D-erythrose 4-phosphate and phosphoenolpyruvate [51]; <14> competitive against phosphoenolpyruvate, noncompetitive against D-erythrose 4-phosphate [20]; <14> DAHP synthase-trp [20]) [20, 51, 55]

diethyl dicarbonate <1> (<1> Phe-sensitive isozyme, pH-dependent, phosphoenolpyruvate protects wild-type and mutants H64G, H207G, H304G [49]) [49]

dihydroxyphenylalanine <1> [13]
iodoacetamide <18> (<18> alkylation, abolishes the dependence on reducing agents [58]) [58]
m-chlorophenylalanine <1> [10]
m-fluorophenylalanine <1> [10]
m-hydroxyphenylalanine <1> [10]
metal chelators <10> [18]
o-chlorophenylalanine <1> [10]
o-fluorophenylalanine <1> [10]
o-hydroxyphenylalanine <1> [10]
p-aminophenylalanine <1> [10]
p-chloromercuribenzoate <1> (<1> complete inhibition at 0.02 mM, reversible by cysteine [1]) [1]
p-fluorophenylalanine <1> [10]
p-hydroxymercuribenzoate <2> [2]
phenylpyruvate <14> (<14> DAHP synthase-tyr [20]) [20]
phosphate <1, 3, 10, 16> (<16> competitive to D-erythrose 4-phosphate, noncompetitive to phosphoenolpyruvate [24]; <3> phosphate buffer destabilizes [4]; <1,10,16> non-competitive with respect to both phosphoenolpyruvate and D-erythrose 4-phosphate [7,18,22]) [4, 7, 18, 22, 24]
phosphoenol 2-oxobutanoate <3> [11]
prephenate <23, 30> (<30> 50% at 0.02 mM [51]; <30> feedback inhibition, noncompetitive against D-erythrose 4-phosphate and phosphoenolpyruvate [51]; <23> isozyme DS-Mn, competitive against D-erythrose 4-phosphate, non-competitive against phosphoenolpyruvate [42]) [42, 51]
tetranitromethane <1> [10]
trinitrobenzene sulfonate <1> [10]
Additional information <1, 5, 30> (<30> no inhibition by aromatic amino acids, folic acid, phenazine 1-carboxylic acid, anthranilic acid, shikimic acid, *p*-aminobenzoic acid and 3-hydroxyanthranilic acid [51]; <1> inhibition mechanism [27]; <5> no inhibition by Tyr and Phe [5]; <25> no inhibition by Tyr and Phe [5]) [5, 27, 51]

Activating compounds
2-mercaptoethanol <1> (<1> activates [1]) [1]
L-Trp <8, 10> (<8,10> activates [16,18]) [16, 18]
N-[phosphomonomethyl]glycine <8> (<8> i.e. glyphosate [15]; <8> induces synthesis of enzyme [15]) [15]
Phe <31> (<31> slightly activating, mutants S187Y, S187F, S187C [54]) [54]
Tyr <29, 31> (<31> mutants S187Y, S187F, S187C [54]; <29> 2.3fold activation of isozyme DS-Mn [50]) [50, 54]
chorismic acid <8, 23> (<8> stimulates [16]; <23> activates isozyme DS-Mn [42]) [16, 42]
dithiothreitol <2, 18> (<18> can partially substitute thioredoxin as activating agent [58]; <2> stimulates [2]) [2, 58]
reduced thioredoxin <18> (<18> required for activity, can be partially substituted by dithiothreitol [58]) [58]

Additional information <1, 18, 32> (<32> not hysteretically activated by dithiols, e.g. DTT [55]; <18> DHS1 is induced 3-5fold by wounding of the leaf and pathogen attack [32]; <1> sulfhydryl groups and possibly a Lys residue may be implicated in activity [10]) [10, 32, 55]

Metals, ions

Cd^{2+} <1> (<1> activates, Tyr-sensitive isozyme [37]) [37]

Co^{2+} <1, 2, 8, 10, 23, 24, 29, 31> (<29> isozyme DS-Co [50]; <31> wild-type and mutants, best activator [53]; <1> 2-2.5fold increase in activity at 0.1 mM [6]; <8> slightly stimulating [16]; <2,10,31> stimulates [2,25,53]; <2> required for full activity [2]; <3> not required, Co^{2+} is inhibitory at concentrations above 0.1 mM [4]; <24> absolute requirement for divalent metal, 70% of the activation with Mg^{2+} [40]; <23> isoenzyme Ds-Co has an absolute requirement for divalent cations, Co^{2+} being best [42]; <6> no stimulation of purified enzyme [12]; <1> no effect [13]) [2, 6, 16, 25, 40, 42, 50, 53]

Cu^{2+} <1> (<1> activates, Tyr-sensitive isozyme [37]; <1> 3 isozymes contain traces of Cu^{2+} [35]) [35, 37]

Fe^{2+} <1, 9> (<1> activates, Tyr-sensitive isozyme [37]; <1> stimulates [28,35]; <9> contains 0.6 mol of iron per mol of enzyme [30]; <1> the 3 isozymes contain 0.2-0.3 mol of iron per mol of enzyme monomer [35]) [28, 30, 35, 37]

Mg^{2+} <1, 1, 8, 10, 18, 24> (<18> stimulating, can partly substitute for Mn^{2+} [58]; <1> activates, Tyr-sensitive isozyme [37]; <8> slightly [16]; <10> stimulates [25]; <24> absolute requirement for divalent metal, Mg^{2+} is most effective [40]) [16, 25, 37, 40, 58]

Mn^{2+} <1, 8, 10, 18, 23, 24, 29> (<1> Tyr-sensitive isozyme, best activator [37]; <1, 8, 10> stimulates [16, 18, 25, 27, 28]; <24> absolute requirement for divalent metal, 15% of the activation with Mg^{2+} [40]; <18, 29> isozyme DS-Mn [50, 58]; <23> isoenzyme DS-Mn is activated 2.6fold by 0.4 mM Mn^{2+} [42]; <18> absolutely required, can partly be substituted by Mg^{2+}, but not by Ca^{2+}, Fe^{2+}, and Co^{2+} [58]) [16, 18, 25, 27, 28, 37, 40, 42, 50, 58]

Ni^{2+} <31> (<31> highly activating [53]; <31> wild-type and mutants [53]) [53]

Zn^{2+} <1> (<1> activates, Tyr-sensitive isozyme [37]; <1> stimulates [28]; <1> 3 isozymes contain variable amounts of zinc [35]) [28, 35, 37]

Additional information <1, 9, 16, 32> (<1> Phe-sensitive isozyme: Mn^{2+}, Co^{2+}, Zn^{2+}, Fe^{3+} have no effect [44]; <1> no effect by Ca^{2+}, Co^{2+}, Cu^{2+}, Fe^{3+}, Mg^{2+} [28]; <9> Tyr-sensitive isozyme, metal factor required [26]; <16> metalloenzyme [22]) [22, 26, 28, 44, 55]

Turnover number (min^{-1})

0.21 <31> (phosphoenolpyruvate, <31> recombinant wild-type, 30°C, pH 6.5 [53]) [53]

1260 <1> (D-erythrose 4-phosphate, <1> Trp-sensitive isozyme, pH 7.0, 25°C [27]) [27]

1260 <1> (phosphoenolpyruvate, <1> Trp-sensitive isozyme, pH 7.0, 25°C [27]) [27]

1770 <1> (phosphoenolpyruvate) [37]

1920 <1> (D-erythrose 4-phosphate, <1> Phe-sensitive isozyme, pH 6.8, 25°C [49]) [49]

1920 <1> (phosphoenolpyruvate, <1> Phe-sensitive isozyme, pH 6.8, 25°C [49]) [49]

4260 <1> (D-erythrose 4-phosphate, <1> Phe-sensitive isozyme, pH 6.8, 25°C [56]) [56]

4260 <1> (phosphoenolpyruvate, <1> Phe-sensitive isozyme, pH 6.8, 25°C [56]) [56]

7300 <1> (D-erythrose 4-phosphate, <1> pH 7.0, 37°C, MW 79000 assumed [7]) [7]

7300 <1> (phosphoenolpyruvate, <1> pH 7.0, 37°C, MW 79000 assumed [7]) [7]

Additional information <31> (<31> mutants [53]) [53]

Specific activity (U/mg)

0.005 <14> (<14> partially purified isozyme synthase-trp [20]) [20]

0.008 <23> (<23> partially purified isoenzyme DS-Mn [42]) [42]

0.026 <4> (<4> partially purified enzyme [5]) [5]

0.056 <14> (<14> partially purified isozyme synthase-tyr [20]) [20]

0.11 <29> (<29> partially purified isozyme DS-Mn, with Mn^{2+} [50]) [50]

0.14 <29> (<29> partially purified isozyme DS-Co, with Co^{2+} [50]) [50]

0.21 <23> (<23> partially purified isozyme DS-Co [42]) [42]

0.27 <16> (<16> partially purified Tyr-sensitive isozyme [23]) [23]

0.4 <1> (<1> partially purified enzyme [1]) [1]

0.8 <32> (<32> recombinant clone M1 in Escherichia coli strain BL21(DE3) [55]) [55]

0.91 <17> (<17> purified enzyme [29]) [29]

1.3 <2> (<2> partially purified enzyme [2]) [2]

2.1 <1> (<1> purified enzyme [6]) [6]

2.3 <10> (<10> purified cytosolic isozyme [25]) [25]

2.4 <32> (<32> partially purified native enzyme [55]) [55]

3.11 <32> (<32> recombinant clone M2 in Escherichia coli strain BL21(DE3) [55]) [55]

4.5 <16> (<16> partially purified Phe-sensitive isozyme [23]) [23]

7.1 <10> (<10> purified enzyme [18]) [18]

7.7 <16> (<16> purified Trp-sensitive isozyme [39]) [39]

7.8 <22> (<22> purified enzyme [41]) [41]

8.1 <16> (<16> purified Trp-sensitive isozyme [23]) [23]

8.24 <9> (<9> purified recombinant Tyr-sensitive isozyme [26]) [26]

8.33 <21> (<21> purified Trp-sensitive isozyme [39]) [39]

10.75 <6> (<6> purified enzyme [12]) [12]

11.02 <22> (<22> purified Trp-sensitive isozyme [39]) [39]

13.1 <1> [28]

16.4 <8> (<8> purified enzyme [16]) [16]

18.5 <3> (<3> purified enzyme [4]) [4]

18.8 <15> (<15> partially purified cytosolic isozyme, + 10 mM Mg^{2+} [40]) [40]

22 <1> (<1> purified recombinant Tyr-sensitive isozyme, with Cd^{2+} [37])
[37]
30-40 <1> (<1> purified Tyr-sensitive isozyme, wild-type and mutant N8K
[52]) [52]
37.5 <3> (<3> purified enzyme [11]) [11]
56 <1> (<1> purified enzyme [7]) [7]
67 <1> (<1> purified enzyme [8]) [8]
82.6 <1> (<1> purified enzyme [13]) [13]
139.3 <31> (<31> recombinant wild-type [53]) [53]
160 <31> (<31> wild-type [54]) [54]
Additional information <1, 9, 31> (<1> metal ions are removed form the
buffer solution via Chelex 100 [56]; <31> mutants, influence of aromatic amino acids on activity [53,54]) [30, 37, 53, 54]

K_m-Value (mM)

0.0026 <22> (D-erythrose 4-phosphate, <22> pH 7.0, 37°C [41]) [41]
0.0032 <31> (phosphoenolpyruvate, <31> wild-type, pH 7.9, 30°C [54]) [54]
0.0043 <31> (phosphoenolpyruvate, <31> recombinant wild-type, 30°C, pH
6.5 [53]) [53]
0.005 <1> (phosphoenolpyruvate, <1> recombinant enzyme, with D-arabinose 5-phosphate as cosubstrate, pH 7.5, 25°C [34]) [34]
0.0053 <1> (phosphoenolpyruvate, <1> recombinant enzyme, with erythrose
4-phosphate as cosubstrate, pH 7.5, 25°C [34]) [34]
0.0058 <1> (phosphoenolpyruvate, <1> pH 7.0, 37°C [7]) [7]
0.0067 <22> (phosphoenolpyruvate, <22> pH 7.0, 37°C [41]) [41]
0.009 <1> (phosphoenolpyruvate, <1> Phe-sensitive isozyme, pH 6.8, 25°C
[49]) [49]
0.01 <1> (phosphoenolpyruvate, <1> recombinant enzyme, with ribose 5-
phosphate as cosubstrate, pH 7.5, 25°C [34]) [34]
0.013 <3> (D-erythrose 4-phosphate, <3> pH 7.4, 37°C [4]) [4]
0.013 <1> (phosphoenolpyruvate, <1> Tyr-sensitive isozyme, 25°C [37]) [37]
0.015 <1, 3> (phosphoenolpyruvate, <3> pH 7.4, 37°C [4]; <1> recombinant
enzyme, with 2-deoxyribose 5-phosphate as cosubstrate, pH 7.5, 25°C [34])
[4, 34]
0.018 <31> (D-erythrose 4-phosphate, <31> wild-type, pH 7.9, 30°C [54])
[54]
0.018 <9> (phosphoenolpyruvate, <9> Phe-sensitive isozyme, pH 6.8, 30°C
[30]) [30]
0.03 <1> (D-arabinose 5-phosphate, <1> recombinant enzyme, pH 7.5, 25°C
[34]) [34]
0.03 <10> (phosphoenolpyruvate, <10> pH 7.0, 25°C [18]) [18]
0.043 <4> (phosphoenolpyruvate, <4> pH 7.8, 25°C [5]) [5]
0.055 <4> (D-erythrose 4-phosphate, <4> pH 7.8, 25°C [5]) [5]
0.07 <10> (D-erythrose 4-phosphate, <10> pH 7.0, 25°C [18]) [18]
0.08 <1> (phosphoenolpyruvate, <1> 37°C, pH 7.0 [9]) [9]
0.0814 <1> (D-erythrose 4-phosphate, <1> Tyr-sensitive isozyme, 25°C [37])
[37]

0.086 <1> (D-erythrose 4-phosphate, <1> Phe-sensitive isozyme, pH 6.8, 25°C [49]) [49]

0.092 <21> (phosphoenolpyruvate, <21> Trp-sensitive isozyme, pH 7.0, 30°C [39]) [39]

0.0965 <1> (D-erythrose 4-phosphate, <1> pH 7.0, 37°C [7]) [7]

0.125 <9> (phosphoenolpyruvate, <9> Tyr-sensitive isozyme, pH 6.8 [26]) [26]

0.13 <9> (D-erythrose 4-phosphate, <9> Phe-sensitive isozyme, pH 6.8, 30°C [30]) [30]

0.141 <1> (D-erythrose 4-phosphate, <1> recombinant enzyme, pH 7.5, 25°C [34]) [34]

0.16 <2> (D-erythrose 4-phosphate, <2> 30°C, pH 7.0 [2]) [2]

0.195 <21> (D-erythrose 4-phosphate, <21> Trp-sensitive isozyme, pH 7.0, 30°C [39]) [39]

0.2 <11> (phosphoenolpyruvate, <11> DAHP synthase-phe, pH 7.0, 37°C [19]) [19]

0.25 <17> (D-erythrose 4-phosphate, <17> Trp-sensitive isozyme, pH 7.2, 37°C [29]) [29]

0.26 <11> (phosphoenolpyruvate, <11> DAHP synthase-tyr, pH 7.0, 37°C [19]) [19]

0.3 <2> (phosphoenolpyruvate, <2> 30°C, pH 7.0 [2]) [2]

0.4 <17, 20> (phosphoenolpyruvate, <17> Trp-sensitive isozyme, pH 7.2, 37°C [29]; <20> isozyme DS II, pH 7.5, 37°C [38]) [29, 38]

0.5 <9> (D-erythrose 4-phosphate, <9> Tyr-sensitive isozyme, pH 6.8 [26]) [26]

0.56 <14> (D-erythrose 4-phosphate, <14> DAHP synthase-trp, pH 7.0, 37°C [20]) [20]

0.77 <11> (D-erythrose 4-phosphate, <11> DAHP synthase-tyr, pH 7.0, 37°C [19]) [19]

0.87 <11> (D-erythrose 4-phosphate, <11> DAHP synthase-phe, pH 7.0, 37°C [19]) [19]

0.9 <1> (D-erythrose 4-phosphate, <1> pH 7.0, 37°C [9]) [9]

1 <14> (D-erythrose 4-phosphate, <14> DAHP synthase-tyr, pH 7.0, 37°C [20]) [20]

1 <14> (phosphoenolpyruvate, <14> DAHP synthase-tyr, pH 7.0, 37°C [20]) [20]

1.1 <20> (D-erythrose 4-phosphate, <20> DS II isozyme, pH 7.5, 37°C [38]) [38]

1.11 <14> (phosphoenolpyruvate, <14> DAHP synthase-trp, pH 7.0, 37°C [20]) [20]

1.2 <1> (D-erythrose 4-phosphate, <1> pH 6.4, 37°C [1]) [1]

1.95 <24> (D-erythrose 4-phosphate, <24> cytosolic isozyme DS-Co, pH 8.6, 37°C [40]) [40]

2 <1> (phosphoenolpyruvate, <1> Phe-sensitive isozyme, pH 6.8, 25°C [56]) [56]

2.5 <24> (D-erythrose, <24> cytosolic isozyme DS-Co, pH 8.6, 37°C [40]) [40]

3 <24> (DL-glyceraldehyde 3-phosphate, <24> cytosolic isozyme DS-Co, pH 8.6, 37°C [40]) [40]

3.3 <10> (D-erythrose 4-phosphate, <10> cytosolic isozyme, pH 8.6 [25]) [25]

3.3 <24> (L-glyceraldehyde, <24> cytosolic isozyme DS-Co, pH 8.6, 37°C [40]) [40]

3.5 <24> (D-glyceraldehyde, <24> cytosolic isozyme DS-Co, pH 8.6, 37°C [40]) [40]

3.5 <1> (phosphoenolpyruvate, <1> pH 6.4, 37°C [1]) [1]

3.6 <24> (glyoxylate, <24> cytosolic isozyme DS-Co, pH 8.6, 37°C [40]) [40]

5.1 <24> (L-erythrose, <24> cytosolic isozyme DS-Co, pH 8.6, 37°C [40]) [40]

6 <1> (ribose 5-phosphate, <1> recombinant enzyme, pH 7.5, 25°C [34]) [34]

6.8 <1> (2-deoxyribose 5-phosphate, <1> recombinant enzyme, pH 7.5, 25°C [34]) [34]

8.4 <24> (D-threose, <24> cytosolic isozyme DS-Co, pH 8.6, 37°C [40]) [40]

8.6 <24> (glycolaldehyde, <24> cytosolic isozyme DS-Co, pH 8.6, 37°C [40]) [40]

13.9 <24> (L-threose, <24> cytosolic isozyme DS-Co, pH 8.6, 37°C [40]) [40]

21 <1> (phosphoenolpyruvate, <1> sensitive to metal ion [56]; <1> Phe-sensitive isozyme, pH 6.8, 25°C [56]) [56]

Additional information <1, 3, 31> (<31> mutants [53, 54]; <1> pH-dependence [37]; <1,3> kinetics [4, 6, 7, 27, 37]) [4, 6, 7, 27, 37, 52-54]

K_i-Value (mM)

0.0009 <9> (Tyr, <9> Tyr-sensitive isozyme, pH 6.8 [26]) [26]

0.005 <14> (Trp, <14> synthase-trp isozyme, versus D-erythrose 4-phosphate, 37°C, pH 7.0 [20]) [20]

0.01 <9> (Phe, <9> Phe-sensitive isozyme, pH 6.8, 30°C [30]) [30]

0.02 <1> (Tyr, <1> pH 7.0, 24°C [6]) [6]

0.023 <14> (Tyr, <14> synthase-tyr isozyme, versus D-erythrose 4-phosphate, 37°C, pH 7.0 [20]) [20]

0.023 <14> (Tyr, <14> synthase-tyr isozyme, versus phosphoenolpyruvate, 37°C, pH 7.0 [20]) [20]

0.04 <14> (Trp, <14> synthase-trp isozyme, versus phosphoenolpyruvate, 37°C, pH 7.0 [20]) [20]

0.27 <9> (Phe, <9> Tyr-sensitive isozyme, pH 6.8 [26]) [26]

1.35 <14> (chorismate, <14> synthase-trp isozyme, versus phosphoenolpyruvate, 37°C, pH 7.0 [20]) [20]

1.35 <14> (phenylpyruvate, <14> synthase-tyr isozyme, versus D-erythrose 4-phosphate, 37°C, pH 7.0 [20]) [20]

2.25 <14> (chorismate, <14> synthase-trp isozyme, versus D-erythrose 4-phosphate, 37°C, pH 7.0 [20]) [20]

2.55 <14> (phenylpyruvate, <14> synthase-tyr isozyme, versus phosphoenolpyruvate, 37°C, pH 7.0 [20]) [20]

22 <16> (phosphate, <16> slope, tryptophan-sensitive isozyme, versus phosphoenolpyruvate, pH 7.4, 37°C [24]) [24]

25 <16> (phosphate, <16> slope, tryptophan-sensitive isozyme, versus D-erythrose 4-phosphate, pH 7.4, 37°C [24]) [24]
54 <16> (phosphate, <16> intercept, tryptophan-sensitive isozyme, versus phosphoenolpyruvate, pH 7.4, 37°C [24]) [24]
Additional information <1, 3, 16> (<3> product inhibition pattern [4]) [4, 9, 22]

pH-Optimum

6.2 <17> (<17> Tris-HCl buffer [29]) [29]
6.2-7.8 <6> [12]
6.4 <1> [1]
6.5 <31> (<31> wild-type and mutants [53]) [53]
6.5-7 <1> [13]
6.8 <1, 9, 14> (<1> wild-type Phe-sensitive isozyme [49]; <1> assay at [31, 44, 56]; <14> DAHP synthase-tyr [20]) [20, 30, 31, 44, 49, 56]
6.8-7.2 <11> (<11> DAHP synthase-phe [19]) [19]
7 <1, 2, 8, 10, 11, 14, 21-23> (<10, 14, 21, 22, 30> assay at [6, 18, 20, 39, 51]; <11> DAHP synthase-tyr [19]; <23> isoenzyme DS-Mn [42]) [2, 6, 7, 9, 16, 18-20, 39, 41, 42, 51]
7-7.4 <21> (<21> Trp-sensitive isozyme [39]) [39]
7-7.5 <1> (<1> recombinant Tyr-sensitive isozyme [37]) [37]
7.1 <10> [18]
7.2 <14, 17> (<14> DAHP synthase-trp [20]; <17> phosphate buffer, Trp-sensitive isozyme [29]) [20, 29]
7.2-8.2 <4> [5]
7.3 <16> (<16> tryptophan-sensitive isozyme [24]) [24]
7.4 <3, 29> (<29> assay at [50]; <3> HEPES buffer [4]) [4, 50]
7.5-8 <24> (<24> chloroplastic isoenzyme DS-Mn [40]) [40]
7.6 <16> (<16> assay at [22]) [22]
8 <1> (<1> above, Phe-sensitive iaozyme mutant C61G [49]) [49]
8.2-8.5 <32> (<32> native enzyme [55]) [55]
8.6 <7, 8, 15, 18, 24> (<7, 8, 15, 18, 24> assay at [14, 40, 58]) [14, 40, 58]
8.8 <23> (<23> isoenzyme DS-Co [42]) [40, 42]
9 <10> (<10> cytosolic isozyme [25]) [25]
9.2 <24> (<24> cytosolic isoenzyme DS-Co [40]) [40]
Additional information <1, 8> (<1> Tyr-sensitive isozyme, pIs: 6.1 and 8.9 [37]; <8> 2 forms with pIs of pH 8.4 and pH 7.8 [16]) [16, 37]

pH-Range

1.5-8.5 <31> (<31> wild-type and mutants [53]) [53]
5.5-8.5 <17> (<17> pH 5.5: about 50% of maximal activity, pH 8.5: about 40% of maximal activity [29]) [29]
5.7-8.5 <1> (<1> pH 5.7: about 25% of maximal activity, pH 8.5: about 30% of maximal activity [7]) [7]
6-7.6 <8> (<8> pH 6.0: about 30% of maximal activity, pH 7.6: about 60% of maximal activity [16]) [16]
6-8.5 <1> (<1> pH 6.0: about 65% of maximal activity, pH 8.5: about 50% of maximal activity [13]) [13]

6.2-7.9 <2> (<2> pH 6.2: about 35% of maximal activity, pH 7.9: about 55% of maximal activity [2]) [2]

6.5-7 <9> [26]

6.5-9 <4> (<4> pH 6.5: about 35% of maximal activity, pH 9.0: about 60% of maximal activity [5]) [5]

Temperature optimum (°C)

24 <1> (<1> assay at [6]) [6]

25 <1, 4, 10, 16> (<1, 4, 10, 16> assay at [5, 18, 22, 28, 31, 44, 49, 56]) [5, 18, 22, 28, 31, 44, 49, 56]

30 <1, 2, 9, 21, 30, 31> (<31> wild-type and mutants [53]; <1, 2, 9, 21, 30> assay at [2, 30, 39, 51, 52]) [2, 30, 39, 51-53]

37 <1, 3, 7, 8, 11, 14, 15, 18, 22, 24, 29> (<1, 3, 7, 8, 11, 14, 15, 18, 22, 24, 29> assay at [1, 4, 7, 9, 14, 19, 20, 40, 41, 50, 58]) [1, 4, 7, 9, 14, 19, 20, 40, 41, 50, 58]

40 <4> [5]

47-49 <8, 24> [40]

52 <32> [55]

Temperature range (°C)

25-50 <31> (<31> wild-type and mutants [53]) [53]

4 Enzyme Structure

Molecular weight

53000 <9> (<9> gel filtration [30]) [30]

66000 <1> (<1> gel filtration [7]) [7]

67000 <11> (<11> DAHP synthase-phe, gel filtration [19]) [19]

69000 <1> (<1> gel filtration [28]) [28]

71000 <1> (<1> Tyr-sensitive isozyme, mutant N8K, gel filtration [52]) [52]

75000 <1> (<1> Tyr-sensitive isozyme, wild-type, gel filtration [52]) [52]

76000 <3> (<3> gel filtration [11]) [11]

97000 <2> (<2> gel filtration [2]) [2]

100000 <3> (<3> gel filtration [4]) [4]

103000 <10> (<10> gel filtration [18]) [18]

105000 <1> (<1> strain JP401, gel filtration [8]) [8]

107000 <21> (<21> Trp-sensitive isozyme, gel filtration [39]) [39]

110000 <1, 8> (<1> strain JP1525 [8]; <1,8> gel filtration [8,16]) [8, 16]

120000 <22> (<22> gel filtration [41]) [41]

135000 <17> (<17> Trp-sensitive isozyme, gel filtration [29]) [29]

136000 <1> (<1> equilibrium sedimentation [13]) [13]

137000 <14> (<14> DAHP synthase-tyr, gel filtration [20]) [20]

140000 <1> (<1> gel filtration and sedimentation equilibrium analysis [13]) [13]

175000 <14> (<14>, DAHP synthase-trp, gel filtration [20]) [20]

200000 <16> (<16> tryptophan-sensitive isozyme, sedimentation equilibrium centrifugation [23]) [23]

251000 <11> (<11> DAHP synthase-tyr, gel filtration [19]) [19]
385000-443000 <24> (<24> gel filtration [40]) [40]
400000 <10> (<10> gel filtration [25]) [25]

Subunits

? <1, 9, 10> (<1> x * 33000, SDS-PAGE [8]; <1> x * 38009-38014, Phe-sensitive isozyme, electron mass spectroscopy and amino acid sequence determination [56]; <1> x * 39000, either a rapid monomer-dimer equilibrium or a
very asymetric shape for the native enzyme, SDS-PAGE [7]; <9> x * 42000,
Tyr-sensitive isozyme, SDS-PAGE [26]; <10> x * 115000, cytosolic isozyme,
SDS-PAGE [25]) [7, 8, 25, 26, 56]
dimer <1, 3, 6, 8, 10, 21, 22> (<1> 2 * 39000, recombinant Trp-sensitive isozyme, SDS-PAGE [28]; <3> 2 * 40000, SDS-PAGE [11]; <10> 2 * 53000, SDS-
PAGE [18]; <8, 16, 21, 22> 2 * 54000, SDS-PAGE [16,39]; <22> 2 * 59000,
SDS-PAGE [41]) [11, 12, 16, 18, 28, 39, 41]
monomer <9> (<9> 1 * 42130, calculation from nucleotide sequence [30])
[30]
tetramer <1, 16, 17> (<17> 4 * 35000, Trp-sensitive isozyme [29]; <1> 4 *
35000, Phe-sensitive isozyme, SDS-PAGE [13]; <16> 4 * 52000, tryptophan-
sensitive isozyme, SDS-PAGE [23]) [13, 23, 29]
Additional information <1, 28> (<1,28> active site structure [47,48]; <28>
subunit structure [48]) [47, 48]

5 Isolation/Preparation/Mutation/Application

Source/tissue

callus <29> [50]
cell culture <7, 8, 10, 19> [15, 25, 33, 40]
cell suspension culture <7, 8, 19> [15, 33, 40]
leaf <15, 18, 29> (<18> DHS1 and DHS2 [32]) [21, 32, 40, 50]
root <10> [18]
seedling <8, 23, 24> [40, 42]
tuber <8> [16, 40]
Additional information <27> (<27> differential expression of the enzyme in
various vegetative and reproductive organs [46]) [46]

Localization

chloroplast <18, 19, 24> (<18> DHS1 and DHS2 [32]; <18,19,24> chloroplastic isoenzyme DS-Mn [33, 40, 58]) [32, 33, 40, 58]
cytosol <7, 8, 10, 19, 24> (<7> isozyme [14]; <19, 8, 24> cytosolic isoenzyme
DS-Co [33,40]) [14, 25, 33, 40]
Additional information <18> (<18> presence of amino-terminal extension
characteristic of chloroplast transit peptides on DHS1 and DHS2 suggests
that both proteins may be targeted to the chloroplast [32]) [32]

Purification

<1> (Tyr-sensitive isozyme, wild-type and mutants from strain AB3257 [52]; Phe-sensitive isozyme, recombinant wild-type from overexpressing strain and recombinant mutants [44, 49]; recombinant from overexpressing strain [28, 34, 35, 37]; 2fold [37]; 5-6fold to homogeneity [28]; 683fold [13]; strain JP1525 [8]; partial [1]; 105fold [7]) [1, 6-8, 13, 28, 34-37, 44, 49, 52]

<2> (170fold [2]) [2]

<3> (173fold [11]) [4, 11]

<4> (partially [5]) [5]

<6> (326fold [12]) [12]

<7> (partial [14]) [14]

<8> (partial [40]; 820fold [16]) [16, 40]

<9> (Phe-sensitive isozyme, 11fold to homogeneity from overproducing strain [30]; Tyr-sensitive isozyme [45]; recombinantly overexpresed Tyr-sensitive isozyme, 7.5fold [26]) [26, 30, 45]

<10> (cytosolic isozyme, 79.8fold to homogeneity [25]; enzyme III [18]) [18, 25]

<14> (partially, 2 isozymes [20]) [20]

<16> (Trp-sensitive isozyme to homogeneity, 2000fold [39]; 2350fold purification of the tryptophan sensitive isozyme, partial purification of the tyrosine- and phenylalanine-sensitive isozyme [23]) [23, 24, 39]

<17> (Trp-sensitive isozyme, 130fold [29]) [29]

<18> (recombinant from Escherichia coli, mature chloroplastic form, purified to homogeneity [58]) [58]

<20> [38]

<21> (Trp-sensitive isozyme to homogeneity, 2380fold, large scale [39]) [39]

<22> (2700fold [41]; Trp-sensitive isozyme to homogeneity, 3750fold [39]) [39, 41]

<23> (partial, isozyme DS-Co and isozyme DS-Mn [42]) [42]

<24> (partial [40]) [40]

<29> (partially, isozymes DS-Co and DS-Mn [50]) [50]

<31> (recombinant His-tagged wild-type and mutants from Escherichia coli [53,54]) [53, 54]

<32> (partial, 19fold, recombinant from Escherichia coli [55]) [55]

Crystallization

<1> (Phe-sensitive isozyme, hanging drop vapour diffusion method, 22°C, with or without inhibitor phenylalanine, at pH 6.3-9.4, 0.1-0.2 M monovalent cations, PEG 1000-4000, 0.01 ml protein solution + 0.3 ml precipitant solution, X-ray diffraction structure determination and analysis [36]) [36]

<9> (Tyr-sensitive isozyme, hanging drop vapour diffusion method, protein solution in 1:1 mixture with precipitant, 0.004 ml, + 1 ml precipitant solution: 1. 50 mM KH_2PO_4, 20% polyethylene glycol 8000 w/v or 2. 0.1 M Tris, pH 9.0, 10 mM $NiCl_2$, 20% polyethylene glycol monomethylester 2000 w/v, room temperature for 3 days, X-ray diffraction structure analysis, usage of cryoprotectant glycerol for investigations [45]) [45]

<28> (Phe-sensitive isozyme, enzyme-Mn^{2+}-2-phosphoglycolate-complexes, hanging drop vapour diffusion method, room temperature, all solutions, except the $MnSO_4$ and the enzyme solution, are treated with Chelex-100 to remove metals, 0.2 mM enzyme subunit solution: 0.37 $MnSO_4$, 4.2 mM 2-phosphoglycolate, 0.1 M Li_2SO_4, 12% PEG 100 w/v, 20% ethanol v/v, 50 mM 1,3-bis[tris(hydroxy-methyl)methylamino]propane buffer, pH 8.7, reservoir solution: 19% PEG 1000, 0.1 M Li_2SO_4, 20% ethanol v/v, 50 mM 1,3-bis[tris(hydroxy-methyl)methylamino]propane buffer, X-ray diffraction structure determination and analysis [48]) [48]

Cloning

<1> (Phe-sensitive isozyme, wild-type and mutants, expression in Escherichia coli strain BL21(DE3) [49]; gene aroG, Phe-sensitive isozyme, overexpression in strain gene aroH, Trp-sensitive isozyme, overexpression in deficient strain AB3248 [28]; gene aroF, Tyr-sensitive isozyme, expression of wild-type and mutants in strain AB3257 [52]; gene aroF, Tyr-sensitive isozyme, overexpression in strain BL21(DE3) [37]; overexpression in BL21(DE3) [34]; overexpression of 3 isozymes [35]) [28, 34, 35, 37, 44, 49, 52]
<9> (ARO3 gene, expression in deficient strain and in Escherichia coli strain JA196 [17]; ARO4 gene, tyrosine-sensitive isozyme, overexpression from plasmid in strain RH1326 [26]) [17, 26]
<18> (expression of mature chloroplastic form, without signal sequence, in Escherichia coli [58]; DHS1 and DHS2, complementation of Saccharomyces cerevisiae mutant strain YBK7 [32]) [32, 58]
<31> (overexpression of His-tagged wild-type and mutants in Escherichia coli [53,54]) [53, 54]
<32> (2 clones M1 and M2, expression in Escherichia coli strain BL21(DE3) and strain AB3248, which is defective in all endogenous isozyme forms, functional complementation of the mutant [55]) [55]

Engineering

$C_{14}5S$ <31> (<31> site-directed mutagenesis, 16% remaining activity compared to the wild-type, 4.6fold increased K_m, 1.6fold decreased k_{cat}, and 13.6fold decreased k_{cat}/K_m for phosphoenolpyruvate [53]) [53]
C328V <1> (<1> oligo-nucleotide mutagenesis, expression in Escherichia coli strains, 20% reduction in the catalytic constant, 2-3fold increase in K_m for the substrates, completely resistant to both spontaneous and Cu^{2+}-catalysed inactivation [44]) [44]
C334S <31> (<31> site-directed mutagenesis, unaltered properties [53]) [53]
C61G <1> (<1> site-directed mutagenesis, highly reduced activity, highly increased K_m for phosphoenolpyruvate, higher pH-optimum than the wild-type [49]) [49]
C61V <1> (<1> oligo-nucleotide mutagenesis, expression in Escherichia coli strains, inactive, does not bind metal ions, resistant to metal attack, no subunit dissociation upon Cu^{2+} treatment [44]) [44]
C67L <31> (<31> site-directed mutagenesis, highly reduced activity, insensitive to inhibition by divalent metal ions [53]) [53]
C67S <31> (<31> site-directed mutagenesis, inactive [53]) [53]

H172G <1> (<1> site-directed mutagenesis, inactive [49]) [49]

H207G <1> (<1> site-directed mutagenesis, reduced activity, increased K_m values for the substrates, reduced k_{cat} [49]) [49]

H268G <1> (<1> site-directed mutagenesis, inactive [49]) [49]

H304G <1> (<1> site-directed mutagenesis, reduced activity, increased K_m values for the substrates, increased k_{cat} [49]) [49]

H64G <1> (<1> site-directed mutagenesis, reduced activity, increased K_m values for the substrates, increased k_{cat} [49]) [49]

H64L <1> (<1> oligo-nucleotide mutagenesis, expression in Escherichia coli strains, unstable to treatment with phosphoenolpyruvate, half-life of about 24 h at 0.4 mM compared to 6 days for the wild-type [44]) [44]

N8K <1> (<1> similar activity and substrate affinities like the wild-type, but insensitive against inhibition by tyrosine, decreased thermostability [52]) [52]

S187A <31> (<31> site-directed mutagenesis, slightly reduced activity [54]) [54]

S187C <31> (<31> site-directed mutagenesis, reduced activity, activation by tyrosine and phenylalanine instead of inhibition like the wild-type [54]) [54]

S187F <31> (<31> site-directed mutagenesis, highly reduced activity, activation by tyrosine and phenylalanine instead of inhibition like the wild-type [54]) [54]

S187Y <31> (<31> site-directed mutagenesis, highly reduced activity, activation by tyrosine and phenylalanine instead of inhibition like the wild-type [54]) [54]

Additional information <1> (<1> deletion mutant of Tr-sensitive isozyme, gene aroF, lacking the first 7 amino acid residues of the N-terminus, mutant is insensitive against inhibition by tyrosine [52]) [52]

Application

agriculture <29> (<29> isozyme DS-Co is a possible target for the herbicide glyphosate, i.e. N-[phosphomonomethyl]glycine [50]) [50]

biotechnology <1> (<1> E. coli strains, overproducing the enzyme, excrete it to the medium, which can also be used as a bioindicator for enhanced carbon commitment into the pathway [43]; <1> increased production of aromatic amino acids in E. coli mutants with modified phosphoenolpyruvate metabolism and enhanced transketolase activity [43]) [43]

6 Stability

pH-Stability

6-8 <2> (<2> 30°C, stable for at least 2 h [2]) [2]

7-7.5 <2> (<2> maximal stability [2]) [2]

Temperature stability

22 <1> (<1> half-life: 1.2 days in absence of phosphoenolpyruvate, half-life: 4 days in presence of 1 mM phosphoenolpyruvate [31]) [31]

25 <1, 18> (<18> $t_{1/2}$: 5 h in absence of DTT [58]; <1> 1 h, in 1,3-bis[tris(hydroxymethyl)methylamino]propane buffer, 10% loss of activity [7]; <1> 2 h, in 3(N-morpholono)propane-sulfonic acid buffer, 50% loss of activity [7]) [7, 58]

30 <2> (<2> pH 6.0-8.0, stable for at least 2 h [2]) [2]

37 <1, 24> (<1> Tyr-sensitive isozyme, wild-type: 50% loss of activity in 80 min in presence of phosphoenolpyruvate, mutant N8K: 50% loss of activity in 20 min in presence of phosphoenolpyruvate [52]; <24> isoenzyme Ds-Co, 95% loss of activity in absence of phosphoenolpyruvate, 1.1 mM phosphoenolpyruvate almost completely protects from inactivation [40]) [40, 52]

50 <2, 4> (<2,4> inactivation above [2,5]) [2, 5]

Additional information <2, 4, 16> (<2,4> phosphoenolpyruvate protects against inactivation [2,5]; <2> D-erythrose 4-phosphate, erythrose and ribose 5-phosphate increase rate of inactivation [2]; <16> phosphoenolpyruvate, D-erythrose 4-phosphate and inorganic phosphate protect the purified enzyme against heat denaturation [24]) [2, 5, 24]

General stability information

<1>, Co^{2+} stabilizes [6]

<1>, Cu^{2+} and Fe^{2+} accelerates subunit dissociation [31]

<1>, metal-catalysed oxidation of the enzyme, the apoenzyme shows an exponentially decrease in activity with a half-life of about 1 day at 22°C, Cu^{2+} and Fe^{2+} accelerated the rate of inactivation and subunit dissociation, phosphoenolpyruvate and EDTA stabilize, mutants are insensitive [44]

<1>, phosphoenolpyruvate stabilizes [7, 44]

<1>, spontaneous inactivation with a net loss of two of the seven thiol groups per subunit is restored by dithiothreitol [31, 44]

<2>, dithiothreitol stabilizes during incubation at 30°C for 15 min [2]

<4>, loss of 50% activity by freezing and thawing the cell extract [5]

<4>, unstable during manipulations such as dialysis, dilution, ammonium sulfate fractionation, chromatography on DEAE-cellulose or Sephadex G-200 [5]

<10>, Mn^{2+} stabilizes during purification [18]

<18>, DTT stabilizes the active protein, no influence on the alkylated enzyme [58]

<21>, overnight dialysis against buffer containing 10 mM EDTA at 4°C results in 85% inactivation [39]

Storage stability

<1>, -15°C, ammonium sulfate and acetone fractions, stable for at least several weeks [1]

<1>, -20°C, 0.1 M potassium phosphate, pH 6.5, 1 mM phosphoenolpyruvate, stable for at least 6 months [8]

<1>, -20°C, in phosphate buffer containing phosphoenolpyruvate, stable for several months [7, 13]

<6>, 0.1 M potassium phosphate, pH 7.4, 1 mM phosphoenolpyruvate, 0.02% NaN_3, purified enzyme, stable for 2-3 months [12]

<9>, -20°C, 50% glycerol, stable for at least 4 weeks [26]

<24>, -80°C, isoenzyme DS-Co, stable for at least several months [40]

References

[1] Srinivasan, P.R.; Sprinson, D.B.: 2-Keto-3-deoxy-D-arabo-heptonic acid 7-phosphate synthetase. J. Biol. Chem., **234**, 716-722 (1954)

[2] Görisch, H.; Lingens, F.: 2-Deoxy-D-arabino-heptulosonate-7-phosphate synthase of Streptomyces aureofaciens Tü 24. Biochim. Biophys. Acta, **242**, 617-629 (1971)

[3] Görisch, H.; Lingens, F.: 3-Deoxy-D-arabino-heptulosonate-7-phosphate synthase of Streptomyces aureofaciens Tü 24. Biochim. Biophys. Acta, **242**, 630-636 (1971)

[4] DeLeo, A.B.; Dayan, J.; Sprinson, D.B.: Purification and kinetics of tyrosine-sensitive 3-deoxy-D-arabino-heptulosonic acid 7-phosphate synthetase from Salmonella. J. Biol. Chem., **248**, 2344-2353 (1973)

[5] Friedrich, C.G.; Schlegel, H.G.: Aromatic amino acid biosynthesis in Alcaligenes eutrophus H16. I. Properties and regulation of 3-deoxy-D-arabino heptulosonate 7-phosphate synthase. Arch. Microbiol., **103**, 133-140 (1975)

[6] Dusha, I.; Dénes, G.: Purification and properties of tyrosine-sensitive 3-deoxy-D-arabino-heptulosonate-7-phosphate synthetase of Escherichia coli K12. Biochim. Biophys. Acta, **438**, 563-573 (1976)

[7] Schoner, R.; Herrmann, K.M.: 3-Deoxy-D-arabino-heptulosonate 7-phosphate synthase. Purification, properties, and kinetics of the tyrosine-sensitive isoenzyme from Escherichia coli. J. Biol. Chem., **251**, 5440-5447 (1976)

[8] Simpson, R.J.; Davidson, B.E.: Studies on 3-deoxy-D-arabinoheptulosonate-7-phosphate synthetase(phe) from Escherichia coli. Purification and subunit structure. Eur. J. Biochem., **70**, 493-500 (1976)

[9] Simpson, R.J.; Davidson, B.E.: Studies on 3-deoxy-D-arabinoheptulosonate-7-phosphate synthetase(phe) from Escherichia coli K12. 2. Kinetic properties. Eur. J. Biochem., **70**, 501-507 (1976)

[10] Simpson, R.J.; Davidson, B.E.: Studies on 3-deoxy-D-arabinoheptulosonate-7-phosphate synthetase(phe) from Escherichia coli K12. 3. Structural studies. Eur. J. Biochem., **70**, 509-516 (1976)

[11] Hu, Ch.-Y.; Sprinson, D.B.: Properties of tyrosine-inhibitable 3-deoxy-D-arabinoheptulosonic acid-7-phosphate synthase from Salmonella. J. Bacteriol., **129**, 177-183 (1977)

[12] Bracher, M.; Schweingruber, M.E.: Purification of the tyrosine inhibitable 3-deoxy-D-arabino-heptulosonate-7-phosphate synthase from Schizosaccharomyces pombe. Biochim. Biophys. Acta, **485**, 446-451 (1977)

[13] McCandliss, R.J.; Poling, M.D.; Herrmann, K.M.: 3-Deoxy-D-arabino-heptulosonate 7-phosphate synthase. Purification and molecular characterization of the phenylalanine-sensitive isoenzyme from Escherichia coli. J. Biol. Chem., **253**, 4259-4265 (1978)

[14] Ganson, R.J.; Jensen, R.A.: The essential role of cobalt in the inhibition of the cytosolic isoenzyme of 3-deoxy-D-arabino-heptulosonate-7-phosphate synthase from Nicotiana silvestris by glyphosate. Arch. Biochem. Biophys., 260, 85-93 (1988)

[15] Pinto, J.E.B.P.; Dyer, W.E.; Weller, S. C.; Herrmann, K.M.: Glyphosate induces 3-deoxy-D-arabino-heptulosonate 7-phosphate synthase in potato (Solanum tuberosum L.) Cells grown in suspension culture. Plant Physiol., 87, 891-893 (1988)

[16] Pinto, J.E.B.P.; Suzich, J.A.; Herrmann, K.M.: 3-Deoxy-D-arabino-heptulosonate 7-phosphate synthase from potato tuber (Solanum tuberosum L.). Plant Physiol., 82, 1040-1044 (1986)

[17] Teshiba, S.; Furter, R.; Niederberger, P.; Braus, G.; Paravicini, G.; Hütter, R.: Cloning of the ARO3 gene of Saccharomyces cerevisiae and its regulation. Mol. Gen. Genet., 205, 353-357 (1986)

[18] Suzich, J.A.; Dean, J.F.D.; Herrmann, K.M.: 3-Deoxy-D-arabino-heptulosonate 7-phosphate synthase from carrot root (Daucus carota) is a hysteretic enzyme. Plant Physiol., 79, 765-770 (1985)

[19] Berry, A.; Johnson, J.L.; Jensen, R.A.: Phenylalanine hydroxylase and isozymes of 3-deoxy-D-arabino-heptulosonate 7-phosphate synthase in relationship to the phylogenetic position of Pseudomonas acidovorans (Ps. Sp. ATCC 11299a). Arch. Microbiol., 141, 32-39 (1985)

[20] Whitaker, R.J.; Fiske, M.J.; Jensen, R.A.: Pseudomonas aeruginosa possesses two novel regulatory isozymes of 3-deoxy-D-arabino-heptulosonate 7-phosphate synthase. J. Biol. Chem., 257, 12789-12794 (1982)

[21] Reinink, M.; Borstlap, A.C.: 3-Deoxy-D-arabino-heptulosonate 7-phosphate synthase from pea leaves: inhibition by L-tyrosine. Plant Sci. Lett., 26, 167-171 (1982)

[22] Nimmo, G.A.; Coggins, J.R.: Some kinetic properties of the tryptophan-sensitive 3-deoxy-D-arabino-heptulosonate 7-phosphate synthase from Neurospora crassa. Biochem. J., 199, 657-665 (1981)

[23] Nimmo, G.A.; Coggins, J.R.: The purification and molecular properties of the tryptophan-sensitive 3-deoxy-D-arabino-heptulosonate 7-phosphate synthase from Neurospora crassa. Biochem. J., 197, 427-436 (1981)

[24] Ip, K.-M.; Doy, C.H.: The kinetic of a purified form of 3-deoxy-D-arabino heptulosonate-7-phosphate synthase (tryptophan) from Neurospora crassa. Eur. J. Biochem., 98, 431-440 (1979)

[25] Suzuki, N.; Sakuta, M.; Shimizu, S.: Purification and characterization of a cytosolic isozyme of 3-deoxy-D-arabino-heptulosonate 7-phosphate synthase from cultured carrot cells. J. Plant Physiol., 149, 19-22 (1996)

[26] Schnappauf, G.; Hartmann, M.; Künzler, M.; Braus, G.H.: The two 3-deoxy-D-arabino-heptulosonate-7-phosphate synthase isoenzymes from Saccharomyces cerevisiae show different kinetic modes of inhibition. Arch. Microbiol., 169, 517-524 (1998)

[27] Akowski, J.P.; Bauerle, R.: Steady state kinetics and inhibitor binding of 3-deoxy-D-arabino-heptulosonate-7-phosphate synthase (tryptophan sensitive) from Escherichia coli. Biochemistry, 36, 15817-15822 (1997)

[28] Ray, J.M.; Bauerle, R.: Purification and properties of tryptophan-sensitive 3-deoxy-D-arabino-heptulosonate-7-phosphate synthase from Escherichia coli. J. Bacteriol., **173**, 1894-1901 (1991)

[29] Tianhui, X.; Chiao, J.S.: Purification and properties of DAHP synthase from Nocardia mediterranei. Biochim. Biophys. Acta, **991**, 1-5 (1989)

[30] Paravicini, G.; Schmidheini, T.; Braus, G.: Purification and properties of the 3-deoxy-D-arabino-heptulosonate-7-phosphate synthase (phenylalanine-inhibitable) of Saccharomyces cerevisiae. Eur. J. Biochem., **186**, 361-366 (1989)

[31] Park, O.K.; Bauerle, R.: Metal-catalyzed oxidation of phenylalanine-sensitive 3-deoxy-D-arabino-heptulosonate-7-phosphate synthase from Escherichia coli: inactivation and destabilization by oxidation of active-site cysteine. J. Bacteriol., **181**, 1636-1642 (1999)

[32] Keith, B.; Dong, X.; Ausubel, F.M.; Fink, G.R.: Differential induction of 3-deoxy-D-arabino-heptulosonate 7-phosphate synthase genes in Arabidopsis thaliana by wounding and pathogenic attack. Proc. Natl. Acad. Sci. USA, **88**, 8821-8825 (1991)

[33] McCue, K.F.; Conn, E.E.: Induction of 3-deoxy-D-arabino-heptulosonate-7-phosphate synthase activity by fungal elicitor in cultures of Petroselinum crispum. Proc. Natl. Acad. Sci. USA, **86**, 7374-7377 (1989)

[34] Sheflyan, G.Y.; Howe, D.L.; Wilson, T.L.; Woodard, R.W.: Enzymatic synthesis of 3-deoxy-D-manno-octulosonate 8-phosphate 3-deoxy-D-altro-octulosonate 8-phosphate, 3,5-dideoxy-D-gluco(manno)-octulosonate 8-phosphate by 3-deoxy-D-arabino-heptulosonate 7-phosphate synthase. J. Am. Chem. Soc., **120**, 11027-11032 (1998)

[35] Stephens, C.M.; Bauerle, R.: Analysis of the metal requirement of 3-deoxy-D-arabino-heptulosonate-7-phosphate synthase from Escherichia coli. J. Biol. Chem., **266**, 20810-20817 (1991)

[36] Shumilin, I.A.; Kretsinger, R.H.; Bauerle, R.: Purification, crystallization, and preliminary crystallographic analysis of 3-deoxy-D-arabino-heptulosonate-7-phosphate synthase from Escherichia coli. Proteins Struct. Funct. Genet., **24**, 404-406 (1996)

[37] Ramilo, C.A.; Evans, J.N.S.: Overexpression, purification, and characterization of tyrosine-sensitive 3-deoxy-D-arabino-heptulosonic acid 7-phosphate synthase from Escherichia coli. Protein Expr. Purif., **9**, 253-261 (1997)

[38] Euverink, G.J.W.; Hessels, G.I.; Franke, C.; Dijkhuizen, L.: Chorismate mutase and 3-deoxy-D-arabino-heptulosonate 7-phosphate synthase of the methylotrophic actinomycete Amycolatopsis methanolica. Appl. Environ. Microbiol., **61**, 3796-3803 (1995)

[39] Walker, G.E.; Dunbar, B.; Hunter, I.S.; Nimmo, H.G.; Coggins, J.R.: Evidence for a novel class of microbial 3-deoxy-D-arabino-heptulosonate-7-phosphate synthase in Streptomyces rimosus and Neurospora crassa. Microbiology, **142**, 1973-1982 (1996)

[40] Doong, R.L.; Gander, J.E.; Ganson, R.J.; Jensen, R.A.: The cytosolic isoenzyme of 3-deoxy-D-arabino-heptulosonate 7-phosphate synthase in Spinacia oleracea and other higher plants: extreme substrate ambiguity and other properties. Physiol. Plant., **84**, 351-360 (1992)

[41] Stuart, F.; Hunter, I.S.: Purification and characterisation of 3-deoxy-D-arabino-heptulosonate-7-phosphate synthase from Streptomyces rimosus. Biochim. Biophys. Acta, **1161**, 209-215 (1993)

[42] Rubin, J.L.; Jensen, R.A.: Differentially regulated isoenzymes of 3-deoxy-D-arabino-heptulosonate-7-phosphate synthase from seedlings of Vigna radiata. Plant Physiol., **79**, 711-718 (1985)

[43] Flores, N.Xiao, J.; Berry, A.; Bolivar, F.; Valle, F.: Pathway engineering for the production of aromatic compouds in Escherichia coli. Nat. Biotechnol., **14**, 620-623 (1996)

[44] Park, O.K.; Bauerle, R.: Metal-catalyzed oxidation of phenylalanine-sensitive 3-deoxy-D-arabino-heptulosonate-7-phosphate synthase from Escherichia coli: inactivation and destabilization by oxidation of active-site cysteines. J. Bacteriol., **181**, 1636-1642 (1999)

[45] Schneider, T.R.; Hartmann, M.; Braus, G.H.: Crystallization and preliminary X-ray analysis of 3-deoxy-D-arabino-heptulosonate-7-phosphate synthase (tyrosine inhibitable) from Saccharomyces cerevisiae. Acta Crystallogr. Sect. D, **55**, 1586-1588 (1999)

[46] Sharma, R.; Jain, M.; Bhatnagar, R.K.; Bhalla-Sarin, N.: Differential expression of DAHP synthase and chorismate mutase in various organs of Brassica juncea and the effect of external factors on enzyme activity. Physiol. Plant., **105**, 739-745 (1999)

[47] Radaev, S.; Dastidar, P.; Patel, M.; Woodard, R.W.; Gatti, D.L.: Structure and mechanism of 3-deoxy-D-manno-octulosonate 8-phosphate synthase. J. Biol. Chem., **275**, 9476-9484 (2000)

[48] Wagner, T.; Shumilin, I.A.; Bauerle, R.; Kretsinger, R.H.: Structure of 3-deoxy-D-arabino-heptulosonate-7-phosphate synthase from Escherichia coli: Comparison of the Mn^{2+}-2-phosphoglycolate and the Pb^{2+}-2-phosphoenolpyruvate complexes and Implications for catalysis. J. Mol. Biol., **301**, 389-399 (2000)

[49] Howe, D.L.; Duewel, H.S.; Woodard, R.W.: Histidine 268 in 3-deoxy-D-arabino-heptulosonic acid 7-phosphate synthase plays the same role as histidine 202 in 3-deoxy-D-manno-octulosonic acid 8-phosphate synthase. J. Biol. Chem., **275**, 40258-40265 (2000)

[50] Jain, M.; Bhalla-Sarin, N.: Effect of glyphosate on the activity of DAHP synthase isozymes in callus cultures of groundnut (Arachis hypogaea L.) selected in vitro. Indian J. Biochem. Biophys., **37**, 235-240 (2000)

[51] Kim, K.-J.: Regulation of 3-deoxy-D-arabinoheptulosonate-7-phosphate (DAHP) synthase of Bacillus sp. B-6 producing phenazine-1-carboxylic acid. J. Biochem. Mol. Biol., **34**, 299-304 (2001)

[52] Jossek, R.; Bongaerts, J.; Sprenger, G.A.: Characterization of a new feedback-resistant 3-deoxy-D-arabino-heptulosonate 7-phosphate synthase AroF of Escherichia coli. FEMS Microbiol. Lett., **202**, 145-148 (2001)

[53] Lin, L.-L.; Liao, H.-F.; Chien, H.R.; Hsu, W.-H.: Identification of essential cysteine residues in 3-deoxy-D-arabino-heptulosonate-7-phosphate synthase from Corynebacterium glutamicum. Curr. Microbiol., **42**, 426-431 (2001)

[54] Liao, H.F.; Lin, L.L.; Chien, H.R.; Hsu, W.H.: Serine 187 is a crucial residue for allosteric regulation of Corynebacterium glutamicum 3-deoxy-D-arabino-heptulosonate-7-phosphate synthase. FEMS Microbiol. Lett., **194**, 59-64 (2001)

[55] Gosset, G.; Bonner, C.A.; Jensen, R.A.: Microbial origin of plant-type 2-keto-3-deoxy-D-arabino-heptulosonate 7-phosphate synthases, exemplified by the chorismate- and tryptophan-regulated enzyme from Xanthomonas campestris. J. Bacteriol., **183**, 4061-4070 (2001)

[56] Parker, E.J.; Bulloch, E.M.M.; Jameson, G.B.; Abell, C.: Substrate deactivation of phenylalanine-sensitive 3-deoxy-D-arabino-heptulosonate 7-phosphate synthase by erythrose 4-phosphate. Biochemistry, **40**, 14821-14828 (2001)

[57] Howe, D.L.; Sundaram, A.K.; Wu, J.; Gatti, D.L.; Woodard, R.W.: Mechanistic insight into 3-deoxy-D-manno-octulosonate-8-phosphate synthase and 3-deoxy-D-arabino-heptulosonate-7-phosphate synthase utilizing phosphorylated monosaccharide analogues. Biochemistry, **42**, 4843-4854 (2003)

[58] Entus, R.; Poling, M.; Herrmann, K.M.: Redox regulation of Arabidopsis 3-deoxy-D-arabino-heptulosonate 7-phosphate synthase. Plant Physiol., **129**, 1866-1871 (2002)

3-Deoxy-8-phosphooctulonate synthase

2.5.1.55

1 Nomenclature

EC number

2.5.1.55

Systematic name

phosphoenolpyruvate:D-arabinose-5-phosphate C-(1-carboxyvinyl)transfer-ase (phosphate-hydrolysing, 2-carboxy-2-oxoethyl-forming)

Recommended name

3-deoxy-8-phosphooctulonate synthase

Synonyms

2-dehydro-3-deoxy-D-octonate-8-phosphate D-arabinose-5-phosphate-lyase (pyruvate-phosphorylating)

2-dehydro-3-deoxyphosphooctonate aldolase

2-keto-3-deoxy-8-phosphooctonic acid synthetase

2-keto-3-deoxy-8-phosphooctonic synthetase

3-deoxy-D-manno-octulosonate 8-phosphate synthetase

3-deoxy-D-manno-octulosonate-8-phosphate synthase

3-deoxy-D-manno-octulosonic acid 8-phosphate synthase

3-deoxy-D-mannooctulosonate-8-phosphate synthetase

3-deoxyoctulosonic 8-phosphate synthetase

8-phospho-2-dehydro-3-deoxy-D-octonate D-arabinose-5-phosphate-lyase (pyruvate-phosporylating)

EC 4.1.2.16 (formerly)

KDO-8-P synthase

KDO-8-P synthetase

KDO8-P

KDO8PS

KDOPS

KDPO synthetase

Kdo8P synthase

KdsA

aldolase, phospho-2-keto-3-deoxyoctonate

phospho-2-dehydro-3-deoxyoctonate aldolase

phospho-2-keto-3-deoxyoctonate aldolase

CAS registry number

9026-96-4

2 Source Organism

<-1> no activity in *Malus sp.* [18]
<1> *Pseudomonas aeruginosa* (pyocyaneus [2]) [1, 2]
<2> *Escherichia coli* (strain B [3,5]; overproducing strain DH5α (pJU1) [9])
    [3, 5-9, 11-13, 15, 16, 21, 23, 24, 29]
<3> *Salmonella typhimurium* (parent strain PR122, and mutant strain PR32, a
    temperature-sensitive lethal mutant that is conditionally defective in 3-
    deoxy-D-manno-octulosonate-8-phosphate synthesis [4]; wild-type strain
    AG701 and temperature-sensitive strain AG701i50 [27]) [4, 27]
<4> *Chlamydia trachomatis* [10]
<5> *Neisseria gonorrhoeae* [14, 26]
<6> *Chlamydia psittaci* (strain 6BC [17]) [17]
<7> *Acinetobacter calcoaceticus* [18]
<8> *Brassica oleracea* (var. italica) [18]
<9> *Daucus carota* [18]
<10> *Cucumis sativus* [18]
<11> *Allium cepa* [18]
<12> *Ipomoea batatas* [18]
<13> *Solanum tuberosum* [18]
<14> *Spinacia oleracea* [18]
<15> *Nicotiana sylvestris* [18]
<16> *Gram-negative bacteria* [19]
<17> *Aquifex aeolicus* [20, 22, 25, 28, 30]
<18> *Actinobacillus actinomycetemcomitans* [22]
<19> *Actinobacillus pleuropneumoniae* [22]
<20> *Arabidopsis thaliana* [22]
<21> *Aquifex aeolicus* [22]
<22> *Bordetella bronchiseptica* [22]
<23> *Bordetella pertussis* [22]
<24> *Campylobacter jejuni* [22]
<25> *Caulobacter crescentus* [22]
<26> *Chlamydia pneumoniae* [22]
<27> *Chlamydia psittaci* [22]
<28> *Chlamydia trachomatis* [22]
<29> *Chlorobium tepidum* [22]
<30> *Escherichia coli* [22]
<31> *Haemophilus influenzae* [22]
<32> *Helicobacter pylori* (J99 [22] (SwissProt-ID: Q9N55)) [22]
<33> *Helicobacter pylori* (26695 [22]; J99 [24]) [22, 24]
<34> *Klebsiella pneumonia* [22]
<35> *Neisseria gonorrhoeae* [22]
<36> *Neisseria meningitidis* [22]
<37> *Pasteurella multocida* [22]
<38> *Pasteurella haemolytica* [22]
<39> *Pisum sativum* [22]

<40> *Porphyromonas gingivalis* [22]
<41> *Pseudomonas aeruginosa* (SwissProt-ID: Q9ZFK4) [22]
<42> *Rickettsia prowazekii* [22]
<43> *Salmonella typhi* [22]
<44> *Shewanella putrefaciens* [22]
<45> *Vibrio cholerae* [22]
<46> *Yersinia pestis* [22]

3 Reaction and Specificity

Catalyzed reaction

phosphoenolpyruvate + D-arabinose 5-phosphate + H$_2$O = 2-dehydro-3-deoxy-D-octonate 8-phosphate + phosphate (<2>, stereochemistry [8]; <2>, reaction pathway involves an acyclic bisphosphate intermediate [9]; <2>, mechanism [11,16]; <2>, steady-state kinetic mechanism, where phosphoenolpyruvate binding precedes that of D-arabinose-5-phosphate, followed by the ordered release of phosphate and 3-deoxy-D-manno-2-octulosonate 8-phosphate [15]; <2>, bisphosphate intermediate formed during the initial step of synthesis may have the pyranose structure with the anomeric phosphate located in the β-configuration [16]; <2>, the synthesis of 2-dehydro-3-deoxy-D-octonate 8-phosphate proceeds through the formation of a linear rather than a cyclic intermediate [21])

Reaction type
condensation

Natural substrates and products
S phosphoenolpyruvate + D-arabinose-5-phosphate + H$_2$O <2, 16, 17> (<2>, the enzyme is involved in KDO biosynthesis before its incorporation into the lipid A precursor [5,13]; <2>, key enzyme in the lipopolysaccharide biosynthesis of gram-negative bacteria [7,13]; <16>, second step in the biosynthesis of 3-deoxy-D-manno-2-octulosonic acid [19]; <2>, 2-dehydro-3-deoxy-D-octonate 8-phosphate is the phosphorylated precursor of 3-deoxy-D-manno-octulosonate, an essential sugar of the liposaccharide of gram-negative bacteria [21]; <17>, the enzyme catalyzes the first commited step in the production of 2-dehydro-3-deoxy-D-octonate, an integral part of the inner core region of the lipopolysaccharide layer in Gram-negative bacteria [22]) [5, 7, 13, 19, 21, 22]
P 2-dehydro-3-deoxy-D-octonate 8-phosphate + phosphate

Substrates and products
S phosphoenolpyruvate + 2-deoxyribose 5-phosphate <2> (Reversibility: ? <2> [29]) [29]
P ?
S phosphoenolpyruvate + D-arabinose 5-phosphate <1-46> (<2,14>, irreversible reaction [11,18]; <1>, absolute specificity for phosphoenolpyruvate and D-arabinose-5-phosphate [2]; <7>, specific for D-arabinose 5-

phosphate [18]; <17>, His185 is necessary for the correct binding of phosphoenolpyruvate and of a catalytic water molecule [25]) (Reversibility: ? <1-46> [1-30]) [1-30]

P 2-dehydro-3-deoxy-D-octonate 8-phosphate + phosphate <1-46> [1-30]

S phosphoenolpyruvate + erythrose 4-phosphate <5, 14, 17> (<15>, 24% of the activity with D-arabinose 5-phosphate [18]; <2,5>, the enzyme does not catalyze the condensation of D-erythrose 4-phosphate and phosphoenolpyruvate to form 3-deoxy-D-ribo-heptulosonate 7-phosphate [26,29]) (Reversibility: ? <5,14,17> [14,18,30]) [14, 18, 30]

P 3-deoxy-D-ribo-heptulosonate 7-phosphate <5> [14]

S phosphoenolpyruvate + ribose 5-phosphate <5, 14> (<15>, 7% of the activity with D-arabinose 5-phosphate [18]) (Reversibility: ? <5,14> [14,18]) [14, 18]

P ?

S Additional information <2> (<2>, in presence of D-erythrose 4-phosphate or D-ribose 5-phosphate the enzyme catalyzes the rapid consumption of approximately 1 mol of phosphoenolpyruvate per active site, after which consumption of phosphoenolpyruvate slows to a negligible but measurable rate [29]) [29]

P ?

Inhibitors

(Z,E)-D-glucophosphoenolpyruvate <16> (<16>, and its carboxylic ester derivatives [19]) [19]

1,10-phenanthroline <2, 17, 27, 33> (<33>, IC50: 0.0293 mM [24]) [22, 24, 30]

2,6-anhydro-3-deoxy-2β-phosphonylmethyl-8-phosphate-D-glycero-D-talo-octonate <2> (<2>, most potent inhibitor [16]) [16]

2,6-pyridine dicarboxylic acid <33> (<33>, IC50: 0.0422 [24]) [24]

2-deoxy-2-fluoro-D-arabinoate-5-phosphate <2> [5]

Cd^{2+} <2, 17> (<2>, 1 mM [5]; <17>, 10 mM [30]) [3, 5, 30]

Co^{2+} <1, 7> (<1>, 1 mM, 22% inhibition [2]) [2, 18]

Cu^{2+} <1, 2, 17, 33> (<1>, 1 mM, 80% inhibition [2]; <2>, 1 mM [5]; <33>, activity is enhanced below 0.001 mM, inhibitory at higher concentrations [24]; <17>, 10 mM [30]) [2, 3, 5, 24, 30]

D-ribose 5-phosphate <2> (<2>, weak, competitive [3]) [3, 5]

EDTA <1, 2, 17, 27, 33> (<33>, IC50: 0.212 mM [24]; <5>, no inhibition [26]; <17>, reactivation by divalent metal ions [30]) [2, 22, 24, 30]

Fe^{2+} <1> (<1>, 1 mM, 30% inhibition [2]) [2]

Hg^{2+} <2> (<2>, reversed by dithiothreitol [3]; <2>, 1 mM [5]) [3, 5]

Mn^{2+} <1, 7, 17> (<1>, 1 mM, 40% inhibition [2]; <17>, 10 mM [30]) [2, 18, 30]

Zn^{2+} <1, 2, 17, 33> (<1>, 1 mM, 34% inhibition [2]; <2>, 1 mM [5]; <33>, mild inhibitor at higher concentrations [24]; <17>, 10 mM [30]) [2, 3, 5, 24, 30]

bromopyruvate <2> (<2>, phosphoenolpyruvate, but not arabinose-5-phosphate protects [11]) [11]

dipicolinic acid <27> [22]

phosphate <1, 2> [2, 5]

Activating compounds

EDTA <2> (<2>, 1 mM, 40% stimulation [24]) [24]

Metals, ions

Ba^{2+} <2> (<2>, 1 mM, 5-10% stimulation [3]) [3]

Cd^{2+} <17, 33> (<17>, in the presence of the metal, the enzyme is asymmetric and appears to alternate catalysis between the active sites located on the other face. In the absence of metal, the asymmetry is lost [20]; <33>, the Zn^{2+} in the enzyme can be quantitatively replaced by Cd^{2+} which increases the observed turnover number and decreases the apparent K_m-value for D-arabinose-5-phosphate by 6.5fold [24]; <17>, the enzyme is most active when the endogenous metal is removed by incubation with EDTA and replaced with Cd^{2+} [25]; <17>, maximal activation below 0.1 mM [30]) [20, 24, 25, 30]

Co^{2+} <33> (<33>, stimulates [24]) [24]

Cu^{2+} <17, 33> (<33>, inhibits at high concentrations, stimulates below 0.001 mM [24]; <17>, maximal activation below 0.1 mM [30]) [24, 30]

Fe^{2+} <17> (<17>, wild-type enzyme contains Zn^{2+} and Fe^{2+} with the ratio Zn^{2+}/Fe^{2+} ranging from 1 to 2 in different preparations. Mutation H185G decreases the ability of the enzyme to bind Fe^{2+}, but not Zn^{2+}. Maximal activity, about 8-10% of the wild-type activity is obtained when the native metal is replaced with Cd^{2+} [25]; <17>, contains 0.2-0.3 equivalents of iron per enzyme subunit. $FeSO_4$ stimulates [30]) [25, 30]

Fe^{3+} <2> (<2>, 1 mM, 5-10% stimulation [3]) [3]

Li^+ <2> (<2>, 1 mM, 5-10% stimulation [3]) [3]

Mg^{2+} <2> (<2>, 1 mM, 5-10% stimulation [3]) [3]

Mn^{2+} <2> (<2>, 1 mM, 5-10% stimulation [3]) [3]

Zn^{2+} <17, 33> (<33>, the enzyme contains one mol of Zn per monomer, apoenzyme is catalytically inactive [24]; <17>, wild-type enzyme contains Zn^{2+} and Fe^{2+} with the ratio Zn^{2+}/Fe^{2+} ranging from 1 to 2 in different preparations [25]; <17>, enzyme contains approximately 0.4 equivalents of zinc per enzyme subunit. Maximal activation below 0.1 mM Zn^{2+} [30]) [24, 25, 30]

Additional information <2, 5> (<2,5>, no metal requirement [3,5,26]; <14>, no stimulation by divalent cations [18]) [3, 5, 18, 26]

Turnover number (min^{-1})

5.7 <32> (phosphoenolpyruvate) [22]

5.7 <32> (phosphoenolpyruvate) [22]

7.2 <2> (2-deoxyribose 5-phosphate, <2>, pH 7.6, 37°C [29]) [29]

20 <32> (D-arabinose 5-phosphate) [22]

20 <32> (phosphoenolpyruvate) [22]

22.8 <17> (D-arabinose 5-phosphate, <17>, pH 7.5, 60°C [28]) [28]

22.8 <17> (phosphoenolpyruvate, <17>, pH 7.5, 60°C [28]) [28]

29.4 <33> (D-arabinose 5-phosphate, <33>, pH 6.5, 25°C, Zn^{2+}-enzyme [24]) [24]

29.4 <33> (phosphoenolpyruvate, <33>, pH 6.5, 25°C, Zn^{2+}-enzyme [24]) [24]

43.2 <3> (phosphoenolpyruvate, <3>, pH 7.4, 25°C, mutant enzyme P145S [27]) [27]

44.4 <3> (D-arabinose 5-phosphate, <3>, pH 7.4, 25°C, mutant enzyme P145S [27]) [27]

52.2 <17> (D-arabinose 5-phosphate, <17>, pH 7.5, 70°C [28]) [28]

52.2 <17> (phosphoenolpyruvate, <17>, pH 7.5, 70°C [28]) [28]

60 <5> (D-arabinose 5-phosphate, <5>, pH 7.4, 37°C [26]) [26]

60 <5> (phosphoenolpyruvate, <5>, pH 7.4, 37°C [26]) [26]

65.1 <33> (D-arabinose 5-phosphate, <33>, pH 6.5, 25°C, Cd^{2+}-enzyme [24]) [24]

65.1 <33> (phosphoenolpyruvate, <33>, pH 6.5, 25°C, Cd^{2+}-enzyme [24]) [24]

88.2 <17> (D-arabinose 5-phosphate, <17>, pH 7.5, 80°C [28]) [28]

88.2 <17> (phosphoenolpyruvate, <17>, pH 7.5, 80°C [28]) [28]

97.8 <3> (D-arabinose 5-phosphate, <3>, pH 7.4, 37°C, mutant enzyme P145S [27]) [27]

105 <3> (D-arabinose 5-phosphate, <3>, pH 7.4, 25°C, wild-type enzyme [27]) [27]

120 <17> (D-arabinose 5-phosphate, <17>, pH 7.5, 90°C [28]) [28]

120 <17> (phosphoenolpyruvate, <17>, pH 7.5, 90°C [28]) [28]

121.2 <3> (phosphoenolpyruvate, <3>, pH 7.4, 37°C, mutant enzyme P145S [27]) [27]

124.8 <3> (phosphoenolpyruvate, <3>, pH 7.4, 25°C, wild-type enzyme [27]) [27]

318.5 <2> (D-arabinose 5-phosphate, <2>, pH 6.5, 25°C [24]) [24]

318.5 <2> (phosphoenolpyruvate, <2>, pH 6.5, 25°C [24]) [24]

336 <3> (D-arabinose 5-phosphate, <3>, pH 7.4, 37°C, wild-type enzyme [27]) [27]

402 <3> (phosphoenolpyruvate, <3>, pH 7.4, 37°C, wild-type enzyme [27]) [27]

408 <2> (D-arabinose 5-phosphate, <2>, pH 7.6, 37°C [29]) [29]

Specific activity (U/mg)

0.038 <1> [1, 2]

3.38 <17> [28]

3.7 <2> [3, 5]

K_m-Value (mM)

0.0028 <3> (phosphoenolpyruvate, <3>, pH 7.4, 25°C, wild-type enzyme [27]) [27]

0.0031 <5> (phosphoenolpyruvate, <5>, pH 7.4, 37°C [26]) [26]

0.0042 <3> (phosphoenolpyruvate, <3>, pH 7.4, 37°C, wild-type enzyme [27]) [27]

0.0044 <3> (D-arabinose 5-phosphate, <3>, pH 7.4, 25°C, wild-type enzyme [27]) [27]

0.00443 <2> (phosphoenolpyruvate, <2>, pH 6.5, 25°C [24]) [24]

0.00593 <33> (D-arabinose 5-phosphate, <33>, pH 6.5, 25°C, Cd^{2+}-enzyme [24]) [24]

0.006 <2> (phosphoenolpyruvate, <2>, pH 7.3, 37°C [3]) [3, 5]

0.00648 <33> (phosphoenolpyruvate, <33>, pH 6.5, 25°C, Cd^{2+}-enzyme [24]) [24]

0.0075 <33> (phosphoenolpyruvate, <33>, pH 6.5, 25°C, Zn^{2+}-enzyme [24]) [24]

0.00773 <2> (D-arabinose 5-phosphate, <2>, pH 6.5, 25°C [24]) [24]

0.008 <17> (D-arabinose 5-phosphate, <17>, pH 7.5, 60°C [28]) [28]

0.0085 <5> (D-arabinose 5-phosphate, <5>, pH 7.4, 37°C [26]) [26]

0.0093 <3> (D-arabinose 5-phosphate, <3>, pH 7.4, 37°C, wild-type enzyme [27]) [27]

0.014 <3> (D-arabinose 5-phosphate, <3>, pH 7.4, 25°C, mutant enzyme P145S [27]) [27]

0.015 <17> (D-arabinose 5-phosphate, <17>, pH 7.5, 70°C [28]) [28]

0.019 <2> (D-arabinose 5-phosphate, <2>, pH 7.6, 37°C [29]) [29]

0.02 <2> (D-arabinose 5-phosphate, <2>, pH 7.3, 37°C [3]) [3, 5]

0.026 <2> (D-arabinose 5-phosphate, <2>, pH 7.3, 37°C [15]) [15]

0.027 <17> (D-arabinose 5-phosphate, <17>, pH 7.5, 80°C [28]) [28]

0.028 <17> (phosphoenolpyruvate, <17>, pH 7.5, at 90°C [28]) [28]

0.038 <17> (phosphoenolpyruvate, <17>, pH 7.5, at 80°C [28]) [28]

0.0385 <33> (D-arabinose 5-phosphate, <33>, pH 6.5, 25°C, Zn^{2+}-enzyme [24]) [24]

0.043 <17> (phosphoenolpyruvate, <17>, pH 7.5, at 60°C or 70°C [28]) [28]

0.05 <2> (2-deoxyribose 5-phosphate, <2>, pH 7.6, 37°C [29]) [29]

0.074 <17> (D-arabinose 5-phosphate, <17>, pH 7.5, 90°C [28]) [28]

0.2 <3> (D-arabinose 5-phosphate, <3>, 28°C, parent strain PR122 [4]) [4]

0.22 <3> (D-arabinose 5-phosphate, <3>, 37°C, parent strain PR122 [4]) [4]

0.23 <3> (D-arabinose 5-phosphate, <3>, 37°C, mutant strain PR32 [4]) [4]

0.26 <3> (phosphoenolpyruvate, <3>, pH 7.4, 25°C, mutant enzyme P145S [27]) [27]

0.28 <3> (D-arabinose 5-phosphate, <3>, pH 7.4, 37°C, mutant enzyme P145S [27]) [27]

0.45 <3> (D-arabinose 5-phosphate, <3>, 42°C, mutant strain PR32 [4]) [4]

0.83 <3> (D-arabinose 5-phosphate, <3>, 42°C, parent strain PR122 [4]) [4]

1.65 <3> (phosphoenolpyruvate, <3>, pH 7.4, 37°C, mutant enzyme P145S [27]) [27]

Additional information <14> [18]

K_i-Value (mM)

0.6 <2> (2,6-anhydro-3-deoxy-2β-phosphonylmethyl-8-phosphate-D-glycero-D-talo-octonate, <2>, pH 7.3, 37°C [16]) [16]

1 <2> (D-ribose 5-phosphate, <2>, pH 7.3, 37°C [3,5]) [3, 5]

pH-Optimum

4-6 <2> (<2>, and a second optimum at pH 9.0 [3,5]) [3, 5]

6.2 <14> [18]

7.2 <1> [1]

9 <2> (<2>, and a second optimum at pH 4.0-6.0 [3,5]) [3, 5]

pH-Range

5.5-10 <14> (<14>, pH 5.5: about 35% of maximal activity, pH 10.0: about 65% of maximal activity, at 50°C [18]) [18]

6.1-9 <1> (<1>, pH 6.1: about 35% of maximal activity, pH 9.0: about 25% of maximal activity [1]) [1]

Temperature optimum (°C)

45 <2> [3]

53 <14> [18]

95 <17> [28]

Temperature range (°C)

37-61 <14> (<14>, 37°C: about 40% of maximal activity, 61°C: about 55% of maximal activity [18]) [18]

4 Enzyme Structure

Molecular weight

90000 <2> (<2>, gel filtration, non-denaturing PAGE [3,5]) [3, 5]

97000 <3> (<3>, wild-type enzyme, gel filtration [27]) [27]

97600 <17> (<17>, gel filtration [28]) [28]

Additional information <3> (<3>, two poorly resolved peaks are detected by gel filtration of the mutant enzyme P145S: 91000 Da and 53000 Da [27]) [27]

Subunits

? <2, 6> (<6>, x * 29348, calculation from nucleotide sequence [17]; <17>, 3 * 29736, electrospray ionization mass spectrometry [28]; <6>, x * 30000, SDS-PAGE [17]; <2>, x * 30842, electrospray mass spectrometry [13]) [13, 17, 28]

tetramer <2> (<2>, the enzyme is a homotetramer in which each monomer has the fold of a $(\beta/\alpha)8$ barrel [21]) [21]

trimer <2, 3, 17> (<2>, 3 * 32000, SDS-PAGE [3,5]) [3, 5, 27, 28]

5 Isolation/Preparation/Mutation/Application

Source/tissue

cell suspension culture <15> (<15>, EE cells) [18]

Purification

<1> [1, 2]

<2> (one-step purification [13]) [3, 13, 21]

<3> (recombinant enzyme from wild-type strain AG701 and temperature-sensitive strain AG701i50 [27]) [27]

<17> (recombinant enzyme [30]) [28, 30]

<33> [24]

Crystallization

<2> (crystallization is performed by vapor diffusion in hanging drops [21]; crystals are grown at 23°C by vapor diffusion in hanging drops, crystal structures of the enzyme in its binary complexes with the substrate phosphoenolpyruvate and with a mechanism-based inhibitor [23]; X-ray structures of the Cd^{2+}/H185G enzyme in its substrate-free form and in complex with phosphoenolpyruvate, and phosphoenolpyruvate plus arabinose 5-phosphate [25]) [12, 21, 23, 25]

<17> (structure of the metal-free and Cd^{2+} forms of the enzyme are determined in the uncomplexed state and in complex with various combinations of phosphoenolpyruvate, arabinose 5-phosphate and erythrose 4-phosphate [20]) [20, 25]

Cloning

<2> (expression in Salmonella typhimurium [6]; expression in Escherichia coli [22]; localization of the kdsA gene with the aid of the physical map of the Escherichia coli chromosome [7]) [6, 7, 13, 22]

<3> (enzyme from wild-type strain AG701 and temperature-sensitive strain AG701i50 expressed in Escherichia coli [27]) [27]

<4> (expression in Escherichia coli [10]) [10]

<6> (expression in Salmonella enterica serovar typhimurium AG701i50 [17]) [17]

<17> (expression in Escherichia coli [28]) [28]

<27> (expression in Escherichia coli [22]) [22]

<33> (expression in Escherichia coli [24]) [24]

Engineering

H185G <17> (<17>, mutation decreases the affinity of the enzyme to bind Fe^{2+}, but not Zn^{2+}. Maximal activity, about 8-10% of the wild-type activity is obtained when the native metal is replaced with Cd^{2+} [25]) [25]

P145S <3> (<3>, natural mutation in the temperature-sensitive strain AG701i50 leads to an increase in K_m-value for both substrates, D-arabinose 5-phosphate and phosphoenolpyruvate, this mutant enzyme also has an altered oligomeric state. Reduced activity, about 35% of the wild-type, between 15 and 30°C. Above 30°C the activity of the mutant enzyme decreases dramatically [27]) [27]

6 Stability

Temperature stability

27 <14> (<14>, unstable in absence of phosphoenolpyruvate [18]) [18]

37 <3> (<3>, half-life of enzyme from mutant strain PR32, without D-arabinose-5-phosphate: 4.5 min. Half-life of enzyme from mutant strain PR32, in presence of 13 mM D-arabinose 5-phosphate: 18 min. Half-life of enzyme from parent strain PR122: 31 min [4]; <3>, 1 h, wild-type enzyme retains full activity, 50% loss of activity of the mutant enzyme P145S [27]) [4, 27]

42 <3> (<3>, half-life of enzyme from mutant strain PR32, without D-arabi-nose-5-phosphate: 2.5 min. Half-life of enzyme from mutant strain PR32, in presence of 13 mM D-arabinose 5-phosphate: 9.5 min. Half-life of enzyme from parent strain PR122: 13 min [4]) [4]

45 <2> (<2>, 15 min, 30% loss of activity [3]) [3]

50 <14> (<14>, stable in presence of phosphoenolpyruvate [18]) [18]

70 <17> (<17>, half-life: 30.3 h [28]) [28]

80 <17> (<17>, half-life: 8.1 h [28]) [28]

90 <17> (<17>, half-life: 1.5 h [28]) [28]

Additional information <3> (<3>, temperature-sensitive lethal mutant of Sal-monella thyphimurium that is conditionally defective in 3-deoxy-D-manno-octulosonate-8-phosphate synthesis due to temperature-sensitive phospho-2-dehydro-3-deoxyoctonate aldolase [4]) [4]

General stability information

<1>, dialysis against 0.01 M Tris-HCl, pH 7.5, containing 0.01 M thioglycerol for 4 h, less than 20% loss of activity [2]

<2>, dialysis against 10 mM EDTA in 50 mM Tris-HCl buffer, pH 7.3, inacti-vates [3]

<2>, repeated freezing and thawing causes loss of activity [5]

Storage stability

<2>, -20°C, 50% loss of activity after 14 days [5]

<2>, -90°C, 0.1 M potassium phosphate buffer, pH 7.2, stable for up to 1 year [5]

References

[1] Levin, D.H.; Racker, E.: Condensation of arabinose 5-phosphate and phos-phorylenol pyruvate by 2-keto-3-deoxy-8-phosphooctonic acid synthetase. J. Biol. Chem., **234**, 2532-2539 (1959)

[2] Levin, D.H.; Racker, E.: 2-Keto-3-deoxy-8-Phosphooctonic acid synthetase. Methods Enzymol., **8**, 216-221 (1966)

[3] Ray, P.: Purification and characterization of 3-deoxy-D-manno-octulosonate 8-phosphate synthetase from Escherichia coli. J. Bacteriol., **141**, 635-644 (1980)

[4] Rick, P.D.; Young, D.A.: Isolation and characterization of a temperature-sensitive lethal mutant of Salmonella thyphimurium that is conditionally defective in 3-deoxy-D-manno-octulosonate-8-phosphate synthesis. J. Bac-teriol., **150**, 447-455 (1982)

[5] Ray, P.H.: 3-Deoxy-D-manno-octulosonate-8-phosphate (KDO-8-P) syntha-se. Methods Enzymol., **83**, 525-530 (1982)

[6] Woisetschläger, M.; Hägenauer, G.: Cloning and characterization of the gene encoding 3-deoxy-D-manno-octulosonate 8-phosphate synthetase from Escherichia coli. J. Bacteriol., **168**, 437-439 (1986)

[7] Woisetschlaeger, M.; Hoedl-Neuhofer, A.; Hoegenauer, G.: Localization of
     the kdsA gene with the aid of the physical map of the Escherichia coli chro-
     mosome. J. Bacteriol., **170**, 5382-5384 (1988)
[8] Dotson, G.D.; Nanjappan, P.; Reily, M.D.; Woodward, R.W.: Stereochemistry
     of 3-deoxyoctulosonate 8-phosphate synthase. Biochemistry, **32**, 12392-
     12397 (1993)
[9] Liang, P.H.; Lewis, J.; Anderson, K.S.: Catalytic mechanism of Kdo8P
     synthase: transient kinetic studies and evaluation of a putative reaction in-
     termediate. Biochemistry, **37**, 1639-16399 (1998)
[10] Wylie, J.L.; Iliffe, E.R.; Wang, L.L.; McClarty, G.: Identification, characteri-
     zation, and developmental regulation of Chlamydia trachomatis 3-deoxy-D-
     manno-octulosonate   (KDO)-8-phosphate synthetase   and   CMP-KDO
     synthetase. Infect. Immun., **65**, 1527-1530 (1997)
[11] Hedstrom, L.; Abeles, R.: 3-Deoxy-D-manno-octulosonate-8-phosphate
     synthase catalyzes the C-O bond cleavage of phosphoenolpyruvate. Bio-
     chem. Biophys. Res. Commun., **157**, 816-820 (1988)
[12] Tolbert, W.D.; Moll, J.R.; Bauerle, R.; Kretsinger, R.H.: Crystallization and
     preliminary crystallographic studies of 3-deoxy-D-manno-octulosonate-8-
     phosphate synthase from Escherichia coli. Proteins Struct. Funct. Genet.,
     24, 407-408 (1996)
[13] Dotson, G.D.; Dua, R.K.; Clemens, J.C.; Wooten, E.W.; Woodard, R.W.:
     Overproduction and one-step purification of Escherichia coli 3-deoxy-D-
     manno-octulosonic acid 8-phosphate synthase and oxygen transfer studies
     during catalysis using isotopic-shifted heteronuclear NMR. J. Biol. Chem.,
     270, 13698-13705 (1995)
[14] Subramaniam, P.S.; Xie, G.; Xia, T.; Jensen, R.A: Substrate ambiguity of 3-
     deoxy-D-manno-octulosonate 8-phosphate synthase from Neisseria gonor-
     rhoeae in the context of its membership in a protein family containing a
     subset of 3-deoxy-D-arabino-heptulosonate 7-phosphate synthases. J. Bac-
     teriol., **180**, 119-127 (1998)
[15] Kohen, A.; Jakob, A.; Baasov, T.: Mechanistic studies of 3-deoxy-D-manno-
     2-octulosonate-8-phosphate synthase from Escherichia coli. Eur. J. Bio-
     chem., **208**, 443-449 (1992)
[16] Baasov, T.; Sheffer-Dee-Noor, S.; Kohen, A.; Jakob, A.; Belakhov, V.: Cataly-
     tic mechanism of 3-deoxy-D-manno-octulosonate-8-phosphate synthase.
     The use of synthetic analogues to probe the structure of the putative reac-
     tion intermediate. Eur. J. Biochem., **217**, 991-999 (1993)
[17] Brabetz, W.; Brade, H.: Molecular cloning, sequence analysis and functional
     characterization of the gene kdsA, encoding 3-deoxy-D-manno-2-octuloso-
     nate-8-phosphate synthase of Chlamydia psittaci 6BC. Eur. J. Biochem., **244**,
     66-73 (1997)
[18] Doong, R.L.; Ahmad, S.; Jensen, R.A.: Higher plants express 3-deoxy-D-
     manno-octulosonate 8-phosphate synthase. Plant Cell Environ., **14**, 113-
     120 (1991)
[19] Coutrot, P.; Dumarcay, S.; Finance, C.; Tabyaoui, M.; Grison, C.: Investiga-
     tion of new potent KDO-8-phosphate synthetase inhibitors. Bioorg. Med.
     Chem. Lett., **9**, 949-952 (1999)

[20] Duewel, H.S.; Radaev, S.; Wang, J.; Woodard, R.W.; Gatti, D.L.: Substrate and metal complexes of 3-deoxy-D-manno-octulosonate-8-phosphate synthase from Aquifex aeolicus at 1.9-A resolution. Implications for the condensation mechanism. J. Biol. Chem., 276, 8393-8402 (2001)

[21] Radaev, S.; Dastidar, P.; Patel, M.; Woodard, R.W.; Gatti, D.L.: Structure and mechanism of 3-deoxy-D-manno-octulosonate 8-phosphate synthase. J. Biol. Chem., 275, 9476-9484 (2000)

[22] Birck, M.R.; Woodard, R.W.: Aquifex aeolicus 3-deoxy-D-manno-2-octulosonic acid 8-phosphate synthase: A new class of KDO 8-P synthase?. J. Mol. Evol., 52, 205-214 (2001)

[23] Asojo, O.; Friedman, J.; Adir, N.; Belokhov, V.; Shonam, Y.; Baasov, T.: Crystal structures of KDOP synthase in its binary complexes with the substrate phosphoenolpyruvate and with a mechanism-based inhibitor. Biochemistry, 40, 6326-6334 (2001)

[24] Krosky, D.J.; Alm, R.; Berg, M.; Carmel, G.; Tummino, P.J.; Xu, B.; Yang, W.: Helicobacter pylori 3-deoxy-D-manno-octulosonate-8-phosphate (KDO-8-P) synthase is a zinc-metalloenzyme. Biochim. Biophys. Acta, 1594, 297-306 (2002)

[25] Wang, J.; Duewel, H.S.; Stuckey, J.A.; Woodard, R.W.; Gatti, D.L.: Function of His185 in Aquifex aeolicus 3-deoxy-D-manno-octulosonate 8-phosphate synthase. J. Mol. Biol., 324, 205-214 (2002)

[26] Sheflyan, G.Y.; Sundaram, A.K.; Taylor, W.P.; Woodard, R.W.: Substrate ambiguity of 3-deoxy-D-manno-octulosonate 8-phosphate synthase from Neisseria gonorrhoeae revisited. J. Bacteriol., 182, 5005-5008 (2000)

[27] Taylor, W.P.; Sheflyan, G.,Y.; Woodard, R.W.: A single point mutation in 3-deoxy-D-manno-octulosonate-8-phosphate synthase is responsible for temperature sensitivity in a mutant strain of Salmonella typhimurium. J. Biol. Chem., 275, 32141-32146 (2000)

[28] Duewel, H.S.; Sheflyan, G.Y.; Woodard, R.W.: Functional and biochemical characterization of a recombinant 3-deoxy-D-manno-octulosonic acid 8-phosphate synthase from the hyperthermophilic bacterium Aquifex aeolicus. Biochem. Biophys. Res. Commun., 263, 346-351 (1999)

[29] Howe, D.L.; Sundaram, A.K.; Wu, J.; Gatti, D.L.; Woodard, R.W.: Mechanistic insight into 3-deoxy-D-manno-octulosonate-8-phosphate synthase and 3-deoxy-D-arabino-heptulosonate-7-phosphate synthase utilizing phosphorylated monosaccharide analogues. Biochemistry, 42, 4843-4854 (2003)

[30] Duewel, H.S.; Woodard, R.W.: A metal bridge between two enzyme families. 3-Deoxy-D-manno-octulosonate-8-phosphate synthase from Aquifex aeolicus requires a divalent metal for activity. J. Biol. Chem., 275, 22824-22831 (2000)

N-Acetylneuraminate synthase 2.5.1.56

1 Nomenclature

EC number

2.5.1.56

Systematic name

phosphoenolpyruvate:N-acetyl-D-mannosamine C-(1-carboxyvinyl)transferase (phosphate-hydrolysing, 2-carboxy-2-oxoethyl-forming)

Recommended name

N-acetylneuraminate synthase

Synonyms

(NANA) condensing enzyme

EC 4.1.3.19 (formerly)

N-acetylneuraminate pyruvate-lyase (pyruvate-phosphorylating)

N-acetylneuraminic acid condensing enzyme

NANA condensing enzyme

NeuAc synthase

synthase, N-acetylneuraminate

CAS registry number

37290-66-7

2 Source Organism

<1> *Neisseria meningitidis* (serogroup B strain M986 [1]; strain PRM102 [1]; strain 60E [3]) [1-3]

<2> *Escherichia coli* (strain K1 [4]; strain K1-M12 [5]) [4, 5, 6, 7]

<3> *Campylobacter jejuni* (NCTC [8]) [8]

<4> *Streptococcus agalactiae* [9]

3 Reaction and Specificity

Catalyzed reaction

phosphoenolpyruvate + N-acetyl-D-mannosamine + H_2O = phosphate + N-acetylneuraminate

Reaction type

condensation

Natural substrates and products

S N-acetyl-D-mannosamine + phosphoenolpyruvate + H_2O <1-4> (<1,4>, the enzyme is involved in sialic acid synthesis [1,9]; <2>, the enzyme is necessary for the synthesis of N-acetylneuraminate which is required for synthesis of the capsular polysaccharide α(2-8)poly-N-acetylneuraminic acid. This polysaccharide is an essential virulence factor of the neuropathogenic bacteria [4]; <3>, the neuB genes appear to be involved in the biosynthesis of at least two distinct surface structures: lipooligosaccharide and flagella [8]) [1, 4, 8, 9]

P ?

Substrates and products

S 2-acetamido-6-azido-2,6-dideoxy-D-mannose + phosphoenolpyruvate + H_2O <1> (Reversibility: ? <1> [3]) [3]

P 5-acetamido-9-azido-3,5,9-trideoxy-D-glycero-D-galacto-2-nonulosonic acid + phosphate <1> [3]

S N-acetyl-D-galactosamine + phosphoenolpyruvate + H_2O <4> (<4>, 15% of the activity with N-acetyl-D-mannosamine [9]) (Reversibility: ? <4> [9]) [9]

P ?

S N-acetyl-D-mannosamine + phosphoenolpyruvate + H_2O <1-4> (<1,2>, specific for [2,4]; <1>, the reaction strongly favors the formation of N-acetylneuraminic acid and appears to be irreversible [2]) (Reversibility: ir <1> [2]; ? <1-4> [1,3-9]) [1-9]

P N-acetylneuraminate + phosphate <1, 2> [2, 4]

S Additional information <4> (<4>, arginine residues are present in the active site and are involved in substrate recognition and binding [9]) [9]

P ?

Inhibitors

$AgNO_3$ <2> (<2>, 1 mM, 61% inhibition [5]) [5]

Cu^{2+} <2, 4> (<2>, 1 mM $CuCl_2$, 29% inhibition [5]; <4>, 10 mM $CuCl_2$, 90% inhibition [9]) [5, 9]

EDTA <1, 4> (<1>, 10 mM, complete inhibition [2]; <4>, 1 or 10 mM, 97% inhibition [9]) [2, 9]

Fe^{2+} <4> (<4>, 1 mM $FeCl_2$, 40% inhibition [9]) [9]

$HgCl_2$ <2> (<2>, at 0.01 mM 85% inhibition, at 0.1 mM complete inhibition [5]) [5]

K^+ <4> (<4>, 1 mM KCl, 92% inhibition [9]) [9]

Li^+ <4> (<4>, 1 mM LiCl, 30% inhibition [9]) [9]

N-bromosuccinimide <2> (<2>, 1 mM, 24% inhibition [5]) [5]

Na^+ <4> (<4>, 1 mM NaCl, 20% inhibition [9]) [9]

PCMB <2> (<2>, 1 mM, complete inhibition [5]) [5]

Zn^{2+} <4> (<4>, 1 mM $ZnCl_2$, 80% inhibition [9]) [9]

hydroxylamine <2> (<2>, 1 mM, 23% inhibition [5]) [5]

iodine <2> (<2>, 1 mM, 30% inhibition [5]) [5]

phenylglyoxal <4> (<4>, N-acetylmannosamine or phosphoenolpyruvate protect [9]) [9]

semicarbazide <2> (<2>, 1 mM, 20% inhibition [5]) [5]

Activating compounds

glutathione <1> (<1>, required for maximal activity [2]) [2]

Metals, ions

Co^{2+} <1, 4> (<1>, at optimal concentration 50% of the activation with Mg^{2+} [2]; <4>, 1 mM stimulates around 2fold, the enzyme is dependent on the presence of metal ions such as Mg^{2+}, Mn^{2+} or Co^{2+} [9]) [2, 9]

Mg^{2+} <1, 2, 4> (<1>, at optimal concentration 50% of the activation with Mg^{2+} [2]; <2>, divalent metal required, optimal activity with Mn^{2+} or Mg^{2+} [4]; <4>, the enzyme is dependent on the presence of metal ions such as Mg^{2+}, Mn^{2+} or Co^{2+} [9]) [2, 4, 9]

Mn^{2+} <1, 2> (<1,2>, divalent metal required [2,4]; <2>, optimal activity with Mn^{2+} or Mg^{2+} [4]; <1>, Mn^{2+} is most effective, optimal concentration: 5 mM [2]; <4>, 1 mM stimulates around 2fold, the enzyme is dependent on the presence of metal ions such as Mg^{2+}, Mn^{2+} or Co^{2+} [9]) [2, 4, 9]

Additional information <2> (<2>, no metal ion requirement [5]) [5]

Specific activity (U/mg)

0.000416 <2> [4]
0.06 <1> [2]
487 <2> [5]

K_m-Value (mM)

0.04 <2> (phosphoenolpyruvate, <2>, pH 8.2, 37°C [5]) [5]
0.042 <1> (phosphoenolpyruvate, <2>, pH 8.2, 37°C [5]) [2]
5.6 <2> (N-acetyl-D-mannosamine, <2>, pH 8.2, 37°C [5]) [5]
6.25 <1> (N-acetyl-D-mannosamine, <2>, pH 8.2, 37°C [5]) [2]

pH-Optimum

7 <4> [9]
7.5 <2> [4]
8.3 <1> (<1>, in Tris buffer [2]) [2]

pH-Range

4-12 <2> (<2>, pH 4.0: about 50% of maximal activity, pH 12.0: about 40% of maximal activity [5]) [5]
6.5-10 <1> (<1>, pH 6.5: about 40% of maximal activity, pH 10.0: about 60% of maximal activity [2]) [2]

Temperature optimum (°C)

35 <2> [5]
45 <2> [4]

Temperature range (°C)

20-60 <2> (<2>, about 60% of maximal activity at 20°C and at 60°C [5]) [5]

4 Enzyme Structure

Molecular weight
   78000 <4> (<4>, gel filtration [9]) [9]
   106000 <2> (<2>, gel filtration [5]) [5]
   135000 <2> (<2>, recombinant protein containing a hexahistidine tag at the
   N-terminus, gel filtration [7]) [7]

Subunits
   dimer <2, 4> (<4>, 2 * 38987, electrospray ionisation mass spectrometry [9];
   <2>, 2 * 52000, SDS-PAGE [5]) [5, 9]
   tetramer <2> (<2>, 4 * 40538, LC-MS analysis [7]) [7]

5 Isolation/Preparation/Mutation/Application

Localization
   cytoplasm <1> [1]

Purification
   <1> [2]
   <2> [4, 5]
   <4> (expression of the native neuB gene product enzyme in E. coli results in
   a product that is prone to proteolysis during purification so the protein is
   tagged with a hexa-histidine tag at its N-terminus and rapidly purified [9])
   [9]

Cloning
   <2> [6]
   <3> (expression of the neuB1 gene in Escherichia coli [8]) [8]
   <4> (overexpression in Escherichia coli [9]) [9]

Application
   synthesis <2> (<2>, development of a novel microbial transformation for the
   synthesis of N-acetyl-D-neuraminic acid using bacterial cells expressing N-
   acetyl-D-glucosamine 2-epimerase and N-acetylneuraminate synthase [6]) [6]

6 Stability

pH-Stability
   4 <2> (<2>, 25°C, 1 h, about 50% loss of activity [5]) [5]
   5 <2> (<2>, 25°C, 1 h, about 40% loss of activity [5]) [5]
   6 <2> (<2>, 25°C, 1 h, about 25% loss of activity [5]) [5]
   7-10 <2> (<2>, 25°C, 1 h, stable [5]) [5]
   11 <2> (<2>, 25°C, 1 h, about 40% loss of activity [5]) [5]
   12 <2> (<2>, 25°C, 1 h, about 60% loss of activity [5]) [5]

Temperature stability

30 <2> (<2>, pH 7.5, 30 min, stable up to [5]) [5]

37 <4> (<4>, thermostable up to [9]) [9]

40 <2> (<2>, pH 7.5, 30 min, about 60% loss of activity [5]) [5]

50 <2, 4> (<2>, pH 7.5, 30 min, about 80% loss of activity [5]; <4>, 30 min, 80-90% loss of activity [9]) [5, 9]

60 <2> (<2>, pH 7.5, 30 min, complete loss of activity [5]) [5]

General stability information

<2>, a specific cleavage by endogenous proteases at Lys280 of the 40000 Da enzyme. Cleavage results in the formation of two inactive fragments of 33000 Da and 7000 Da. The fragmentation is associated with a significant change of the enzyme from a tetrameric to trimeric form, and alterations in both secondary and native quarternary structures [7]

<4>, expression of the native neuB gene product enzyme in E. coli results in a product that is prone to proteolysis during purification so the protein is tagged with a hexa-histidine tag at its N-terminus and rapidly purified [9]

References

[1] Masson, L.; Holbein, B.E.: Physiology of sialic acid capsular polysaccharide synthesis in serogroup B Neisseria meningitidis. J. Bacteriol., **154**, 728-736 (1983)

[2] Blacklow, R.S.; Warren, L.: Biosynthesis of sialic acids by Neisseria meningitidis. J. Biol. Chem., **237**, 3520-3526 (1962)

[3] Brossmer, R.; Rose, U.: Enzymic synthesis of 5-acetamido-9-azido-3,5,9-trideoxy-D-glycero-D-galacto-2-nonulosonic acid, a 9-azido-9-deoxy derivative of N-acetylneuraminic acid. Biochem. Biophys. Res. Commun., **96**, 1282-1289 (1980)

[4] Vann, W.F.; Tavarez, J.J.; Crowley, J.; Vimr, E.; Silver, R.P.: Purification and characterization of the Escherichia coli K1 neuB gene product N-acetylneuraminic acid synthetase. Glycobiology, **7**, 697-701 (1997)

[5] Komaki, E.; Ohta, Y.; Tsukada, Y.: Purification and characterization of N-acetylneuraminate synthase from Escherichia coli K1-M12. Biosci. Biotechnol. Biochem., **61**, 2046-2050 (1997)

[6] Tabata, K.; Koizumi, S.; Endo, T.; Ozaki, A.: Production of N-acetyl-D-neuraminic acid by coupling bacteria expressing N-acetyl-D-glucosamine 2-epimerase and N-acetyl-D-neuraminic acid synthetase. Enzyme Microb. Technol., **30**, 327-333 (2002)

[7] Hwang, T.-S.; Hung, C.-H.; Teo, C.-F.; Chen, G.-T.; Chang, L.-S.; Chen, S.-F.; Chen, Y.-J.; Lin, C.-H.: Structural characterization of Escherichia coli sialic acid synthase. Biochem. Biophys. Res. Commun., **295**, 167-173 (2002)

[8] Linton, D.; Karlyshev, A.V.; Hitchen, P.G.; Morris, H.R.; Dell, A.; Gregson, N.A.; Wren, B.W.: Multiple N-acetyl neuraminic acid synthetase (neuB) genes in Campylobacter jejuni: identification and characterization of the gene in-

volved in sialylation of lipo-oligosaccharide. Mol. Microbiol., **35**, 1120-1134 (2000)

[9] Suryanti, V.; Nelson, A.; Berry, A.: Cloning, over-expression, purification, and characterization of N-acetylneuraminate synthase from Streptococcus agalactiae. Protein Expr. Purif., **27**, 346-356 (2003)

N-Acylneuraminate-9-phosphate synthase 2.5.1.57

1 Nomenclature

EC number

2.5.1.57

Systematic name

phosphoenolpyruvate:N-acyl-D-mannosamine-6-phosphate 1-(2-carboxy-2-oxoethyl)transferase

Recommended name

N-acylneuraminate-9-phosphate synthase

Synonyms

DmSAS <1> [10]

EC 4.1.3.20 (formerly)

N-acetylneuraminate-9-phosphate lyase

N-acetylneuraminate-9-phosphate sialic acid 9-phosphate synthase

N-acetylneuraminate-9-phosphate synthetase

N-acylneuraminate-9-phosphate pyruvate-lyase (pyruvate-phosphorylating)

neu5ac-9-P synthase <6> [8]

neuac-9-phosphate synthase <8> [6]

sialic acid 9-phosphate synthase

CAS registry number

9031-58-7

2 Source Organism

<1> *Drosophila melanogaster* (sialic acid synthase, single genomic clone [10] (SwissProt-ID: Q9VG74)) [10]

<2> *Drosophila melanogaster* (sialic acid synthase cDNA sequence [10]) [10]

<3> *Escherichia coli* (neuB gene product amino acid sequence accession. no. [8-10]) [8-10]

<4> *Homo sapiens* (SwissProt-ID: Q9NR45) [9, 10]

<5> *Homo sapiens* (human [3]) [3]

<6> *Mus musculus* (mouse, neuB gene product accession. no. [8]) [8]

<7> *Ovis aries* (sheep [3]) [3]

<8> *Rattus norvegicus* (rat [3-6,9]; Sprague-Dawley [4,6]) [3-6, 9]

<9> *Sus scrofa* (pig [1,2]; hog [3]; porcine [7]) [1-3, 7]

3 Reaction and Specificity

Catalyzed reaction

phosphoenolpyruvate + N-acyl-D-mannosamine 6-phosphate + H_2O = N-acylneuraminate 9-phosphate + phosphate

Reaction type

condensation

phosphorylation

Natural substrates and products

S   N-acetyl-D-mannosamine 6-phosphate + phosphoenolpyruvate + H_2O <1, 3, 5-9> (<8> final steps of the biosynthetic pathway of N-acetylneuraminic acid [5]; <8> sialic acid metabolism, biosynthesis of N-acetyl-D-neuraminic acid via its 9-phosphate [6]; <9> N-acetylneuraminic acid biosynthesis [7]; <1> sialic acid biosynthesis [10]) [1-10]

P   N-acetyl-neuraminate 9-phosphate + phosphate

Substrates and products

S   N-acetyl-D-mannosamine 6-phosphate + phosphoenolpyruvate + H_2O <1, 3, 5-9> (Reversibility: ? <1,3,5-9> [1-10]) [1-10]

P   N-acetyl-neuraminate 9-phosphate + phosphate <1, 3, 5-9> [1-10]

S   N-acyl-D-mannosamine 6-phosphate + phosphoenolpyruvate + H_2O <5, 7, 9> (Reversibility: ? <5,7,9> [3]) [3]

P   N-acylneuraminic acid 9-phosphate + phosphate <5, 7, 9> [3]

S   N-acyl-D-mannosamine 6-phosphate + phosphoenolpyruvate + H_2O <9> (Reversibility: ? <9> [3]) [3]

P   sialic acid 9-phosphate + phosphate <9> [3]

S   N-glycolyl-D-mannosamine 6-phosphate + phosphoenolpyruvate + H_2O <5, 7, 9> (Reversibility: ? <5,7,9> [1-3]) [1-3]

P   N-glycolyl-neuraminate 9-phosphate + phosphate <9> [2]

S   D-mannose 6-phosphate + phosphoenolpyruvate + H_2O <1> (Reversibility: ? <1> [10]) [10]

P   2-keto-3-deoxy-D-glycero-D-galactononoic acid + phosphate <1> [10]

Inhibitors

5,5'-dithio-bis(2-nitrobenzoic acid) <8> [9]

EDTA <9> [2, 3]

N-ethylmaleimide <8> [9]

NaF <8> [5]

Metals, ions

Co^{2+} <9> [2, 2]

Fe^{2+} <8, 9> (<8> absolute requirement for divalent cations [9]) [3, 9]

Mg^{2+} <8, 9> (<8> absolute requirement for divalent cations [9]; <9> maximal enzyme activity in presence of 0.02 M [2]; <9> essential, 0.01 M required for maximal activity [3]) [2, 3, 9]

Mn^{2+} <8, 9> (<8> absolute requirement for divalent cations [9]) [2, 3, 9]

Ni^{2+} <8, 9> (<8> absolute requirement for divalent cations [9]) [2, 3, 9]

Zn^{2+} <9> [3]

Additional information <9> (<8,9> Ca^{2+} or Cu^{2+} does not substitute for Mg^{2+} [2,3,9]; <9> Al^{3+} does not substitute for Mg^{2+} [3]) [2, 3, 9]

Specific activity (U/mg)

0.0013 <9> (<9> smooth muscle cells [7]) [7]

0.0015 <9> (<9> endothelial cells [7]) [7]

0.2 <9> [1]

0.26 <5> [3]

0.505 <8> [9]

0.83 <9> [2, 3]

K_m-Value (mM)

0.035 <8> (N-acetyl-D-mannosamine 6-phosphate) [9]

0.1 <8> (phosphoenolpyruvate) [9]

0.21 <8> (N-acetyl-D-mannosamine 6-phosphate, <8> in presence of NaF [5]) [5]

0.25 <8> (N-acetyl-D-mannosamine 6-phosphate) [5]

0.69 <9> (N-acetyl-D-mannosamine 6-phosphate) [2, 3]

1.6 <9> (N-glycolyl-D-mannosamine 6-phosphate) [2, 3]

pH-Optimum

7.8 <9> [2, 3]

4 Enzyme Structure

Molecular weight

41000 <1> (<1> SDS-PAGE [10]) [10]

75000 <8> (<8> gel filtration [9]) [9]

80000 <6> (<6> gel filtration [8]) [8]

Subunits

dimer <6, 8> (<8> 2 + 37000, SDS-PAGE [9]; <6> 2 * 50000, SDS-PAGE [8]; <6> 2 * 40000, amino acid sequence [8]) [8, 9]

5 Isolation/Preparation/Mutation/Application

Source/tissue

SCHNEIDER-2 cell <1> [10]

adrenal gland <9> [3]

aorta <9> [7]

blood <5> [3]

brain <8, 9> [3, 5]

colon <8> [4]

colonic mucosa <8> [4, 6]

embryo <1> [10]

endothelial cell <9> [7]

erythrocyte <5, 8> [3, 5]
heart <9> [3]
intestinal mucosa <8> [5]
kidney <8> [3, 5, 6]
larva <1> [10]
liver <8> [3, 5, 6, 9]
lung <8> [3, 5]
muscle <8, 9> [3, 5]
ovary <9> [3]
pancreas <8> [5]
salivary gland <8> [5, 9]
skin <8> [3]
smooth muscle <9> [7]
spleen <8> [3, 5]
submaxillary gland <5, 7, 9> [1-3]
testis <8> [3]
thymus <8> [5]

Localization
cytosol <6, 8> [5, 6, 8, 9]

Purification
<6> (recombinant enzyme [8]) [8]
<8> [9]
<9> [1-3]

Cloning
<1> (cloned and expressed in either bacterial or baculoviral expression systems [10]) [10]
<2> [10]
<3> [8]
<4> [9, 10]
<6> (cDNA cloned by PCR, transiently expressed in HeLa cells [8]) [8]

6 Stability

General stability information
<8>, protected from inactivation by presence of phosphoenolpyruvate [9]
<9>, purified enzyme is unstable [2]
<9>, purified enzyme is unstable to freezing and thawing [3]

Storage stability
<8>, 4°C, crude extract retains 60% of enzyme activity when stored for 1 week [9]
<8>, 4°C, purified enzyme loses activity completely after 1 day in phosphate buffer [9]

<8>, 4°C, purified enzyme, 50% of enzyme activity remains after 2 days in phosphate buffer containing 5 mM MgCl$_2$, enzyme activity is lost entirely after 3 days [9]
<9>, -18°C, crude extract can be stored [3]
<9>, -20°C, crude extract stable for at least 1 month [2]
<9>, 0°C, loses approximately 25% of the activity during a 24-h period [2]
<9>, 4°C, purified enzyme loses approximately half of its activity in 3 days [3]

References

[1] Roseman, S.; Jourdian, G.W.; Watson, D.; Rood, R.: Enzymatic synthesis of sialic acid 9-phosphates. Biochemistry, **47**, 958-961 (1961)
[2] Watson, D.; Jourdian, G.W.; Roseman, S.: N-Acylneuraminic (sialic) acid 9-phosphate synthetase. Methods Enzymol., **8**, 201-205 (1966)
[3] Watson, D.; Jourdian, G.W.; Roseman, S.: The sialic acids. VIII. Sialic acid 9-phosphate synthetase. J. Biol. Chem., **241**, 5627-5636 (1966)
[4] Corfield, A.P.; Rainey, J.B.; Clamp, J.R.; Wagner, S.A.: Changes in the activity of the enzymes involved in sialic acid metabolism in isolated rat colonic mucosal cells on administration of azoxymethane. Biochem. Soc. Trans., **11**, 766-767 (1983)
[5] Van Rinsum, J.; Van Dijk, W.; Hooghwinkel, G.J.M.; Ferwerda, W.: Subcellular localization and tissue distribution of sialic acid-forming enzymes. N-Acetylneuraminate-9-phosphate synthase and N-acetylneuraminate 9-phosphatase. Biochem. J., **223**, 323-328 (1984)
[6] Corfield, A.P.; Clamp, J.R.; Wagner, S.A.: The metabolism of sialic acids in isolated rat colonic mucosal cells. Biochem. J., **226**, 163-174 (1985)
[7] Hagemeier, C.; Vischer, P.; Buddecke, E.: Studies on the biosynthesis of N-acetylneuraminic acid in cultured arterial wall cells. Biochem. Soc. Trans., **14**, 1173-1174 (1986)
[8] Nakata, D.; Close, B.E.; Colley, K.J.; Matsuda, T.; Kitajima, K.: Molecular cloning and expression of the mouse N-acetylneuraminic acid 9-phosphate synthase which does not have deaminoneuraminic acid (KDN) 9-phosphate synthase activity. Biochem. Biophys. Res. Commun., **273**, 642-648 (2000)
[9] Chen, H.; Blume, A.; Zimmermann-Kordmann, M.; Reutter, W.; Hinderlich, S.: Purification and characterization of N-acetylneuraminic acid-9-phosphate synthase from rat liver. Glycobiology, **12**, 65-71 (2002)
[10] Kim, K.; Lawrence, S.M.; Park, J.; Pitts, L.; Vann, W.F.; Betenbaugh, M.J.; Palter, K.B.: Expression of a functional Drosophila melanogaster N-acetylneuraminic acid (Neu5Ac) phosphate synthase gene: Evidence for endogenous sialic acid biosynthetic ability in insects. Glycobiology, **12**, 73-83 (2002)

Protein farnesyltransferase   2.5.1.58

1 Nomenclature

EC number
   2.5.1.58

Systematic name
   farnesyl-diphosphate:protein-cysteine farnesyltransferase

Recommended name
   protein farnesyltransferase

Synonyms
   CAAX farnesyltransferase
   FPT
   FTase
   PFT
   RAS farnesyltransferase
   farnesyltransferase
   farnesyltransferase, farnesyl pyrophosphate-protein
   farnesyltransferase, protein
   farnsesyl protein transferase
   prenylprotein transferase
   prenyltransferase
   protein cysteine farnesyltransferase
   protein farnesyltransferase
   protein prenyltransferase

CAS registry number
   131384-38-8

2 Source Organism

<1> *Homo sapiens* (mammalian HeLa cells [10]) [7, 8, 10, 25]
<2> *Saccharomyces cerevisiae* [2, 7, 9, 18]
<3> *Rattus norvegius* [1, 3-7, 11, 14-17, 19, 23-25, 27]
<4> *Toxoplasma gondii* (tachyzoites of the RH strain were grown and maintained in Cercopithecus aethiops kidney cells [10]) [10]
<5> *Plasmodium falciparum* (protozoa malaria parasite, infected erythrocytes [26]) [26]
<6> *Trypanosoma brucei brucei* (EATRO 140 [11]) [11, 12, 13]
<7> *Trypanosoma cruzi* (Tulahuen, allelic form I of the α-subunit [13]) [13]

<8> *Trypanosoma cruzi* (Tulahuen, allelic form II of the α-subunit [13]) [13]
<9> *Trypanosoma cruzi* (Tulahuen, β-subunit [13]) [13]
<10> *Leishmania major* (α-subunit [13]) [13]
<11> *Leishmania major* (β-subunit [13]) [13]
<12> *Leishmania amazonensis* (strain LV78 [13]) [13]
<13> *Bos taurus* [20]
<14> *Pisum sativum* [21]
<15> *Nicotiana tabacum* (BY-2 cells [21]) [21]
<16> *Lycopersicon esculentum* (4-week-old hydroponically grown tomato plants, VFNT cherry [22]) [22]

3 Reaction and Specificity

Catalyzed reaction

farnesyl diphosphate + protein-cysteine = S-farnesyl protein + diphosphate
(This enzyme, along with geranylgeranyltransferase types I, EC 2.5.1.59 and II, EC 2.5.1.60, constitutes the protein prenyltransferase family of enzymes. Catalyses the formation of a thioether linkage between the C-1 of an isoprenyl group and a cysteine residue fourth from the C-terminus of the protein. These protein acceptors have the C-terminal sequence CA1A2X, where the terminal residue, X, is preferably serine, methionine, alanine or glutamine; leucine makes the protein a substrate for EC 2.5.1.59. The enzymes are relaxed in specificity for A1, but cannot act if A2 is aromatic. Substrates of prenyltransferases include Ras, Rho, Rab, other Ras-related small GTP-binding proteins, γ-subunits of heteotrimeric G-proteins, nuclear lamins, centromeric proteins and many proteins involved in visual signal transduction. A zinc metalloenzyme that requires Mg^{2+} for activity; <2,3> farnesyl diphosphate binds the enzyme in a two step process that may involve an enzyme conformational change, the enzyme-substrate complex then rapidly reacts with the peptide substrate to form a product, and product release is the rate-limiting step in catalysis [1, 2]; <2, 3> kinetic mechanism [1, 2, 4, 5, 15, 16, 25]; <2,3> enzyme constitutes the protein prenyltransferase family of enzymes [2,5,9,25]; <3> farnesyl diphosphate binds exclusively to the β subunit [7]; <3> both subunits would appear to have a role in binding and orienting the leaving group of farnesyl diphosphate approximately for catalysis [23]; <2> Enzyme preferentially farnesylates CaaX sequences ending in methionine, cysteine or serine. Enzyme also attaches geranylgeranyl to some CaaX sequences ending in methionine, leucine and cysteine. Substrate overlap may occur between the Saccharomyces cerevisae farnesyl transferase and geranylgeranyltransferase types I in vivo [9])

Natural substrates and products

S farnesyl diphosphate + protein-cysteine <1-3, 14-16> (<1-3> process necessary for the subcellular localisation of substrate to the plasma membrane [3-5, 7-9, 15, 16, 23]; <1-3> process required for the transforming activity of oncogenic variants of Ras, making enzyme a prime target for

anticancer therapeutics [3-5, 7, 8, 9]; <3, 4> enzyme responsible for cata-
lysing isoprene lipid modifications [3, 4, 7, 10]; <14, 15> potential role for
plant enzyme in cell cycle regulation [21]; <16> a possible role for the
enzyme during fruit development [22]; <3> A series of steps modifies
the C-terminus of Ras. Enzyme catalyses the first of these steps by transferring the farnesyl group from farnesyl diphosphates to the cysteine side
chain in the CAAX sequence motif at the C-terminus of Ras [4,23]; <5>
potential role for cell cycle [26]) [1-5, 8-9, 15, 16, 21-27]

P S-farnesyl protein + diphosphate

Substrates and products

S 10-benzyloxy-3-methyl-2-decen-1-ol-diphosphate + N-dansyl-Gly-Cys-Val-
Leu-Ser <3> (Reversibility: ? <3> [4]) [4]

P diphosphate + N-dansyl-Gly-S-(10-benzyloxy-3-methyl-2-hexenyl)-Cys-
Val-Leu-Ser

S 6-benzyloxy-3-methyl-2-hexen-1-ol-diphosphate + N-dansyl-Gly-Cys-Val-
Leu-Ser <3> (Reversibility: ? <3> [4]) [4]

P diphosphate + N-dansyl-Gly-S-(6-benzyloxy-3-methyl-2-hexenyl)-Cys-
Val-Leu-Ser

S 7-benzyloxy-3-methyl-2-hepten-1-ol-diphosphate + N-dansyl-Gly-Cys-
Val-Leu-Ser <3> (Reversibility: ? <3> [4]) [4]

P diphosphate + N-dansyl-Gly-S-(7-benzyloxy-3-methyl-2-heptenyl)-Cys-
Val-Leu-Ser

S 8-benzyloxy-3-methyl-2-octen-1-ol-diphosphate + N-dansyl-Gly-Cys-Val-
Leu-Ser <3> (Reversibility: ? <3> [4]) [4]

P diphosphate + N-dansyl-Gly-S-(8-benzyloxy-3-methyl-2-octenyl)-Cys-Val-
Leu-Ser

S 9-benzyloxy-3-methyl-2-nonen-1-ol-diphosphate + N-dansyl-Gly-Cys-Val-
Leu-Ser <3> (Reversibility: ? <3> [4]) [4]

P diphosphate + N-dansyl-Gly-S-(9-benzyloxy-3-methyl-2-nonenyl)-Cys-Val-
Leu-Ser

S farnesyl diphosphate + GST-CIIL <2> (Reversibility: ? <2> [18]) [18]

P diphosphate + S-farnesyl-GST-Cys-Ile-Ile-Ser

S farnesyl diphosphate + GST-Cys-Ile-Ile-Ser <2> (Reversibility: ? <2>
[18]) [18]

P diphosphate + S-farnesyl-GST-Cys-Ile-Ile-Ser

S farnesyl diphosphate + N-dansyl-Gly-Cys-Val-Leu-Ser <3> (<3> the best
substrate [4]) (Reversibility: ? <3> [4]) [4]

P diphosphate + N-dansyl-Gly-S-farnesyl-Cys-Val-Leu-Ser

S farnesyl diphosphate + Ras-Cys-Val-Ile-Met <3, 6, 13, 16> (<3, 16> best
substrate [22, 24]) (Reversibility: ? <3, 6, 13, 16> [11, 20, 22, 24]) [11, 20,
22, 24]

P diphosphate + Ras-S-farnesyl-Cys-Val-Ile-Met

S farnesyl diphosphate + protein-cysteine <1-16> (<3, 13, 14> posttranslational lipid modification in which a 15-carbon farnesyl isoprenoid is
linked via a thioether bond to specific cysteine residues of proteins, the
reactive cysteine is located in the C-terminal Ca1a2X motif in which C is

the modified cysteine, a1 and a2 are often an aliphatic residue, and X is Ser, Met, Ala or Gln [3-5, 7, 19-21, 25, 27]; <2, 3, 14> prenylation, farnesylation, substrates are Ras, nuclear lamins, transducin γ subunit, protein substrate motif: Cys-alipatic-aliphatic-X, X: M, S, Q, A, F [1, 2, 6, 11, 14-19, 21]; <3> substrates are cellular proteins such as Ras at a cysteine residue near their carboxy-terminus [3]; <3> substrate are: Ras superfamiliy and Ras-related proteins for example Rap, Rab, Rac and Ral, large G proteins, nuclear lamins, rhodopsin kinase and the retinal cGMP phosphodiesterase [24,25]; <2,3> prefered CaaX-substrate: CAIM [9,27]; <4> substrate: Biotin lamin-B protein [10]; <3,6> Ha-Ras-CVLS, which is a good substrate for mammalian enzyme, not farnesylated by the parasite enzyme, only the mammalian enzyme substrate RAS1-CVIM farnesylated by parasite enzyme [11]; <6-11> enzyme prefers Gln and Met at the X position but not Ser, Thr or Cys, which are good substrates for mammalian enzyme [12, 13]; <13> substrate: Ras-CVIM and GST-PEP [20]; <16> the specific CAQQ motif of the ANJ1 protein has a low affinity for enzyme in vitro, enzyme specifically farnesylated the RasB-CVIM peptide substrate [22]; <3> enzyme prefers linear residues in the X position of the CaaX box over branched residues, enzyme prefers Gln, Met and Ser at the X position [24]) (Reversibility: ir <3> [1]; ? <2-16> [2-27]) [1-27]

P diphosphate + S-farnesyl protein <1-16> [1-27]
S farnesyl monophosphate + Gly-Cys-Val-Leu-Ser <3> (Reversibility: ? <3> [16,17]) [16, 17]
P phosphate + Gly-S-farnesyl-Cys-Val-Leu-Ser
S geranylgeranyl diphosphate + N-dansyl-Gly-Cys-Val-Leu-Ser <3> (Reversibility: ? <3> [4]) [4]
P diphosphate + N-dansyl-Gly-S-geranylgeranyl-Cys-Val-Leu-Ser
S geranylgeranyl diphosphate + protein-cysteine <3> (Reversibility: ? <3> [4]) [4]
P diphosphate + S-geranylgeranylcysteinyl protein
S Additional information <3> (<3> not: geranylgeranyl diphosphate [4,7]; <3> although farnesyl diphosphate and geranylgeranyl diphospahte bind competitively, the geranyl geranyl diphosphate is not transferred efficiently to the protein substrate by enzyme [7]) [4, 7]
P ?

Inhibitors

(+)-6-(camphorquinone-10-sulfonamido)-hexanoic acid <1> [8]
(-)-6-(camphorquinone-10-sulfonamido)-hexanoic acid <1> [8]
2-oxododecanal <1> [8]
4,4-(biphenyldiglyoxaldehyde) <1> [8]
5,9,13-trimethyl-8,12-tetradecadiene-2,3-dione <1> (<1> 0.093 mM causes a 94% reduction in enzyme activity after 30 min [8]) [8]
5,9-dimethyl-8-decene-2,3-dione <1> (<1> 0.017 mM causes a 62% reduction in enzyme activity after 30 min [8]) [8]
Cys-Val-Phe-Cys <3> [24]
Cys-Val-Phe-Met <3> [24]

Cys-Val-Phe-Met <3> [6]

Cys-Val-Phe-Phe <3> [24]

Cys-Val-Phe-aminohexanoate <3> [24]

EDTA <13, 16> [20, 22]

FTI-276 <3, 6> (<3,6> about 2fold more potent on parasite enzyme than on mammalian enzyme [11]) [11]

FTI-277 <3, 5, 6> (<3,6> inhibition with potencies 10fold higher on parasite enzyme than on mammalian enzyme, however growth of bloodstream Trypanosoma brucei completely inhibited with 0.001 mM [11]) [11, 26]

GGTI-297 <3, 6> (<3,6> inhibition with potencies 10fold higher on parasite enzyme than on mammalian enzyme [11]) [11]

GGTI-298 <3, 5, 6> (<3,6> inhibition with potencies 10fold higher on parasite enzyme than on mammalian enzyme, however growth of bloodstream Trypanosoma brucei cmpletely inhibited with 0.005 mM [11]) [11, 26]

KKTKSKCVIM <3> [25]

KTSCVAM <1, 4> (<1,4> 40% inhibition of farnesylation in Toxoplasma gondii, 100% inhibition of enzyme in HeLa cells [10]) [10]

KTSCVFM <1, 4> (<1,4> 80% inhibition of farnesylation in Toxoplasma gondii and 80% inhibition of enzyme in HeLa cells [10]) [10]

KTSCVIA <1, 4> (<1,4> 50% inhibition of farnesylation in Toxoplasma gondii and no inhibition of enzyme in HeLa cells [10]) [10]

KTSCVIF <1, 4> (<1,4> no inhibition of farnesylation in Toxoplasma gondii, 80% inhibition of enzyme in HeLa cells [10]) [10]

KTSSVIM <1, 4> (<1,4> 80% inhibition of farnesylation in Toxoplasma gondii, 100% inhibition of enzyme in HeLa cells [10]) [10]

L-745,631 <3, 6> (<3,6> similarly potent on parasite enzyme and on mammalian enzyme [11]) [11]

L-penicillamine-Ile-Ile-Met <3> [24]

L-penicillamine-Val-Ile-Ala <3> [24]

L-penicillamine-Val-Ile-Cys <3> [24]

L-penicillamine-Val-Ile-Gln <3> [24]

L-penicillamine-Val-Ile-Met <3> [24]

L-penicillamine-Val-Ile-Phe <3> [24]

L-penicillamine-Val-Ile-aminohexanoate <3> [24]

L-penicillamine-Val-Ile-homoserine <3> [24]

L-penicillamine-Val-Val-Met <3> [24]

Lys-Lys-Cys-Val-Ile-Met <13> (<13> 50% inhibition at 0.0025 mM [20]) [20]

NH_2-KTKCVFM <3> [25]

SCH-44342 <3, 6> (<3,6> 2 orders of magnitude less potent on parasite enzyme than on mammalian enzyme [11]) [11]

dehydroascorbic acid 6-palmitate <1> [8]

farnesyltransferase inhibitors I and II <4> (<4> peptidomimemtic inhibitors, inhibitor I: 50% inhibition at 270 nM, inhibitor II: 50% inhibition at 0.97 nM [10]) [10]

manumycin <15> (<15> inhibition at concentration as low as 0.0025 mM, inhibitor most likely to affect only the processes that involve protein isoprenylation [21]) [21]

phenylglyoxal <1> [8]
sodium deoxycholate <13> [20]
Additional information <1-5> (<3> two primary modes of inhibition: blocking the peptide substrate site and blocking the exit groove [5]; <3> hexapeptide containing a switch of leucine for methionine at the COOH terminus or a switch of cysteine for serine at the prenylation site is essentially inactive in competing for enzyme activity [6]; <1> higher concentrations of the α-dicarbonyl compound results in more rapid and more extensive inactivation [8]; <1,4> not: KTSSVIM [10]; <2,3> enzyme inhibited by peptides containing the C-terminal CAAX sequence [18,25]; <3> tetrapeptides with sequence L-penicillamine-Val-Ile-X and Cys-Val-Phe-X [24]) [5, 6, 8, 10, 18, 24, 26]

Activating compounds
Additional information <3, 13> (<3> infection of Sf9 cells with the recombinant baculovirus containing DNA sequences encoding both enzyme activity results in a dramatic enhancement of enzyme activity [6]; <13> n-octyl β-D-glycopyranoside stimulates enzyme activity even at concentrations of 0.5% [20]) [6, 20]

Metals, ions
Cd^{2+} <3> (<3> substitution of the active site zinc with cadmium increases the affinity of the peptide substrate and decreases the rate constant for the chemical step [15]) [15]
Co^{2+} <3> (<3> evidence for a catalytically relevant interaction between the metal ion and the protein substrate in the enzyme [7]) [6, 7]
Mg^{2+} <2-4, 13, 16> (<2-4, 13, 16> required [2, 4, 5, 9, 10, 18, 20, 22]; <3> appears to coordinate the diphosphate moiety of farnesyl diphosphate [4, 15, 16]; <3> magnesium is not required for formation of the thioether product but its presence increases the single-turnover rate constant by several orders of magnitude at saturating enzyme and substrate concentrations [15]; <3> the magnesium affinity of enzyme increases with pH with a pK_a of 7.4, presumably reflecting the deprotonation of the farnesyl diphosphate to enhance magnesium coordination [16]) [2, 4, 5, 9, 10, 15, 16, 18, 20, 22]
Zn^{2+} <1-4, 6-11, 13, 16> (<2-4, 13, 16> required [2-4, 7, 9, 10, 16, 19, 20, 22]; <3> the zinc ion activates the cysteine thiolate for nucleophilic attack on the C_1 atom of the farnesyl diphosphate substrate [5]; <3, 13> zinc metalloenzyme [6, 14-17, 20]; <3> enzyme contains a stoichiometric amount of zinc [6]; <1> a single zinc ion bound to the β subunit, near the subunit interface, which marks the location of the active site [7]; <1,4> optimal activity in the presence of 0.025 mM [10]; <3, 6-11> zinc ion is coordinated by three residues in the β subunit: Asp-297, Cys-299, and H-362 and a water molecule [7, 13, 14, 23, 27]; <3> zinc plays a major catalytic role in the mechanism of enzyme, zinc seems to activate the cysteine thiol of protein substrate for attack at C-1 of the isoprenoid substrate [14]) [2-7, 9, 10, 13-16, 19, 20, 22, 23, 27]
Additional information <3, 4> (<3, 4> not: Ca^{2+} [6,10]; <3> Co^{2+} [6]; <3> a metal-assisted nucleophile is involved in the catalytic mechanism of enzyme [15, 16]) [6, 10, 15, 16]

Turnover number (min^{-1})

50.4 <3> (7-benzyloxy-3-methyl-2-hepten-1-ol-diphosphate) [4]
900 <3> (geranylgeranyl diphosphate) [4]
1500 <3> (10-benzyloxy-3-methyl-2-decen-1-ol-diphosphate) [4]
1620 <3> (8-benzyloxy-3-methyl-2-octen-1-ol-diphosphate) [4]
2580 <3> (9-benzyloxy-3-methyl-2-nonen-1-ol-diphosphate) [4]
2640 <3> (6-benzyloxy-3-methyl-2-hexen-1-ol-diphosphate) [4]
6000 <3> (farnesyl diphosphate) [4]
Additional information <1, 3> (<3> comparison of k_{cat} of wild-type and mutant enzymes [14]; <3> comparison of k_{cat} of metal-substituted enzymes, k_{cat}
for product formation decreases for C^3-fluoromethyl farnesyl diphosphate
substrates, paralleling the number of fluorines at the C^3-methyl position
[15]; <3> comparison of k_{cat}-value of substrates with the structure Lys-Lys-
Ser-Ser-Cys-Val-Ile-X for for farnesyltransferase and geranylgeranyltransferase [24]; <1,3> comparison of k_{cat} of mutant enzymes [25]) [14, 15, 16, 20, 24,
25]

Specific activity (U/mg)

0.0085 <3> [6]
0.0207 <13> [20]
0.08 <16> [22]
13.5 <6> [11]

K$_m$-Value (mM)

0.0051 <2> (GST-CIIS) [18]
0.0081 <2> (farnesyl diphosphate) [18]
0.046 <3> (farnesyl diphosphate) [4]
0.056 <3> (8-benzyloxy-3-methyl-2-octen-1-ol-diphosphate) [4]
0.089 <3> (10-benzyloxy-3-methyl-2-decen-1-ol-diphosphate) [4]
0.091 <3> (9-benzyloxy-3-methyl-2-nonen-1-ol-diphosphate) [4]
0.13 <3> (7-benzyloxy-3-methyl-2-hepten-1-ol-diphosphate) [4]
0.45 <3> (geranylgeranyl diphosphate) [4]
0.89 <3> (6-benzyloxy-3-methyl-2-hexen-1-ol-diphosphate) [4]
Additional information <1, 3, 13> (<3> correlation between K_m and analogue hydrophobicity and length [4]; <3> wild-type enzyme and the H362E
mutant possess similar K_m for H-Ras [14]; comparison of K_M-value of substrates with the structure Lys-Lys-Ser-Ser-Cys-Val-Ile-X for for farnesyltransferase and geranylgeranyltransferase [24]; <1,3> comparison of K_m of mutant
enzymes [25]) [4, 14, 20, 24, 25]

K$_i$-Value (mM)

0.00008 <3> (Cys-Val-Phe-Met) [24]
0.00012 <3> (NH$_2$-KTKCVFM) [25]
0.0003 <3> (L-penicillamine-Val-Ile-Gln) [24]
0.00052 <3> (L-penicillamine-Ile-Ile-Met) [24]
0.0008 <3> (L-penicillamine-Val-Ile-Cys) [24]
0.00082 <3> (L-penicillamine-Val-Val-Met) [24]
0.0011 <3> (L-penicillamine-Val-Ile-Met) [24]

0.0011 <3> (L-penicillamine-Val-Ile-Met) [24]
0.0013 <3> (KKSKTKCVIM) [25]
0.0017 <3> (L-penicillamine-Val-Ile-homoserine) [24]
0.0028 <3> (Cys-Val-Phe-Phe) [24]
0.0029 <3> (Cys-Val-Phe-Cys) [24]
0.0034 <3> (L-penicillamine-Val-Ile-aminohexanoate) [24]
0.0045 <3> (L-penicillamine-Val-Ile-Phe) [24]
0.0053 <3> (L-penicillamine-Val-Ile-Ala) [24]
0.0059 <3> (Cys-Val-Phe-aminohexanoate) [24]
Additional information <3> (<3> comparison of K_i-values of tetrapeptide inhibitors L-penicillamine-Val-Ile-X and Cys-Val-Phe-X for farnesyltransferase and geranylgeranyltransferase [24]) [24]

pH-Optimum
6.5-7 <16> [22]
7.2 <13> [20]
7.8 <2> [18]
7.9 <2> [9]
8 <4> [10]

pH-Range
6-8.4 <4> (<4> pH 6.0 and 8.4: about 80% of maximal activity, the enzyme activity is completely abolished at pH 4.0 [10]) [10]
Additional information <3> (<3> pH-dependence of the chemical step of product formation: the reaction is characterized by two ionizations, drastic differences in the pH dependence of farnesyl diphosphate and farnesyl monophosphate, the magnesium affinity of enzyme increases with pH with a pK_a of 7.4 [16]) [16]

Temperature optimum (°C)
30 <16> [22]
36 <4> [10]
50 <2, 13> [18, 20]

4 Enzyme Structure

Molecular weight
90000 <2> [18]
100000 <13, 16> (<13,16> gel filtration [20,22]) [20, 22]

Subunits
heterodimer <1, 2, 3, 6, 13, 16> (<2,3> $\alpha\beta$, 1 * 48000 + 1 * 46000 [2,6,7]; <3> SDS-PAGE [6]; <6> $\alpha\beta$, 1 * 61000 + 1 * 65000, SDS-PAGE [11]; <2> 1 * 43000 + 1 * 34000 [18]; <3> $\alpha\beta$, 1 * 49000 + 1 * 46000, SDS-PAGE [19]; <3> $\alpha\beta$, 1 * 50000 + 1 * 44000, SDS-PAGE, 1 * 46200 + 1 * 487000, mass spectroscopy [25]) [2, 6, 7, 11, 18-20, 22, 27]
Additional information <3, 14> (<3> the secondary structures of α subunit include 15 α helices, three short 3,10 helices and a β strand, the β subunit

202

contains 14 α helices, seven short 3,10 helices and three β strands [7]; <3> the α-subunit plays a direct role in the catalytic reaction in addition to its role in the stabilization of the β-subunit [19]; <14> α-subunit of enzyme plays a role in binding isoprenyl substrates [21]) [7, 19, 21]

5 Isolation/Preparation/Mutation/Application

Source/tissue
amastigote <10, 11> [13]
bloodstream form <6> [12]
brain <3> [6, 19]
epimastigote <7-9> [13]
fruit <16> (<16> decreased activity approximately 10fold between young and mature green fruits, increased activity during fruit development in red ripe fruit [22]) [22]
leaf <16> (<16> low activity [22]) [22]
procyclic form <6> [11, 12, 13]
promastigote <10, 11> [13]
root <16> [22]
stem <16> (<16> high activity in tissues from young plants [22]) [22]
testis <13> [20]
trypomastigote <7-9> [13]
Additional information <5, 16> (<16> no significant difference in enzyme activity between dark-grown and light-grown seedlings [22]; <5> appearance during specific stages of Plasmodium falciparum differentiation [26]) [22, 26]

Purification
<2> [18]
<2> (recombinant enzyme [9]) [9, 18]
<3> [19]
<3> (from recombinant baculovirus-infected Sf9 cells [6]) [6]
<3> (recombinant enzyme [25]) [25]
<4> [10]
<6> [11, 12]
<13> [20]
<16> (recombinant enzyme [22]) [22]

Crystallization
<3> (X-ray crystal structure, modelling indicates that geranylgeranyl diphosphate adopts a different conformation than the farnesyl chain of farnesyl diphosphate [4]) [4]
<3> (a new crystalline form of enzyme for the mutant truncated 10 residues at the C-terminus [25]) [25]
<3> (at 2.25 Angstrom, zinc occurs at a junction between a hydrophilic surface groove near the subunit interface: peptide binding site, and a deep lipophilic cleft in the β subunit lined with aromatic residues: farnesyl diphosphate binding site [27]) [7, 27]

<3> (farnesyl diphosphate and peptide substrate can be accommodated in the hydrophobic active-site barrel, with the sole charged residue inside the barrel, Arg202 of the β-subunit, forming a salt bridge with the negatively charged carboxy terminus of peptide substrate [23]) [23]

<3> (the isoprenoid moiety of farnsyl diphophate binds in an extended conformation in a hydrophobic cavity of the β-subunit of the enzyme, and the diphosphate moiety binds to a positively charged cleft at the top of this cavity near the subunit interface [3]) [3]

<3> (two crystal structures of enzyme complex: one containing farnesylated Ras peptide product alone and a complex that contains both the farnesylated peptide and an additional farnesyl diphosphate substrate [5]) [5]

Cloning

<1> (enzyme produced using an Sf9 cell overexpression [25]) [25]

<3> (a novel translationally coupled two-cistron expression system for enzyme in Escherichia coli [25]) [25]

<3> (coexpression of α and β subunits of enzyme with a recombinant baculovirus [23]) [23]

<3> (enzyme produced using an Sf9 cell overexpression [3,5,27]) [3, 5, 27]

<3> (wild-type enzyme and mutant enzymes expressed in Escherichia coli [14,17]) [14, 17]

<6> (expression in bloodstream and insect stage parasites, and expression in Escherichia coli [12]) [12]

<2, 3, 14, 16> (expression in Escherichia coli [4,9,15,22]) [4, 9, 15, 21, 22]

<3, 1> (coexpression of α and β subunits of enzyme by Sf9 cells infected with a recombinant baculovirus [6]) [6, 24, 8]

<6-11> (coexpression of α and β subunits of enzyme by Sf9 cells infected with a recombinant baculovirus [13]) [13]

Engineering

C309A <2> (<2> lower k_{cat} than the wild-type enzyme [7]) [7]

D109N <2> (<2> loss of affinity of the enzyme for its protein substrate [7]) [7]

D200N <1> (<1> decrease of protein substrate affinity without affecting the affinity for farnesyl diphosphate substrates [7]) [7]

D297A <3> (<3> β-subunit, 200fold decrease in k_{cat} [14]) [14]

D297N <3> (<3> β-subunit, 200fold decrease in k_{cat} [14]) [14]

D307A <2> (<2> lower k_{cat} than the wild-type enzyme [7]) [7]

E256A <2> (<2> 130fold higher K_m for the farnesyl diphosphate substrate [7]) [7]

G249V <1> (<1> decrease in the affinity of both protein and farnesyl diphosphate substrates [7]) [7]

G259V <2> (<2> loss of affinity of the enzyme for its protein substrate [7]) [7]

G328S <2> (<2> loss of affinity of the enzyme for its protein substrate [7]) [7]

G349S <1> (<1> decrease of protein substrate affinity without affecting the affinity for farnesyl diphosphate substrates [7]) [7]

H248A <3> (<3> β-subunit, mutation no effect on substrate binding affinity, the farnesylation rate constant is 10fold decreased in comparison of wild-type, the rate constant by chemical step using farnesyl monophosphate 5fold decreased [17]) [17]

H362A <3> (<3> β-subunit, 50fold decrease in k_{cat} [14]) [14]

H362Q <3> (<3> β-subunit, 15fold decrease in k_{cat} [14]) [14]

H362Q <3> (<3> β-subunit, 500fold decrease in k_{cat} [14]) [14]

H363A <2> (<2> lower k_{cat} than the wild-type enzyme [7]) [7]

K164A <3> (<3> α-subunit, mutation no effect on substrate binding affinity, the farnesylation rate constant is 30fold decreased in comparison of wild-type enzyme, the rate constant by chemical step using farnesyl monophosphate unaffected [17]) [17]

K164N <3> (<3> mutation abolishes enzyme activity [7,19]) [7, 19]

N199D <3> (<3> mutation reduces enzyme activity [7,19]) [7, 19]

R172E <3> (<3> mutation reduces enzyme activity [7,19]) [7, 19]

R211Q <2> (<2> lower k_{cat} than the wild-type enzyme [7]) [7]

W203H <3> (<3> mutation reduces enzyme activity [7,19]) [7, 19]

Y166F <3> (<3> mutation reduces enzyme activity [7,19]) [7, 19]

Y300F <3> (<3> β-subunit, mutation has no effect on substrate binding affinity, the farnesylation rate constant is 500fold decreased in comparison of wild-type enzyme, the rate constant by chemical step using farnesyl monophosphate 300fold decreased. Transition state for farnesylation is stabilized by interactions between the α-phosphate of the isoprenoid substrate and the side chains of Y300β [17]) [17]

Y310F <2> (<2> lower k_{cat} than the wild-type enzyme [7]) [7]

Additional information <1, 3> (<1> H248β, R291β, K294β, W303β involved with the binding and utilization of the farnesyl diphosphate substrate, mutations in R202β affect the binding of the protein substrate [7]; <3> all five mutant enzymes bind farnesyl diphosphate with similar affinity to that of the wild-type enzyme, indicating that the targeted residues neither directly nor indirectly influence the farnesyl diphosphate binding site, only the wild-type enzyme able to bind zinc, while all five of the mutant enzymes lose this ability [14]; <3> mutations have a little effect on the pH and magnesium dependence: side chains K164α, H248β and Y300β not function either as general acid-base catalyst or as magnesium ligands [17]; <3> deletion of 51 amino acids at the NH$_2$ terminus of α subunit: activity the same as wild-type enzyme, deletion of 106 amino acids at the NH$_2$ terminus: complete loss of activity, suggesting that residues between 51 and 106 are important for activity, deletion of 5 amino acids at the COOH terminus reduces α-subunit activity to about 50% of activity of wild-type enzyme, removal of 20 amino acids at the COOH terminus: complete loss of activity [19]; <3> three β-subunit C-terminal truncation mutants, deletion of 5, 10 and 14 residues, the overall catalytic efficiency of enzyme decreases gradually with increasing C-terminal deletion [25]) [7, 14, 17, 19, 25]

Application

medicine <1-5, 7-12> (<1, 3, 4> prime target for development of anticancer therapeutics [1-5, 7-12, 23]; <1-3> evidence that inhibitors of enzyme could be effective therapeutic agents in treatment of many human cancers [2-5, 7, 9, 14, 15, 25]; <1> possibility of development of specific inhibitors for parasitic protozoa [10]; <8-12> enzyme from trypanosomatid parasites are target of anti-trypanosomatid agents because inhibitors of this enzyme are highly toxic to these parasites compared to mammalian cells [13]; <5> enzyme a target in developing drug therapy for malaria [26]) [1-5, 7-23, 25-27]

6 Stability

pH-Stability

4 <4> (<4> the enzyme activity completely abolishes [10]) [10]

Organic solvent stability

Triton X-100 <13> (<13> and Zwittergent 3-14, no effect on the activity of the enzyme [20]) [20]

Additional information <3, 13, 16> (<13> n-octyl β-D-glycopyranoside stimulates even at concentrations of 0.5% [20]; <3, 13, 16> EDTA-treatment decreases activity [6,20,22]) [6, 20, 22]

General stability information

<3>, all activity is lost, when the subunits are dissociated chemically [19]

<3>, the heterodimer cannot be dissociated unless it is denatured, and each individual subunit appears to be unstable in solution, the number of hydrogen bonds found in the enzyme subunit interface may explain the unusual stability of the heterodimer [7]

References

[1] Furfine, E.S; Leban, J.J.; Landavazo, A.; Moomaw, J.F.; Casey, P.J.: Protein farnesyltransferase: kinetics of farnesyl pyrophosphate binding and product release. Biochemistry, **34**, 6857-6862 (1995)

[2] Casey, P.J.; Seabra, M.C.: Protein prenyltransferases. J. Biol. Chem., **271**, 5289-5292 (1996)

[3] Long, S.B.; Casey, P.J.; Beese, L.S.: Cocrystal structure of protein farnesyltransferase complexed with a farnesyl diphosphate substrate. Biochemistry, **37**, 9612-9618 (1998)

[4] Micali, E.; Chehade, K.A.; Isaacs, R.J.; Andres, D.A.; Spielmann, H.P.: Protein farnesyltransferase isoprenoid substrate discrimination is dependent on isoprene double bonds and branched methyl groups. Biochemistry, **40**, 12254-12265 (2001)

[5] Long, S.B.; Casey, P.J.; Beese, L.S.: Reaction path of protein farnesyltransferase at atomic resolution. Nature, **419**, 645-650 (2002)

[6] Chen, W.J.; Moomaw, J.F.; Overton, L.; Kost, T.A.; Casey, P.J.: High level expression of mammalian protein farnesyltransferase in a baculovirus system. J. Biol. Chem., **286**, 9675-9680 (1993)

[7] Park, H.W.; Beese, L.S.: Protein farnesyltransferase. Curr. Opin. Struct. Biol., **7**, 873-880 (1997)

[8] Okolotowicz, K.J.; Lee, W.-J.; Hartman, R.F.; Kim, A.Y.; Ottersberg, S.R.; Robinson, D.E., Jr.; Lefler, S.R.; Rose, S.D.: Inactivation of protein farnesyltransferase by active-site-targeted dicarbonyl compounds. Arch. Pharmacol., **334**, 194-202 (2001)

[9] Caplin, B.E.; Hettich, L.A.; Marshall, M.S.: Substrate characterization of the Saccharomyces cerevisiae protein farnesyltransferase and type-I protein geranylgeranyltransferase. Biochim. Biophys. Acta, **1205**, 39-48 (1994)

[10] Ibrahim, M.; Azzouz, N.; Gerold, P.; Schwarz, R.T.: Identification and characterisation of Toxoplasma gondii protein farnesyltransferase. Int. J. Parasitol., **31**, 1489-1497 (2001)

[11] Yokoyama, K.; Trobridge, P.; Buckner, F.S.; Van Voorhis, W.C.; Stuart, K.D.; Gelb, M.H.: Protein farnesyltransferase from Trypanosoma brucei. A heterodimer of 61- and 65-kDa subunits as a new target for antiparasite therapeutics. J. Biol. Chem., **273**, 26497-26505 (1998)

[12] Buckner, F.S.; Yokoyama, K.; Nguyen, L.; Grewal, A.; Erdjument-Bromage, H.; Tempst, P.; Strickland, C.L.; Xiao, L.; Van Voorhis, W.C.; Gelb, M.H.: Cloning, heterologous expression, and distinct substrate specificity of protein farnesyltransferase from Trypanosoma brucei. J. Biol. Chem., **275**, 21870-21876 (2000)

[13] Buckner, F.S.; Eastman, R.T.; Nepomuceno-Silva, J.L.; Speelmon, E.C.; Myler, P.J.; Van Voorhis, W.C.; Yokoyama, K.: Cloning, heterologous expression, and substrate specificities of protein farnesyltransferases from Trypanosoma cruzi and Leishmania major. Mol. Biochem. Parasitol., **122**, 181-188 (2002)

[14] Fu, H.-W.; Beese, L.S.; Casey, P.J.: Kinetic analysis of zinc ligand mutants of mammalian protein farnesyltransferase. Biochemistry, **37**, 4465-4472 (1998)

[15] Huang, C.-c.; Hightower, K.E.; Fierke, C.A.: Mechanistic studies of rat protein farnesyltransferase indicate an associative transition state. Biochemistry, **39**, 2593-2602 (2000)

[16] Saderholm, M.J.; Hightower, K.E.; Fierke, C.A.: Role of metals in the reaction catalyzed by protein farnesyltransferase. Biochemistry, **39**, 12398-12405 (2000)

[17] Pickett, J.S.; Bowers, K.E.; Hartman, H.L.; Fu, H.W.; Embry, A.C.; Casey, P.J.; Fierke, C.A.: Kinetic studies of protein farnesyltransferase mutants establish active substrate conformation. Biochemistry, **42**, 9741-9748 (2003)

[18] Gomez, R.; Goodman, L.E.; Tripathy, S.K.; O'Rourke, E.; Manne, V.; Tamanoi, F.: Purified yeast protein farnesyltransferase is structurally and functionally similar to its mammalian counterpart. Biochem. J., **289**, 25-31 (1993)

[19] Andres, D.A.; Goldstein, J.L.; Ho, Y.K.; Brown, M.S.: Mutational analysis of α-subunit of protein farnesyltransferase. Evidence for a catalytic role. J. Biol. Chem., **268**, 1383-1390 (1993)

[20] Ryu, K.-Y.; Baik, Y.-J.; Yang, C.-H.: Purification and characterization of far-nesyl protein transferase from bovine testis. J. Biochem. Mol. Biol., 28, 197-203 (1995)

[21] Qian, D.; Zhou, D.; Ju, R.; Cramer, C.L.; Yang, Z.: Protein farnesyltransferase in plants: molecular characterization and involvement in cell cycle control. Plant Cell, 8, 2381-2394 (1996)

[22] Schmitt, D.; Callan, K.; Gruissem. W.: Molecular and biochemical character-ization of tomato farnesyl-protein transferase. Plant Physiol., 112, 767-777 (1996)

[23] Dunten, P.; Kammlott, U.; Crowther, R.; Weber, D.; Palermo, R.; Birktoft, J.: Protein farnesyltransferase: Structure and implications for substrate bind-ing. Biochemistry, 37, 7907-7912 (1998)

[24] Roskoski, R., Jr.; Ritchie, P.: Role of the carboxyterminal residue in peptide binding to protein farnesyltransferase and protein geranylgeranyltransfer-ase. Arch. Biochem. Biophys., 356, 167-176 (1998)

[25] Wu, Z.; Demma, M.; Strickland, C.L.; Syto, R.; Le, H.V.; Windsor, W.T.; We-ber, P.C.: High-level expression, purification, kinetic characterization and crystallization of protein farnesyltransferase β-subunit C-terminal mutants. Protein Eng., 12, 341-348 (1999)

[26] Chakrabarti, D.; Da Silva, T.; Barger, J.; Paquette, S.; Patel, H.; Patterson, S.; Allen, C.M.: Protein Farnesyltransferase and protein prenylation in Plasmo-dium falciparum. J. Biol. Chem., 277, 42066-42073 (2002)

[27] Park, H.-W.; Boduluri, S.R.; Moomaw, J.F.; Casey, P.J.; Beese, L.S.: Crystal structure of protein farnesyltransferase at 2.25 Angstrom resolution. Science, 275, 1800-1804 (1997)

Protein geranylgeranyltransferase type I 2.5.1.59

1 Nomenclature

EC number
2.5.1.59

Systematic name
geranylgeranyl-diphosphate:protein-cysteine geranyltransferase

Recommended name
protein geranylgeranyltransferase type I

Synonyms
CAAX geranylgeranyltransferase
GGTase-I
GGTaseI
PGGT
PGGT-I
PGGTaseI
geranylgeranyl protein transferase type I
geranylgeranyltransferase-1
geranylgeranyltransferase I
protein geranylgeranyltransferase
protein geranylgeranyltransferase I

CAS registry number
135371-29-8

2 Source Organism

<1> *Saccharomyces cerevisiae* (strains YOT433-2C, YOT434-2B, YOT435-1A, YOT436-3C, YOT437-2C [11]) [1, 7-11]
<2> *Bos taurus* [2, 3, 9]
<3> *Rattus norvegicus* (enzyme produced in Sf9cells using baculovirus system [4]) [4, 5, 12, 14]
<4> *Homo sapiens* (enzyme produced in Sf9cells using baculovirus system [6]) [6]
<5> *Schizosaccharomyces pombe* [13]

3 Reaction and Specificity

Catalyzed reaction

geranylgeranyl diphosphate + protein-cysteine = S-geranylgeranyl-protein + diphosphate (<1-3, 5> this enzyme, along with protein farnesyltransferase, EC 2.5.1.58 and protein geranylgeranyltransferase type II, EC 2.5.1.60, constitutes the protein prenyltransferase family of enzymes [1, 3-5, 7, 8, 10, 12, 13]; <1> catalyses the formation of a thioether linkage between the C-1 atom of the geranylgeranyl group and a cysteine residue fourth from the C-terminus of the protein, these protein acceptors have the C-terminal sequence CA1A2X, where the terminal residue, X, is preferably leucine, but serine, methionine, alanine or glutamine makes the protein a substrate for EC 2.5.1.58, the enzymes are relaxed in specificity for A1, but cannot act if A2 is aromatic, known targets of this enzyme include most γ-subunits of heterotrimeric G proteins and Ras-related GTPases such as members of the Ras and Rac/Rho families [1,10]; <2> enzyme and geranylgeranyl diphosphate produce a stable binary complex and the tightly bound geranylgeranyl diphosphate is transferred to a peptide acceptor, without the equilibration of enzyme-bound geranylgeranyl diphosphate with free geranylgeranyl diphosphate [2,3]; <2,3> random sequential mechanism [3,4]; <1> an ordered binding mechanism for enzyme where geranylgeranyl diphosphate adds before peptide [8])

Natural substrates and products

S geranylgeranyl diphosphate + protein-cysteine <1-3, 5> (<2> enzyme catalyses the transfer of the 20-carbon prenyl group from geranylgeranyl diphosphate to the cysteine residue near the C-termini of a variety of eukaryotic proteins [3, 5, 7, 8, 11]; <1, 5> this prenylation is necessary for many proteins to interact with membrane localized at proper intracellular sites [11,13]; <1> yeast enzyme is essential for yeast cell growth [11]) [1-3, 5, 7-11, 13]

P S-geranylgeranyl-protein + diphosphate

Substrates and products

S Ser-Ser-Cys-Ile-Leu-Leu + geranylgeranyl diphosphate <2> (Reversibility: ? <2> [2]) [2]

P diphosphate + Ser-Ser-S-geranylgeranyl-Cys-Ile-Leu-Leu

S farnesyl diphosphate + (biotin-CONH-$(CH_2)_5$-CO-)-NPFREKKFFCAIL <2> (<2> ECB-γ6 [2]; <2> biotin-γ6 [3]) (Reversibility: ? <2> [2,3]) [2, 3]

P diphosphate + (biotin-CONH-$(CH_2)_5$-CO-)-NPFREKKFF-S-farnesyl-CAIL

S farnesyl diphosphate + Ki-Ras4B <4> (Reversibility: ? <4> [6]) [6]

P diphosphate + S-geranylgeranyl-Ki-Ras4B

S geranylgeranyl diphosphate + (biotin-CONH-$(CH_2)_5$-CO-)-ALEPPE-TEPKRKCCIF <2> (<2> ECB-G25K, 42% activity in comparison of protein substrate motif: CAA-leucine [2]) (Reversibility: ? <2> [2,3]) [2, 3]

P diphosphate + (biotin-CONH-$(CH_2)_5$-CO-)-ALEPPETEPKRK-S-geranyl-geranyl-CCIF

S geranylgeranyl diphosphate + (biotin-CONH-$(CH_2)_5$-CO-)-GTPRASNRS-
CAIS <2> (<2> ECB-laminB(S), 5% activity in comparison of protein
substrate motif: CAA-leucine [2]) (Reversibility: ? <2> [2]) [2]

P diphosphate + (biotin-CONH-$(CH_2)_5$-CO-)-GTPRASNRS-S-geranylgera-
nyl-CAIS

S geranylgeranyl diphosphate + (biotin-CONH-$(CH_2)_5$-CO-)-NPFREKKFF-
CAIL <2> (<2> ECB-γ6, preferred substrate [2]; <2> biotin-γ6, preferred
substrate [3]) (Reversibility: ? <2> [2,3]) [2, 3]

P diphosphate + (biotin-CONH-$(CH_2)_5$-CO-)-NPFREKKFF-S-geranylgera-
nyl-CAIL

S geranylgeranyl diphosphate + Arg-Arg–Cys-Val-Leu-Leu <3> (Reversibility: ? <3> [12]) [12]

P diphosphate + Arg-Arg-S-geranylgeranyl-Cys-Val-Leu-Leu

S geranylgeranyl diphosphate + GST-CDC42 <2> (<2> recombinant glutathione S-transferase fusion protein of CDC42Hs with a C-terminal Cys-
Cys-Ile-Phe sequence [3]) (Reversibility: ? <2> [3]) [3]

P diphosphate + GST-S-geranylgeranyl-CDC42

S geranylgeranyl diphosphate + Ki-Ras4A <4> (Reversibility: ? <4> [6]) [6]

P diphosphate + S-geranylgeranyl-Ki-Ras4A

S geranylgeranyl diphosphate + Leu-Pro-Cys-Val-Val-Met <3> (Reversibility: ? <3> [12]) [12]

P diphosphate + Leu-Pro-S-geranylgeranyl-Cys-Val-Val-Met

S geranylgeranyl diphosphate + Lys-Lys-Cys-Ile-Ile-Met <3> (Reversibility:
? <3> [12]) [12]

P diphosphate + Lys-Lys-S-geranylgeranyl-Cys-Ile-Ile-Met

S geranylgeranyl diphosphate + N-Ras <4> (Reversibility: ? <4> [6]) [6]

P diphosphate + S-geranylgeranyl-N-Ras

S geranylgeranyl diphosphate + Phe-Phe-Cys-Ala-Ile-Leu <3> (Reversibility: ? <3> [12]) [12]

P diphosphate + Phe-Phe-S-geranylgeranyl-Cys-Ala-Ile-Leu

S geranylgeranyl diphosphate + Ras-Cys-Val-Leu-Leu <3, 4> (Reversibility:
? <3,4> [4,6]) [4, 6]

P diphosphate + Ras-S-geranylgeranyl-Cys-Val-Leu-Leu

S geranylgeranyl diphosphate + S-geranylgeranyl-Ki-Ras4B <4> (<4> both
the polylysine and the carboxy-terminal methionine are important for
geranylgeranylation of this substrate [6]) (Reversibility: ? <4> [6]) [6]

P diphosphate + S-geranylgeranyl-Ki-Ras4B

S geranylgeranyl diphosphate + dansyl-Cys-Ile-Ile-Leu <1> (Reversibility: ?
<1> [7]) [7]

P diphosphate + dansyl-S-geranylgeranyl-Cys-Ile-Ile-Leu

S geranylgeranyl diphosphate + gluthathione S-transferase-GCVKIKKCVIL
<1, 2> [9]

P diphosphate + ?

S geranylgeranyl diphosphate + protein-cysteine <1-5> (<1, 3> prenylation,
substrates are Rho, Rac, most trimeric G protein γ subunits [1, 4, 6-11];
<1-5> enzyme requires that protein substrates contain a Cys residue
fourth from the C terminus, protein substrate motif: Cys-aliphatic-alipha-

tic-X [1, 4, 6-10, 12, 13]; <1, 2, 4> X is Leu [6, 7, 9, 10, 12-14]; <1> X is Leu or Phe [1, 8]; <2, 3> X is Leu, Phe or Ser, preferred substrate: Leu [2, 3, 4]; <2> high concentrations of peptide substrates results in lost of specificity [3]; <3> the Rab protein Rab8-GTPase, which ends with a Cys-Val-Leu-Leu motif is able to serve as a substrate for either geranylgeranyl transferase I and II, but modified predominantly by geranylgeranyl transferase II, Y78D mutation selective prevents prenylation of Rab8 by geranylgeranyl transferase II, but not geranylgeranyltransferaseI [14]) (Reversibility: ? <1-5> [1-13]) [1-14]

P S-geranylgeranyl-protein + diphosphate <1-5> [1-14]

S Additional information <3-5> (<4> not: N-Ras [6]; <3> some of the 18 analysed hexapeptides effieciently farnsylated by enzyme [12]; <5> requires geranylgeranylation of rho1p for the correct function of enzyme [13]) [6, 12, 13]

P ?

Inhibitors

1-phosphono-(E,E,E)-geranylgeraniol <1, 2> (<2> competitive to geranylgeranyl diphosphate, noncompetitive to GST-CDC42 [3]; <1> competitive to geranylgeranyl diphosphate, potent substrate inhibition to dansyl-Gly-Cys-Ile-Ile-Leu [8]) [3, 8]

3-aza-geranylgeranyl-diphosphate <3> (<3> competitive inhibitor with respect to geranylgeranyl diphosphate, non-competitive to Cys-Val-Phe-Leu [4]) [4]

Cys-3-(aminomethyl)benzoic acid-Leu <1> (<1> noncompetitive to geranylgeranyl diphosphate, competitive to dansyl-Gly-Cys-Ile-Ile-Leu [8]) [8]

Cys-Val-Phe-Leu <3> (<3> noncompetitive inhibitor with respect to geranylgeranyl diphosphate, 50% inhibition at 0.0001 mM, competitive to Cys-Val-Phe-Leu [4]) [4]

GGTI-297 <3> [14]

GGTI-298 <3> [14]

NPFREKKFFCAIL <2> (<2> biotin-γ6, substrate inhibition at high peptide concentration [3]) [3]

PD-083176 <2> (<2> noncompetitive to geranylgeranyl diphosphate, competitive to GST-CDC42, modest inhibitor [3]) [3]

Thr-Lys-Cys-Val-Ile-Leu <3> (<3> potent competitor, 50% inhibition at 0.001 mM [4]) [4]

Thr-Lys-Cys-Val-Ile-Met <3> (<3> potent competitor, 50% inhibition at 0.008 mM [4]) [4]

diethyl dicarbonate <1> (<1> 80% loss of activity at 5 mM [10]) [10]

phenylglyoxal <1> (<1> 80% loss of activity, inactivation of inhibition in the presence of geranylgeranyl diphosphate [10]) [10]

Additional information <3> (<3> TKSer-Val-Ile-Leu inactive as competitor [4]) [4]

Metals, ions

Cd^{2+} <1-3> (<1-3> the zinc in enzyme can be replaced by Cd^{2+} [1,3,5]; <2> 80% of maximal activity in the presence of 0.5 mM alone [3]; <3> Cd-substituted enzyme has altered specificities with regard to utilization of both peptide and isoprenoid substrates [5]) [1, 3, 5]

Mg^{2+} <1-3> (<1,3> no requirement [1,5]; <1> requirement, optimal activity at 0.5 mM in presence of 0.1 mM Zn^{2+} [7]; <1> required, optimal activity of EDTA-treated enzyme at 0.5 mM, Mg^{2+} probably activates the diphosphate leaving group of enzyme, and is essential for transferring the bound peptide substrate [9]; <2> 20% of maximal activity in the presence of 0.5 mM alone [3]) [1, 3, 5, 7]

Mn^{2+} <2> (<2> 60% of maximal activity in the presence of 0.5 mM alone [3]) [3]

Zn^{2+} <1-3> (<1,3> A zinc metalloenzyme. The Zn^{2+} is required for peptide, but not for isoprenoid, substrate binding [1,5]; <2> required, but no activity in presence of Zn^{2+} alone, only when either 0.1 mM Mg^{2+}, Mn^{2+} or Cd^{2+} is additionally present [3]; <1> requirement, optimal activity of EDTA-treated enzyme at 0.01 mM in presence of 1 mM Mg^{2+} [7,9]; <3> enzyme contains a single zinc atom [4]; <2> enzyme contains tightly bound Zn^{2+} which is required for its catalytic activity [9]) [1, 3-5, 7, 9]

Additional information <2> (<2> not: Co^{2+} [3]; <2> metals not needed for complex formation between enzyme and geranylgeranyl diphosphate, but addition of prenyl acceptor only results in prenyl transfer in presence of Mg^{2+} and Zn^{2+} [3]) [3]

Turnover number (min^{-1})

0.342 <2> (GST-CDC42, <2> prenyl donor: geranylgeranyl diphosphate [3]) [3]

0.68 <4> (N-Ras) [6]

0.84 <4> (Ras-Cys-Val-Leu-Leu) [6]

1.2 <3> (Ras-Cys-Val-Leu-Leu) [4]

1.2 <3> (geranylgeranyl diphosphate) [4]

1.68 <2> ((biotin-CONH-$(CH_2)_5$-CO-)-NPFREKKFFCAIL, <2> prenyl donor: geranylgeranyl diphosphate [3]) [3]

4 <4> (Ki-Ras4A) [6]

4.6 <4> (Ki-Ras4B) [6]

6 <2> ((biotin-CONH-$(CH_2)_5$-CO-)-NPFREKKFFCAIL, <2> prenyl donor: farnesyl diphosphate [3]) [3]

7.2 <2> ((biotin-CONH-$(CH_2)_5$-CO-)-NPFREKKFFCAIL, <2> prenyl donor: geranylgeranyl diphosphate [3]) [3]

20.4 <1> (geranylgeranyl diphosphate) [8]

Additional information <1, 2> (<2> comparison of k_{cat} of CAAX-substrates with X is Leu, Phe or Ser [3]; <1> comparison of k_{cat} of wild-type and mutant enzyme [10]) [3, 10]

Specific activity (U/mg)

0.0059 <2> [2]

0.007 <3> [4]

Additional information <1> [9]

Kₘ-Value (mM)

0.000003 <3> (geranylgeranyl diphosphate, <3> pH 7.7, 30°C [4]) [4]

0.00001 <1> (geranylgeranyl diphosphate, <1> pH 7.5, 37°C [9]) [9]

0.00002 <3> (Arg-Arg-Cys-Val-Leu-Leu, <3> pH 7.6, 30°C [12]) [12]

0.00013 <3> (Lys-Lys-Cys-Ile-Ile-Met, <3> pH 7.6, 30°C [12]) [12]

0.0002 <2> (Ser-Ser-Cys-Ile-Leu-Leu, <2> pH 7.4, 30°C [2]) [2]

0.00027 <2> ((biotin-CONH-(CH₂)₅-CO-)-NPFREKKFFCAIL, <2> pH 7.0, 30°C, prenyl donor: geranylgeranyl diphosphate [3]) [3]

0.0009 <4> (Ras-Cys-Val-Leu-Leu, <4> pH 7.5, 37°C [6]) [6]

0.00094 <1> (geranylgeranyl diphosphate, <1> pH 7.5, 30°C [8]) [8]

0.001 <1> (geranylgeranyl diphosphate, <1> pH 7.5, 30°C [7]) [7]

0.0012 <3> (Ras-Cys-Val-Leu-Leu, <3> pH 7.7, 30°C [4]) [4]

0.00143 <1> (gluthathione S-transferase-GCVKIKKCVIL, <1> pH 7.5, 37°C [9]) [9]

0.0018 <1> (dansyl-Gly-Cys-Ile-Ile-Leu, <1> pH 7.5, 30°C [8]) [8]

0.002 <3> (Lys-Pro-Cys-Val-Val-Met, <3> pH 7.6, 30°C [12]) [12]

0.002 <3> (Phe-Phe-Cys-Ala-Ile-Leu, <3> pH 7.6, 30°C [12]) [12]

0.0021 <4> (N-Ras, <4> pH 7.5, 37°C [6]) [6]

0.0024 <1> (dansyl-Gly-Cys-Ile-Ile-Leu, <1> pH 7.5, 30°C [7]) [7]

0.0078 <2> ((biotin-CONH-(CH₂)₅-CO-)-NPFREKKFFCAIL, <2> pH 7.0, 30°C, prenyl donor: farnesyl diphosphate [3]) [3]

0.0088 <4> (Ki-Ras4A, <4> pH 7.5, 37°C [6]) [6]

0.012 <4> (Ki-Ras4B, <4> pH 7.5, 37°C [6]) [6]

0.016 <2> (GST-CDC42, <2> pH 7.0, 30°C, prenyl donor: geranylgeranyl diphosphate [3]) [3]

0.025 <2> ((biotin-CONH-(CH₂)₅-CO-)-NPFREKKFFCAIL, <2> pH 7.0, 30°C, prenyl donor: geranylgeranyl diphosphate [3]) [3]

Additional information <1, 2> (<2> comparison of Kₘ of CAAX-substrates with X is Leu, Phe or Ser [3]; <1> comparison of Kₘ of wild-type and mutant enzyme [10]; <3> comparison of Kₘ of 18 hexapeptides as potential substrates of the farnesyltransferase and geranylgeranyl transferaseI, recognition of overlapping sequences by farnesyltransferase and geranylgeranyltransferaseI [12]) [3, 10]

Kᵢ-Value (mM)

0.000015 <3> (3-aza-geranylgeranyl-diphosphate, <3> with respect to geranylgeranyl diphosphate [4]) [4]

0.000022 <3> (3-aza-geranylgeranyl-diphosphate, <-1> with respect to Ras-Cys-Val-Leu-Leu [4]) [4]

0.00005 <3> (Cys-Val-Phe-Leu, <3> with respect to Ras-Cys-Val-Leu-Leu [4]) [4]

0.000065 <3> (Cys-Val-Phe-Leu, <3> with respect to geranylgeranyl diphosphate [4]) [4]

0.0002 <1> (1-phosphono-(E,E,E)-geranylgeraniol, <1> with respect to geranylgeranyl diphosphate [8]) [8]

0.00072 <2> (1-phosphono-(E,E,E)-geranylgeraniol, <2> with respect to geranylgeranyl diphosphate [3]) [3]

0.00086 <2> (1-P-GGOH, <2> with respect to GST-CDC42 [3]) [3]

0.0082 <2> (PD-083176, <2> with respect to GST-CDC42 [3]) [3]

0.0087 <2> (PD-083176, <2> with respect to geranylgeranyl diphosphate [3]) [3]

0.078 <1> (Cys-3-(aminomethyl)benzoicacid-Leu, <1> with respect to dansyl-Gly-Cys-Ile-Ile-Leu [8]) [8]

0.14 <1> (Cys-3-(aminomethyl)benzoicacid-Leu, <1> with respect to geranylgeranyl diphosphate [8]) [8]

Additional information <1> (<1> K_i for varied concentrations of dansyl-Gly-Cys-Ile-Ile-Leu [8]) [8]

pH-Optimum

6.9-7.3 <1> [9]

7.5 <1> [7]

pH-Range

7-8.5 <1> [7]

4 Enzyme Structure

Subunits

heterodimer <1, 2> (<1> $\alpha\beta$, 1 * 48000 + 1 * 43000 [1]; <2> 1 * 48000 + 1 * 40000 [2]; <1> $\alpha\beta$, 1 * 34000 + 1 * 42000 [7]) [1, 2, 7]

Posttranslational modification

side-chain modification <1> (<1> chemical modification with phenylglyoxal of arginine residue: 80% loss of activity in 30 min, chemical modification with diethyl dicarbonate of histidine residue: 80% loss of activity at 5 mM [10]) [10]

5 Isolation/Preparation/Mutation/Application

Source/tissue

brain <2, 3> [2, 3, 12]

Localization

cytosol <3> [12]

Purification

<1> (recombinant enzyme [7,9]) [7, 9]

<1> (recombinant wild-type and mutant enzymes [10,11]) [10, 11]

<2> [3]

<2> (peptide affinity chromatography of enzyme on SSCILL-Sepharose [2]) [2]

<3> (partial [12]) [12]

<3> (recombinant enzyme [4]) [4]

<4> (recombinant enzyme [6]) [6]

Cloning

<1> (α-subunit encoded by RAM2 and β-subunit encoded by CDC43 translationally coupled by overlapping the RAM-CDC43 stop-start codons and by locating a ribosome-binding site near the 3' end of RAM2, recombinant enzyme overproduced in Escherichia coli [7,8]) [7, 8]

<1> (α-subunit encoded by RAM2 cloned to the pFC vector, which has an chloroamphenicol resistance gene and β-subunit encoded by CAL1 cloned to the pFlag vector, which has an ampicillin resistance gene, these two recombinant enzymes are cotransformed and expressed in Escherichia coli [9,10]) [9, 10]

<1> (wild-type and mutant enzymes expressed in Escherichia coli [11]) [11]

<3> (coexpression of two subunits of enzyme by Sf9 cells infected with recombinant baculovirus [4]) [4, 5, 14]

<5> (identification of the cwp1+ gene, which encodes the α-subunit of enzyme, coexpression of cwp1p and cwg2p, β-subunit, in Escherichia coli [13]) [13]

Engineering

D140N <1> (<1> α-subunit, increased K_m [10]) [10]

H145D <1> (<1> α-subunit, increased K_m [10]) [10]

H216D <1> (<1> β-subunit, K_M of geranylgeranyl diphosphate increased 12fold [10]) [10]

N282A <1> (<1> α-subunit, increased K_M [10]) [10]

R166I <1> (<1> β-subunit, K_M of geranylgeranyl diphosphate increased 29fold, no forming of geranygeranyl diphosphate [10]) [10]

Additional information <1> (<1> comparison of substrate specificity, using GST-CAIL and geranylgeranyl diphosphate: best substrate of wild-type enzyme, using GST-CVIM and geranylgeranyl diphosphate: best substrate of mutant β H216D, using GST-CAIL and farnesyl diphosphate: best substrate of mutant βR166I [10]; <1> seven different mutations in CAL1/CDC43 gene: cal1-1 and cdc43-2 to cdc43-7, all of mutants possess reduced activity and none show temperature-sensitive enzymatic activities, but all of them show temperature-sensitive growth phenotypes; cal1-1 mutation preferentially affects Rho1p geranylgeranylation, and it can be suppressed by Rho1p overexpression, the cdc43-5 mutation in the same gene causes accumulation of Cdc42p in soluble form and can be suppressed by Cdc42p overproduction [11]; <5> Cwg2-1 mutant has a defect in the actin organization, this mutant is identified as a single nucleotide change causing an A202T substitution in a residue conserved among the β-subunit of enzyme, the deletion of cwg2 causes cell death [13]) [10, 11, 13]

Application

medicine <3> (<3> S-prenylated enzymes are interesting targets for noncytotoxic anticancer agents [12]) [12]

6 Stability

pH-Stability
8.6 <1> (<1> at pH 8.6 no enzyme activity is detected [9]) [9]

Organic solvent stability
Additional information <1> (<1> no enzyme activity in the presence of EDTA [9]) [9]

References

[1] Casey, P.J.; Seabra, M.C.: Protein prenyltransferases. J. Biol. Chem., **271**, 5289-5292 (1996)

[2] Yokoyama, K.; Gelb, M.H.: Purification of a mammalian protein geranylgeranyltransferase. Formation and catalytic properties of an enzyme-geranylgeranyl pyrophosphate complex. J. Biol. Chem., **268**, 4055-4060 (1993)

[3] Yokoyama, K.; McGeady, P.; Gelb, M.H.: Mammalian protein geranylgeranyltransferase-I: Substrate specificity, kinetic mechanism, metal requirements, and affinity labeling. Biochemistry, **34**, 1344-1354 (1995)

[4] Zhang, F.L.; Moomaw, J.F.; Casey, P.J.: Properties and kinetic mechanism of recombinant mammalian protein geranylgeranyltransferase type I. J. Biol. Chem., **269**, 23465-23470 (1994)

[5] Zhang, F.L.; Casey, P.J.: Influence of metal ions on substrate binding and catalytic activity of mammalian protein geranylgeranyltransferase type-I. Biochem. J., **320**, 925-932 (1996)

[6] Zhang, F.L.; Kirschmeier, P.; Carr, D.; James, L.; Bond, R.W.; Wang, L.; Patton, R.; Windsor, W.T.; Syto, R.; Zhang, R.; Bishop, W.R.: Characterization of Ha-Ras, N-Ras, Ki-Ras4A, and Ki-Ras4B as in vitro substrates for farnesyl protein transferase and geranylgeranyl protein transferase type I. J. Biol. Chem., **272**, 10232-10239 (1997)

[7] Stirtan, W.G.; Poulter, C.D.: Yeast protein geranylgeranyltransferase type-I: overproduction, purification, and characterization. Arch. Biochem. Biophys., **321**, 182-190 (1995)

[8] Stirtan, W.G.; Poulter, C.D.: Yeast protein geranylgeranyltransferase type-I: steady-state kinetics and substrate binding. Biochemistry, **36**, 4552-4557 (1997)

[9] Kim, H.; Koo, S.H.; Choi, H.J.; Yang, C.H.: Characterization of yeast geranylgeranyl transferase type I expressed in Escherichia coli. Mol. Cell., **6**, 602-608 (1996)

[10] Kim, H.; Yang, C.H.: Active site determination of yeast geranylgeranyl protein transferase type I expressed in Escherichia coli. Eur. J. Biochem., **265**, 105-111 (1999)

[11] Ohya, Y.; Caplin, B.E.; Qadota, H.; Tibbetts, M.F.; Anraku, Y.; Pringle, J.R.; Marshall, M.S.: Mutational analysis of the β-subunit of yeast geranylgeranyl transferase I. Mol. Gen. Genet., **252**, 1-10 (1996)

[12] Boutin, J.A.; Marande, W.; Goussard, M.; Loynel, A.; Canet, E.; Fauchere, J.L.: Chromatographic assay and peptide substrate characterization of partially purified farnesyl- and geranylgeranyltransferases from rat brain cytosol. Arch. Biochem. Biophys., **354**, 83-94 (1998)

[13] Arellano, M.; Coll, P.M.; Yang, W.; Duran, A.; Tamanol, F.; Perez, P.: Characterization of the geranylgeranyl transferase type I from Schizosaccharomyces pombe. Mol. Microbiol., **29**, 1357-1367 (1998)

[14] Wilson, A.L.; Erdman, R.A.; Castellano, F.; Maltese, W.A.: Prenylation of Rab8 GTPase by type I and type II geranylgeranyl transferases. Biochem. J., **333**, 497-504 (1998)

Protein geranylgeranyltransferase type II 2.5.1.60

1 Nomenclature

EC number

2.5.1.60

Systematic name

geranylgeranyl-diphosphate,geranylgeranyl-diphosphate:protein-cysteine ger-anyltransferase

Recommended name

protein geranylgeranyltransferase type II

Synonyms

GGTase-II

GGTaseII

PGGT-II

REP/GGTase

Rab GG transferase

Rab geranylgeranyl transferase

Rab geranylgeranyltransferase

RabGGTase

geranylgeranyltransferase type II

protein geranylgeranyltransferase type-II

CAS registry number

135371-29-8

2 Source Organism

<1> *Saccharomyces cerevisiae* (strain JRY1550 [16]) [1, 11, 15, 16]

<2> *Bos taurus* [2]

<3> *Rattus norvegius* [3-13]

<4> *Homo sapiens* [13]

<5> *Lycopersicon esculentum* [15]

<6> *Nicotiana tabacum* [15]

<7> *Mus musculus* [14]

3 Reaction and Specificity

Catalyzed reaction

2 geranylgeranyl diphosphate + protein-cysteine = 2 S-geranylgeranyl-protein
+ 2 diphosphate (<1, 3> enzyme is unique in the protein prenyltransferase
family, consisting protein farnesyltransferase, EC 2.5.1.58, protein geranylgeranyltransferase type I, EC 2.5.1.59 and Rab geranylgeranyltransferase [1, 2, 5,
8, 11, 16]; <3> mechanism, recombinant enzyme/REP-1 complex catalyzes
the geranylgeranylation of both adjacent cysteines in Rab1A, Rab3A and
Rab5A [4]; <3> mechanism [5, 6]; <3> mechanism, phosphoisoprenoid acts
both as a substrate and as a sensor governing the kinetics of protein protein
interactions in the double prenylation reaction [7]; <3> mechanism, geranylgeranyl diphosphate first acts as an allosteric regulator of enzyme, second
acts as the phosphoisoprenoid donor in the prenylation reaction and third
senses the completion of catalysis and concomitantly triggers substrate release by increasing the dissociation rate of the product [8]; <1,3> mammalian enzyme have revealed two alternative pathways for the assembly of catalytic complex. The classical pathway: the Rab protein forms a stable complex
with REP. Enzyme covalently attaches the geranylgeranyl groups to two C-
terminal cysteines of the Rab protein. Upon prenylation, the Rab/REP complex remains tightly associated with the enzyme until binding of a new molecule of isoprenoid decreases its affinity for the prenylated complex. The prenylated complex dissociated from enzyme and delivers Rab protein to its target membrane. In the alternative pathway, enzyme and REP form a tight complex in the presence of the phosphoisoprenoid. This complex is catalytically
competent and can recruit and prenylate Rab protein. [11,12]; <1> the yeast
enzyme follows only the classical pathway [11]; <3> mechanism of enzyme/
REP-complex formation [12])

Natural substrates and products

S geranylgeranyl diphosphate + protein-cysteine <1-3, 7> (<1-3> reaction
is critical for membrane localization of Rab proteins and for their interaction with soluble regulatory proteins [2, 11]; <2> the Rab protein Rab8-
GTPase, which end with a Cys-Val-Leu-Leu motif able to serve as a substrate for either geranylgeranyl transferase I and II, but modified predominantly by either geranylgeranyl transferase II in vivo [2]; <3, 7> enzyme crucial for membrane association and function of Rab proteins in
intracellular vesicular trafficking [5, 14, 15]; <3> this posttranslational
modification is essential for the biological activity of Rab proteins
[8,10]) [2, 5, 8, 10, 11, 14, 15]

P S-geranylgeranyl-protein + diphosphate

Substrates and products

S geranylgeranyl diphosphate + Rab1A <3> (<3> substrate motif: -XXCC
[3, 4]) (Reversibility: ? <3> [3,4]) [3, 4]

P ?

S geranylgeranyl diphosphate + Rab1B <2> (Reversibility: ? <2> [2]) [2]

P ?

S geranylgeranyl diphosphate + Rab3A <3> (<3> substrate motif: -XCXC [3,4]) (Reversibility: ? <3> [3,4]) [3, 4]

P ?

S geranylgeranyl diphosphate + Rab5A <3> (<3> substrate motif: -CCXX [4]) (Reversibility: ? <3> [4]) [4]

P ?

S geranylgeranyl diphosphate + Rab8 <2> (<2> the Rab protein Rab8-GTPase, which end with a Cys-Val-Leu-Leu motif able to serve as a substrate for either geranylgeranyl transferase I and II, but modified predominantly by either geranylgeranyl transferase II in vitro, but the mutation of Y78D prevents Rab8 from serving as a substrate [2]) (Reversibility: ? <2> [2]) [2]

P ?

S geranylgeranyl diphosphate + Ypt1p <1> (<1> yeast enzyme catalyses the prenylation of Ypt1p in the presence of an escort protein, Msi4p [16]) (Reversibility: ? <1> [16]) [16]

P ?

S geranylgeranyl diphosphate + protein-cysteine <1-7> (<1, 2> enzyme attaches geranylgeranyl groups to two C-terminal cysteines in Ras-related GTPases of a single family, the Rab family, Ypt/Sec4 in lower eukaryotes, that terminate in XXCC, XCXC and CCXX motifs, reaction only if Rab is bound to a carrier protein termed Rab escort protein, REP [1-7,12,15,16]; <1,2> Rab protein forms a stable complex with Rab escort protein, REP [1-12]; <3> enzyme catalyses the transfer of two 20-carbon geranylgeranyl groups from geranylgeranyl diphosphate onto C-terminal cysteine residues of Rab-proteins [9,10,15]; <4> GDP-bound form of Rab3 is preferred substrate of enzyme [13]; <7> substrate are Rab3 and Rab4 [14]; <1,5,6> substrate are three tomato cDNAs endcoding Rab-like proteins: LeRab1A, B and C with mutant and wild-type prenylation motif, LeRab1B is the best substrate for the plant enzyme, but LeRab1A is the best substrate for yeast enzyme [15]) (Reversibility: ? <1-7> [1-16]) [1-16]

P S-geranylgeranyl-protein + diphosphate <1-7> [1-16]

S Additional information <1, 3> (<3> farnesyl diphosphate and geranylgeranyl diphosphate bind to enzyme with very similar rate constant, but the binding mechanism of farnesyl diphosphate differs from that of geranylgeranyl diphosphate, shortening of the isoprenoid chain leads to a drastic decreases in affinity [6]; <3> the affinity of the mono-prenylated Rab7/REP-1 complex is very close to that observed for double prenylated Rab7/REP-1 complex binding to enzyme [7]; <1,3> yeast enzyme binding of farnesyl diphosphate is as weaker as of geranylgeranyl diphosphate, the length of isoprenoids has an influence on their affinity for enzyme, but unlike the mammalian enzyme, yeast enzyme binds prenylated and unprenylated Yptp/Mrs6p complex with similar affinities, phosphoisoprenoids do not influence the affinity of Mrs6p for yeast enzyme [11]; <4> the C-terminal cysteines of Rab3a contribute to the interaction with enzyme [13]) [6, 7, 11]

P ?

Inhibitors

KCl <3> (<3> 50% inhibition at 100 mM [3]) [3]

NaCl <3> (<3> 50% inhibition at 100 mM [3]) [3]

NaOAc <3> (<3> 50% inhibition at 100 mM [3]) [3]

Zn^{2+} <1, 3> (<3> in the presence of saturating concentrations of $MgCl_2$ 100% inhibition at 0.2 mM [3]; <1> at concentrations above 0.01 mM [16]) [3, 16]

Additional information <5-7> (<7> perillyl alcohol inhibits insulin induced activation of enzyme, but not GGTI⁻298 and α-hxdroxyfarnesylphosphonic acid, PD 98056 inhibits the ability of insulin to increase the amount of geranylgeranylated Rab3 and Rab4 proteins [14]; <5,6> plant enzyme is inhibited by mutant Rab lacking a prenylation consence sequence [15]) [14, 15]

Activating compounds

Nonidet P-40 <1, 5, 6> (<5,6> the plant enzyme activity is enhanced by detergent [15]; <1> in assay requirement [16]) [15, 16]

insulin <7> (<7> promotes the phosphorylation of the α subunit of enzyme without an effect on β subunit, and activation of enzyme, insulin increases the amount of geranylgeranylated Rab3 and Rab4 proteins [14]) [14]

Metals, ions

Mg^{2+} <1, 3> (<3> requirement, K_M 0.5 mM [3]; <3> no requirement [6]; <1> requirement, optimal concentration at 8 mM [16]) [3, 6, 16]

Zn^{2+} <1, 3> (<3> the location of the zinc ion in the β subunit, where it is coordinated by Asp238β, Cys240β and His290β, the fourth ligand is residue His2α [5]; <1> requirement, optimal concentration at 0.008 mM, zinc content: 0.7-0.12 mol/mol enzyme [16]) [5, 16]

Turnover number (min⁻¹)

Additional information <3> (<3> double prenylation the rate for the first prenyl transfer reaction is 0.00026/min, while the second prenyl transfer reaction is 4fold slower [7]) [7]

Specific activity (U/mg)

0.0082 <1> [16]

0.03 <3> [3]

0.098 <3> [9]

K_m-Value (mM)

0.00002 <3> (Rab3A, <3> pH 7.2, 37°C [3]) [3]

0.00002 <3> (geranylgeranyl diphosphate, <3> pH 7.2, 37°C [3]) [3]

0.0011 <1> (Yptlp, <1> pH 7.3, 30°C [16]) [16]

0.0016 <1> (geranylgeranyl diphosphate, <1> pH 7.3, 30°C [16]) [16]

Additional information <1, 3, 4> (<1,3> yeast enzyme displays significantly lower affinity for its reaction product than its mammalian counterpart [11]; <4> comparison of K_M of Rab3a proteins [13]) [11, 13]

pH-Optimum

7-7.5 <1> [16]

4 Enzyme Structure

Molecular weight
100000 <3> [9]

Subunits
heterodimer <1, 3> (<1,3> $\alpha\beta$, 1 * 60000 + 1 * 38000 [1,3]; <3> $\alpha\beta$ [10]; <1,3> $\alpha\beta$, but structural differences between enzyme of higher and lower eukaryotes [11]) [1, 9-11]
Additional information <1, 3> (<1,3> yeast enzyme does not possess the immunoglobulin-like domain and a leucine-rich repeat domain found in mammalian enzyme [11]) [11]

5 Isolation/Preparation/Mutation/Application

Source/tissue
adipocyte <7> [14]
brain <3, 4> [3, 13]
fibroblast <7> [14]

Purification
<1> (recombinant enzyme [16]) [16]
<3> (component B [3]) [3]
<3> (recombinant enzyme [9,10]) [9, 10]

Crystallization
<3> [10]
<3> (at 2.0 Angstrom resolution, the β subunit forms an α-α barrel that contains most of the residues in the active site, the α subunit consists of a helical domain, an immunoglobulin-like domain and a leucine-rich repeat domain, the N-terminal α subunit binds to the active site in the β subunit [5]) [5]
<3> (at 2.7 Angstrom, crystal structure of REP-1 in complex with farnesyl-loaded enzyme, contact between the two proteins is formed through domain II of REP-1 and the α subunit of the enzyme [12]) [12]

Cloning
<1> (expression in Escherichia coli [11]) [11]
<1> (expression in Escherichia coli [16]) [16]
<3> (development of a two plasmid expression system in Escherichia coli that achieves large quantities of enzyme [9]) [9]
<3> (expression in SF21 cells [6-8]) [6-8]
<3> (recombinant enzyme produced in baculovirus and Escherichia coli system [10]) [10]
<3> (recombinant enzyme produced in insect cells [4]) [4]
<4> (cloning of the β subunit of enzyme by means of the yeast two-hybrid system [13]) [13]

References

[1] Casey, P.J. and Seabra, M.C.: Protein prenyltransferases. J. Biol. Chem., **271**, 5289-5292 (1996)

[2] Wilson, A.L.; Erdman, R.A.; Castellano, F.; Maltese, W.A.: Prenylation of Rab8 GTPase by type I and type II geranylgeranyl transferases. Biochem. J., **333**, 497-504 (1998)

[3] Seabra, M.C.; Goldstein, J.L.; Sudhof, T.C.; Brown, M.S.: Rab geranylgeranyl transferase. A multisubunit enzyme that prenylates GTP-binding proteins terminating in Cys-X-Cys or Cys-Cys. J. Biol. Chem., **267**, 14497-14503 (1992)

[4] Farnsworth, C.C.; Seabra, M.C.; Ericsson, L.H.; Gelb, M.H.; Glomset, J.A.: Rab geranylgeranyl transferase catalyses the geranylgeranylation of adjacent cysteines in the small GTPases Rab1A, Rab3A, and Rab5A. Proc. Natl. Acad. Sci. USA, **91**, 11963-11967 (1994)

[5] Zhang, H.; Seabra, M.C.; Deisenhofer, J.: Crystal structure of Rab geranylgeranyltransferase at 2.0 Å resolution. Structure Fold Des., **8**, 241-251 (2000)

[6] Thomä, N.H.; Iakovenko, A.; Owen, D.; Scheidig, A.S.; Waldmann, H.; Goody, R.S.; Alexandrov, K.: Phosphoisoprenoid binding specificity of geranylgeranyltransferase type II. Biochemistry, **39**, 12043-12052 (2000)

[7] Thomä, N.H.; Niculae, A.; Goody, R.S.; Alexandrov, K.: Double prenylation by RabGGTase can proceed without dissociation of the mono-prenylated intermediate. J. Biol. Chem., **276**, 48631-48636 (2001)

[8] Thomä, N.H.; Iakovenko, A.; Kalinin, A.; Waldmann, H.; Goody, R.S.; Alexandrov, K.: Allosteric regulation of substrate binding and product release in geranylgeranyltransferase type II. Biochemistry, **40**, 268-274 (2001)

[9] Kalinin, A.; Thomä, N.H.; Iakovenko, A.; Heinemann, I.; Rostkova, E.; Constantinescu, A.T.; Alexandrov, K.: Expression of mammalian geranylgeranyltransferase type-II in Escherichia coli and its application for in vitro prenylation of Rab proteins. Protein Expr. Purif., **22**, 84-91 (2001)

[10] Rak, A.; Niculae, A.; Kalinin, A.; Thomä, N.H.; Sidorovitch, V.; Goody, R.S.; Alexandrov, K.: In vitro assembly, purification, and crystallization of the Rab geranylgeranyl transferase:substrate complex. Protein Expr. Purif., **25**, 23-30 (2002)

[11] Dursina, B.; Thomä, N.H.; Sidorovitch, V.; Niculae, A.; Iakovenko, A.; Rak, A.; Albert, S.; Ceacareanu, A.C.; Kolling, R.; Herrmann, C.; Goody, R.S.; Alexandrov, K.: Interaction of yeast Rab geranylgeranyl transferase with its protein and lipid substrates. Biochemistry, **41**, 6805-6816 (2002)

[12] Pylypenko, O.; Rak, A.; Reents, R.; Niculae, A.; Sidorovitch, V.; Cioaca, M.D.; Bessolitsyna, E.; Thoma, N.H.; Waldmann, H.; Schlichting, I.; Goody, R.S.; Alexandrov, K.: Structure of Rab escort protein-1 in complex with Rab geranylgeranyltransferase. Mol. Cell., **11**, 483-494 (2003)

[13] Johannes, L.; Perez, F.; Laran-Chich, M.P.; Henry, J.P.; Darchen, F.: Characterization of the interaction of the monomeric GTP-binding protein Rab3a with geranylgeranyl transferase II. Eur. J. Biochem., **239**, 362-368 (1996)

[14] Goalstone, M.L.; Leitner, J.W.; Golovchenko, I.; Stjernholm, M.R.; Cormont, M.; Le Marchand-Brustel, Y.; Draznin, B.: Insulin promotes phosphorylation and activation of geranylgeranyltransferase II. Studies with geranylgeranylation of rab-3 and rab-4. J. Biol. Chem., **274**, 2880-2884 (1999)

[15] Loraine, A.E.; Yalovsky, S.; Fabry, S.; Gruissen, W.: Tomato Rab1A homologs as molecular tools for studying Rab geranylgeranyl transferase in plant cells. Plant Physiol., **110**, 1337-1347 (1996)

[16] Witter, D.J.; Poulter, C.D.: Yeast geranylgeranyltransferase type-II: steady state kinetic studies of the recombinant enzyme. Biochemistry, **35**, 10454-10463 (1996)

Hydroxymethylbilane synthase

1 Nomenclature

EC number
2.5.1.61

Systematic name
porphobilinogen:(4-[2-carboxyethyl]-3-[carboxymethyl]pyrrol-2-yl)methyl-transferase (hydrolysing)

Recommended name
hydroxymethylbilane synthase

Synonyms
(HMB)-synthase <8> [15]
EC 4.3.1.8 (formerly)
HMB-S <8, 13, 14> [12, 14, 18, 22]
HMBS <14> [16, 23, 29]
PBG-D <13, 16, 17> [30, 31, 33]
PBG-deaminase <8-11, 16> [8, 25, 28, 30]
PBGD <8> [22]
UPGI-S <4, 8> [5, 10]
URO-S <13> [18]
porphobilinogen ammonia-lyase (polymerizing)
porphobilinogen deaminase <3, 5, 8, 9, 13-15, 17> [2-5, 7, 11, 13, 15, 17, 18, 22-27, 31-33]
pre-uroporphyrinogen synthase
preuroporphyrinogen synthetase <3> [7]
synthase, uroporphyrinogen I
urogenI synthase <4> [10]
uroporphyrinogen I synthase <8, 13> [5, 6, 18]
uroporphyrinogen I synthetase <1-3, 8, 12, 13> [1-3, 9, 13, 28]
uroporphyrinogen synthase
uroporphyrinogen synthetase <3> [3]

CAS registry number
9036-47-9
9074-91-3

2 Source Organism

<1> *Mus musculus* [1]
<2> *Spinacia oleracea* (spinach) [1, 3, 5, 9, 18, 31]

<3> *Rhodopseudomonas sphaeroides* (ATCC 11167 [2]; N.C.I.B. 8253 [3, 7]) [2, 3, 5, 7, 9, 31]

<4> *Triticum aestivum* (wheat) [3, 9, 10]

<5> *Arabidopsis thaliana* (Columbia [32]) [32]

<6> *Bos taunus* [3, 5]

<7> *Glycine max* (soya-bean) [3, 26]

<8> *Homo sapiens* [4-6, 12, 15, 18, 21-23, 25, 28, 31]

<9> *Euglena gracilis* [8, 11, 23, 31]

<10> *Propionibacterium shermanii* [8]

<11> *Gallus gallus* [8]

<12> *Chlorella regularis* (S-50 [9]) [9, 18, 31]

<13> *Rattus norvegicus* (male Wistar rat [13]; Sprague-Dawley [14]; Wistar albino rat [18]; male Chbb Thom rat [33]) [13, 14, 18, 23, 31, 33]

<14> *Escherichia coli* (K12, JA 200/pLC 41-4 [17]; TG1 [19]; wild-type TG' recO, mutants K55Q, K59Q, and K55Q-K59Q [24]; K12, mutant PO1562 [29]) [16-20, 23, 24, 27, 29, 31]

<15> *Pisum sativum* (pea) [26, 31]

<16> *Saccharomyces cerevisiae* (wild-type D273-10B, mutant B231 [30]) [30, 31]

<17> *Scenedesmus obliquus* (wild-type D3, Sammlung von Algenkulturen, Pflanzenphysiologisches Institiut Universität Göttingen, Germany [31]) [31]

3 Reaction and Specificity

Catalyzed reaction

4 porphobilinogen + H_2O = hydroxymethylbilane + 4 NH_3 (The enzyme works by stepwise addition of pyrrolylmethyl groups until a hexapyrrole is present at the active centre. The terminal tetrapyrrole is then hydrolysed to yield the product, leaving a cysteine-bound dipyrrole on which assembly continues. In the presence of a second enzyme, EC 4.2.1.75 uroporphyrinogen-III synthase, which is often called cosynthase, the product is cyclized to form uroporphyrinogen-III. If EC 4.2.1.75 is absent, the hydroxymethylbilane cyclizes spontaneously to form uroporphyrinogen I; <9-11>, mechanism [8])

Reaction type

elimination (of NH_3, C-N bond cleavage)

Natural substrates and products

S   porphobilinogen <1-17> [1-33]

P   ?

S   Additional information <8, 9, 13> (<8>, occurrence of multiple forms of the enzyme. These isomers correspond to the enzyme-substrate intermediates: mono-, di-, tri-, and tetrapyrrole [6]; <8,9,13>, third enzyme in the heme biosynthetic pathway [18, 22, 23, 25]; <9>, enzyme from chloroplasts is a nuclear-encoded protein, contains a transit-peptide part for transport across the organelle membrans [23]; <9>, in presence of a second enzyme, EC 4.2.1.75, uroporphyrinogen-III synthase, often called

co-synthase, the product is cyclized to form uroporphyrinogen-III [11])
[6, 11, 18, 22, 23, 25]

P ?

Substrates and products

S porphobilinogen + H_2O <1-17> (<1-17>, in the presence of uroporphyrinogen III cosynthase, EC 4.2.1.75 [1-33]) (Reversibility: ?, <1-17> [1-33])
[1-33]

P uroporphyrinogen III <1-17> (<1>, uroporphyrinogen III cosynthase is
more heat-labile than porphobilinogen deaminase. Heat-treatment of the
porphobilinogen deaminase/uroporphyrinogen III cosynthase enzyme
system forms only 2-4% of uroporphyrinogen III [1]) [1-33]

S porphobilinogen + H_2O <1-17> (<4,8>, PBG [4, 10]; <9>, stoichiometry
of enzyme reaction [11]) (Reversibility: ?, <1-17> [1-33]) [1-33]

P hydroxylmethylbilane + NH_3 <1-17> (<1, 2>, uroporphyrinogen [1]; <1,
3, 12>, uroporphyrinogen I [1, 3, 9]; <3>, at pH 8.2 [3]; <8>, HMB [15];
<5>, 1-hydroxymethylbilane, highly unstable [32]) [1-33]

Inhibitors

2-bromoporphobilinogen <5> [32]

2-methylopsopyrroledicarboxylic acid <8> [4]

5,5'-dithiobis(2-nitrobenzoic acid) <13> (<13>, 5 mM, pH 8.0, 50% inhibition [33]) [33]

Ag^{2+} <5> (<5>, 1 mM $AgCl_2$, 90% inhibition [32]) [32]

Al^{3+} <13> [14]

Ba^{2+} <13> [14]

Ca^{2+} <3, 8, 15> (<3>, $CaCl_2$ [2]; <8>, 25 mM $CaCl_2$, 90% inhibition [4];
0.1 mM $CaCl_2$, 80% inhibition [6]; <15>, $CaCl_2$, weak [26]) [2-4, 6, 26]

Cd^{2+} <5, 8, 13> (<8>, 12 mM, complete inhibition [4]; <13>, strong [14];
<5>, 0.1 mM $CdCl_2$, 55% inhibition [32]) [4, 14, 32]

Co^{2+} <5, 13, 15> (<15>, $CoCl_2$, strong [26]; <5>, 0.1 mM $CoCl_2$, 95% inhibition [32]) [14, 26, 32]

Cr^{3+} <13> (<13>, strong [14]) [14]

Cu^{2+} <5, 8, 13> (<8>, 1 mM $CuSO_4$, 87% inhibition [6]; <13>, strong [14];
<5>, 1 mM $CuSO_4$, 66% inhibition [32]) [6, 14, 32]

Fe^{2+} <3, 8, 15> (<3>, 0.1 mM $FeSO_4$ [2]; <8>, 1 mM $FeSO_4$, 58% inhibition
[6]; <15>, $FeSO_4$, strong [26]) [2, 6, 26]

Fe^{3+} <3, 8, 13> (<3>, 0.1 mM $FeCl_3$ [2]; <8>, 1 mM $FeCl_3$, 62% inhibition [6];
<13>, strong [14]) [2, 6, 14]

Hg^{2+} <3-5, 8, 13> (<3>, $HgCl_2$, strong [2]; <8>, 0.0004 mM $HgCl_2$, 80% inhibition [6]; <4>, inhibition by mercury derivates of porphobilinogen, i.e.
PBG-Hg, and opsopyrroledicarboxylic acid, i.e. OPD-Hg, is enhanced by porphobilinogen [10]; <4>, opsopyrroledicarboxylic acid enhances inhibition by
$HgCl_2$ [10]; <13>, strong [14]; <5>, 0.1 mm $HgCl_2$, 76% inhibition [32]) [2, 6,
10, 14, 32]

K^+ <5> (<5>, at high concentrations [32]) [32]

Mg^{2+} <3, 8, 13> (<3>, $MgCl_2$ [2]; <8>, 25 mM, 90% inhibition [4]; <8>,
50 mM $MgCl_2$, 55% inhibition [6]; <15>, $MgCl_2$, weak [26]) [2-4, 6, 14, 26]

Mn^{2+} <3, 13, 15> (<3,15>, $MnCl_2$ [2, 26]; <15>, strong inhibition at submillimolar concentrations [26]) [2, 14, 26]

N-ethylmaleimide <3, 4, 8, 12, 13> (<8>, 1mM 38% inhibition [6]; <12>, 1 mM, 59% inhibition [9]; <13>, 5 mM, pH 8.0, 70% inhibition [33]) [2, 4, 6, 9, 10, 33]

NH_4^+ <3, 5, 13, 15> [3, 18, 26, 32]

Na^+ <5, 13, 15> (<15>, NaCl greater than 300 mM [26]; <5>, at high concentrations [32]) [14, 26, 32]

$NaBH_4$ <14> (<14>, partially inactivates [24]) [24]

Ni^{2+} <13> [14]

Pb^{2+} <13> (<13>, strong [14]) [14]

$PbNO_3$ <8> (<8>, 0.005 mM, 35% inhibition [6]) [6]

UO_2^+ <13> [14]

VO_2^+ <13> [14]

Zn^{2+} <5, 13, 15> (<13>, strong [14]; <15>, $ZnCl_2$, strong [26]; <5>, 1 mM $ZnCl_2$, 78% inhibition [32]) [14, 26, 32]

aniline <3> [3]

coproporphyrinogen <8> [28]

glycerol <8> (<8>, 15% inhibits 40% of enzyme activity [4]) [4]

hydrazine <3> [3]

hydroxylamine <3, 5> [3, 32]

iodine <3> [2]

isoporphobilinogen <8> [4]

methoxyamine <3> [3]

methylamine <3> [3]

opsopyrroledicarboxylic acid <3, 4> (<3>, 1 mM, 40% inhibition [3]; <4>, i.e. OPD, competitive [10]) [3, 10]

p-chloromercuribenzoate <3, 8, 12> (<3>, strong [2]; <8>, 0.002 mM, 74% inhibition [6]; <12>, 0.025 mM, 62% inhibition [9]) [2, 6, 9]

p-hydroxymercuribenzoate <4, 8> (<4>, i.e. PHMB, inhibition is enhanced by porphobilinogen [10]) [4, 10]

protoporphyrin IX <13> (<13>, enzyme activity decreases with increasing concentrations, reaching a 33% inhibition at the vivo concentration of porphyrin, 0.00185 mM, and 0.014 mM porphobilinogen [33]) [33]

protoporphyrinogen <8> (<8>, inhibits both erythroid and lymphoblast forms of the enzyme [28]) [28]

pyridoxal 5'-phosphate <14> (<14>, partially inactivates [24]) [24]

Additional information <3-5, 13> (<3>, mechanism of inhibition [3]; <4>, effects of sulfhydryl reagents [10]) [3, 10, 14, 32, 33]

Turnover number (min^{-1})

15 <8> (porphobilinogen) [6]

Specific activity (U/mg)

0.00117 <8> [22]

0.0058 <12> [9]

0.0128 <8> (<8>, 40 kDa enzyme form [22]) [22]

0.0183 <8> (<8>, 42 kDa enzyme form [22]) [22]

0.038 <8> (<8>, A form [6]) [6]
0.039 <8> (<8>, B form [6]) [6]
0.04 <8> [21]
0.0402 <8> [25]
0.073 <5> [32]
0.0813 <15> [26]
0.1 <9> [11]
0.113 <17> (<17>, at pH 7.4 [31]) [31]
0.45 <3> [3]
Additional information <1, 3, 5, 8, 13, 14, 16> (<8>, enzyme activity is tem-
perature-dependent [6]; <8>, enzyme activity is sharply dependent on pH
[21]; <8>, enzyme activity in lymphoblasts from variegate porphyria subjects
[28]; <16>, porphobilinogenase system [30]) [1-6, 13, 14, 17, 21, 22, 24, 28-
30, 32, 33]

K_m-Value (mM)

0.0011 <13> (porphobilinogen, <13>, 37°C, pH 8.0 [33]) [33]
0.006 <8> (porphobilinogen, <8>, A and B form [6]) [6]
0.0081 <8> (porphobilinogen, <8>, lymphoblast enzyme from control subjects [28]) [28]
0.0089 <8> (porphobilinogen) [25]
0.012 <16> (porphobilinogen, <16>, wild type D273-10B [30]) [30]
0.013 <3> (hydroxymethylbilane) [2]
0.017 <5, 13> (porphobilinogen, <13>, 0.1 M phosphate buffer, pH 7.5 [18])
[18, 32]
0.0195 <16> (porphobilinogen, <16>, mutant B231 [30]) [30]
0.02 <3> (porphobilinogen, <3>, 0.1 mM porphobilinogen and 5 mM $NaBH_4$
[2]) [2]
0.037 <1> (porphobilinogen, <1>, 60°C [1]) [1]
0.04 <3> (porphobilinogen) [3]
0.046 <8> (porphobilinogen, <8>, porphobilinogen deaminase/uroporphyri-
nogen III cosynthase enzyme system [4]) [4]
0.05 <4> (porphobilinogen) [9]
0.051 <8> (hydroxymethylbilane, <8>, porphobilinogen deaminase/uropor-
phyrinogen III cosynthase enzyme system [4]) [4]
0.072 <2> (porphobilinogen) [9]
0.077 <8> (hydroxymethylbilane) [4]
0.079 <17> (porphobilinogen, <17>, 37°C, pH 7.4 [31]) [31]
0.085 <12> (hydroxymethylbilane) [9]
0.089 <12> (porphobilinogen) [9]
0.13 <8> (porphobilinogen) [4]
Additional information <8, 9, 13, 14, 16, 17> (<13>, K_m value decreases with
decreasing pH on the acid side of the pH optimum [13]; <9>, kinetics [11])
[4, 6, 11, 13-15, 17, 24, 28-31, 33]

pH-Optimum

6.9 <17> [31]
7.2-7.3 <13> [14]

7.4 <8, 12, 14> (<8>, phosphate buffer [4]; <12>, both in phosphate buffer and Tris buffer [9]; <14>, SeMet-labelled enzyme [29]) [4, 9, 29]

7.5 <13> (<13>, at 37°C [18]) [18]

7.6 <3, 8> (<3>, phosphate buffer and Tris buffer [2]) [2, 21]

7.7-8.5 <5> [32]

7.8-8 <3> (<3>, 0.1-0.4 mM porphobilinogen [3]) [3]

7.8-8.2 <3> [9]

7.9-8.2 <15> [26]

8 <5> [32]

8-8.2 <13> [33]

8.2 <4, 8> (<8>, Tris-HCl buffer [4]; <8>, A and B form [6]) [4, 6, 9]

Additional information <8, 17> [12, 31]

pH-Range

5-9 <8> (<8>, pH 5.0: unstable below, pH 9.0 stable above [4]) [4]

6-8 <3> (<3>, pH 6.0: enzyme is rapidly and irreversibly inactivated below, pH 8.0: about 50% of maximal activity [2]) [2]

6-9 <8, 9, 12> (<8>, pH 6.0: no activity below, pH 9.0: 50% of maximal activity [6]; <9>, pH 6.0: 40% of maximal activity, pH 9.0: 35% of maximal activity [11]) [6, 9, 11]

6-10 <5> (<5>, pH 6.0: about 15% of maximal activity, pH 10.0: 25% of maximal activity [32]) [32]

6-10.6 <13> (<13>, no activity below pH 6.0 and above pH 10.6 [33]) [33]

6.2-8.4 <13> (<13>, pH 6.2: 50% of maximal activity, pH 8.4: 30% of maximal activity [14]) [14]

6.3-7.5 <17> (<17>, pH 6.3 and pH 7.5: 85% of maximal activity [31]) [31]

6.5-9 <15> (<15>, pH 6.5: 25% of maximal activity, pH 9.0: 50% of maximal activity [26]) [26]

6.5-9.2 <3> (<3>, 50% of maximal activity at pH 6.5 and pH 9.2 [3]) [3]

Additional information <9, 14, 17> (<9>, stoichiometry and kinetic mechanism of the enzyme reaction at different pH values [11]) [11, 17, 31]

Temperature optimum (°C)

37 <3> (<3>, assay at [2, 3]) [2, 3]

45 <8> (<8>, activity twice that at 37°C [4]) [4]

60 <13> (<13>, maximal activity [33]) [33]

65 <12> (<12>, 24% above the activity at 37°C [9]) [9]

Temperature range (°C)

20-80 <12> (<12>, 10% of maximal activity at 20°C, 8% of maximal activity at 80°C [9]) [9]

4 Enzyme Structure

Molecular weight

25000 <8> (<8>, gel filtration [4]) [4]

31000 <5> (<5>, gel filtration [32]) [32]

33000 <17> (<17>, gel filtration [31]) [31]
35000 <3> (<3>, gel filtration [3]) [3]
35000-36000 <12> (<12>, gel filtration [9]) [9]
35500 <3> (<3>, gel filtration [2]) [2]
36000 <8> (<8>, gel filtration, fraction A and B [5]; <8>, gel filtration, A form [6]) [5, 6]
38000 <2, 8, 13> (<8>, gel filtration, B form [6]; <2>, sucrose density gradient centrifugation [9]; <13>, gel filtration [33]) [6, 9, 33]
39000 <3> (<3>, sucrose density gradient centrifugation [2]) [2]
40000 <2> (<2>, gel filtration [9]) [9]
41000 <9, 13> (<9,13>, gel filtration [11, 18]) [11, 18]
44000 <8> (<8>, gel filtration [21]) [21]

Subunits

monomer <3, 5, 8, 9, 12-15> (<3>, 1 * 36000, SDS-PAGE in the presence of β-mercaptoethanol [2, 3]; <3>, 1 * 36000, SDS-PAGE in the presence and in the absence of 2-mercaptoethanol [3]; <5>, 1 * 35000, SDS-PAGE [32]; <8>, 1 * 38000, SDS-PAGE, isozyme B1 [5]; <8>, 1 * 39500, SDS-PAGE, isozymes B2 and B3 [5]; <8>, 1 * 37000, SDS-PAGE, A and B form [6]; <8>, 1 * 42000, SDS-PAGE [22]; <8>, 1 * 40000, SDS-PAGE [22]; <8>, 1 * 41000, SDS-PAGE [25]; <8>, 1 * 41200, SDS-PAGE, erythrocyte enzyme [28]; <15>, 1 * 45000, SDS-PAGE [26]; <9>, 1 * 36927, calculation from sequence of cDNA [23]; <12>, 1 * 33000, calculation from amino acid residues [9]; <13>, 1 * 43000, SDS-PAGE [13]; <14>, 1 * 39100, SDS-PAGE [17]; <14>, 1 * 34268, calculation from sequence of amino acid [29]; <14>, 1 * 34000, calculation from crystal structure, asymmetric unit [27]; <14>, enzyme structure [16]) [1-3, 5, 6, 9, 13, 16, 17, 22, 23, 25-29, 32]
Additional information <8> (<8>, the two enzyme forms can each be separated into three differently charged subforms by Mono Q chromatography [22]) [22]

5 Isolation/Preparation/Mutation/Application

Source/tissue

callus <6, 7> (<6>, tissue [5]) [3, 5, 26]
erythrocyte <8, 13> [4-6, 12, 13, 15, 21]
germ <4> [3, 9, 10]
harderian gland <13> [33]
leaf <2, 5, 15> [3, 26, 32]
liver <6, 13> [3, 13, 14, 18]
lymphoblast <8> (<8>, variegate porphyria lymphoblasts [28]) [22, 25, 28]
spleen <1, 13> (<13>, 5 multiple enzyme forms from spleen of phenylhydrazine-treated male rats [13]) [1, 13]

Localization

chloroplast <9, 15> [23, 26]
cytosol <8> [15]

Purification

<1> (partial [1]) [1]
<3> [2, 3]
<5> [32]
<8> (partial [4]; 3 isoenzymes [5]; 5 forms, A: native enzyme, B-E: isomeres corresponding to the enzyme-substrate intermediates [6]; 2 forms [22]) [4-6, 21, 22, 25]
<9> [11]
<12> [9]
<13> [14, 18, 33]
<14> (SeMet-labelled enzyme, i.e. [SeMet] HMBS, and wild-type enzyme [29]) [17, 29]
<15> (dye-ligand affinity chromatography [26]) [26]
<17> [31]

Crystallization

<14> (SeMet-labelled enzyme [29]) [27, 29]

Cloning

<9> [23]
<14> (SeMet-labelled enzyme [29]) [29]

Engineering

K55Q <14> (K55Q mutant [24]) [24]
K55Q/K59Q <14> (K55Q-K59Q mutant, lower specific activity than the wild-type enzyme [24]) [24]
K59Q <14> (K59Q mutant, lower specific activity than the wild-type enzyme [24]) [24]
Additional information <14> (all, six, methionine residues are replaced by SeMet [29]) [29]

Application

medicine <8> (<8> a HPLC method for simultaneous determination of porphobilinogen deaminase and uroporphobilinogen III synthase activity. The estimation of porphobilinogen deaminase activity is widely used for the diagnosis of acute intermittent porphyria where the abnormality of the enzyme is the primary genetic defect [12]) [12]

6 Stability

pH-Stability

6-9 <12> (stable between [9]) [9]

Temperature stability

56 <8> (<8>, 2h, no loss of activity [6]) [6]
60 <3, 8> (<3>, during early stages of purification enzyme may be heated with no loss of activity. Inactivation is quite rapid above [2]; <8>, 2 h, 10% loss of activity [6]) [2, 6]

65 <8, 12> (<8>, 15 min, no loss of activity [4]; <12>, 1 h, little loss of activity [9]) [4, 9]

70 <5, 13> (<5,13>, 10 min, heat-stable up to 70°C [32, 33]) [32, 33]

75 <12> (<12>, 13% loss of activity after 1h [9]) [9]

80 <5> (<5>, 10 min, 40% loss of activity [32]) [32]

Additional information <8> [4, 6]

Oxidation stability

<8>, photooxidation of the enzyme in the presence of methylene blue, 0.03%, and Rose bengal, 0.03%, over a 5 min period, inhibits 90% and 60% of the enzyme activity, respectively [4]

General stability information

<3>, enzyme is most stable in 0.3 M LiBr, 0.1 M 2-mercaptoethanol, 30% glycerol, 0.15 M Tris chloride, pH 8.0 to 9.0, or 0.3 M $(NH_4)_2SO_4$ [2]

<8>, phosphate stabilizes [4]

Storage stability

<3>, -20°C, stable for over 1 year [3]

<3>, 0°C, stable in solution at pH 7.5 and 9.5 for 3 weeks [3]

<3>, stored in 0.1 M Tris chloride, pH 8.5, containing 30% glycerol and 0.01 M 2-mercaptoethanol. Under these conditions 50% loss of activity after 3 weeks [2]

<8>, -20°C, forms A to E, stable for up to 18 months in 10 mM potassium phosphate buffer, pH 8.0, containing 0.2 mM dithioerythritol [6]

<8>, -15°C, purified enzyme rapidly loses its activity [4]

<8>, 4°C, purified enzyme stable for more than a month [4]

<8>, erythrocytes lysates stored in 1 mM potassium phosphate, pH 7.6, containing 0.05% Trition X-100, 1 mM dithiothreitol and 1 mM MgCl₂, no or little loss of activity after 1 year [6]

<8>, purified enzyme loses its activity when it is frozen [5]

<9>, 0-4°C, less than 5% loss of activity over 10 days [11]

<15>, -20°C or 4°C, unstable, loss of activity after several days [26]

<15>, -20°C, crude plastide lysate is stable for at least 2 weeks [26]

References

[1] Levin, E.Y.; Coleman, D.L.: The enzymatic conversion of porphobilinogen to uroporphyrinogen catalyzed by extracts of hematopoietic mouse spleen. J. Biol. Chem., **242**, 4248-4253 (1967)

[2] Jordan, P.M.; Shemin, D.: Purification and properties of uroporphyrinogen I synthetase from Rhodopseudomonas spheroides. J. Biol. Chem., **248**, 1019-1024 (1973)

[3] Davies, R.C.; Neuberger, A.: Polypyrroles formed from porphobilinogen and amines by uroporphyrinogen synthetase of Rhodopseudomonas spheroides. Biochem. J., **133**, 471-492 (1973)

[4] Frydman, R.B.; Feinstein, G.: Studies on porphobilinogen deaminase and uroporphyrinogen III cosynthase from human erythrocytes. Biochim. Biophys. Acta, **350**, 358-373 (1974)

[5] Miyagi, K.; Kaneshima, M.; Kawakami, J.; Nakada, F.; Petryka, Z.J.; Watson, C.J.: Uroporphyrinogen I synthetase from human erythrocytes: Separation, purification, and properties of isoenzymes. Proc. Natl. Acad. Sci. USA, **76**, 6172-6176 (1979)

[6] Anderson, P.M.; Desnick, R.J.: Purification and properties of uroporphyrinogen I synthase from human erythrocytes. J. Biol. Chem., **255**, 1993-1999 (1980)

[7] Jordan, P.M.; Berry, A.: Preuroporphyrinogen, a universal intermediate in the biosynthesis of uroporphyrinogen III. FEBS Lett., **112**, 86-88 (1980)

[8] Battersby, A.R.; Fookes, C.J.R.; Matcham, G.W.J.; McDonald, E.: Biosynthesis of the pigments of life: formation of the macrocycle. Nature, **285**, 17-19 (1980)

[9] Shioi, Y.; Nagamine, M.; Kuroki, M.; Sasa, T.: Purification by affinity chromatography and properties of uroporphyrinogen I synthetase from Chlorella regularis. Biochim. Biophys. Acta, **616**, 300-309 (1980)

[10] Russell, C.S.; Rockwell, P.: The effects of sulfhydryl reagents on the activity of wheat germ uroporphyrinogen I synthase. FEBS Lett., **116**, 199-202 (1980)

[11] Williams, D.C.; Morgan, G.S.; McDonald, E.; Battersby, A.R.: Purification of porphobilinogen deaminase from Euglena gracilis and studies of its kinetics. Biochem. J., **193**, 301-310 (1981)

[12] Wright, D.J.; Lim, C.K.: Simultaneous determination of hydroxymethylbilane synthase and uroporphyrinogen III synthase in erythrocytes by high-performance liquid chromatography. Biochem. J., **213**, 85-88 (1983)

[13] Williams, D.C.: Characterization of the multiple forms of hydroxymethylbilane synthase from rat spleen. Biochem. J., **217**, 675-683 (1984)

[14] Farmer, D.J.; Hollebone, B.R.: Comparative inhibition of hepatic hydroxymethylbilane synthase by both hard and soft metal cations. Can. J. Biochem. Cell Biol., **62**, 49-54 (1984)

[15] Brown, R.C.; Elder, G.H.; Urquhart, A.J.: Purification of hydroxymethylbilane synthase from human erythrocytes. Biochem. Soc. Trans., **13**, 1227-1228 (1985)

[16] Helliwell, J.R.; Nieh, Y.P.; Raftery, J.; Cassata, A.; Habash, J.; Carr, P.D.; Ursby, T.; Wulff, M.; Thompson, A.W.; Niemann, A.C.; Haedener, A.: Time-resolved structures of hydroxymethylbilane synthase (Lys59Gln mutant) as it is loaded with substrate in the crystal determined by Laue diffraction. J. Chem. Soc. Faraday Trans., **94**, 2615-2622 (1998)

[17] Hart, G.J.; Abell, C.; Battersby, A.R.: Purification, N-terminal amino acid sequence and properties of hydroxymethylbilane synthase (porphobilinogen deaminase) from Escherichia coli. Biochem. J., **240**, 273-276 (1986)

[18] Mazzetti, M.B.; Tomio, J.M.: Purification and some properties of rat liver uroporphyrinogen I synthase. Anal. Asoc. Quim. Argent., **76**, 207-215 (1988)

[19] Miller, A.D.; Hart, G.J.; Packman, L.C.; Battersby, A.R.: Evidence that the pyrromethane cofactor of hydroxymethylbilane synthase (porphobilinogen deaminase) is bound to the protein through the sulphur atom of cysteine-242. Biochem. J., **254**, 915-918 (1988)

[20] Beifuss, U.; Hart, G.J.; Miller, A.D.; Battersby, A.R.: ^{13}C-N.M.R. Studies on the pyrromethane cofactor of hydroxymethylbilane synthase. Tetrahedron Lett., **29**, 2591-2594 (1988)

[21] Smythe, E.; Williams, D.C.: A simple rapid purification scheme for hydroxymethylbilane synthase from human erythrocytes. Biochem. J., **251**, 237-241 (1988)

[22] Lannfelt, L.; Wetterberg, L.; Lilius, L.; Thunell, S.; Joernvall, H.; Pavlu, B.; Wielburski, A.; Gellerfors, P.: Porphobilinogen deaminase in human erythrocytes. Scand. J. Clin. Lab. Invest., **49**, 677-684 (1989)

[23] Sharif, A.; Smith, A.G.; Abell, C.: Isolation and characterisation of cDNA clone for chlorophyll synthesis enzyme from Euglena gracilis. Eur. J. Biochem., **184**, 353-359 (1989)

[24] Haedener, A.; Alefounder, P.; Hart, G.J.; Abell, C.; Battersby, A.R.: Investigation of putative active-site lysine residues in C. Biochem. J., **271**, 487-491 (1990)

[25] Corrigall, A.V.; Meissner, P.N.; Kirsch, R.E.: Purification of human erythrocyte porphobilinogen deaminase. S. Afr. Med. J., **80**, 294-269 (1991)

[26] Spano, A.J.; Timko, M.P.: Isolation, characterization and partial amino acid sequence of a chloroplast-localized porphobilinogen deaminase from pea (Pisum sativum L.). Biochim. Biophys. Acta, **1076**, 29-36 (1991)

[27] Jordan, P.M; Warren, M.J.; Mgbeje, B.I.A.; Wood, S.P.; Cooper, J.B.; Louie, G.; Brownlie, P.; Lambert, R.; Blundell, T.L.: Crystallization and preliminary X-ray investigation of Escherichia coli porphobilinogen deaminase.. J. Mol. Biol., **224**, 269-271 (1992)

[28] Meissner, P.; Adams, P.; Kirsch, R.: Allosteric inhibition of human lymphoblast and purified porphobilinogen deaminase by protoporphyrinogen and coproporphyrinogen. J. Clin. Invest., **91**, 1436-1444 (1993)

[29] Haedener, A.; Matzinger, P.K.; Malashkevich, V.N.; Louie, G.V.; Wood, S.P.; Oliver, P.; Alefounder, P.R.; Pitt, A.R.; Abell, C.; Battersby, A.R.: Purification, characterization, crystallisation and X-ray analysis of selenomethionine-labelled hydroxymethylbilane synthase from Escherichia coli. Eur. J. Biochem., **211**, 615-624 (1993)

[30] Araujo, L.S.; Lombardo, M.E.; Batlle, A.M.D.C.: Inhibition of porphobilinogenase by porphyrins in Saccharomyces cerevisiae. Int. J. Biochem., **26**, 1377-1381 (1994)

[31] Juknat, A.A.; Doernemann, D.; Senger, H.: Purification and kinetic studies on a porphobilinogen deaminase from the unicellular green alga Scenedesmus obliquus. Planta, **193**, 123-130 (1994)

[32] Jones, R.M.; Jordan, P.M.: Purification and properties of porphobilinogen deaminase from Arabidopsis thaliana. Biochem. J., **299**, 895-902 (1994)

[33] Cardalda, C.A.; Juknat, A.A.; Princ, F.G.; Batlle, A.: Rat harderian gland porphobilinogen deaminase: characterization studies and regulatory action of protoporphyrin IX. Arch. Biochem. Biophys., **347**, 69-77 (1997)

Chlorophyll synthase 2.5.1.62

1 Nomenclature

EC number
2.5.1.62

Systematic name
chlorophyllide-a:phytyl-diphosphate phytyltransferase

Recommended name
chlorophyll synthase

Synonyms
chlorophyll a synthase
chlorophyll synthetase
synthase, chlorophyll

CAS registry number
9077-08-1

2 Source Organism

<1> *Avena sativa* (recombinant enzyme expressed in E. coli [1]) [1, 3, 5, 10]
<2> *Synechocystis sp.* (PCC 6803 [2]) [2]
<3> *Helianthus annuus* (L., line 2-24, albina mutant and wild type line 3629 [4]) [4]
<4> *Triticum aestivum* (L. cv. Walde [6]; L. cv. Kosack [7]) [6, 7, 11, 12]
<5> *Capsicum annuum* [8]
<6> *Narcissus pseudomnarcissus* [9]
<7> *Secale cereale* [10]

3 Reaction and Specificity

Catalyzed reaction
chlorophyllide a + phytyl diphosphate = chlorophyll a + diphosphate (<1>, ping-pong mechanism [1])

Reaction type
polyprenyl-group transfer

Substrates and products

S chlorophyllide a + farnesyl diphosphate <1, 5> (Reversibility: ? <1, 5> [3, 8]) [3, 8]

P 3-farnesylchlorophyllide a + diphosphate

S chlorophyllide a + geranylgeranyl diphosphate <1, 2, 4, 5> (<2>, 42% of the activity with phythyl diphosphate [2]) (Reversibility: ? <1, 2, 4, 5> [2, 3, 6, 8]) [2, 3, 6, 8]

P geranylgeranylchlorophyllide a + diphosphate

S chlorophyllide a + phytyl diphosphate <1, 2> (<2>, marked preference of phytyl diphosphate as substrate over geranylgeranyl diphosphate [2]) (Reversibility: ? <1,2> [2,3]) [2, 3]

P chlorophyll a + diphosphate

S chlorophyllide a + tetraprenyl diphosphate <1> (<1>, ping-pong mechanism. Tetraprenyl diphosphate must bind to the enzyme as the first substrate and esterification occurs when the pre-loaded enzyme meets the second substrate [1]) (Reversibility: ? <1> [1]) [1]

P diphosphate + 3-tetraprenylchlorophyllide a

S pheophorbide a + geranylgeranyl diphosphate <4> (<4> Mg- and Zn-complexes are good substrates, the Co-, Cu- and Ni-complexes are neither substrates nor competitive inhibitors [11]; <4> no esterification of synthetic zinc-pheophorbide a derivatives with the stereochemistry of chlorophyllide a' [12]) (Reversibility: ? <4> [11]) [11]

P ?

S Additional information <1, 2, 4> (<2>, no activity with bacteriochlorophyllide a [2]; <1>, phytol, farnesyl, geranylgeranniol and its monophosphate derivative are incorporated into chlorophyll only in the presence of ATP [3]; <4> hydrogen bonding or electronic interaction involving the carbomethoxy group is not essential for substrate binding [12]) [2, 3, 12]

P ?

Inhibitors

N-phenylmaleimide <1> (<1>, since the wild-type enzyme and all other Cys-mutants with the exception of the mutant C304A are inhibited, it is concluded that the inhibitor binds to a non-essential Cys residue to abolish activity [5]) [5]

diacetyl <1> [5]

Metals, ions

Mg^{2+} <1> (<1>, required, reconstitutes activity after treatment with EDTA [5]) [5]

Mn^{2+} <1> (<1>, partially reconstitutes activity after treatment with EDTA [5]) [5]

Zn^{2+} <1> (<1>, partially reconstitutes activity after treatment with EDTA [5]) [5]

5 Isolation/Preparation/Mutation/Application

Source/tissue

cotyledon <3> (<3>, of wild-type line 3629 [4]) [4]

fruit <5> [8]

leaf <1, 3, 4, 7> (<3>, white leaf of line 2-24, albino mutant [4]; <1,7>, nearly normal or partially reduced activity in bleached leaf tissue compared with green primary leaves [10]) [4, 6, 7, 10]

Localization

chloroplast envelope <5> [8]

chromoplast <6> (<6>, peripheral membrane protein which releases its product into the membrane [9]) [9]

chromoplast membrane <5> [8]

etioplast prolamellar body <4> (<4>, enzyme is present in an inactive state [6]; <4>, chlorophyll synthetase activity is relocated from transforming prolamellar bodies to developing thylakoids during irradiation of dark-grown Triticum aestivum [7]) [6, 7]

plastid <1, 7> (<1,7>, heat-bleached, ribosome-deficient plastids [10]) [10]

thylakoid <4> (<4>, chlorophyll synthetase activity is relocated from transforming prolamellar bodies to developing thylakoids during irradiation of dark-grown Triticum aestivum [7]) [7]

Cloning

<1> (expression in Escherichia coli [5]) [5]

<2> (amplified by the polymerase chain reaction and cloned into T7 RNA polymerase-based expression plasmid, heterologous expression of chlG in Escherichia coli [2]) [2]

Engineering

C109A <1> (<1>, mutant enzyme exhibits nearly no enzymatic activity, N-phenylmaleimide results in an additional decrease of activity [5]) [5]

C130A <1> (<1>, mutant enzyme shows reduced activity and sensitivity to N-phenylmaleimide [5]) [5]

C137A <1> (<1>, mutant enzyme is not impaired in enzymatic activity and shows the same inhibition by N-phenylmaleimide as the wild-type enzyme [5]) [5]

C262A <1> (<1>, mutant enzyme is not impaired in enzymatic activity and shows the same inhibition by N-phenylmaleimide as the wild-type enzyme [5]) [5]

C304A <1> (<1>, as active as the wild-type enzyme, mutant enzyme is not inhibited by N-phenylmaleimide [5]) [5]

D147A <1> (<1>, mutant enzyme without activity [1]) [1]

D150A <1> (<1>, mutant enzyme without activity [1]) [1]

D154A <1> (<1>, mutant enzyme without activity [1]) [1]

N146A <1> (<1>, mutant enzyme without activity [1]) [1]

R151A <1> (<1>, mutant enzyme with 35% of the activity compared to wild-type enzyme [5]) [5]

R161A <1> (<1>, mutant enzyme without activity [1,5]) [1, 5]
R161H <1> (<1>, mutant enzyme with 1% activity compared to wild type enzyme [1]) [1]
R161K <1> (<1>, mutant enzyme with 34% activity compared to wild type enzyme [1]) [1]
R284A <1> (<1>, mutant enzyme shows wild-type activity [5]) [5]
R91A <1> (<1>, mutant enzyme without activity [5]) [5]
Additional information <1> (<1>, deletion of the presequence yields a protein with full activity, even further deletion of the N-terminus including amino acid residues 1-87 results in a core protein that is still enzymatically active. Deletion of the 88 N-terminal residues yields a protein without enzymatic activity. At the C-terminus, only one residue H378 can be deleted without loss of activity, while deletion of S77 together with H378, and all shorter sequences show no activity [5]) [5]

References

[1] Schmid, H.C.; Rassadina, V.; Oster, U.; Schoch, S.; Rüdiger, W.: Pre-loading of chlorophyll synthase with tetraprenyl diphosphate is an obligatory step in chlorophyll biosynthesis. Biol. Chem., **383**, 1769-1778 (2002)
[2] Oster, U.; Bauer, C.E.; Rüdiger, W.: Characterization of chlorophyll a and bacteriochlorophyll a synthases by heterologous expression in Escherichia coli. J. Biol. Chem., **272**, 9671-9676 (1997)
[3] Rüdiger, W.; Benz, J.; Guthoff. C.: Detection and partial characterization of activity of chlorophyll synthetase in etioplast membranes. Eur. J. Biochem., **109**, 193-200 (1980)
[4] Vezitskii, A.Y.; Lezhneva, L.A.; Scherbakov, R.A.; Rassadina, V.V.; Averina, N.G.: Activity of chlorophyll synthetase and chlorophyll b reductase in the chlorophyll-deficient plastome mutant of sunflower. Russ. J. Plant Physiol., **46**, 502-506 (1999)
[5] Schmid, H.C.; Oster, U.; Kogel, J.; Lenz, S.; Rüdiger, W.: Cloning and characterisation of chlorophyll synthase from Avena sativa. Biol. Chem., **382**, 903-911 (2001)
[6] Lindsten, A.; Welch, C.J.; Schoch, S.; Ryberg, A.; Rüdiger, W.; Sundqvist, C.: Chlorophyll synthetase is latent in well preserved prolamellar bodies of etiolated wheat. Physiol. Plant., **80**, 277-285 (1990)
[7] Lindsten, A.; Wiktorsson, B.; Ryberg, M.; Sundqvist, C.: Chlorophyll synthetase activity is relocated from transforming prolamellar bodies to developing thylakoids during irradiation of dark-grown wheat. Physiol. Plant., **88**, 29-36 (1993)
[8] Dogbo, O.; Bardat, F.; Camara, B.: Terpenoid metabolism in plastids: activity, localization and substrate specificity of chlorophyll synthetase in Capsicum annuum plastids. Physiol. Veg., **22**, 75-82 (1984)
[9] Kreuz, K.; Kleinig, H.: Chlorophyll synthetase in chlorophyll-free chromoplasts. Plant Cell Rep., **1**, 40-42 (1981)

[10] Hess, W.R.; Blank-Huber, M.; Fieder, B.; Boerner, T.; Ruediger, W.: Chlorophyll synthetase and chloroplast tRNA^{Glu} are present in heat-bleached, ribosome-deficient plastids. J. Plant Physiol., **139**, 427-430 (1992)

[11] Helfrich, M.; Ruediger, W.: Various metallopheophorbides as substrates for chlorophyll synthetase. Z. Naturforsch. C, **47**, 231-238 (1992)

[12] Helfrich, M.; Schoch, S.; Lempert, U.; Cmiel, E.; Ruediger, W.: Chlorophyll synthetase cannot synthesize chlorophyll a'. Eur. J. Biochem., **219**, 267-275 (1994)

Adenosyl-fluoride synthase

1 Nomenclature

EC number
2.5.1.63

Systematic name
S-adenosyl-L-methionine:fluoride adenosyltransferase

Recommended name
adenosyl-fluoride synthase

Synonyms
5'-fluorodeoxyadenosine synthase
SAM fluorinase
fluorinase
fluorinase, S-adenosyl-L-methionine

CAS registry number
438583-16-5

2 Source Organism

<1> *Streptomyces cattleya* [1, 2]

3 Reaction and Specificity

Catalyzed reaction
S-adenosyl-L-methionine + fluoride = 5'-deoxy-5'-fluoroadenosine + L-methionine

Reaction type
adenosyl group transfer
halogenation

Natural substrates and products
S   S-adenosyl-L-methionine + fluoride <1> (<1> involved in production of toxic fluoroacetate [1]) [1]
P   5'-deoxy-5'-fluoroadenosine + L-methionine

Substrates and products
S   S-adenosyl-L-methionine + fluoride <1> (Reversibility: ? <1> [1,2]) [1, 2]
P   5'-deoxy-5'-fluoroadenosine + L-methionine

Inhibitors

S-adenosyl-L-homocysteine <1> (<1> competent, competitive [1]) [1]
sinefungin <1> (<1> weak [1]) [1]

Activating compounds

EDTA <1> (<1> 1 mM increases enzyme activity by 25% [1]) [1]

Metals, ions

Additional information <1> (<1> Mg^{2+} at 1 mM does not affect enzyme activity [1]) [1]

K_m-Value (mM)

0.42 <1> (S-adenosyl-L-methionine) [1]
8.56 <1> (fluoride) [1]

K_i-Value (mM)

0.029 <1> (S-adenosyl-homocysteine) [1]

Temperature optimum (°C)

55 <1> [1]

4 Enzyme Structure

Molecular weight

180000 <1> (<1> gel filtration [1]) [1]

Subunits

hexamer <1> (<1> 6 * 32000, SDS-PAGE and electrospray mass spectrometry [1]) [1]

5 Isolation/Preparation/Mutation/Application

Purification

<1> [1]

References

[1] Schaffrath, C.; Deng, H.; O'Hagan, D.: Isolation and characterisation of 5'-fluorodeoxyadenosine synthase, a fluorination enzyme from Streptomyces cattleya. FEBS Lett., **547**, 111-114 (2003)
[2] O'Hagan, D.; Schaffrath, C.; Cobb, L.; Hamilton, J.T.G.; Murphy, C.D.: Biosynthesis of an organofluorine molecule. Nature, **416**, 279 (2002)

2-Succinyl-6-hydroxy-2,4-cyclohexadiene-1-carboxylate synthase

2.5.1.64

1 Nomenclature

EC number

2.5.1.64

Systematic name

isochorismate:2-oxoglutarate:cyclodienyltransferase (decarboxylating, pyruvate-forming)

Recommended name

2-succinyl-6-hydroxy-2,4-cyclohexadiene-1-carboxylate synthase

Synonyms

2-succinyl-6-hydroxy-2,4-cyclohexadiene-1-carboxylate synthase-α-ketoglutarate decarboxylase 2-succinyl-6-hydroxy-2,4-cyclohexadiene-1-carboxylic acid synthase
6-hydroxy-2-succinylcyclohexa-2,4-diene-1-carboxylate synthase
SHCHC synthase

CAS registry number

122007-88-9

2 Source Organism

<1> *Escherichia coli* (bifunctional enzyme 2-succinyl-6-hydroxy-2,4-cyclohexadiene-1-carboxylic acid synthase and α-ketoglutarate decarboxylase [1]; strain JRG511 [3]) [1, 3]
<2> *Bacillus subtilis* [2]
<3> *Synechocystis sp.* (PCC 6803 [4]) [4]

3 Reaction and Specificity

Catalyzed reaction

2-oxoglutarate + isochorismate = (1S,6R)-6-hydroxy-2-succinylcyclohexa-2,4-diene-1-carboxylate + pyruvate + CO_2

Reaction type

acyl group transfer

Natural substrates and products

S 2-oxoglutarate + isochorismate <1, 2, 3> (<1,2> the enzyme is involved in the pathway for the biosynthesis of menaquinone [1,2]; <1> early enzyme in menaquinone biosynthesis [3]; <3> menD codes for 2-succinyl-6-hydroxyl-2.4-cyclohexadiene-1-carboxylate synthase. menD- mutants lack phyloquinone and are sensitive to high light intensities. In menD- mutants the ratio of photosystem I to photosystem II is reduced relative to wild-type. The lower growth rate and high-light sensitivity of the menD-mutants are attributed to a lower content of photosystem I per cell [4]) (Reversibility: ? <1, 2, 3> [1, 2, 3]) [1, 2, 3, 4]

P (1S,6R)-6-hydroxy-2-succinylcyclohexa-2,4-diene-1-carboxylate + pyruvate + CO_2

Substrates and products

S 2-oxoglutarate + isochorismate <1, 2, 3> (<1> two reactions are identified in the formation of (1S,6R)-6-hydroxy-2-succinylcyclohexa-2,4-diene-1-carboxylate. These are the decarboxylation of 2-oxoglutarate, which results in the formation of succinic semialdehyde-thiamine diphosphate anion, and the addition of succinic semialdehyde-thiamine diphosphate anion to isochorismate with the elimination of the pyruvoyl moiety [1]; <2> the 2-oxoglutarate decarboxalase activity involved is distinct from that of the E1 of the KGDH complex [2]) (Reversibility: ? <1,2,3> [1,2„3,4]) [1, 2, 3, 4]

P (1S,6R)-6-hydroxy-2-succinylcyclohexa-2,4-diene-1-carboxylate + pyruvate + CO_2 <1-3> [1-4]

Specific activity (U/mg)

Additional information <1> (<1> convenient HPLC assay for the study of the overall synthesis of *o*-succinylbenzoic acid from isochorismate or other substrates, i.e. combined activity of 2-succinyl-6-hydroxy-2,4-cyclohexadiene-1-carboxylate synthase, *o*-succinylbenzoic acid synthase and putative decarboxylase. The method also has been adapted to separate measurement of 2-succinyl-6-hydroxy-2,4-cyclohexadiene-1-carboxylate synthase plus decarboxylase activity [3]) [3]

4 Enzyme Structure

Subunits

? <1> (<1> x * 69000, a single gene encodes 2-succinyl-6-hydroxy-2,4-cyclohexadiene-1-carboxylic acid synthase and a-ketoglutarate decarboxylase, calculation from nucleotide sequence [1]) [1]

5 Isolation/Preparation/Mutation/Application

Source/tissue

cell extract <2> [2]

References

[1] Palaniappan, C.; Sharma, V.; Hudspeth, M.E.; Meganathan, R.: Menaquinone (vitamin K_2) biosynthesis: evidence that the Escherichia coli menD gene encodes both 2-succinyl-6-hydroxy-2,4-cyclohexadiene-1-carboxylic acid synthase and a-ketoglutarate decarboxylase activities. J. Bacteriol., **174**, 8111-8118 (1992)

[2] Palaniappan, C.; Taber, H.; Meganathan, R.: Biosynthesis of o-succinylbenzoic acid in Bacillus subtilis: identification of menD mutants and evidence against the involvement of the a-ketoglutarate dehydrogenase complex. J. Bacteriol., **176**, 2648-2653 (1994)

[3] Popp, J. L.; Berliner, C.; Bentley, R.: Vitamin K (menaquinone) biosynthesis in bacteria: high-performance liquid chromatography assay of the overall synthesis of o-succinylbenzoic acid and 2-succinyl-6-hydroxy-2,4-cyclohexa-diene-1-carboxylic acid synthase. Anal. Biochem., **178**, 306-310 (1989)

[4] Johnson, T.W.; Naithani, S.; Stewart, C. Jr.; Zybailov, B.; Jones, A.D.; Golbeck, J.H.: Chitnis, P.R.: The menD and menE homologs code for 2-succinyl-6-hydroxyl-2,4-cyclohexadiene-1-carboxylate synthase and O-succinylbenzoic acid-CoA synthase in the phylloquinone biosynthetic pathway of Synechocystis sp. PCC 6803. Biochim. Biophys. Acta, **1557**, 67-76 (2003)

Aspartate transaminase

1 Nomenclature

EC number

2.6.1.1

Systematic name

L-aspartate:2-oxoglutarate aminotransferase

Recommended name

aspartate transaminase

Synonyms

2-oxoglutarate-glutamate aminotransferase

AAT

AST

AspAT <1, 3, 7, 12, 35, 37, 39, 42> [8, 54, 55, 59, 60-63, 67, 69]

AspT

GOT <9> [15]

GOT (enzyme)

L-aspartate transaminase

L-aspartate-2-ketoglutarate aminotransferase

L-aspartate-2-oxoglutarate aminotransferase

L-aspartate-2-oxoglutarate-transaminase

L-aspartate-α-ketoglutarate transaminase

L-aspartic aminotransferase

aminotransferase, aspartate

aspartate α-ketoglutarate transaminase

aspartate aminotransferase

aspartate-2-oxoglutarate transaminase

aspartate:2-oxoglutarate aminotransferase

aspartic acid aminotransferase

aspartic aminotransferase

aspartyl aminotransferase

glutamate oxaloacetate transaminase

glutamate-oxalacetate aminotransferase

glutamate-oxalate transaminase

glutamic oxalic transaminase

glutamic-aspartic aminotransferase

glutamic-aspartic transaminase

glutamic-oxalacetic transaminase

glutamic-oxaloacetic transaminase

oxaloacetate transferase

oxaloacetate-aspartate aminotransferase
transaminase A
Additional information <3, 37> (<3,37> aspartate aminotransferases are divided into 2 subgroups: subgroup Ia, inclusive the enzyme of Escherichia coli, posses Arg292 for substrate binding and show a specificity for acidic substrates only, subgroup Ib, inclusive the enzyme of Thermus thermophilus, can utilize acidic as well as neutral substrates [61]; <3> EC 2.6.1.57 may be converted to EC 2.6.1.1 by controlled proteolysis with subtilisin, the 2 enzymes are encoded by 2 different genes with high sequence homology [1]) [1, 61]

CAS registry number
9000-97-9

2 Source Organism

<1> *Gallus gallus* (cytosolic and mitochondrial isozymes [53,63]; different molecular forms: α, β, γ, δ, ε [44]) [10, 12, 28, 44, 53, 63]
<2> *Pseudomonas striata* (strain IFO 12996 [14]) [14]
<3> *Escherichia coli* (isoenzymes A and B [3]; mutant V39L [31]; mutant K258A [29]; B [3,43]) [1, 3, 11, 29-32, 43, 58, 61, 67, 72]
<4> *Trichomonas vaginalis* (Bushby strain [2]) [2]
<5> *Sus scrofa* (purified cytosolic isozyme [60]; cytosolic and mitochondrial isozyme [19]; mitochondrial isozyme [27]) [4, 5, 9, 10, 13, 19, 27, 38, 60]
<6> *Oryza sativa* (cv. Koganemasari [6,7]; 2 isoenzymes: AAT-1, AAT-2 [6,7]) [6, 7]
<7> *Haloferax mediterranei* (strain R-4, ATCC 33500 [8,49]) [8, 49]
<8> *Lupinus angustifolius* (cv. Uniharvest [51]; 2 cytosolic isozymes: AAT-P1 and AAT-P2 [51]) [51]
<9> *Trichoderma viride* [15]
<10> *Saccharomyces cerevisiae* [52]
<11> *Oryctolagus cuniculus* (cytosolic and mitochondrial isozyme [16]) [16]
<12> *Rattus norvegicus* (male Wistar rats [56,60]; purified cytosolic isozyme [56]; purified premature mitochondrial isozyme pmAAT [74]; mitochondrial isozyme mAspAT [54,68]; cytosolic and mitochondrial isozyme [17,60]) [17, 54, 56, 60, 68, 74]
<13> *Bos taurus* (ox [22]) [18, 22]
<14> *Rhizobium japonicum* (strain 392 [20]) [20]
<15> *Glycine max* (3 isozymes: 2 cytosolic, 1 mitochondrial [20]) [20]
<16> *Leptospira michotii* (2 isozymes A and B [21]) [21]
<17> *Panicum maximum* (Jacq. var. trichoglume Eyles [23]; 2 isozymes [23]) [23]
<18> *Ovis aries* (cytosolic and mitochondrial isozymes [24]; cytosolic isozyme [39]) [24, 39]
<19> *Sulfolobus solfataricus* [25]

<20> *Canis familiaris* (Mongrel dogs [26]; cytosolic and mitochondrial isozymes [26]) [26]

<21> *Methanobacterium thermoautotrophicum* (strains FTF-INRA, DSM 3012 [33]) [33]

<22> *Rhodotorula minuta* (cytosolic and mitochondrial isozyme [34]) [34]

<23> *Torulopsis candida* (cytosolic and mitochondrial isozyme [34]) [34]

<24> *Methanococcus aeolicus* [35]

<25> *Eleusine corocana* (isoenzymes: AspAT-1, AspAT-2, AspAT-3 [36]) [36]

<26> *Bacillus sp.* (strain YM-2 [37]) [37]

<27> *dolphin* (cytosolic and mitochondrial isozyme [40]) [40]

<28> *Atriplex spongiosa* (2 major isozymes: Asp-DEAE-1 and Asp-DEAE-2 [41]) [41]

<29> *Daucus carota* (cytosolic form I, and mitochondrial forms II and III [42]) [42]

<30> *Arabidopsis thaliana* (3 genes encoding 3 different isozymes: asp1 is mitochondrial, asp3 is amyloplastidic and asp2 is cytosolic [64]; 2 genotypes [45]) [45, 64]

<31> *Medicago sativa* (2 cytosolic isozymes: AAT-1 and AAT-2, the latter has the 3 forms AAT-2a, AAT-2b, AAT-2c [46]) [46]

<32> *Chlamydomonas reinhardtii* (strain 6145c [47]) [47]

<33> *Methanobacterium thermoformicicum* (strain SF-4 [33,48]; DSM 6457 [33]) [33, 48]

<34> *Avena sativa* (var. Lodi [50]; 2 isozymes [50]) [50]

<35> *Panicum miliaceum* (at least 3 isozymes: inducible cytosolic cAspAT and mitochondrial mAspAT, and a constitutive minor plastidic pAspAT [55]) [55]

<36> *Homo sapiens* (cytosolic isozyme [57]) [57]

<37> *Thermus thermophilus* (strain HB8 [59,61]) [59, 61]

<38> *Mus musculus* [60]

<39> *Phormidium lapideum* [62]

<40> *Ruditapes philippinarum* (clam, mollusca: Bivalvia [65]) [65]

<41> *Mugil auratus* (grey mullet, Osteichthyes [66]) [66]

<42> *Pseudoalteromonas haloplanktis* (strain TAC 125 [69]) [69]

<43> *Lupinus albus* (2 isozymes: AAT1 and AAT2 [70,73]; cv. Estoril [70,73]) [70, 73]

<44> *Hyalomma dromedarii* (4 isozymes: AST I , AST II, AST III, AST IV [71]; camel tick [71]) [71]

<45> *Rattus norvegicus* (mature mitochondrial isozyme mAAT [74]) [74]

3 Reaction and Specificity

Catalyzed reaction

L-aspartate + 2-oxoglutarate = oxaloacetate + L-glutamate (<3, 37> active site structure, substrate recognition mechanism [61]; <3> intermediate formation [58]; <3> mechanism [29, 58]; <1, 3-5, 8, 11, 12, 19, 32, 37, 39, 43> bi bi ping pong reaction kinetic [2, 5, 16, 17, 25, 28, 31, 47, 51, 59, 62, 63, 70])

Reaction type
amino group transfer

Natural substrates and products
S L-aspartate + 2-oxoglutarate <1-45> (<43> enzyme plays an important role in numerous metabolic processes [70]; <12> anti-enzyme antibodies selectively inhibit the uptake of oleate in 3T3-L1 adipocytes by 31-55%, but only poorly in fibroblasts, the antibodies have no effect on uptake of 2-deoxyglucose and octanoate [54]; <31> key enzyme in assimilation of C and N compounds [46]; <27> role in energy metabolism of diving animals [40]; <5> enzyme plays a central role in nitrogen metabolism of cells [13]; <2> enzyme plays the most important role in amino acid metabolism [14]; <25> function in C_4 acid pathway [36]) (Reversibility: ? <1-45> [1-74]) [1-74]

P oxaloacetate + L-glutamate <1-45> [1-74]

Substrates and products
S 2,4-diaminobutyric acid + 2-oxoglutarate <12> (Reversibility: ? <12> [17]) [17]

P ?

S 2-aminohexanedioic acid + 2-oxoglutarate <12> (<12> i.e. α-aminoadipic acid [17]) (Reversibility: ? <12> [17]) [17]

P 2-oxohexanedioic acid + L-glutamate

S L-2-amino-4-methoxy-4-oxobutanoic acid + 2-oxoglutarate <43> (<43> isozymes AAT1 and AAT2, the latter shows lower activity [70]) (Reversibility: ? <43> [70]) [70]

P 4-methoxy-2,4-dioxobutanoic acid + L-glutamate

S L-2-amino-4-methoxy-4-oxobutanoic acid + oxaloacetate <43> (<43> isozymes AAT1 and AAT2, low activity [70]) (Reversibility: ? <43> [70]) [70]

P 4-methoxy-2,4-dioxobutanoic acid + L-aspartate

S L-alanine + 2-oxoglutarate <4, 9, 37> (<9> 7.7% of the activity with aspartate [15]) (Reversibility: r <37> [61]; ? <4, 9> [2,15]) [2, 15, 61]

P pyruvate + L-glutamate <37> [61]

S L-aspartate + 2-oxobutyrate <39> (<39> 0.4% activity compared to 2-oxoglutarate [62]) (Reversibility: ? <39> [62]) [62]

P oxaloacetate + 2-aminobutyrate

S L-aspartate + 2-oxoglutarate <1-45> (<39> 45% activity with L-aspartate compared to L-glutamate [62]; <39> 2-oxoglutarate is the best acceptor substrate [62]; <37> enzyme can also act on neutral amino acid substrates due to a substrate-binding pocket with a more flexible conformation, Lys109 is the major determinant for the acidic substrate specificity [61]; <28> isozyme Asp-DEAE-1 performs preferably the reverse reaction, isozyme the forward reaction [41]; <9> best amino acid donor [15]; <12> best amino acid donor of cytosolic enzyme [17]; <3> 30% of the activity with glutamate [43]; <33> 25.2% of the activity with glutamate [48]; <2> 34% of the activity with glutamate [14]; <19> 48% of the activity with cysteine sulfinic acid [25]; <8> absolute specificity for aspartate and 2-oxoglutarate and for glutamate and oxaloacetate [51]) (Reversibility: r

<3-6, 8, 10, 17, 22, 23, 25, 27, 28, 30, 35-37, 39, 43> [1, 2, 6, 23, 27, 29-31, 34, 36, 40, 41, 51, 52, 55, 57-59, 61, 62, 64, 67, 68, 70, 72, 73]; ? <1-3, 5, 7, 9, 11-16, 18-21, 24, 26, 29-34, 38, 40-42, 44, 45> [3-5, 7-22, 24-26, 28, 32, 33, 35, 37-39, 42-50, 53, 54, 56, 60, 63, 65, 66, 69, 71, 74]) [1-74]

P oxaloacetate + L-glutamate <1-45> [1-74]

S L-aspartate + 2-oxoisocaproate <39> (<39> 1.0% activity compared to 2-oxoglutarate [62]) (Reversibility: ? <39> [62]) [62]

P oxaloacetate + L-leucine

S L-aspartate + 4-hydroxyphenylpyruvate <23> (<23> 1.48% of the activity with 2-oxoglutarate [34]) (Reversibility: ? <23> [34]) [34]

P oxaloacetate + L-tyrosine

S L-aspartate + phenylpyruvate <23> (<39> no activity [62]; <23> 7.12% of the activity with 2-oxoglutarate [34]) (Reversibility: ? <23> [34]) [34]

P oxaloacetate + L-phenylalanine

S L-aspartate + pyruvate <39> (<39> 1.9% activity compared to 2-oxoglutarate [62]) (Reversibility: ? <39> [62]) [62]

P oxaloacetate + L-alanine

S L-cysteic acid + 2-oxoglutarate <12> (<43> no activity [70]; <12> i.e. 2-amino-2-sulfopropionate [17]) (Reversibility: ? <12> [17]) [17]

P 2-oxo-3-sulfopropionate + L-glutamate

S L-cysteine + 2-oxoglutarate <9, 21, 33> (<9> 26.2% of the activity with L-aspartate [15]) (Reversibility: ? <9, 21, 33> [15, 33]) [15, 33]

P 2-oxopropionate + L-glutamate

S L-cysteine sulfinic acid + 2-oxoglutarate <2, 6, 10, 12, 16, 19, 26, 33, 39> (<39> 27% activity compared to L-glutamate [62]; <16> only isozyme B, 27% activity compared to L-aspartate [21]; <10> 48% of the activity with L-glutamate [52]; <2> 21% of the activity with L-glutamate [14]; <6> 63% of the activity with aspartate [6]; <12> best amino acid donor for mitochondrial enzyme [17]) (Reversibility: ir <26> [37]; ? <2, 6, 10, 12, 16, 19, 33, 39> [6, 14, 17, 21, 25, 48, 52, 62]) [6, 14, 17, 21, 25, 37, 48, 52, 62]

P 2-oxo-3-sulfinopropionic acid + L-glutamate

S L-erythro-3-hydroxyaspartate + 2-oxoglutarate <3> (<3> high stabilization of the quinonoid intermediate formed by L-erythro-3-hydroxyaspartate and pyridoxal 5'-phosphate at the active site via Tyr70, kinetic analysis [58]; <3> wild-type, no activity with mutant T70F [58]) (Reversibility: ? <3> [58]) [58]

P erythro-3-hydroxy-2-oxoaspartate + L-glutamate

S L-glutamate + 2-oxoglutarate <2, 3, 6, 19, 21, 27, 33, 39> (<2, 3, 6, 27, 33, 39> L-glutamate is best amino acid donor [6, 14, 40, 43, 48, 62]; <19> 25% of the activity with cysteine sulfinic acid [25]) (Reversibility: ? <2, 3, 6, 19, 21, 27, 33, 39> [6, 14, 25, 33, 40, 43, 48, 52, 62]) [6, 14, 25, 33, 40, 43, 48, 62]

P 2-oxoglutarate + L-glutarate

S L-methionine + 2-oxoglutarate <3, 9, 33, 39> (<39> 0.1% activity compared to L-glutamate [62]; <9> 93% of the activity with aspartate [15]; <3, 33> 1.0% of the activity with glutamate [43, 48]) (Reversibility: ? <3, 9, 33, 39> [15, 43, 48, 62]) [15, 43, 48, 62]

P L-glutamate + 4-methylsulfanyl-2-oxobutyric acid
S L-phenylalanine + 2-oxoglutarate <2-4, 6, 9, 22, 27, 32> (<32> low activity [47]; <9> 7.7% of the activity with L-aspartate [15]; <2> 6.9% of the
   activity with glutamate [14]) (Reversibility: ir <27> [40]; ? <2-4, 6, 9, 22,
   32> [1, 2, 6, 14, 15, 34, 43, 47]) [1, 2, 6, 14, 15, 34, 40, 43, 47]
P 2-oxo-3-phenylpropionic acid + L-glutamate <2-4, 6, 9, 22, 27> (<2-4, 6,
   9, 22> i.e. phenylpyruvate [1, 2, 6, 14, 15, 34, 43]) [1, 2, 6, 14, 15, 34, 40,
   43]
S L-serine + 2-oxoglutarate <9> (<9> 7.7% of the activity with aspartate
   [15]) (Reversibility: ? <9> [15]) [15]
P 3-hydroxy-2-oxopropionic acid + L-glutamate
S L-tryptophan + 2-oxoglutarate <3, 4, 12, 24, 26> (<26> 0.1% of the activity with aspartate [37]; <24> 0.08% of the activity with aspartate [35];
   <3> 11% of the activity with glutamate [43]) (Reversibility: ? <3, 4, 12,
   24, 26> [1, 2, 17, 35, 37, 43]) [1, 2, 17, 35, 37, 43]
P 3-indole-2-oxopropionic acid + L-glutamate
S L-tyrosine + 2-oxoglutarate <2-4, 9, 24, 27, 32> (<32> low activity [47];
   <3> 1.5 of the activity with glutamate [43]; <2> 1.2% of the activity with
   glutamate [14]; <24> 0.04% of the activity with aspartate [35]; <9> 84.6%
   of the activity with aspartate [15]) (Reversibility: ir <27> [40]; ? <2-4, 9,
   24, 32> [1, 2, 14, 15, 35, 43, 47]) [1, 2, 14, 15, 35, 40, 43, 47]
P 3-(4-hydroxyphenyl)-2-oxopropionic acid + L-glutamate <2-4, 9, 24, 27>
   (<2-4, 9, 24> i.e. 4-hydroxyphenylpyruvate [1, 2, 14, 15, 35, 43]) [1, 2, 14,
   15, 35, 40, 43]
S glycine + 2-oxoglutarate <9> (<9> 7.7% of the activity with aspartate
   [15]) (Reversibility: ? <9> [15]) [15]
P glyoxylate + L-glutamate
S Additional information <2-6, 10, 22, 23, 32, 33, 39, 43> (<43> poor substrates are: glyoxylic acid, pyruvate, succinate, maleate, L-asparagine, L-
   glutamine, L-alanine, glycine [70]; <43> no activity with L-2-amino-5-
   methoxy-5-oxopentanoic acid, fumarate, hydroxypyruvate, glutarate, L-
   proline, L-arginine, and L-lysine [70]; <39> indole-3-pyruvate, p-hydro-
   xy-phenylpyruvate, and 2-oxo-n-caproate are no substrates [62]; <4> no
   activity with L-cysteine, L-isoleucine, L-leucine, L-valine, L-glutamine,
   glycine, L-ornithine, L-lysine, L-arginine [2]; <2-6, 10, 22, 23, 32, 33, 39,
   43> substrate specificity [2-4, 6, 14, 34, 47, 48, 52, 62, 70]) [2-4, 6, 14, 34,
   47, 48, 52, 62, 70]
P ?

Inhibitors

2-methyl-DL-aspartate <37> (<37> binds to the pyridoxal 5'-phosphate form
of the enzyme, formation of an external aldimine complex [61]) [61]
2-oxoglutaconate <5, 12, 38> (<12> cytosolic and mitochondrial isozymes
[60]; <5, 12> 30% inhibition at 25°C, pH 7.2, 10 mM, after 2 h [60]; <5, 12>
2-oxoglutarate protects against inhibition, L-glutamate enhances the inhibitory effect [60]; <5> binds the active site of the pyridoxal 5'-phosphate en-

zyme form, cytosolic isozyme [60]; <38> injection into mice leads to inhibition of the kidney enzyme [60]) [60]

2-oxoglutaconic acid dimethyl ester <5> (<5> pyridoxal 5'-phosphate enzyme form, cytosolic isozyme [60]; <5> 15% inhibition at 25°C, pH 7.0, 10 mM, after 2 h [60]) [60]

2-oxoglutarate <4, 8, 12, 13, 18, 27, 33> (<12> competitive to cysteine sulfinic acid or aspartate [17]; <4,27> substrate inhibition at high concentration [2, 40]; <13> 1-3 mM, at pH 6.0, not at pH 8.0 [18]; <18> cytoplasmic enzyme inhibited above 0.25 mM, mitochondrial enzyme not [24]; <27> oxoglutarate production [40]; <33> competitive to L-aspartate [48]; <8> product and substrate inhibition [51]) [2, 17, 18, 24, 40, 48, 51]

CN⁻ <15, 32> (<15> cytosolic isoform, at 5 mM [20]) [20, 47]

Ca^{2+} <44> (<44> AST II [71]) [71]

Hg^{2+} <9> [15]

KCl <7> (<7> 1.3 M [49]) [49]

L-2-amino-5-methoxy-5-oxopentanoic acid <43> (<43> only isozyme AAT2, competitive against L-asparate [70]) [70]

L-ascorbate <32> [47]

L-aspartate <12> (<12> weak, competitive against 2-oxoglutarate [17]) [17]

L-cycloserine <15> (<15> cytosolic isozyme, weak inhibition only after 24 h incubation, 1 mM [20]) [20]

L-cysteic acid <43> (<43> competitive [70]) [70]

L-cysteine sulfinic acid <12> (<12> weak, competitive to 2-oxoglutarate [17]) [17]

L-glutamate <4, 8, 27, 43> (<43> isozyme AAT1 and AAT2, product inhibition [73]; <4> competitive [2]; <27> forward reaction, substrate L-aspartate [40]; <8> product inhibition, noncompetitive to 2-oxoglutarate and competitive against aspartate [51]) [2, 40, 51, 73]

L-histidine <24> [35]

L-serine O-sulfate <12> (<12> inhibition of cytosolic enzyme, no inhibition of mitochondrial enzyme [17]) [17]

Mg^{2+} <44> (<44> AST II [71]) [71]

N-5'-phosphopyridoxyl L-aspartate <5> (<5> cofactor analogue binds covalently to the enzyme [27]) [27]

N-ethylmaleimide <5> (<5> alkylation of cysteine residues [60]) [60]

NaCl <7> (<7> 1.3 M [49]) [49]

Ni^{2+} <44> (<44> AST II [71]) [71]

Zn^{2+} <44> (<44> AST II [71]) [71]

adipate <33> (<4> no inhibition [2]; <33> moderately [48]) [48]

aminoguanidine <12> (<12> competes with the enzyme for pyridoxal 5'-phosphate, forms complexes with pyridoxal 5'-phosphate, 70% enzyme inhibition at 1 mM [56]) [56]

chymoptrypsin <41> (<41> cytosolic isozyme at 1 mg/ml, no inhibition of mitochondrial isozyme [66]) [66]

fumarate <33, 43> (<43> competitive [70]; <33> weak [48]) [48, 70]

γ-acetylenic GABA <12> [17]

glutarate <33, 43> (<43> isozyme AAT1: competitive [70]; <33> weak [48])
[48, 70]
hadacidin <28> (<28> i.e. N-formylhydroxyaminoacetic acid [41]) [41]
isonicotinic acid hydrazine <14, 15> (<14> complete inhibition at 0.3 mM,
partially reversible by pyridoxyl 5'-phosphate [20]; <15> cytosolic isozyme,
only after 24 h incubation, 1 mM [20]) [20]
malate <27, 28, 43> (<43> competitive [70]; <27> isozymes Asp-DEAE-1 and
Asp-DEAE-2 [40]; <28> oxoglutarate production, oxaloacetate production
[41]) [40, 41, 70]
maleate <33, 36, 37, 43> (<37> binds noncovalently to the pyridoxal 5'-phos-
phate form of the enzyme [61]; <36> 78% inhibition of cytosolic isozyme at
4 mM, competitive to 2-oxoglutarate [57]) [48, 57, 61, 70]
methylacetimidate <32> (<32> complete inhibition at 50 mM [47]) [47]
oxaloacetate <8, 27, 43> (<43> isozyme AAT1 and AAT2, product inhibition
[73]; <8> isozymes AAT-P1 and AAT-P2 [51]; <8,27> substrate inhibition
[40,51]) [40, 51, 73]
p-chloromercuribenzoate <14, 15> (<14> complete inhibition at 0.15 mM
[20]; <15> cytosolic isozyme, 50% inhibition after 10 min at 30°C, 1 mM
[20]) [20]
p-hydroxymercuribenzoate <32> [47]
phosphate <18> (<18> inhibition of cofactor binding to the apoenzyme, cytosolic and mitochondrial isozymes [24]) [24]
potassium phosphate <18> [24]
proteinase K <41> (<41> both cytosolic and mitochondrial isozymes [66])
[66]
sodium mersalyl <15> (<15> cytosolic isozyme, 40% inhibition after 60 min
at 5 mM and 30°C [20]) [20]
subtilisin <41> (<41> only cytosolic isozyme [66]) [66]
succinate <9, 27, 43> (<43> competitive [70]; <27> forward reaction [40])
[15, 40, 70]
thiosemicarbazide <32> [47]
trypsin <41> (<41> weak inhibition of mitochondrial isozyme [66]) [66]
Additional information <12, 13, 15, 32, 40> (<40> no inhibition by Cd^{2+},
Pb^{2+} and Cu^{2+} [65]; <15> cytosolic isozyme, no inhibition by L-cysteine
and EDTA [20]; <12> no inhibition by dipropylacetic acid, vinyl-GABA [17];
<13> no inhibition by 4-aminobutyric acid [18]; <32> no substrate inhibi-
tion [47]) [17, 18, 20, 47, 65]

Cofactors/prosthetic groups
pyridoxal 5'-phosphate <1, 3, 5, 6, 7, 10, 12-14, 18, 19, 22, 23, 26, 32, 37-39,
42> (<1, 3, 5, 6, 7, 10, 12, 13, 18, 19, 22, 23, 32, 37-39> a pyridoxal 5'-phos-
phate protein [1, 3, 4, 6, 22, 24, 25, 28, 29, 34, 39, 47, 49, 52, 53, 56, 59, 60, 62,
74]; <37, 39> enzyme-bound [59, 62]; <1,5> bound to the active site [53, 60];
<18> 2.1mol per mol of enzyme [39]; <3, 7, 32> 2 mol of pyridoxal phos-
phate per mol of enzyme [3, 47, 49]; <18> 2.1 mol of pyridoxal phosphate per
mol of enzyme [39]; <10> 1.7-1.9 mol per mol of enzyme [52]; <19, 26> 1
mol of pyridoxal phosphate per mol of subunit [25, 37]; <6> isozyme AAT-1:

0.87 mol per mol of subunit, isozyme AAT-2: 0.93 mol per mol of subunit [6];
<5> dependent on [4]) [1, 3, 4, 6, 11, 12, 18, 20, 22, 24, 25, 28, 29, 34, 37, 39,
44, 47, 49, 52, 53, 56, 59, 60, 62, 69, 74]
pyridoxamine 5'-phosphate <1, 3, 19> (<1,3,19> reverse reaction [25,28,29];
<19> 1 mol per mol of subunit [25]) [25, 28, 29]

Activating compounds

Additional information <12, 45> (<12,45> cytosolic Hsp70, from bovine
brain, protein exclusively binds to the mitochondrial isozyme, regulatory
function [74]; <12> intramitochondrial chaperone homologues GroEL and
GroES can facilitate the folding of nascent premature mitochondrial isoform
pmAspAT in rabbit reticulocyte lysate under conditions where it otherwise
would not, GroEL alone inhibits the import into mitochondria, which is reversed by GroES [68]) [68, 74]

Metals, ions

Mn^{2+} <44> (<44> activation, isozyme AST II [71]) [71]
Additional information <15> (<15> cytosolic isozyme, no metal ion requirement [20]) [20]

Turnover number (min^{-1})

240 <1> (2-oxoglutarate, <1> recombinant deletion mutant, pH 7.5, 25°C
[63]) [63]
240 <1> (L-aspartate, <1> recombinant deletion mutant, pH 7.5, 25°C [63])
[63]
5340 <1> (2-oxoglutarate, <1> recombinant C166S mutant, pH 7.5, 25°C
[63]) [63]
5340 <1> (L-aspartate, <1> recombinant C166S mutant, pH 7.5, 25°C [63])
[63]
5400 <1> (2-oxoglutarate, <1> recombinant premature isozyme pmAspAT,
pH 7.5, 25°C [63]) [63]
5400 <1> (L-aspartate, <1> recombinant premature isozyme pmAspAT, pH
7.5, 25°C [63]) [63]
5760 <1> (2-oxoglutarate, <1> recombinant C166A mutant, pH 7.5, 25°C
[63]) [63]
5760 <1> (L-aspartate, <1> recombinant C166A mutant, pH 7.5, 25°C [63])
[63]
6060 <39> (L-glutamate, <39> pH 8.0, 45°C [62]) [62]
7200 <37> (2-oxoglutarate, <37> wild-type, pH 8.0, 25°C [61]) [61]
7200 <37> (L-aspartate, <37> wild-type, pH 8.0, 25°C [61]) [61]
10560 <30> (2-oxoglutarate, <30> isozyme AAT5, pH 8.0, 25°C [64]) [64]
11220 <1> (2-oxoglutarate, <1> recombinant isozyme mAspAT, pH 7.5, 25°C
[63]) [63]
11220 <1, 39> (L-aspartate, <1> recombinant isozyme mAspAT, pH 7.5, 25°C
[63]; <39> pH 8.0, 45°C [62]) [62, 63]
12300 <30> (2-oxoglutarate, <30> isozyme AAT1, pH 8.0, 25°C [64]) [64]
13020 <30> (2-oxoglutarate, <30> isozyme AAT2, pH 8.0, 25°C [64]) [64]
15540 <3> (L-aspartate, <3> wild-type, pH 8.4, 25°C [72]) [72]

16740 <30> (oxaloacetate, <30> isozyme AAT5, pH 8.0, 25°C [64]) [64]
19140 <30> (oxaloacetate, <30> isozyme AAT1, pH 8.0, 25°C [64]) [64]
19310 <18> (2-oxoglutarate, <18> pH 7.4, 25°C [39]) [39]
19310 <18> (L-aspartate, <18> pH 7.4, 25°C [39]) [39]
34440 <30> (oxaloacetate, <30> isozyme AAT2, pH 8.0, 25°C [64]) [64]
Additional information <3> (<3> mutants [72]) [72]

Specific activity (U/mg)
0.061 <44> (<44> partially purified isozyme AST II [71]) [71]
0.11 <39> (<39> cell extract [62]) [62]
0.24 <23> [34]
0.576 <15> (<15> partially purified cytosolic isozyme [20]) [20]
1.53 <7> (<7> purified enzyme [49]) [49]
2.4 <22> [34]
2.5 <23> (<23> purified enzyme [34]) [34]
8.44 <33> (<33> purified enzyme [48]) [48]
12.92 <42> (<42> recombinant enzyme in Escherichia coli cells [69]) [69]
18.8 <1> (<1> purified mitochondrial isozyme from liver [53]) [53]
21.8 <21> (<21> purified enzyme [33]) [33]
22.6 <1> (<1> purified mitochondrial isozyme from heart [53]) [53]
45.15 <2> (<2> purified enzyme [14]) [14]
69.4 <14> (<14> partially purified enzyme [20]) [20]
74 <1> (<1> purified cytosolic isozyme from liver [53]) [53]
83.5 <1> (<1> purified cytosolic isozyme from heart [53]) [53]
84 <19, 32> (<19,32> purified enzyme [25,47]) [25, 47]
100 <27> (<27> purified cytosolic isozyme [40]) [40]
105 <8, 25> (<8> partially purified enzyme [51]; <25> purified enzyme [36])
[36, 51]
108 <3> (<3> mutant V39L [31]) [31]
112.1 <37> (<37> purified recombinant enzyme [59]) [59]
120 <34> [50]
130 <31> (<31> purified cytosolic isozyme AAT-2 form 2a [46]) [46]
136 <13> (<13> partially purified enzyme [18]) [18]
137.7 <42> (<42> purified recombinant enzyme [69]) [69]
140 <31> (<31> purified cytosolic isozyme AAT-2 form 2c [46]) [46]
150 <1> (<1> purified, mitochondrial isozyme [12]) [12]
151.5 <18> (<18> mitochondrial isozyme [24]) [24]
156 <20> (<20> mitochondrial fraction [26]) [26]
165 <17, 30> (<30> purified recombinant isozyme AAT5 [64]; <17> synthesis of aspartate [23]) [23, 64]
170 <5> (<5> purified enzyme [38]) [38]
172 <4> (<4> purified enzyme [2]) [2]
178 <35> (<35> purified pAspAT, γ-form [55]) [55]
182 <17> (<17> forward reaction [23]) [23]
200 <3> (<3> purified enzyme [3]) [3]
202 <12> (<12> purified mitochondrial isozyme [17]) [17]
210 <3> (<3> purified apoenzyme [3,43]) [3, 43]

215 <12> (<12> purified cytosolic isozyme [17]) [17]
217 <18> (<18> purified enzyme [39]) [39]
220 <26> [37]
221 <39> (<39> purified enzyme [62]) [62]
228 <27> (<27> purified mitochondrial isozyme [40]) [40]
231 <11> (<11> purified mitochondrial isozyme [16]) [16]
232 <3> (<3> purified recombinant enyme [11]) [11]
240 <31> (<31> purified cytosolic isozyme AAT-2 form 2b [46]) [46]
252 <35> (<35> purified pAspAT, α-form [55]) [55]
292.3 <23> [34]
322 <35> (<35> purified pAspAT, β-form [55]) [55]
341 <20> (<20> supernatant fraction [26]) [26]
390 <6> (<6> purified isozyme AAT-1 [6]) [6]
403.4 <22> (<22> purified enzyme [34]) [34]
443 <11> (<11> purified cytosolic isozyme [16]) [16]
450 <6> (<6> purified isozyme AAT-2 [6]) [6]
454 <30> (<30> recombinant purified isozyme AAT1 [64]) [64]
512 <35> (<35> purified mAspAT [55]) [55]
546.5 <16> (<16> purified isozyme B [21]) [21]
671 <35> (<35> purified cAspAT, α-form [55]) [55]
805.6 <16> (<16> purified isozyme A [21]) [21]
953 <35> (<35> purified cAspAT, γ-form [55]) [55]
1035 <35> (<35> purified cAspAT, β-form [55]) [55]
2934 <30> (<30> recombinant purified isozyme AAT2 [64]) [64]
Additional information <2, 3, 6, 7, 13-15, 18, 30, 36, 41-43> (<30> activity of recombinant isozymes from different E. coli host strains [64]) [3, 6, 8, 14, 18, 20, 39, 43, 57, 64, 66, 67, 69, 70]

K_m-Value (mM)

0.00025 <3> (pyridoxyl 5'-phosphate, <3> pH 7.6, 25°C [1]) [1]
0.007 <1> (2-oxoglutarate, <1> recombinant deletion mutant, pH 7.5, 25°C [63]) [63]
0.01 <3> (oxaloacetate, <3> 25°C, pH 8.0 [3]) [3]
0.02 <8> (oxaloacetate, <8> isozyme AAT-P2, pH 7.6, 25°C [51]) [51]
0.023 <35> (oxaloacetate, <35> isozyme pAspAT, γ-form, pH 8.0, 25°C [55]) [55]
0.027 <31> (oxaloacetate, <31> isozyme AAT-2, including forms 2a, 2b and 2c, pH 8.0 [46]) [46]
0.03 <35> (oxaloacetate, <35> isozyme pAspAT, α-form, pH 8.0, 25°C [55]) [55]
0.031 <31> (oxaloacetate, <31> isozyme AAT-1, pH 8.0 [46]) [46]
0.032 <39> (oxaloacetate, <39> pH 8.0, 45°C [62]) [62]
0.033 <35> (oxaloacetate, <35> isozyme pAspAT, β-form, pH 8.0, 25°C [55]) [55]
0.043 <6> (oxaloacetate, <6> isoenzyme AAT-1 [6]) [6]
0.044 <32> (oxaloacetate, <32> pH 7.8, 25°C [47]) [47]
0.045 <28> (oxaloacetate, <28> isozyme Asp-DEAE-2, pH 8.0 [41]) [41]

0.048 <6> (oxaloacetate, <6> isoenzyme AAT-2 [6]) [6]

0.049 <17> (oxaloacetate, <17> pH 8.0, 25°C [23]) [23]

0.056 <35> (oxaloacetate, <35> isozyme mAspAT, pH 8.0, 25°C [55]) [55]

0.062 <4> (2-oxoglutarate, <4> 30°C [2]) [2]

0.065 <35> (oxaloacetate, <35> isozyme cAspAT, α-form, pH 8.0, 25°C [55]) [55]

0.07 <35> (2-oxoglutarate, <35> isozyme pAspAT, γ-form, pH 8.0, 25°C [55]) [55]

0.071 <3> (2-oxoglutarate, <3> pH 7.6, 25°C [1]) [1]

0.075 <35> (2-oxoglutarate, <35> isozyme pAspAT, β-form, pH 8.0, 25°C [55]) [55]

0.08 <1, 28> (2-oxoglutarate, <1> recombinant mutant C166S, pH 7.5, 25°C [63]; <28> isozyme Asp-DEAE-1, pH 8.0 [41]) [41, 63]

0.08 <35> (oxaloacetate, <35> isozyme cAspAT, γ-form, pH 8.0, 25°C [55]) [55]

0.085 <35> (oxaloacetate, <35> isozyme cAspAT, β-form, pH 8.0, 25°C [55]) [55]

0.087 <35> (2-oxoglutarate, <35> isozyme pAspAT, α-form, pH 8.0, 25°C [55]) [55]

0.088-0.095 <1> (2-oxoglutarate, <1> pH 7.4, 30°C [44]) [44]

0.09 <1> (L-aspartate, <1> recombinant deletion mutant, pH 7.5, 25°C [63]) [63]

0.093 <18> (2-oxoglutarate, <18> cytosolic isozyme, pH 7.4, 25°C [24]) [24]

0.098 <18> (2-oxoglutarate, <18> mitochondrial isozyme, pH 7.4, 25°C [24]) [24]

0.1 <8, 28> (oxaloacetate, <8> isozyme AAT-P1, pH 7.6, 25°C [51]; <28> isozyme Asp-DEAE-1, pH 8.0 [41]) [41, 51]

0.105 <6> (2-oxoglutarate, <6> isoenzyme AAT-2, pH 7.8, 37°C [6]) [6]

0.11 <6, 17> (2-oxoglutarate, <17> pH 8.3, 25°C [23]; <6> isoenzyme AAT-1, pH 7.8, 37°C [6]) [6, 23]

0.12 <28> (2-oxoglutarate, <28> isozyme Asp-DEAE-2, pH 8.0 [41]) [41]

0.13 <35> (2-oxoglutarate, <35> isozyme mAspAT, pH 8.0, 25°C [55]) [55]

0.14 <35> (2-oxoglutarate, <35> isozyme cAspAT, β-form, pH 8.0, 25°C [55]) [55]

0.17 <13, 35> (2-oxoglutarate, <35> isozyme cAspAT, α-form, pH 8.0, 25°C [55]; <13> pH 7.4, 25°C [18]) [18, 55]

0.2 <8, 25, 35, 39> (2-oxoglutarate, <39> pH 8.0, 45°C [62]; <35> isozyme cAspAT, γ-form, pH 8.0, 25°C [55]; <8> isozyme AAT-P2, pH 7.6, 25°C [51]; <25> isozyme AspAT-3, 25°C, pH 8.0 [36]) [36, 51, 55, 62]

0.2 <25> (oxaloacetate, <25> isozyme AspAT-1, 25°C, pH 8.0 [36]) [36]

0.23 <25> (2-oxoglutarate, <25> isozyme AspAT-1, 25°C, pH 8.0 [36]) [36]

0.24 <3> (2-oxoglutarate, <3> 25°C, pH 8.0 [3]) [3]

0.25 <1> (L-aspartate, <1> recombinant premature isozyme pmAspAT, pH 7.5, 25°C [63]) [63]

0.26 <8> (2-oxoglutarate, <8> isozyme AAT-P1, pH 7.6, 25°C [51]) [51]

0.29 <1> (L-aspartate, <1> recombinant mutant C166A, pH 7.5, 25°C [63]) [63]

0.3 <1, 11> (2-oxoglutarate, <1> recombinant isozyme mAspAT, pH 7.5, 25°C [63]; <11> cytosolic isozyme, pH 8.0, 37°C [16]) [16, 63]

0.32 <25> (oxaloacetate, <25> isozyme AspAT-3, 25°C, pH 8.0 [36]) [36]

0.33 <1> (2-oxoglutarate, <1> recombinant premature isozyme pmAspAT, pH 7.5, 25°C [63]) [63]

0.33 <1> (L-aspartate, <1> recombinant mutant C166S, pH 7.5, 25°C [63]) [63]

0.36 <1> (L-aspartate, <1> recombinant isozyme mAspAT, pH 7.5, 25°C [63]) [63]

0.37 <3> (oxaloacetate, <3> pH 7.6, 25°C [1]) [1]

0.39 <31> (2-oxoglutarate, <31> isozyme AAT-2, including forms 2a, 2b and 2c, pH 8.0 [46]) [46]

0.4 <18> (L-aspartate, <18> mitochondrial isozyme, pH 7.4, 25°C [24]) [24]

0.5 <28> (L-aspartate, <28> isozyme Asp-DEAE-2, pH 8.0 [41]) [41]

0.51 <20> (L-aspartate, <20> mitochondrial isozyme [26]) [26]

0.55 <31, 32> (2-oxoglutarate, <31> isozyme AAT-1, pH 8.0 [46]; <32> pH 7.8, 25°C [47]) [46, 47]

0.59 <3> (2-oxoglutarate, <3> wild-type, pH 8.4, 25°C [72]) [72]

0.67 <44> (2-oxoglutarate, <44> isozyme AST II, 37°C, pH 7.5 [71]) [71]

0.75 <28> (L-glutamate, <28> isozyme Asp-DEAE-2, pH 8.0 [41]) [41]

0.8 <1> (2-oxoglutarate, <1> recombinant mutant $C_{16}6A$, pH 7.5, 25°C [63]) [63]

0.85 <28> (L-aspartate, <28> isozyme Asp-DEAE-1, pH 8.0 [41]) [41]

0.96 <4> (L-aspartate, <4> 30°C [2]) [2]

1.2 <31> (L-aspartate, <31> isozyme AAT-1, pH 8.0 [46]) [46]

1.3 <3, 35> (L-aspartate, <35> isozyme mAspAT, pH 8.0, 25°C [55]; <3> 25°C, pH 8.0 [3]) [3, 55]

1.43 <3> (L-tyrosine, <3> pH 7.6, 25°C [1]) [1]

1.5 <31> (L-aspartate, <31> isozyme AAT-2, including forms 2a, 2b and 2c, pH 8.0 [46]) [46]

1.6 <28> (L-glutamate, <28> isozyme Asp-DEAE-1, pH 8.0 [41]) [41]

1.7 <20> (2-oxoglutarate, <20> cytosolic isozyme [26]) [26]

1.7 <25, 37> (L-aspartate, <37> wild-type, pH 8.0, 25°C [61]; <25> isozyme AspAT-3, 25°C, pH 8.0 [36]) [36, 61]

1.9 <3, 11> (L-aspartate, <3> wild-type, pH 8.4, 25°C [72]; <11> mitochondrial isozyme, pH 8.0, 37°C [16]) [16, 72]

2 <13> (L-aspartate, <13> pH 7.4, 25°C [18]) [18]

2.06 <36> (2-oxoglutarate, <36> cytosolic isozyme, pH 6.8, 37°C [57]) [57]

2.1-2.9 <1> (L-aspartate, <1> pH 7.4, 30°C [44]) [44]

2.17 <3> (L-phenylalanine) [1]

2.2 <20> (2-oxoglutarate, <20> mitochondrial isozyme [26]) [26]

2.2 <8, 35> (L-aspartate, <35> isozyme pAsp-AT, β-form, pH 8.0, 25°C [55]; <8> isozyme AAT-P1, pH 7.6, 25°C [51]) [51, 55]

2.3 <17, 35> (L-aspartate, <35> isozyme pAsp-AT, γ-form, pH 8.0, 25°C [55]; <17> pH 8.3, 25°C [23]) [23, 55]

2.4 <37> (2-oxoglutarate, <37> wild-type, pH 8.0, 25°C [61]) [61]

2.4 <6> (L-aspartate, <6> isoenzyme AAT-1, pH 7.8, 37°C [6]) [6]

2.53 <32> (L-aspartate, <32> pH 7.8, 25°C [47]) [47]

2.6 <26> (2-oxoglutarate, <26> pH 8.0, 50°C [37]) [37]

2.6 <8, 35> (L-aspartate, <35> isozyme pAsp-AT and cAspAT, α-form, pH 8.0, 25°C [55]; <8> isozyme AAT-P2, pH 7.6, 25°C [51]) [51, 55]

3 <18, 26, 35> (L-aspartate, <18> cytosolic isozyme, pH 7.4, 25°C [24]; <35> isozyme cAspAT, γ-form, pH 8.0, 25°C [55]; <26> pH 8.0, 50°C [37]) [24, 37, 55]

3.2 <35> (L-aspartate, <35> isozyme cAspAT, β-form, pH 8.0, 25°C [55]) [55]

3.7 <6> (L-aspartate, <6> isoenzyme AAT-2, pH 7.8, 37°C [6]) [6]

3.88 <32> (L-glutamate, <32> pH 7.8, 25°C [47]) [47]

3.9 <11> (L-aspartate, <11> cytosolic isozyme, pH 8.0, 37°C [16]) [16]

4.1 <43> (L-2-amino-4-methoxy-4-oxobutanoic acid, <43> isozyme AAT2, 25°C [70]) [70]

4.4 <3> (L-aspartate) [1]

5 <39> (L-aspartate, <39> pH 8.0, 45°C [62]) [62]

5 <3, 6> (L-tryptophan, <3> pH 7.6, 25°C [1]; <6> L-glutamate, isoenzyme AAT-1 [6]) [1, 6]

5.7 <39> (L-glutamate, <39> pH 8.0, 45°C [62]) [62]

6.5 <20> (L-aspartate, <20> cytosolic isozyme [26]) [26]

6.9 <11> (2-oxoglutarate, <11> mitochondrial isozyme, pH 8.0, 37°C [16]) [16]

7.5 <25> (L-aspartate, <25> isozyme AspAT-1, 25°C, pH 8.0 [36]) [36]

7.5 <35> (L-glutamate, <35> isozyme mAspAT, pH 8.0, 25°C [55]) [55]

12 <8> (L-glutamate, <8> isozyme AAT-P2, pH 7.6, 25°C [51]) [51]

12.2 <25> (L-glutamate, <25> isozyme AspAT-3, 25°C, pH 8.0 [36]) [36]

13 <17, 35> (L-glutamate, <35> isozyme cAspAT, γ-form, pH 8.0, 25°C [55]; <17> pH 8.0, 25°C [23]) [23, 55]

13.2 <6> (L-glutamate, <6> isoenzyme AAT-2 [6]) [6]

14 <35> (L-glutamate, <35> isozyme cAspAT, β-form, pH 8.0, 25°C [55]) [55]

15 <3> (L-glutamate, <3> 25°C, pH 8.0 [3]) [3]

15.1 <44> (L-aspartate, <44> isozyme AST II, 37°C, pH 7.5 [71]) [71]

17 <35> (L-glutamate, <35> isozyme cAspAT, α-form, pH 8.0, 25°C [55]) [55]

17.4 <25> (L-glutamate, <25> isozyme AspAT-1, 25°C, pH 8.0 [36]) [36]

18 <35> (L-glutamate, <35> isozyme pAspAT, γ-form, pH 8.0, 25°C [55]) [55]

18.5 <31> (L-glutamate, <31> isozyme AAT-1, pH 8.0 [46]) [46]

19.4 <31> (L-glutamate, <31> isozyme AAT-2, including forms 2a, 2b and 2c, pH 8.0 [46]) [46]

20 <35> (L-glutamate, <35> isozyme pAspAT, β-form, pH 8.0, 25°C [55]) [55]

20.3 <43> (L-2-amino-4-methoxy-4-oxobutanoic acid, <43> isozyme AAT1, 25°C [70]) [70]

22 <8> (L-glutamate, <8> isozyme AAT-P1, pH 7.6, 25°C [51]) [51]

22.5 <36> (L-aspartate, <36> cytosolic isozyme, pH 6.8, 37°C [57]) [57]

32 <35> (L-glutamate, <35> isozyme pAspAT, α-form, pH 8.0, 25°C [55]) [55]

Additional information <1, 3, 7, 10, 12, 14-16, 18, 27, 30, 34, 36, 37, 42, 43> (<3> mutants [72]; <42> recombinant enzyme, high temperature dependence of K_m [69]; <37> recombinant enzyme [59]; <3,37> wild-type and mutants [58, 61]; <1, 3, 18, 30, 36, 37> kinetics [24,53,57-59,61,64,72]) [17, 20, 21, 24, 35, 40, 49, 50, 52, 53, 57-59, 61, 64, 69, 72, 73]

K_i-Value (mM)

0.005 <43> (oxaloacetate, <43> isozyme AAT1, competitive versus 2-oxoglutarate [73]) [73]

0.0195 <18> (phosphate, <18> mitochondrial isozyme, pH 7.4, 37°C [24]) [24]

0.024 <43> (oxaloacetate, <43> isozyme AAT2, noncompetitive versus L-aspartate [73]) [73]

0.026 <43> (oxaloacetate, <43> isozyme AAT2, competitive versus 2-oxoglutarate [73]) [73]

0.03 <43> (oxaloacetate, <43> isozyme AAT1, noncompetitive versus L-aspartate [73]) [73]

0.032 <18> (phosphate, <18> cytosolic isozyme, pH 7.4, 37°C [24]) [24]

0.96 <12> (2-oxoglutarate, <12> cytosolic isozyme, pH 8.6, versus cysteinic acid [17]) [17]

1.21 <43> (maleate, <43> isozyme AAT2, versus 2-oxoglutarate, 25°C [70]) [70]

1.4 <43> (maleate, <43> isozyme AAT1, versus 2-oxoglutarate, 25°C [70]) [70]

1.72 <36> (maleate, <36> cytosolic isozyme, pH 6.8, 37°C [57]) [57]

1.9 <12> (2-oxoglutarate, <12> mitochondrial isozyme, pH 8.6, versus L-aspartate [17]) [17]

2 <5> (2-oxoglutaconate, <5> cytosolic isozyme, 25°C, pH 7.2 [60]) [60]

2.5 <12> (2-oxoglutarate, <12> mitochondrial isozyme, pH 8.6, versus cysteinic acid [17]) [17]

2.8 <4> (L-glutamate, <4> 30°C, competitive versus L-aspartate [2]) [2]

5.1 <12> (L-cysteinic acid, <12> cytosolic isozyme, pH 8.6, versus 2-oxoglutarate [17]) [17]

6.78 <43> (fumarate, <43> isozyme AAT2, 25°C, versus 2-oxoglutarate [70]) [70]

6.8 <4> (L-glutamate, <4> 30°C, competitive versus 2-oxoglutrate [2]) [2]

8.5 <43> (maleate, <43> isozyme AAT2, versus L-aspartate, 25°C [70]) [70]

8.86 <43> (succinate, <43> isozyme AAT2, 25°C, versus 2-oxoglutarate [70]) [70]

10.5 <43> (L-glutamate, <43> isozyme AAT1, competitive versus L-aspartate [73]) [73]

10.58 <43> (maleate, <43> AAT1, versus L-aspartate, 25°C [70]) [70]

10.66 <43> (succinate, <43> isozyme AAT1, 25°C, versus 2-oxoglutarate [70]) [70]

11.1 <43> (L-glutamate, <43> isozyme AAT1, noncompetitive versus 2-oxoglutarate [73]) [73]

15.05 <43> (glutarate, <43> isozyme AAT1, 25°C, versus 2-oxoglutarate [70]) [70]

16 <43> (L-glutamate, <43> isozyme AAT2, noncompetitive versus 2-oxoglutarate [73]) [73]

16.29 <43> (L-cysteinic acid, <43> isozyme AAT1, 25°C, versus 2-oxoglutarate [70]) [70]

16.84 <43> (succinate, <43> isozyme AAT2, 25°C, versus L-aspartate [70]) [70]

17 <12> (2-oxoglutarate, <12> cytosolic isozyme, pH 8.6, versus L-aspartate [17]) [17]

20.73 <43> (L-cysteinic acid, <43> isozyme AAT2, 25°C, versus 2-oxoglutarate [70]) [70]

21 <12> (L-aspartate, <12> cytosolic isozyme, pH 8.6, versus 2-oxoglutarate [17]) [17]

24.59 <43> (fumarate, <43> isozyme AAT1, 25°C, versus 2-oxoglutarate [70]) [70]

25.85 <43> (L-2-amino-5-methoxy-5-oxopentanoic acid, <43> isozyme AAT2, 25°C, versus 2-oxoglutarate [70]) [70]

26 <12> (L-cysteinic acid, <12> mitochondrial isozyme, pH 8.6, versus 2-oxoglutarate [17]) [17]

26.55 <43> (glutarate, <43> isozyme AAT1, 25°C, versus L-aspartate [70]) [70]

30 <4> (2-oxoglutarate, <4> 30°C, substrate inhibition [2]) [2]

30 <43> (L-glutamate, <43> isozyme AAT2, competitive versus L-aspartate [73]) [73]

33.72 <43> (fumarate, <43> isozyme AAT2, 25°C, versus L-aspartate [70]) [70]

36.89 <43> (succinate, <43> isozyme AAT1, 25°C, versus L-aspartate [70]) [70]

44.12 <43> (fumarate, <43> isozyme AAT1, 25°C, versus L-aspartate [70]) [70]

97.63 <43> (L-cysteinic acid, <43> isozyme AAT1, 25°C, versus L-aspartate [70]) [70]

150.8 <43> (L-cysteinic acid, <43> isozyme AAT2, 25°C, versus L-aspartate [70]) [70]

263 <12> (aspartate, <12> mitochondrial isozyme, pH 8.6, versus 2-oxoglutarate [17]) [17]

pH-Optimum

6.3-7.3 <27> (<27> mitochondrial enzyme [40]) [40]

6.5-8.5 <18> [24]

6.8 <36> (<36> cytosolic isozyme [57]) [57]

6.8-8.3 <28> (<28> formation of oxaloacetate, isoenzyme Asp-DEAE-1 [41]) [41]

7 <12, 45> (<12,45> assay at [74]) [74]

7-8 <6> (<6> forward reaction, isoenzymes AAT-1 and AAT-2 [6]) [6]

7-9.5 <8> [51]

7.2 <26> [37]

7.3-8.3 <27> (<27> cytosolic enzyme [40]) [40]

7.4 <1> (<1> assay at [44]) [44]

7.4-7.6 <1> [44]

7.5 <1, 15, 16, 34, 44> (<44> isozyme AST II [71]; <1> recombinant enzyme, assay at [63]; <16> isozymes A and B [21]; <15> cytosolic isozyme [20]) [20, 21, 50, 63, 70, 71]

7.5-8.5 <3, 43> (<43> isozyme AAT1 [73]) [3, 43, 73]

7.6 <18> [39]

7.6-7.9 <7> [49]

7.8 <7, 32> (<7> assay at [8]) [8, 47]

8 <1, 3, 11, 14, 17, 25, 28, 30, 31, 35, 37, 39, 41, 42> (<1, 3, 11, 30, 35, 37, 41, 42> assay at [16, 53, 55, 59, 64, 66, 67, 69]; <17> reverse reaction [23]; <25> formation of glutamate [36]; <3> formation of aspartate [1]; <28> oxaloacetate formation, isoenzyme Asp-DEAE-2 [41]; <31> isoenzyme AAT-1, isoenzyme AAT-2 [46]) [1, 16, 20, 23, 36, 41, 46, 53, 55, 59, 62, 64, 66, 67, 69]

8-8.3 <2> (<2> 2 reaction methods [14]) [14]

8-8.5 <6> (<6> reverse reaction, isoenzyme AAT-1 [6]) [6]

8-8.8 <6> (<6> reverse reaction, isoenzyme AAT-2 [6]) [6]

8-8.9 <8, 13> [18, 51]

8-9 <10> [52]

8.3 <17> (<17> forward reaction [23]) [23]

8.3-8.6 <2> [14]

8.5 <3, 19, 21, 25, 33> (<19> assay at [25]; <3> substrates L-phenylalanine or L-tryptophan [1]; <25> L-aspartate formation [36]) [1, 25, 33, 36, 48]

8.5-9.5 <43> (<43> isozyme AAT2 [73]) [73]

8.5-10 <12> (<12> cytosolic enzyme [17]) [17]

8.7 <12> (<12> mitochondrial enzyme [17]) [17]

9 <3, 9> (<3> substrate L-tyrosine [1]) [1, 15]

pH-Range

5-8.8 <39> [62]

5.5-9 <16> (<16> isozyme A [21]) [21]

5.5-9.5 <19> (<19> no definite pH optimum [25]) [25]

5.5-10 <16> (<16> isozyme B [21]) [21]

6.3-9 <15> (<15> about 60% of activity maximum at pH 6.3 and pH 9.0 [20]) [20]

6.4-7.2 <36> (<36> cytosolic isozyme [57]) [57]

6.5-8 <17> (<17> pH 6.5: 63% of activity maximum of oxaloacetate formation, 50% of activity maximum of aspartate formation, pH 8.0: activity maximum [23]) [23]

6.8-8.3 <27> (<27> pH 6.8: 76% of activity maximum, pH 7.3-8.3: activity maximum, supernatant enzyme [40]) [40]

7-9 <14> (<14> pH 7.0: about 45% of activity maximum, pH 9.0: about 40% of activity maximum [20]) [20]

7-11 <33> [48]

7.4-8.7 <12> (<12> pH 7.4: 95% of activity maximum, pH 8.7: activity maximum, mitochondrial enzyme [17]) [17]

7.4-10 <12> (<12> pH 7.4: 60% of activity maximum, pH 8.5-10.0: activity maximum, cytosolic enzyme [17]) [17]

Temperature optimum (°C)

25 <1, 3, 7, 13, 18, 25, 30, 35, 37, 41, 43> (<1, 3, 7, 13, 25, 30, 35, 37, 41, 43> assay at [8, 11, 12, 18, 36, 55, 59, 63, 64, 66, 67, 70]) [8, 11, 12, 18, 36, 39, 55, 59, 63, 64, 66, 67, 70]

30 <1, 2, 16, 34, 42> (<1,2,16,34> assay at [14, 21, 44, 50]) [14, 21, 44, 50, 69]

34 <42> (<42> optimal temperature of recombinant enzyme in Escherichia coli in vivo [69]) [69]

37 <11, 12, 32, 36, 44, 45> (<11, 12, 36, 44, 45> assay at [16, 57, 71, 74]) [16, 47, 57, 71, 74]

55 <7> (<7> 0.7 M KCl [8]) [8]

60 <19, 21, 33> (<19> assay at [25]) [25, 33, 48]

65 <7, 43> (<7> 3.5 M KCl [49]; <7> 1.6-3.4 M KCl [8]) [8, 49, 73]

70 <21, 26> (<21> strain FTF-INRA [33]) [33, 37]

80 <39> [62]

95 <19> (<19> above [25]) [25]

Temperature range (°C)

25-95 <19> [25]

35-63 <42> (<42> half-maximal activity at 35°C and 63°C [69]) [69]

40-80 <7> (<7> 40°C: about 40% of activity maximum, 80°C: about 65% of activity maximum [8]) [8]

45-85 <39> (<39> 44% of maximum activity at 45°C, 80% of maximum activity at 85°C [62]) [62]

4 Enzyme Structure

Molecular weight

52000 <44> (<44> isozyme AST II, gel filtration [71]) [71]

57000 <33> (<33> gel filtration, gradient PAGE [48]) [48]

64200-68200 <7> (<7> gel filtration, PAGE, sedimentation equilibrium [49]) [49]

80000 <2, 25, 31> (<2> meniscus depletion sedimentation equilibrium method [14]; <25> gel filtration [36]; <31> isozymes AAT-2a, 2b and 2c, gel filtration [46]) [14, 36, 46]

80000-84000 <3> (<3> gel filtration, meniscus depletion sedimentation equilibrium method [43]) [43]

82000 <3> (<3> PAGE [1]) [1]

84000 <3, 12> (<3> meniscus depletion sedimentation equilibrium method [3]; <12> cytosolic isozyme, gel filtration [17]) [3, 17]

85000 <39> (<39> gel filtration [62]) [62]

86000 <12, 18> (<12> mitochondrial isozyme, gel filtration [17]; <18> cytosolic isozyme, gel filtration [39]) [17, 39]

87100 <18> (<18> sedimentation equilibrium analysis [24]) [24]
88000 <6, 22, 43> (<43> isozyme AAT1, gel filtration [73]; <6> isozyme
AAT-1 [6]; <22> mitochondrial isozyme [34]; <6,22> gel filtration [6,34])
[6, 34, 73]
88900 <18> (<18> cytosolic isozyme, sedimentation equilibrium [39]) [39]
90000 <13> (<13> analytical ultracentrifugation [18]) [18]
90000-92000 <10> (<10> gel filtration, sedimentation equilibrium analysis
[52]) [52]
91000 <11> (<11> mitochondrial isozyme, sedimentation equilibrium analysis [16]) [16]
92000 <16, 43> (<43> isozyme AAT2, gel filtration [73]; <16> isozymes A
and B, PAGE, gel filtration [21]) [21, 73]
93150 <5> (<5> amino acid sequence [4]) [4]
94000 <6, 11> (<6> isozyme AAT-2, gel filtration [6]; <11> cytosolic isozyme, sedimentation equilibrium analysis [16]) [6, 16]
95000 <23> (<23> cytosol, gel filtration [34]) [34]
96000 <8> (<8> isozyme AAT-P2, gel filtration [51]) [51]
97000-98000 <26> (<26> gel filtration, sedimentation equilibrium [37]) [37]
100000 <4, 14, 15, 17, 22> (<4, 14, 15, 17, 22> gel filtration [2, 20, 23, 34]) [2,
20, 23, 34]
103000 <13> (<13> gel filtration [18]) [18]
105000 <8> (<8> gel filtration, isozyme AAT-P2 [51]) [51]
108000 <19> (<19> gel filtration [25]) [25]
111000 <29> (<29> native PAGE [42]) [42]
112000 <43> (<43> isozyme AAT1, native PAGE [73]) [73]
130000 <34> (<34> sedimentation equilibrium centrifugation [50]) [50]
136000 <43> (<43> isozyme AAT2, native PAGE [73]) [73]
138000 <32> (<32> gel filtration [47]) [47]
162000 <24> (<24> gel filtration [35]) [35]
180000 <21, 33> (<21,33> gel filtration [33]) [33]

Subunits
? <12, 30, 42> (<12> x * 43000, SDS-PAGE [54]; <42> x * 43776, recombinant
enzyme, DNA sequence determination and mass spectrometry [69]; <30> x *
44000-45000, isoymes AAT1, AAT2 and AAT5, SDS-PAGE [64]; <12> x *
47000, premature mitochondrial isozyme pmAspAT, SDS-PAGE [68]) [54, 64,
68, 69]
dimer <2-8, 10-13, 17, 19, 22, 23, 25, 26, 29, 31, 32, 39, 43> (<7> 2 * 32500,
SDS-PAGE [49]; <25> 2 * 40000, SDS-PAGE [36]; <31> 2 * 40000, isozymes
AAT-2a, 2b and 3c, SDS-PAGE [46]; <12> 2 * 41000, cytosolic isozyme, SDS-
PAGE [17]; <17,26,39> 2 * 42000, SDS-PAGE [23,37,62]; <3> 2 * 42000-45000,
SDS-PAGE [1,11]; <3> 2 * 43000, SDS-PAGE [3,43]; <12> 2 * 43400, mito-
chondrial isozyme, SDS-PAGE [17]; <4,22> 2 * 44000, mitochondria, SDS-
PAGE [2,34]; <6> 2 * 44000, isozyme AAT-1, SDS-PAGE [6]; <10> 2 * 45000,
SDS-PAGE [52]; <8> AAT-P2 und AAT-P2 [51]; <11> 2 * 45400, mitochon-
drial isozyme, SDS-PAGE [16]; <22,23> 2 * 46000, cytosol, SDS-PAGE [34];
<11> 2 * 46300, cytosolic isozyme [16]; <5> 2 * 46344, amino acid sequence

[4]; <13> 2 * 46500, SDS-PAGE [22]; <5> 2 * 46470, SDS-PAGE [38]; <6> 2 *
47000, AAT-2, SDS-PAGE [6]; <43> 2 * 52200, isozyme AAT1, SDS-PAGE [73];
<19> 2 * 53000, SDS-PAGE [25]; <43> 2 * 54800, isozyme AAT2, SDS-PAGE
[73]; <32> 2 * 65000, SDS-PAGE [47]) [1-4, 6, 11, 14, 16, 17, 22, 23, 25, 34, 36-
38, 42, 43, 46, 47, 49, 51, 52, 62, 73]
monomer <33, 44> (<33,44> 1 * 50000, SDS-PAGE [48,71]) [48, 71]
tetramer <21, 33> (<21,33> 4 * 43000, SDS-PAGE [33]) [33]
Additional information <42> (<42> structure modeling, subunit organization [69]) [69]

5 Isolation/Preparation/Mutation/Application

Source/tissue
3T3-L1 cell <12> (<12> proliferating fibroblasts, differentiation to adipocytes, plasma membrane [54]) [54]
adipocyte <12> (<12> differentiated from 3T3-L1 fibroblasts [54]) [54]
brain <12, 13> [17, 18]
bran <6> [6, 7]
bundle sheath <25, 28> (<28> isozyme Asp-DEAE-2 [41]; <25> AspAT-3
[36]) [36, 41]
cell suspension culture <29> [42]
digestive gland <40> [65]
fibroblast <12> (<12> 3T3-L1 cells, low activity [54]) [54]
foot <40> [65]
gill <40> [65]
heart <1, 5, 13, 20> [4, 5, 9, 12, 13, 19, 22, 26, 38, 53, 60, 63]
hypha <16> [21]
kidney <12, 38> [60]
leaf <6, 17, 25, 28, 30, 31, 35, 43> (<43> isozymes AAT1 and AAT2 [70,73];
<30> cytosolic, mitochondrial and amyloplastidic isozymes [64]; <35>
green, greening and etiolated: all isozymes, low content in etiolated leaves
[55]; <17> mesophyll [23]; <31> low activity [46]) [6, 23, 36, 41, 46, 55, 64,
70, 73]
liver <1, 11, 12, 18, 41, 45> (<12,45> premature and mature mitochondrial
isozyme [74]) [16, 24, 39, 44, 53, 54, 56, 66, 74]
mesocotyl <35> (<35> cytosolic and mitochondrial isozymes, low content
[55]) [55]
mesophyll <17, 25, 28> (<28> isozyme Asp-DEAE-1 [41]; <25> AspAT-1
[36]) [23, 36, 41]
muscle <27> [40]
mycelium <9, 16> [15, 21]
placenta <36> [57]
plasma <41> [66]
red muscle <41> [66]
root <6, 31, 35> (<35> cytosolic isozyme, low content [55]; <31> low activity
[46]) [6, 46, 55]

root nodule <8, 31> [46, 51]
seedling <34> [50]
soft body part <40> [65]
stem <6, 31> (<31> low activity [46]) [6, 46]
zygote <44> (<44> 3-days-old eggs [71]) [71]

Localization

amyloplast <30> (<30> isozyme AAT5 [64]) [64]
cell wall <16> (<16> isozymes A and B, major part [21]) [21]
chloroplast <34, 35> (<35> isozyme pAspAT, multiple forms [55]; <34> only
form I occurs in chloroplast enriched fraction [50]) [50, 55]
cytosol <1, 5, 6, 8, 11-16, 18, 20, 22, 23, 27, 29-31, 36, 41> (<30> isozyme
AAT2 [64]; <16> isozymes A and B [21]; <15> 2 isozymes [20]; <1, 13, 18,
31, 35> multiple catalytically active forms of the cytosolic enzyme [18, 39, 46,
53, 55]; <1, 5, 11, 12, 20, 22, 23, 27, 29, 36, 41> cytosolic isoenzyme [5, 16, 19,
26, 34, 40, 42, 53, 56, 57, 60, 63, 66]; <6> isoenzyme AAT-1 mainly in cytosol
[7]; <16> isoenzyme A [21]) [5, 7, 16-22, 24, 26, 28, 34, 39, 40, 42, 44, 46, 51,
53, 55-57, 60, 63, 64, 66]
membrane <6, 12> (<6> isozyme AAT-2, very low activity of isoenzyme
AAT-1 [7]) [7, 54]
mitochondrion <1, 5, 6, 11-13, 15, 16, 18, 20, 22, 23, 27, 29, 30, 34, 35, 41, 45>
(<45> mature isozyme mAAT [74]; <30> isozyme AAT1 [64]; <1,12> mature
mAspAT and premature pmAspAT forms of the isozyme [63, 68]; <1, 35>
multiple catalytically active forms of mitochondrial isozyme [53, 55]; <16>
isozyme B [21]; <1, 5, 11, 12, 15, 20, 22, 27, 29, 41> mitochondrial isoenzyme
[5, 10, 12, 16, 19, 20, 26, 27, 34, 40, 42, 53, 54, 60, 66, 74]; <34> form I and II
occurs in chloroplast-enriched fraction [50]; <6> isoenzyme AAT-2: major
isoenzyme in mitochondria and peroxisomes [7]; <29> form II and III in
poorly resolved plastid and mitochondrial fraction [42]) [5, 7, 9, 10, 12, 16,
17, 19-22, 24, 26, 27, 34, 38, 40, 42, 50, 53-55, 60, 63, 64, 66, 68, 74]
peroxisome <6> (<6> AAT-2: major isoenzyme in mitochondria and peroxisomes, very low activity of AAT-1 [7]) [7]
plasma membrane <12> [54]
plastid <29> (<29> form II and III in poorly resolved plastid and mitochondrial fraction [42]) [42]
Additional information <12, 45> (<12,45> cytosolic and mitochondrial isozymes bind differently to the 70 kDa heat shock protein Hsp70, which regu-
lates the folding and aggregation process of polypeptides, sequence and distribution of Hsp70-binding regions allow the classification of isozymes into a
phylogenetic tree, overview [74]) [74]

Purification

<1> (recombinant mAspAT, pmAspAT and mutants from Escherichia coli
[63]; cytosolic isozyme [53]; mitochondrial isozyme [12,53]) [12, 53, 63]
<2> [14]
<3> (recombinant wild-type and mutants from overexpressing strain MG204
[72]; mutant V39L [31]; isoenzymes A and B [3]; recombinantly overexpressed enzyme [11,30]) [1, 3, 11, 30, 31, 43, 72]

<4> (200-1000fold, copurification with the aromatic amino acid aminotrans-
ferase 2.6.1.57 [2]) [2]
<5> (large-scale [38]) [38]
<6> (2 isoenzymes: AAT-1, AAT-2 [6]) [6]
<7> [8, 49]
<8> (isoenzymes: AAT-P1 and AAT-P2, partial [51]) [51]
<9> (partial [15]) [15]
<10> [52]
<11> (mitochondrial and cytosolic isozymes [16]) [16]
<12> (mitochondrial isozyme [54]; mitochondrial and cytosolic isozymes
[17]) [17, 54]
<13> (380fold [18]) [18, 22]
<14> (partial, 92fold [20]) [20]
<15> (1 cytosolic isozyme, partial, 120fold [20]) [20]
<16> (2 forms: A and B [21]) [21]
<17> (2 isozymes [23]) [23]
<18> (mitochondrial isozyme, 57.3fold [24]; cytosolic isozyme [39]) [24, 39]
<19> (840fold [25]) [25]
<20> (partial, cytosolic, 171fold, and mitochondrial, 78fold, isozymes [26])
[26]
<21> [33]
<22> (2 isozymes [34]) [34]
<23> (mitochondrial isoform [34]) [34]
<24> (partial [35]) [35]
<25> (isoenzymes: AspAT-1, AspAT-2, AspAT-3 [36]) [36]
<26> [37]
<27> (cytosolic and mitochondrial isozymes [40]) [40]
<28> (isoenzymes: Asp-DEAE-1 and Asp-DEAE-2 [41]) [41]
<29> (cytosolic form I, and mitochondrial forms II and III [42]) [42]
<30> (recombinant isozymes AAT1, AAT2 and AAT5 from Escherichia coli
[64]) [45, 64]
<31> (cytosolic isozyme: multiple forms AAT-2a, AAT-2b, AAT-2c [46]) [46]
<32> (400fold [47]) [47]
<33> [48]
<34> (isoenzyme I and II [50]) [50]
<35> (isozymes mAsp-AT, cAsp-AT, and pAspAT from young green leaves
[55]) [55]
<36> (partial, 404fold, cytosolic isozyme [57]) [57]
<37> (recombinant from overproducing Escherichia coli AB1157 cells, 66fold
[59]) [59]
<39> (2450fold [62]) [62]
<42> (partial purification of native enzyme, purification of recombinant en-
zyme from Escherichia coli [69]) [69]
<43> (isozymes AAT1 and AAT2 [70,73]) [70, 73]
<44> (isozyme AST II, 30.25fold [71]) [71]

Renaturation

<3> (wild-type and mutants, unfolding by guanidinium hydrochloride and renaturing kinetics, fluorescent spectrometric analysis [67]; reversible dissociation and unfolding of the dimeric enzyme by guanidine hydrochloride [30]) [30]

<5> (reconstitution of holoenzyme after purification of apoenzyme with the inhibitor N-5'-phosphopyridoxyl L-aspartate [27]) [27]

<7> (reconstitution of holoenzyme after purification of apoenzyme with pyridoxal 5'-phosphate [8]; thermal denaturation is not reversible [8,49]) [8, 49]

<12> (guanidinium hydrochloride denatured premature form pmAspAT cannot refold at 30°C, but refolds rapidly in presence of the intramitochondrial chaperone homologues GroEL and GroES [68]) [68]

<19> (reconstitution of holoenzyme with pyridoxal 5'-phosphate or pyridoxamine 5'-phosphate [25]) [25]

Crystallization

<1> (purified cytosolic enzyme, variation of crystallization conditions resulting in different crystal types, influence of various divalent metal ions, dioxane and non-ionic detergent β-octylglucoside, structure analysis [28]; vapour diffusion method with hanging drops, protein solution: 10 mg/ml, 50mM sodium phosphate, pH 7.5, 8-24% polyethylene glycol 2000-20000, reservoir solution: 8-24% polyethylene glycol 2000-20000, equal amounts, room temperature, X-ray structure analysis [12]) [12, 28]

<2> (concentrated protein solution, potassium phosphate 0,01 M, pH 7.2, 0.01 mM pyridoxal 5'-phosphate, ammonium sulfate [14]) [14]

<3> (hanging drop vapour diffusion, crystals of mutants P195A and P138A/ P195A in complex with maleate, protein solution: 20 mg/ml, 10 mM potassium phosphate, 0.01 mM pyridoxal 5'-phosphate, 2 mM EDTA, 0.5 mM DTT, reservoir solution: 1.9-2.1 M ammonium sulfate, 0.2 M 2-methylmorpholine, 2% polyethylene glycol 400, mutant P138A/P195A is also crystallized in complex with maleate, 20 mM, X-ray diffraction structure analysis [67]; vapour diffusion technique, protein solution: 10 mg/ml, 10 mM potassium phosphate,pH 7.0, 2.5 mM pyridoxal 5'-phosphate, 2.5% polyethylene glycol w/v, 4°C, 7-10 days [43]; purified mutant V39L, hanging drop method, room temperature, protein solution: 12.5 mg/ml, 20 mM sodium phosphate, pH 7.5, 30 mM maleate, 0.4 M sodium chloride, 12.5% w/v polyethylene glycol 4000, reservoir solution: 23-25% w/v polyethylene glycol 4000, 0.8 M sodium chloride, 20 mM sodium phosphate, pH 7.5, X-ray diffraction structure determination [31]; purified mutant K258A, batch method, protein solution: 5.2 mg/ml, 50 mM potassium phosphate, pH 7.5, 53% ammonium sulfate, room temperature, 1 week, preparation of crystals with heavy atom derivatives by soaking crystals for 24 or 48 h, structure analysis [29]; sitting drop method, protein solution: 30 mg/ml, 10 mM potassium phosphate, pH 7.0, 0.01 mM pyridoxal 5'-phosphate, 1 mM EDTA, 2 mM succinate, 0.3 mM azide, reservoir solution: same composition as protein solution, contains no protein but polyethylene glycol 4000, 11-17% w/v, 20°C, stepwise addition of protein solution, X-ray diffraction analysis [11]) [11, 29, 31, 43, 67]

<5> (forced dialysis lead to crystal formation or pure protein, 8.5 mg/ml, in solution for microdiffusion, 53% $(NH_4)_2SO_4$ and pH 6.0-7.0, 4°C, 2 months [38]; crystals of enzyme-inhibitor complex, sitting drop vapour diffusion method, 4°C, protein solution: 7 mg/ml, polyethylene glycol 4000 11% w/v, 50 mM potassium phosphate, pH 7.5, reservoir solution: polyethylene glycol 4000 22% w/v, 50 mM potassium phosphate, pH 7.5, X-ray structure analysis [27]; vapour diffusion method with hanging drops, crystals from solution: 10 mg/ml protein, polyethylene glycol 4000 16-17% w/v, 10 mM sodium phosphate, pH 7.5, injection of seed crystal into the hanging drop of chicken mitochondrial aspartate aminotransferase [10]; vapour diffusion method with hanging drops, crystals from solution: polyethylene glycol 4000 17% w/v, 10 mM glycine/NaOH, pH 9.1, microspectrophotometric equilibrium and kinetic analysis [9]; crystals of cytosolic enzyme, or enzyme in complex with substrates, N-methylpyridoxal, and inhibitors, depression-plate and sandwich-box techniques, protein solution: 35-40 mg/ml, pH 5.4, 40 mM sodium acetate, 4% polyethylene glycol 4000-6000, with or without substrate, cofactor or inhibitor compounds, reservoir solution: 40 mM sodium acetate, pH 5.4, 8% polyethylene glycol 4000-6000, crystals appear within 1 week, spectroscopic analysis [5]) [5, 9, 10, 13, 27, 38]

<6> (concentrated isoenzyme AAT-2 solution, 10 mM potassium phosphate, pH 7.0, 1 mM 2-oxoglutarate, 2 mM KOH, 0.001 mM pyridoxal 5'-phosphate, 0.01% 2-mercaptoethanol, dialysis against distilled water at 4°C for 3 days [6]; concentrated isoenzyme AAT-1 solution, 10 mM Tris-HCl, pH 8.0, 1 mM 2-oxoglutarate, 2 mM KOH, 0.01 pyridoxal 5'-phosphate, 0.01% 2-mercaptoethanol, polyethylene glycol 6000 8% w/v, 4°C, 6 days [6]) [6]

<10> [52]

<13> (vapour diffusion method, 10 mM NaOH-glycine, pH 9.1, polyethylene glycol 4000 18% w/v, 1 week, X-ray analysis [22]) [22]

<20> [26]

Cloning

<1> (expression of wild-type mAspAT and pmAspAT and mutants in Escherichia coli [63]) [63]

<3> (overexpression of wild-type enzyme and mutants from plasmid in Escherichia coli strain MG204, which lacks endogenous enzyme activity [72]; overexpression from plasmid in strain JM103 [11]) [11, 32, 72]

<12> (cloning of a liver cDNA fragment, nucleotides 1-419, into a plasmid for production of cRNA probes [54]) [54]

<30> (AAT1, AAT2, and AAT5, overexpression in Escherichia coli strains [64]) [64]

<35> (cloning and DNA-sequencing of cDNA encoding the plastidic isozyme pAspAT, DNA and amino acid sequence analysis, [55]) [55]

<37> (gene aspC, expression of wild-type and K109 mutants in Escherichia coli BL21(DE3) [61]; cloning, DNA and amino acid sequence determination and analysis, overexpression in Escherichia coli AB1157 [59]) [59, 61]

<42> (isolation of genomic DNA and construction of a genetic library, cloning of gene PhaspC, DNA and amino acid sequence determination, overexpression in Escherichia coli strain TY103 [69]) [69]

Engineering

C166A <1> (<1> site-directed mutagenesis of isozyme mAspAT, decreased ability to undergo transition from the open to the closed conformation essential for the reaction mechanism, reduced reactivity with DTNB [63]) [63]

C166S <1> (<1> site-directed mutagenesis of isozyme mAspAT, decreased ability to undergo transition from the open to the closed conformation essential for the reaction mechanism, reduced reactivity with DTNB [63]) [63]

E265K <3> (<3> site-directed mutagenesis, reduced k_{cat}, reduced K_m for L-aspartate and 2-oxoglutarate [72]) [72]

E265Q <3> (<3> site-directed mutagenesis, reduced k_{cat}, reduced K_m for L-aspartate and 2-oxoglutarate [72]) [72]

K109S <37> (<37> site-directed mutagenesis, loss of activity towards acidic substrates, increased activity towards the neutral substrate alanine, increase in pKa value [61]) [61]

K109V <37> (<37> site-directed mutagenesis, loss of activity towards acidic substrates, increased activity towards the neutral substrate alanine [61]) [61]

K258A <3> (<3> exchange of active site lysine for alanine, binds still pyridoxal 5'-phosphate or pyridoxamine 5'-phosphate, but not covalently, the latter forms when bound to the enzyme a covalent stable ketimine with 2-oxoglutarate and effects a β-decarboxylation of oxaloacetate, followed by transamination to give the pyridoxal 5'-phosphate aldimine of alanine as a final product [29]) [29]

K68E <3> (<3> site-directed mutagenesis, reduced k_{cat}, reduced K_m for L-aspartate and increased K_m for 2-oxoglutarate [72]) [72]

K68E/E265K <3> (<3> site-directed mutagenesis, charge exchange in the conserved intrasubunit salt bridge, reduced k_{cat}, reduced K_m for L-aspartate and 2-oxoglutarate [72]) [72]

K68M <3> (<3> site-directed mutagenesis, reduced k_{cat}, reduced K_m for L-aspartate and increased K_m for 2-oxoglutarate [72]) [72]

K68M/E265Q <3> (<3> site-directed mutagenesis, introduction of a neutral amino acid pair in the conserved intrasubunit salt bridge, reduced k_{cat}, reduced K_m for L-aspartate and 2-oxoglutarate [72]) [72]

P138A <3> (<3> oligonucleotide-directed mutagenesis, no significant effect, unaltered compared to the wild-type [67]) [67]

P138A/P195A <3> (<3> oligonucleotide-directed mutagenesis, [67]) [67]

P195A <3> (<3> oligonucleotide-directed mutagenesis, no significant effect, unaltered compared to the wild-type [67]) [67]

T70F <3> (<3> nearly no activity with L-erythro-3-hydroxyaspartate, in opposite to the wild-type, altered reaction kinetics due to inability to stabilize the reaction quinonoid intermediate [58]) [58]

V39L <3> (<3> crystal structure analysis [31]) [31]

Additional information <1> (<1> construction of deletion mutant Δ^3-11mAspAT of isozyme mAspAT, enhanced thermostability, reduced k_{cat} and K_m, enhanced reactivity of Cys166 with DTNB [63]) [63]

6 Stability

pH-Stability
2 <12> (<12> irreversibly destroyed below [17]) [17]
3.3 <12> (<12> 1 h, 10% loss of activity of cytoplasmic enzyme, 40% loss of activity of mitochondrial enzyme [17]) [17]
5 <19> (<19> irreversible inactivation [25]) [25]
6-9 <19> (<19> stable [25]) [25]
6-11 <26> (<26> 37°C, 1 h, stable [37]) [37]
12 <12> (<12> irreversibly destroyed above [17]) [17]

Temperature stability
37 <9, 26> (<9> relatively stable [15]; <26> 1 h, pH 6.0-11.0, stable [37]) [15, 37]
40 <21> (<21> pH 7.5, 1 h, gradual loss of activity above [33]) [33]
50 <7, 9, 12, 42-44> (<43> isozyme AAT1 $t_{1/2}$: 3.6 min, isozyme AAT2 $t_{1/2}$: 41.2 min [73]; <44> isozyme AST II, 15 min, stable [71]; <42> $t_{1/2}$: 6.8 min [69]; <7> under physiological or nearly physiological conditions, 3.5 M KCl, pH 7.8, stable for several h [8]; <9> 30 min, 20% loss of activity [15]; <12> 15 min, 10% loss of activity for the mitochondrial isozyme, 80% loss of activity with aspartate for the cytosolic isozyme [17]) [8, 15, 17, 69, 71, 73]
50-65 <42> (<42> recombinant enzyme, transition temperature [69]) [69]
55 <3, 29> (<3> 10 min, 48% loss of activity with tyrosine, 36% loss of activity with phenylalanine [1]; <29> 30 min, isoenzyme I: completely stable, isoenzyme II: 31% loss of activity, isoenzyme III: 67% loss of activity [42]) [1, 42]
60 <1, 9, 32> (<1> 15 min, stable [44]; <9> 30 min, 75% loss of activity [15]; <32> 5 min, more than 90% loss of activity [47]) [15, 44, 47]
65 <43> (<43> inactivation, isozymes AAT 1 and 2 [73]) [73]
65.5 <42> (<42> recombinant enzyme, melting temperature [69]) [69]
70 <1, 9, 12, 21, 33> (<1> $t_{1/2}$ wild-type of mature mitochondrial isozyme: 2.7 min [63]; <33> pH 7.5, 1 h, stable up to [33]; <21> pH 7.5, 1 h, enzyme loses 65% of activity [33]; <9> 30 min, 80% loss of activity [15]; <12> 15 min, mitochondrial enzyme completely destroyed [17]; <33> pH 7.5, 1 h, stable up to [48]) [15, 17, 33, 48, 63]
75 <37, 39> (<37> stable up to [59,62]) [59, 62]
78.5 <7> (<7> denaturation [49]) [49]
80 <26, 39, 44> (<44> isozyme AST II, loss of 89% activity [71]; <39> 10 min, 75% remaining activity [62]; <26> 20 min, 35% loss of activity [37]) [37, 62, 71]
90 <19> (<19> 6 h, stable [25]) [25]
100 <19> (<19> $t_{1/2}$: 2 h [25]) [25]

Additional information <1, 7, 12, 30, 37, 39> (<1> the premature form of the mitochondrial isozyme is more thermostable than the mature form [63]; <7,39> 2-oxoglutarate protects against thermal inactivation [8,62]; <37> high thermostability, extra prolyl residues, located at the surface of the protein, are involved [59]; <37> thermal denaturation is irreversible [59]; <12> cysteine sulfinic acid protects against heat inactivation [17]; <30> no difference in thermostability between plants of genotypes acclimated at 4 different thermo-periods [45]; <7> thermal denaturation is not reversible, denaturation temperature of holoenzyme is 78.5°C [49]) [17, 44, 45, 49, 59, 62, 63]

General stability information

<7>, inactivation at 1.3 M NaCl and KCl [49]

<8>, complete loss of activity after freezing and thawing [51]

<12>, cysteinic acid protects the cytosolic isoform against thermal denaturation, not aspartate or 2-oxoglutarate [17]

<12>, enzyme is stabilized by the intramitochondrial chaperone homologues GroEL and GroES [68]

<14>, labile enzyme is stabilized by pyridoxyl 5'-phosphate and L-aspartate [20]

<32>, 2-oxoglutarate does not protect against heat inactivation [47]

<32>, addition of glycerol, DTT and pyridoxal 5'-phosphate to buffers, required for stabilization during purification [47]

<32>, aspartate, 10 mM, or pyridoxal 5'-phosphate, 0.2 mM protects against heat inactivation [47]

<7, 39>, 2-oxoglutarate stabilizes against thermal inactivation [8, 62]

Storage stability

<3>, -70°C, 0.7-0.9 mg/ml protein in 20 to 40 mM sodium phosphate buffer, pH 7.0, 0.01 mM pyridoxal phosphate, 1 mM 2-oxoglutarate [31]

<3>, 4°C, 0.01 M potassium phosphate buffer, pH 7.0, 0.001 mM pyridoxal 5'-phosphate, 75% ammonium sulfate, stable for several months [3]

<3>, 4°C, stable for several weeks [1]

<17>, -80°C, stable for at least 1 year [23]

<18>, -20°C, protein concentration up to 2 mg/ml, 30-40% loss of activity per month [39]

<26>, -20°C, protein concentration above 10 mg/ml [37]

<28>, 0°C, stable for at least several h in solution [41]

<33>, -20°C or 4°C, stable for at least 2 weeks [48]

<36>, 4°C, partially purified cytosolic isozyme, stable for 1 month, gradual decrease of activity afterwards [57]

References

[1] Mavrides, C.; Orr, W.: Multispecific aspartate and aromatic amino acid aminotransferases in Escherichia coli. J. Biol. Chem., **250**, 4128-4133 (1975)

[2] Lowe, P.N.; Rowe, A.F.: Aspartate: 2-oxoglutarate aminotransferase from trichomonas vaginalis. Identity of aspartate aminotransferase and aromatic amino acid aminotransferase. Biochem. J., 232, 689-695 (1985)

[3] Yagi, T.; Kagamiyama, H.; Nozaki, M.; Soda, K.: Glutamate-aspartate transaminase from microorganisms. Methods Enzymol., 113, 83-89 (1985)

[4] Ovchinnikov, Y.A.; Egorov, T.A.; Aldanova, N.A.; Feigina, M.Y.; Lipkin, V.M.; Abdulaev, N.G.; Grishin, E.V.; Kiselev, A.P.; Modyanov, M.N.; Braunstein, A.E.; Polyanovsky, O.L.; Nosikov, V.V.: The complete amino acid sequence of cytoplasmic aspartate aminotransferase from pig heart. FEBS Lett., 29, 31-34 (1973)

[5] Metzler, C.M.; Rogers, P.H.; Arone, A.; Martin, D.S.; Metzler, D.E.: Investigation of crystalline enzyme-substrate complexes of pyridoxal phosphate-dependent enzymes. Methods Enzymol., 62, 551-558 (1979)

[6] Yagi, T.; Sako, M.; Moriuti, S.; Shounaka, M.; Masaki, K.; Yamamoto, S.: Purification and characterization of aspartate aminotransferase isoenzymes from rice bran. Biosci. Biotechnol. Biochem., 57, 2074-2080 (1993)

[7] Yagi, T.; Masaki, K.; Yamamoto, S.: Distribution and cellular localization of aspartate aminotransferase isoenzymes in rice. Biosci. Biotechnol. Biochem., 57, 2081-2084 (1993)

[8] Muriana, F.J.G.; Alvarez-Ossorio, M.C.; Relimpio, A.M.: Further thermal characterization of an aspartate aminotransferase from a halophilic organism. Biochem. J., 298, 465-470 (1994)

[9] Mozzarelli, A.; Ottonello, S.; Rossi, G.L.; Fasella, P.: Catalytic activity of aspartate aminotransferase in the crystal. Equilibrium and kinetic analysis. Eur. J. Biochem., 98, 173-179 (1979)

[10] Eichele, G.; Ford, G.C.; Jansonius, J.N.: Crystallization of pig mitochondrial aspartate aminotransferase by seeding with crystals of the chicken mitochondrial isoenzyme. J. Mol. Biol., 135, 513-516 (1979)

[11] Kamitori, S.; Hirotsu, K.; Higuchi, T.; Kondo, K.; Inoue, K.; Kuramitsu, S.; Kagamiyama, H.; Higuchi, Y.; Yasuoka, N.; Kusunoki, M.; Matsuura, Y.: Overproduction and preliminary X-ray characterization of aspartate aminotransferase from Escherichia coli. J. Biochem., 101, 813-816 (1987)

[12] Gehring, H.; Christen, P.; Eichele, G.; Glor, M.; Jansonius, J.N.; Reimer, A.-S.; Smit, J.D.G.; Thaller, C.: Isolation, crystallization and preliminary crystallographic data of aspartate aminotransferase from chicken heart mitochondria. J. Mol. Biol., 115, 97-101 (1977)

[13] Arone, A.; Rogers, P.H.; Schmidt, J.; Han, C.-N.; Harris, C.M.; Metzler, D.E.: Preliminary crystallographic study of aspartate: 2-oxoglutarate aminotransferase from pig heart. J. Mol. Biol., 112, 509-513 (1977)

[14] Yagi, T.; Toyosato, M.; Soda, K.: Crystalline aspartate aminotransferase from Pseudomonas striata. FEBS Lett., 61, 34-37 (1976)

[15] Eze, L.C.; Echetebu, C.O.: Some properties of aspartate and alanine aminotransferases from Trichoderma viride. J. Gen. Microbiol., 120, 523-527 (1980)

[16] Kuramitsu, S.; Inoue, K.; Kondo, K.; Aki, K.; Kagamiyama, H.: Aspartate aminotransferase isozymes from rabbit liver. Purification and properties. J. Biochem., 97, 1337-1345 (1985)

[17] Recasens, M.; Benezra, R.; Mandel, P.: Cysteine sulfinate aminotransferase and aspartate aminotransferase isoenzymes of rat brain. Purification, characterization, and further evidence for identity. Biochemistry, 19, 4583-4589 (1980)

[18] Krista, M.L.; Fonda, M.L.: Beef brain cytoplasmic aspartate aminotransferase. Purification, kinetics, and physical properties. Biochim. Biophys. Acta, 309, 83-96 (1973)

[19] Barra, D.; Bossa, F.; Doonan, S.; Fahmy, H.M.A.; Hughes, G.J.; Martini, F.; Petruzzelli, R.; Wittmann-Liebold, B.: The cytosolic and mitochondrial aspartate aminotransferases from pig heart. A comparison of their primary structures, predicted secondary structures and some physical properties. Eur. J. Biochem., 108, 405-414 (1980)

[20] Ryan, E.; Bodley, F.; Fottrell, P.F.: Purification and characterization of aspartate aminotransferases from soybean root nodules and Rhizobium japonicum. Phytochemistry, 11, 957-963 (1972)

[21] Romestant, M.; Jerebzoff, S.; Noaillac-Depeyre, J.; Gas, N.; Dargent, R.: Aspartate aminotransferase isoenzymes in Leptosphaeria michotii. Properties and intracellular location [published erratum appears in Eur J Biochem 1989 Jul 1;182(3):737]. Eur. J. Biochem., 180, 153-159 (1989)

[22] Capasso, S.; Garzillo, A.M.; Marino, G.; Mazzarella, L.; Pucci, P.; Sannia, G.: Mitochondrial bovine aspartate aminotransferase. Preliminary sequence and crystallographic data. FEBS Lett., 101, 351-354 (1979)

[23] Numazawa, T.; Yamada, S.; Hase, T.; Sugiyama, T.: Aspartate aminotransferase from Panicum maximum Jacq. var. trichoglume Eyles, a C_4 plant: purification, molecular properties, and preparation of antibody. Arch. Biochem. Biophys., 270, 313-319 (1989)

[24] Orlacchio, A.; Campos-Cavieres, M.; Pashev, I.; Mun, E.A.: Some kinetic and other properties of the isoenzymes of aspartate aminotransferase isolated from sheep liver. Biochem. J., 177, 583-593 (1979)

[25] Marino, G.; Nitti, G.; Arnone, M.I.; Sannia, G.; Gambacorta, A.; De Rosa, M.: Purification and characterization of aspartate aminotransferase from the thermoacidophilic archaebacterium Sulfolobus solfataricus. J. Biol. Chem., 263, 12305-12309 (1988)

[26] Aoki, K.; Calva, E.; Fonseca, L.: Crystallization and immunological identification of dog heart aspartate aminotransferase isoenzymes. J. Biochem., 72, 511-519 (1972)

[27] Izard, T.; Fol, B.; Paupit, R.A.; Jansonius, J.N.: Trigonal crystals of porcine mitochondrial aspartate aminotransferase. J. Mol. Biol., 215, 341-344 (1990)

[28] Malashkevich, V.N.; Sinitzina, N.I.: New crystal form of cytosolic chicken aspartate aminotransferase suitable for high-resolution X-ray analysis. J. Mol. Biol., 221, 61-63 (1991)

[29] Smith, D.L.; Almo, S.C.; Toney, M.D.; Ringe, D.: 2.8-A-resolution crystal structure of an active-site mutant of aspartate aminotransferase from Escherichia coli. Biochemistry, 28, 8161-8167 (1989)

[30] Herold, M.; Kirschner, K.: Reversible dissociation and unfolding of aspartate aminotransferase from Escherichia coli: characterization of a monomeric intermediate. Biochemistry, 29, 1907-1913 (1990)

[31] Jäger, J.; Köhler, E.; Tucker, P.; Sauder, U.; Housley-Markovic, Z.; Fothering-ham, I.; Edwards, M.; Hunter, M.; Kirschner, K.; Jansonius, J.N.: Crystalliza-tion and preliminary X-ray studies of an aspartate aminotransferase mutant from Escherichia coli. J. Mol. Biol., 209, 499-501 (1989)

[32] Fotheringham, I.; Dacey, S.; Taylor, P.; Smith, T.; Hunter, M.; Finlay, M.; Primrose, S.; Parker, D.; Edwards, M.: The cloning ans sequence analysis of the aspC and tyrB genes from Escherichia coli K12. Comparison of the primary structures of the aspartate aminotransferase and aromatic amino-transferase of E. coli with those of the pig aspartate aminotransferase iso-enzymes. Biochem. J., 234, 592-604 (1986)

[33] Tanaka, T.; Yamamoto, S.; Taniguchi, M.; Hayashi, H.; Kuramitsu, S.; Kaga-miyama, H.; Oi, S.: Further studies on aspartate aminotransferase of ther-mophilic methanogens by analysis of general properties, bound cofactors, and subunit structures. J. Biochem., 112, 811-815 (1992)

[34] Yagi, T.; Shounaka, M.; Yamamoto, S.: Distribution of aspartate aminotrans-ferase activity in yeasts, and purification and characterization of mitochon-drial and cytosolic isoenzymes from Rhodotorula minuta [corrected] [pub-lished erratum appears in J Biochem (Tokyo) 1990 Dec;108(6):1070]. J. Bio-chem., 107, 151-159 (1990)

[35] Xing, R.; Whitman, W.B.: Characterization of amino acid aminotrans-ferases of Methanococcus aeolicus. J. Bacteriol., 174, 541-548 (1992)

[36] Taniguchi, M.; Sugiyama, T.: Aspartate aminotransferase from Eleusine cor-acana, a C_4 plant: purification, characterization, and preparation of anti-body. Arch. Biochem. Biophys., 282, 427-432 (1990)

[37] Sung, M.-H.; Tanizawa, K.; Tanaka, H.; Kuramitsu, S.; Kagamiyama, H.; Soda, K.: Purification and characterization of thermostable aspartate ami-notransferase from a thermophilic Bacillus species. J. Bacteriol., 172, 1345-1351 (1990)

[38] Barra, D.; Bossa, F.; Doonan, S.; Fahmy, H.M.A.; Martini, F.; Hughes, G.J.: Large-scale purification and some properties of the mitochondrial aspar-tate aminotransferase from pig heart. Eur. J. Biochem., 64, 519-526 (1976)

[39] Campos-Cavieres, M.; Munn, E.A.: Purification and some properties of cy-toplasmic aspartate aminotransferase from sheep liver. Biochem. J., 135, 683-693 (1973)

[40] Owen, T.G.; Hochachka, P.W.: Purification and properties of dolphin muscle aspartate and alanine transaminases and thier possible roles in the energy metabolism of diving mammals. Biochem. J., 143, 541-553 (1974)

[41] Hatch, M.D.: Separation and properties of leaf aspartate aminotransferase and alanine aminotransferase isoenzymes operative in the C_4 pathway of photosynthesis. Arch. Biochem. Biophys., 156, 207-214 (1973)

[42] Turano, F.J.; Wilson, B.J.; Matthews, B.F.: Rapid purification and thermo-stability of the cytoplasmic aspartate aminotransferase from carrot suspen-sion cultures. Plant Physiol., 97, 606-612 (1991)

[43] Yagi, T.; Kagamiyama, H.; Motosugi, K.; Nozaki, M.; Soda, K.: Crystalliza-tion and properties of aspartate aminotransferase from Escherichia coli B. FEBS Lett., 100, 81-84 (1979)

[44] Quiroga, C.; Imperial, S.; Busquets, M.; Cortes, A.: Comparison of some of the properties of the holoenzymes and apoenzymes of the molecular forms of chicken liver cytoplasmic aspartate aminotransferase. Biochem. Soc. Trans., 19, 74S (1991)

[45] Potvin, C.; Simon, J.-P.; Blanchard, M.-H.: Thermal properties of NAD malate dehydrogenase and glutamate oxaloacetate transaminase in two genotypes pf Arabidopsis thaliana (Cruciferae) from contrasting environments. Plant Sci. Lett., 31, 35-47 (1983)

[46] Griffith, S.M.; Vance, C.P.: Aspartate aminotransferase in alfalfa root nodules. I. Purification and partial characterization. Plant Physiol., 90, 1622-1629 (1989)

[47] Lain-Guelbenzu, B.; Munoz-Blanco, J.; Cardenas, J.: Purification and properties of L-aspartate aminotransferase of Chlamydomonas reinhardtii. Eur. J. Biochem., 188, 529-533 (1990)

[48] Tanaka, T.; Tokuda, T.; Tachibana, A.; Taniguchi, M.; Oi, S.: Purification and some properties of aspartate aminotransferase of Methanobacterium thermoformicicum SF-4. Agric. Biol. Chem., 54, 625-631 (1990)

[49] Muriana, F.J.G.; Alvarez-Ossorio, M.C.; Relimpio, A.: Purification and characterization of aspartate aminotransferase from the halophile archaebacterium Haloferax mediterranei. Biochem. J., 278, 149-154 (1991)

[50] Reed, R.E.; Hess, J.L.: Partial purification and characterization of aspartate aminotransferases from seedling oat leaves. J. Biol. Chem., 250, 4456-4461 (1975)

[51] Reynolds, P.H.S.; Boland, M.J.; Farnden, K.J.F.: Enzymes of nitrogen metabolism in legume nodules: partial purification and properties of the aspartate aminotransferases from lupine nodules. Arch. Biochem. Biophys., 209, 524-533 (1981)

[52] Yagi, T.; Kagamiyama, H.; Nozaki, M.: Aspartate: 2-oxoglutarate aminotransferase from bakers yeast: crystallization and characterization. J. Biochem., 92, 35-43 (1982)

[53] Shrawder, E.J.; Martinez-Carrion, M.: Simultaneous isolation and characterization of chicken supernatant and mitochondrial isoenzymes of aspartate transmainase. J. Biol. Chem., 248, 2140-2146 (1973)

[54] Zhou, S.L.; Stump, D.; Kiang, C.L.; Isola, L.M.; Berk, P.D.: Mitochondrial aspartate aminotransferase expressed on the surface of 3T3-L1 adipocytes mediates saturable fatty acid uptake. Proc. Soc. Exp. Biol. Med., 208, 263-270 (1995)

[55] Taniguchi, M.; Kobe, A.; Kato, M.; Sugiyama, T.: Aspartate aminotransferase isozymes in Panicum miliaceum L., an NAD-malic enzyme-type C_4 plant: comparison of enzymatic properties primary structures, and expression patterns. Arch. Biochem. Biophys., 318, 295-306 (1995)

[56] Okada, M.; Ayabe, Y.: Effects of aminoguanidine and pyridoxal phosphate on glycation reaction of aspartate aminotransferase and serum albumin. J. Nutr. Sci. Vitaminol., 41, 43-50 (1995)

[57] Vessal, M.; Taher, M.: Partial purification and kinetic properties of human placental cytosolic aspartate transaminase. Comp. Biochem. Physiol. B, 110, 431-437 (1995)

[58] Hayashi, H.; Kagamiyama, H.: Reaction of aspartate aminotransferase with L-erythro-3-hydroxyaspartate: involvement of Tyr70 in stabilization of the catalytic intermediates. Biochemistry, 34, 9413-9423 (1995)

[59] Okamoto, A.; Kato, R.; Masui, R.; Yamagishi, A.; Oshima, T.; Kuramitsu, S.: An aspartate aminotransferase from an extremely thermophilic bacterium, Thermus thermophilus HB8. J. Biochem., 119, 135-144 (1996)

[60] Kato, Y.; Asano, Y.; Makar, T.K.; Cooper, A.J.: Irreversible inactivation of aspartate aminotransferase by 2-oxoglutaconic acid and its dimethyl ester. J. Biochem., 120, 531-539 (1996)

[61] Nobe, Y.; Kawaguchi, S.; Ura, H.; Nakai, T.; Hirotsu, K.; Kato, R.; Kuramitsu, S.: The novel substrate recognition mechanism utilized by aspartate aminotransferase of the extreme thermophile Thermus thermophilus HB8. J. Biol. Chem., 273, 29554-29564 (1998)

[62] Kim, H.; Ikegami, K.; Nakaoka, M.; Yagi, M.; Shibata, H.; Sawa, Y.: Characterization of aspartate aminotransferase from the cyanobacterium Phormidium lapideum. Biosci. Biotechnol. Biochem., 67, 490-498 (2003)

[63] Azzariti, A.; Vacca, R.A.; Giannattasio, S.; Merafina, R.S.; Marra, E.; Doonan, S.: Kinetic properties and thermal stabilities of mutant forms of mitochondrial aspartate aminotransferase. Biochim. Biophys. Acta, 1386, 29-38 (1998)

[64] Wilkie, S.E.; Warren, M.J.: Recombinant expression, purification, and characterization of three isoenzymes of aspartate aminotransferase from Arabidopsis thaliana. Protein Expr. Purif., 12, 381-389 (1998)

[65] Blasco, J.; Puppo, J.: Effect of heavy metals (Cu, Cd and Pb) on aspartate and alanine aminotransferase in Ruditapes philippinarum (Mollusca: Bivalvia). Comp. Biochem. Physiol. C, 122, 253-263 (1999)

[66] Petrovic, S.; Semencic, L.; Ozretic, B.; Krajnovic-Ozretic, M.: Selective determination of fish aspartate aminotransferase isoenzymes by their differential sensitivity to proteases. Comp. Biochem. Physiol. B, 124, 209-214 (1999)

[67] Birolo, L.; Malashkevich, V.N.; Capitani, G.; De Luca, F.; Moretta, A.; Jansonius, J.N.; Marino, G.: Functional and structural analysis of cis-proline mutants of Escherichia coli aspartate aminotransferase. Biochemistry, 38, 905-913 (1999)

[68] Mattingly, J.R., Jr.; Yanez, A.J.; Martinez-Carrion, M.: The folding of nascent mitochondrial aspartate aminotransferase synthesized in a cell-free extract can be assisted by GroEL and GroES. Arch. Biochem. Biophys., 382, 113-122 (2000)

[69] Birolo, L.; Tutino, M.L.; Fontanella, B.; Gerday, C.; Mainolfi, K.; Pascarella, S.; Sannia, G.; Vinci, F.; Marino, G.: Aspartate aminotransferase from the Antarctic bacterium Pseudoalteromonas haloplanktis TAC 125. Cloning, expression, properties, and molecular modelling. Eur. J. Biochem., 267, 2790-2802 (2000)

[70] Martins, L.L.; Mourato, M.P.; de Varennes, A.: Effects of substrate structural analogues on the enzymatic activities of aspartate aminotransferase isoenzymes. J. Enzyme Inhib., 16, 251-257 (2001)

[71] Mohamed, T.M.: Purification and characterization of aspartate aminotrans-ferase from developing embryos of the camel tick Hyalomma dromedarii. Exp. Appl. Acarol., **25**, 231-244 (2001)

[72] Deu, E.; Koch, K.A.; Kirsch, J.F.: The role of the conserved Lys68:Glu265 intersubunit salt bridge in aspartate aminotransferase kinetics: multiple forced covariant amino acid substitutions in natural variants. Protein Sci., **11**, 1062-1073 (2002)

[73] Martins, M.L.; Mourato, M.P.; de Varennes e Mendonca, A.P.: Characteriza-tion of aspartate aminotransferase isoenzymes from leaves of Lupinus albus L. cv Estoril. J. Biochem. Mol. Biol., **35**, 220-227 (2002)

[74] Artigues, A.; Crawford, D.L.; Iriarte, A.; Martinez-Carrion, M.: Divergent Hsc70 binding properties of mitochondrial and cytosolic aspartate amino-transferase. Implications for their segregation to different cellular compart-ments. J. Biol. Chem., **273**, 33130-33134 (1998)

Alanine transaminase 2.6.1.2

1 Nomenclature

EC number
2.6.1.2

Systematic name
L-alanine:2-oxoglutarate aminotransferase

Recommended name
alanine transaminase

Synonyms
ALT
GPT
L-alanine aminotransferase
L-alanine transaminase
L-alanine-α-ketoglutarate aminotransferase
alanine aminotransferase
alanine-α-ketoglutarate aminotransferase
alanine-pyruvate aminotransferase
aminotransferase, alanine
β-alanine aminotransferase
glutamic acid-pyruvic acid transaminase
glutamic-alanine transaminase
glutamic-pyruvic aminotransferase
glutamic-pyruvic transaminase
pyruvate transaminase
pyruvate-alanine aminotransferase
pyruvate-glutamate transaminase

CAS registry number
9000-86-6

2 Source Organism

<1> *Atriplex spongiosa* [9]
<2> *Lycopersicon esculentum* (tomato, 2 alanine aminotransferases, soluble and Triton X-100 extractable [10]) [10]
<3> *Sus scrofa* [11]
<4> *Chlamydomonas reinhardtii* [13]
<5> *Candida maltosa* [14]

<6> *Panicum miliaceum* [15]
<7> *Hordeum vulgare* (4 isoforms [16]) [16]
<8> *Lagenorhynchus obliquidens* (dolphin [6]) [6]
<9> *Leptosphaeria michotii* (fungi [1]) [1]
<10> *Bombyx mori* [2]
<11> *Bos taurus* [3, 4]
<12> *Schistocerca gregaria* [5]
<13> *Oryctolagus cuniculus* [17]
<14> *Rattus norvegicus* (Sprague-Dawley female rats [18]) [7, 18, 24]
<15> *Homo sapiens* (high alanine aminotransferase activity is a marker of risk for type 2 diabetis [25]) [7, 12, 19, 25]
<16> *Canis familiaris* [7]
<17> *Felis catus* [7]
<18> *Trichoderma viride* (soil fungus [8]) [8]
<19> *Zea mays* [20, 21]
<20> *Ruditapes philippinarum* (mollusca: bivalvia [22]) [22]
<21> *Pyrococcus furiosus* [23]

3 Reaction and Specificity

Catalyzed reaction

L-alanine + 2-oxoglutarate = pyruvate + L-glutamate (<2, 4, 13> ping pong bi bi mechanism [10, 13, 17]; <3> ping pong type reaction [11])

Reaction type

amino group transfer

Natural substrates and products

S L-alanine + 2-oxoglutarate <13, 15> (<13> important function in β-alanine metabolism [17]; <15> mitochondrial isoenzyme shows higher affinity for L-alanine, mainly functions in the direction of conversion of alanine to pyruvate whereas the cytosolic isoenzyme would function in both directions [12]) [12, 17]

P pyruvate + L-glutamate

Substrates and products

S 2-oxoglutarate + 5-aminopentanoate <13> (<13> 121% of the activity with β-alanine [17]) (Reversibility: ? <13> [17]) [17]

P glutamate + 5-oxopentanoate <13> [17]

S 2-oxoglutarate + 6-aminohexanoate <13> (<13> 8% of the activity with β-alanine [17]) (Reversibility: ? <13> [17]) [17]

P glutamate + 6-oxohexanoate <13> [17]

S 2-oxoglutarate + DL-3-aminoisobutanoate <13> (<13> 51% of the activity with β-alanine [17]) (Reversibility: ? <13> [17]) [17]

P glutamate + 3-oxoisobutanoate <13> [17]

S 4-aminobutanoate + 2-oxoglutarate <13> (<13> at 131% of the activity with β-alanine [17]) (Reversibility: ? <13> [17]) [17]

P 4-oxobutanoate + glutamate <13> [17]
S L-alanine + 2-oxobutyrate <5> (<5> 10.2% of the activity with 2-oxoglutarate [14]) (Reversibility: ? <5> [14]) [14]
P pyruvate + 2-aminobutanoate <5> [14]
S L-alanine + 2-oxoglutarate <1-21> (<7, 19> highly specific for alanine and 2-oxoglutarate [16, 20]; <14> with 2-oxoglutarate specific for L-alanine [7]; <3> D-alanine at 2% of the activity with L-alanine [11]; <14> glutamate-glyoxylate aminotransferase is identical with alanine-2-oxoglutarate aminotransferase [7]; <9> no activity with L-aspartate, cysteine sulfinate, 2-aminobutanoate and cyclic amino acids [1]; <7> cannot use glyoxylate as keto donor [16]; <2> rate of the forward reaction is 4 times higher than the rate of the reverse reaction, no activity with aspartate, glycine, threonine, leucine, methionine, arginine, phenylalanine, tyrosine, tryptophan, glyoxylate and 2-oxo-3-methylvalerate [10]) (Reversibility: r <1, 2, 5, 6, 19, 21> [9, 10, 14, 15, 20, 23]; ? <3, 4, 7-18, 20> [1-11, 12, 13, 16, 17, 22]) [1-17, 20, 22, 23]
P pyruvate + L-glutamate <1-21> [1-17, 20, 22, 23]
S L-glutamate + 2-oxobutyrate <5> (<5> 32.4% of the activity with pyruvate [14]) (Reversibility: ? <5> [14]) [14]
P 2-oxoglutarate + 2-aminobutanoate <5> [14]
S aspartate + 2-oxoglutarate <21> (<21> 14% of activity with glutamate and pyruvate [23]) (Reversibility: ? <21> [23]) [23]
P malate + L-glutamate <21> [23]
S aspartate + pyruvate <21> (<21> 10% of activity with alanine and 2-oxoglutarate [23]) (Reversibility: ? <21> [23]) [23]
P malate + alanine <21> [23]
S β-alanine + glyoxylic acid <13> (<13> 7% of the activity with 2-oxoglutarate [17]) (Reversibility: ? <13> [17]) [17]
P malonic semialdehyde + glycine <13> [17]
S β-alanine + pyruvate <13> (<13> 1% of the activity with 2-oxoglutarate [17]) (Reversibility: ? <13> [17]) [17]
P malonic semialdehyde + alanine <13> [17]
S glycine + 2-oxoglutarate <18> (<18> 24% of activity with DL-alanine [8]) (Reversibility: ? <18> [8]) [8]
P glyoxylate + glutamate <18> [8]
S glyoxylate + alanine <4, 14> (<14> 70% of activity with glutamate [7]) (Reversibility: ir <14> [7]; ? <4> [13]) [7, 13]
P glycine + pyruvate <4, 14> [7, 13]
S glyoxylate + glutamate <4, 14> (Reversibility: ir <14> [7]; ? <4> [13]) [7, 13]
P glycine + 2-oxoglutarate <4, 14> [7, 13]
S glyoxylate + glutamine <14> (<14> 42% of activity with glutamate [7]) (Reversibility: ? <14> [7]) [7]
P glycine + 2-oxosuccinamate <14> [7]
S glyoxylate + methionine <14> (<14> 37% of activity with glutamate [7]) (Reversibility: ? <14> [7]) [7]
P glycine + 4-methylsulfanyl-2-oxobutanoate <14> [7]

S methionine + 2-oxoglutarate <18> (<18> 75% of activity with DL-alanine [8]) (Reversibility: ? <18> [8]) [8]

P 4-methylsulfanyl-2-oxobutanoate + glutamate <18> [8]

S serine + 2-oxoglutarate <18> (<18> 22% of activity with DL-alanine [8]) (Reversibility: ? <18> [8]) [8]

P 3-hydroxy-2-oxopropanoate + glutamate <18> [8]

S tyrosine + 2-oxoglutarate <18> (<18> 36.5% of activity with DL-alanine [8]) (Reversibility: ? <18> [8]) [8]

P 3-(4-hydroxyphenyl)-2-oxopropanoate + glutamate <18> [8]

Inhibitors

1,10-phenanthroline <4> (<4> 2 mM, 51% inhibition [13]) [13]

2-oxoglutarate <13> (<13> substrate inhibition at neutral pH, not seen at pH 8.8 [17]) [17]

3-chloro-L-alanine <19> (<19> 0.1 mM, 78% inhibition [21]) [21]

6-azauracil <13> (<13> 1.5 mM, 50% inhibition [17]) [17]

Ag^+ <11> [3]

Al^{3+} <11> [3]

Cd^{2+} <19, 20> (<19> 0.1 mM, complete inhibition [21]; <20> 0.2 mM, 39% inhibition of alanine aminotransferase activity measured in soft body and gills, 0.6 mM, 46% inhibition [22]) [21, 22]

Co^{2+} <11> [3]

Cu^{2+} <11, 20> (<11> reversed by L-cysteine and pyridoxal phosphate [3]; <20> inhibition of alanine aminotransferase activity in gills [22]) [3, 22]

D-alanine <3> (<3> competitive vs. L-alanine [11]) [11]

EDTA <11> [3]

Hg^{2+} <5, 11, 14, 18> (<18> 93% inhibition [8]; <5> 0.1 mM, 81% inhibition [14]) [3, 7, 8, 14]

KCN <14> [7]

L-ascorbic acid <4> (<4> 2 mM, 25% inhibition [13]) [13]

Pb^{2+} <11, 20> (<11> reversed by L-cysteine and pyridoxal phosphate [3]; <20> 0.7 mM, 58% inhibition of alanine aminotransferase activity measured in soft body, 0.35 mM, 32% inhibition [22]) [3, 22]

Zn^{2+} <11, 19> (<19> 1 mM, 75% inhibition [21]) [3, 21]

aminooxyacetate <3, 4, 5, 7> (<3> 0.0004 mM, 85% inhibition of mitochondrial alanine aminotransferase [11]; <4> 1 mM, complete inhibition [13]; <5> 2 mM, complete inhibition [14]; <7> 1 mM, 98% inhibition [16]) [11, 13, 14, 16]

citric acid <11> [3]

cycloserine <3, 19> (<3> 1 mM, 10 mM, 50 mM, 10%, 52% and 100% inhibition respectively [11]; <19> 0.1 mM, 98% inhibition [21]) [11, 21]

fumaric acid <11> [3]

glutarate <7> (<7> 10 mM, 94% inhibition [16]) [16]

hydroxylamine <4, 5, 10, 12, 14> (<10> activity recovered by pyridoxal 5'-phosphate [2]; <14> 1 mM, 96% inhibition [7]; <4> 2 mM, 96% inhibition [13]; <5> 2 mM, 73% inhibition [14]) [2, 5, 7, 13, 14]

isonicotinic hydrazide <14> [7]

malate <1> (<1> 5 mM, 25% inhibition [9]) [9]
maleic acid <11> [3]
oxalic acid <11> [3]
p-chloromercuribenzoate <10, 12> (<10> reactivation by dithiothreitol [2])
[2, 5]
p-hydroxymercuribenzoate <4, 5> (<4> 1 mM, complete inhibition [13]) [13,
14]
p-hydroxymercuribenzoate <5, 19> (<5> 1 mM, 87% inhibition [14]; <19>
0.1 mM, 72% inhibition [21]) [14, 21]
phenelzine <14> (<14> monoamine oxidase inhibitor used as antidepres-
sant/antipanic drug, maximum inhibition, i.e. 32%, of brain alanine aminotransferase at 15 mg/kg [24]) [24]
phenylhydrazine <12> [5]
semicarbazide <7, 14> (<14> 1 mM, 40% inhibition [7]; <7> 10 mM, 97%
inhibition [16]) [7, 16]
succinate <18> (<18> competitive inhibitior [8]) [8]
vigabatrin <15> (<15> structural analogue to γ-amino butyric acid, anti-epi-
lepsy drug, 1 mM, approx. 80% inhibition in vitro, in vivo alanine aminotransferase activity is reduced 30-40% 1-2h after administration [19]) [19]
Additional information <13> (<13> not inhibited by 6-azauridine or 6-
azauridine 5'-phosphate [17]) [17]

Cofactors/prosthetic groups
pyridoxal 5'-phosphate <4, 5, 13, 19> (<4,5> a pyridoxal phosphate protein,
2 mol pyridoxal phosphate per mol of enzyme [13,14]; <13> 1 mol of pyridoxal 5'-phosphate per mol of dimer [17]) [13, 14, 17, 20]
Additional information <7> (<7> anaerobically induced isoform does not
require the presence of pyridoxal 5'-phosphate to retain its activity [16]) [16]

Specific activity (U/mg)
1.39 <19> [21]
1.58 <13> [17]
3.55 <3> [11]
18.66 <4> [13]
35.5 <8> [6]
120 <2> [10]
127.5 <19> [20]
156 <5> (<5> reverse reaction [14]) [14]
158 <21> (<21> native enzyme [23]) [23]
213 <3> [11]
243 <21> (<21> recombinant enzyme [23]) [23]
360 <9> [1]
387.9 <6> [15]
480 <14> [7]
594 <5> (<5> forward reaction [14]) [14]
2231 <7> [16]

K$_m$-Value (mM)

0.0058 <11> (2-oxoglutarate, <11> pH 7.5, 35°C [3]) [3]

0.022 <1> (pyruvate, <1> pH 7.5, 22°C, isoenzyme 3 [9]) [9]

0.028 <1> (2-oxoglutarate, <1> pH 7.5, 22°C, isoenzyme 3 [9]) [9]

0.04 <1> (pyruvate, <1> pH 7.5, 22°C, isoenzyme 2 [9]) [9]

0.05 <4> (2-oxoglutarate, <4> pH 7.3, 40°C [13]) [13]

0.06 <7> (pyruvate, <7> pH 8.0, 23°C [16]) [16]

0.09 <1> (2-oxoglutarate, <1> pH 7.5, 22°C, isoenzyme 2 [9]) [9]

0.09 <2> (pyruvate, <2> pH 7.25, 37°C [10]) [10]

0.15 <6> (2-oxoglutarate, <6> pH 7.5, 25°C [15]) [15]

0.16 <5> (pyruvate, <5> pH 7.5, 25°C [14]) [14]

0.19 <15> (2-oxoglutarate, <15> pH 7.8, 30°C, mitochondrial enzyme [12]) [12]

0.2 <9> (2-oxoglutarate) [1]

0.22 <15> (2-oxoglutarate, <15> pH 7.8, 30°C, cytosolic enzyme [12]) [12]

0.23 <10> (2-oxoglutarate) [2]

0.24 <4> (pyruvate, <4> pH 7.3, 40°C [13]) [13]

0.25 <1> (alanine, <1> pH 7.5, 22°C, isoenzyme 1 [9]) [9]

0.28 <2> (2-oxoglutarate, <2> pH 7.25, 37°C [10]) [10]

0.3 <1> (2-oxoglutarate, <1> pH 7.5, 22°C, isoenzyme 2 [9]) [9]

0.33 <6> (pyruvate, <6> pH 7.5, 25°C [15]) [15]

0.37 <5> (2-oxoglutarate, <5> pH 7.5, 25°C [14]) [14]

0.4 <3> (2-oxoglutarate, <3> pH 7.8, 30°C, mitochondrial alanine amino-transferase [11]) [11]

0.4 <11> (pyruvate, <11> pH 7.5 mitochondrial alanine aminotransferase [4]) [4]

0.4 <3> (pyruvate, <3> pH 7.8, 30°C, mitochondrial alanine aminotransfer-ase [11]) [11]

0.45 <8> (2-oxoglutarate, <8> pH 7.3, 37°C [6]) [6]

0.45 <1> (pyruvate, <1> pH 7.5, 22°C, isoenzyme 1 [9]) [9]

0.5 <11> (pyruvate, <11> pH 7.5, cytosolic alanine aminotransferase [4]) [4]

0.51 <13> (4-aminobutanoate, <13> pH 8.8, 37°C, cosubstrate 2-oxoglutarate [17]) [17]

0.52 <4> (L-glutamate, <4> pH 7.3, 40°C [13]) [13]

0.54 <13> (2-oxoglutarate, <13> pH 8.8, 37°C [17]) [17]

0.55 <1> (glutamate, <1> pH 7.5, 22°C, isoenzyme 1 [9]) [9]

0.59 <11> (L-alanine, <11> pH 7.5, 35°C [3]) [3]

0.7 <11> (2-oxoglutarate, <11> pH 7.5, mitochondrial and cytosolic alanine aminotransferase [4]) [4]

0.8 <1> (glutamate, <1> pH 7.5, 22°C, isoenzyme 3 [9]) [9]

0.87 <8> (pyruvate, <8> pH 7.3, 37°C [6]) [6]

0.9 <21> (2-oxoglutarate, <21> pH 7.2, 80°C, native enzyme [23]) [23]

1 <7> (glutamate, <7> pH 8.0, 23°C [16]) [16]

1.1 <21> (2-oxoglutarate, <21> pH 7.2, 80°C, recombinant enzyme [23]) [23]

1.15 <1> (glutamate, <1> pH 7.5, 22°C, isoenzyme 2 [9]) [9]

1.4 <13> (2-oxoglutarate, <13> pH 8.8, 37°C, cosubstrate β-alanine [17]) [17]

1.6 <12> (2-oxoglutarate) [5]

1.85 <18> (2-oxoglutarate, <18> pH 8.6, 37°C [8]) [8]

2 <3> (alanine, <3> pH 7.8, 30°C, mitochondrial alanine aminotransferase [11]) [11]

2.3 <2> (glutamate, <2> pH 7.25, 37°C [10]) [10]

2.6 <5> (L-alanine, <5> pH 7.5, 25°C [14]) [14]

2.7 <4> (L-alanine, <4> pH 7.3, 40°C [13]) [13]

2.8 <2, 21> (alanine, <2> pH 7.25, 37°C [10]; <21> pH 7.2, 80°C, native enzyme [23]) [10, 23]

3 <1> (alanine, <1> pH 7.5, 22°C, isoenzyme 2 [9]) [9]

3.1 <1> (alanine, <1> pH 7.5, 22°C, isoenzyme 3 [9]) [9]

3.2 <21> (L-alanine, <21> pH 7.2, 80°C, recombinant enzyme [23]) [23]

3.9 <13> (β-alanine, <13> pH 8.8, 37°C, cosubstrate 2-oxoglutarate [17]) [17]

4.3 <11, 21> (glutamate, <11> pH 7.5, cytosolic alanine aminotransferase [4]; <21> pH 7.2, 80°C, native enzyme [23]) [4, 23]

4.7 <21> (glutamate, <21> pH 7.2, 80°C, recombinant enzyme [23]) [23]

5 <7> (2-oxoglutarate, <7> pH 8.0, 23°C [16]) [16]

5 <6> (glutamate, <6> pH 7.5, 25°C [15]) [15]

5.1 <11, 15> (L-alanine, <11> pH 7.5, mitochondrial alanine aminotransferase [4]; <15> pH 7.8, 30°C, mitochondrial enzyme [12]) [4, 12]

5.4 <21> (pyruvate, <21> pH 7.2, 80°C, native enzyme [23]) [23]

5.6 <21> (pyruvate, <21> pH 7.2, 80°C, recombinant enzyme [23]) [23]

6.2 <10> (alanine) [2]

6.3 <5> (L-glutamate, <5> pH 7.5, 25°C [14]) [14]

6.6 <11> (L-glutamate, <11> pH 7.5, mitochondrial alanine aminotransferase [4]) [4]

6.67 <6> (alanine, <6> pH 7.5, 25°C [15]) [15]

6.85 <9> (L-alanine) [1]

8.2 <8> (L-alanine, <8> pH 7.3, 37°C [6]) [6]

8.3 <18> (DL-alanine, <18> pH 8.6, 37°C [8]) [8]

15 <8> (glutamate, <8> pH 7.3, 37°C [6]) [6]

17 <7> (alanine, <7> pH 8.0, 23°C [16]) [16]

19.05 <12> (L-alanine) [5]

21 <15> (alanine, <15> pH 7.8, 30°C, cytosolic enzyme [12]) [12]

30.3 <11> (L-alanine, <11> pH 7.5, cytosolic alanine aminotransferase [4]) [4]

32 <3> (glutamate, <3> pH 7.8, 30°C, mitochondrial alanine aminotransferase [11]) [11]

K_i-Value (mM)

17 <18> (succinate, <18> pH 8.6, 37°C [8]) [8]

pH-Optimum

6-7.5 <5> [14]

6.5-7.5 <6> (<6> forward and reverse reaction [15]) [15]

6.5-7.8 <21> (<21> native and recombinant enzyme [23]) [23]

7.2-7.8 <8> [6]

7.3 <4> [13]

7.3-7.5 <2> [10]

7.4 <12> (<12> in phosphate buffer [5]) [5]
7.5 <1, 9, 11> [1, 3, 9]
7.5-8.5 <3> [11]
7.6 <10> [2]
7.8 <11> [4]
8-8.3 <7> (<7> with alanine and 2-oxoglutarate [16]) [16]
8.3 <12> (<12> in borate buffer [5]) [5]
8.6 <18> [8]
8.8 <13> [17]

pH-Range
3.5-9.5 <3> (<3> active over whole range [11]) [11]
5.5-8.2 <21> (<21> native and recombinant enzyme [23]) [23]
5.5-9.5 <5> (<5> approx. 70% of maximal activity at pH 5.5, approx. 50% of maximal activity at pH 9.5: [14]) [14]
6.3-7.8 <8> (<8> approx. 40% of maximal activity at pH 6.3, 75% of maximal activity at pH 6.8 [6]) [6]
6.4-8.7 <1> (<1> at least 50% of maximal activity at pH 6.4 and 8.7 [9]) [9]
7-7.5 <2> (<2> rapid decrease of activity below pH 7 and above pH 7.5 [10]) [10]

Temperature optimum (°C)
23 <2> (<2> assay at [10]) [10]
30 <3> (<3> assay at [11]) [11]
35 <11> [3]
37 <2, 8, 14-17> (<2,8,14-17> assay at [6,7,10]) [6, 7, 10]
50 <4> [13]
55 <5> [14]
Additional information <21> (<21> above 95°C, native and recombinant enzyme [23]) [23]

Temperature range (°C)
25-55 <5> (<5> strong decrease above [14]) [14]
30-95 <21> (<21> native and recombinant enzyme [23]) [23]

4 Enzyme Structure

Molecular weight
69000 <10> [2]
80000 <3> (<3> gel filtration [11]) [11]
92000 <11> [3]
93400 <21> (<21> gel filtration [23]) [23]
95000 <13, 19> (<13> gel filtration [17]; <19> gel filtration [20]) [17, 20]
97000 <7> (<7> gel filtration [16]) [16]
99000 <5> (<5> gel filtration [14]) [14]
100000 <2> (<2> gel filtration [10]) [10]
102000 <6> (<6> gel filtration [15]) [15]

105000 <4> (<4> gel filtration [13]) [13]
110000 <9> [1]
112000 <14> (<14> sucrose density gradient centrifugation [7]) [7]
112000-118000 <15-17> (<15-17> gel filtration [7]) [7]
115000 <14> (<14> gel filtration [7]) [7]
118000 <13> (<13> sucrose density gradient centrifugation [17]) [17]

Subunits

dimer <4, 5, 6, 7, 9, 13, 19, 21> (<7> 2 * 50000, SDS-PAGE [16]; <6> 2 *
50000, SDS-PAGE [15]; <4> 2 * 45000, SDS-PAGE [13]; <13> 2 * 48000,
SDS-PAGE [17]; <5> 2 * 52000, SDS-PAGE [14]; <19> 2 * 50000, SDS-PAGE
[20]; <21> 2 * 46000, SDS-PAGE [23]) [1, 13-17, 20, 23]
monomer <3> (<3> 1 * 75000, SDS-PAGE [11]) [11]

5 Isolation/Preparation/Mutation/Application

Source/tissue

brain <11, 14> (<11> cortex [4]) [4, 24]
digestive gland <20> [22]
erythrocyte <11> (<11> stroma [3]) [3]
flight muscle <12> [5]
foot <20> [22]
fruit <2> [10]
gill <20> [22]
kidney <3> (<3> cortex [11]) [11]
leaf <1, 6, 19> [9, 15, 20]
liver <13-17> [7, 12, 17, 18, 25]
mammary gland <14> (<14> increase in alanine aminotransferase activity
during lactation and weaning when compared to virgin rats [18]) [18]
midgut <10> [2]
muscle <8, 14> (<14> sceletal muscle [18]) [6, 18]
mycelium <18> [8]
root <7> [16]
soft body part <20> [22]

Localization

mitochondrion <3, 12> [5, 11]
soluble <2> [10]

Purification

<1> (ammonium sulfate, DEAE-cellulose, 3 isoforms [9]) [9]
<2> (ammonium sulfate, DEAE-Sephadex, Sephadex G-200, partial purifica-
tion [10]) [10]
<3> (Sephadex G-150, DEAE-cellulose, ammonium sulfate, isoelectric focus-
ing [11]) [11]
<4> [13]
<5> [14]

<6> (2 minor forms: AlaAT-1, AlaAT-3, 1 major form, AlaAT-2, ammonium sulfate, DEAE-cellulose, Sephacryl S-200, phenyl-Sepharose [15]) [15]
<7> (anaerobically induced isoform, ammonium sulfate, gel filtration, Q-Sepharose, Mono P, Mono Q [16]) [16]
<8> [6]
<9> [1]
<10> [2]
<11> (ammonium sulfate, Sephadex G-150, DEAE-Sephadex, partial purification [4]; DEAE-Sephadex [3]) [3, 4]
<12> [5]
<13> (heat treatment, ammonium sulfate, DEAE-Sepharose, hydroxylapatite, Sephacryl S-200 [17]) [17]
<14> (heat treatment, ammonium sulfate, DEAE-cellulose, isoelectric focusing, Sephadex G-150, hydroxylapatite, DEAE-cellulose [7]) [7]
<15> (ammonium sulfate, DEAE-Sepharose, partial purification of cytosolic and mitochondrial isoenzymes [12]) [12]
<18> (partial [8]) [8]
<19> (ammonium sulfate, hydroxylapatite, preparative electrophoresis [21]) [20, 21]
<21> (native enzyme, Q-Sepharose, phenyl-Sepharose, hydroxyapatite, S-Sepharose, Mono Q, Superdex 200, recombinant enzyme, heat treatment, Mono Q [23]) [23]

Cloning
<21> (expression in Escherichia coli [23]) [23]

6 Stability

pH-Stability
6.5 <2> (<2> strong loss of activity below [10]) [10]
7.5 <5> (<5> 40°C, 30 min, 50% loss of activity [14]) [14]
8.5-9.5 <5> (<5> more stable at alkaline pH than at neutral pH [14]) [14]
9 <5> (<5> 40°C, 30 min, 5% loss of activity [14]) [14]

Temperature stability
22 <1> (<1> isoenzymes 2 and 3, 10-20% loss of activity after 3 h [9]) [9]
40 <5> (<5> 30 min, 50% loss of activity at pH 7.5, 5% loss of activity at pH 9.0 [14]) [14]
50 <2, 4> (<2> quickly inactivated, 100 mM alanine gives slight protection [10]; <4> 50% loss of activity after 10 min [13]) [10, 13]
60 <4> (<4> more than 85% loss of activity after 5 min [13]) [13]
Additional information <4> (<4> pyridoxal 5-phosphate, L-alanine or 2-oxoglutarate protect against heat inactivation [13]) [13]

General stability information
<2>, during the purification procedure the enzyme needs protective substances, i.e. glycerol, 2-mercaptoethanol and alanine [10]

<2>, uneffected by repeated freezing and thawing [10]
<3>, K$^+$ or L-alanine stabilizes [11]
<3>, enzyme does not tolerate freezing at any stage after sonication of the isolated mitochondria [11]
<4>, pyridoxal 5'-phosphate, L-alanine or 2-oxoglutarate protects against heat inactivation [13]

Storage stability

<1>, 0°C, stable for several h in solution [9]
<2>, -20°C, stable for several months [10]
<7>, -20°C, 10% glycerol, half-life of 2 weeks [16]
<13>, -25°C, 10 mM potassium phosphate, pH 7.5, 1 mM EDTA, 2 mM 2-mer-captoethanol, 0.04 mM pyridoxal 5'-phosphate, a few days, 50% loss of activity [17]
<13>, 4°C, 80% saturated solution of ammonium sulfate, 10 days, 40% loss of activity [17]
<14>, -20°C, 50 mM potassium phosphate buffer, pH 7.5, 5 mM 2-mercaptoethanol, 5% glycerol, stable for at least 5 weeks [7]
<14>, 0-5°C, 50 mM potassium phosphate buffer, pH 7.5, 5 mM 2-mercaptoethanol, 5% glycerol, stable for 2 weeks [7]

References

[1] Abou-Zeid, A.M.; Jerebzoff, S.; Jerebzoff-Quintin, S.: Alanine aminotransferase in Leptosphaeria michotii. Isolation, properties and cyclic variations of its activity in relation to polypeptide level. Physiol. Plant., **82**, 401-408 (1991)
[2] Nakamura, M.; Horie, Y.: Partial purification and properties of alanine and aspartate aminotransferases in the midgut of the silkworm, Bombyx mori (Lepidoptera:Bombycidae). Appl. Entomol. Zool., **21**, 236-243 (1986)
[3] Agarwal, K.: Purification and properties of L-alanine 2-oxoglutarate aminotransferase of bovine erythrocytes stroma. Indian J. Biochem. Biophys., **22**, 102-106 (1985)
[4] Ruscak, M.; Orlicky, J.; Zubor, V.; Hager, H.: Alanine aminotransferase in bovine brain: purification and properties. J. Neurochem., **39**, 210-216 (1982)
[5] Mane, S.D.; Mehrotra, K.N.: Alanine aminotransferase: properties and distribution in the desert locust, Schistocerca gregaria. Insect Biochem., **7**, 419-426 (1977)
[6] Owen, T.G.; Hochachka, P.W.: Purification and properties of dolphin muscle aspartate and alanine transaminases and thier possible roles in the energy metabolism of diving mammals. Biochem. J., **143**, 541-553 (1974)
[7] Noguchi, T.; Takada, Y.; Kido, R.: Glutamate-glyoxylate aminotransferase in rat liver cytosol. Purification, properties and identity with alanine-2-oxoglutarate aminotransferase. Hoppe-Seyler's Z. Physiol. Chem., **358**, 1533-1542 (1977)

[8] Eze, L.C.; Echetebu, C.O.: Some properties of aspartate and alanine amino-transferases from Trichoderma virie. J. Gen. Microbiol., **120**, 523-527 (1980)

[9] Hatch, M.D.: Separation and properties of leaf aspartate aminotransferase and alanine aminotransferase isoenzymes operative in the C_4 pathway of photosynthesis. Arch. Biochem. Biophys., **156**, 207-214 (1973)

[10] Rech, J.; Crouzet, J.: Partial purification and initial studies of the tomato L-alanine:2-oxoglutarate aminotransferase. Biochim. Biophys. Acta, **350**, 392-399 (1974)

[11] De Rosa, G.; Burk, T.L.; Swick, R.W.: Isolation and characterization of mi-tochondrial alanine aminotransferase from porcine tissue. Biochim. Biophys. Acta, **567**, 116-124 (1979)

[12] Gubern, G.; Imperial, S.; Busquets, M.; Cortes, A.: Partial characterization of the alanine aminotransferase isoenzymes from human liver. Biochem. Soc. Trans., **18**, 1288-1289 (1990)

[13] Lain-Guelbenzu, B.; Cardenas, J.; Munoz-Blanco, J.: Purification and prop-erties of L-alanine aminotransferase from Chlamydomonas reinhardtii. Eur. J. Biochem., **202**, 881-887 (1991)

[14] Umemura, I.; Yanagiya, K.; Komatsubara, S.; Sato, T.; Tosa, T.: Purification and some properties of alanine aminotransferase from Candida maltosa. Biosci. Biotechnol. Biochem., **58**, 283-287 (1994)

[15] Son, D.; Jo, J.; Sugiyama, T.: Purification and characterization of alanine aminotransferase from Panicum miliaceum leaves. Arch. Biochem. Bio-phys., **289**, 262-266 (1991)

[16] Good, A.G.; Muench, D.G.: Purification and properties of β-alanine amino-transferase from rabbit liver. Plant Physiol., **99**, 1520-1525 (1992)

[17] Tamaki, N.; Aoyama, H.; Kubo, K.; Ikeda, T.; Hama, T.: Purification and properties of β-alanine aminotransferase from rabbit liver. J. Biochem., **92**, 1009-1017 (1982)

[18] Desantiago, S.; Aleman, G.; Hernandez-Montes, H.: Alanine aminotransfer-ase activity in mammary tissue, muscle and liver of dam rat during lacta-tion and weaning. Arch. Med. Res., **27**, 443-448 (1996)

[19] Richens, A.; McEwan, J.R.; Deybach, J.C.; Mumford, J.P.: Evidence for both in vivo and in vitro interaction between vigabatrin and alanine transami-nase. Br. J. Clin. Pharmacol., **43**, 163-168 (1997)

[20] Orzechowski, S.; Socha-Hanc, J.; Paszkowski, A.: Purification and proper-ties of alanine aminotransferase from maize (Zea mays L.) leaves. Acta Phy-siol. Plant., **21**, 323-330 (1999)

[21] Orzechowski, S.; Socha-Hanc, J.; Paszkowski, A.: Alanine aminotransferase and glycine aminotransferase from maize (Zea mays L.) leaves. Acta Bio-chim. Pol., **46**, 447-457 (1999)

[22] Blasco, J.; Puppo, J.: Effect of heavy metals (Cu, Cd and Pb) on aspartate and alanine aminotransferase in Ruditapes philippinarum (Mollusca: Bival-via). Comp. Biochem. Physiol. C, **122**, 253-263 (1999)

[23] Ward, D.E.; Kengen, S.W.; van Der Oost, J.; de Vos, W.M.: Purification and characterization of the alanine aminotransferase from the hyperthermophi-lic Archaeon pyrococcus furiosus and its role in alanine production. J. Bac-teriol., **182**, 2559-2566 (2000)

[24] Tanay, V.A.; Parent, M.B.; Wong, J.T.; Paslawski, T.; Martin, I.L.; Baker, G.B.: Effects of the antidepressant/antipanic drug phenelzine on alanine and alanine transaminase in rat brain. Cell. Mol. Neurobiol., 21, 325-339 (2001)

[25] Vozarova, B.; Stefan, N.; Lindsay, R.S.; Saremi, A.; Pratley, R.E.; Bogardus, C.; Tataranni, P.A.: High alanine aminotransferase is associated with decreased hepatic insulin sensitivity and predicts the development of type 2 diabetes. Diabetes, 51, 1889-1895 (2002)

Cysteine transaminase

2.6.1.3

1 Nomenclature

EC number

2.6.1.3

Systematic name

L-cysteine:2-oxoglutarate aminotransferase

Recommended name

cysteine transaminase

Synonyms

CGT

L-cysteine aminotransferase

L-cysteine-2-oxoglutarate aminotransferase

aminotransferase, cysteine

cysteine aminotransferase

Additional information <1> (<1> cytosolic cysteine aminotransferase, EC 2.6.1.3, is identical with cytosolic aspartate aminotransferase, EC 2.6.1.1 [2]) [2]

CAS registry number

9030-32-4

2 Source Organism

<1> *Rattus norvegicus* [1, 2]

3 Reaction and Specificity

Catalyzed reaction

L-cysteine + 2-oxoglutarate = mercaptopyruvate + L-glutamate

Reaction type

amino group transfer

Natural substrates and products

S L-cysteine + 2-oxoglutarate <1> (<1> important part of enzyme together with β-mercaptopyruvate sulfurtransferase in the regulation of thiosulfate formation [1]) (Reversibility: ? <1> [1]; r <1> [2]) [1, 2]

P L-glutamate + β-mercaptopyruvate <1> [1, 2]

Substrates and products

S L-alanine 3-sulfinic acid + 2-oxoglutarate <1> (<1> at pH 8.0 [2]) (Reversibility: ? <1> [2]) [2]

P pyruvate 3-sulfinic acid + L-glutamate <1> [2]

S L-cysteine + 2-oxoglutarate <1> (Reversibility: ? <1> [1]; r <1> [2]) [1, 2]

P L-glutamate + β-mercaptopyruvate <1> [1, 2]

Inhibitors

L-aspartate <1> [2]

α-methyl-DL-cysteine <1> (<1> inhibits transamination of cysteine [1]) [1]

maleate <1> [1]

sodium phosphate buffer <1> [1]

Cofactors/prosthetic groups

pyridoxal 5'-phosphate <1> (<1> pyridoxal phosphate protein [1]) [1]

Specific activity (U/mg)

2.085 <1> [1]

K_m-Value (mM)

0.06 <1> (2-oxoglutarate) [2]

22.2 <1> (L-cysteine) [2]

pH-Optimum

8.2 <1> [1]

9.7 <1> [2]

pH-Range

7-11 <1> (<1> about 60% of activity maximum at pH 7 and 11 [1]) [1]

Temperature optimum (°C)

37 <1> (<1> assay at [1]) [1]

4 Enzyme Structure

Molecular weight

74000 <1> (<1> gel filtration [2]) [2]

83500 <1> (<1> gel filtration [1]) [1]

Subunits

monomer <1> (<1> 1 * 84000, SDS-PAGE [1]) [1]

5 Isolation/Preparation/Mutation/Application

Source/tissue

liver <1> [1, 2]

Localization

cytosol <1> [2]

Purification

<1> (TEAE-cellulose, isoelectric focusing [1]) [1]

6 Stability

Temperature stability

Additional information <1> (<1> 1 h at room temperature, complete inactivation [1]) [1]

Storage stability

<1>, -20°C, 68% inactivated after 32 days [1]

References

[1] Thibert, R.J.; Schmidt, D.E.: Purification and partial characterization of cysteine-glutamate transaminase from rat liver. Can. J. Biochem., **55**, 958-964 (1977)
[2] Akagi, R.: Purification and characterization of cysteine aminotransferase from rat liver cytosol. Acta Med. Okayama, **36**, 187-197 (1982)

Glycine transaminase 2.6.1.4

1 Nomenclature

EC number
2.6.1.4

Systematic name
glycine:2-oxoglutarate aminotransferase

Recommended name
glycine transaminase

Synonyms
GlyAT <5, 6> [5, 6]
L-glutamate:glyoxylate aminotransferase
aminotransferase, glycine
glutamate-glycine transaminase <2> [2]
glutamate:glyoxylate aminotransferase <3> [3]
glutamic-glycine transaminase <1> [1]
glutamic-glyoxylic transaminase
glycine aminotransferase
glyoxylate-glutamate aminotransferase
glyoxylate-glutamic transaminase

CAS registry number
9032-99-9

2 Source Organism

<1> *Rattus norvegicus* [1]
<2> *Homo sapiens* [2]
<3> *Spinacia oleracea* (var. Longstanding Bloomsdale [3]) [3]
<4> *Lactobacillus plantarum* [4]
<5> *Zea mays* (var. Duet [5]; 5 isoforms with varying substrate specificities [5])
[5]
<6> *Rhodopseudomonas palustris* (No. 7 [6]) [6]

3 Reaction and Specificity

Catalyzed reaction
glycine + 2-oxoglutarate = glyoxylate + L-glutamate

Reaction type
amino group transfer

Natural substrates and products
S glyoxylate + L-glutamate <1-3, 5, 6> (<6> involved in the supply of glycine for extracellular porphyrin production [6]; <2> may be involved in
control of human oxalate synthesis by providing alternate pathway for
glyoxylate metabolism [2]) (Reversibility: ? <5, 6> [5, 6]; ir <1-3> [1-3])
[1-3, 5, 6]

P glycine + 2-oxoglutarate <1-3, 5, 6> [1-3, 5, 6]

Substrates and products
S glyoxylate + L-alanine <2, 3, 5> (<1> no activity [1]) (Reversibility: ir
<2,3> [2,3]; ? <5> [5]) [2, 3, 5]

P glycine + pyruvate

S glyoxylate + L-glutamate <1-3, 5, 6> (<6> highly specific for glyoxylate
and L-glutamate as substrates [6]; <2> best substrate [2]) (Reversibility:
? <5,6> [5,6]; ir <1-3> [1-3]) [1-3, 5, 6]

P glycine + 2-oxoglutarate <1-3, 5, 6> [1-3, 5, 6]

S glyoxylate + L-glutamine <2> (Reversibility: ir <2> [2]) [2]

P glycine + 2-oxoglutaramate

S Additional information <1, 2, 5> (<5> substrate specificities of isoforms
[5]; <1,2> substrate specificity [1,2]; <2> poor substrates are arginine or
methionine [2]; <1> no substrates are D-amino acid isomers [1]; <2> no
activity with serine, valine, aspartate, histidine, phenylalanine, tyrosine,
isoleucine [2]; <1> no activity with ammonia, glutathione, L-ornithine
or 4-aminobutanoate [1]) [1, 2, 5]

P ?

Inhibitors
3-chloro-L-alanine <5> [5]

Cd^{2+} <5> (<5> 85% inhibition at 1 mM [5]) [5]

Cu^{2+} <2> (<2> strong [2]) [2]

L-alanine <5> (<5> competitive [5]) [5]

L-serine <6> (<6> weak, 20% inhibition at 15 mM [6]) [6]

PCMB <2> (<1> no inhibition [1]) [2]

Zn^{2+} <5> (<5> weak, 25% inhibition at 1 mM [5]) [5]

aminooxyacetate <5> [5]

ammonium chloride <6> (<6> 50% inhibition at 15 mM [6]) [6]

glycine <6> (<6> weak, 30% inhibition at 15 mM [6]) [6]

hydroxylamine <1> (<1> 50% inhibition at 5 mM [1]; <2> no inhibition [2])
[1]

isonicotinic acid hydrazide <1> (<1> glyoxylate partially protects [1]; <2>
no inhibition [2]) [1]

maleic acid <1> (<1> weak [1]; <2> no inhibition [2]) [1]

phenylhydrazine <2> (<2> pyridoxal phosphate restores activity [2]) [2]

Additional information <2, 3, 5, 6> (<6> no inhibition by aspartate [6]; <6> no substrate inhibition [6]; <5> *p*-hydroxymercuribenzoate and cycloserine are poor inhibitors [5]; <3> no inhibition by phosphate [3]; <2> no inhibition by CN⁻ [2]) [2, 3, 5, 6]

Cofactors/prosthetic groups

pyridoxal 5'-phosphate <1, 2, 4-6> (<2,4> required [2,4]; <2> tightly bound to apoenzyme [2]; <1,2> addition to enzyme system mostly without effect [1,2]) [1, 2, 4-6]

Metals, ions

Mg^{2+} <4> (<4> requirement [4]) [4]

Turnover number (min⁻¹)

2808 <6> (L-glutamate, <6> pH 7.5, 30°C [6]) [6]
2892 <6> (glyoxylate, <6> pH 7.5, 30°C [6]) [6]

Specific activity (U/mg)

0.13 <1> (<1> partially purified enzyme [1]) [1]
0.185 <2> (<2> purified enzyme [2]) [2]
2.4 <3> (<3> peroxisomes [3]) [3]
3.25 <6> (<6> purified enzyme [6]) [6]

K_m-Value (mM)

2 <2> (glutamate, <2> pH 7.7, 37°C [2]) [2]
2 <2> (glyoxylate, <2> pH 7.7, 37°C [2]) [2]
2.75 <3, 4> (glutamate, <4> pH 8.0, 35-37°C [4]; <3> pH 7.0, 25°C [3]) [3, 4]
3.75 <6> (glyoxylate, <6> pH 7.5, 30°C [6]) [6]
4.6 <1> (glutamate, <1> pH 7.3 [1]) [1]
6.2 <6> (L-glutamate, <6> pH 7.5, 30°C [6]) [6]
6.25 <3, 4> (glyoxylate, <4> pH 8.0, 35-37°C [4]; <3> pH 7.0, 25°C [3]) [3, 4]
8.3 <1> (glyoxylate, <1> pH 7.3 [1]) [1]

pH-Optimum

7 <3, 6> [3, 6]
7.3 <1, 2> [1, 2]
7.5 <5> (<5> assay at [5]) [5]
8 <4> (<4> phosphate or veronal buffers [4]) [4]
Additional information <3> (<3> pI: 5.8 [3]) [3]

pH-Range

5-7.5 <6> (<6> 80% activity at pH 5.0 and pH 7.5 [6]) [6]
6.2-8.2 <1> (<1> about half-maximal activity at pH 6.2 and pH 8.2 [1]) [1]
6.3-8.5 <2> (<2> about half-maximal activity at pH 6.3 and about 60% of maximal activity at pH 8.5 [2]) [2]

Temperature optimum (°C)

25 <3> (<3> assay at [3]) [3]
30 <5> (<5> assay at [5]) [5]
35-37 <4> [4]
37 <2> (<2> assay at [2]) [2]
40-45 <6> [6]

4 Enzyme Structure

Molecular weight
   37000 <4> (<4> gel filtration [4]) [4]
   45500 <6> (<6> gel filtration [6]) [6]
   50000 <5> (<5> 5 isoforms, gel filtration [5]) [5]

Subunits
   monomer <6> (<6> 1 * 42000, SDS-PAGE [6]) [6]

5 Isolation/Preparation/Mutation/Application

Source/tissue
   heart <1> (<1> limited activity [1]) [1]
   kidney <1> [1]
   leaf <3, 5> [3, 5]
   liver <1, 2> [1, 2]
   spleen <1> [1]
   Additional information <1, 6> (<6> activity depends on growth phase [6];
   <1> tissue distribution, very low activity in lung, testes or diaphragm [1]) [1,
   6]

Localization
   peroxisome <3> [3]
   Additional information <3> (<3> subcellular distribution [3]) [3]

Purification
   <1> (partial, about 30fold [1]) [1]
   <2> (290fold [2]) [2]
   <4> (84fold [4]) [4]
   <5> (partial, 5 isoforms [5]) [5]
   <6> (191fold to homogeneity [6]) [6]

6 Stability

pH-Stability
   9 <6> (<6> inactivation [6]) [6]

Temperature stability
   60 <1> (<1> 5 min, stable [1]) [1]
   80 <6> (<6> inactivation [6]) [6]

Storage stability
   <6>, -20°C, fairly stable for 8 months [6]

References

[1] Nakada, H.I.: Glutamic-glycine transaminase from rat liver. J. Biol. Chem., **239**, 468-471 (1964)

[2] Thompson, J.S.; Richardson, K.E.: Isolation and characterization of a glutamate-glycine transaminase from human liver. Arch. Biochem. Biophys., **117**, 599-603 (1966)

[3] Rehfeld, D.W.; Tolbert, N.E.: Aminotransferases in peroxisomes from spinach leaves. J. Biol. Chem., **247**, 4803-4811 (1972)

[4] Galas, E.; Florianowicz, T.: L-glutamate-glyoxylate aminotransferase in Lactobacillus plantarum. Acta Microbiol. Pol., **7**, 243-252 (1975)

[5] Orzechowski, S.; Socha-Hanc, J.; Paszkowski, A.: Alanine aminotransferase and glycine aminotransferase from maize (Zea mays L.) leaves. Acta Biochim. Pol., **46**, 447-457 (1999)

[6] Yamaguchi, H.; Ohtani, M.; Amachi, S.; Shinoyama, H.; Fujii, T.: Some properties of glycine aminotransferase purified from Rhodopseudomonas palustris No. 7 concerning extracellular porphyrin production. Biosci. Biotechnol. Biochem., **67**, 783-789 (2003)

Tyrosine transaminase

1 Nomenclature

EC number
2.6.1.5

Systematic name
L-tyrosine:2-oxoglutarate aminotransferase

Recommended name
tyrosine transaminase

Synonyms
L-phenylalanine 2-oxoglutarate aminotransferase
L-tyrosine aminotransferase
L-tyrosine-2-oxoglutarate aminotransferase
aminotransferase, tyrosine
glutamic phenylpyruvic aminotransferase
glutamic-hydroxyphenylpyruvic transaminase
phenylalanine aminotransferase
phenylalanine transaminase
phenylalanine-α-ketoglutarate transaminase
phenylpyruvate transaminase
phenylpyruvic acid transaminase
tyrosine aminotransferase
tyrosine-2-ketoglutarate aminotransferase
tyrosine-2-oxoglutarate aminotransferase
tyrosine-α-ketoglutarate aminotransferase
tyrosine-α-ketoglutarate transaminase
Additional information <4> (<4> the mitochondrial enzyme may be identical with EC 2.6.1.1 and with EC 2.6.1.57 [1]) [1]

CAS registry number
9014-55-5

2 Source Organism

<1> *Crithidia fasciculata* [17]
<2> *Agama agama* (rainbow lizard [18]) [18]
<3> *Drosophila hydei* [19]

<4> *Rattus norvegicus* (4 enzyme forms from kidney, heart and brain [6]; 4
forms, one of them is probably identical to mitochondrial L-aspartate
aminotransferase EC 2.6.1.1 [20]) [1-12, 15, 20, 22, 24-26]
<5> *Rana temporaria* (frog [16]) [16, 20]
<6> *Trichoderma viride* [13]
<7> *Anchusa officinalis* [14]
<8> *Trypanosoma cruzi* (epimastigote [21,27]) [21, 27]
<9> *Escherichia coli* [28]
<10> *Felis catus* [23]
<11> *Gallus gallus* (chicken [29]) [29]

3 Reaction and Specificity

Catalyzed reaction
L-tyrosine + 2-oxoglutarate = 4-hydroxyphenylpyruvate + L-glutamate

Reaction type
amino group transfer

Natural substrates and products
  S L-tyrosine + 2-oxoglutarate <5, 7> (<7> first step in tyrosine pathway
     leading to the formation of rosmarinic acid (α-O-caffeoyl-3,4-dihydroxy-
     phenyllactic acid) [14]) (Reversibility: ? <5, 7> [14, 16]) [14, 16]
  P 4-hydroxyphenylpyruvate + L-glutamate <5, 7> [14, 16]

Substrates and products
  S (2-aminophenyl)alanine + 2-oxoglutarate <4> (<4> 10% of the activity
     that with L-tyrosine [12]) (Reversibility: ? <4> [12]) [12]
  P 3-(4-aminophenyl)-2-oxopropanoate + L-glutamate <4> [12]
  S (3-fluorophenyl)alanine + 2-oxoglutarate <4> (<4> 4% of the activity that
     with L-tyrosine [12]) (Reversibility: ? <4> [12]) [12]
  P 3-(3-fluorophenyl)-2-oxopropanoate + L-glutamate <4> [12]
  S (4-chlorophenyl)alanine + 2-oxoglutarate <4> (<4> 17% of the activity
     with L-tyrosine [1]) (Reversibility: ? <4> [1]) [1]
  P 3-(3-chlorophenyl)-2-oxopropanoate + L-glutamate <4> [1]
  S (4-fluorophenyl)alanine + 2-oxoglutarate <4> (<4> 7.9% of the activity
     that with L-tyrosine [12]) (Reversibility: ? <4> [12]) [12]
  P 3-(4-fluorophenyl)-2-oxopropanoate + L-glutamate <4> [12]
  S 3,4-dihydroxyphenylalanine + 2-oxoglutarate <4> (<4> 85% of the activ-
     ity than with L-tyrosine [1]) (Reversibility: ? <4> [1,12]) [1, 12]
  P 3-(3,4-dihydroxyphenyl)-2-oxopropanoate + L-glutamate <4> [1, 12]
  S 3-aminotyrosine + 2-oxoglutarate <4> (<4> 16% of the activity than with
     L-tyrosine [12]) (Reversibility: ? <4> [12]) [12]
  P 3-(3-amino-4-hydroxyphenyl)-2-oxopropanoate + L-glutamate <4> [12]
  S 3-iodotyrosine + 2-oxoglutarate <4> (<4> as effective as L-tyrosine [12];
     <4> 84% of the activity than with L-tyrosine [1]) (Reversibility: ? <4>
     [1,12]) [1, 12]

P 3-(4-hydroxy-3-iodophenyl)-2-oxopropanoate + L-glutamate <4> [1, 12]
S 3-methoxytyrosine + 2-oxoglutarate <4, 4> (<4> 22% of the activity with
   L-tyrosine [1]) (Reversibility: ? <4> [1]) [1]
P 3-(3-methoxy-4-hydroxyphenyl)-2-oxopropanoate + L-glutamate <4> [1]
S 4-hydroxyphenylpyruvate + L-aspartate <4> (<4> 35% of the activity
   with L-glutamate [1]) (Reversibility: r <4> [1]) [1]
P L-tyrosine + oxaloacetate <4> [1]
S L-asparagine + 2-oxoglutarate <4> (<4> 15% of the activity with L-tyro-
   sine [1]) (Reversibility: ? <4> [1]) [1]
P 2-oxosuccinamate + L-glutamate <4> [1]
S L-aspartate + 2-oxoglutarate <4, 7> (<7> isoenzymes TAT-2 and TAT-3,
   best substrate for isoenzyme TAT-3 [14]; <4> higher activity than with L-
   tyrosine [1]) (Reversibility: ? <4,7> [1,14]) [1, 14]
P oxaloacetate + L-glutamate <4, 7> [1, 14]
S L-cysteine + 2-oxoglutarate <4> (<4> higher activity than with L-tyrosine
   [1]) (Reversibility: ? <4> [1]) [1]
P 3-mercapto-2-oxopropanoate + L-glutamate <4> [1]
S L-ethionine + 2-oxoglutarate <4> (<4> 17% of the activity than with L-
   tyrosine [1]) (Reversibility: ? <4> [1]) [1]
P 4-ethylsulfanyl-2-oxobutanoate + L-glutamate <4> [1]
S L-phenylalanine + 2-oxoglutarate <1, 4, 6> (<4> higher rate than with L-
   tyrosine [1]; <4> lower activity than with L-tyrosine [12]; <6> no activity
   [13]) (Reversibility: ? <1,4,6> [1, 6, 12, 13, 15, 17]) [1, 6, 12, 13, 15, 17]
P phenylpyruvate + L-glutamate <1, 4, 6> [1, 6, 12, 13, 15, 17]
S L-tyrosine + 2-oxoglutarate <1-9> (<4> maximal activity with L-tyrosine
   [12]; <7> isoenzymes TAT-1, TAT-2 and TAT-3 show a pronounced preference for L-tyrosine over other aromatic amino acids [14]; <2,6> highly
   specific for: L-tyrosine [13,18]; <7> isoenzyme TAT-1: oxaloacetate and 2-
   oxoglutarate utilized equally well as amino acceptor, isoenzyme TAT-2:
   oxaloacetate is most effective, isoenzyme TAT-3: 2-oxoglutarate is most
   effective [14]; <5> specific for 2-oxoglutarate as the amino group acceptor
   [16]) (Reversibility: r <4> [1]; ? <1-3,5-9> [2-29]) [1-29]
P 4-hydroxyphenylpyruvate + L-glutamate <1-8> [1-29]
S L-tyrosine + oxaloacetate <1, 4-7> (<5, 6> no activity [13, 16]) (Reversibility: r <4> [1]; ? <1,5-7> [13, 14, 16, 17]) [1, 13, 14, 16, 17]
P 4-hydroxyphenylpyruvate + L-aspartate [1, 13, 14, 16, 17]
S methionine + 2-oxoglutarate <4> (<4> 20% of the activity than with L-
   tyrosine [1]) (Reversibility: ? <4> [1]) [1]
P 4-methylsulfanyl-2-oxobutanoate + L-glutamate <4> [1]
S tryptophan + 2-oxoglutarate <4, 6> (<4> 62% of the activity than with L-
   tyrosine [1]; <6> no activity [13]) (Reversibility: ? <4> [1, 6, 12, 15]) [1,
   6, 12, 13, 15]
P 3-indole-2-oxopropanoate + L-glutamate <4, 6> [1, 6, 12, 13, 15]

Inhibitors

(4-hydroxyphenyl)acetic acid <4> (<4> 74% inhibition with tyrosine at
3 mM and (4-hydroxyphenyl)acetic acid at 12 mM [12]) [12]

2-oxoglutarate <3> (<3> substrate inhibition at high concentration [19]) [19]
3,4-dihydroxyphenylalanine <4> [12]
3,4-dihydroxyphenyllactate <7> (<7> strong inhibition of all the three TAT isoenzymes [14]) [14]
3-(3,4-dihydroxyphenyl)-2-methylalanine <4> [12]
3-methoxydopamine <4> [12]
3-methoxytyrosine <4> [12]
5-hydroxyindole acetic acid <4> (<4> 55% inhibition with tyrosine at 3 mM and 5-hydroxyindole acetic acid at 12 mM [12]) [12]
5-hydroxytryptophan <4> (<4> 30% inhibition with tyrosine at 3 mM and 5-hydroxytryptophan at 12 mM [12]) [12]
Ag^{2+} <2> [18]
Ca^{2+} <2> [18]
D-aspartate <4> (<4> weak inhibition [1]) [1]
D-tyrosine <4> (<4> 9% inhibition with tyrosine at 3 mM and D-tyrosine at 12 mM [12]) [12]
Hg^{2+} <2> [18]
L-aspartate <4> (<4> with L-tyrosine and 2-oxoglutarate or oxaloacetate as substrates [1]) [1]
L-glutamate <2, 4> (<2> competitive inhibition [18]) [6, 18]
Mn^{2+} <2> [18]
Zn^{2+} <2> [18]
α-methyl-L-aspartate <4> (<4> with L-tyrosine and 2-oxoglutarate or oxaloacetate as substrates [1]) [1]
α-methyltyrosine <4> [12]
aminooxyacetate <5> (<5> probably acts reacting with the cofactor [16]) [16]
β-methyl-L-aspartate <4> (<4> with L-tyrosine and 2-oxoglutarate or oxaloacetate as substrates [1]) [1]
canaline <4, 5> (<5> probably acts reacting with the cofactor [16]) [15, 16]
dihydroxymandelic acid <4> (<4> 65% inhibition with tyrosine at 3 mM and dihydroxymandelic acid at 3 mM [12]) [12]
dihydroxyphenylacetic acid <4> (<4> 88% inhibition with tyrosine at 3 mM and dihydroxyphenylacetic acid at 3 mM [12]) [12]
dopamine <4> (<4> 100% inhibition with tyrosine at 3 mM and dopamine at 12 mM [12]) [12]
homogentisate <5> (<5> half-maximal inhibition with 15 micromolar [16]) [16]
homovanillic acid <4> [12]
indole-3-acetic acid <4> (<4> 42% inhibition with tyrosine at 3 mM and indole-3-acetic acid at 12 mM [12]) [12]
indole-3-butyric acid <4> (<4> 71% inhibition with tyrosine at 3 mM and indole-3-butyric acid at 12 mM [12]) [12]
indole-3-propionic acid <4> (<4> 48% inhibition with tyrosine at 3 mM and indole-3-propionic acid at 12 mM [12]) [12]
norepinephrine <4> (<4> competitive inhibition [12]) [12]
normetanephrine <4> [12]

phenylacetic acid <4> (<4> 30% inhibition with tyrosine at 3 mM and phenylacetic acid at 12 mM [12]) [12]
phenylethylamine <4> (<4> 80% inhibition with tyrosine at 3 mM and phenylethylamine at 12 mM [12]) [12]
rosmarinic acid <7> (<7> isoenzymes TAT-2 and TAT-3 [14]) [14]
serotonin <4> (<4> 37% inhibition with tyrosine at 3 mM and serotonin at
12 mM [12]) [12]
tyramine <4> (<4> 13% inhibition with tyrosine at 3 mM and tyramine at
12 mM [12]) [12]
vanillylmandelic acid <4> (<4> 51% inhibition with tyrosine at 3 mM and
vanillylmandelic acid at 3 mM [12]) [12]

Cofactors/prosthetic groups

pyridoxal 5'-phosphate <4, 8> (<4> two mol coenzyme per mol of enzyme
[3]; <8> relative position of the cofactor at the active site determined by X-
ray crystallography [27]) [3, 27]
pyridoxamine 5'-phosphate <4> (<4> can replace pyridoxal 5'-phosphate
[6]) [6]

Activating compounds

adrenaline <5> (<5> activation is independent of the concentration of tyrosine, maximal activation achieved with 7-10 micromolar of adrenaline [16])
[16]
dopamine <5> (<5> slight activation [16]) [16]
metanephrine <5> (<5> slight activation [16]) [16]
noradrenaline <5> (<5> slight activation [16]) [16]

Specific activity (U/mg)

0.155 <4> [1]
6.06 <8> [21]
203 <5> (<5> isoenzyme I [16]) [16]
290.4 <4> [3]
400 <4> [11]
416 <4> (<4> enzyme expressed in Escherichia coli [15]) [15]
616 <4> [2]
655 <4> [5]
1200 <4> (<4> from rats stimulated with corticosterone and dexamethasone
[7]) [7]
Additional information <2, 4, 7, 11> (<4> activities of different isoenzymes
from liver, kidney and heart [6]; <4> effect of glucagon, cyclic nucletides,
insulin and hydrocortisone on activity [6]; <4> effect of induction with hydrocortisone on activity [12]; <7> activities for different isoenzymes [14];
<4> effect of estradiol and ovariectomy on activity [22]; <10> effect of vita-
min B-6 deficiency on activity [23]; <4> effect of induction with dexamethasone on activity [24]; <4> effect of induction with cAMP on activity [25];
<4> effect of acute and chronic ethanol administration on synthesis and ac-
tivity of TAT [26]; <11> effect of chronic heat stress on activity [29]) [6, 8, 11,
12, 14, 15, 18, 22-26, 29]

K$_m$-Value (mM)

0.25 <6> (2-oxoglutarate, <6> pH 7.6, 37°C [13]) [13]

0.45 <7> (L-tyrosine, <7> pH 9.2, 40°C, isoenzyme TAT-3 [14]) [14]

0.5 <6> (L-tyrosine, <6> pH 7.6, 37°C [13]) [13]

0.5 <8> (pyruvate, <8> pH 7.0, 37°C, tyrosine as co-substrate [21]) [21]

0.7 <4> (2-oxoglutarate, <4> pH 7.2, 37°C [15]) [15]

0.9 <4> (2-oxoglutarate, <4> pH 7.6, 37°C, enzyme form I [6]) [6]

1 <4> (2-oxoglutarate, <4> pH 7.6, 37°C, enzyme form III [6]) [6]

1.1 <4> (2-oxoglutarate, <4> pH 7.6, 37°C, enzyme form II [6]) [6]

1.2 <4> (2-oxoglutarate, <4> pH 7.6, 37°C, enzyme form IV [6]) [6]

1.4 <4> (L-tyrosine, <4> pH 7.6, 37°C, enzyme form II [6]) [6]

2.1 <10> (L-tyrosine, <10> pH 7.6, 37°C [23]) [23]

3.4 <4> (L-tyrosine, <4> pH 7.2, 37°C [15]) [15]

5 <7> (L-tyrosine, <7> pH 9.2, 40°C, isoenzyme TAT-2 [14]) [14]

6.8 <8> (L-tyrosine, <8> pH 7.0, 37°C, pyruvate as co-substrate [21]) [21]

15 <4> (L-tyrosine, <4> pH 7.9, 37°C [12]) [12]

16 <8> (oxaloacetate, <8> pH 7.0, 37°C, tyrosine as co-substrate [21]) [21]

17.9 <8> (phenylalanine, <8> pH 7.0, 37°C, pyruvate as co-substrate [21]) [21]

20 <7> (L-tyrosine, <7> pH 9.2, 40°C, isoenzyme TAT-1 [14]) [14]

21.4 <8> (tryptophan, <8> pH 7.0, 37°C, pyruvate as co-substrate [21]) [21]

26 <4> (3,4-dihydroxyphenylalanine, <4> pH 7.9, 37°C [12]) [12]

30 <4> (tryptophan, <4> pH 7.9, 37°C [12]) [12]

38 <8> (2-oxoglutarate, <8> pH 7.0, 37°C, tyrosine as co-substrate [21]) [21]

60 <4> (tryptophan, <4> pH 7.2, 37°C [15]) [15]

80 <4> (phenylalanine, <4> pH 7.9, 37°C [12]) [12]

Additional information <2-5> (<5> non-Michaelis complex kinetic [16]) [18-20]

K$_i$-Value (mM)

20.3 <4> (L-glutamate, <4> pH 7.6, 37°C, isozyme I from kidney, L-tyrosine as substrate [6]) [6]

26.8 <4> (L-glutamate, <4> pH 7.6, 37°C, isozyme IV from liver, L-tyrosine as substrate [6]) [6]

27.4 <4> (L-glutamate, <4> pH 7.6, 37°C, isozyme III from liver, L-tyrosine as substrate [6]) [6]

31.3 <4> (L-glutamate, <4> pH 7.6, 37°C, isozyme II from liver, L-tyrosine as substrate [6]) [6]

Additional information <2> (<2> inhibition due to L-glutamate [18]) [18]

pH-Optimum

6.8 <3> [19]

7 <8> [21]

7.6 <2, 6> [13, 18]

7.7 <4> (<4> recombinant enzyme expressed in Escherichia coli [15]) [15]

9 <7> (<7> L-tyrosine as substrate, isoenzyme TAT-1 [14]) [14]

Additional information <7> (<7> optimum pH for the isoenzyme TAT-2 is above pH 9.6 [14]) [14]

pH-Range

6-8 <6> [13]

8.8-9.4 <7> (<7> L-tyrosine as substrate, isoenzyme TAT-3 [14]) [14]

Additional information <4> [6]

Temperature optimum (°C)

37 <4-6> (assay at [1, 5, 7, 11, 13, 16]) [1, 5, 7, 11, 13, 16]

40 <7> (assay at [14]) [14]

4 Enzyme Structure

Molecular weight

85000 <5> (<5> gel filtration, isoforms I and II [16]) [16]

90000 <4> (<4> isoenzymes: TAT-I, TAT-II, TAT-III, gel filtration [8]; <4> amino acid composition and gel filtration [1]) [1, 8]

91000 <8> (<8> gel filtration [21]) [21]

100000 <1> [17]

100000 <4> (<4> gel filtration [3]) [3]

104500 <4> (<4> ultracentrifugation [7]) [7]

105000 <4> (<4> glycerol gradient centrifugation [3]) [3]

107000 <4> (<4> gel filtration, different isoenzymes [6]) [6]

110000 <4, 6> (<4> recombinant expressed in Escherichia coli, ultracentrifugation [15]; <6> gel filtration [13]) [13, 15]

110500 <4> (<4> sucrose density gradient sedimentation [4]) [4]

114000 <4> (<4> sedimentation analysis [1]) [1]

130000 <2> [18]

150000 <4> (<4> recombinant expressed in Escherichia coli, native gradient PAGE [15]) [15]

160000 <4> (<4> recombinant expressed in Escherichia coli, gel filtration [15]) [15]

180000 <7> (<7> isoenzyme TAT-1, gel filtration [14]) [14]

220000 <7> (<7> isoenzymes TAT-2 and TAT-3, gel filtration [14]) [14]

Subunits

dimer <4, 8> (<4> 2 * 50000, SDS-PAGE [3,15]; <4> 2 * 53000, SDS-PAGE [4]; <4> 2 * 44000, isoenzymes TAT-I and TAT-II, SDS-PAGE [8]; <8> 2 * 45000, SDS-PAGE [21]) [3, 4, 8, 15, 21]

heterodimer <1> (<1> 1 * 50000 + 1 * 48000 [17]) [17]

monomer <3> (<3> 1 * ?, SDS-PAGE [19]) [19]

oligomer <4> (<4> ? * 50000, SDS-PAGE [5]; <4> x * 52000, SDS-PAGE [7]) [5, 7]

tetramer <7> (<7> 4 * 43000, isoenzyme TAT-1, SDS-PAGE [14]; <7> 4 * 56000, isoenzyme TAT-2, SDS-PAGE [14]) [14]

Posttranslational modification

proteolytic modification <4> (<4> liver isoform I is converted into functional lower molecular weight isoenzymes by a lysosomal convertase [9]) [9]

Additional information <1, 4, 8> (<4> no measurable amount of fucose, mannose, galactose, N-acetylglucose, N-acetylgalactose nor sialic acid [7]; <1,8> not a glycoprotein [17,21]) [7, 17, 21]

5 Isolation/Preparation/Mutation/Application

Source/tissue
cell suspension culture <7, 8> [14, 21]
heart <4> [6]
kidney <4> [6]
liver <2, 4, 5> (<4> cultured hepatocytes [24,25]) [1-12, 15, 16, 18, 20, 24, 25]
salivary gland <3> [19]

Localization
cytoplasm <2, 4, 5> [7, 8, 10, 18, 20]
mitochondrion <4> [1]

Purification
<1> (partial [17]) [17]
<2> (partial, using ammonium sulfate precipitation and several chromato-graphic steps [18]) [18]
<3> (partial [19]) [19]
<4> (partial, using Sephadex chromatography and isoelectrofocusing [1]; 4 enzyme forms separated by isoelectrofocusing and hydroxyapatite chromato-graphy, one of them is probably identical to mitochondrial L-aspartate ami-notransferase EC 2.6.1.1 [20]; separation of 4 enzyme forms: I, II, III, IV, using hydroxyapatite chromatography [6]; partial purification of enzyme ex-pressed in Saccharomyces cerevisiae and Escherichia coli using affinity chro-matography [15]; high yield procedure using a thermal inactivation of the lysosomal converting factor that generates two additional, lower molecular weight forms [2]; native and one of the modified forms [4]; 3 enzyme forms isolated using pyridoxamine-affinity column chromatography [5]; isoen-zymes TAT-I and TAT-II [8]; partial, using pyridoxamine-affinity column cheomatography [11]) [1-8, 11, 15, 20]
<5> (partial purification of two isoenzymes, using several chromatographic steps [16]) [16, 20]
<6> (partial, using gel filtration [13]) [13]
<7> (three isoenzymes TAT-1, TAT-2 and TAT-3, isolated using anion-ex-change chromatography and electrofocusing [14]) [14]
<8> (partial, using chromatography on DEAE-cellulose and gel filtration [21]) [21]
<10> (partial [23]) [23]

Crystallization
<4> (spontaneous crystallization when enzyme concentration is about 10 mg protein per ml [1]) [1]
<8> (structure determined at 2.5 A resolution [27]) [27]

<9> (low resolution structure of the enzyme bound to pyridoxal 5'-phosphate [28]) [28]

Cloning

<4> (expression of rat liver enzyme in Saccharomyces cerevisiae and Escherichia coli [15]) [15]

6 Stability

Temperature stability

50 <7> (<7> 15 min, 50% loss of activity, isoenzyme TAT-3 [14]) [14]
55 <7> (<7> 15 min, 50% loss of activity, isoenzyme TAT-2 [14]) [14]
65 <7, 8> (<7> 15 min, 50% loss of activity, isoenzyme TAT-1 [14]; <8> complete inactivation after 5 min [21]) [14, 21]
Additional information <2, 4> (<2,4> pyridoxal 5'-phosphate protects against thermal inactivation [15,18]; <2> 2-oxoglutarate protects against thermal inactivation [18]; <4> the apoenzyme is thermally unstable: $t_{1/2}$ of 2 min at 55°C [15]) [15, 18]

General stability information

<2>, 2-oxoglutarate protects against thermal inactivation [18]
<2>, sensitive to SH inactivation [18]
<2, 4>, pyridoxal 5'-phosphate protects against thermal inactivation [15, 18]

Storage stability

<4>, -20°C, 50 mM sodium phosphate, pH 6.5, 2 mM EDTA, 1 mM 2-mercaptoethanol, 2.5 mM 2-oxoglutarate [5]
<4>, -20°C, 50% v/v glycerol, 25 mM phosphate buffer, pH 7.6, 0.1 mM pyridoxal 5'-phosphate, 0.5 mM 2-oxoglutarate, 1 mM DTT, stable for several months [7]

References

[1] Miller, J.E.; Litwack, G.: Purification, properties, and identity of liver mitochondrial tyrosine aminotransferase. J. Biol. Chem., 246, 3234-3240 (1971)
[2] Hargrove, J.L.; Granner, D.K.: Purification of the native form of tyrosine aminotransferase from rat liver. Anal. Biochem., 104, 231-235 (1980)
[3] Lee, K.L.; Roberson, L.E.; Kenney, F.T.: Properties of tyrosine aminotransferase from rat liver. Anal. Biochem., 95, 188-193 (1979)
[4] Hargrove, J.L.; Granner, D.K.: Physical properties, limited proteolysis, and acetylation of tyrosine aminotransferase from rat liver. J. Biol. Chem., 256, 8012-8017 (1981)
[5] Donner, P.; Wagner, H.; Kröger, H.: Tyrosine aminotransferase from rat liver, a purification in three steps. Biochem. Biophys. Res. Commun., 80, 766-772 (1978)

[6] Iwasaki, Y.; Lamar, C.; Danenberg, K.; Pitot, H.C.: Studies on the induction and repression of enzymes in rat liver. Characterization and metabolic regulation of multiple forms of tyrosine aminotransferase. Eur. J. Biochem., **34**, 347-357 (1973)

[7] Belarbi, A.; Bollack, C.; Befort, N.; Beck, J.P.; Beck, G.: Purification and characterization of rat liver tyrosine aminotransferase. FEBS Lett., **75**, 221-225 (1977)

[8] Roewekamp, W.; Sekeris, C.E.: Purification and subunit structure of tyrosine aminotransferase from rat liver cytosol. FEBS Lett., **73**, 225-228 (1977)

[9] Gohda, E.; Pitot, H.C.: Purification and characterization of a factor catalyzing the conversion of the multiple forms of tyrosine aminotransferase from rat liver. J. Biol. Chem., **255**, 7371-7379 (1980)

[10] Johnson, R.W.; Roberson, L.E.; Kenney, F.T.: Regulation of tyrosine aminotransferase in rat liver. X. Characterization and interconversion of the multiple enzyme forms. J. Biol. Chem., **248**, 4521-4527 (1973)

[11] Miller, J.V.; Cuatrecasas, P.; Thompson, E.B.: Purification of tyrosine aminotransferase by affinity chromatography. Biochim. Biophys. Acta, **276**, 407-415 (1972)

[12] Jacoby, G.A.; La Du, B.N.: Studies on the specificity of tyrosine-α-ketoglutarate transaminase. J. Biol. Chem., **239**, 419-424 (1964)

[13] Echetebu, C.O.: Some properties of tyrosine aminotransferase from Trichoderma viride. J. Gen. Microbiol., **128**, 2735-2738 (1982)

[14] De-Eknamkul, W.; Ellis, B.E.: Purification and characterization of tyrosine aminotransferase activities from Anchusa officinalis cell cultures. Arch. Biochem. Biophys., **257**, 430-438 (1987)

[15] Dietrich, J.B.; Lorber, B.; Kern, D.: Expression of mammalian tyrosine aminotransferase in Saccharomyces cerevisiae and Escherichia coli. Purification to homogeneity and characterization of the enzyme overproduced in the bacteria. Eur. J. Biochem., **201**, 399-407 (1991)

[16] Ohisalo, J.J.; Andersson, S.M.; Pispa, J.P.: Partial purification and properties of frog liver tyrosine aminotransferase. Biochem. J., **163**, 411-417 (1977)

[17] Rege, A.A.: Purification and characterization of a tyrosine aminotransferase from Crithidia fasciculata. Mol. Biochem. Parasitol., **25**, 1-9 (1987)

[18] Echetebu, C.O.; Ifem, F.M.; Echetebu, Z.O.: Hepatic tyrosine aminotransferase from the rainbow lizard Agama agama: purification and some properties. Biochimie, **69**, 223-230 (1987)

[19] Belew, K.; Brady, T.: Tyrosine aminotransferase from Drosophila hydei salivary glands: characterization, purification and antibody production. Insect Biochem., **11**, 239-245 (1981)

[20] Ohisalo, J.J.; Pispa, J.P.: Heterogeneity of hepatic tyrosine aminotransferase. Separation of the multiple forms from rat and frog liver by isoelectro focussing and hydroxyapatite column chromatography and their partial characterization. Acta Chem. Scand., **B30**, 491-500 (1976)

[21] Montemartini, M.; Santome, J.A.; Cazzulo, J.J.; Nowick, C.: Purification and partial structural and kinetic characterization of tyrosine aminotransferase from epimastigotes of Trypanosoma cruzi. Biochem. J., **292**, 901-906 (1993)

[22] Presch, I.; Birnbacher, R.; Herkner, K.; Lubec, G.: The effect of estradiol and ovariectomy on tyrosine hydroxylase, tyrosine aminotransferase and phenylalanine hydroxylase. Life Sci., **60**, 479-484 (1997)

[23] Bai, S.C.; Rogers, Q.R.; Wong, D.L.; Sampson, D.A.; Morris, J.G.: Vitamin B$_6$ deficiency and level of dietary protein affect hepatic tyrosine aminotransferase activity in cats. J. Nutr., **128**, 1995-2000 (1998)

[24] Nimi, S.; Yamaguchi, T.; Hayakawa, T.: Effect of dexamethasone pretreatment on the dexamethasone-dependent induction of tyrosine aminotransferase activity in primary cultured rat hepatocytes. Biol. Pharm. Bull., **21**, 1009-1012 (1998)

[25] Pickering, C.S.; Watkins, R.H.; Dickson, A.J.: Rat primary hepatocytes and H4 hepatoma cells display differential sensitivity to cyclic AMP at the level of expression of tyrosine aminotransferase. Biochem. Biophys. Res. Commun., **252**, 764-769 (1998)

[26] Donohue, T.M., Jr.; Drey, M.L.; Zetterman, R.K.: Contrasting effects of acute and chronic ethanol administration on rat liver tyrosine aminotransferase. Alcohol, **15**, 141-146 (1998)

[27] Blankenfeldt, W.; Nowicki, C.; Montemartini-Kalisz, M.; Kalisz, H.M.; Hecht, H.J.: Crystal structure of Trypanosoma cruzi tyrosine aminotransferase: substrate specificity is influenced by cofactor binding mode. Protein Sci., **8**, 2406-2417 (1999)

[28] Ko, T.P.; Wu, S.P.; Yang, W.Z.; Tsai, H.; Yuan, H.S.: Crystallization and preliminary crystallographic analysis of the Escherichia coli tyrosine aminotransferase. Acta Crystallogr. Sect. D, **55 (Pt 8)**, 1474-1477 (1999)

[29] Tabiri, H.Y.; Sato, K.; Takahashi, K.; Toyomizu, M.; Akiba, Y.: Hepatic tyrosine aminotransferase activity is affected by chronic heat stress and dietary tyrosine in broiler chickens. Br. Poult. Sci., **43**, 629-634 (2002)

Leucine transaminase   2.6.1.6

1 Nomenclature

EC number
2.6.1.6

Systematic name
L-leucine:2-oxoglutarate aminotransferase

Recommended name
leucine transaminase

Synonyms
L-leucine aminotransferase
leucine 2-oxoglutarate transaminase
leucine aminotransferase
leucine-α-ketoglutarate transaminase

CAS registry number
9030-37-9

2 Source Organism

<1> *Amaranthus dubius* [5]
<2> *Glycine max* (soybean, var. Kali tur [3]) [3]
<3> *Rattus norvegicus* (rat, strain Wistar [2]) [1, 2, 4]

3 Reaction and Specificity

Catalyzed reaction
L-leucine + 2-oxoglutarate = 4-methyl-2-oxopentanoate + L-glutamate

Reaction type
amino group transfer

Natural substrates and products
S L-leucine + 2-oxoglutarate <2, 3> (<3> leucine metabolism and hence of
cholesterogenesis and ketogenesis, not present in fetal rat liver, but after
birth activity appears and increases rapidly [1]) (Reversibility: ? <2, 3>
[1-3]) [1-3]
P 4-methyl-2-oxopentanoate + L-glutamate <2, 3> [1-3]
S L-methionine + 2-oxoglutarate <3> (Reversibility: ? <3> [2]) [2]
P 4-methylsulfanyl-2-oxobutanoate + L-glutamate

Substrates and products

S L-leucine + 2-oxoglutarate <2, 3> (Reversibility: r <2,3> [1-4]) [1-4]
P 4-methyl-2-oxopentanoate + L-glutamate <2, 3> [1-3]
S L-leucine + 4-methyl-2-oxopentanoate <2> (Reversibility: ? <2> [3]) [3]
P 4-methyl-2-oxopentanoate + L-leucine
S L-leucine + oxaloacetate <3> (Reversibility: ? <3> [1]) [1]
P 4-methyl-2-oxopentanoate + L-aspartate
S L-methionine + 2-oxoglutarate <3> (<3> mitochondrial isoenzyme [4])
(Reversibility: r <3> [2,4]) [2, 4]
P 4-methylsulfanyl-2-oxobutanoate + L-glutamate
S Additional information <3> (<3> specific for leucine, isoleucine or va-
line, no other amino acid would serve as an amino donor, pyruvate is no
amino acid acceptor [1]) [1]
P ?

Inhibitors

2-mercaptoethanol <3> (<3> slight inhibition [1]) [1]
adipic acid <2> [3]
fumaric acid <2> [3]
glutaric acid <2> [3]
hydroxylamine <2> (<2> inhibits the aldehyde form of the enzyme while the
amino form is found to be inert [3]) [3]
maleate <2> [3]
maleic acid <2> [3]
p-chloromercuribenzoate <3> (<3> 1.0 mM, complete inhibition [1]) [1]
phenylhydrazine <2> [3]
potassium cyanide <2> [3]
sodium bisulfite <2> [3]
succinic acid <2> [3]
thiosemicarbazide <2> [3]

Cofactors/prosthetic groups

pyridoxal 5'-phosphate <3> (<3> K_m 0.004 mM [1]) [1]

Specific activity (U/mg)

1.033 <3> [1]
7.4 <2> (<2> isoenzyme LAT I [3]) [3]
8.7 <2> (<2> isoenzyme LAT II [3]) [3]

K_m-Value (mM)

0.065 <3> (2-oxoglutarate, <3> pH 8.2, 37°C [1]) [1]
1.67 <2> (L-leucine, <2> isozyme LAT II, pH 9, 37°C [3]) [3]
2.3 <2> (2-oxoglutarate, <2> isozyme LAT II, pH 9, 37°C [3]) [3]
2.5 <2> (L-leucine, <2> isozyme LAT I, pH 9, 37°C [3]) [3]
3.3 <2> (2-oxoglutarate, <2> isozyme LAT I, pH 9, 37°C [3]) [3]
25 <3> (L-leucine, <3> pH 8.2, 37°C [1]) [1]

pH-Optimum
8.7 <1, 3> [1, 2, 5]
8.9 <2> [3]

4 Enzyme Structure

Molecular weight
65000 <3> (<3> gel filtration [2]) [2]
68000 <2> (<2> gel filtration, isoenzyme LAT I [3]) [3]
93300 <2> (<2> gel filtration, isoenzyme LAT II [3]) [3]

5 Isolation/Preparation/Mutation/Application

Source/tissue
leaf <1> [5]
liver <3> [1, 2, 4]
seedling <2> [3]

Localization
cytoplasm <3> (<3> predominantly found in mitochondrion [1]) [1]
microsome <3> [1]
mitochondrion <3> [1, 2, 4]
nucleus <3> [1]

Purification
<1> (partially, 2 isoenzymes, LAT I and LAT II [5]) [5]
<2> (partially, 2 isoenzymes, LAT I and LAT II [3]) [3]
<3> (partially [2]) [1, 2, 4]

6 Stability

Temperature stability
45 <3> (<3> quite stable on heating for 1 h [2]) [2]

Organic solvent stability
guanidine hydrochloride <1> (<1> shows sensitivity towards [5]) [5]
urea <1> (<1> shows sensitivity towards [5]) [5]

Storage stability
<2>, 4°C, concentrated enzyme, activity is stable for 2 months [3]

References

[1] Aki, K.; Ogawa, K.; Ichihara, A.: Transaminases of branched chain amino acids. IV. Purification and properties of two enzymes from rat liver. Biochim. Biophys. Acta, **159**, 276-284 (1968)

[2] Ikeda, T.; Konishi, Y.; Ichihara, A.: Transaminase of branched chain amino acids. XI. Leucine (methionine) transaminase of rat liver mitochondria. Biochim. Biophys. Acta, **445**, 622-631 (1976)

[3] Pathre, U.; Singh, A.K.; Viswanathan, P.N.; Sane, P.V.: Purification and properties of leucine aminotransferase from soybean seedlings. Phytochemistry, **26**, 2913-2917 (1987)

[4] Korpela, T.K.: Purification of branched-chain-amino-acid aminotransferase from pig heart. Methods Enzymol., **166**, 269-274 (1988)

[5] Singh, A.K.; Pathre, U.; Sane, P.V.: Purification and characterization of leucine aminotransferase from green leaves of Amaranthus dubius. Biochem. Physiol. Pflanz., **187**, 337-345 (1991)

Kynurenine-oxoglutarate transaminase 2.6.1.7

1 Nomenclature

EC number

2.6.1.7

Systematic name

L-kynurenine:2-oxoglutarate aminotransferase

Recommended name

kynurenine-oxoglutarate transaminase

Synonyms

KAT <16, 17, 18> [19, 20, 21]

KAT I <19> [22]

KAT II <19> [22]

L-kynurenine aminotransferase

aminotransferase, kynurenine

kynurenine 2-oxoglutarate transaminase

kynurenine aminotransferase

kynurenine pyruvate aminotransferase <1> (<1> i.e. KAT-I [25]) [25]

kynurenine transaminase (cyclizing)

kynurenine-2-oxoglutarate aminotransferase <1> (<1> i.e. KAT-II [25]) [25]

Additional information <1, 16> (<1> enzyme from mitochondria may be identical with EC 2.6.1.39 [18]; <1> not identical with EC 2.6.1.39 [16,17]; <16> identical to kidney soluble cysteine conjugate β-lyase, EC 2.6.1.64, also referred to as glutamine transaminase K, EC 2.6.1.15, K stands for kidney [19]; <1> brain KAT I is identical with glutamine transferase K [27]) [16, 17, 18, 19, 27]

CAS registry number

9030-38-0

2 Source Organism

<1> *Rattus norvegicus* (male [15]; adult albino [3,15]; male Wistar [18]; Sprague-Dawley [7]; Donryu [5,6]; isoenzyme 2 [5]; KAT I and KAT II [27]; KAT II accounts for more than 70% of kynurenate production under physiological conditions [31]; brain KAT I [35]) [1, 3-7, 13, 15-18, 23, 24, 25, 26, 27, 30, 31, 35, 36]

<2> *Drosophila sordidula* [12]

<3> *Neurospora crassa* [2]

<4> *Hansenula schneggii* (yeast [8-10,14]) [8-10, 14]
<5> *Saccharomyces cerevisiae* [9]
<6> *Saccharomyces marxianus* [9]
<7> *Pichia polymorpha* (yeast [9]) [9]
<8> *Debaryomyces hansenii* (yeast [9]) [9]
<9> *Pseudomonas fluorescens* [11]
<10> *Saccharomyces fragilis* (low activity [9]) [9]
<11> *Debariomyces globosus* (low activity [9]) [9]
<12> *Cryptococcus albidus* (low activity [9]) [9]
<13> *Candida rugosa* (low activity [9]) [9]
<14> *Tricasporon cutaneum* (low activity [9]) [9]
<15> *Schwanniomyces occidentalis* (low activity [9]) [9]
<16> *Rattus norvegicus* (identical to kidney soluble cysteine conjugate β-lyase also referred to as glutamine transaminase K, K stands for kidney [19]) [19]
<17> *Rattus norvegicus* (UniProt-ID S619603) [20]
<18> *Rattus norvegicus* (brain KAT I [21,34]) [21, 34]
<19> *Homo sapiens* (kynurenine aminotransferase I and II, i.e. KAT I and KAT II, in elderly Down syndrome patients [22]) [22, 28, 32, 37]
<20> *Penaeus monodon* (black tiger prawn [29]) [29]
<21> *Aedes aegypti* (yellow fever mosquito [33]) [33]
<22> *Rattus norvegicus* (cytosolic KAT I [37]) [37]

3 Reaction and Specificity

Catalyzed reaction

L-kynurenine + 2-oxoglutarate = 4-(2-aminophenyl)-2,4-dioxobutanoate + L-glutamate (<4> mechanism [10]; <1> transamination of the amino acid to a transient keto acid intermediate which spontaneously undergoes ring closure [3])

Reaction type

amino group transfer

Natural substrates and products

S L-kynurenine + 2-oxoglutarate <1, 4> (<1,4> inducible enzyme of tryptophan catabolism [3,8]; <4> initial reaction in quinaldine pathway of kynurenine catabolism [8]) [3, 8]
P kynurenic acid + L-glutamate

Substrates and products

S 3,5-diiodo-L-tyrosine + 2-oxoglutarate <1> (<1> slightly better than L-kynurenine [3]) (Reversibility: ? <1> [3]) [3]
P 3-(2,5-diiodo-4-hydroxyphenyl)-2-oxopropanoate + L-glutamate <1> [3]
S L-2-aminoheptane-1,7-dioate + 2-oxoglutarate <1> (<1> at 39.7% of the activity with 2-aminohexane-1,6-dioate [3]) (Reversibility: ? <1> [3]) [3]
P 2-oxoheptane-1,7-dioate + L-glutamate <1> [3]

S L-2-aminohexane-1,6-dioate + 2-oxoglutarate <4> (<4> at about 60% the rate of L-kynurenine [9]) (Reversibility: ? <4> [9]) [9]

P 2-oxohexane-1,6-dioate + L-glutamate <4> [9]

S L-3-hydroxykynurenine + 2-oxoglutarate <1-4, 17> (<4> at about 45% the rate of L-kynurenine [9]; <1> no activity with D-isomer [3]; <1> KAT I, no good substrate [25]) (Reversibility: ir <1> [18]; ? <2-4> [2-4, 6, 9, 12, 16, 20]) [2-4, 6, 9, 12, 16, 18, 20, 25]

P xanthurenic acid + L-glutamate <1-4, 17> [2-4, 6, 9, 12, 16, 18, 20, 25]

S L-4-chloro-kynurenine + 2-oxoglutarate <1> (Reversibility: ? <1> [23]) [23]

P L-7-chloro-kynurenic acid + L-glutamate <1> [23]

S L-5-chloro-kynurenine + 2-oxoglutarate <1> (Reversibility: ? <1> [23]) [23]

P L-6-chloro-kynurenic acid + L-glutamate <1> [23]

S L-histidine + 2-oxoglutarate <1> (Reversibility: ? <1> [16]) [16]

P 3-(1H-imidazol-4-yl)-2-oxopropanoate + L-glutamate <1> [16]

S L-kynurenine + 2-oxoadipate <17, 19, 21> (<17> 91% of activity with 2-oxoglutarate [20]; <21> 25% of activity with pyruvate [33]) (Reversibility: ? <17, 19, 21> [20, 28, 33]) [20, 28, 33]

P L-kynurenic acid + 2-aminoadipate <17, 19, 21> [20, 28, 33]

S L-kynurenine + 2-oxobutyrate <3, 4, 17> (<3, 4> less effective than 2-oxoglutarate [2, 8]; <17> 28% of activity with 2-oxoglutarate [20]) (Reversibility: ? <3, 4, 17> [2, 8, 20]) [2, 8, 20]

P kynurenic acid + 2-aminobutanoate <3, 4, 17> [2, 8, 20]

S L-kynurenine + 2-oxoglutarate <1-19, 21> (<1, 4> best substrates [8, 9, 16, 18]; <1> isoenzyme 2, no activity with L-tyrosine, L-phenylalanine, L-tryptophan, 5-hydroxy-L-tryptophan and L-aspartate [5]; <1,4> broad specificity [8, 15]; <3> no activity with D-kynureinine [2]; <16> soluble kidney/brain kynurenine aminotransferase has both amino transferase and cysteine conjugate β-lyase, i.e., glutamine transaminase K activities [19]; <17> enzyme has both kynurenine aminotransferase and α-aminoadipate aminotransferase activity [20]; <18> enzyme has both kynurenine aminotransferase and glutamine transaminase K activity [21]; <19> KAT I shows far higher activity with 2-oxoisohexanoate than with 2-oxoglutarate; [22]; <1> KAT II shows equal activity with 2-oxoglutarate and pyruvate, KAT I shows lower activity with 2-oxoglutarate than with pyruvate, ratio of KAT I to KAT II in normal brain is approx. 1/4 [27]; <21> 31% of activity with pyruvate [33]) (Reversibility: ir <1, 4> [8, 18]; ? <2-3, 5-9, 16, 17, 18, 19, 21> [1-7, 9-17, 19, 20, 21, 33]) [1-22, 27, 33]

P kynurenic acid + L-glutamate <1-19, 21> (<4-8> via o-aminobenzoylpyruvic acid which is immediately converted to the intramolecularly dehydrated and cyclized form kynurenic acid [9]) [1-22, 27, 33]

S L-kynurenine + 2-oxohexanedioate <1, 4> (Reversibility: ? <1, 4> [3, 9, 18]) [3, 9, 18]

P kynurenic acid + 2-aminohexanedioate <1, 4> [3, 9, 18]

S L-kynurenine + 2-oxohexanoate <4> (Reversibility: ? <4> [8]) [8]

P kynurenic acid + norleucine <4> [8]

S L-kynurenine + 2-oxoisohexanoate <1, 4, 19> (<1> less effective than 2-
oxoglutarate [1]; <19> KAT I shows far higher activity with 2-oxoisohex-
anoate than with 2-oxoglutarate [22]; <19> heart KAT, low activity [28])
(Reversibility: ? <1, 4, 19> [1, 8, 9, 22]) [1, 8, 9, 22, 28]

P kynurenic acid + 2-aminoisohexanoate <1, 4, 19> [1, 8, 9, 22, 28]

S L-kynurenine + 2-oxoisopentanoate <3> (<3> poor substrate [2]) (Reversibility: ? <3> [2]) [2]

P kynurenic acid + 2-aminoisopentanoate <3> [2]

S L-kynurenine + 2-oxopentanoate <3, 4> (<3,4> less effective than 2-oxo-
glutarate [2,8]) (Reversibility: ? <3,4> [2,8]) [2, 8]

P kynurenic acid + norvaline <3, 4> [2, 8]

S L-kynurenine + 3-indole-2-oxopropanoate <3, 4> (Reversibility: ? <3,4>
[2,9]) [2, 9]

P kynurenic acid + L-tryptophan <3, 4> [2, 9]

S L-kynurenine + 3-methyl-2-oxopentanoate <3> (<3> poor substrate [2])
(Reversibility: ? <3> [2]) [2]

P kynurenic acid + L-isoleucine <3> [2]

S L-kynurenine + 3-phenyl-2-oxopropanoate <3, 4> (Reversibility: ? <3,4>
[2,9]) [2, 9]

P kynurenic acid + phenylalanine <3, 4> [2, 9]

S L-kynurenine + 4-methyl-2-oxovalerate <17> (<17> 5.9% of activity with
2-oxoglutarate [20]) (Reversibility: ? <17> [20]) [20]

P L-kynurenic acid + 2,4-dimethylvalerate <17> [20]

S L-kynurenine + 4-methylsulfanyl-2-oxobutyrate <4> (Reversibility: ? <4>
[8,9]) [8, 9]

P kynurenic acid + L-methionine <4> [8, 9]

S L-kynurenine + glyoxylate <1> (<1> poor substrate [18]) (Reversibility: ?
<1> [18]) [18]

P kynurenic acid + glycine <1> [18]

S L-kynurenine + oxaloacetate <1, 21> (<1> less effective than 2-oxogluta-
rate [18]; <21> 90% of activity with pyruvate [33]) (Reversibility: ? <1,
21> [1, 18, 33]) [1, 18, 33]

P kynurenic acid + L-aspartate <1, 21> [1, 18, 33]

S L-kynurenine + phenylpyruvate <17> (<17> 43% of activity with 2-oxo-
glutarate [20]) (Reversibility: ? <17> [20]) [20]

P phenylalanine + L-glutamate <17> [20]

S L-kynurenine + pyruvate <1, 3, 16, 17, 18, 19, 20, 21> (<1> less effective
than 2-oxoglutarate [1, 3, 16, 18]; <16> 100fold higher maximal velocity
than with 2-oxoglutarate [19]; <17> 3% of activity with 2-oxoglutarate
[20]; <19> KAT I shows far higher activity with 2-oxoisohexanoate and
pyruvate than with 2-oxoglutarate [22]; <19> best substrate of heart KAT
[28]; <19> placental KAT I, higher activity than with 2-oxoglutarate [32];
<21> best amino acceptor [33]) (Reversibility: ? <1, 3, 16, 17, 18, 19, 20,
21> [1, 3, 6, 16, 18, 19, 20, 21, 22, 29, 33]) [1-3, 6, 16, 18, 19, 20, 21, 22, 28,
29, 32, 33]

P kynurenic acid + L-alanine <1, 3, 16, 17, 18, 19, 20, 21> [1-3, 6, 16, 18, 19,
20, 21, 22, 28, 29, 32, 33]

S   L-leucine + 2-oxoglutarate <4> (<4> at about 85% the rate of L-kynure-
nine [9]) (Reversibility: ? <4> [9]) [9]
P   4-methyl-2-oxopentanoate + L-glutamate <4> [9]
S   L-methionine + 2-oxoglutarate <1, 4> (<4> at about 36% the rate of L-
kynurenine [9]; <1> 12% the rate of L-kynurenine [3]) (Reversibility: ?
<1,4> [3,9]) [3, 9]
P   4-methylsulfanyl-2-oxobutanoate + glutamate <1, 4> [3, 9]
S   L-norleucine + 2-oxoglutarate <1> (<1> at about 30% the rate of L-kynur-
enine [3]) (Reversibility: ? <1> [3]) [3]
P   2-oxohexanoate + L-glutamate <1> [3]
S   L-phenylalanine + 2-oxoglutarate <1> (Reversibility: ? <1> [3]) [3]
P   phenylpyruvate + L-glutamate <1> [3]
S   L-tryptophan + 2-oxoglutarate <1, 4> (<4> at about 60% the rate of L-
kynurenine [9]; <1> 16% the rate of L-kynurenine [3]; <1> poor substrate
[16]) (Reversibility: ? <1,4> [3, 9, 16]) [3, 9, 16]
P   3-indole-2-oxopropanoate + L-glutamate <1, 4> [3, 9, 16]
S   L-tyrosine + 2-oxoglutarate <1> (Reversibility: ? <1> [3,16]) [3, 16]
P   3-(4-hydroxyphenyl)-2-oxopropanoate + L-glutamate <1> [3, 16]
S   L-vinylglycine <4> (<4> deamination at 1% the rate of kynurenine, sui-
cide substrate [10]) (Reversibility: ? <4> [10]) [10]
P   2-oxobutyrate + NH$_3$ <4> [10]
S   L-vinylglycine + 1-propanethiol <4> (<4> γ-addition reaction in the pres-
ence of alkanethiols [10]) (Reversibility: ? <4> [10]) [10]
P   S-n-propylhomocysteine <4> [10]
S   L-vinylglycine + 2-oxobutyrate <4> (Reversibility: ? <4> [10]) [10]
P   2-aminobutyrate + 2-oxo-3-butenoate <4> [10]
S   L-vinylglycine + ethanethiol <4> (<4> γ-addition reaction in the presence
of alkanethiols [10]) (Reversibility: ? <4> [10]) [10]
P   ethionine <4> (<4> i.e. S-ethylhomocysteine [10]) [10]
S   L-vinylglycine + methanethiol <4> (<4> γ-addition reaction in the pres-
ence of alkanethiols [10]) (Reversibility: ? <4> [10]) [10]
P   methionine <4> [10]

Inhibitors

2-aminoadipate <1> [3, 18]
2-oxoadipate <1> [3, 18]
2-oxoglutarate <1> (<1> substrate inhibition [3,18]) [3, 18]
3-hydroxykyunerine <1> (<1> 2 mM, 50% inhibition of KAT I activity [25];
<1> 0.6 mM, 54% inhibition of KAT I, 88% inhibition of KAT II [27]) [25, 27]
3-methyl-2-benzothiazolone hydrazone hydrochloride <4> (<4> 0.1 mM,
complete inhibition, reversible by pyridoxal phosphate [8]) [8]
4-carboxy-3-hydroxyphenylglycine <1> (<1> inhibition of brain KAT II [31])
[31]
4-chloromercuribenzoate <1, 4> (<4> 1 mM, 78% inhibition [8]) [8, 9, 18]
Co^{2+} <1> [3, 18]
Cu^{2+} <1> [18]

EGTA <1> (<1> strong inhibiton at pH 6.8 and Ca^{2+} concentrations below
0.5 mM [7]) [7]
Fe^{3+} <1> [18]
Hg^{2+} <1, 4> (<1> strong inhibiiton [18]; <4> 0.1 mM, 89% inhibition [8]) [3,
8, 9, 18]
KCN <1> (<1> complete inhibition [3]) [3]
L-2-amino-4-phosphonobutyric acid <1> (<1> inhibition of brain KAT II
[31]) [31]
L-4-chlorokynurenine <1> [23]
L-5-chlorokynurenine <1> [23]
L-aspartate <1> (<1> 0.72 mM, 50% inhibition of brain KAT I, 0.073 mM,
50% inhibition of brain KAT II [36]) [36]
L-cycloserine <4> (<4> 0.1 mM, 88% inhibition, reversible by pyridoxal
phosphate [8]) [8]
L-cysteate <1> (<1> 5.38 mM, 50% inhibition of brain KAT I, 1.55 mM, 50%
inhibition of brain KAT II [36]) [36]
L-cysteine <1> (<1> inhibition of KAT I [25]) [25]
L-cysteine sulfinate <1> (<1> 0.002 mM, 50% inhibition of brain KAT II [36])
[36]
L-glutamate <1> (<1> 1.3 mM, 50% inhibition of brain KAT I, 0.79 mM, 50%
inhibition of brain KAT II [36]) [36]
L-homocysteate <1> (<1> 5.0 mM, 50% inhibition of brain KAT I, 2.1 mM,
50% inhibition of brain KAT II [36]) [36]
L-homocysteine sulfinate <1> (<1> 1.59 mM, 50% inhibition of brain KAT II
[36]) [36]
L-kynurenine <1, 3> (<1> substrate inhibition above 6 mM [3,4,18]; <3> in-
hibition above 5 mM [2]) [2-4, 18]
L-penicillamine <4> (<4> 0.1 mM, 62% inhibition, reversible by pyridoxal
phosphate [8]) [8]
L-phenylalanine <1, 19> (<19> inhibition of KAT I [22]; <1> strong inhibition of KAT I [25]; <1> strong inhibition of KAT I [27]) [22, 25, 27, 32]
L-serine-O-phosphate <1> (<1> inhibition of brain KAT II [31]) [31]
L-tryptophan <19> (<19> inhibition of KAT I [22]; <1> strong inhibition of
KAT I [27]) [22, 27, 28, 32]
N-ethylmaleimide <4> [8, 9]
Ni^{2+} <1> [3]
S-dichlorovinyl-L-cysteine <1> (<1> strong inhibition of KAT I [25]; <1>
2 mM, 90% inhibition of KAT I and KAT II [27]) [25, 27]
Sn^{2+} <1> [3]
Zn^{2+} <1> [3, 18]
adipate <1, 4> (<1> strong inhibition [18]; <4> 10 mM, 39% inhibition [8])
[3, 8, 16, 18]
α-aminoadipate <1> (<1> strong inhibition of KAT II [27]) [27]
anthranilate <1> (<1> weak inhibition [18]) [18]
azelaic acid <1> [16]
decanoic acid <1> [16]
dicarboxylic acids <1, 4> (<1> C-2 to C-10 [3]) [3, 8, 9]

diethylglutaric acid <1> [16]
dimethylglutaric acid <1> [16]
glutamate <1> (<1> weak inhibition [18]) [18]
glutamine <1, 19> (<19> inhibition of KAT I [22]; <1> strong inhibition of KAT I, phenylpyruvate, 2-oxo-4-methiolbutyrate or 2-oxo-isocaproate protect [25]; <1> strong inhibition of KAT I [27]; <19> 0.08 mM; 50% inhibition of placental KAT I [32]) [22, 25, 27, 32]
hydroxylamine <1, 4> (<1> complete inhibition [3]; <4> 0.1 mM, complete inhibition after 30 min [8]) [3, 8]
iodoacetate <4> (<4> 1 mM, 64% inhibition [8]) [8, 9]
isonicotinic acid hydrazide <1> (<1> anti-tuberculosis drug, enzyme is inhibited in animals fed with a diet containing isonicotinic acid hydrazide and by injection of the drug [24]) [24]
kynurenate <1> (<1> weak inhibition [18]) [18]
nicotinate <1> (<1> weak inhibition [18]) [18]
phenylhydrazine <4> (<4> 0.1 mM, 81% inhibiiton, reversible by pyridoxal phosphate [8]) [8]
phosphate <1> [3]
pimelate <1, 4> (<4> 10 mM, 53% inhibition [8]) [8, 16]
pyridoxal 5'-phosphate <3> (<3> at high concentrations [2]) [2]
quinolinic acid <1> (<1> strong inhibiton at pH 6.8 and Ca^{2+} concentrations below 0.5 mM [7]) [7]
quisqualate <1> (<1> 0.52 mM, 50% inhibition of KAT II, 2 mM, 80% inhibition [27]; <1> inhibition of brain KAT II [31]) [27, 31]
sulfate <1> [3]
sulfate or phosphate esters of steroids <1> [3]
trans-pyrrolidine-2,4-dicarboxylate <1> (<1> inhibition of brain KAT II [31]) [31]
xanthurenate <1> (<1> weak inhibition [18]) [18]
Additional information <1, 4, 19> (<1> not inhibited by antibodies against 2-aminoadipate transaminase, [17]; <1> not inhibited by aspartate, tyrosine, phenylalanine [5]; <1> not inhibited by EGTA [7]; <4> not inhibited by EDTA, 2,2'-dipyridyl [8, 9]; <1> not inhibited by quinolinate [7, 18]; <1> not inhibited by oxaloacetic acid, 3-methylglutaric acid [16]; <1> not inhibited by glycerol, DTT, up to 10%, tryptophan, 3-hydroxyanthranilate, picolinate, NAD^+, NADH, $NADP^+$, NADPH, serotonin, tryptamine, 5-hydroxytryptophan, indole-3-acetate, organic acids of tricarboxylic acid cycle, glyoxylate, pyruvate, glutamine, Mg^{2+}, Ca^{2+}, Mn^{2+}, monovalent cations [18]; <1> not inhibited by 2-aminoadipate [16]; <1> not inhibited by L-3-hydroxykynurenine at higher concentrations [3]; <19> KAT II is not inhibited by glutamine, tryptophan and phenylalanine [22]; <1> KAT I is not inhibited by quisqualate [27]; <1> brain KAT I is not inhibited by quiqualate, 4-carboxy-3-hydroxyphenylglycine, L-2-amino-4-phosphonobutyric acid, L-serine-O-phosphate, and trans-pyrrolidine-2,4-dicarboxylate [31]; <1> brain KAT I is not inhibited by L-cysteine sulfinate and L-homocysteine sulfinate [36]) [3, 5, 7-9, 16-18, 22, 27, 31, 36]

Cofactors/prosthetic groups

pyridoxal 5'-phosphate <1-4> (<4> required for activity, 2 mol per mol holoenzyme [8]; <3> K_m-value: 0.001 mM [2]; <1> K_m-value: 0.00085 mM [3];
<4> K_m-value: 0.0013 mM [8]) [1-5, 8, 9, 12, 14, 18]
Additional information <3> (<3> pyridoxamine 5'-phosphate is not a cofactor [2]) [2]

Activating compounds

2-mercaptoethanol <1> (<1> activation [3]) [3]
L-tryptophan <19> (<19> 2 mM, 84% increase in activity of heart KAT at pH
8.0 [28]) [28]
pyridoxamine 5'-phosphate <4> (<4> activation [8]) [8]

Metals, ions

$CaCl_2$ <1> (<1> approx. 4fold increase in activity at pH 6.4 [7]) [7]
Additional information <1> (<1> not activated by chloride salts of Mg^{2+},
Mn^{2+}, Na^+, K^+ or NH_4^+ [7]) [7]

Specific activity (U/mg)

0.13 <1> (<1> hydroxykynurenine [18]; <1> isoenzyme 2 [5]) [18, 5]
0.17 <4> [10]
0.36 <16> [19]
1.5 <1, 3> (<1,3> L-kynurenine [2,18]) [2, 18]
1.56 <17> [20]
2.65 <1> [3, 15]
3.5 <1> [17]
4.22 <1> [16]
7.8 <1> [5]
16.9 <4> [8, 9]
Additional information <20> (<20> prawns at intermoult show significant
higher KAT activity compared to prawns from premoult [29]) [29]

K_m-Value (mM)

0.00085 <1> (pyridoxal 5'-phosphate, <1> pH 6.5, 37°C [3]) [3]
0.0013 <4> (pyridoxal 5'-phosphate, <4> pH 8.0, 37°C [8]) [8]
0.0032 <4> (pyridoxamine 5'-phosphate, <4> pH 8.0, 37°C [8]) [8]
0.01 <1> (2-oxoadipate) [18]
0.013 <1> (2-oxoglutarate, <1> cosubstrate hydroxykynurenine [18]) [18]
0.02 <1> (2-oxoglutarate, <1> cosubstrate L-kynurenine [16,18]) [16, 18]
0.024 <1> (2-oxoglutarate, <1> pH 9.5, 37°C, KAT I [27]) [27]
0.045 <17> (2-oxoglutarate, <17> pH 8.0, 37°C, recombinant enzyme, cosubstrate L-kynurenine [20]) [20]
0.053 <1> (pyruvate, <1> pH 9.5, 37°C, KAT I [27]) [27]
0.083 <16> (L-kynurenine, <16> pH 8.0, 37°C, recombinant enzyme, cosubstrate 2-oxoglutarate [19]) [19]
0.096 <18> (2-oxoglutarate, <18> pH 8.0, 37°C, recombinant mitochondrial
enzyme [21]) [21]
0.1 <18> (pyruvate, <18> E61G mutant KAT I from brain [34]) [34]
0.2 <18> (L-kynurenine, <18> E61G mutant KAT I from brain [34]) [34]

0.25 <20> (L-kynurenine, <20> pH 8.0. 30°C [29]) [29]

0.62 <19> (2-oxoglutarate, <19> pH 8.0, 37°C, heart KAT, at 5 mM L-kynurenine [28]) [28]

0.7 <3> (2-oxoglutarate, <3> pH 7.5, 37°C, cosubstrate L-kynurenine [2]) [2]

0.7 <3> (pyruvate, <3> pH 7.5, 37°C, cosubstrate L-kynurenine [2]) [2]

0.717 <1> (pyruvate, <1> pH 7.0, 37°C, KAT II [27]) [27]

0.879 <1> (2-oxoglutarate, <1> pH 7.0, 37°C, KAT II [27]) [27]

0.95 <17> (L-kynurenine, <17> pH 8.0, 37°C, recombinant enzyme [20]) [20]

1 <1> (2-oxoglutarate, <1> pH 6.5, 37°C [3]) [3]

1 <4> (2-oxoisocaproate, <4> pH 8.0, 37°C, cosubstrate L-kynurenine [8]) [8,
9]

1.1 <4> (2-oxoglutarate, <4> pH 8.0, 37°C, cosubstrate L-kynurenine [8]) [8,
9]

1.15 <19> (pyruvate, <19> pH 8.0, 37°C, heart KAT, at 5 mM L-kynurenine
[28]) [28]

1.2 <16> (L-kynurenine, <16> pH 8.0, 37°C, recombinant enzyme, cosubstrate pyruvate [19]) [19]

1.2 <4> (β-phenylpyruvate, <4> pH 8.0, 37°C, cosubstrate L-kynurenine [8])
[8, 9]

1.2 <18> (pyruvate, <18> recombinant brain KAT I [34]) [34]

1.36 <17> (L-3-hydroxykynurenine, <17> pH 8.0, 37°C, recombinant enzyme
[20]) [20]

1.4 <1> (L-kynurenine, <1> pH 6.5, 37°C [3]) [3]

1.4 <18> (L-kynurenine, <18> recombinant brain KAT I [34]) [34]

1.5 <3> (L-kynurenine, <3> pH 7.5, 37°C [2]) [2]

1.6 <1> (L-kynurenine, <1> pH 6.5, 37°C, enzyme from supernatant [6]) [6]

1.63 <18> (pyruvate, <18> pH 8.0, 37°C, recombinant mitochondrial enzyme
[21]) [21]

1.7 <16> (L-kynurenine, <16> pH 8.0, 37°C, cosubstrate pyruvate [19]) [19]

1.9 <4> (L-kyrunenine, <4> pH 8.0, 37°C [8]) [8]

2 <1> (L-3-hydroxykynurenine, <1> pH 6.5, 37°C [3]) [3]

2.2 <4> (L-kynurenine) [14]

2.5 <1> (L-3-hydroxykynurenine, <1> pH 7.2, 37°C, kidney mitochondria
[4]) [4]

2.5 <1> (L-kynurenine, <1> pH 6.5, 37°C, mitochondrial enzyme [6]) [6]

2.5 <1> (L-kynurenine, <1> pH 7.2, 37°C, kidney supernatant [4]) [4]

2.6 <4> (L-tryptophan, <4> pH 8.0, 37°C, cosubstrate 2-oxoglutarate [8]) [8,
9]

2.7 <4> (L-vinylglycine, <4> pH 8.0. 25°C [10]) [10]

3 <1> (L-kynurenine, <1> pH 7.2, 37°C, liver supernatant [4]) [4]

3.4 <1> (L-kynurenine, <1> pH 7.2, 37°C, kidney mitochondria [4]) [4]

4 <1> (L-kynurenine) [16]

4.3 <1> (L-kynurenine) [18]

5 <1> (L-3-hydroxykynurenine, <1> pH 7.2, 37°C, liver supernatant [4]) [4]

5 <1> (L-kynurenine, <1> pH 7.2, 37°C, liver mitochondria [4]) [4]

5 <4> (L-leucine, <4> pH 8.0, 37°C, cosubstrate 2-oxoglutarate [8]) [8, 9, 14]

5 <4> (L-methionine, <4> pH 8.0, 37°C, cosubstrate 2-oxoglutarate [8]) [8, 9, 14]

5.3 <1> (L-3-hydroxykynurenine, <1> pH 7.2, 37°C, kidney supernatant [4]) [4]

5.6 <17> (L-glutamate, <17> pH 8.0, 37°C, α-aminoadipate aminotransferase activity [20]) [20]

6.3 <1> (L-3-hydroxykynurenine, <1> pH 7.2, 37°C, liver mitochondria [4]) [4]

20 <19> (L-kynurenine, <19> pH 8.0, 37°C, heart KAT, cosubstrate 5 mM pyruvate [28]) [28]

27 <19> (L-kynurenine, <19> pH 8.0, 37°C, heart KAT, cosubstrate 5 mM 2-oxoglutarate [28]) [28]

K_i-Value (mM)

9.5 <4> (pimelate) [8]

13 <4> (adipate) [8]

pH-Optimum

6-6.5 <1> [5, 6, 13]

6.5 <1> (<1> with L-kynurenine as substrate [3]) [3, 18]

7-7.5 <1> (<1> KAT II [27]) [27]

7-8 <19> (<19> KAT II from brain [22]) [22]

7.5 <1, 3> (<1> with L-α-aminoadipate as substrate [3]) [2, 3]

8 <2, 4, 20> [8, 12, 29]

8-9 <19> (<19> heart enzyme [28]) [28]

8.3 <1> (<1> KAT I [25]) [25]

8.5 <21> (<21> optima at pH 8.5 and 10.0 [33]) [33]

8.8 <19> (<19> total placental KAT activity [32]) [32]

9-9.5 <1> (<1> KAT I [27]) [27]

9.6-10 <19> (<19> KAT I from brain, sharp optimum [22]) [22]

9.8 <19> (<19> placental KAT I activity [32]) [32]

10 <21> (<21> optima at pH 8.5 and 10.0 [33]) [33]

Additional information <1, 4> (pI: 4.95 [8]; <1> pI: 5.9 [16]; pI: 6.56 [3,15]) [3, 8, 15, 16]

pH-Range

6.2-9.2 <3> (<3> approx. half-maximal activity at pH 6.2, approx. 60% of maximal activity at pH 9.2 [2]) [2]

6.5-8.8 <1> (<1> maximal activity at pH 6.5, approx. half-maximal activity at pH 8.8 [6]) [6]

Temperature optimum (°C)

37 <1, 3, 4> (<1,3,4> assay at [2-6,9,15-18]) [2-6, 9, 15-18]

47 <2> [12]

60 <4, 21> (<21> recombinant enzyme [33]) [8, 33]

Temperature range (°C)

0-45 <20> (<20> linear increase in activity [29]) [29]

20-60 <4> (<4> linear increase of activity [8]) [8]

4 Enzyme Structure

Molecular weight

85000 <1> (<1> gel filtration, sucrose density gradient centrifugation [3]) [3, 15]

90000-100000 <16, 17> (<16> gel filtration [19]; <17> gel filtration [20]) [19, 20]

91000 <1> (<1> gel filtration [16]) [16]

97000 <21> (<21> truncated recombinant enzyme, gel filtration [33]) [33]

98000 <4> (<4> sedimentation equilibrium [8,9]) [8, 9]

100000 <1> (<1> isoenzyme 2, sucrose density gradient centrifugation [5]) [5]

102000 <4> (<4> native PAGE [9]) [9]

106000 <4> (<4> gel filtration [8,9]) [8, 9]

Additional information <1, 4> (<1,4> amino acid composition [8,16]) [8, 16]

Subunits

? <21> (<21> x * 53490, deduced from nucleotide sequence [33]; <21> x * 53000, SDS-PAGE [33]) [33]

dimer <1, 4, 16, 17, 21> (<1> 2 * 44000, SDS-PAGE [16]; <1> 2 * 45000, SDS-PAGE in the absence of 2-mercaptoethanol [3,15]; <1> 2 * 45500, SDS-PAGE in the presence of 2-mercaptoethanol [3]; <1> 2 * 49500, SDS-PAGE [18]; <4> 2 * 52000, SDS-PAGE [8]; <16> 2 * 47000, SDS-PAGE [19]; <17> 2 * 45000, SDS-PAGE [20]; <21> 2 * 48000, truncated recombinant enzyme, SDS-PAGE [33]) [3, 8, 15, 16, 18, 19, 20, 33]

5 Isolation/Preparation/Mutation/Application

Source/tissue

brain <1, 16, 18, 19> (<19> frontal and temporal cortex [22]; <1> KAT I and KAT II [27]; <1> KAT I activity in glial cells of striatum, in astrocytes of the hippocampus and in glial cells of the temporal lobe [30,35]) [19, 21, 22, 27, 30, 35]

frontal lobe <19> [22]

ganglion <20> (<20> highest activity in cerebral ganglion, low activity in thoracic and abdominal ganglion [29]) [29]

heart <1, 19, 20> (<1> KAT 1, low activity [25]; <20> low activity [29]) [25, 28, 29]

hemolymph <20> [29]

kidney <1, 16, 17> (<1> KAT I and KAT II, major activity is performed by KAT II, KAT I contributes to 1/15 of total kynurenine aminotransferase activity [25]) [4, 15-17, 19, 20, 25]

liver <1> (<1> KAT I [25]) [4, 15, 18, 25]

lung <1> (<1> KAT 1, low activity [25]) [25]

muscle <1> (<1> KAT 1, low activity [25]) [25]

pancreas <1> (<1> KAT I, low activity [25]) [25]

placenta <19> (<19> KAT I [32]) [32]

small intestine <1> (<1> KAT 1, low activity [25]) [25]

spleen <1> (<1> KAT 1, low activity [25]) [25]
stomach <1> (<1> KAT 1, low activity [25]) [25]
testis <1> (<1> KAT I, low activity [25]) [25]

Localization

cytosol <16, 22> (<22> KAT I, the enzyme exists in a cytosolic and in a mitochondrial form because of the presence of 2 different KAT I mRNAs coding for a protein respectively with and without leader sequence targeting the protein into mitochondria [37]) [19, 37]
endoplasmic reticulum <1> (<1> KAT I of astroglial cells is associated with ribosomes of the endoplasmic reticulum [30]) [30]
lysosome <1> (<1> 10% of liver total activity at pH 6.5 [6]) [6]
mitochondrial inner membrane <1> [18]
mitochondrial matrix <1> [26]
mitochondrion <1, 18, 21> (<1> 40% of kidney total activity [4]; <1> 80% of liver total activity [4]; <18> cDNA encodes for a kynurenine aminotransferase that is identical to a soluble form but carries an additional stretch of 32 amino acids resembling a mitochondrial import sequence [21]; <21> deduced amino acid sequence contains a mitochondrial leader sequence [33]) [3-7, 15, 18, 21, 26, 33]
soluble <1, 16, 17> (<1> 60% of kidney total activity [4]; <1> 12% of liver total activity at pH 6.5 [6]; <1> brain KAT I [35]) [3-6, 15, 16, 19, 20, 35]
Additional information <1> (<1> subcellular distribution [4,6]) [4, 6]

Purification

<1> (partial purification of kynurenine apotransaminase [1]; acid precipitation, ammonium sulfate, DEAE-cellulose, hydroxylapatite [3]; DEAE-cellulose, heat treatment, hydroxyapatite, CM-Sephadex C-25, Sephadex G-100 [5]; KAT I [25]; DEAE-Sepharose, ammonium sulfate, Sephadex G-25, partial purification of KAT I and KAT II [27]) [1, 3, 5, 15-18, 25, 27]
<4> (polyethyleneimine, DEAE-cellulose, Bio-Gel P-300, hydroxyapatite, crystallization [8]) [8, 9, 14]
<16> (DEAE-Sepharose, phenyl-Sepharose, Superose 12, Mono Q [19]) [19]
<17> (ammonium sulfate, DEAE-Sepharose, phenyl-Sepharose, Mono Q, Mono P [20]) [20]
<21> (truncated recombinant enzyme, DEAE-Sepharose, native PAGE [33]) [33]

Crystallization

<4> [8, 9]

Cloning

<16> (transient expression in HEK-293 cells [19]) [19]
<17> (transient expression in HEK-293 cells [20]; expression of wild-type and E61G mutant enzyme in Escherichia coli [34]) [20, 34]
<18> (transient expression in HEK-293 cells [21]) [21]
<19> (expression of KAT I in Cos-1 cells [37]) [37]
<21> (expression of full length enzyme and truncated version lacking the mitochondrial leader sequence in Sf9 insect cells [33]) [33]

Engineering

E61G <18> (<18> 40-60% of wild-type activity, K_m for L-kynurenine and pyruvate is reduced, missense mutation found in spontaneously hypertensive rats [34]) [34]

6 Stability

pH-Stability

4.5 <1> (<1> 75 min, inactivation, $t_{1/2}$: 30 min [17]) [17]

5 <1> (<1> and below, inactivation even in the presence of pyridoxal phosphate [3]) [3]

6.5-7.5 <1> (<1> most stable [3]) [3]

Temperature stability

37 <1> (<1> phosphate buffer, pH 6.3, inactivation within a few min, pyridoxal phosphate restores activity, only slow inactivation at pH 7.4, 2-oxoglutarate protects [1]) [1]

40 <1> (<1> and above, 5 min, decreasing activity [3]) [3]

50 <4> (<4> and below, 5 min stable between pH 6.0 and pH 10.0 [8,9]; <1> in 20% glycerol, 0.1 mM DTT, 2 mM pyridoxal phosphate, 25 mM imidazole-HCl buffer, pH 7, 30 min stable [18]) [8, 9, 18]

65 <4> (<4> above, rapid inactivation [8]) [8]

70 <1> (<1> 5 min, complete inactivation [3]; <1> 20 min, inactivation [18]) [3, 18]

General stability information

<1>, alcohol fractionation inactivates [1]

<1>, ammonium sulfate fractionation inactivates [1]

<1>, dialysis against phosphate buffer inactivates, 2-oxoglutarate protects to some extent [1]

<1>, dialysis of the apoenzyme against Tris-HCl or imidazole-HCl buffer inactivates [3]

<1>, freeze-thawing inactivates [1]

<1>, slightly acidic buffer solutions inactivate, pyridoxal phosphate or pyridoxamine phosphate restores activity, 2-oxoglutarate protects [1]

<1>, stable in phosphate buffer supplemented with pyridoxal phosphate [3]

<4>, repeated freeze-thawing inactivates enzyme irriversible in dilute solution, below 0.1 mg/ml [8]

Storage stability

<1>, -20°C, in 10 mM phosphate buffer, pH 7, 50% glycerol, at least 2 months [18]

<4>, -20°C, at protein concentrations above 0.6 mg/ml, 0.02 mM pyridoxal phosphate, 0.01% mercaptoethanol, several weeks, no loss of activity [8, 9]

<4>, 4°C, crystalline enzyme as suspension in 70% saturated ammonium sulfate, pH 7.2, more than 6 months [8, 9]

References

[1] Mason, M.: Kynurenine transaminase of rat kidney: a study of coenzyme dissociation. J. Biol. Chem., **227**, 61-68 (1957)

[2] Jakoby, W.B.; Bonner, D.M.: Kynurenine transaminase from Neurospora. J. Biol. Chem., **221**, 689-695 (1956)

[3] Tobes, M.C.: Kynurenine-oxoglutarate aminotransferase from rat kidney. Methods Enzymol., **142**, 217-224 (1987)

[4] De Antoni, A.; Costa, C.; Allegri, G.: Studies on the kynurenine aminotransferase activity in rat liver and kidney. Hoppe-Seyler's Z. Physiol. Chem., **357**, 1707-1712 (1976)

[5] Noguchi, T.; Minatogawa, Y.; Okuno, E.; Nakatani, M.; Morimoto, M.; Kido, R.: Purification and characterization of kynurenine-2-oxoglutarate aminotransferase from the liver, brain and small intestine of rats. Biochem. J., **151**, 399-406 (1975)

[6] Nakatani, M.; Morimoto, M.; Noguchi, T.; Kido, R.: Subcellular distribution and properties of kynurenine transaminase in rat liver. Biochem. J., **143**, 303-310 (1974)

[7] Mason, M.: Effects of calcium ions and quinolinic acid on rat kidney mitochondrial kynurenine aminotransferase. Biochem. Biophys. Res. Commun., **60**, 64-69 (1974)

[8] Asada, Y.; Sawa, Y.; Tanizawa, K.; Soda, K.: Purification and characterization of yeast L-kynurenine aminotransferase with broad substrate specificity. J. Biochem., **99**, 1101-1110 (1986)

[9] Tanizawa, K.; Asada, Y.; Soda, K.: L-Kynurenine transaminase from Hansenula schneggii. Methods Enzymol., **113**, 90-95 (1985)

[10] Asada, Y.; Tanizawa, K.; Yonaha, K.; Soda, K.: Deamination and γ-addition reactions of vinylglycine catalyzed by yeast kynurenine aminotransferase, and suicidal inactivation of the enzyme during its processing. Agric. Biol. Chem., **52**, 2873-2878 (1988)

[11] Miller, I.L.; Tsuchida, M.; Adelberg, E.A.: J. Biol. Chem., **203**, 205-211 (1953)

[12] Real, M.D.; Ferre, J.: Analysis of kynurenine transaminase activity in Drosophila by high performance liquid chromatography. Insect Biochem., **21**, 647-652 (1991)

[13] Kido, R.; Noguchi, T.: Subcellular distribution and properties of rat liver kynurenine transaminase. Acta Vitaminol. Enzymol., **29**, 307-309 (1975)

[14] Yonaha, K.; Moriguchi, M.; Hirasawa, T.; Yamamoto, T.; Soda, K.: Occurence of kynurenine aminotransferase in extracts of yeasts. Bull. Inst. Chem. Res., **53**, 315-318 (1975)

[15] Hartline, R.A.: Kynurenine aminotransferase from kidney supernatant. Methods Enzymol., **113**, 664-673 (1985)

[16] Mawal, M.R.; Mukhopadhyay, A.; Deshmukh, D.R.: Purification and properties of kynurenine aminotransferase from rat kidney. Biochem. J., **279**, 595-599 (1991)

[17] Mawal, M.R.; Deshmukh, D.R.: α-Aminoadipate and kynurenine aminotransferase activities from rat kidney. Evidence for separate identity. J. Biol. Chem., **266**, 2573-2575 (1991)

[18] Takeuchi, F.; Otsuka, H.; Shibata, Y.: Purification, characterization and identification of rat liver mitochondrial kynurenine aminotransferase with α-aminoadipate aminotransferase. Biochim. Biophys. Acta, **743**, 323-330 (1983)

[19] Alberati-Giani, D.; Malherbe, P.; Kohler, C.; Lang, G.; Kiefer, V.; Lahm, H.W.; Cesura, A.M.: Cloning and characterization of a soluble kynurenine aminotransferase from rat brain: identity with kidney cysteine conjugate β-lyase. J. Neurochem., **64**, 1448-1455 (1995)

[20] Buchli, R.; Alberati-Giani, D.; Malherbe, P.; Kohler, C.; Broger, C.; Cesura, A.M.: Cloning and functional expression of a soluble form of kynurenine/α-aminoadipate aminotransferase from rat kidney. J. Biol. Chem., **270**, 29330-29335 (1995)

[21] Malherbe, P.; Alberati-Giani, D.; Koehler, C.; Cesura, A.M.: Identification of a mitochondrial form of kynurenine aminotransferase/glutamine transaminase K from rat brain. FEBS Lett., **367**, 141-144 (1995)

[22] Baran, H.; Cairns, N.; Lubec, B.; Lubec, G.: Increased kynurenic acid levels and decreased brain kynurenine aminotransferase I in patients with Down syndrome. Life Sci., **58**, 1891-1899 (1996)

[23] Varasi, M.; Della Torre, A.; Heidempergher, F.; Pevarello, P.; Speciale, C.; Guidetti, P.; Wells, D.R.; Schwarcz, R.: Derivatives of kynurenine as inhibitors of rat brain kynurenine aminotransferase. Eur.J.Med.Chem., **31**, 11-21 (1996)

[24] Shibata, K.; Marugami, M.; Kondo, T.: In vivo inhibition of kynurenine aminotransferase activity by isonicotinic acid hydrazide in rats. Biosci. Biotechnol. Biochem., **60**, 874-876 (1996)

[25] Okuno, E.; Nishikawa, T.; Nakamura, M.: Kynurenine aminotransferases in the rat. Localization and characterization. Adv. Exp. Med. Biol., **398**, 455-464 (1996)

[26] Cesura, A.M.; Alberati-Giani, D.; Buchli, R.; Broger, C.; Kohler, C.; Vilbois, F.; Lahm, H.W.; Heitz, M.P.; Malherbe, P.: Molecular characterization of kynurenine pathway enzymes: 3-Hydroxyanthranilic-acid dioxygenase and kynurenine aminotransferase. Adv. Exp. Med. Biol., **398**, 477-483 (1996)

[27] Guidetti, P.; Okuno, E.; Schwarcz, R.: Characterization of rat brain kynurenine aminotransferases I and II. J. Neurosci. Res., **50**, 457-465 (1997)

[28] Baran, H.; Amann, G.; Lubec, B.; Lubec, G.: Kynurenic acid and kynurenine aminotransferase in heart. Pediatr. Res., **41**, 404-410 (1997)

[29] Meunpol, O.; Hall, M.R.; Kapoor, V.: Partial characterization and distribution of kynurenine aminotransferase activity in the Black Tiger prawn (Penaeus monodon). Comp. Biochem. Physiol. B, **120B**, 139-143 (1998)

[30] Knyihar-Csillik, E.; Okuno, E.; Vecsei, L.: Effects of in vivo sodium azide administration on the immunohistochemical localization of kynurenine aminotransferase in the rat brain. Neurosci., **94**, 269-277 (1999)

[31] Battaglia, G.; Rassoulpour, A.; Wu, H.Q.; Hodgkins, P.S.; Kiss, C.; Nicoletti, F.; Schwarcz, R.: Some metabotropic glutamate receptor ligands reduce ky-

nurenate synthesis in rats by intracellular inhibition of kynurenine aminotransferase II. J. Neurochem., **75**, 2051-2060 (2000)

[32] Milart, P.; Urbanska, E.M.; Turski, W.A.; Paszkowski, T.; Sikorski, R.: Kynurenine aminotransferase I activity in human placenta. Placenta, **22**, 259-261 (2001)

[33] Fang, J.; Han, Q.; Li, J.: Isolation, characterization, and functional expression of kynurenine aminotransferase cDNA from the yellow fever mosquito, Aedes aegypti(1). Insect Biochem. Mol. Biol., **32**, 943-950 (2002)

[34] Kwok, J.B.J.; Kapoor, R.; Gotoda, T.; Iwamoto, Y.; Iizuka, Y.; Yamada, N.; Isaacs, K.E.; Kushwaha, V.V.; Church, W.B.; Schofield, P.R.; Kapoor, V.: A missense mutation in kynurenine aminotransferase-1 in spontaneously hypertensive rats. J. Biol. Chem., **277**, 35779-35782 (2002)

[35] Csillik, A.; Knyihar, E.; Okuno, E.; Krisztin-Peva, B.; Csillik, B.; Vecsei, L.: Effect of 3-nitropropionic acid on kynurenine aminotransferase in the rat brain. Exp. Neurol., **177**, 233-241 (2002)

[36] Kocki, T.; Luchowski, P.; Luchowska, E.; Wielosz, M.; Turski, W.A.; Urbanska, E.M.: L-cysteine sulphinate, endogenous sulphur-containing amino acid, inhibits rat brain kynurenic acid production via selective interference with kynurenine aminotransferase II. Neurosci. Lett., **346**, 97-100 (2003)

[37] Mosca, M.; Croci, C.; Mostardini, M.; Breton, J.; Malyszko, J.; Avanzi, N.; Toma, S.; Benatti, L.; Gatti, S.: Tissue expression and translational control of rat kynurenine aminotransferase/glutamine transaminase K mRNAs. Biochim. Biophys. Acta, **1628**, 1-10 (2003)

2,5-Diaminovalerate transaminase

1 Nomenclature

EC number
2.6.1.8

Systematic name
2,5-diaminopentanoate:2-oxoglutarate aminotransferase

Recommended name
2,5-diaminovalerate transaminase

Synonyms
aminotransferase, diamino acid
diamino acid aminotransferase
diamino-acid transaminase

CAS registry number
9030-39-1

2 Source Organism

<1> *Escherichia coli* [1]
<2> *Aspergillus fumigatus* [1]
<3> *Mus musculus* [1]

3 Reaction and Specificity

Catalyzed reaction
2,5-diaminopentanoate + 2-oxoglutarate = 5-amino-2-oxopentanoate + L-glutamate

Reaction type
amino group transfer

Natural substrates and products
S   L-2,5-diaminopentanoate + 2-oxoglutarate <1, 2> [1]
P   5-amino-2-oxopentanoate + L-glutamate

Substrates and products
S   2,4-diaminoglutarate + 2-oxoglutarate <1-3> (Reversibility: ? 1-3<> [1])
    [1]
P   4-amino-2-oxobutanoate + L-glutamate

S   L-2,5-diaminopentanoate + 2-oxoglutarate <1, 2> (Reversibility: ? <1,2>
    [1]) [1]
P   5-amino-2-oxopentanoate + L-glutamate
S   Additional information <1-3> (<1-3> ω-amino monocarboxylic acids,
    e.g. 5-aminopentanoate 4-aminobutanoate, L-aspartate may be transami-
    nated with 2-oxoglutarate by crude enzyme extracts [1]) [1]
P   ?

Inhibitors
streptomycin <1, 3> (<1,3> weak, with 2,4-diaminoglutarate as substrate [1])
[1]

Cofactors/prosthetic groups
pyridoxal 5'-phosphate <1> [1]

Temperature optimum (°C)
30 <1-3> (<1-3> assay at [1]) [1]

5   Isolation/Preparation/Mutation/Application

Source/tissue
brain <3> [1]
liver <3> [1]

Purification
<1> (partial [1]) [1]
<2> (partial [1]) [1]
<3> (partial [1]) [1]

References

[1] Roberts, E.: Studies of transamination. Arch. Biochem. Biophys., **48**, 395-401
    (1954)

Histidinol-phosphate transaminase

1 Nomenclature

EC number
2.6.1.9

Systematic name
L-histidinol-phosphate:2-oxoglutarate aminotransferase

Recommended name
histidinol-phosphate transaminase

Synonyms
HisC
IAP transaminase
L-histidinol phosphate aminotransferase
aminotransferase, histidinol phosphate
glutamic-imidazoleacetol phosphate transaminase
histidine:imidazoleacetol phosphate transaminase
histidinol phosphate aminotransferase
histidinol-phosphate aminotransferase
imidazoleacetol phosphate transaminase
imidazolylacetolphosphate aminotransferase
imidazolylacetolphosphate transaminase

CAS registry number
9032-98-8

2 Source Organism

<1> *Neurospora sp.* ($C_{14}1$-T1710, C84, $C_{14}1$ [1]) [1]
<2> *Salmonella typhimurium* (derepressed histidine auxotroph [2,5]; pyridoxin auxotroph strain [8]) [2-6, 8]
<3> *Bacillus subtilis* [7]
<4> *Nicotiana tabacum* [9]
<5> *Zymomonas mobilis* [10]
<6> *Escherichia coli* [11, 12, 13]

3 Reaction and Specificity

Catalyzed reaction

L-histidinol phosphate + 2-oxoglutarate = 3-(imidazol-4-yl)-2-oxopropyl phosphate + L-glutamate (<2, 3> ping-pong bi bi mechanism [3, 7]; <5> mechanism [10])

Reaction type

amino group transfer

Natural substrates and products

S 3-(imidazol-4-yl)-2-oxopropyl phosphate + L-glutamate <2, 6> (<2> eighth step in histidine biosynthesis [2]; <6> histidine pathway [11]) [2, 11]

P L-histidinol phosphate + 2-oxoglutarate

Substrates and products

S 3-(imidazol-4-yl)-2-oxopropyl phosphate + L-2-aminoadipate <1> (<1> 40% of the activity with L-glutamate [1]) (Reversibility: r <1> [1]) [1]

P L-histidinol phosphate + 2-oxoadipate

S 3-(imidazol-4-yl)-2-oxopropyl phosphate + L-arginine <1> (<1> 33% of the activity with L-glutamate [1]) (Reversibility: r <1> [1]) [1]

P L-histidinol phosphate + 2-oxo-5-guanidinopentanoate

S 3-(imidazol-4-yl)-2-oxopropyl phosphate + L-glutamine <1> (<1> 4% of the activity with L-glutamate [1]) (Reversibility: r <1> [1]) [1]

P L-histidinol phosphate + 4-carbamoyl-2-oxobutanoate

S 3-(imidazol-4-yl)-2-oxopropyl phosphate + L-histidine <1> (<1> 30% of the activity with L-glutamate [1]) (Reversibility: r <1> [1]) [1]

P L-histidinol phosphate + 3-(1H-imidazol-4-yl)-2-oxopropanoate

S L-histidine + 2-oxoglutarate <5> (Reversibility: ? <5> [10]) [10]

P 3-(imidazol-4-yl)-2-oxopropanoate + L-glutamate

S L-histidinol phosphate + 2-oxo-5-guanidinopentanoate <1> (Reversibility: r <1> [1]) [1]

P 3-(imidazol-4-yl)-2-oxopropyl phosphate + L-arginine

S L-histidinol phosphate + 2-oxoadipate <1> (<1> 70% of the activity with 2-oxoglutarate [1]) (Reversibility: r <1> [1]) [1]

P 3-(imidazol-4-yl)-2-oxopropyl phosphate + 2-aminoadipate

S L-histidinol phosphate + 2-oxoglutarate <1-3, 5, 6> (Reversibility: r <1> [1]; ? <2, 3, 5, 6> [2, 4, 5, 7, 10, 11, 13]) [1, 2, 4, 5, 7, 10, 11, 13]

P 3-(imidazol-4-yl)-2-oxopropyl phosphate + L-Glu <1, 6> (<1,6> i.e. imidazoleacetol phosphate + L-Glu [1,11]) [1, 11]

S L-histidinol phosphate + 4-hydroxyphenylpyruvate <5, 5> (Reversibility: ? <5> [10]) [10]

P 3-(imidazol-4-yl)-2-oxopropyl phosphate + L-tyrosine

S L-histidinol phosphate + oxaloacetate <5> (Reversibility: ? <5> [10]) [10]

P 3-(imidazol-4-yl)-2-oxopropyl phosphate + L-aspartate

S L-histidinol phosphate + phenylpyruvate <5> (Reversibility: ? <5> [10]) [10]

P   3-(imidazol-4-yl)-2-oxopropyl phosphate + L-phenylalanine
S   L-histidinol phosphate + pyruvate <5> (Reversibility: ? <5> [10]) [10]
P   3-(imidazol-4-yl)-2-oxopropyl phosphate
S   L-phenylalanine + 2-oxoglutarate <3, 5> (Reversibility: ? <3,5> [7,10]) [7, 10]
P   phenylpyruvate + L-glutamate
S   L-tryptophan + 2-oxoglutarate <3> (Reversibility: ? <3> [7]) [7]
P   3-indole-2-oxopropanoate + L-glutamate
S   L-tyrosine + 2-oxoglutarate <5> (Reversibility: ? <5> [10]) [10]
P   3-(4-hydroxyphenyl)-2-oxopropanoate + L-glutamate
S β-chloro-L-alanine <2> (Reversibility: ? <2> [3]) [3]
P   pyruvate + NH_3 + Cl^- <2> [3]
S   leucine + 2-oxoglutarate <5> (Reversibility: ? <5> [10]) [10]
P   4-methyl-2-oxopentanoate + L-glutamate
S   Additional information <1> (<1> no substrate: oxaloacetate, pyruvate [1]) [1]
P   ?

Inhibitors
Co^{2+} <1> (<1> 0.004 M, 80% inhibition [1]) [1]
Cu^{2+} <1> (<1> 0.004 M, 80% inhibition [1]) [1]
L-glutamate <3> (<3> competitive to tyrosine [7]) [7]
hydroxylamine <1> (<1> 0.004 M, 80% inhibition [1]) [1]
iodoacetate <1> (<1> 0.004 M, 95% inhibition [1]) [1]
semicarbazide <1> (<1> 0.004 M, 80% inhibition [1]) [1]

Cofactors/prosthetic groups
pyridoxal 5'-phosphate <1-3, 5> (<1-3> a pyridoxal phosphate protein [1-3, 5, 7]; <2> sequence of pyridoxal phosphate binding site [3]; <1> K_m: 0.0012 mM [1]; <3> 0.0022 mM [7]; <2> 1 pyridoxal phosphate group associated with each molecule of enzyme [2]; <2> 2 mol of pyridoxal phosphate per mol of enzyme [5]; <5> removal of cofactor triggers irreversible denaturation [10]) [1-3, 5, 7, 8, 10]

Activating compounds
ω-methylpyridoxal phosphate <1> (<1> can replace pyridoxal phosphate, K_m: 0.004 mM [1]) [1]

Specific activity (U/mg)
0.01266 <4> (<4> protein extract after expression in wild-type Escherichia coli DH5α [9]) [9]
7 <5> (<5> 2-oxoglutarate + histidinol-phosphate [10]) [10]
7.11 <5> (<5> 4-hydroxyphenylpyruvate + histidinol-phosphate [10]) [10]
11.6 <5> (<5> 2-oxoglutarate + phenylalanine [10]) [10]
11.6 <5> (<5> 2-oxoglutarate + tyrosine [10]) [10]
18.24 <5> [10]
40 <2> [6]
52.8 <2> [5]

60.3 <1> [1]
325 <3> [7]
1885 <2> [2]

K$_m$-Value (mM)
0.15 <3> (L-histidinol phosphate) [7]
0.17 <5> (L-histidinol phosphate, <5> pH 7, 37°C [10]) [10]
0.71 <3> (L-tyrosine) [7]
1.1 <1> (L-histidinol phosphate, <1> pH 8.1, 37°C [1]) [1]
1.2 <1> (2-oxoglutarate, <1> pH 8.1, 37°C [1]) [1]
2.5 <3> (L-phenylalanine) [7]
3.39 <5> (tyrosine, <5> pH 7, 37°C [10]) [10]
7.1 <3> (2-oxoglutarate) [7]
43.48 <5> (phenylalanine, <5> pH 7, 37°C [10]) [10]

pH-Optimum
7.1 <3> [7]
8 <5> [10]
8.1 <1> (<1> diphosphate buffer [1]) [1]
8.4 <1> (<1> diethanolamine buffer [1]) [1]
8.5 <1> (<1> triethanolamine buffer [1]) [1]

pH-Range
7.5-9 <1> (<1> 7.5: about 85% of activity maximum, 9.0: about 50% of activity maximum, diphosphate buffer [1]) [1]

Temperature optimum (°C)
37 <1, 2> (<1,2> assay at [1,2,5]) [1, 2, 5]

4 Enzyme Structure

Molecular weight
33000 <3> (<3> gel filtration [7]) [7]
37000 <2> (<2> sedimentation equilibrium [8]) [8]
45400 <4> (<4> calculated from nucleotide sequence [9]) [9]
53000 <4> (<4> SDS-PAGE [9]) [9]
59000 <2> (<2> equilibrium centrifugation, pyridoxal phosphate analysis, spectrophotometric assay, phosphate analysis, viscometry [2]) [2]
60000 <6> (<6> dynamic light scattering experiments [13]) [13]
74000 <2> (<2> sedimentation equilibrium measurement [5]) [5]
85000 <5> (<5> gel filtration [10]) [10]

Subunits
dimer <2, 5, 6> (<2> 2 * 25300-37000, nonidentical, equilibrium centrifugation in guanidine-HCl, sedimentation velocity in guanidine-HCl, viscosity analysis in guanidine-HCl [4]; <2> 2 * 37000, SDS-PAGE [5]; <5> 2 * 40000, homodimer, SDS-PAGE [10]; <6> homodimer, crystallization experiments

[11]; <6> homodimer, dynamic light scattering experiments [13]) [4, 5, 10, 11, 13]

Additional information <2> (<2> marked tendency to polymerize upon ageing [2]) [2]

5 Isolation/Preparation/Mutation/Application

Localization
chloroplast <4> [9]

Purification
<1> (C141-T1710, C141 [1]) [1]
<2> (use of ω-aminoalkylagaroses, 15fold purification in one step [6]; from derepressed histidine auxotrophs, 95% pure [2]) [2, 4-6]
<3> (220fold [7]) [7]
<5> (recombinant protein [10]) [10]
<6> (wild-type and mutant enzymes [12]) [12, 13]

Renaturation
<2> (holoenzyme and apoenzyme can be constituted in vitro with pyridoxal 5'-phosphate [8]) [8]

Crystallization
<2> [5]
<6> (native and complexed with L-histidinol phosphate or N-(5'-phospho-pyridoxyl)-L-glutamate, hanging drop vapor diffusion method [11]; hanging drop vapor diffusion method [13]) [11, 13]

Cloning
<4> [9]
<5> [10]
<6> [12, 13]

Engineering
K214A <6> (<6> increased pKa [12]) [12]
N157A <6> (<6> increased pKa [12]) [12]
N157A/R335L <6> (<6> increased pKa [12]) [12]
R335L <6> (<6> increased pKa [12]) [12]

6 Stability

pH-Stability
6 <2> (<2> unstable below [2]) [2]
6-8.5 <3> (<3> stable overnight [7]) [7]

Temperature stability

37 <2> (<2> holoenzyme and apoenzyme are stable for 15 min [8]) [8]
45 <2> (<2> holoenzyme loses 15% activity after 5 min, apoenzyme loses 70% activity after 5 min [8]) [8]
50 <5> (<5> no loss of activity for 2 min [10]) [10]
55 <3, 5> (<3> 4 min, more than 90% loss of activity, 10 min, in presence of pyridoxal 5'-phosphate, 75% loss of activity [7]; <5> enzyme loses activity when incubated above 55°C for 2 min [10]) [7, 10]

Storage stability

<1>, -20°C, stable for several months [1]
<3>, 0°C, 1 month, 30% loss of activity [7]
<5>, -20°C over night and thawing on ice: 98% loss of activity [10]
<5>, -80°C no loss of activity after several months when stored in 10% glycerol [10]
<5>, -80°C over night and thawing on ice: 58% loss of activity [10]

References

[1] Ames, B.N.; Horecker, B.L.: Biosynthesis of histidine: imidazoleacetol phosphate transaminase. J. Biol. Chem., **220**, 113-128 (1956)
[2] Martin, R.G.; Goldberger, R.F.: Imidazolylacetolphosphate:L-glutamate aminotransferase. Purification and physical properties. J. Biol. Chem., **242**, 1168-1174 (1967)
[3] Hsu, L.C.; Okamoto, M.; Snell, E.E.: L-Histidinol phosphate aminotransferase from Salmonella typhimurium. Kinetic behavior and sequence at the pyridoxal-P binding site. Biochimie, **71**, 477-489 (1989)
[4] Martin, R.G.; Voll, M.J.; Appella, E.: Imidazolylacetolphosphate:L-glutamate aminotransferase. Composition and substructure. J. Biol. Chem., **242**, 1175-1181 (1967)
[5] Henderson, G.B.; Snell, E.E.: Crystalline L-histidinol phosphate aminotransferase from Salmonella typhimurium. Purification and subunit structure. J. Biol. Chem., **248**, 1906-1911 (1973)
[6] Henderson, G.B.; Shaltiel, S.; Snell, E.E.: ω-aminoalkylagaroses in the purification of L-histidinol-phosphate aminotransferase. Biochemistry, **13**, 4335-4338 (1974)
[7] Weigent, D.A.; Nester, E.W.: Purification and properties of two aromatic aminotransferases in Bacillus subtilis. J. Biol. Chem., **251**, 6974-6980 (1976)
[8] Roberts, J.H.; Levin, A.P.: Imidazolylacetolphosphate aminotransferase. Properties of the apoprotein produced in a pyridoxine auxotroph of Salmonella typhimurium. J. Biol. Chem., **248**, 77408-7753 (1973)
[9] El Malki, F.; Jacobs, M.: Molecular characterization and expression study of a histidine auxotrophic mutant (his1-) of Nicotiana plumbaginifolia. Plant Mol. Biol., **45**, 191-199 (2001)

[10] Gu, W.; Zhao, G.; Eddy, C.; Jensen, R.A.: Imidazole acetol phosphate amino-transferase in Zymomonas mobilis: molecular genetic, biochemical, and evolutionary analyses. J. Bacteriol., 177, 1576-1584 (1995)

[11] Haruyama, K.; Nakai, T.; Miyahara, I.; Hirotsu, K.; Mizuguchi, H.; Hayashi, H.; Kagamiyama, H.: Structures of Escherichia coli histidinol-phosphate aminotransferase and its complexes with histidinol-phosphate and N-(5'-phosphopyridoxyl)-L-glutamate: double substrate recognition of the enzyme. Biochemistry, 40, 4633-4644 (2001)

[12] Mizuguchi, H.; Hayashi, H.; Miyahara, I.; Hirotsu, K.; Kagamiyama, H.: Characterization of histidinol phosphate aminotransferase from Escherichia coli. Biochim. Biophys. Acta, 1647, 321-324 (2003)

[13] Sivaraman, J.; Li, Y.; Larocque, R.; Schrag, J.D.; Cygler, M.; Matte, A.: Crystal structure of histidinol phosphate aminotransferase (HisC) from Escherichia coli, and its covalent complex with pyridoxal-5'-phosphate and l-histidinol phosphate. J. Mol. Biol., 311, 761-776 (2001)

D-Aspartate transaminase

1 Nomenclature

EC number
2.6.1.10 (deleted, included in EC 2.6.1.21)

Recommended name
D-aspartate transaminase

Acetylornithine transaminase 2.6.1.11

1 Nomenclature

EC number
2.6.1.11

Systematic name
N^2-acetyl-L-ornithine:2-oxoglutarate aminotransferase

Recommended name
acetylornithine transaminase

Synonyms
ACOAT
AOTA
DapATase
EC 2.6.1.69 (deleted, incorporated)
N-acetylornithine aminotransferase
N-acetylornithine-δ-transaminase
N^2-acetyl-L-ornithine:2-oxoglutarate 5-aminotransferase
N^2-acetylornithine 5-aminotransferase
N^2-acetylornithine 5-transaminase
SOAT
acetylornithine 5-aminotransferase
acetylornithine aminotransferase
acetylornithine δ-transaminase
acetylornithine transaminase
aminotransferase, acetylornithine
succinyldiaminopimelate transferase
succinylornithine aminotransferase

CAS registry number
9030-40-4

2 Source Organism

<1> *Klebsiella aerogenes* (one arginine-inducible form and one arginine-repressible form [4]) [4]
<2> *Escherichia coli* (wild-type and arginine-inducible form [5]; Escherichia coli W [2, 5]; ornithine δ transaminase activity (EC 2.6.1.13) in Escherichia coli is due to acetylornithine δ transaminase and not to a separate

specific ornithine δ transaminase [2]; one arginine inducible form and one arginine-repressible form [3]) [1-3, 5]
<3> *Pseudomonas aeruginosa* [6]
<4> *Pseudomonas cepacia* (NCTC 10743 [7]) [7]

3 Reaction and Specificity

Catalyzed reaction
N^2-acetyl-L-ornithine + 2-oxoglutarate = N-acetyl-L-glutamate 5-semialdehyde + L-glutamate

Reaction type
amino group transfer

Natural substrates and products
S   N-acetyl-L-glutamic γ-semialdehyde + L-glutamate <2, 3> (fourth step in biosynthesis of arginine, arginine-repressible biosynthetic enzyme) (Reversibility: ? <2, 3> [2, 3, 6]) [2, 3, 6]
P N^2-acetyl-L-ornithine + 2-oxoglutarate
S N^2-acetyl-L-ornithine + 2-oxoglutarate <1, 3, 4> (arginine degradation, arginine-inducible catabolic enzyme) (Reversibility: ? <1, 3, 4> [4, 6, 7]) [4, 6, 7]
P   N-acetyl-L-glutamate 5-semialdehyde + L-glutamate

Substrates and products
S   L-ornithine + 2-oxoglutarate <4> (Reversibility: ? <4> [7]) [7]
P   ?
S N^2-acetyl-L-ornithine + 2-oxohexanedioic acid <1, 4> (i.e. 2-ketoadipate) (Reversibility: ? <1,4> [4,7]) [4, 7]
P   N-acetyl-L-glutamate 5-semialdehyde + 2-aminohexanedioic acid (i.e. 2-aminoadipic acid)
S N^2-succinyl-L-ornithine + 2-oxoglutarate <4> (Reversibility: ? <4> [7]) [7]
P N^2-succinylglutamate semialdehyde + glutamate <4> [7]
S N^α-acetyl-L-ornithine + 2-oxoglutarate <1-3> (Reversibility: r <2> [1, 5]) [1-6]
P   N-acetyl-L-glutamate 5-semialdehyde + L-glutamate <2, 3> [1, 2, 5, 6]
S   Additional information <1-3> (purified enzyme also has activity of EC 2.6.1.13, <1-3> [2, 4, 6]; not: putrescine, <1> [4]; 4-aminobutyrate, <1> [4]) [2, 4, 6]
P   ?

Inhibitors
Cu^{2+} <2> [1]
L-ornithine <2> [2]
p-chloromercuribenzoate <2> [1]
Additional information <2> (<2> not: EDTA [1]) [1]

Cofactors/prosthetic groups

pyridoxal 5'-phosphate <1, 2> (<1,2> a pyridoxal phosphate protein [1, 4]; <2> inactive unless pyridoxal 5-phosphate is supplied [1]; <2> K_m: 0.0017 mM [1]) [1, 4]

Metals, ions

Ni^{2+} <2> (<2> stimulates [1]) [1]

Specific activity (U/mg)

34.6 <1> [4]
Additional information <2> [1, 5]

K_m-Value (mM)

0.34 <2> (N^2-acetyl-L-ornithine) [1]
0.48 <2> (N^2-acetyl-L-ornithine, <2> wild type enzyme [2]) [2]
0.7 <1> (2-oxoglutarate, <1> cosubstrate acetylornithine [4]) [4]
1.1 <1, 2> (N^2-acetyl-L-ornithine, <2> cosubstrate 2-oxoglutarate, arginine inducible and arginine repressible enzyme [3]) [3, 4]
1.25 <4> (N^2-succinyl-L ornithine) [7]
1.3 <2> (N^2-acetyl-L-ornithine, <2> arginine-repressible enzyme [3]) [3]
1.54 <2> (N^2-acetyl-L-ornithine, <2> arginine-inducible enzyme [2]) [2]
2.5 <2> (2-oxoglutarate) [1]
3.1 <2> (N^2-acetyl-L-ornithine, <2> arginine-inducible enzyme [3]) [3]
6.3 <4> (N^2-acetyl-L-ornithine) [7]
100 <4> (L-ornithine) [7]
Additional information <2, 3> [5, 6]

pH-Optimum

8 <2> [2]
8.1 <1, 2> (<1> biosynthetic enzyme [4]) [1, 4]
8.2 <4> (<4> N^2-acetyl-L-ornithine [7]) [7]
8.5 <3> (<3> N^2-acetyl-L-ornithine [6]) [6]
8.6 <4> (<4> N^2-succinyl-L-ornithine [7]) [7]
8.7 <1> (<1> catabolic enzyme [4]) [4]

pH-Range

6.5-10 <1, 3> (<3> pH 6.5: about 30% of activity maximum, pH 10.0: about 45% of activity maximum [6]; <1> active over the range, catabolic enzyme [4]) [4, 6]
7-9 <2> (<2> about 35% of activity maximum at pH 7.0 and 9.0 [2]) [2]
7.1-9.1 <2> (<2> pH 7.1: about 70% of activity maximum, pH 9.1: about 15% of activity maximum [1]) [1]

Temperature optimum (°C)

37 <2, 4> (<2,4> assay at [1,2,7]) [1, 2, 7]

4 Enzyme Structure

Molecular weight

59000 <1> (<1> arginine-inducible catabolic enzyme, gel filtration [4]) [4]
61000 <2> (<2> gel filtration, arginine-inducible enzyme [3,5]) [3, 5]
105000 <3> (<3> thin-layer gel filtration, sucrose gradient centrifugation [6])
[6]
119000 <2> (<2> gel filtration, arginine-repressible enzyme [3, 5]; sedimentation equilibrium studies [5]) [3, 5]

Subunits

? <2> (<2> x * 31000, wild-type and arginine-inducible form are composed
of 31000 MW subunits, which are products of 2 different structural genes,
high-speed sedimentation equilibrium in 6 M guanidine-HCl, SDS-PAGE
[5]) [5]
dimer <1, 3> (<1> 2 * 27000, SDS-PAGE [4]; <3> 2 * 55000, SDS-PAGE [6])
[4, 6]

5 Isolation/Preparation/Mutation/Application

Purification

<1> (arginine-inducible form, partial purification of arginine-repressible
form [4]) [4]
<2> (partial [1]) [1, 2]
<3> (copurification of EC 2.6.1.11 and 2.6.1.13 [6]) [6]

Crystallization

<2> [5]

6 Stability

Storage stability

<3>, -20°C, partially purified enzyme is stable for at least 3 months [6]

References

[1] Albrecht, A.M.; Vogel, H.J.: Acetylornithine δ-transaminase. Partial purification and repression behavior. J. Biol. Chem., **239**, 1872-1876 (1964)
[2] Billheimer, J.T.; Carnevale, H.N.; Leisinger, T.; Eckardt, T.; Jones, E.E.: Ornithine δ-transaminase activity in Escherichia coli: its identity with acetylornithine δ-transaminase. J. Bacteriol., **127**, 1315-1323 (1976)
[3] Billheimer, J.T.; Jones, E.E.: Inducible and repressible acetylornithine δ-transaminase in Escherichia coli: different proteins. Arch. Biochem. Biophys., **161**,
647-651 (1974)

[4] Friedrich, B.; Friedrich, C.G.; Magasanik, B.: Catabolic N^2-acetylornithine 5-aminotransferase of Klebsiella aerogenes: control of synthesis by induction, catabolite repression, and activation by glutamine synthetase. J. Bacteriol., **133**, 686-691 (1978)

[5] Billheimer, J.T.; Shen, M.Y.; Carnevale, H.N.; Horton, H.R.; Jones, E.E.: Isolation and characterization of acetylornithine δ-transaminase of wild-type Escherichia coli W. Comparison with arginine-inducible acetylornithine δ-transaminase. Arch. Biochem. Biophys., **195**, 401-413 (1979)

[6] Voellmy, R.; Leisinger, T.: Dual role for N-2-acetylornithine 5-aminotransferase from Pseudomonas aeruginosa in arginine biosynthesis and arginine catabolism. J. Bacteriol., **122**, 799-809 (1975)

[7] Vander Wauven, C.; Stalon, V.: Occurrence of succinyl derivatives in the catabolism of arginine in Pseudomonas cepacia. J. Bacteriol., **164**, 882-886 (1985)

Alanine-oxo-acid transaminase

2.6.1.12

1 Nomenclature

EC number
2.6.1.12

Systematic name
L-alanine:2-oxo-acid aminotransferase

Recommended name
alanine-oxo-acid transaminase

Synonyms
L-alanine-α-keto acid aminotransferase
alanine-keto acid aminotransferase
alanine-oxo acid aminotransferase
aminotransferase, alanine-keto acid
leucine-alanine transaminase

CAS registry number
9030-41-5

2 Source Organism

<1> *Brucella abortus* [1]
<2> *Pseudomonas sp.* (enzyme F-II [2]) [2]

3 Reaction and Specificity

Catalyzed reaction
L-alanine + a 2-oxo acid = pyruvate + an L-amino acid

Reaction type
amino group transfer

Substrates and products
S L-alanine + 2-oxobutanoate <2> (Reversibility: r <2> [2]) [2]
P pyruvate + 2-aminobutanoate <2> [2]
S L-alanine + 2-oxoglutarate <2> (<2> at 2% of the activity with 2-oxobutanoate [2]) (Reversibility: r <2> [2]) [2]
P pyruvate + L-glutamate <2> [2]

S L-alanine + 2-oxohexanoate <2> (<2> at 81% of the activity with 2-oxo-butanoate [2]) (Reversibility: r <2> [2]) [2]

P pyruvate + 2-aminohexanoate <2> [2]

S L-alanine + 2-oxoisohexanoate <2> (<2> at 8% of the activity with 2-ox-obutanoate [2]) (Reversibility: r <2> [2]) [2]

P pyruvate + L-leucine <2> [2]

S L-alanine + 2-oxoisopentanoate <2> (<2> at 10% of the activity with 2-oxobutanoate [2]) (Reversibility: r <2> [2]) [2]

P pyruvate + L-valine <2> [2]

S L-alanine + 2-oxopentanoate <2> (<2> at 108% of the activity with 2-oxobutanoate [2]) (Reversibility: r <2> [2]) [2]

P pyruvate + L-norvaline <2> [2]

S L-alanine + glyoxylate <2> (<2> 53% of the activity with 2-oxobutanoate [2]) (Reversibility: ? <2> [2]) [2]

P pyruvate + glycine <2> [2]

S L-alanine + oxaloacetate <2> (<2> at 16% of the activity with 2-oxobutanoate [2]) (Reversibility: r <2> [2]) [2]

P pyruvate + L-aspartate <2> [2]

S pyruvate + L-2-aminobutanoate <2> (<2> preferred amino donor [2]) (Reversibility: r <2> [2]) [2]

P L-alanine + 2-oxobutanoate <2> [2]

S pyruvate + L-aspartate <2> (<2> 4% of the activity with L-2-aminobu-tanoate [2]) (Reversibility: ? <2> [2]) [2]

P L-alanine + oxaloacetate <2> [2]

S pyruvate + L-glutamate <2> (<2> 4% of the activity with L-2-aminobu-tanoate [2]) (Reversibility: r <2> [2]) [2]

P L-alanine + 2-oxoglutarate <2> [2]

S pyruvate + L-isoleucine <1, 2> (<2> 6% of the activity with L-2-amino-butanoate [2]) (Reversibility: ? <1,2> [1,2]) [1, 2]

P L-alanine + 3-methyl-2-oxopentanoate <1, 2> [1, 2]

S pyruvate + L-leucine <1, 2> (<2> 16% of the activity with L-2-aminobu-tanoate [2]) (Reversibility: r <2> [2]; ? <1> [1]) [1, 2]

P alanine + 4-methyl-2-oxopentanoate <1, 2> (<1,2> i.e. 2-oxoisohexanoate [1,2]) [1, 2]

S pyruvate + L-norleucine <1, 2> (<2> 31% of the activity with L-2-amino-butanoate [2]) (Reversibility: ? <1,2> [1,2]) [1, 2]

P L-alanine + 2-oxohexanoate <1, 2> [1, 2]

S pyruvate + L-norvaline <2> (<2> 84% of the activity with L-2-aminobu-tanoate [2]) (Reversibility: r <2> [2]) [2]

P L-alanine + 2-oxopentanoate <2> [2]

S pyruvate + L-phenylalanine <1, 2> (<2> 4% of the activity with L-2-ami-nobutanoate [2]) (Reversibility: ? <1,2> [1,2]) [1, 2]

P L-alanine + 2-oxo-3-phenylpropanoate <1, 2> [1, 2]

S pyruvate + L-threonine <2> (<2> 3% of the activity with L-2-aminobu-tanoate [2]) (Reversibility: ? <2> [2]) [2]

P L-alanine + 3-hydroxy-2-oxobutanoate <2> [2]

S pyruvate + L-valine <2> (<2> 8% of the activity with L-2-aminobutano-
ate [2]) (Reversibility: r <2> [2]) [2]

P L-alanine + 3-methyl-2-oxobutanoate <2> [2]

Cofactors/prosthetic groups
pyridoxal 5'-phosphate <1> (<1> a pyridoxal phosphate protein [1]) [1]

pH-Optimum
8 <2> (<2> assay at [2]) [2]

Temperature optimum (°C)
30 <2> (<2> assay at [2]) [2]

5 Isolation/Preparation/Mutation/Application

Purification
<2> (enzyme F-II [2]) [2]

References

[1] Altenbern, R.A.; Housewright, R.D.: Transaminase in smooth Brucella abor-
tus, strain 19. J. Biol. Chem., **204**, 159-167 (1953)

[2] Koide, Y.; Honma, M.; Shimomura, T.: L-Alanine-α-ketoacid aminotransferase in Pseudomonas sp. Agric. Biol. Chem., **41**, 781-784 (1977)

Ornithine-oxo-acid transaminase

2.6.1.13

1 Nomenclature

EC number
2.6.1.13

Systematic name
L-ornithine:2-oxo-acid aminotransferase

Recommended name
ornithine-oxo-acid transaminase

Synonyms
L-ornithine 5-aminotransferase
L-ornithine aminotransferase
L-ornithine:α-ketoglutarate δ-aminotransferase
OAT
aminotransferase, ornithine-keto acid
ornithine 5-aminotransferase
ornithine δ-transaminase
ornithine transaminase
ornithine-2-oxoacid aminotransferase
ornithine-α-ketoglutarate aminotransferase
ornithine-keto acid aminotransferase
ornithine-keto acid transaminase
ornithine-ketoglutarate aminotransferase
ornithine-oxo acid aminotransferase
ornithine:α-oxoglutarate transaminase

CAS registry number
9030-42-6

2 Source Organism

<1> *Erwinia aroidea* [24]
<2> *Ancyclostoma ceylanicum* (hookworm parasite [4]) [4]
<3> *Nippostrongylus brasiliensis* [4]
<4> *Bos taurus* [5, 7]
<5> *Corynebacterium sepedonicum* [8]
<6> *Fasciola hepatica* [9]
<7> *Schistosoma mansoni* [9, 10]
<8> *Cucurbita pepo* (squash [11]) [11]

<9>   *Aspergillus nidulans* [13]
<10> *Bacillus sphaericus* (IFO3525 [24]) [14, 15, 24]
<11> *Raphanus sativus* (radish [16]) [16]
<12> *Manduca sexta* (tobacco hornworm [23]) [23]
<13> *Staphylococcus aureus* [24]
<14> *Rattus norvegicus* (Wistar [21]) [1, 2, 12, 17-22, 25, 26, 27, 28, 29, 39, 40]
<15> *Bacillus brevis* (gramicidin S-producing [3]) [3]
<16> *Homo sapiens* (patients with hyperonithinemia [32,39]; hereditary en-
      zyme deficiency [40]) [21, 25, 32, 39, 40, 41, 42, 43]
<17> *Mus musculus* (OF-1 Swiss mouse [34]) [17, 27, 34, 35]
<18> *Gallus gallus* [6]
<19> *Neurospora crassa* [29]
<20> *Sus scrofa* [30]
<21> *Plasmodium falciparum* (3D7 [31]) [31]
<22> *Bacillus sp.* (YM-2 [33]) [33]
<23> *Brassica juncea* [36]
<24> *Arabidopsis thaliana* [37, 38]
<25> *Oryza sativa* (rice, cv. Taichung Native 1 [44]) [44]

3 Reaction and Specificity

Catalyzed reaction

L-ornithine + a 2-oxo acid = L-glutamate 5-semialdehyde + an L-amino acid
(<16,22> mechanism [33,40])

Reaction type

amino group transfer

Natural substrates and products

S L-ornithine + 2-oxoglutarate <10, 14, 16, 25> (<14> metabolism of pro-
       line to ornithine [1]; <16> urea cycle [40]; <25> ornithine pathway [44])
       [1, 24, 40, 44]
P Δ^1-pyrroline-5-carboxylate + L-glutamate

Substrates and products

S D-ornithine + 2-oxoglutarate <22> (<22> 6% the rate of L-ornithine [33])
       (Reversibility: ? <22> [33]) [33]
P glutamate + Δ^1-pyrroline-5-carboxylate
S L-lysine + 2-oxoglutarate <10> (<10> weak [24]) (Reversibility: ? <10>
       [24]) [24]
P glutamate + 2-amino-7-oxooctanoic acid
S L-ornithine + 2-oxoglutarate <1-17, 19-22, 24, 25> (<14> exchanges the
       pro-S hydrogen on the δ-carbon atom of ornithine exclusively, also trans-
       fers the α amino group of glutamate, kinetics of the half reactions be-
       tween the enzyme and both amino acids [22]; <14> D-ornithine is not
       accepted as substrate [1]) (Reversibility: r <14> [18]; ? <1-17, 19-22, 24,
       25> [1-17, 19-31, 33, 37-40, 44]) [1-31, 33, 37-40, 44]

P Δ^1-pyrroline-5-carboxylate + L-glutamate <10, 12, 14, 16, 19> (<14> equilibrium constant: 71 [1]; <12, 14> DL-pyrroline-5-carboxylate [18, 23]; <10, 14, 16, 19, 22> glutamate-γ semialdehyde spontaneously cyclizes to form Δ1-pyrroline-5-carboxylate [12, 24, 29, 33, 40]) [1, 12, 14, 18, 23, 24, 26, 29, 34, 40]

S L-ornithine + glyoxylate <10, 14, 18, 22> (<10, 22> weak [14, 33]) (Reversibility: ? <10, 14, 18, 22> [1, 6, 14, 29, 33]) [1, 6, 14, 29, 33]

P Δ^1-pyrroline-5-carboxylate + glycine

S L-ornithine + pyruvate <7, 10, 14, 18, 22> (<7> considerably less effective than 2-oxoglutarate [10]; <10> weak [14]) (Reversibility: ? <7, 10, 14, 18, 22> [1, 6, 10, 14, 29, 33]) [1, 6, 10, 14, 29, 33]

P Δ^1-pyrroline-5-carboxylate + L-alanine

S N-acetylornithine + 2-oxoglutarate <21> (Reversibility: ? <21> [31]) [31]

P glutamate + N-acetyl-glutamate-γsemialdehyde

S β-lysine + 2-oxoglutarate <10> (<10> weak [14,24]) (Reversibility: ? <10> [14, 24]) [14, 24]

P glutamate + 3-amino-7-oxooctanoic acid

S ornithine + oxaloacetate <7, 9, 10, 18, 22> (<10, 22> weak [14, 33]) (Reversibility: ? <7, 9, 10, 18, 22> [6, 10, 13, 14, 29, 33]) [6, 10, 13, 14, 29, 33]

P Δ^1-pyrroline-5-carboxylate + L-aspartate

S Additional information <8-10, 14> (<9, 10, 14> specific for ornithine [1, 13, 24]; <8, 10> specific for 2-oxoglutarate [11, 24]; <10, 14> inactive as amino group acceptor: 2-oxohexanoate [1, 14]; <10, 14> inactive as amino group acceptor: 2-oxopentanoate [1, 14]; <10, 14> inactive as amino group acceptor: 2-oxobutanoate [1, 14]; <10, 14> inactive as amino group acceptor: 2-oxoisohexanoate [1, 14]; <10, 14> inactive as amino group acceptor: 2-oxoisopentanoate [1, 14]; <14> inactive as amino group acceptor: oxaloacetate [1]; <10> inactive as amino group acceptor: 3-phenylpyruvate [14]; <9, 10> inactive as amino group donor: L-lysine [13, 14]; <9,10> inactive as amino group donor: L-arginine [13, 14]; <9> inactive as amino group donor: 4-aminobutanoate [13]; <9, 10> inactive as amino group donor: D-ornithine [13, 14]; <9, 10> inactive as amino group donor: putrescine [13, 14]; <10> inactive as amino group donor: D-lysine [14]; <10> inactive as amino group donor: L-citrulline [14]; <10> inactive as amino group donor: glycine [14]; <10> inactive as amino group donor: L-alanine [14]; <10> inactive as amino group donor: L-aspartate [14]; <10> inactive as amino group donor: L-valine [14]; <10> inactive as amino group donor: L-leucine [14]; <10> inactive as amino group donor: L-2-aminobutanoate [14]; <10> inactive as amino group donor: β-alanine [14]; <10> inactive as amino group donor: taurine [14]; <10> inactive as amino group donor: butylamine [14]; <10> inactive as amino group donor: cadaverine [14]) [1, 11, 13, 14, 24]

P ?

Inhibitors

2-aminooxyacetic acid <14, 16> (<14> almost complete inhibition at 0.2 mM [40]) [40]

2-oxobutanoate <14> [1]

2-oxoglutarate <7, 14> (<7,14> α-keto acid acceptor, inhibitor at high concentrations [1,10]) [1, 10]

2-oxohexanoate <14> [1]

2-oxoisopentanoate <14> [1]

3-methyl-2-benzothiazolone hydrazone hydrochloride <10> (<10> fully reversed by addition of pyridoxal phosphate [24]) [24]

4-amino-5-hexynoic acid <11> (<11> synonym: AHA, irreversible, pretreatment with ornithine protects [16]) [16]

4-aminobutanoate <14> [1, 26]

5,5'-dithiobis(2-nitrobenzoic acid) <14> [26]

5-aminopentanoate <14> (<14> 5-aminovaleric acid [28]) [1, 26, 28]

5-fluoromethylornithine <14, 16, 17> (<14,17> specific irreversible inhibitor in vivo and in vitro [17]; <16> competitive [40]) [17, 40]

DL-norvaline <14> (<14> 44% inhibition at 100 mM [26]) [1, 26]

L-arginine <10> (<10> fully reversed by addition of pyridoxal phosphate, 28% inhibition at 10 mM after 20 min [24]) [24]

L-canaline <4, 12, 14, 16, 21> (<14> i.e. 2-amino-4-(aminooxy)-n-butanoic acid, irreversible, forms oxime-type compound with pyridoxal 5'-phosphate [28]; <12> binds covalently to enzyme forming a canaline-pyridoxal phosphate oxime, dialysis against free pyridoxal phosphate slowly reactivates the enzyme as the oxime is replaced by pyridoxal phosphate [23]; <21> aminooxy analogue of ornithine [31]; <16> irreversible [40]) [7, 23, 28, 31, 40, 41]

L-canavanine <14> (<14> complete inhibition at 0.025 M [1]; <14> 23% inhibition at 20 mM [26]) [1, 26]

L-cycloserine <10> (<10> fully reversed by addition of pyridoxal phosphate [24]) [24]

L-isoleucine <14> (<14> inhibitory action largely depends on the branched structure of the hydrophobic residue [18]; <14> 19% inhibition at 50 mM [26]) [18, 26]

L-leucine <10, 14> (<10> fully reversed by addition of pyridoxal phosphate [24]; <14> inhibitory action largely depends on the branched structure of the hydrophobic residue [18]; <14> 17% inhibition at 50 mM [26]) [1, 18, 24, 26]

L-lysine <10> (<10> 35% inhibition at 10 mM after 20 min [24]) [24]

L-ornithine <9> (<9> slight inhibition above 40 mM [13]) [13]

L-valine <4, 10, 14> (<10> fully reversed by addition of pyridoxal phosphate, 35% inhibition at 10 mM after 20 min [24]; <4,14> competitive with L-ornithine [5,18]; <14> inhibitory action largely depends on the branched structure of the hydrophobic residue [18]; <14> 33% inhibition at 50 mM [26]) [1, 5, 18, 24, 26]

N-ethylmaleimide <12> (<12> complete inactivation at 0.1 and 1 mM after 20 min [23]) [23]

N^2-acetyl-L-ornithine <14> (<14> competitive with L-ornithine [28]) [28]

NH_2OH <8> [11]

NaCN <8> [11]

β-lysine <10> (<10> 80% inhibition at 10 mM after 20 min [24]) [24]

butyrate <14> (<14> 27% inhibition at 25 mM [26]) [26]

cadaverine <10, 14> (<10> fully reversed by addition of pyridoxal phosphate, 57% inhibition at 10 mM after 20 min [24]) [1, 24]

caproate <14> (<14> 34% inhibition at 25 mM [26]) [26]

carboxylmethoxyamine <10> (<10> fully reversed by addition of pyridoxal phosphate [24]) [24]

gabaculine <10, 11, 16, 17, 25> (<10> synonym: 5-amino-1,3-cyclohexadienyl carboxylate, irreversible inactivation of the holo form, but not the apo form [15]; <11> almost complete irreversible inhibition at 0.001 mM within a few hours, pretreatment with ornithine protects [16]; <25> 75% inhibition at 0.05 mM [44]) [15, 16, 34, 41, 44]

glyoxylate <7, 14> (<7,14> α-keto acid acceptor, inhibitor at high concentrations [1,10]) [1, 10]

hexanoate <14> [26]

hydroxylamine <10, 14> (<14> reversible, noncompetitive [28]) [24, 28]

iodoacetamide <14> [26]

isocaproate <14> (<14> 41% inhibition at 25 mM [26]) [26]

isohexanoate <14> [26]

isoleucine <7> [1, 10]

isonicotinylhydrazide <8> [11]

isopentanoate <14> [26]

isovalerate <14> (<14> 41% inhibition at 25 mM [26]) [26]

norvaline <14> (<14> inhibitory action largely depends on the branched structure of the hydrophobic residue [18]) [18]

ornithine <7, 9, 14> (<9> concentration above 40 mM, slight inhibition [13]; <7,14> at high concentrations [1,10]) [1, 10, 13]

oxaloacetate <7> [10]

p-chloromercuribenzoate <14> [26]

p-chloromercuriphenylsulfonic acid <18> [6]

phenylhydrazine <10> (<10> fully reversed by addition of pyridoxal phosphate [24]) [24]

putrescine <10, 14> (<10> fully reversed by addition of pyridoxal phosphate, 63% inhibition at 10 mM after 20 min [24]) [1, 24]

pyridoxal 5'-phosphate <14> (<14> in excess: partial inactivation, 70% inactivation at 0.75 and 7.5 mmol/g enzyme [12]) [12]

pyruvate <7, 14> (<7,14> α-keto acid acceptor, inhibitor at high concentrations [1,10]) [1, 10]

serine <7> [10]

valerate <14> (<14> 38% inhibition at 25 mM [26]) [26]

Additional information <6, 7, 16, 18, 25> (<18> inhibition by high substrate concentration [6]; <6,7> by substrate analogs and pyridoxal phosphate inhibitors [9]; <16> overview: inhibitors [40]; <25> no inhibitor: cycloheximide [44]) [6, 9, 40, 44]

Cofactors/prosthetic groups

pyridoxal 5'-phosphate <4, 8-10, 12, 14, 16, 17, 22> (<8> stimulates purified enzyme [11]; <10> activates [24]; <9> enzyme is dependent on pyridoxal 5'-phosphate [13]; <4> K_m: 0.01 mM [7]; <16> K_m: 0.0018 mM [25]; <1, 4, 10, 12, 14> a pyridoxal phosphate protein [5, 23, 24, 28]; <12, 14> 4 mol of pyridoxal phosphate per mol of enzyme [23, 26]; <4, 10, 22> 1 mol per mol of enzyme [5, 24, 33]; <10> pyridoxal 5'-phosphate is bound through an aldimine linkage to the ε-amino group of a lysine residue [24]) [5, 7, 11, 13, 23-26, 28, 33, 34, 40]

Turnover number (min^{-1})

2 <12> (2-oxoglutarate) [23]
2 <12> (L-ornithine) [23]

Specific activity (U/mg)

0.000184 <16> (<16> healthy patients [32]) [32]
0.0011 <16> (<16> native chorionic villi [39]) [39]
0.0011-0.0072 <16> (<16> fibroblast [39]) [39]
0.0026-0.228 <20> (<20> depending on gestational age [30]) [30]
0.0052 <16> (<16> cultured chorionic villi [39]) [39]
0.12 <14> [17]
0.184 <14> (<14> liver [39]) [39]
0.184 <14> (<14> skeletal muscle [39]) [39]
0.184 <16> (<16> healthy patients [32]) [32]
0.21 <14> [1]
0.55 <13> [24]
0.67 <10> [24]
1.38 <11> [16]
1.6 <12> [23]
16 <14> [25]
19.08 <17> [35]
33 <16> [25]
53.6 <10> [14]
76 <22> [33]
Additional information <14> [2]

K$_m$-Value (mM)

0.191 <14> (2-oxoglutarate, <14> gut, pH 7.5 [21]) [21]
0.23 <14> (2-oxoglutarate, <14> kidney, pH 7.5 [21]) [21]
0.235 <14> (2-oxoglutarate, <14> liver, pH 7.5 [21]) [21]
0.25 <14> (2-oxoglutarate, <14> kidney, pH 7.5 [21]) [21]
0.28 <14> (2-oxoglutarate, <14> pH 7.1, 37°C [1]) [1]
0.337 <14> (2-oxoglutarate, <14> AH 130 cells, pH 7.5 [21]) [21]
0.5 <4> (2-oxoglutarate) [5]
0.56 <21> (2-oxoglutarate, <21> pH 7.4, 37°C [31]) [31]
0.67 <14> (2-oxoglutarate, <14> rhinopharyngeal tumor, pH 7.5 [21]) [21]
0.7 <17> (2-oxoglutarate, <17> pH 7.8, 37°C [27]) [27]
1.03 <18> (2-oxoglutarate) [6]

1.1 <14> (2-oxoglutarate, <14> pH 8, 37°C [25]) [25]
1.1 <14> (L-ornithine, <14> pH 8, 37°C [25]) [25]
1.113 <14> (L-ornithine, <14> gut, pH 7.5 [21]) [21]
1.136 <14> (2-oxoglutarate, <14> well differentiated gastroadenocarcinoma, pH 7.5 [21]) [21]
1.282 <14> (L-ornithine, <14> well differentiated gastroadenocarcinoma, pH 7.5 [21]) [21]
1.4 <14> (DL-pyrroline-5-carboxylate) [18]
1.4 <14> (L-ornithine, <14> kidney, pH 7.5 [21]) [21]
1.4 <14> (L-ornithine, <14> liver, pH 7.5 [21]) [21]
1.53 <21> (N-acetylornithine, <21> pH 7.4, 37°C [31]) [31]
1.7 <14> (L-ornithine, <14> rhinopharyngeal tumor, pH 7.5 [21]) [21]
1.7 <14> (ornithine, <14> with pyruvate as cosubstrate, pH 7.1, 37°C [1]) [1]
1.8 <10> (L-ornithine, <10> pH 8.5, 37°C [14]) [14]
1.8 <16> (L-ornithine, <16> pH 8, 37°C [25]) [25]
2 <14> (L-ornithine, <14> AH 130 cells, pH 7.5 [21]) [21]
2 <22> (L-ornithine, <22> 37°C [33]) [33]
2.3 <12> (L-ornithine, <12> pH 8, 37°C [23]) [23]
2.4 <14> (2-oxoglutarate, <14> poorly differentiated gastroadenocarcinoma, pH 7.5 [21]) [21]
2.7 <16> (2-oxoglutarate, <16> pH 8, 37°C [25]) [25]
2.7 <14> (L-ornithine, <14> pH 7.8, 37°C [27]) [27]
2.8 <14> (ornithine, <14> with 2-oxoglutarate as cosubstrate, pH 7.1, 37°C [1]) [1]
2.9 <17> (L-ornithine, <17> pH 8, 37°C [35]) [35]
2.9 <14> (ornithine, <14> with glyoxylate as cosubstrate, pH 7.1, 37°C [1]) [1]
3 <14> (2-oxoglutarate, <14> pH 7.8, 37°C [27]) [27]
3.2 <12> (2-oxoglutarate, <12> pH 8, 37°C [23]) [23]
3.275 <14> (L-ornithine, <14> poorly differentiated gastroadenocarcinoma, pH 7.5 [21]) [21]
3.3 <14> (L-ornithine, <14> kidney, pH 7.5 [21]) [21]
3.6 <17> (2-oxoglutarate, <17> pH 8, 37°C [35]) [35]
3.95 <21> (L-ornithine, <21> pH 7.4, 37°C [31]) [31]
4.6 <10> (2-oxoglutarate, <10> pH 8.5, 37°C [14]) [14]
4.8 <17> (L-ornithine, <17> pH 7.8, 37°C [27]) [27]
5.78 <4> (L-ornithine) [5]
5.79 <2> (L-ornithine) [4]
6.1 <22> (2-oxoglutarate, <22> 37°C [33]) [33]
6.2 <9> (2-oxoglutarate, <9> pH 7.6, 37°C [13]) [13]
6.2 <9> (L-ornithine, <9> pH 7.6, 37°C [13]) [13]
6.31 <18> (L-ornithine) [6]
6.7 <14> (glyoxylate, <14> pH 7.1, 37°C [1]) [1]
22 <14> (pyruvate, <14> pH 7.1, 37°C [1]) [1]
25 <14> (L-glutamate) [18]
43.4 <2> (2-oxoglutarate) [4]
87 <24> (L-ornithine, <24> salt stress conditions [38]) [38]

91 <24> (L-ornithine, <24> normal conditions [38]) [38]
Additional information <4, 8, 14> (<14> aggregation results in increase of
K_m [20]; <14> K_m for ornithine increases below pH 7.5, K_m for 2-oxogluta-
rate decreases below pH 7 [26]) [7-11, 18, 20, 26]

K_i-Value (mM)

0.00043 <14> (hydroxylamine, <14> pH 8, 37°C [28]) [28]
0.000492 <21> (L-canaline, <21> pH 7.4, 37°C [31]) [31]
0.0017 <14> (5-aminopentanoate, <14> pH 8, 37°C [28]) [28]
0.0088 <14> (N^2-L-acetyl ornithine, <14> pH 8, 37°C [28]) [28]
0.07 <14> (5-fluoromethylornithine, <14> 37°C, pH 8 [17]) [17]
2 <14> (L-valine, <14> forward reaction, 37°C, pH 8 [18]) [18]
9.5 <4> (L-valine) [5]
20 <14> (L-valine, <14> reverse reaction, 37°C, pH 8 [18]) [18]

pH-Optimum

7.3 <18> [6]
7.4 <2> [4]
7.5 <17> [27]
7.6 <9> [13]
7.8 <14> [27]
8 <4, 5, 7, 14, 16> [5, 8, 10, 18, 25]
8-8.1 <4> [7]
8-8.5 <6, 7> [9]
8.15 <14> [12]
8.5 <3, 10> [4, 14]

pH-Range

7-9 <16> (<16> 7.0: about 50% of activity maximum, 9.0: about 60% of activity maximum [25]) [25]

Temperature optimum (°C)

37 <12, 14, 16, 17> (<12, 14, 16, 17> assay at [1, 12, 20, 23, 25, 27, 28]) [1, 12,
20, 23, 25, 27, 28]
45 <6> [9]
55 <7> [9, 10]
70 <22> [33]

4 Enzyme Structure

Molecular weight

45000 <14, 16> (<14> SDS-PAGE [27]; <16> mature protein [40]) [27, 40]
46000-47000 <17> (<17> SDS-PAGE [27]) [27]
48000 <8> (<8> gel filtration [11]) [11]
63000 <18> (<18> gel filtration [6]) [6]
74000 <4> [7]
80000-85000 <10> (<10> sedimentation equilibrium method, gel filtration
[24]) [24]

82000 <22> (<22> gel filtration [33]) [33]
88000 <4> (<4> gel filtration [5]) [5]
90000 <15> (<15> gel filtration [3]) [3]
92000 <17> (<17> gel filtration [35]) [35]
109000 <21> (<21> gel filtration [31]) [31]
132000 <14> (<14> equilibrium ultracentrifugation [2]) [2]
148000-152000 <12> (<12> nondenaturing PAGE, gel filtration [23]) [23]
161000 <14> (<14> equilibrium sedimentation analysis [26]) [26]
177000 <16> (<16> sucrose density gradient centrifugation [25]) [25]
180000 <5> [8]
256000 <14> (<14> equilibrium sedimentation [19]) [19]

Subunits

? <14, 17> (<14> x * 45000, SDS-PAGE [27]; <17> x * 46000-47000, SDS-PAGE [27]; <14> decreasing pH or raising ionic strength enhances aggregation of the enzyme [20]) [20, 27]

dimer <4, 10, 15, 17, 21, 22> (<4> 2 * 42000, SDS-PAGE [5]; <15> [3]; <10> 2 * 41000, SDS-PAGE [24]; <21> 2 * 45000, SDS-PAGE, homodimer [31]; <22> 2 * 41000, SDS-PAGE [33]; <17> 2 * 46000, SDS-PAGE [35]) [3, 5, 24, 31, 33, 35]

hexamer <14, 16> (<14> 6 * 45000, SDS-PAGE and equilibrium sedimentation in presence of 6 M guanidine-HCl [19]; <16> α_6 hexameric molecule, crystallization experiments [42]) [19, 42]

monomer <18> (<18> 1 * 62000, SDS-PAGE [6]) [6]

tetramer <12, 14, 16> (<14> 4 * 33000, empirical weight determined from the values for the thiol and half-cystine content [2]; <14> 4 * 43000, SDS-PAGE [26]; <12> 4 * 36000, SDS-PAGE [23]; <16> 4 * 44000, SDS-PAGE [25]) [2, 23, 25, 26]

5 Isolation/Preparation/Mutation/Application

Source/tissue

Yoshida AH-130 cell <14> [21]
brain <14, 17> [17, 29]
chorionic villus <16> [39]
ciliary body <4> [7]
cotyledon <8, 11> [11, 16]
eye <4> [7]
fetus <17, 20> (<20> small intestine [30]) [30, 35]
fibroblast <16> [39]
gastric adenocarcinoma cell <16> (<16> well and poorly differentiated [21]) [21]
gut <14> [21]
heart <14> [29]
intestine <16> [40]
iris <4> [7]

kidney <14, 17> [21, 26, 29, 34, 40]
leaf <11, 23, 25> [16, 36, 44]
liver <4, 14, 16, 17> [1, 2, 5, 6, 12, 17-22, 25, 27, 29, 35, 39, 40]
lymphoblast <16> [25]
lymphocyte <16> [32]
nephron <17> [34]
retina <16> [40]
skeletal muscle <14> [29, 39]
spleen <14> [29]

Localization
mitochondrial matrix <16> [40]
mitochondrion <14, 16> [20, 29, 39, 40]
polysome <14, 17> [27]

Purification
<4> (partial [7]) [5, 7]
<5> [8]
<6> [9]
<7> [10]
<8> [11]
<9> (partial [13]) [13]
<10> (homogeneity [15]) [14, 15, 24]
<12> [23]
<14> (partial [1,17]) [1, 2, 12, 19, 22, 25, 26, 27]
<16> (partial [25]) [25, 42, 43]
<17> (homogeneity [35]) [27, 35]
<18> [6]
<21> [31]
<22> (homogeneity [33]) [33]
<23> (partial [36]) [36]

Crystallization
<5> [8]
<10> [14, 24]
<14> (hanging drop vapor diffusion method [19]) [2, 19, 26]
<16> (complexed to gabaculine and L-canaline, hanging drop vapor diffusion method [41]; hanging drop vapor diffusion method [42]; complexed with 5-fluoromethylornithine, hanging drop vapor diffusion method [43]) [25, 40, 41, 42, 43]
<17> [27]

Cloning
<16> (fusion with maltose binding protein [43]) [41, 42, 43]
<21> (calmodulin-binding peptide fusion protein [31]) [31]
<24> [37, 38]

6 Stability

pH-Stability

6.8-8 <14> (<14> equally stable at pH 6.8 and pH 8.0 [26]) [26]

Temperature stability

65 <5> (<5> stable up to [8]) [8]

66 <14> (<14> pH 7.5, 15 min, 85% loss of activity, 60 min, 96% loss of activity [26]) [26]

General stability information

<5>, pyridoxal 5'-phosphate slightly increases heat stability [8]

<7>, pyridoxal 5'-phosphate is required during solubilization and purification [10]

<8>, stability enhanced by presence of DTT and glycerol [11]

<12>, 2-mercaptoethanol or DTT stabilizes [23]

<14>, activity of the purified enzyme is preserved by addition of pyridoxal 5'-phosphate [1]

<14>, pyridoxal 5'-phosphate stabilizes [1]

Storage stability

<12>, -60°C, long-term storage is best as a pellet under liquid-saturated ammonium sulfate [23]

<12>, 1°C, sharp loss in activity, especially in absence of free pyridoxal phosphate and mercaptans [23]

<14>, -15°C, pH 6-8, 0.01 M potassium phosphate buffer, about 20% loss of activity after 4 months [26]

<14>, -15°C, rat liver homogenate in sucrose or phosphate solution, stable for some weeks [1]

<16>, -20°C, 50 mM potassium phosphate buffer, pH 8, 5 mM 2-oxoglutarate, 0.02 mM pyridoxal phosphate, 90-100% retention of activity after 1 month [25]

References

[1] Strecker, H.J.: Purification and properties of rat liver ornithine Δ-transaminase. J. Biol. Chem., **240**, 1225-1230 (1965)

[2] Peraino, C.; Bunville, L.G.; Tahmisian, T.N.: Chemical, physical, and morphological properties of ornithine aminotransferase from rat liver. J. Biol. Chem., **244**, 2241-2249 (1969)

[3] Kanda, M.; Hori, K.; Kurotsu, T.; Saito, Y.: Purification and properties of ornithine and N-acetylornithine aminotransferases from gramicidin S-producing Bacillus brevis. Enzymes Depend. Pyridoxal Phosphate Other Carbonyl Compd. Cofactors, Proc. Int. Symp. Vitamin B6 Carbonyl Catal., Meeting Date 1990 (Fukui, T., ed.), Pergamon, Oxford, **8**, 241-243 (1991)

[4] Sharma, V.; Katiyar, J.C.; Ghatak, S.; Shukla, O.P.: Ornithine aminotransferase of hookworm parasites. Indian J. Parasitol., **12**, 237-242 (1988)

[5] Park, K.S.: Characterization of bovine liver ornithine aminotransferase. Han'guk Saenghwa Hak Hoe Chi, **18**, 351-360 (1985)

[6] Kim, G.S.; Park, J.O.; Park, K.S.: Studies on the ornithine aminotransferase from chicken liver. Han'guk Saenghwa Hak Hoe Chi, **18**, 109-115 (1985)

[7] Shiono, T.; Hayasaka, S.; Mizuno, K.: Partial purification and certain properties of ornithine ketoacid aminotransferases in the ciliary body and iris and the retinal pigment epithelium of the bovine eye. Exp. Eye Res., **32**, 475-490 (1981)

[8] Yasuda, M.; Toyama, S.; Yonaha, K.; Soda, K.: Purification and crystallization of L-ornithine:α-ketoglutarate Δ-aminotransferase from Corynebacterium sepedonicum. Bull. Inst. Chem. Res., **58**, 366-370 (1980)

[9] Goldberg, M.; Flescher, E.; Gold, D.; Lengy, J.: Ornithine-Δ-transaminase from the liver fluke, Fasciola hepatica and the blood fluke, Schistosoma mansoni: a comparative study. Comp. Biochem. Physiol. B, **65**, 605-613 (1980)

[10] Goldberg, M.; Flescher, E.; Lengy, J.: Schistosoma mansoni: partial purification and properties of ornithine-δ-transaminase. Exp. Parasitol., **47**, 333-341 (1979)

[11] Lu, T.-S.; Mazelis, M.: L-ornithine 2-oxoacid aminotransferase from squash (Curcubita pepo) cotyledons. Purification and properties. Plant Physiol., **55**, 502-506 (1975)

[12] Peraino, C.: Functional properties of ornithine-ketoacid aminotransferase from rat liver. Biochim. Biophys. Acta, **289**, 117-127 (1972)

[13] Stevens, L.; Heaton, A.: Induction, partial purification and properties of ornithine transaminase from Aspergillus nidulans. Biochem. Soc. Trans., **1**, 749-751 (1973)

[14] Yasuda, M.; Misono, H.; Soda, K.; Yonaha, K.; Toyama, S.: Purification and crystallization of L-ornithine:α-ketoglutarate δ-aminotransferase from Bacillus sphaericus. FEBS Lett., **105**, 209-212 (1979)

[15] Yasuda, M.; Toyama, S.; Rando, R.R.; Esaki, N.; Tanizawa, K.; Soda, K.: Irreversible inactivation of L-ornithine: α-ketoglutarate Δ-aminotransferase by gabaculine. Agric. Biol. Chem., **44**, 3005-3006 (1980)

[16] Hervieu, F.; Le Dily, F.; Le Saos, J.; Billard, J.-P.; Huault, C.: Inhibition of plant ornithine aminotransferase by gabaculine and 4-amino-5-hexynoic acid. Phytochemistry, **34**, 1231-1234 (1993)

[17] Daune, G.; Gerhart, F.; Seiler, N.: 5-Fluoromethylornithine, an irreversible and specific inhibitor of L-ornithine:2-oxo-acid aminotransferase. Biochem. J., **253**, 481-488 (1988)

[18] Matsuzawa, T.: Characteristics of the inhibition of ornithine-δ-aminotransferase by branched-chain amino acids. J. Biochem., **75**, 601-609 (1974)

[19] Marcovic-Housley, Z.; Kania, M.; Lustig, A.; Vincent, M.G.; Jansonius, J.N.; John, R.A.: Quaternary structure of ornithine aminotransferase in solution and preliminary crystallographic data. Eur. J. Biochem., **162**, 345-350 (1987)

[20] Boernke, W.E.; Stevens, F.J.; Peraino, C.: Effects of self-association of ornithine aminotransferase on its physicochemical characteristics. Biochemistry, **20**, 115-121 (1981)

[21] Matsuzawa, T.; Sugimoto, N.; Sobue, M.; Ishiguro, I.: Ornithine oxoacid aminotransferase found in AH 130 ascites hepatoma cells. Biochim. Biophys. Acta, **714**, 356-360 (1982)

[22] Williams, J.A.; Bridge, G.; Fowler, L.J.; John, R.A.: The reaction of ornithine aminotransferase with ornithine. Biochem. J., **201**, 221-225 (1982)

[23] Rosenthal, G.A.; Dahlman, D.L.: Interaction of L-canaline with ornithine aminotransferase of the tobacco hornworm, Manduca sexta (Sphingidae). J. Biol. Chem., **265**, 868-873 (1990)

[24] Yasuda, M.; Tanizawa, K.; Misono, H.; Toyama, S.; Soda, K.: Properties of crystalline L-ornithine: α-ketoglutarate δ-aminotransferase from Bacillus sphaericus. J. Bacteriol., **148**, 43-50 (1981)

[25] Ohura, T.; Kominami, E.; Tada, K.; Katunuma, N.: Crystallization and properties of human liver ornithine aminotransferase. J. Biochem., **92**, 1785-1792 (1982)

[26] Kalita, C.; Kerman, J.D.; Strecker, H.J.: Preparation and properties of ornithine-oxo-acid aminotransferase of rat kidney. Comparison with the liver enzyme. Biochim. Biophys. Acta, **429**, 780-797 (1976)

[27] Burcham, J.M.; Giometti, C.S.; Tollaksen, S.L.; Peraino, C.: Comparison of rat and mouse ornithine aminotransferase with respect to molecular properties and regulation of synthesis. Arch. Biochem. Biophys., **262**, 501-507 (1988)

[28] Kito, K.; Sanada, Y.; Katunuma, N.: Mode of inhibition of ornithine aminotransferase by L-canaline. J. Biochem., **83**, 201-206 (1978)

[29] Cooper, A.J.L.: Glutamate-ornithine transaminases. Methods Enzymol., **113**, 76-79 (1985)

[30] Dekaney, C.M.; Wu, G.; Jaeger, L.A.: Ornithine aminotransferase messenger RNA expression and enzymatic activity in fetal porcine intestine. Pediatr. Res., **50**, 104-109 (2001)

[31] Gafan, C.; Wilson, J.; Berger, L.C.; Berger, B.J.: Characterization of the ornithine aminotransferase from Plasmodium falciparum. Mol. Biochem. Parasitol., **118**, 1-10 (2001)

[32] Heinanen, K.; Nanto-Salonen, K.; Leino, L.; Pulkki, K.; Heinonen, O.; Valle, D.; Simell, O.: Gyrate atrophy of the choroid and retina: lymphocyte ornithine-d-aminotransferase activity in different mutations and carriers. Pediatr. Res., **44**, 381-385 (1998)

[33] Jhee, K.H.; Yoshimura, T.; Esaki, N.; Yonaha, K.; Soda, K.: Thermostable ornithine aminotransferase from Bacillus sp. YM-2: purification and characterization. J. Biochem., **118**, 101-108 (1995)

[34] Levillain, O.; Diaz, J.J.; Reymond, I.; Soulet, D.: Ornithine metabolism along the female mouse nephron: localization of ornithine decarboxylase and ornithine aminotransferase. Pflugers Arch., **440**, 761-769 (2000)

[35] Lim, S.N.; Rho, H.W.; Park, J.W.; Jhee, E.C.; Kim, J.S.; Kim, H.R.: A variant of ornithine aminotransferase from mouse small intestine. Exp. Mol. Med., **30**, 131-135 (1998)

[36] Madan, S.; Nainawatee, H.S.: Properties of partially purified leaf ornithine aminotransferase of Brassica juncea under salt stress. Acta Physiol. Plant., **22**, 444-450 (2000)

[37] Roosens, N.H.; Bitar, F.A.; Loenders, K.; Angenon, G.; Jacobs, M.: Overex-
     pression of ornithine-δ-aminotransferase increases proline biosynthesis
     and confers osmotolerance in transgenic plants. Mol. Breed., **9**, 73-80
     (2002)
[38] Roosens, N.H.C.J.; Thu, T.T.; Iskandar, H.M.; Jacobs, M.: Isolation of the
     ornithine-d-aminotransferase cDNA and effect of salt stress on its expres-
     sion in Arabidopsis thaliana. Plant Physiol., **117**, 263-271 (1998)
[39] Roschinger, W.; Endres, W.; Shin, Y.S.: Characteristics of L-ornithine: 2-ox-
     oacid aminotransferase and potential prenatal diagnosis of gyrate atrophy
     of the choroid and retina by first trimester chorionic villus sampling. Clin.
     Chim. Acta, **296**, 91-100 (2000)
[40] Seiler, N.: Ornithine aminotransferase, a potential target for the treatment
     of hyperammonemias. Curr. Drug Targets, **1**, 119-153 (2000)
[41] Shah, S.A.; Shen, B.W.; Brunger, A.T.: Human ornithine aminotransferase
     complexed with L-canaline and gabaculine: structural basis for substrate
     recognition. Structure, **5**, 1067-1075 (1997)
[42] Shen, B.W.; Hennig, M.; Hohenester, E.; Jansonius, J.N.; Schirmer, T.: Crys-
     tal structure of human recombinant ornithine aminotransferase. J. Mol.
     Biol., **277**, 81-102 (1998)
[43] Storici, P.; Capitani, G.; Muller, R.; Schirmer, T.; Jansonius, J.N.: Crystal
     structure of human ornithine aminotransferase complexed with the highly
     specific and potent inhibitor 5-fluoromethylornithine. J. Mol. Biol., **285**,
     297-309 (1999)
[44] Yang, C.-W.; Kao, C.H.: Importance of ornithine-δ-aminotransferase to pro-
     line accumulation caused by water stress in detached rice leaves. Plant
     Growth Regul., **27**, 189-192 (1999)

Asparagine-oxo-acid transaminase

1 Nomenclature

EC number
2.6.1.14

Systematic name
L-asparagine:2-oxo-acid aminotransferase

Recommended name
asparagine-oxo-acid transaminase

Synonyms
aminotransferase, asparagine-keto acid
asparagine keto-acid transaminase
asparagine oxoacid aminotransferase
asparagine-pyruvate transaminase

CAS registry number
9030-43-7

2 Source Organism

<1> *Rattus norvegicus* [1, 2, 4]
<2> *Mus musculus* [3]

3 Reaction and Specificity

Catalyzed reaction
L-asparagine + a 2-oxo acid = 2-oxosuccinamate + an amino acid (<1> ping-pong mechanism [2])

Reaction type
amino group transfer

Natural substrates and products
S   glyoxylate + amino acid <1> [2]
P   ?

Substrates and products
S   2-oxosuccinamic acid + L-2-aminobutanoic acid <1, 2> (Reversibility: r <1> [2]; ? <2> [1,3]) [1-3]
P   L-asparagine + 2-oxobutanoate

S 2-oxosuccinamic acid + L-alanine <1, 2> (Reversibility: ? <1,2> [1-3]) [1-3]

P L-asparagine + pyruvate

S 2-oxosuccinamic acid + L-cysteine <1> (Reversibility: ? <1> [2]) [2]

P L-asparagine + 3-mercapto-2-oxopropanoate

S 2-oxosuccinamic acid + L-ethionine <1> (Reversibility: ? <1> [1]) [1]

P L-asparagine + 4-ethylsulfanyl-2-oxobutanoate

S 2-oxosuccinamic acid + L-glutamine <1, 2> (Reversibility: ? <1,2> [1-3]) [1-3]

P L-asparagine + ?

S 2-oxosuccinamic acid + L-histidine <1> (Reversibility: ? <1> [1]) [1]

P L-asparagine + 3-(1H-imidazol-4-yl)-2-oxopropanoate

S 2-oxosuccinamic acid + L-homocysteine <2> (Reversibility: ? <2> [3]) [3]

P L-asparagine + 4-mercapto-2-oxobutanoate

S 2-oxosuccinamic acid + L-homoserine <1> (Reversibility: ? <1> [2]) [2]

P L-asparagine + 4-hydroxy-2-oxobutanoate

S 2-oxosuccinamic acid + L-leucine <1, 2> (Reversibility: ? <1,2> [1-3]) [1-3]

P L-asparagine + 4-methyl-2-oxopentanoate

S 2-oxosuccinamic acid + L-methionine <1, 2> (Reversibility: r <1,2> [2,3]; ? <1> [1]) [1-3]

P L-asparagine + 4-methylsulfanyl-2-oxobutanoate

S 2-oxosuccinamic acid + L-methylcysteine <1> (Reversibility: ? <1> [2]) [2]

P L-asparagine + ?

S 2-oxosuccinamic acid + L-norleucine <1> (Reversibility: ? <1> [1]) [1]

P L-asparagine + 2-oxohexanoate

S 2-oxosuccinamic acid + L-norvaline <1, 2> (Reversibility: ? <1,2> [1,3]) [1, 3]

P L-asparagine + 2-oxopentanoate

S 2-oxosuccinamic acid + L-phenylalanine <1> (Reversibility: ? <1> [1,2]) [1, 2]

P L-asparagine + 2-oxo-3-phenylpropanoate

S 2-oxosuccinamic acid + L-serine <1> (Reversibility: ? <1> 1,2[]) [1, 2]

P L-asparagine + 3-hydroxy-2-oxopropanoate

S 2-oxosuccinamic acid + L-tryptophan <1> (Reversibility: ? <1> [1]) [1]

P L-asparagine + 3-indole-2-oxopropanoate

S 2-oxosuccinamic acid + L-tyrosine <1> (Reversibility: ? <1> [1]) [1]

P L-asparagine + 3-(4-hydroxyphenyl)-2-oxopropanoate

S 2-oxosuccinamic acid + aspartate <1> (Reversibility: ? <1> [2]) [2]

P L-asparagine + 2-oxo-propane-1,3-dioate (<1> i.e. 2-oxosuccinate [2])

S L-asparagine + 2-oxobutanoate <1> (<1> weak [2]) (Reversibility: r <1> [2]) [2]

P 2-oxosuccinamate + 2-aminobutanoate

S L-asparagine + 2-oxoisohexanoate <2> (Reversibility: ? <2> [3]) [3]

P 2-oxosuccinamate + 2-aminoisohexanoate

S L-asparagine + 2-oxosuccinamate <1> (Reversibility: ? <1> [2]) [2]

P 2-oxosuccinamate + L-asparagine
S L-asparagine + 4-hydroxy-2-oxoglutarate <2> (Reversibility: ? <2> [3])
[3]
P 2-oxosuccinamate + 2-amino-4-hydroxyglutarate
S L-asparagine + 4-methylsulfanyl-2-oxobutanoate <1, 2> (<2> weak [3])
(Reversibility: r <1,2> [2, 3]) [2, 3]
P 2-oxosuccinamate + L-methionine
S L-asparagine + glyoxylate <1, 2> (Reversibility: ? <1,2> [2,3]) [2, 3]
P 2-oxosuccinamate + glycine
S L-asparagine + pyruvate <1, 2> [2, 3]
P 2-oxosuccinamate + L-alanine
S glyoxylate + L-asparagine <1> (Reversibility: ? <1> [2]) [2]
P ? + 2-oxosuccinamate
S phenylalanine + pyruvate <1> (Reversibility: ? <1> [4]) [4]
P phenylpyruvate + L-alanine
S phenylpyruvate + L-asparagine <1, 2> (Reversibility: ? <1,2> [2,3]) [2, 3]
P L-phenylalanine + 2-oxosuccinamate
S phenylpyruvate + L-glutamine <2> (Reversibility: ? <2> [3]) [3]
P L-phenylalanine + 2-oxoglutaramate
S phenylpyruvate + L-homoserine <2> (Reversibility: ? <2> [3]) [3]
P phenylalanine + 4-hydroxy-2-oxobutanoate
S phenylpyruvate + methionine <2> (Reversibility: ? <2> [3]) [3]
P L-phenylalanine + 4-methylsulfanyl-2-oxobutanoate
S Additional information <1> (<1> no or very low activity with: L-valine,
L-isoleucine, L-alloisoleucine, L-tertiary-leucine, L-2-aminophenylacetic
acid, L-glutamic acid, L-aspartic acid [1]; <1> the following 2-oxo acids
are inactive at concentrations of 20 mM: 2-oxoglutarate, oxaloacetate, 2-
oxoisopentanoate, L-2-oxo-3-methylvalerate, 3-sulfopyruvate, 3-hydroxy-
2-oxobutanoate, trimethylpyruvate [2]) [1, 2]
P ?

Inhibitors

$(NH_4)_2SO_4$ <2> (<2> at high concentration [3]) [3]
$AgNO_3$ <2> [3]
D-cycloserine <2> (<2> weak [3]) [3]
$HgCl_2$ <2> [3]
KCN <2> [3]
L-2-aminobutanoate <1> (<1> weak, asparagine/pyruvate reaction [2]) [2]
L-alanine <1> (<1> weak, asparagine/pyruvate reaction [2]) [2]
L-aspartate <1, 2> (<2> weak, asparagine/pyruvate reaction [3]) [2, 3]
L-cysteine <2> [3]
L-glutamate <2> (<2> weak, asparagine/pyruvate reaction [3]) [3]
L-glutamine <1, 2> (<2> asparagine/pyruvate reaction [3]) [2, 3]
L-methionine-(SR)-sulfoxime <1> (<1> asparagine/pyruvate reaction [2]) [2]
N-ethylmaleimide <2> [3]
S-methyl-L-cysteine <1> (<1> asparagine/pyruvate reaction [2]) [2]
aminooxyacetate <2> [3]

glycine <1, 2> (asparagine/pyruvate reaction [3]) [2, 3]
homoserine <1, 2> (<1> DL-isomer [2]; <1,2> asparagine/pyruvate reaction [2, 3]) [2, 3]
hydroxylamine <2> [3]
leucine <1, 2> (<1> weak [2]; <1,2> asparagine/pyruvate reaction [2, 3]) [2, 3]
methionine <1, 2> (<1> L-isomer [2]; <1,2> asparagine/pyruvate reaction [2,3]) [2, 3]
norvaline <2> (<2> asparagine/pyruvate reaction [3]) [3]
p-chloromercuribenzoate <2> [3]
phenylalanine <1, 2> [3, 4]
serine <1, 2> (<1,2> asparagine/pyruvate reaction [2,3]) [2, 3]
Additional information <2> (<2> strong substrate and product inhibition [3]) [3]

Cofactors/prosthetic groups
pyridoxal 5'-phosphate <1> (<1> a pyridoxal phosphate protein [1]) [1]

Specific activity (U/mg)
0.25 <1> [2]
5.08 <2> [3]
Additional information <1, 2> (<1> similar activity in glycylglycine, ammediol and Tris/HCl buffer [2]; <2> highest activity at pH 8.5 in N,N-bis(2-hydroxyethyl)glycine buffer and 4-(2-hydroxyethyl)-1-piperazinepropane sulfonic acid buffer, low activity in borate buffer and glycylglycine buffer [3]) [2, 3]

K_m-Value (mM)
0.03 <2> (L-methionine, <2> cosubstrate phenylpyruvate [3]) [3]
0.077 <1> (glyoxylate, <1> cosubstrate asparagine [2]) [2]
0.35 <1> (4-methylsulfanyl-2-oxobutanoate, <1> cosubstrate L-asparagine [2]) [2]
0.43 <2> (L-glutamine, <2> cosubstrate phenylpyruvate [3]) [3]
0.45 <1> (phenylpyruvate, <1> cosubstrate L-asparagine [2]) [2]
1 <2> (L-asparagine, <2> cosubstrate phenylpyruvate [3]) [3]
1.4 <2> (L-homoserine, <2> cosubstrate phenylpyruvate [3]) [3]
2.5 <1> (2-oxosuccinamate, <1> cosubstrate L-asparagine [2]) [2]
4.8 <2> (L-alanine, <2> cosubstrate 2-oxosuccinamate [3]) [3]
5.1 <1> (pyruvate, <1> cosubstrate L-asparagine [2]) [2]
6.5 <2> (L-norvaline, <2> cosubstrate 2-oxosuccinamate [3]) [3]
6.8 <2> (L-methionine, <2> cosubstrate 2-oxosuccinamate) [3]
21.4 <2> (L-glutamine, <2> cosubstrate 2-oxosuccinamate [3]) [3]
24.7 <2> (L-homocysteine, <2> cosubstrate 2-oxosuccinamate [3]) [3]
28.7 <2> (L-leucine, <2> cosubstrate 2-oxosuccinamate [3]) [3]
28.8 <2> (2-aminoisobutanoate, <2> cosubstrate 2-oxosuccinamate [3]) [3]
100 <1> (2-oxobutanoate, <1> cosubstrate asparagine [2]) [2]

pH-Optimum
8.2 <1> (<1> assay at [1]) [1]
8.5-8.6 <1> [2]

Temperature optimum (°C)
37 <1> (<1> assay at [1,2]) [1, 2]

4 Enzyme Structure

Molecular weight
70800 <2> (<2> mouse mitochondrial enzyme [3]) [3]

5 Isolation/Preparation/Mutation/Application

Source/tissue
liver <1, 2> [1-4]
Additional information <1> (<1> no activity in rat brain and skeletal muscle [1]) [1]

Localization
cytosol <2> [3]
mitochondrion <1, 2> [3, 4]

Purification
<1> [2, 4]
<2> [3]

References

[1] Meister, A.; Fraser, P.E.: Enzymatic formation of L-asparagine by transamination. J. Biol. Chem., **210**, 37-43 (1954)
[2] Cooper, A.J.L.: Asparagine transaminase from rat liver. J. Biol. Chem., **252**, 2032-2038 (1977)
[3] Maul, D.M.; Schuster, S.M.: Kinetic properties and characteristics of mouse liver mitochondrial asparagine aminotransferase. Arch. Biochem. Biophys., **251**, 585-593 (1986)
[4] hongo, S.; Ito, H.; Takeda, M.; Sato, T.: Identity of rat liver mitochondrial asparagine-pyruvate transaminase with phenylalanine-pyruvate transaminase. Enzyme, **36**, 232-238 (1987)

Glutamine-pyruvate transaminase

2.6.1.15

1 Nomenclature

EC number
2.6.1.15

Systematic name
L-glutamine:pyruvate aminotransferase

Recommended name
glutamine-pyruvate transaminase

Synonyms
aminotransferase, glutamine-keto acid
γ-glutaminyltransferase
glutaminase II
glutamine transaminase
glutamine transaminase L
glutamine-α-keto acid transamidase
glutamine-α-keto acid transaminase
glutamine-keto acid aminotransferase
glutamine-oxo acid aminotransferase
glutamine-oxo-acid transaminase

CAS registry number
9030-44-8

2 Source Organism

<1> *Rattus norvegicus* (male Sprague-Dawley [1,2,5]) [1-5]
<2> *Bos taurus* [6]
<3> *Rhizobium etli* [7]

3 Reaction and Specificity

Catalyzed reaction
L-glutamine + pyruvate = 2-oxoglutaramate + L-alanine

Reaction type
amino group transfer

Natural substrates and products

S L-glutamine + α-keto acid <1> (Reversibility: ? <1> [2, 5]) [2, 5]

P α-ketoglutarate + amino acid <1> [2, 5]

S L-glutamine + pyruvate <1> (Reversibility: ? <1> [2]) [2]

P 2-oxoglutaramate + L-alanine <1> [2]

Substrates and products

S 2-oxoglutaramate + L-alanine <1> (<1> poor substrate [2]) (Reversibility: ? <1> [2]) [2]

P L-glutamine + pyruvate <1> [2]

S 2-oxoglutaramate + L-asparagine <1> (<1> poor substrate [2]) (Reversibility: ? <1> [2]) [2]

P L-glutamine + 2-oxosuccinamate <1> [2]

S 2-oxoglutaramate + L-cysteine <1> (<1> poor substrate [2]) (Reversibility: ? <1> [2]) [2]

P L-glutamine + 3-mercapto-2-oxopropanoate <1> [2]

S 2-oxoglutaramate + L-ethionine <1> (Reversibility: ? <1> [2]) [2]

P L-glutamine + 4-ethylsulfanyl-2-oxobutanoate <1> [2]

S 2-oxoglutaramate + L-glutamic acid 4-benzylester <1> (<1> poor substrate [2]) (Reversibility: ? <1> [2]) [2]

P L-glutamine + 2-oxopentane-1,5-dioic acid benzyl ester <1> [2]

S 2-oxoglutaramate + L-homoserine <1> (Reversibility: ? <1> [2]) [2]

P L-glutamine + 4-hydroxy-2-oxobutanoate <1> [2]

S 2-oxoglutaramate + L-methionine <1> (<1> best amino group donor [2]) (Reversibility: ? <1> [2,5]) [2, 5]

P L-glutamine + 4-methylsulfanyl-2-oxobutanoate <1> [2, 5]

S 2-oxoglutaramate + L-methionine (SR)-sulfoxide <1> (Reversibility: ? <1> [2]) [2]

P L-glutamine + 4-methylsulfinyl-2-oxobutanoate <1> [2]

S 2-oxoglutaramate + L-methionine sulfone <1> (Reversibility: ? <1> [2]) [2]

P L-glutamine + 4-methylsulfonyl-2-oxobutanoate <1> [2]

S 2-oxoglutaramate + L-phenylalanine <1> (<1> poor substrate [2]) (Reversibility: ? <1> [2]) [2]

P L-glutamine + phenylpyruvate <1> [2]

S 2-oxoglutaramate + L-serine <1> (<1> poor substrate [2]) (Reversibility: ? <1> [2]) [2]

P L-glutamine + 3-hydroxy-2-oxopropanoate <1> [2]

S 2-oxoglutaramate + glycine <1> (<1> poor substrate [2]) (Reversibility: ? <1> [2]) [2]

P L-glutamine + glyoxylate <1> [2]

S L-albizziin + glyoxylate <1> (<1> S-carbamoyl-L-cysteine or O-carbamoyl-L-serine can replace albizziin [4]) (Reversibility: ? <1> [3,4]) [3, 4]

P glycine + 2-imidazolidone-5-hydroxy-5-carboxylic acid <1> (<1> undergoes non-enzymatic dehydration to 2-imidazolinone-5-carboxylate [3]) [3, 4]

S L-glutamine + 2-oxoglutaramate <1> (Reversibility: ? <1> [1,2]) [1, 2]

P 2-oxoglutaramate + L-glutamine <1> [1, 2]

S L-glutamine + 3-hydroxypyruvate <1> (Reversibility: ? <1> [2]) [2]

P 2-oxoglutaramate + L-serine <1> [2]

S L-glutamine + 3-mercaptopyruvate <1> (Reversibility: ? <1> [1,2]) [1, 2]

P 2-oxoglutaramate + L-cysteine <1> [1, 2]

S L-glutamine + 4-methylsulfanyl-2-oxobutanoate <1> (Reversibility: ? <1> [2]) [2]

P 2-oxoglutaramate + L-methionine <1> [2]

S L-glutamine + α-keto acid <1, 3> (Reversibility: ? <1,3> [2, 4, 5, 7]) [2, 4, 5, 7]

P α-ketoglutarate + amino acid <1, 3> [2, 4, 5, 7]

S L-glutamine + pyruvate <1> (Reversibility: ? <1> [2]) [2]

P 2-oxoglutaramate + L-alanine <1> [2]

S S-(3-aminopropyl)cysteine + pyruvate <2> (Reversibility: ? <2> [6]) [6]

P S-(3-aminopropyl)thiopyruvic acid + amino acid <2> [6]

S γ-glutamylhydrazones of 2-oxo acid <1> (<1> γ-glutamylhydrazones of 4-ethylsulfanyl-2-oxobutanoate, glyoxylate, 2-oxoisovalerate, transfers the 2-amino group of the glutamyl moiety to the oxo acid moiety, the reaction rates are higher than those of the corresponding glutamine α-keto acid transaminations [3]) (Reversibility: ? <1> [3,5]) [3, 5]

P L-amino acid + 2-hydroxy-tetrahydro-6-pyridazinone-3-carboxylic acid <1> [3, 5]

S glyoxylate + 4-glutamylmethylamide <1> (Reversibility: ? <1> [2,5]) [2, 5]

P glycine + ? <1> [2, 5]

S glyoxylate + DL-homocysteine <1> (<1> poor substrate [2]) (Reversibility: ? <1> [2]) [2]

P glycine + 2-oxo-4-thiobutanoate <1> [2]

S glyoxylate + L-2-amino-4-oxo-5-chloropentanoic acid <1> (<1> poor substrate [2]) (Reversibility: ? <1> [2]) [2]

P glycine + 5-chloro-2,4-dioxopentanoate <1> [2]

S glyoxylate + L-alanine <1> (<1> poor substrate [2]) (Reversibility: ? <1> [2]) [2]

P glycine + pyruvate <1> [2]

S glyoxylate + L-asparagine <1> (<1> poor substrate [2]) (Reversibility: ? <1> [2]) [2]

P glycine + 2-oxosuccinamate <1> [2]

S glyoxylate + L-cysteine <1> (<1> poor substrate [2]) (Reversibility: ? <1> [2,5]) [2, 5]

P glycine + 3-mercapto-2-oxopropanoate <1> [2, 5]

S glyoxylate + L-ethionine <1> (Reversibility: ? <1> [2,5]) [2, 5]

P glycine + 4-ethylsulfanyl-2-oxobutanoate <1> [2, 5]

S glyoxylate + L-glutamine <1> (<1> best substrate [1]) (Reversibility: ? <1> [1-5]) [1-5]

P glycine + 2-oxoglutaramate <1> (<1> 2-oxoglutaramate cyclizes spontaneously to form 2-pyrrolidone-5-hydroxy-5-carboxylate [4]) [1-5]

S glyoxylate + L-homoserine <1> (<1> poor substrate [2]) (Reversibility: ? <1> [2,5]) [2, 5]

P glycine + 4-hydroxy-2-oxobutanoate <1> [2, 5]

S glyoxylate + L-methionine <1> (Reversibility: ? <1> [1,2,5]) [1, 2, 5]

P glycine + 4-methylsulfanyl-2-oxobutanoate <1> [1, 2, 5]

S glyoxylate + L-methionine (SR)-sulfoxide <1> (Reversibility: ? <1> [2,5]) [2, 5]

P glycine + ? <1> [2, 5]

S glyoxylate + L-methionine (SR)-sulfoximine phosphate <1> (<1> poor substrate [2]) (Reversibility: ? <1> [2,5]) [2, 5]

P glycine + ? <1> [2, 5]

S glyoxylate + L-methionine sulfone <1> (Reversibility: ? <1> [2,5]) [2, 5]

P glycine + 4-methylsulfonyl-2-oxobutanoate <1> [2, 5]

S glyoxylate + L-phenylalanine <1> (<1> poor substrate [2]) (Reversibility: ? <1> [2,5]) [2, 5]

P glycine + phenylpyruvate <1> [2, 5]

S glyoxylate + L-serine <1> (<1> poor substrate [2]) (Reversibility: ? <1> [2]) [2]

P glycine + 3-hydroxy-2-oxopropanoate <1> [2]

S glyoxylate + S-methyl-L-cysteine <1> (<1> poor substrate [2]) (Reversibility: ? <1> [2,5]) [2, 5]

P glycine + 3-methylsulfanyl-2-oxopropanoate <1> [2, 5]

S glyoxylate + glutamic acid 4-methylester <1> (Reversibility: ? <1> [2,5]) [2, 5]

P glycine + 2-oxoglutarate 4-methylester <1> [2, 5]

S pyruvate + L-4-glutamylmethylamide <1> (Reversibility: ? <1> [2]) [2]

P L-alanine + ? <1> [2]

S pyruvate + L-glutamine <1> (<1> pyruvate cannot be replaced by its hydrazone, oxime or semicarbazone derivatives [3]; <1> L-glutamine cannot be replaced by glycine [2]) (Reversibility: ? <1> [2,3,5]) [2, 3, 5]

P L-alanine + 2-oxoglutaramate <1> [2, 3, 5]

S pyruvate + L-methionine <1> (<1> poor substrate [2]) (Reversibility: ? <1> [2]) [2]

P L-alanine + 4-methylsulfanyl-2-oxobutanoate <1> [2]

S pyruvate + L-methionine (SR)-sulfoxide <1> (<1> poor substrate [2]) (Reversibility: ? <1> [2]) [2]

P L-alanine + 4-methylsulfinyl-2-oxobutanoate <1> [2]

S pyruvate + L-methionine (SR)-sulfoximine phosphate <1> (<1> poor substrate [2]) (Reversibility: ? <1> [2]) [2]

P L-alanine + ? <1> [2]

S pyruvate + L-methionine sulfone <1> (<1> poor substrate [2]) (Reversibility: ? <1> [2]) [2]

P L-alanine + 4-methylsulfonyl-2-oxobutanoate <1> [2]

S pyruvate + L-phenylalanine <1> (Reversibility: ? <1> [2,3,5]) [2, 3, 5]

P L-alanine + phenylpyruvate <1> [2, 3, 5]

S selenium-(3-aminopropyl)selenocysteine + pyruvate <2> (Reversibility: ? <2> [6]) [6]

P selenium-(3-aminopropyl)selenopyruvic acid + amino acid <2> [6]

Inhibitors

2-oxoisopentanoate <1> (<1> transamination between glutamine and glyoxylate [3]) [3]

3-mercaptopyruvate <1> [2]

4-methylsulfanyl-2-oxobutanoate <1> [2]

L-γ-glutamylhydrazide <1> (<1> with γ-glutamylhydrazone of glyoxylate as substrate, weak [3]) [3]

L-glutamine <1> (<1> with γ-glutamylhydrazone of glyoxylate as substrate [3]) [3]

aminoxyacetate <1> (<1> soluble fraction [1]) [1]

carbonyl reagents <1> [2]

cycloserine <1> (<1> soluble fraction [1]) [1]

glyoxylate <1> (<1> at high concentration, transamination between albizziin and glyoxylate [3]) [3]

hydroxylamine <1> (<1> soluble fraction [1]) [1, 2]

isonicotinic acid hydrazide <1> [2]

sodium borohydride <1> [2]

Specific activity (U/mg)

0.013 <3> [7]

0.017 <1> [4]

0.113 <1> (<1> mitochondrion [1]) [1]

0.467 <1> [5]

5 <1> (<1> soluble fraction [1]) [1, 2]

K_m-Value (mM)

1.9 <1> (L-methionine, <1> pH 8.4, 37°C [2]) [2]

2 <1> (L-glutamine, <1> pH 8.4, 37°C [2]) [2]

2 <1> (L-methionine (SR)-sulfoxide, <1> pH 8.4, 37°C [2]) [2]

3.2 <1> (4-methylsulfanyl-2-oxobutanoate, <1> pH 8.4, 37°C [2]) [2]

4.4 <1> (L-glutamic acid γ-methylester, <1> pH 8.4, 37°C [2]) [2]

4.8 <1> (3-mercaptopyruvate, <1> pH 8.4, 37°C [2]) [2]

6 <1> (L-glutamine, <1> pH 8.4, 37°C [3]) [3]

8 <1> (3-hydroxypyruvate, <1> pH 8.4, 37°C [2]) [2]

8 <1> (glyoxylate, <1> pH 8.4, 37°C [2]) [2]

10.5 <1> (pyruvate, <1> pH 8.4, 37°C [2]) [2]

23 <1> (γ-glutamylhydrazone of glyoxylate, <1> pH 8.4, 37°C [3]) [3]

50 <1> (2-oxoglutaramate, <1> pH 8.4, 37°C [2]) [2]

pH-Optimum

8.2-8.5 <1> [2]

pH-Range

7.6-8.8 <1> (<1> about half-maximum activity at pH 7.6 and about 90% of maximum activity at pH 8.8) [2, 2]

Temperature optimum (°C)

37 <1> (<1> assay at [1-5]) [1-5]

4 Enzyme Structure

Molecular weight
    110000 <1> (<1> PAGE [2]) [1, 2]

5 Isolation/Preparation/Mutation/Application

Source/tissue
    liver <1, 2> [1-6]

Localization
    cytosol <1> [5, 1]
    mitochondrion <1> [1]

Purification
    <1> (DE-52 chromatography, hydroxyapatite chromatography [5]) [5]
    <1> (DE-52 chromatography, hydroxyapatite chromatography for soluble
    fraction, DE-52 chromatography for mitochondrion fraction [1]) [1]
    <1> (ammonium sulfate fractionation , DE-52 chromatography [2]) [2]

6 Stability

Temperature stability
    60 <1> (<1> in crude extract inactivation within 10 min, in the presence of
    pyruvate, glyoxylate or phenylpyruvate 1 h stability, 25% loss of activity with-
    in 3 h [2]) [2]

General stability information
    <1>, 2-mercaptoethanol and α-keto acids stabilize [2]

Storage stability
    <1>, -20°C, in 30-50 mM potassium phosphate buffer, pH 7.2, 1 mM 2-mer-
    captoethanol, at least 7 weeks [2]
    <1>, 0-5°C, in 30-50 mM potassium phosphate buffer, pH 7.2, 1 mM 2-mer-
    captoethanol, little or no loss of activity for 1-2 weeks, in the absence of
    mercaptoethanol inactivation within 1-2 days [2]
    <1>, 24-26°C, 20% loss of activity of purified enzyme in 3 days [2]
    <1>, 4°C, 10 mM potassium phosphate, pH 7.2, 10% glycerol, 1 mM 2-mer-
    captoethanol, 0.01 mM pyridoxal 5'-phosphate, 3 months [1, 2]

References

[1] Cooper, A.J.L.; Meister, A.: Glutamine transaminase L from rat liver. Methods
    Enzymol., **113**, 338-343 (1985)

[2] Cooper, A.J.L.; Meister, A.: Isolation and properties of highly purified glutamine transaminase. Biochemistry, 11, 661-671 (1972)

[3] Cooper, A.J.L.; Meister, A.: Activity of liver glutamine transaminase toward L-γ-glutamyl hydrazones of α-keto acids. J. Biol. Chem., 248, 8489-8498 (1973)

[4] Cooper, A.J.L.; Meister, A.: Action of liver glutamine transaminase and L-amino acid oxidase on several glutamine analogs. Preparation and properties of the 4-S, O, and NH analogs of α-ketoglutaramic acid. J. Biol. Chem., 248, 8499-8505 (1973)

[5] Cooper, A.J.L.; Meister, A.: Isolation and properties of a new glutamine transaminase form rat kidney. J. Biol. Chem., 249, 2554-2561 (1974)

[6] Coccia, R.; Foppoli, C.; Blarzino, C.; De Marco, C.; Pensa, B.: Transamination of some sulphur- or selenium-containing amino acids by bovine liver glutamine transaminase. Physiol. Chem. Phys. Med. NMR, 24, 313-321 (1992)

[7] Duran, S.; Calderon, J.: Role of the glutamine transaminase-ω-amidase pathway and glutaminase in glutamine degradation in Rhizobium etli. Microbiology, 141, 589-595 (1995)

Glutamine-fructose-6-phosphate transaminase (isomerizing)

2.6.1.16

1 Nomenclature

EC number

2.6.1.16

Systematic name

L-glutamine:D-fructose-6-phosphate isomerase (deaminating)

Recommended name

glutamine-fructose-6-phosphate transaminase (isomerizing)

Synonyms

2-amino-2-deoxy-D-glucose-6-phosphate ketol-isomerase
EC 5.3.1.19 (formerly)
GFAT
L-glutamine fructose 6-phosphate transamidase
glucosamine 6-phosphate synthase
glucosamine 6-phosphate synthetase
glucosamine phosphate isomerase (glutamine-forming)
glucosamine synthase
glucosamine-6-phosphate isomerase (glutamine-forming)
glucosamine-6-phosphate synthetase
glutamine-fructose 6-phosphate amidotransferase
glutamine-fructose 6-phosphate aminotransferase
glutamine:fructose-6-phosphate aminotransferase
hexosephosphate aminotransferase
isomerase, glucosamine phosphate (glutamine-forming)

CAS registry number

9030-45-9

2 Source Organism

<1> *Gallus gallus* [15]
<2> *Saccharomyces cerevisiae* [18]
<3> *Escherichia coli* (3000 Hfr strain (ATCC25257) [24]; HB101 strain [30]) [1, 8, 13, 14, 19, 23, 24, 26, 27, 29, 30]
<4> *Neurospora crassa* [1]
<5> *Rattus norvegicus* (rat [1, 2, 4, 5, 9, 16, 21]) [1, 2, 4, 5, 9, 16, 21, 25, 28]
<6> *Bos taurus* (bovine [3]) [3, 17]
<7> *Phaseolus aureus* [20]

<8> *Homo sapiens* [11, 22]
<9> *Mus musculus* (mouse [12]) [5, 12]
<10> *Salmonella typhimurium* (LT2 [6,7]) [6, 7]
<11> *Pseudomonas aeruginosa* [8]
<12> *Arthrobacter aurescens* [8]
<13> *Glomerella cingulata* [1]
<14> *Helminthosporium sativum* [1]
<15> *Neurospora tetrasperma* [1]
<16> *Penicillium sp.* [1]
<17> *Dactylium dendroides* [1]
<18> *Aspergillus parasiticus* [1]
<19> *Penicillium notatum* [1]
<20> *Aspergillus flavus* [1]
<21> *Bacillus thuringiensis* [8]
<22> *Blastocladiella emersonii* [10]
<23> *Candida albicans* [13]
<24> *Bacillus pumilus* [13]

3 Reaction and Specificity

Catalyzed reaction

L-glutamine + D-fructose 6-phosphate = L-glutamate + D-glucosamine 6-phosphate (<1> rapid equilibrium random mechanism [15])

Reaction type

amino group transfer

Natural substrates and products

S L-glutamine + D-fructose 6-phosphate <2, 3, 8> (<8> enzyme plays a key role in induction of insulin resistance in cultured cells [11]; <8> importance for N-acetylglucosamine synthesis in human liver [22]) (Reversibility: ? <2, 3, 8> [11, 18, 22, 27]) [11, 18, 22, 27]

P L-glutamate + D-glucosamine 6-phosphate <2, 3, 8> [11, 18, 22, 27]

Substrates and products

S D-fructose 6-phosphate <3> (<3> isomerase activity studied over C-terminal D-fructose 6-phosphate binding domain constituted by residues 241 to 608 [30]) (Reversibility: r <3> [30]) [30]

P D-glucose 6-phosphate <3> [30]

S L-glutamine + D-fructose 6-phosphate <1-24> (<2-5,7> specific for: L-glutamine, [1,18,20]; <3> mechanism proposed [29]) (Reversibility: ir <3-5> [1, 19]) [1-22]

P L-glutamate + D-glucosamine 6-phosphate <1-24> [1-22]

S Additional information <7> (<7> asparagine is not a substrate [20]) [20]

P ?

Inhibitors

1,1'-dithiodiformamidine <3, 10-12, 21> (<3, 10-12, 21> irreversible inhibition [7, 8]) [7, 8]

1,2-anhydrohexitol 6-phosphate <3> (<3> mixture of the four diastereoisomers. Irreversible inactivation. D-fructose 6-phosphate and 2-amino-2-deoxyglucitol protect, L-glutamine does not [26]) [26]

2-amino-2-deoxyglucitol 6-phosphate <3> (<3> competitive with respect to D-fructose 6-phosphate [23, 24]) [23, 24]

4,4'-dithiodipyridine <5> (<5> inactivation reversed by dithiothreitol. Competitive with respect to L-glutamine. Non-competitive with respect to D-fructose 6-phosphate [28]) [28]

4-glutamylhydroxamate <10> [7]

5,5'-dithionitrobenzoic acid <3, 5, 11, 12, 21> (<3, 10-12, 21> irreversible inhibition [7, 8]; <5> inactivation reversed by dithiothreitol [28]) [7, 8, 28]

6,6'-dithiodinicotinic acid <3, 10-12, 21> (<3, 10-12, 21> irreversible inhibition [7,8]) [7, 8]

6-diazo-5-oxo-L-norleucine <3-6, 10> (<6> competitive with respect to L-glutamine [3]) [1, 3, 7, 14, 16, 17, 19, 28]

D-glucitol 6-phosphate <3> (<3> competitive with respect to D-fructose 6-phosphate [26]) [26]

DL-Δ^1-pyrroline-5-carboxylate <3> (<3> competitive with respect to L-glutamine [23]) [23]

L-2,3-diaminopropanoic acid <3, 23, 24> [13]

L-α-glycerophosphate <5> (<5> weak [9]) [9]

N-ethylmaleimide <2, 3, 10-12, 21> (<3,10-12,21> irreversible inhibition [7,8]; <2> 78% inhibition at 1 mM [18]) [7, 8, 18]

N-iodoacetylglucosamine 6-phosphate <3> (<3> D-fructose 6-phosphate protects [24]) [24]

N^3-(4-methoxyfumaroyl)-L-2,3-diaminopropanoic acid <3, 10> (<10> L-glutamine and some analogs protect [7]) [7, 14]

N^3-bromoacetyl-L-2,3-diaminopropanoic acid <3, 23, 24> (<3, 23, 24> competitive with respect to L-glutamine [13]) [13]

N^3-chloroacetyl-L-2,3-diaminopropanoic acid <3, 23, 24> (<3, 23, 24> competitive with respect to L-glutamine [13]) [13]

N^3-fumaramoyl-L-2,3-diaminopropanoic acid <3, 10> [7, 14]

N^3-fumaroyl-L-2,3-diaminopropanoic acid <3, 10> [7, 14]

N^3-fumaroylcarboxyamido-L-2,3-diaminopropionic acid <10> (<10> competitive with L-glutamine [6]) [6, 7]

N^3-iodoacetyl-L-2,3-diaminopropanoic acid <3, 23, 24> (<3, 23, 24> competitive with respect to L-glutamine [13]) [13]

N^4-(4-methoxyfumaroyl)-L-2,4-diaminobutanoic acid <10> [7]

UDP-N-acetylglucosamine <2, 5-9> (<9> weak [12]; <6> partial [17]; <7> competitive inhibitor with respect to D-fructose 6-phosphate [20]; <6> competitive with respect to D-fructose 6-phosphate, non-competitive with respect to L-glutamine [3]; <2> 40% inhibition at 1 mM [18]; <8> feed-back inhibition [22]) [3, 12, 17, 18, 20, 22, 28]

UDP-glucose <9> [12]

albizziin <10> [7]

anticapsin <3, 10-12, 21> (<3, 11, 12, 21> L-glutamine protects, irreversible inhibition [8]) [7, 8]

azaserine <3-5, 10> (<3,5> weak [1]) [1, 7]

dihydroxyacetone <5> (<5> weak [9]) [9]

dihydroxyacetone phosphate <5> (<5> weak [9]) [9]

D-fructose 1,6-diphosphate <5> (<5> weak [9]) [9]

glyceraldehyde 3-phosphate <5> (<5> 50% inhibition at 0.2 mM [9]) [9]

glycolaldehyde <5> (<5> weak [9]) [9]

glyoxal <5> (<5> 50% inhibition at 0.03 mM [9]) [9]

iodoacetamide <3, 10-12, 21> (<3, 10-12, 21> irreversible inhibition [7,8]) [7, 8, 19]

iodoacetate <10> [7]

mercuric chloride <2> (<2> 84% inhibition at 1 mM [18]) [18]

methylglyoxal <5> (<5> 50% inhibition at 0.01 mM, non competitive [9]; <5> inhibits preincubated enzyme less profoundly than the untreated enzyme [21]) [9, 21]

p-chloromercuribenzoate <2, 9> (<2> 84% inhibition at 0.1 mM [18]) [12, 18]

p-hydroxymercuribenzoate <10> [7]

pyridoxamine-5'-phosphate <5> [2]

tolbutamide <5> (<5> 80% inhibition at 2 mg/ml [4]) [4]

Cofactors/prosthetic groups

Additional information <3-5> (<3-5> no cofactor requirement [1]) [1, 5]

Activating compounds

chondroitin sulfate B <5> [4]

dithiothreitol <2> (<2> 1 mM, activates 1.8fold [18]) [18]

Additional information <5> (<5> small acidic compound devoid of SH groups thermolabile at 100°C dialyzable, not precipitable with $HClO_4$ easily separated from proteins in a Sephadex G-25 column [5]) [5]

Specific activity (U/mg)

0.0047 <1> [15]

0.00783 <2> [18]

0.013 <5> [2]

0.0144 <7> [20]

0.16 <5> [16]

0.24 <3> [24]

0.866 <5> [28]

1.2 <11> [8]

3.5 <22> [10]

3.6 <21> [8]

4 <3> [8]

4.66 <1> [15]

7.63 <3> [19]

8 <12> [8]

34.8 <8> [22]

K$_m$-Value (mM)

0.2 <6> (L-glutamine, <6> pH 7.4, 37°C [17]) [17]

0.225 <12> (L-glutamine, <12> pH 7.5, 37°C [8]) [8]

0.3 <1, 10> (L-glutamine, <10> pH 7.5, 37°C [7]; <1> pH 6.3, 38°C [15]) [7, 15]

0.38 <5> (D-fructose 6-phosphate, <5> pH 7.7, 37°C [1]) [1]

0.4 <5> (D-fructose 6-phosphate) [28]

0.4 <3> (L-glutamine, <3> pH 7.5, 37°C [8,19]) [8, 19]

0.5 <1> (D-fructose 6-phosphate, <1> pH 6.3, 38°C [15]) [15]

0.5 <7> (L-glutamine, <7> pH 6.5, 30°C [20]) [20]

0.6 <21> (L-glutamine, <21> pH 7.5, 37°C [8]) [8]

0.65 <3> (L-glutamine, <3> pH 7.0, 37°C [1]) [1]

0.75 <4> (D-fructose 6-phosphate, <4> pH 7.0, 37°C [1]) [1]

0.8 <5> (L-glutamine) [28]

0.85 <11> (L-glutamine, <11> pH 7.5, 37°C [8]) [8]

1.6 <5> (L-glutamine, <5> pH 7.7, 37°C [1]) [1]

2 <3> (D-fructose 6-phosphate, <3> pH 7.0, 37°C [1]; <3> pH 7.5, 37°C [19]) [1, 19]

3.8 <3, 7> (D-fructose 6-phosphate, <7> pH 6.5, 30°C [20]; <3> pH 7.2, 37°C, isomerization catalyzed by C-terminal domain [30]) [20, 30]

16 <3> (D-glucose 6-phosphate, <3> pH 7.2, 37°C, isomerization catalyzed by C-terminal domain [30]) [30]

K$_i$-Value (mM)

0.0001 <23> (N^3-bromoacetyl-L-2,3-diaminopropanoic acid, <23> pH 6.5, 25°C [13]) [13]

0.00011 <11> (anticapsin, <11> pH 7.5, 37°C [8]) [8]

0.000125 <23> (N^3-iodoacetyl-L-2,3-diaminopropanoic acid, <23> pH 6.5, 25°C [13]) [13]

0.00027 <10> (N^3-(4-methoxyfumaroyl)-L-2,3-diaminopropanoic acid, <10> pH 7.5, 37°C [7]) [7]

0.00028 <10> (anticapsin, <10> pH 7.5, 37°C [7]) [7]

0.00035 <3> (N^3-(4-methoxyfumaroyl)-L-2,3-diaminopropanoic acid, <3> pH 7.5, 37°C [14]) [14]

0.00053 <12> (anticapsin, <12> pH 7.5, 37°C [8]) [8]

0.00056 <3> (anticapsin, <3> pH 7.5, 37°C [8]) [8]

0.0006 <23> (N^3-chloroacetyl-L-2,3-diaminopropanoic acid, <23> pH 6.5, 25°C [13]) [13]

0.00075 <21> (anticapsin, <21> pH 7.5, 37°C [8]) [8]

0.001 <6> (6-diazo-5-oxo-L-norleucine, <6> pH 7.4, 37°C [17]) [17]

0.0013 <7> (UDP-N-acetylglucosamine, <7> pH 6.5, 30°C [20]) [20]

0.0014 <3> (6-diazo-5-oxo-L-norleucine, <3> pH 7.5, 37°C [14]) [14]

0.0025 <5> (6-diazo-5-oxo-L-norleucine) [28]

0.0028 <3> (6-diazo-5-oxo-L-norleucine, <3> pH 7.5, room temperature [19]) [19]

0.0075 <10> (6-diazo-5-oxo-L-norleucine, <10> pH 7.5, 37°C [7]) [7]

0.008 <10> (N^3-fumaroyl-L-2,3-diaminopropanoic acid, <10> pH 7.5, 37°C [7]) [7]

0.0183 <5> (UDP-N-acetylglucosamine) [28]

0.019 <10> (N^3-fumaramoyl-L-2,3-diaminopropanoic acid, <10> pH 7.5, 37°C [7]) [7]

0.0193 <3> (2-amino-2-deoxyglucitol 6-phosphate, <3> pH 7.5, 25°C [24]) [24]

0.023 <3> (DL-Δ^1-pyrroline-5-carboxylate) [23]

0.04 <6> (UDP-N-acetylglucosamine, <6> pH 7.4, 37°C [17]) [17]

0.0551 <3> (N^3-fumaroyl-L-2,3-diaminopropanoic acid, <3> pH 7.5, 37°C [14]) [14]

0.088 <3> (2-amino-2-deoxyglucitol 6-phosphate, <3> takes into account 2.2% of the open form present in a solution of D-fructose 6-phosphate [23]) [23]

0.095 <10> (4-glutamylhydroxamate, <10> pH 7.5, 37°C [7]) [7]

0.125 <10> (N^3-fumaroylcarboxyamido-L-2,3-diaminopropionic acid, <10> pH 7.5, 37°C [6]) [6]

0.77 <10> (azaserine, <10> pH 7.5, 37°C [7]) [7]

1.5 <10> (albizziin, <10> pH 7.5, 37°C [7]) [7]

2.46 <3> (D-glucitol 6-phosphate, <3> pH 7.2 [26]) [26]

12.6 <3> (iodoacetamide, <3> pH 7.5, room temperature [19]) [19]

pH-Optimum

6.1-6.4 <1> [15]

6.2-6.7 <7> (<7> potassium phosphate buffer [20]) [20]

6.5-7.5 <6> [17]

6.7 <4> [1]

7 <2> (<2> potassium phosphate buffer [18]) [18]

7-8 <5> [16]

7.2 <3> (<3> TES buffer [19]) [19]

7.4 <6> [3]

7.5 <5> [28]

7.7 <3, 5> (<3,5> assay at [1]) [1]

7.7-7.9 <5> [1]

7.9 <3> [1]

pH-Range

6-8.5 <6> [17]

Temperature optimum (°C)

30 <7> (<7> assay at [20]) [20]

37 <3, 5, 6> (<3,5,6> assay at [1,8,9,17]) [1, 8, 9, 17]

4 Enzyme Structure

Molecular weight

170000-193000 <3> (<3> gel filtration [19]) [19]

280000 <5> (<5> gel filtration [28]) [28]

300000 <7> (<7> ultrafiltration [20]) [20]
340000 <5> (<5> gel filtration, preincubated enzyme [21]) [21]
350000 <6> (<6> gel filtration [3]) [3, 17]
410000 <5> (<5> gel filtration, untreated enzyme [21]) [21]

Subunits

? <8> (<8> ? * 77000, SDS-PAGE [11]) [11]
dimer <3> (<3> 2 * 70800, SDS-PAGE [19]; <3> 2 * 70000, calculated by means of crystalline structure determined by X ray diffraction [29]) [19, 29]
tetramer <5> (<5> 4 * 75000, gel filtration, SDS-PAGE [28]) [28]
Additional information <8> (<8> The fact that the 77 kDa protein expressed in E. coli is sensitive to inhibition by UDP-N-acetylglucosamine is consistent with the idea that mammalian GFAT is comprised of four identical subunits [11]) [11]

5 Isolation/Preparation/Mutation/Application

Source/tissue

ascites tumor cell <9> [5]
carcinoma cell <5> [5]
cartilage <1> [15]
fibroblast <5> [25]
liver <5, 8, 9, 13-20> [1, 2, 4, 5, 9, 16, 21, 22, 28]
seed <7> [20]
small intestine <9> [12]
sublingual gland <9> [12]
thyroid gland <6> [3]
trachea <6> [17]
zoospore <22> [10]

Crystallization

<3> (crystal structure of intact protein [29]) [29]
<3> (two crystal complexes of the isomerase domain with D-glucose 6-phosphate and 2-amino-2-deoxyglucitol 6-phosphate [27]) [27]

Cloning

<8> (expressed in Escherichia coli [11]) [11]

6 Stability

Temperature stability

37 <2> (<2> 90 min, complete loss of activity without stabilizing agent. Dithiothreitol, D-fructose 6-phosphate, D-glucose 6-phosphate or L-glutamine, 1 mM, increase thermal stability at 37°C [18]) [18]

General stability information

<2>, complete loss of activity after treatment with ammonium sulfate, 30-50% saturation [18]

<2>, dithiothreitol, D-fructose 6-phosphate, D-glucose 6-phosphate or L-glutamine, 1 mM, increases thermal stability at 37°C [18]

<2>, freezing and thawing has no effect on activity [18]

<3>, inactivated by freezing and thawing [1]

<5>, can be concentrated by precipitation with 2.3 M ammonium sulfate with some concomitant loss of activity [16]

<5>, extremely unstable during purification [28]

<5>, very unstable [1]

<6>, glucose 6-phosphate, 12 mM, required as protective agent during purification [17]

<8>, isopropanol, 1%, stabilizes during purification [22]

<22>, sucrose needed as stabilizing agent [10]

<3-5>, inactivated when exposed to a solution of pH 5.5 or below [1]

Storage stability

<3>, 0-2°C, potassium phosphate buffer, pH 7.0, 0.06 M L-glutamine, 0.01 M EDTA, stable for up to 3 or 4 weeks [1]

<3>, 5°C, potassium phosphate buffer, pH 7.5, 1 mM EDTA, 1 mM dithiothreitol, 0.5 mM L-glutamine, 0.05 mM D-glucose 6-phosphate, 600 mM sucrose, stable for few weeks [8]

<4>, -18°C, little loss of activity after 1 week for lyophilized material after calcium phosphate gel treatment [1]

<5>, -80°C, 3 months [28]

<5>, 4°C or -15°C, 8 days, 50% activity loss [16]

<5>, 4°C or 15°C, 50% loss of activity after 8 days [16]

<7>, in liquid N_2, stable for up to 2 weeks [20]

<8>, -25°C, 1 month, 15% activity loss, ammonium sulfate precipitated enzyme [22]

<10>, 4°C, 25 mM potassium phosphate buffer, pH 7.5, 1 mM EGTA, 2 mM dithiothreitol, 500 mM sucrose, stable for up to 2 or 3 weeks [6]

<22>, -20°C, 0.6 M sucrose, 60% loss of activity after 45 days [10]

References

[1] Ghosh, S.; Blumenthal, H.J.; Davidson, E.; Roseman, S.: Glucosamine metabolism.V. Enzymatic synthesis of glucosamine 6-phosphate. J. Biol. Chem., 235, 1265-1273 (1960)

[2] Gryder, R.M.; Pogell, B.M.: Further studies on glucosamine 6-phosphate synthesis by rat liver enzymes. J. Biol. Chem., 235, 558-562 (1960)

[3] Trujillo, J.L.; Gan, J.C.: Purification and some kinetic properties of bovine thyroid gland L-glutamine D-fructose-6-phosphate amidotransferase. Int. J. Biochem., 5, 515-523 (1974)

[4] Malathy, K.; Kurup, P.A.: Inhibition of hexosephosphate aminotransferase activity in vitro by tolbutamide. Indian J. Biochem. Biophys., **9**, 310-312 (1972)

[5] Ikeda, Y.; Tsuiki, S.: Occurrence and properties of a natural activator of L-glutamine:D-fructose 6-phosphate amidotransferase. Sci. Rep. Res. Inst. Tohoku Univ. Ser. C Med., **19**, 53-63 (1972)

[6] Chmara, H.; Andruszkiewicz, R.; Borowski, E.: Inactivation of glucosamine-6-phosphate synthetase from Salmonella typhimurium LT 2 SL 1027 by N β-fumarylcarboxyamido-L-2,3-diamino-propionic acid. Biochem. Biophys. Res. Commun., **120**, 865-872 (1984)

[7] Chmara, H.; Andruszkiewicz, R.; Borowski, E.: Inactivation of glucosamine-6-phosphate synthethase from Salmonella typhimurium LT2 by fumarolyl diaminopropanoic acid derivatives, a novel group of glutamine analogs. Biochim. Biophys. Acta, **870**, 357-366 (1985)

[8] Chmara, H.; Zähner, H.: The inactivation of glucosamine synthetase from bacteria by anticapsin, the C-terminal epoxyamino acid of the antibiotic tetaine. Biochim. Biophys. Acta, **787**, 45-52 (1984)

[9] Kikuchi, H.; Ikeda, Y.; Tsuiki, S.: Inhibition of L-glutamine:D-fructose-6-phosphate aminotransferase by methylglyoxal. Biochim. Biophys. Acta, **289**, 303-310 (1972)

[10] Norrman, J.; Myers, R.B.; Giddings, T.H.; Cantino, E.C.: Partial purification of L-glutamine:D-fructose 6-phosphate aminotransferase from zoospores of Blastocladiella emersonii. Biochim. Biophys. Acta, **302**, 173-177 (1973)

[11] McKnight, G.L.; Mudri, S.L.; Mathewes, S.L.; Traxinger, R.R.; Marshall, S.; Sheppard, P.O.; O'Hara, P.J.: Molecular cloning, cDNA sequence, and bacterial expression of human glutamine:fructose-6-phosphate amidotransferase. J. Biol. Chem., **267**, 25208-25212 (1992)

[12] Hosoi, K.; Kobayashi, S.; Ueha, T.: Affinity adsorption of L-glutamine D-fructose-6-phosphate aminotransferase to Sepharose coupled with p-chloromercuribenzoate. Biochem. Biophys. Res. Commun., **85**, 558-563 (1978)

[13] Milewski, S.; Chmara, H.; Andruszkiewicz, R.; Borowski, E.: N^3-haloacetyl derivatives of L-2,3-diaminopropanoic acid: novel inactivators of glucosamine-6-phosphate synthase. Biochim. Biophys. Acta, **1115**, 225-229 (1992)

[14] Badet, B.; Vermoote, P.; Le Goffic, F.: Glucosamine synthetase from Escherichia coli: kinetic mechanism and inhibition by N^3-fumaroyl-L-2,3-diaminopropionic derivatives. Biochemistry, **27**, 2282-2287 (1988)

[15] Calcagno, M.; Levy, J.A.; Arrambide, E.; Mizraji, E.: L-glutamine:D-fructose 6-phosphate amidotransferase of chick cartilage. Evidence for a random mechanism. Enzymologia, **41**, 175-182 (1971)

[16] Zalkin, H.: Glucosamine-6-phosphate synthase. Methods Enzymol., **113**, 278-281 (1985)

[17] Ellis, D.B.; Sommar, K.M.: Biosynthesis of respiratory tract musins. II. Control of hexosamine metabolism by L-glutamine D-fructose 6-phosphate aminotransferase. Biochim. Biophys. Acta, **267**, 105-112 (1972)

[18] Moriguchi, M.; Yamamoto, K.; Kawai, H.; Tochikura, T.: The partial purification of by L-glutamine D-fructose 6-phosphate amidotransferase from Baker's yeast. Agric. Biol. Chem., **40**, 1655-1656 (1976)

[19] Badet, B.; Vermoote, P.; Haumont, P.-Y.; Lederer, F.; Le Goffic, F.: Glucosamine synthetase from Escherichia coli: purification, properties, and glutamine-utilizing site location. Biochemistry, 26, 1940-1948 (1987)

[20] Vessal, M.; Hassid, W.Z.: Partial purification and properties of L-glutamine D-fructose 6-phosphate amidotransferase from Phaseolus aureus. Plant Physiol., 49, 977-981 (1972)

[21] Kikuchi, H.; Tsuiki, S.: Stabilization of glucosaminephosphate synthase from rat liver by hexose 6-phosphates. Properties and interconversion of two molecular forms. Biochim. Biophys. Acta, 422, 231-240 (1976)

[22] Kikuchi, H.; Tsuiki, S.: Glucosaminephosphate synthase of human liver. Biochim. Biophys. Acta, 422, 241-246 (1976)

[23] Badet-Denisot, M.-A.; Leriche, C.; Massiere, F.; Badet, B.: Nitrogen transfer in E. coli glucosamine-6P synthase. Investigations using substrate and bisubstrate analogs. Bioorg. Med. Chem. Lett., 5, 815-820 (1995)

[24] Bearne, S.L.: Active site-directed inactivation of Escherichia coli glucosamine-6-phosphate synthase. Determination of the fructose 6-phosphate binding constant using a carbohydrate-based inactivator. J. Biol. Chem., 271, 3052-3057 (1996)

[25] Hebert, L.F., Jr.; Daniels, M.C.; Zhou, J.; Crook, E.D.; Turner, R.L.; Simmons, S.T.; Neidigh, J.L.; Zhu, J.-S.; Baron, A.D.; McClain, D.A.: Overexpression of glutamine:fructose-6-phosphate amidotransferase in transgenic mice leads to insulin resistance. J. Clin. Invest., 98, 930-936 (1996)

[26] Leriche, C.; Badet-Denisot, M.A.; Badet, B.: Affinity labeling of Escherichia coli glucosamine-6-phosphate synthase with a fructose 6-phosphate analog. Evidence for proximity between the N-terminal cysteine and the fructose-6-phosphate-binding site. Eur. J. Biochem., 245, 418-422 (1997)

[27] Teplyakov A,; Galya O.; Badet-Denisot, M.; Badet B.: The mechanism of sugar phosphate isomerization by glucosamine 6-phosphate synthase. Protein Sci., 8, 596-602 (1999)

[28] Huynh, Q.K.; Gulve, E.A.; Dian, T.: Purification and characterization of glutamine:fructose 6-phosphate amidotransferase from rat liver. Arch. Biochem. Biophys., 379, 307-313 (2000)

[29] Teplyakov, A.; Obmolova, G.; Badet, B.; Badet-Denisot, M.-A.: Channeling of ammonia in glucosamine-6-phosphate synthase. J. Mol. Biol., 313, 1093-1102 (2001)

[30] Todorova, R.: Isomerase activity of the C-terminal fructose-6-phosphate binding domain of glucosamine-6-phosphate synthase from Escherichia coli. J. Enzyme Inhib., 16, 373-380 (2001)

Succinyldiaminopimelate transaminase 2.6.1.17

1 Nomenclature

EC number
2.6.1.17

Systematic name
N-succinyl-L-2,6-diaminoheptanedioate:2-oxoglutarate aminotransferase

Recommended name
succinyldiaminopimelate transaminase

Synonyms
DAP-AT
N-succinyl-L-diaminopimelic-glutamic transaminase
aminotransferase, succinyldiaminopimelate
succinyldiaminopimelate aminotransferase

CAS registry number
9030-46-0

2 Source Organism

<-4> no activity in *pig heart* [1]
<-3> no activity in *Brewer's yeast* [1]
<-2> no activity in *Streptococcus sp.* [1]
<-1> no activity in *Lactobacillus casei* [1]
<1> *Escherichia coli* [1-5]
<2> *Azotobacter vinelandii* [1]
<3> *Rhodopseudomonas sphaeroides* [1]
<4> *Alcaligenes faecalis* [1]
<5> *Aerobacter aerogenes* [1]
<6> *Micrococcus lysodeikticus* [1]
<7> *Lactobacillus arabinosus* [1]
<8> *Nostoc muscorum* [1]
<9> *Mycobacterium smegmatis* [5]
<10> *Bordetella pertussis* [6]

3 Reaction and Specificity

Catalyzed reaction
N-succinyl-L-2,6-diaminoheptanedioate + 2-oxoglutarate = N-succinyl-2-L-amino-6-oxoheptanedioate + L-glutamate (a pyridoxal-phosphate protein)

Reaction type
amino group transfer

Natural substrates and products
S   N-succinyl-2-amino-6-oxoheptanedioate + L-glutamate <1-8> (<1> a key enzyme in the bacterial pathway to L-lysine [3]) [1, 3]
P   N-succinyl-L-2,6-diaminoheptanedioate + 2-oxoglutarate

Substrates and products
S   N-succinyl-2-amino-6-oxoheptanedioate + L-glutamate <1-10> (<1> completely specific towards L-glutamate [1]; <1> i.e. N-succinyl-α-amino-ε-ketopimelic acid [3]) (Reversibility: r <1-10> [1-6]) [1-6]
P   N-succinyl-L-2,6-diaminoheptanedioate + 2-oxoglutarate <1-8> [1]
S   Additional information <1, 9> (<1> structural requirements for substrate recognition [3]; <1,9> enzyme is not identical to N-acetylornithine aminotransferase [5]) [3, 5]
P   ?

Inhibitors
$(NH_4)_2SO_4$ <1> [1]
2-(N-(succinylamino))-6-hydrazinoheptane-1,7-dioic acid <1> (<1> very potent slow-binding inhibitor, synthesis [3]) [3]
2-(N-carbobenzoxy-amino)-6-hydrazinoheptane-1,7-dioic acid <1> (<1> very potent slow-binding inhibitor, synthesis [3]) [3]
2-oxoglutarate <1> (<1> at high concentrations, inhibition in both directions [1]) [1]
carbenicillin <1> [4]
chloramphenicol <1> [4]
hydroxylamine <1> [1]
tetracycline <1> [4]
Additional information <1> (<1> synthesis of hydrazino product analogues which are potent slow tight-binding inhibitors, kinetics [4]) [4]

Cofactors/prosthetic groups
pyridoxal 5'-phosphate <1> (<1> a pyridoxal phosphate protein [1]) [1, 3]

Specific activity (U/mg)
5.33 <1> (<1> pH 7.4 [1]) [1]

K_m-Value (mM)
0.18 <1> (N-succinyl-2-amino-6-oxoheptanedioate, <1> 30°C, pH 8.0 [3]) [3]
0.5 <1> (N-succinyl-2-amino-6-oxoheptanedioate, <1> pH 7.4 [1]) [1]
1.21 <1> (L-glutamate, <1> 30°C, pH 8.0 [3]) [3]
5.2 <1> (L-glutamate, <1> pH 7.4 [1]) [1]

K$_i$-Value (mM)

0.000004 <1> (2-(N-(succinylamino))-6-hydrazinoheptane-1,7-dioic acid, <1> 30°C, pH 8.0 [3]) [3]

0.000054 <1> (2-(N-carbobenzoxy-amino)-6-hydrazinoheptane-1,7-dioic acid, <1> 30°C, pH 8.0 [3]) [3]

pH-Optimum

7.4 <1> (<1> assay at [1]) [1]

8 <1> [1]

pH-Range

6.5-9 <1> (<1> pH 6.5: about 25% of maximum activity, pH 9: about 60% of maximum activity [1]) [1]

4 Enzyme Structure

Molecular weight

82000 <1> (<1> gel filtration [3]) [3]

Subunits

dimer <1> (<1> 2 * 39900, SDS-PAGE, MALDI-TOF mass spectra [3]) [3]

Additional information <1> (<1> contains multiple imperfect-hexapeptide-repeat units [2]) [2]

5 Isolation/Preparation/Mutation/Application

Purification

<1> [1, 3]

References

[1] Peterkofsky, B.; Gilvarg, C.: N-succinyl-L-diaminopimelic-glutamic transaminase. J. Biol. Chem., **236**, 1432-1438 (1961)

[2] Vaara, M.: Eight bacterial proteins, including UDP-N-acetylglucosamine acyltransferase (LpxA) and three other transferases of Escherichia coli, consist of a six-residue periodicity theme. FEMS Microbiol. Lett., **97**, 249-254 (1992)

[3] Cox, R.J.; Sherwin, W.A.; Lam, L.K.P.; Vederas, J.C.: Synthesis and evaluation of novel substrates and inhibitors of N-succinyl-LL-diaminopimelate aminotransferase (DAP-AT) from Escherichia coli. J. Am. Chem. Soc., **118**, 7449-7460 (1996)

[4] Cox, R.J.; Schouten, J.A.; Stentiford, R.A.; Wareing, K.J.: Peptide inhibitors of N-succinyl diaminopimelic acid aminotransferase (DAP-AT): a novel class of antimicrobial compounds. Bioorg. Med. Chem. Lett., **8**, 945-950 (1998)

[5] Cox, R.J.; Wang, P.S.H.: Is N-acetylornithine aminotransferase the real N-succinyl-LL-diaminopimelate aminotransferase in Escherichia coli and Mycobacterium smegmatis?. J.Chem. Soc. Perkin Trans.I, 1, 2006-2008 (2001)
[6] Fuchs, T.M.; Schneider, B.; Krumbach, K.; Eggeling, L.; Gross, R.: Characterization of a Bordetella pertussis diaminopimelate (DAP) biosynthesis locus identifies dapC, a novel gene coding for an N-succinyl-L,L-DAP aminotransferase. J. Bacteriol., 182, 3626-3631 (2000)

β-Alanine-pyruvate transaminase   2.6.1.18

1 Nomenclature

EC number
2.6.1.18

Systematic name
L-alanine:3-oxopropanoate aminotransferase

Recommended name
β-alanine-pyruvate transaminase

Synonyms
aminotransferase, β-alanine-pyruvate
β-alanine-α-alanine transaminase <1> [1]
β-alanine-pyruvate aminotransferase

CAS registry number
9030-47-1

2 Source Organism

<1> *Pseudomonas fluorescens* (ATCC 11250 [1]) [1]
<2> *Phaseolus vulgaris* (wax bean, var. Kinghorn [2]) [2]
<3> *Bacillus cereus* (K-22 [3]) [3]
<4> *Rattus norvegicus* [4]
<5> *Burkholderia cepacia* (ATCC 25416 [5]) [5]

3 Reaction and Specificity

Catalyzed reaction
L-alanine + 3-oxopropanoate = pyruvate + β-alanine

Reaction type
amino group transfer

Natural substrates and products
  S β-alanine + pyruvate <5> (<5> enzyme may be involved in pyrimidine base catabolism [5]) [5]
  P malonic semialdehyde + L-alanine

Substrates and products

S β-alanine + 2-oxobutanoate <1, 3> (<3> 2-oxo-n-butyric acid [3]; <1> at 9% of the activity with pyruvate [1]; <3> at 64% of the activity with pyruvate [3]) (Reversibility: ? <1,3> [1,3]) [1, 3]

P malonic semialdehyde + 2-aminobutanoate <1, 3> [1, 3]

S β-alanine + 2-oxohexanoate <1> (<1> at 6% of the activity with pyruvate [1]) (Reversibility: ? <1> [1]) [1]

P malonic semialdehyde + L-norleucine <1> [1]

S β-alanine + 2-oxoisohexanoate <1> (<1> at 4% of the activity with pyruvate [1]) (Reversibility: ? <1> [1]) [1]

P malonic semialdehyde + L-isoleucine <1> [1]

S β-alanine + 2-oxoisopentanoate <3> (<3> at 13% of the activity with pyruvate [3]) (Reversibility: ? <3> [3]) [3]

P malonic semialdehyde + L-valine <3> [3]

S β-alanine + 2-oxopentanoate <3> (<3> at 27% of the activity with pyruvate [3]) (Reversibility: ? <3> [3]) [3]

P malonic semialdehyde + L-norvaline <3> [3]

S β-alanine + acetoacetate <1> (<1> at 4% of the activity with pyruvate [1]) (Reversibility: ? <1> [1]) [1]

P malonic semialdehyde + 3-aminobutanoate <1> [1]

S β-alanine + glyoxylic acid <1> (<1> at 17% of the activity with pyruvate [1]) (Reversibility: ? <1> [1]) [1]

P malonic semialdehyde + glycine <1> [1]

S β-alanine + oxaloacetic acid <3> (<3> at 115% of the activity with pyruvate [3]) (Reversibility: ? <3> [3]) [3]

P malonic semialdehyde + L-aspartate <3> [3]

S β-alanine + pyruvate <1-3> (<1,2,3> i.e. 3-aminopropanoate + 2-oxopropanoate [1-3]) (Reversibility: r <1> [1]; ? <2,3> [2,3]) [1-3]

P malonic semialdehyde + L-alanine <1-3> [1-3]

S pyruvate + 2-aminoethylphosphonic acid <3> (<3> at 28% of the activity with β-alanine [3]) (Reversibility: ? <3> [3]) [3]

P L-alanine + 2-oxoethylphosphonic acid <3> [3]

S pyruvate + 2-aminoisobutanoate <1> (<1> at 4% of the activity with β-alanine [1]) (Reversibility: ? <1> [1]) [1]

P L-alanine + 2-oxo-3-methylpropanoate <1> [1]

S pyruvate + 4-aminobutanoate <1, 3> (<1> at 118% of the activity with β-alanine [1]; <3> at 43% of the activity with β-alanine [3]) (Reversibility: ? <1,3> [1,3]) [1, 3]

P L-alanine + succinic semialdehyde <1, 3> [1, 3]

S pyruvate + 5-aminopentanoate <1> (<1> at 17% of the activity with β-alanine [1]) (Reversibility: ? <1> [1]) [1]

P L-alanine + 5-oxopentanoate <1> [1]

S pyruvate + 6-aminohexanoate <1, 3> (<1> at 134% of the activity with β-alanine [1]; <3> at 9% of the activity with β-alanine [3]) (Reversibility: ? <1,3> [1,3]) [1, 3]

P L-alanine + 6-oxohexanoate <1, 3> [1, 3]

S pyruvate + DL-3-aminoisobutanoate <1> (<1> at 160% of the activity with β-alanine [1]) (Reversibility: ? <1> [1]) [1]
P L-alanine + 3-oxoisobutanoate <1> [1]
S pyruvate + L-2,4-diaminobutanoate <1> (<1> at 4% of the activity with β-alanine [1]) (Reversibility: ? <1> [1]) [1]
P L-alanine + ?
S pyruvate + L-lysine <1> (<1> at 7% of the activity with β-alanine [1]) (Reversibility: ? <1> [1]) [1]
P L-alanine + ?
S pyruvate + L-ornithine <1> (<1> at 1% of the activity with β-alanine [1]) (Reversibility: ? <1> [1]) [1]
P L-alanine + ?
S pyruvate + glycine <1> (<1> at 15% of the activity with β-alanine [1]) (Reversibility: ? <1> [1]) [1]
P L-alanine + glyoxylate <1> [1]
S pyruvate + propylamine <3> (<3> at 37% of the activity with β-alanine [3]) (Reversibility: ? <3> [3]) [3]
P L-alanine + propanal <3> [3]
S pyruvate + taurine <1> (<1> at 8% of the activity with β-alanine [1]) (Reversibility: ? <1> [1]) [1]
P L-alanine + 2-oxoethanesulfonic acid <1> [1]

Inhibitors

cysteine <3> (<3> 0.5 mM, 14% inhibition [3]) [3]
glutathione <3> (<3> 0.5 mM, 12% inhibition [3]) [3]
monoiodoacetic acid <3> (<3> 0.5 mM, 24% inhibition [3]) [3]
p-chloromercuribenzoate <1, 3> (<1> 0.1 mM, 70% inhibition [1]; <3> 0.5 mM, 71% inhibition [3]) [1, 3]

Cofactors/prosthetic groups

pyridoxal 5'-phosphate <3> (<3> a pyridoxal phosphate protein, 40% stimulation on addition of the cofactor even after the enzyme has previously been incubated with saturating amounts of pyridoxal phosphate [3]) [3]

Activating compounds

pyridoxamine 5'-phosphate <1> (<1> shows 20% of pyridoxal 5'-phosphate activation [1]) [1]

Specific activity (U/mg)

56.6 <1> [1]

K_m-Value (mM)

1.1 <3> (β-alanine, <3> pH 10.0, 35°C, cosubstrate pyruvate [3]) [3]
14 <1> (β-alanine, <1> pH 8.0, 35°C, cosubstrate pyruvate [1]) [1]
62 <1> (pyruvate, <1> pH 8.0, 35°C, cosubstrate β-alanine [1]) [1]

pH-Optimum

7.2-7.4 <2> (<2> assay at [2]) [2]
9.2 <1> [1]
10 <3> [3]

pH-Range

7-11 <1> (<1> approx. 35% of maximal activity at pH 7.0, approx. 50% of maximal activity at pH 11.0 [1]) [1]

Temperature optimum (°C)

33 <2> (<2> assay at [2]) [2]
35 <1, 3> (<1> assay at [1]) [1, 3]

5 Isolation/Preparation/Mutation/Application

Source/tissue

cotyledon <2> [2]
culture condition:(NH$_4$)$_2$SO$_4$-grown cell <5> (<5> carbon source succinate or glucose [5]) [5]
culture condition:5-methylcytosine-grown cell <5> (<5> carbon source glucose [5]) [5]
culture condition:β-L-alanine-grown cell <5> (<5> carbon source glucose [5]) [5]
culture condition:β-aminoisobutyric acid-grown cell <5> (<5> carbon source glucose [5]) [5]
culture condition:cytosine-grown cell <5> (<5> carbon source glucose [5]) [5]
culture condition:dihydrothymine-grown cell <5> (<5> carbon source glucose [5]) [5]
culture condition:dihyrouracil-grown cell <5> (<5> carbon source glucose [5]) [5]
culture condition:thymine-grown cell <5> (<5> carbon source glucose [5]) [5]
culture condition:uracil-grown cell <5> (<5> carbon source glucose, 10fold increase in β-alanine-pyruvate tranasaminase activity compared to ammonium sulfate-grown cells [5]) [5]
kidney <4> [4]
liver <4> (<4> expression increases with increasing dietary protein levels [4]) [4]
Additional information <4> (<4> no activity in brain [4]) [4]

Purification

<1> (heat treatment, alumina C-γ gel, protamine sulfate, ammonium sulfate [1]) [1]
<3> (K-22, partial purification [3]) [3]

6 Stability

pH-Stability

6.5-10.5 <3> (<3> stable [3]) [3]

Temperature stability

40 <3> (<3> stable up to [3]) [3]

Storage stability

<1>, -20°C, several weeks, no loss of activity [1]

References

[1] Hayashi, O.; Nishizuka, Y.; Tatibana, M.; Takeshita, M.; Kuno, S.: Enzymatic studies on the metabolism of β-alanine. J. Biol. Chem., **236**, 781-790 (1961)
[2] Stinson, R.A.; Spencer, M.S.: β alanine aminotransferase (s) from a plant source. Biochem. Biophys. Res. Commun., **34**, 120-127 (1969)
[3] Nakano, Y.; Tokunaga, H.; Kitaoka, S.: Two ω-amino acid transaminases from Bacillus cereus. J. Biochem., **81**, 1375-1381 (1977)
[4] Ito, S.; Ohyama, T.; Kontani, Y.; Matslida, K.; Sakata, S.F.; Tamaki, N.: Influence of dietary protein levels on β-alanine aminotransferase expression and activity in rats. J. Nutr. Sci. Vitaminol., **47**, 275-282 (2001)
[5] West, T.P.: Role of cytosine deaminase and β-alanine-pyruvate transaminase in pyrimidine base catabolism by Burkholderia cepacia. Antonie Leeuwenhoek, **77**, 1-5 (2000)

4-Aminobutyrate transaminase

1 Nomenclature

EC number
2.6.1.19

Systematic name
4-aminobutanoate:2-oxoglutarate aminotransferase

Recommended name
4-aminobutyrate transaminase

Synonyms
4-aminobutyrate aminotransferase
4-aminobutyrate-2-ketoglutarate aminotransferase
4-aminobutyrate-2-oxoglutarate aminotransferase
4-aminobutyrate-2-oxoglutarate transaminase
4-aminobutyric acid 2-ketoglutaric acid aminotransferase
4-aminobutyric acid aminotransferase
GABA aminotransferase
GABA transaminase
GABA transferase
GABA-2-oxoglutarate aminotransferase
GABA-2-oxoglutarate transaminase
GABA-α-ketoglutarate aminotransferase
GABA-α-ketoglutarate transaminase
GABA-α-ketoglutaric acid transaminase
GABA-α-oxoglutarate aminotransferase
GABA-oxoglutarate aminotransferase
GABA-oxoglutarate transaminase
aminobutyrate aminotransferase
aminobutyrate transaminase
aminotransferase, aminobutyrate
β-alanine aminotransferase
β-alanine transaminase
β-alanine-oxoglutarate aminotransferase
β-alanine-oxoglutarate transaminase
γ-aminobutyrate aminotransaminase
γ-aminobutyrate transaminase
γ-aminobutyrate-α-ketoglutarate aminotransferase
γ-aminobutyrate-α-ketoglutarate transaminase
γ-aminobutyrate:α-oxoglutarate aminotransferase
γ-aminobutyric acid aminotransferase

γ-aminobutyric acid pyruvate transaminase
γ-aminobutyric acid transaminase
γ-aminobutyric acid-2-oxoglutarate transaminase
γ-aminobutyric acid-α-ketoglutarate transaminase
γ-aminobutyric acid-α-ketoglutaric acid aminotransferase
γ-aminobutyric transaminase
glutamate-succinic semialdehyde transaminase
Additional information (<4> may be identical with EC 2.6.1.22 [30])

CAS registry number
9037-67-6

2 Source Organism

<1> *Mus musculus* (Swiss albino [1]; two isozymes, mitochondrial GABA-T and synaptosomal GABA-T [6]) [1, 3, 4, 6, 7, 20]
<2> *Sus scrofa* (expressed in E. coli [27, 45]) [2, 5, 7, 8, 16, 17, 27, 35, 37, 38, 40, 42, 44, 45, 50]
<3> *Oryctolagus cuniculus* [7, 12, 13]
<4> *Rattus norvegicus* (weanling [43]) [7, 9, 10, 14, 18, 19, 21, 30, 31, 34, 36, 41, 43, 49]
<5> *Ovis aries* [15]
<6> *Homo sapiens* [7, 9, 11, 32, 39, 47]
<7> *Schistocerca gregaria* (locust [15]) [15]
<8> *Euglena gracilis* (streptomycin-bleached mutant strain [29]) [29]
<9> *Neurospora crassa* (wild-type strain 3a6A [23]) [23]
<10> *Candida guilliermondii var. membranaefaciens* (Y43 [22]) [22]
<11> *Pseudomonas sp.* (F-126 [25]) [25]
<12> *Pseudomonas fluorescens* [19, 24, 26, 33]
<13> *Streptomyces griseus* [28]
<14> *Rhizobium leguminosarum bv. viciae* (VF39 [46]) [46]
<15> *Nicotiana tabacum* [48]
<16> *Meriones unguiculatus* (Mongolian gerbil [49]) [49]
<17> *Bos taurus* [51]

3 Reaction and Specificity

Catalyzed reaction
4-aminobutanoate + 2-oxoglutarate = succinate semialdehyde + L-glutamate
(<1, 4, 10, 12> mechanism [19, 20, 22]; <6> γ-aminobutyric acid transaminase and β-alanine transaminase are identical [47])

Reaction type
amino group transfer

Natural substrates and products

S   4-aminobutanoate + 2-oxoglutarate <1-4, 6, 8, 11, 12, 14> (<1> involved in γ-aminobutyrate metabolism [1]; <1-4, 6, 8, 11> key-reaction of γ-aminobutyrate(GABA)-shunt or bypass [7, 8, 17, 25, 29]; <2> involved in β-alanine metabolism [8]; <12> inducible enzyme [24]; <14> involved in 4-aminobutanoate metabolism via the GABA shunt pathway [46]) [1, 7, 8, 17, 24, 25, 29, 46]

P   4-oxobutanoate + L-glutamate

Substrates and products

S   (1R,4S)-4-amino-2-cyclopentene-1-carboxylic acid + 2-oxoglutarate <2> (<2> analogue of 4-aminobutanoate, vigabatrin [42]) (Reversibility: ? <2> [42]) [42]

P   ? + L-glutamate

S   (4R)-4-amino-1-cyclopentene-1-carboxylic acid + 2-oxoglutarate <2> (<2> analogue of 4-aminobutanoate, vigabatrin [42]) (Reversibility: ? <2> [42]) [42]

P   4-oxo-1-cyclopentene1-carboxylic acid + L-glutamate

S   (S)-4-amino-4,5-dihydro-2-thiophenecarboxylic acid + 2-oxoglutarate <2> (<2> mechanism-based inactivator that partly undergoes inactivation [37]) (Reversibility: ? <2> [37]) [37]

P   4-oxo-4,5-dihydro-2-thiophenecarboxylic acid + L-glutamate

S   3-aminoisobutanoate + 2-oxoglutarate <1, 2> (<1> transamination at 55% the rate of 4-aminobutanoate [1]; <2> transamination at 14% the rate of 4-aminobutanoate [8]) (Reversibility: r <1, 2> [1, 7, 8]) [1, 7, 8]

P   L-glutamate + 3-oxoisobutanoate

S   4-aminobutanoate + 2-oxoglutarate <1-14, 16, 17> (<10, 13> best substrate [22, 28]; <1> specific for 2-oxoglutarate [1]; <15> no substrate: 2-oxoglutarate [48]) (Reversibility: r <1-13> [1-31]; ? <14, 16, 17> [46, 51]) [1-31, 45, 46, 48, 49, 51]

P   4-oxobutanoate + L-glutamate <1-4, 6, 8, 10-13> (<1> i.e. succinic semialdehyde + L-Glu [1]) [1-13, 19-22, 24-29, 30, 31, 46]

S   4-aminobutanoate + glyoxylate <11> (<11> at 16% of the activity of 2-oxoglutarate [25]) (Reversibility: r <11> [25]) [25]

P   4-oxobutanoate + glycine

S   4-aminobutanoate + pyruvate <11, 15> (<11> at 7% of the activity of 2-oxoglutarate [25]; <15> not using 2-oxoglutarate [48]) (Reversibility: r <11> [25]; ? <15> [48]) [25, 48]

P   4-oxobutanoate + alanine

S   5-aminopentanoate + 2-oxoglutarate <1, 2, 10, 13> (<2> transamination at 85% the rate of 4-aminobutanoate [8]; <13> transamination at 60% the rate of 4-aminobutanoate [28]; <1> transamination at 48% the rate of 4-aminobutanoate [1]; <10> transamination at 45% the rate of 4-aminobutanoate [22]; <12> not [24]) [1, 2, 7, 8, 22, 28]

P   L-glutamate + 5-oxopentanoate

S 6-aminohexanoate + 2-oxoglutarate <1, 2, 10, 13> (<13> transamination at 29% the rate of 4-aminobutanoate [28]; <10> transamination at 27% the rate of 4-aminobutanoate [22]; <12> not [24]) [7, 22, 28]

P L-glutamate + 6-oxohexanoate

S D-lysine + 2-oxoglutarate <11, 13> (<11, 13> poor amino donor [25, 28]) (Reversibility: r <11,13> [25, 28]) [25, 28]

P L-glutamate + 5-aminopentanol

S DL-3-amino-1-cyclopentene-1-carboxylic acid + 2-oxoglutarate <2> (<2> analogue of 4-aminobutanoate, vigabatrin [42]) (Reversibility: ? <2> [42]) [42]

P 3-oxo-1-cyclopentene-1-carboxylic acid + L-glutamate

S DL-3-hydroxy-4-aminobutanoate + 2-oxoglutarate <1, 2> (<2> transamination at 20% the rate of 4-aminobutanoate [8]) (Reversibility: r <1, 2> [1, 7, 8]) [1, 7, 8]

P L-glutamate + 3-hydroxy-4-oxobutanoate

S DL-ornithine + 2-oxoglutarate <11, 13> (<11,13> poor amino donor [25,28]; <12> not [24]) (Reversibility: r <11> [13]) [25, 28]

P L-glutamate + 4-methyl-2-oxopentanoate

S β-alanine + 2-oxoglutarate <1-4, 6, 9, 10> (<1,2> effective amino group donor [1, 8]; <10> poor substrate [22]; <11-13> not [24, 25, 28]) (Reversibility: r <1-4, 6, 9, 10> [1, 2, 7-12, 17, 18, 22, 23, 31, 47]) [1, 2, 7-12, 17, 18, 22, 23, 31]

P malonic semialdehyde + L-glutamate <2, 4, 9> [17, 18, 23, 30, 31, 47]

S cadaverine + 2-oxoglutarate <11> (<11> poor amino donor [25]) (Reversibility: r <11> [25]) [25]

P L-glutamate + ?

S hypotaurine + 2-oxoglutarate <11> (<11> poor amino donor [25]; <13> not [28]) (Reversibility: r <11> [25]) [25]

P L-glutamate + 2-oxoethanesulfinic acid

S putrescine + 2-oxoglutarate <11> (<11> poor amino donor [25]) (Reversibility: r <11> [25]) [25]

P L-glutamate + 4-aminobutanol

S Additional information <1, 9-12> (<1,9,12> no amino acceptors are oxaloacetate, pyruvate [1, 23, 24]; <1> no amino acceptors are 2-oxobutyrate, phenylpyruvate, α-ketoadipate [1]; <12> no amino acceptors are ketomalonic acid, α-ketoisovaleric acid [24]; <12> aspartate, L-2,4-diaminoglutarate, alanine, D-glutamate or malonic semialdehyde cannot replace L-glutamate or succinic semialdehyde in the reverse reaction [24]; <10> no substrates are L-aminovalerate, L-norleucine, straight α-amino acids, diamino acids [22]; <11> no substrates are taurine, 3-aminopropane sulfonate, glycine [25]; <12> no substrate: lysine [24]; <1, 2, 12> overview [1, 8, 24]) [1, 8, 22-25]

P ?

Inhibitors

(1R,4S)-4-amino-2-cyclopentene-1-carboxylic acid <2> (<2> analogue of 4-aminobutanoate, vigabatrin [42]) [42]

(1S,4R)-4-amino-2-cyclopentene-1-carboxylic acid <2> (<2> analogue of 4-aminobutanoate, vigabatrin [42]) [42]

(4R)-4-amino-1-cyclopentene-1-carboxylic acid <2> (<2> analogue of 4-aminobutanoate, vigabatrin [42]) [42]

(4S)-4-amino-1-cyclopentene-1-carboxylic acid <2> (<2> analogue of 4-aminobutanoate, vigabatrin [42]) [42]

(S)-4-amino-4,5-dihydro-2-thiophenecarboxylic acid <2> (<2> mechanism-based inactivator, reacts via aromatization mechanism [37]) [37]

2,4-diaminobutanoate <4> (<4> kinetics [10]; <5,7> not [15]) [10]

2-aminobutanoate <10> [22]

2-oxoadipic acid <1> [4]

2-oxoglutarate <2-4, 10> [8, 12, 14, 22]

2-thiouracil <3> (<3> weak [12]) [12]

3-chloro-4-aminobutanoate <1> [6]

3-mercaptopropionic acid <1> [4]

3-methyl-2-benzothiazolone hydrazone hydrochloride <11> [25]

3-phenyl-4-aminobutanoate <1> [6]

4-amino-2-fluorobutanoate <12> (<12> reversible, competitive to 4-aminobutanoate [33]) [33]

4-amino-hex-5-enoic acid <4> (<4> substrate analogue, irreversible, in vitro and in vivo [19]) [19]

4-aminohex-5-ynoic acid <12> (<12> irreversible, in vitro and in vivo, kinetics [26]) [26]

5-diazouracil <3> (<3> weak [12]) [12]

5-fluorouracil <3, 4> (<3> weak [12]) [12, 31]

5-iodouracil <3> (<3> 84% inhibition at 1 mM [12]) [12]

5-nitrouracil <3> (<3> weak [12]) [12]

5-thiouracil <3> (<3> weak [12]) [12]

6-azathymine <1> [6]

6-azauracil <1> (<1> 63% inhibition at 1 mM, reversible by dialysis, not by pyridoxal phosphate addition [6]) [6]

ADP <1> (<1> reversible [3]) [3]

ATP <1> (<1> reversible [3]) [3]

Ba^{2+} <1> (<1> order of decreasing inhibitory potency: Hg^{2+}, Cd^{2+}, Cu^{2+}, Co^{2+}, Ba^{2+}, Sr^{2+}, Ni^{2+}, Mn^{2+}, Ca^{2+}, Mg^{2+} [4]) [4]

Ca^{2+} <1> (<1> order of decreasing inhibitory potency: Hg^{2+}, Cd^{2+}, Cu^{2+}, Co^{2+}, Ba^{2+}, Sr^{2+}, Ni^{2+}, Mn^{2+}, Ca^{2+}, Mg^{2+} [4]) [4]

Cd^{2+} <1> (<1> order of decreasing inhibitory potency: Hg^{2+}, Cd^{2+}, Cu^{2+}, Co^{2+}, Ba^{2+}, Sr^{2+}, Ni^{2+}, Mn^{2+}, Ca^{2+}, Mg^{2+} [4]) [4]

Co^{2+} <1> (<1> order of decreasing inhibitory potency: Hg^{2+}, Cd^{2+}, Cu^{2+}, Co^{2+}, Ba^{2+}, Sr^{2+}, Ni^{2+}, Mn^{2+}, Ca^{2+}, Mg^{2+} [4]) [4]

Cu^{2+} <1, 11, 13> (<1> order of decreasing inhibitory potency: Hg^{2+}, Cd^{2+}, Cu^{2+}, Co^{2+}, Ba^{2+}, Sr^{2+}, Ni^{2+}, Mn^{2+}, Ca^{2+}, Mg^{2+} [4]) [4, 25, 28]

D-cycloserine <3, 11, 13> [12, 25, 28]

D-penicillamine <11, 13> [25, 28]

DL-3-amino-1-cyclopentene-1-carboxylic acid <2> (<2> analogue of 4-aminobutanoate, vigabatrin [42]) [42]

DL-cysteine <11> [25]

DL-trans-4-amino-2-cyclopentene-1-carboxylic acid <2> (<2> analogue of 4-aminobutanoate, vigabatrin [42]) [42]

GDP <1> (<1> reversible [3]) [3]

$HgCl_2$ <1, 11-13> (<1> strong, 50% inhibition at 0.007 mM [4]; <12> 24% inhibition at 0.05 mM, pyridoxal 5'-phosphate protects [33]) [4, 25, 28, 33]

KCN <12> [24]

Mg^{2+} <1> (<1> order of decreasing inhibitory potency: Hg^{2+}, Cd^{2+}, Cu^{2+}, Co^{2+}, Ba^{2+}, Sr^{2+}, Ni^{2+}, Mn^{2+}, Ca^{2+}, Mg^{2+} [4]) [4]

Mn^{2+} <1> (<1> order of decreasing inhibitory potency: Hg^{2+}, Cd^{2+}, Cu^{2+}, Co^{2+}, Ba^{2+}, Sr^{2+}, Ni^{2+}, Mn^{2+}, Ca^{2+}, Mg^{2+} [4]) [4]

Ni^{2+} <1> (<1> order of decreasing inhibitory potency: Hg^{2+}, Cd^{2+}, Cu^{2+}, Co^{2+}, Ba^{2+}, Sr^{2+}, Ni^{2+}, Mn^{2+}, Ca^{2+}, Mg^{2+} [4]) [4]

SH-group reagents <1, 2, 11, 13> (<2> 2-oxoglutarate protects [2]; <11> 2-mercaptoethanol or DTT reactivates [25]) [2, 4, 25, 28]

Sr^{2+} <1> (<1> order of decreasing inhibitory potency: Hg^{2+}, Cd^{2+}, Cu^{2+}, Co^{2+}, Ba^{2+}, Sr^{2+}, Ni^{2+}, Mn^{2+}, Ca^{2+}, Mg^{2+} [4]) [4]

acetic acid <1> [4]

adipic acid <1> [4]

α-alanine <10> [22]

aminooxyacetate <1, 3, 4, 11, 13, 15> (<4> kinetics [10]; <3> 99% inhibition at 1 mM [12]; <15> 80% inhibition at 2 mM [48]) [1, 3, 10, 12, 25, 28, 48]

β-alanine <10> (<3> not [12]) [22]

branched-chain fatty acids <4, 6> [9]

butyric acid <1, 10> [4, 22]

carbonyl reagents <11, 13> [25, 28]

cis-3-aminocyclohex-4-ene-1-carboxylic acid <2> (<2> conformationally rigid analogue of vigabatrin, mechanism [38]) [38]

cycloserine <3> (<3> 90% inhibition at 1 mM [12]) [12]

dioxan <1> (<1> 5% v/v [4]) [4]

divalent metal ions <1> (<1> with decreasing efficiency: Hg^{2+}, Cd^{2+}, Co^{2+}, Ba^{2+}, Sr^{2+}, Ni^{2+}, Mn^{2+}, Ca^{2+}, Mg^{2+} [4]) [4]

ethanol <1, 4> (<1> 10% v/v, weak [4]; <4> in presence of disulfiram, i.e. N,N,N',N'-tetraethylthiuram disulfide [43]) [4, 43]

ethanolamine O-sulfate <4> (<4> active-site directed, ir, in vitro and in vivo, kinetics [21]) [21]

gabaculine <1, 3-5, 7, 11, 13, 15> (<1> i.e. 5-amino-1,3-cyclohexadienylcarboxylate, ir, kinetics [20]; <1> not its tert-butylcarbamate derivative [20]; <3> 98% inhibition at 1 mM [12]; <15> 80% inhibition at 2 mM [48]) [12, 15, 18, 20, 25, 28, 48]

glutamic acid <1> [4]

glyoxylate <3> (<3> weak [12]) [12]

hydrazine <1, 11, 13> [4, 25, 28]

hydroxylamine <1, 11-13> [4, 24, 25, 28]

lysyl reagents <2> (<2> 2-oxoglutarate protects [2]) [2]

methanol <1> (<1> 10% v/v, weak [4]) [4]

monoiodoacetate <13> (<11> not [25]) [28]

muscimol <5, 7> (<5,7> i.e. 5-(aminomethyl)-3-isoxazolol [15]) [15]
oxalacetate <3> [12]
p-chloromercuribenzoate <1, 11-13> (<1> strong [4]) [4, 25, 28, 33]
phenylhydrazine <11, 13> [25, 28]
pimelic acid <1> [4]
propionic acid <1, 10> [4, 22]
pyruvate <3> [12]
succinic semialdehyde <4, 15> (<4> substrate inhibition [14]) [14, 48]
vigabatrin <2, 4> (<4> γ-vinyl GABA, anticonvulsant, induces spontaneous release of 4-aminobutanoate [34]; <2> mechanism [37]) [34, 38]
Additional information <3, 5, 7, 11, 13> (<3> not inhibitory: 5-aminouracil [12]; <3> no inhibition by 6-azauridine, 6-azauridine 5'-phosphate, uracil, (iso)orotic acid, cytosine, thymine, dihydrothymine, 2-thiocytosine, thiourea [12]; <11,13> no inhibition by chelating agents, non-substrate L- or D-amino acids, metal ions [25, 28]; <7> no inhibition by 3-aminopropane-1-sulfonic acid, isoguvacine (i.e. 1,2,3,4-tetrahydro-1-methyl-3-pyridine carboxylic acid), baclofen (i.e. β-(aminomethyl)-4-chlorobenzenepropanoic acid), bicuculline, picrotoxin, Schistocerca gregaria: antiserum against sheep enzyme [15]) [12, 15, 25, 28]

Cofactors/prosthetic groups

pyridoxal 5'-phosphate <1, 2, 4, 5, 7, 9, 11-13> (<2> requirement, bound to lysine-residue 330 [17,45]; <13> 2 mol per mol enzyme [28]; <12> no activation by addition of pyridoxal phosphate, but sensitive to carbonyl reagents [24]) [1, 2, 15, 17-21, 23, 25-28, 30, 31, 45]

Turnover number (min^{-1})

792 <5> (4-aminobutanoate) [15]

Specific activity (U/mg)

1.8-3.5 <2> (<2> 37°C [8]) [8]
2.5 <4> (<4> 38°C, pH 8.6 [9]) [9]
3.4 <12> [24]
4 <6> [11]
4.1 <5> [15]
4.36 <7> [15]
4.9 <6> (<4> 38°C, pH 8.6 [9]) [9]
5 <1> (<1> 37°C, pH 8.0 [1]) [1]
9.64 <13> [28]
10 <4> (<4> 38°C, pH 8.5 [14]) [14]
18 <17> (<17> pH 8.4, 25°C [51]) [51]
18.2 <2> [27]
46.8 <11> [25]
52.33 <10> [22]
170 <4> (<4> pH 7.3, 37°C, liver enzyme [41]) [41]
260 <4> (<4> pH 7.3, 37°C, kidney enzyme [41]) [41]
600 <4> (<4> pH 7.3, 37°C, brain enzyme [41]) [41]
Additional information <6> (<6> stable-isotope dilution assay [47]) [47]

K$_m$-Value (mM)

0.1 <2> ((4R)-4-amino-1-cyclopentene-1-carboxylic acid) [42]

0.1 <17> (α-ketoglutarate, <17> pH 8.4, 25°C [51]) [51]

0.11 <6> (2-oxoglutarate, <6> 25°C, pH 8.4 [39]) [39]

0.13-0.27 <1-3> (2-oxoglutarate, <1> 37°C, pH 8.0 [1]; <3> 37°C, pH 7.3 [12]) [1, 12, 27]

0.22 <5> (2-oxoglutarate, <5> pH 8.5 [15]) [15]

0.24 <15> (pyruvate, <15> pH 8.2 [48]) [48]

0.27 <7> (2-oxoglutarate, <7> pH 8.5 [15]) [15]

0.4 <6> (4-aminobutanoate) [11]

0.7 <2> (2-oxoglutarate, <2> 37°C [8]) [8]

0.79 <7> (4-aminobutanoate, <7> pH 8.5 [15]) [15]

1 <6> (2-oxoglutarate) [11]

1 <17> (2-oxoglutarate, <17> pH 8.4, 25°C [51]) [51]

1.05 <4> (β-alanine) [30]

1.1 <1, 2> (4-aminobutanoate, <1> 37°C, pH 8.0 [1]; <2> 37°C [8]) [1, 8, 27]

1.2 <15> (4-aminobutanoate, <15> pH 8.2 [48]) [48]

1.27 <6> (4-aminobutanoate, <6> 25°C, pH 8.4 [39]) [39]

1.3-1.5 <2> (2-oxoglutarate) [5]

1.5 <10> (2-oxoglutarate) [22]

1.6 <2> ((4R)-4-amino-1-cyclopentene-1-carboxylic acid) [42]

1.7 <6> (4-aminobutanoate, <6> pH 8.0, 37°C [47]) [47]

2.2-2.3 <5, 10> (4-aminobutanoate, <7> pH 8.5 [15]) [15, 22]

2.3 <2> (DL-3-amino-1-cyclopentene-1-carboxylic acid) [42]

2.72 <4> (L-3-aminoisobutanoate) [30]

2.8 <11> (2-oxoglutarate, <11> pH 8.0, 37°C [25]) [25]

3.3 <13> (4-aminobutanoate, <13> pH 8.0, 37°C [28]) [28]

3.3-4.8 <2, 11> (4-aminobutanoate, <11> pH 8.0, 37°C [25]) [5, 25]

4 <4> (4-aminobutanoate, <4> 38°C, pH 8.5 [14]) [14]

4.4 <6> (β-alanine, <6> pH 8.0, 37°C [47]) [47]

5.5 <4> (2-oxoglutarate, <4> 38°C, pH 8.5 [14]) [14]

6.5 <4> (4-aminobutanoate, <4> mitochondrial enzyme; pH 8.0 [10]) [10]

8.3 <13> (2-oxoglutarate, <13> pH 8.0, 37°C [28]) [28]

18 <3> (β-alanine) [12]

53 <4> (4-aminobutanoate, <4> synaptosomal enzyme, pH 8.0 [10]) [10]

Additional information <4, 6> (<4> kinetic study [14]; <4> pH-dependence of kinetic data [14]) [9, 14, 18]

K$_i$-Value (mM)

0.00018 <4> (2-oxoglutarate, <4> 38°C, pH 8.5 [14]) [14]

0.002 <10> (butyric acid) [22]

0.003 <10> (propionic acid) [22]

0.0044 <4> (ethanolamine O-sulfate, <4> 37°C, pH 8.4 [21]) [21]

0.0066 <4> (succinic semialdehyde, <4> 38°C, pH 8.5 [14]) [14]

0.008 <4> (2,4-diaminobutanoate, <4> synaptosomal enzyme, pH 8.0 [10]) [10]

0.013 <4> (2,4-diaminobutanoate, <4> cytoplasmic enzyme, pH 8.0 [10]) [10]

0.013 <1> (3-mercaptopropionic acid) [4]

0.033 <10> (2-aminobutanoate) [22]

0.0355 <10> (β-alanine) [22]

0.039 <10> (α-alanine) [22]

0.052 <2> ((1R,4S)-4-amino-2-cyclopentene-1-carboxylic acid) [42]

0.44 <12> (4-amino-2-fluorobutanoate, <12> pH 8.6, 37°C [33]) [33]

0.6 <2> (DL-3-amino-1-cyclopentene-1-carboxylic acid) [42]

1 <6> (vigabatrin, <6> pH 8.4, 25°C [39]) [39]

1.2 <2> ((4R)-4-amino-1-cyclopentene-1-carboxylic acid) [42]

2.7 <2> ((1S,4R)-4-amino-2-cyclopentene-1-carboxylic acid) [42]

3.7 <1> (ATP) [3]

38 <2> (DL-trans-4-amino-2-cyclopentene-1-carboxylic acid) [42]

72 <2> ((4S)-4-amino-1-cyclopentene-1-carboxylic acid) [42]

pH-Optimum

7 <1> (<1> mitochondrial enzyme [6]) [6]

7-8 <9> (<9> mitochondrial enzyme [23]) [23]

7.5-8.5 <13> [28]

7.8-8 <10> (<10> at 35°C [22]) [22]

8 <1> (<1> synaptosomal enzyme [6]) [6]

8.1 <1> [1]

8.5 <4, 6> [9, 14]

8.5-8.6 <4> [9]

8.5-9 <11> [25]

8.6 <6> [11]

8.6-9 <12> (<12> at 25°C [24]) [24]

8.7 <2> (<2> isozyme II [8]) [5, 8]

8.8 <2> (<2> isozyme I [8]) [8]

Additional information <2, 4, 5, 7, 10, 12> (<5> pI: 5.5 [15]; <10> pI: 5.7 [22]; <2> pI: 5.9 [5]; <2> pI: 5.9 and 6.35 (isozyme II), pI: 6.1 and 6.3 (isozyme I) [8]; <7> pI: 6.7 [15]; <4> pI: 6.8 [14]) [5, 8, 14, 15, 22, 24]

pH-Range

5.5 <13> (<13> no activity below [28]) [28]

6.5-11 <11> (<11> no activity above or below [25]) [25]

6.9-8.8 <1> (<1> about half-maximal activity at pH 6.9 and 8.8 [1]) [1]

7.4-9 <4> (<4> about half-maximal activity at pH 7.4 and about 90% of maximal activity at pH 9 [14]) [14]

8-9.2 <4, 12> (<4,12> about half-maximal activity at pH 8 [14, 24]; <4> about 50% of maximal activity at pH 9.2 [14]; <12> about 60% of maximal activity at pH 9.2 [24]) [14, 24]

11 <13> (<13> and above, no activity [28]) [28]

Temperature optimum (°C)

30-35 <9> [23]

45 <4, 10> [14, 22]

50 <6, 13> [11, 28]
60 <11> [25]

Temperature range (°C)
20-40 <9> (<9> about half-maximal activity at 20°C and 40°C [23]) [23]
20-50 <13> (<13> linear increase, above 55°C rapid decrease [28]) [28]
25-45 <10> (<10> linear increase, above 45°C rapid decrease [22]) [22]
30-55 <4> (<4> about half-maximal activity at 30°C and 55°C [14]) [14]
30-60 <11> (<11> linear increase, above 65°C rapid decrease [25]) [25]
35-70 <6> (<6> about half-maximal activity at 35°C and 70°C [11]) [11]

4 Enzyme Structure

Molecular weight
97000 <5, 7> (<5,7> gel filtration [15]) [15]
100000 <13> (<13> gel filtration [28]) [28]
103000 <4> (<4> PAGE [14]) [14]
105000 <2, 4> (<2,4> gel filtration [5,8,14]) [5, 9, 14]
107000 <4, 10> (<4> calculated from amino acid composition [14]; <10> gel filtration [22]) [14, 22]
109000 <1> (<1> high speed sedimentation equilibrium centrifugation [1]) [1]
110000 <2> (<2> liver enzyme, gel filtration [8]) [8]
110100 <2> (<2> liver enzyme, crystallographic data [16]) [16]
176000-178000 <11> (<11> gel filtration, sedimentation equilibrium centrifugation [25]) [25]
Additional information <4> (<4> compared to liver enzyme, mature form of brain enzyme has an additional peptide at the N-terminus which may be cleaved by liver mitochondrial extract [41]) [41]

Subunits
? <4, 16, 17> (<4,16> x * 50000, SDS-PAGE [49]; <17> x * 52000, SDS-PAGE, x * 53950, deduced from gene sequence [51]) [49, 51]
dimer <1, 2, 4-6, 13> (<2> brain enzyme, SDS- or SDS/urea-PAGE [5]; <5> 2 * 50000, SDS-PAGE [15]; <2> α_2, liver enzyme, crystallization data [16]; <13> 2 * 50000, SDS-PAGE [28]; <1> 1 * 53000 + 1 * 58000, SDS-PAGE [4]; <2> 2 * 55000, liver enzyme, SDS-PAGE [8]; <6> 2 * 57000, SDS-PAGE [9]; <4> 2 * 57000, SDS-PAGE with or without DTT or urea [9,14]; <4> 2 * 53300, SDS-PAGE, brain enzyme, 2 * 52700, SDS-PAGE, liver enzyme [41]) [4, 5, 8, 9, 14-16, 22, 28, 41]
Additional information <2, 13> (<13> amino acid composition [28]; <2> crosslinking with bifunctional sulfhydryl reagent DMDS at one mol per enzyme dimer [35]; <2> three different enzyme species isolated are not isozymes but products of proteolysis differing by 3,7,12 residues, Na₂EDTA inhibits N-terminal cleavage during preparation [44]) [28, 35, 44]

5 Isolation/Preparation/Mutation/Application

Source/tissue

brain <1-6, 16, 17> (<4, 16> immunohistochemical study on distribution [49]) [1-7, 9-11, 14, 15, 17, 19-21, 27, 39, 42, 44, 45, 49, 50, 51]

ganglion <7> (<7> supraoesophageal [15]) [15]

leukocyte <6> [47]

liver <2-4> [7, 8, 12, 13, 16, 18, 19, 30, 31]

mycelium <9> [23]

thymus <4, 6> (<6> biochemical and histochemical data to distribution, young to elderly men [32]; <4> mostly arteries and partly veins of, strong decrease of activity with age [36]) [32, 36]

Localization

cytoplasm <10> [22]

mitochondrion <1, 2, 4, 6-8, 15> [1, 3, 6, 7, 10, 11, 14, 15, 27, 29, 48]

soluble <11-13> [24, 25, 28]

synaptosome <1, 4> [6, 10]

Purification

<1> (partial [3]) [1, 3]

<2> (cationic (I) and anionic (II) form [8]; expressed in E. coli [27]) [8, 27]

<3> [13]

<4> [14]

<5> [15]

<6> [9, 11]

<7> [15]

<10> [22]

<11> (F-126 [25]) [25]

<12> (partial [24]) [24]

<13> [28]

<15> (partial [48]) [48]

Renaturation

<3> (enzyme inhibited by 6-azauracil is reactivated by dialysis but not by addititon of pyridoxal 5'-phosphate [12]) [12]

<11> (after inhibition by carbonyl reagents or sulfhydryl reagents enzyme may be reactivated by incubation with pyridoxal 5-phosphate or thiol reagents [25]) [25]

Crystallization

<2> [40]

<2, 11> [7, 16, 25]

Cloning

<2> (cDNA clone isolation and sequence determination [17]; brain cDNA, expressed in Escherichia coli strain BL21(DE3)pLysS transformed with expression vector pETG1.5 [27]) [17, 27]

<14> [46]

<17> [51]

Engineering

K330R <2> (<2> no catalytic activity, no pyridoxal 5'-phosphate covalently linked to protein [45]) [45]

Additional information <2> (<2> C-terminal mutant lacking 5 amino acids, no interference with kinetical parameters or functional properties but change in stability of dimeric structure at acidic pH [50]) [50]

6 Stability

pH-Stability

6-10 <11, 13> (<11,13> 15 min stable, 37°C [25,28]) [25, 28]

6.5-7.5 <12> (<12> below and above, rapid inactivation [24]) [24]

Temperature stability

35 <4> (<4> $t_{1/2}$: 3 h [14]) [14]

37 <11> (<11> 15 min stable, pH 6-10 [25]) [25]

42 <10> (<10> up to, at least 10 min stable [22]) [22]

45 <13> (<13> up to, at least 10 min stable, pH 8 [28]) [28]

46 <10> (<10> up to, at least 10 min stable in the presence of pyridoxal phosphate [22]) [22]

50 <4, 10, 13> (<4> 3 h, 80% loss of activity [14]; <10> up to, at least 10 min stable in the presence of 2-oxoglutarate [22]; <13> up to, stable [28]; <13> above, 10 min, inactivation [28]) [14, 22, 28]

50-60 <6> (<6> slow decrease of activity, rapid inactivation above 60°C [11]) [11]

55 <4, 6> (<4> rapid inactivation [9]; <6> slow inactivation [9]; <4> 3 h, inactivation [14]) [9, 14]

60 <11, 13> (<11> up to, at least 15 min stable, pH 8 [25]; <13> 10 min stable in the presence of pyridoxal phosphate [28]) [25, 28]

65 <11> (<11> above, 15 min, inactivation [25]) [25]

70 <11> (<11> 15 min stable in the presence of pyridoxal phosphate [25]) [25]

Additional information <10, 11, 13> (<10> 2-oxoglutarate protects against heat inactivation, better than pyridoxal phosphate [22]; <11,13> pyridoxal phosphate protects against heat inactivation [25, 28]) [22, 25, 28]

General stability information

<1>, freeze-thawing slightly inactivates [1]

<10>, 2-oxoglutarate protects against heat inactivation, better than pyridoxal phosphate [22]

<11>, pyridoxal phosphate protects against heat inactivation [25]

Storage stability

<1>, -20°C, protected with 2-aminoethylisothiouronium bromide hydrobromide and pyridoxal phosphate, several months [1]

<12>, -20°C, several months [24]

References

[1] Schousboe, A.; Wu, J.Y.; Roberts, E.: Purification and characterization of the 4-aminobutyrate–2,ketoglutarate transaminase from mouse brain. Biochemistry, **12**, 2868-2873 (1973)

[2] Choi, S.Y.; Kim, D.S.: Catalytic and structural properties of 4-aminobutyrate aminotransferase. Han'guk Saenghwa Hak Hoe Chi, **24**, 508-514 (1991)

[3] Tunnicliff, G.; Youngs, T.L.: Competitive inhibition of mouse brain γ-aminobutyrate aminotransferase by ATP. Proc. Soc. Exp. Biol. Med., **192**, 11-15 (1989)

[4] Schousboe, A.; Wu, J.-Y.; Roberts, E.: Subunit structure and kinetic properties of 4-aminobutyrate-2-ketoglutarate transaminase purified from mouse brain. J. Neurochem., **23**, 1189-1195 (1974)

[5] Bloch-Tardy, M.; Rolland, B.; Gonnard, P.: Pig brain 4-aminobutyrate 2-ketoglutarate transaminase. Purification, kinetics and physical properties. Biochimie, **56**, 823-832 (1974)

[6] Buu, N.T.; van Gelder, N.M.: Differences in biochemical properties of γ-aminobutyric acid aminotransferase from synaptosome-enriched and cytoplasmic mitochondria-enriched subcellular fractions of mouse brain. Can. J. Physiol. Pharmacol., **52**, 674-680 (1974)

[7] Cooper, A.J.L.: Glutamate-γ-aminobutyrate transaminase. Methods Enzymol., **113**, 80-82 (1985)

[8] Buzenet, A.M.; Fages, C.; Bloch-Tardy, M.; Gonnard, P.: Purification and properties of 4-aminobutyrate 2-ketoglutarate aminotransferase from pig liver. Biochim. Biophys. Acta, **522**, 400-411 (1978)

[9] Maître, M.; Ciesielski, L.; Cash, C.; Mandel, P.: Comparison of the structural characteristics of the 4-aminobutyrate:2-oxoglutarate transaminases from rat and human brain, and of their affinities for certain inhibitors. Biochim. Biophys. Acta, **522**, 385-399 (1978)

[10] Tunnicliff, G.; Ngo, T.T.; Rojo-Ortega, J.M.; Barbeau, A.: The inhibition by substrate analogues of γ-aminobutyrate aminotransferase from mitochondria of different subcellular fractions of rat brain. Can. J. Biochem., **55**, 479-484 (1977)

[11] Cash, C.; Maître, M.; Ciesielski, L.; Mandel, P.: Purification and partial characterisation of 4-aminobutyrate 2-ketoglutarate transaminase from human brain. FEBS Lett., **47**, 199-203 (1974)

[12] Tamaki, N.; Kubo, K.; Aoyama, H.; Funatsuka, A.: Inhibitory effect of 6-azauracil on purified rabbit liver 4-aminobutyrate aminotransferase. J. Biochem., **93**, 955-959 (1983)

[13] Tamaki, N.; Aoyama, H.; Kubo, K.; Ikeda, T.; Hama, T.: Purification and properties of β-alanine aminotransferase from rabbit liver. J. Biochem., **92**, 1009-1017 (1982)

[14] Maître, M.; Ciesielski, L.; Cash, C.; Mandel, P.: Purification and studies on some properties of the 4-aminobutyrate: 2-oxoglutarate transaminase from rat brain. Eur. J. Biochem., **52**, 157-169 (1975)

[15] Jeffery, D.; Rutherford, D.M.; Witzman, P.D.J.; Lunt, G.G.: Purification and partial characterization of 4-aminobutyrate:2-oxoglutarate aminotransferase from sheep brain and locust ganglia. Biochem. J., 249, 795-799 (1988)

[16] Markovic-Housley, Z.; Schirmer, T.; Fol, B.; Jansonius, J.N.; De Biase, D.; John, R.A.: Crystallization and preliminary X-ray analysis of γ-aminobutyric acid transaminase. J. Mol. Biol., 214, 821-823 (1990)

[17] Kwon, O.-S.; Park, J.; Churchich, J.E.: Brain 4-aminobutyrate aminotransferase. Isolation and sequence of a cDNA encoding the enzyme. J. Biol. Chem., 267, 7215-7216 (1992)

[18] Kaneko, M.; Fujimoto, S.; Kikugawa, M.; Tamaki, N.: Irreversible inhibition of D-3-aminoisobutyrate-pyruvate aminotransferase by gabaculine. FEBS Lett., 276, 115-118 (1990)

[19] Lippert, B.; Metcalf, B.W.; Jung, M.J.; Casara, P.: 4-Amino-hex-5-enoic acid, a selective catalytic inhibitor of 4-aminobutyric-acid aminotransferase in mammalian brain. Eur. J. Biochem., 74, 441-445 (1977)

[20] Rando, R.R, Bangerter, F.W.: The irreversible inhibition of mouse brain γ-aminobutyric acid (GABA)-α-ketoglutaric acid transaminase by gabaculine. J. Am. Chem. Soc., 98, 6762-6764 (1976)

[21] Fowler, L.J.; John, R.A.: Active-site-directed irreversible inhibition of rat brain 4-aminobutyrate aminotransferase by ethanolamine O-sulphate in vitro and in vivo. Biochem. J., 130, 569-573 (1972)

[22] Der Garabedian, P.A.; Lotti, A.-M.; Vermeersch, J.J.: 4-Aminobutyrate:2-oxoglutarate aminotransferase from Candida. Purification and properties. Eur. J. Biochem., 156, 589-596 (1986)

[23] Aurich, H.: Über die β-Alanin-α-Ketoglutarat-Transaminase aus Neurospora crassa. Hoppe-Seyler's Z. Physiol. Chem., 326, 25-33 (1961)

[24] Scott, E.M.; Jacoby, W.B.: Soluble γ-aminobutyric-glutamic transaminase from Pseudomonas fluorescens. J. Biol. Chem., 234, 932-936 (1959)

[25] Yonaha, K.; Toyama, S.: γ-Aminobutyrate:α-ketoglutarate aminotransferase from Pseudomonas sp. F^-126: purification, crystallization, and enzymologic properties. Arch. Biochem. Biophys., 200, 156-164 (1980)

[26] Jung, M.J.; Metcalf, B.W.: Catalytic inhibition of γ-aminobutyric acid - α-ketoglutarate transaminase of bacterial origin by 4-aminohex-5-ynoic acid, a substrate analog. Biochem. Biophys. Res. Commun., 67, 301-306 (1975)

[27] Park, J.; Osei, Y.D.; Churchich, J.E.: Isolation and characterization of recombinant mitochondrial 4-aminobutyrate aminotransferase. J. Biol. Chem., 268, 7636-7639 (1993)

[28] Yonaha, K.; Suzuki, K.; Toyama, S.: 4-Aminobutyrate:2-oxoglutarate aminotransferase of Streptomyces griseus: purification and properties. Eur. J. Biochem., 146, 101-106 (1985)

[29] Tokunaga, M.; Nakano, Y.; Kitaoka, S.: Subcellualr localization of the GABA-shunt enzymes in Euglena gracilis strain Z. J. Protozool., 26, 471-473 (1979)

[30] Tamaki, N.; Fujimoto, S.; Mizota, C.; Kikugawa, M.: Identity of β-alanine-oxo-glutarate aminotransferase and L-β-aminoisobutyrate aminotransferase in rat liver. Biochim. Biophys. Acta, 925, 238-240 (1987)

[31] Kaneko, M.; Kontani, Y.; Kikugawa, M.; Tamaki, N.: Inhibition of D-3-aminoisobutyrate-pyruvate aminotransferase by 5-fluorouracil and α-fluoro-β-alanine. Biochim. Biophys. Acta, **1122**, 45-49 (1992)

[32] Cavallotti, D.; Artico, M.; De Santis, S.; Cavallotti, C.: Occurrence of γ-aminobutyric acid-transaminase activity in nerve fibers of human thymus. Hum. Immunol., **60**, 1072-1079 (1999)

[33] Tunnicliff, G.; Crites, G.J.: Chemical inactivation of bacterial GABA aminotransferase. Biochem. Mol. Biol. Int., **46**, 43-54 (1998)

[34] Wu, Y.; Wang, W.; Richerson, G.B.: GABA transaminase inhibition induces spontaneous and enhances depolarization-evoked GABA efflux via reversal of the GABA transporter. J. Neurosci., **21**, 2630-2639 (2001)

[35] Sung, B.K.; Kim, Y.T.: Structural arrangement for functional requirements of brain recombinant 4-aminobutyrate aminotransferase. J. Biochem. Mol. Biol., **33**, 43-48 (2000)

[36] Cavalotti, C.; Artico, M.; De Santis, S.: Occurence of GABA transaminase in the thymus gland of juvenile and aged rats. Eur. J. Histochem., **43**, 293-299 (1999)

[37] Fu, M.; Nikolic, D.; Van Breemen, R.B.; Silverman, R.B.: Mechanism of inactivation of γ-aminobutyric acid aminotransferase by (S)-4-amino-4,5-dihydro-2-thiophenecarboxylic acid. J. Am. Chem. Soc., **121**, 7751-7759 (1999)

[38] Choi, S.; Storici, P.; Schirmer, T.; Silverman, R.B.: Design of a conformationally restricted analogue of the antiepilepsy drug Vigabatrin that directs its mechanism of inactivation of γ-aminobutyric acid aminotransferase. J. Am. Chem. Soc., **124**, 1620-1624 (2002)

[39] Jeon, S.G.; Bahn, J.H.; Jang, J.S.; Park, J.; Kwon, O.S.; Cho, S.W.; Choi, S.Y.: Human brain GABA transaminase tissue distribution and molecular expression. Eur. J. Biochem., **267**, 5601-5607 (2000)

[40] Storici, P.; Capitani, G.; De Biase, D.; Moser, M.; John, R.A.; Jansonius, J.N.; Schirmer, T.: Crystal structure of GABA-aminotransferase, a target for antiepileptic drug therapy. Biochemistry, **38**, 8628-8634 (1999)

[41] Kontani, Y.; Sakata, S.F.; Matsuda, K.; Ohyama, T.; Sano, K.; Tamaki, N.: The mature size of rat 4-aminobutyrate aminotransferase is different in liver and brain. Eur. J. Biochem., **264**, 218-222 (1999)

[42] Qiu, J.; Pingsterhaus, J.M.; Silverman, R.B.: Inhibition and substrate activity of conformationally rigid vigabatrin analogues with γ-aminobutyric acid aminotransferase. J. Med. Chem., **42**, 4725-4728 (1999)

[43] Kontani, Y.; Kawasaki, S.; Kaneko, M.; Matsuda, K.; Sakata, S.F.; Tamaki, N.: Inhibitory effect of ethanol administration on β-alanine-2-oxoglutarate aminotransferase (GABA aminotransferase) in disulfiram-pretreated rats. J. Nutr. Sci. Vitaminol., **44**, 165-176 (1998)

[44] Koo, Y.K.; Nandi, D.; Silverman, R.B.: The multiple active enzyme species of γ-aminobutyric acid aminotransferase are not isozymes. Arch. Biochem. Biophys., **374**, 248-254 (2000)

[45] Kim, Y.T.; Song, Y.H.; Churchich, J.E.: Recombinant brain 4-aminobutyrate aminotransferases overexpression, purification, and identification of Lys-330 at the active site. Biochim. Biophys. Acta, **1337**, 248-256 (1997)

[46] Prell, J.; Boesten, B.; Poole, P.; Priefer, U.B.: The Rhizobium leguminosarum bv. viciae VF39 γ-aminobutyrate (GABA) aminotransferase gene (gabT) is induced by GABA and highly expressed in bacteroids. Microbiology, **148**, 615-623 (2002)

[47] Schor, D.S.; Struys, E.A.; Hogema, B.M.; Gibson, K.M.; Jakobs, C.: Development of a stable-isotope dilution assay for γ-aminobutyric acid (GABA) transaminase in isolated leukocytes and evidence that GABA and β-alanine transaminases are identical. Clin. Chem., **47**, 525-531 (2001)

[48] Van Cauwenberghe, O.R.; Shelp, B.J.: Biochemical characterization of partially purified gaba:pyruvate transaminase from Nicotiana tabacum. Phytochemistry, **52**, 575-581 (1999)

[49] Kang, T.C.; Park, S.K.; Bahn, J.H.; Chang, J.S.; Cho, S.W.; Choi, S.Y.; Won, M.H.: Comparative studies on the GABA-transaminase immunoreactivity in rat and gerbil brains. Mol. Cells, **11**, 321-325 (2001)

[50] Sung, B.K.; Cho, J.J.; Kim, Y.T.: Functional expression and characterization of C-terminal mutant of 4-aminobutyrate aminotransferase. J. Biochem. Mol. Biol., **32**, 181-188 (1999)

[51] Jeon, S.G.; Bahn, J.H.; Jang, J.S.; Jang, S.H.; Lee, B.R.; Lee, K.S.; Park, J.; Kang, T.C.; Won, M.H.; Kim, H.B.; Kwo, O.S.; Cho, S.W.; Choi, S.Y.: Molecular cloning and functional expression of bovine brain GABA transaminase. Mol. Cells, **12**, 91-96 (2001)

Tyrosine-pyruvate transaminase

2.6.1.20

1 Nomenclature

EC number
2.6.1.20 (deleted)

Recommended name
tyrosine-pyruvate transaminase

D-Alanine transaminase 2.6.1.21

1 Nomenclature

EC number
2.6.1.21

Systematic name
D-alanine:2-oxoglutarate aminotransferase

Recommended name
D-alanine transaminase

Synonyms
D-alanine aminotransferase
D-aspartate transaminase
D-aspartic aminotransferase
aminotransferase, D-alanine

CAS registry number
37277-85-3

2 Source Organism

<1> *Bacillus sphaericus* [4, 5, 7, 9]
<2> *Bacillus subtilis* [1, 2, 5]
<3> *Pisum sativum* (var Alaska [3]) [3]
<4> *Rhizobium japonicum* [6]
<5> *Bacillus sp.* (YM-1, thermophile [8]) [8, 9, 10, 11, 12, 13, 14, 15, 16]

3 Reaction and Specificity

Catalyzed reaction
D-alanine + 2-oxoglutarate = pyruvate + D-glutamate (<4> ping pong mechanism [6]; <5> proposed reaction mechanism [16])

Reaction type
amino group transfer

Natural substrates and products
S D-alanine + 2-oxoglutarate <4> (<4> function of the enzyme is probably the provision of D-amino acids for cell-wall synthesis [6]) [6]
P pyruvate + D-glutamate

Substrates and products

S D-2-aminobutanoate + pyruvate <1-3> (Reversibility: r <2> [1]; ? <1,3> [3,7]) [1, 3, 7]

P 2-oxobutanoate + D-alanine <1-3> [1, 3, 7]

S D-alanine + 2-oxobutanoate <2> (Reversibility: r <2> [1]) [1]

P pyruvate + 2-aminobutanoate <2> [1]

S D-alanine + 2-oxoglutarate <1-5> (<2> D-alanine formation is favoured [1]; <5> very low activity with: D-histidine, D-phenylalanine, D-arginine and D-lysine, amino acceptors 2-oxoglutarate and pyruvate [11]) (Reversibility: r <1-5> [1, 3, 6, 7, 8]) [1, 3, 6, 7, 8, 11]

P pyruvate + D-glutamate <1-5> [1, 3, 6, 7, 8, 11]

S D-alanine + 2-oxoisovalerate <5> (Reversibility: ? <5> [10]) [10]

P pyruvate + D-valine <5> [10]

S D-alanine + pyruvate <2> (Reversibility: ? <2> [2]) [2]

P pyruvate + D-alanine <2> [2]

S D-aminobutanoate + 2-oxoglutarate <1-3> (<1> 97% of the activity with D-alanine [7]) (Reversibility: ? <1-3> [1, 3, 7]) [1, 3, 7]

P oxobutanoate + D-glutamate <1-3> [1, 3, 7]

S D-arginine + 2-oxoglutarate <1> (<1> 6% of the activity with D-alanine [7]) (Reversibility: ? <1> [7]) [7]

P ? + D-glutamate <1> [7]

S D-asparagine + 2-oxoglutarate <1, 2, 5> (<1> 31% of the activity with D-alanine [7]; <5> 38% of activity with D-alanine [11]) (Reversibility: ? <1, 2, 5> [1, 7, 11]) [1, 7, 11]

P 2-oxosuccinamate + D-glutamate <1, 2, 5> [1, 7, 11]

S D-asparagine + pyruvate <1, 2, 5> (Reversibility: ? <1,2,5> [1,7,11]) [1, 7, 11]

P 2-oxosuccinamate + D-alanine <1, 2, 5> [1, 7, 11]

S D-aspartate + 2-oxoglutarate <1, 2> (<1> 30% of the activity with D-alanine [7]) (Reversibility: ? <1,2> [1,7]) [1, 7]

P 2-oxosuccinate + D-glutamate <1, 2> [1, 7]

S D-aspartate + pyruvate <1-3> (Reversibility: ? <1-3> [1,3,7]) [1, 3, 7]

P 2-oxosuccinate + D-alanine <1-3> [1, 3, 7]

S D-ethionine + 2-oxoglutarate <1> (<1> 76% of the activity with D-alanine [7]) (Reversibility: ? <1> [7]) [7]

P 4-ethylsulfanyl-2-oxobutanoate + D-glutamate <1> [7]

S D-glutamate + 2-oxobutanoate <1, 2> (Reversibility: ? <1,2> [1,7]) [1, 7]

P 2-oxoglutarate + 2-aminobutanoate <1, 2> [1, 7]

S D-glutamate + 2-oxoglutarate <2, 4, 5> (Reversibility: ? <2, 4, 5> [1, 6, 11]) [1, 6, 11]

P 2-oxoglutarate + D-glutamate <2, 4, 5> [1, 6, 11]

S D-glutamate + pyruvate <1-3, 5> (Reversibility: r <1-3, 5> [1, 3, 7, 11]) [1, 3, 7, 11]

P 2-oxoglutarate + D-alanine <1-3, 5> [1, 3, 7, 11]

S D-glutamine + 2-oxoglutarate <1, 5> (<1> 52% of the activity with D-alanine [7]; <5> 25% of activity with d-alanine [11]) (Reversibility: ? <1,5> [7,11]) [7, 11]

P 4-carbamoyl-2-oxobutanoate + D-glutamate <1, 5> [7, 11]

S D-glutamine + pyruvate <5> (Reversibility: ? <5> [11]) [11]

P 4-carbamoyl-2-oxobutanoate + D-alanine <5> [11]

S D-histidine + 2-oxoglutarate <1> (<1> 6% of the activity with D-alanine [7]) (Reversibility: ? <1> [7]) [7]

P 3-(1H-imidazol-4-yl)-2-oxopropanoate + L-glutamate <1> [7]

S D-leucine + 2-oxoglutarate <1> (<1> 8% of the activity with D-alanine [7]) (Reversibility: ? <1> [7]) [7]

P 4-methyl-2-oxopentanoate + D-glutamate <1> [7]

S D-lysine + pyruvate <1, 3> (<1, 3> weak activity [3,7]) (Reversibility: ? <1,3> [3, 7]) [3, 7]

P ? + D-alanine <1, 3> [3, 7]

S D-methionine + 2-oxoglutarate <1-3, 5> (<2,3> weak activity [1,3]; <1> 61% of the activity with D-alanine [7]; <5> 30% of activity with d-alanine [11]) (Reversibility: ? <1-3, 5> [1, 3, 7, 11]) [1, 3, 7, 11]

P 4-methylsulfanyl-2-oxobutanoate + D-glutamate <1-3, 5> [1, 3, 7, 11]

S D-methionine + pyruvate <1-3, 5> (Reversibility: ? <1-3, 5> [1, 3, 7, 11]) [1, 3, 7, 11]

P 4-methylsulfanyl-2-oxobutanoate + L-alanine <1-3, 5> [1, 3, 7, 11]

S D-norleucine + 2-oxoglutarate <1, 5> (<1> 46% of the activity with D-alanine [7]; <5> 7% of activity with D-alanine [11]) (Reversibility: ? <1,5> [7,11]) [7, 11]

P 2-oxopentanoate + D-glutamate <1, 5> [7, 11]

S D-norleucine + pyruvate <5> (Reversibility: ? <5> [11]) [11]

P 2-oxopentanoate + D-alanine <5> [11]

S D-norvaline + 2-oxoglutarate <1, 2> (<1> 83% of the activity with D-alanine [7]) (Reversibility: ? <1,2> [1,7]) [1, 7]

P 2-oxopentanoate + D-glutamate <1, 2> [1, 7]

S D-norvaline + pyruvate <1, 2> (Reversibility: ? <1,2> [1,7]) [1, 7]

P 2-oxopentanoate + D-alanine <1, 2> [1, 7]

S D-ornithine + 2-oxoglutarate <1, 2, 5> (<2> weak activity [1]; <1> 16% of activity with D-alanine [7]; <5> 65 of activity with D-alanine [11]) (Reversibility: ? <1, 2, 5> [1, 7, 11]) [1, 7, 11]

P ? + D-glutamate <1, 2, 5> [1, 7, 11]

S D-ornithine + pyruvate <1, 2, 5> (Reversibility: ? <1, 2, 5> [1, 7, 11]) [1, 7, 11]

P ? + L-alanine <1, 2, 5> [1, 7, 11]

S D-phenylalanine + 2-oxoglutarate <1, 3> (<1> 32% of the activity with D-alanine [7]) (Reversibility: ? <1,3> [3,7]) [3, 7]

P phenylpyruvate + D-glutamate <1, 3> [3, 7]

S D-phenylalanine + pyruvate <1, 3> (<,3> weak activity [3,7]) (Reversibility: ? <1,3> [3,7]) [3, 7]

P phenylpyruvate + D-alanine <1, 3> [3, 7]

S D-serine + 2-oxoglutarate <2, 5> (<2> weak activity [1]; <5> 8% of activity with d-alanine [11]) (Reversibility: ? <2,5> [1,11]) [1, 11]

P 3-hydroxy-2-oxopropanoate + D-glutamate <2, 5> [1, 11]

S D-serine + pyruvate <5> (Reversibility: ? <5> [11]) [11]

P   3-hydroxy-2-oxopropanoate + D-alanine <5> [11]
S   D-theanine + 2-oxoglutarate <1> (<1> 40% of the activity with D-alanine
       [7]) (Reversibility: ? <1> [7]) [7]
P   ? + D-glutamate <1> [7]
S   D-tryptophan + 2-oxoglutarate <1, 3> (<1> 4% of the activity with D-
       alanine [7]) (Reversibility: ? <1,3> [3,7]) [3, 7]
P   3-indole-2-oxopropanoate + D-glutamate <1, 3> [3, 7]
S   D-tryptophan + pyruvate <1, 3> (<1> weak activity [7]) (Reversibility: ?
       <1,3> [3,7]) [3, 7]
P   3-indole-2-oxopropanoate + D-alanine <1, 3> [3, 7]
S   D-valine + 2-oxoglutarate <5> (<5> 5% of activity with D-alanine [11])
       (Reversibility: ? <5> [11]) [11]
P   2-oxoisovalerate + D-glutamate <5> [11]
S   D-valine + pyruvate <5> (Reversibility: ? <5> [11]) [11]
P   2-oxoisovalerate + D-alanine <5> [11]
S β-chloro-D-alanine <1> (<1> loss in enzyme activity during β-elimina-
       tion [4]) (Reversibility: ? <1> [4]) [4]
P   pyruvate + ammonia + Cl⁻ <1> [4]

Inhibitors

2-amino-3-butenoate <1, 2> (<1,2> i.e. vinylglycine, little if any inactivation
in absence of 2-oxoglutarate as cosubstrate, in presence of 2-oxoglutarate
pseudo-first order kinetics inactivation, inactivation mechanism [5]; <2> D-
alanine in a 10fold excess over vinylglycin affords a 70% protection against
inactivation [5]) [5]
3-methyl-2-benzothiazolone hydrazone hydrochloride <1> [7]
$CaCl_2$ <5> (<5> 45 mM, 50% inhibition [16]) [16]
D-2-aminobutanoate <2> (<2> inhibition of D- and L-transamination of D-
alanine to 2-oxoglutarate, weak [2]) [2]
D-asparagine <2> (<2> inhibition of D- and L-transamination of D-alanine to
α-ketoglutarate, weak [2]) [2]
D-aspartate <2> (<2> inhibition of D- and L-transamination of D-alanine to
2-oxoglutarate, weak [2]) [2]
D-cycloserine <1, 3> [3, 7]
D-cysteine <2> (<2> 0.1 mM, 60% inhibition [2]) [2]
D-glutamate <2> (<2> competitive to D-alanine, noncompetitive to 2-oxoglu-
tarate [2]) [2]
D-methionine <2> (<2> inhibition of D- and L-transamination of D-alanine
to 2-oxoglutarate, weak [2]) [2]
D-oxamycin <2> (<2> 0.000025 mM, 50% inhibition [2]) [2]
D-penicillamine <1> (<1> 1 mM, 73% inhibition [7]) [7]
D-serine <2> (<2> inhibition of D- and L-transamination of D-alanine to 2-
oxoglutarate [2]) [2]
DL-α-methylserine <2> (<2> inhibition of D- and L-transamination of D-ala-
nine to α-ketoglutarate, weak [2]) [2]
$HgCl_2$ <1> (<1> 0.1 mM, complete inhibition [7]) [7]
KCl <5> (<5> 80 mM, 50% inhibition [16]) [16]

L-cysteine <2> (<2> 2 mM, 49% inhibition [2]) [2]
L-oxamycin <2> (<2> 0.001 mM, 50% inhibition [2]) [2]
L-penicillamine <1> (<1> 1 mM, 28% inhibition [7]) [7]
L-serine <2> (<2> 10 mM, 22% inhibition [2]) [2]
N-ethylmaleimide <1> (<1> 1 mM, 50% inhibition [7]) [7]
NH$_4$Cl <5> (<5> 80 mM, 50% inhibition [16]) [16]
Na$_2$SO$_4$ <5> (<5> 40 mM, 50% inhibition [16]) [16]
NaCl <5> (<5> 75 mM, 50% inhibition [16]) [16]
Tris-Cl <5> (<5> 100 mM, 50% inhibition [16]) [16]
aminooxyacetic acid <2> (<2> 0.0001 mM, 50% inhibition [2]) [2]
aminoxyacetate <1> (<1> 0.01 mM, 90% inhibition [7]) [7]
β-chloro-D-alanine <1> (<1> catalyzes α,β elimination to yield pyruvate, chloride and ammonia, loss of activity during β-elimination in presence of D-alanine and 2-oxoglutarate, a potent competitive inhibitor vs. D-alanine [4]; <1> 0.025, 56% inhibition [7]) [4, 7]
cysteine <2> (<2> inhibition of D- and L-transamination of D-alanine to 2-oxoglutarate [2]) [2]
hydroxylamine <1-3> (<2> 0.1 mM, 91% inhibition [2]; <3> 1 mM, 74% inhibition [3]; <1> 0.01 mM, 61% inhibition [7]) [2, 3, 7]
maleate <2> (<2> 12 mM, 30% inhibition at pH 7.1, no inhibition at pH 8.3 [2]) [2]
oxamycin <2> (<2> D-isomer by a factor of 40 more inhibitory than L-isomer [2]) [2]
p-chloromercuribenzoate <4> (<4> 4 mM, 36% inhibition, D-alanine and pyridoxal 5'-phosphate protect to some extent [6]) [6]
phenylhydrazine <1, 3> (<3> 1 mM; 30% inhibition [3]; <1> 2 mM, 85% inhibition [7]) [3, 7]
succinate <2> (<2> 12 mM, 18% inhibition at pH 7.1, no inhibition at pH 8.3 [2]) [2]
Additional information <2, 5> (<2> not inhibited by malic acid, KCN, iodoacetic acid, p-chloromercuribenzoate and HgCl$_2$ [2]; <5> L201A and L201W mutant enzymes lose their activity by incubation with D-alanine with biphasic kinetics [8]) [2, 8]

Cofactors/prosthetic groups

pyridoxal 5'-phosphate <1-3, 5> (<1,2> a pyridoxal phosphate protein [5,7]; <3> addition of pyridoxal phosphate to the final preparation slightly stimulates [3]; <1> 2 mol of pyridoxal 5'-phosphate per mol of enzyme [7]; <2> holoenzyme contains 1 mol of vitamin B6 per molecule of enzyme, 25-50% of maximal activity in the absence of pyridoxal 5-phosphate [1]; <5> wild-type, L201A and L201W mutant enzyme: 0.89, 0.78 and 0.56 mol per mol of subunit, respectively [8]) [1, 3, 5, 7, 8]

Activating compounds

pyridoxamine phosphate <2> [1]
Additional information <2> (<2> not activated by pyridoxamine [1]) [1]

Turnover number (min^{-1})

7.2 <5> (D-alanine, <5> pH 8.0, 37°C, L201W mutant enzyme [8]) [8]

30 <5> (D-alanine, <5> pH 8.5, 50°C, Y31A mutant enzyme [10]) [10]

78 <5> (D-alanine, <5> pH 8.5, 50°C, E32A mutant enzyme [10]) [10]

342 <5> (D-alanine, <5> pH 8.0, 37°C, L201A mutant enzyme [8]) [8]

804 <5> (D-alanine, <5> pH 8.5, 50°C, E32Q mutant enzyme [10]) [10]

5058 <5> (D-alanine, <5> pH 8.5, 50°C, K35A mutant enzyme [10]) [10]

5136 <5> (D-alanine, <5> pH 8.5, 50°C, E32D mutant enzyme [10]) [10]

7872 <5> (D-alanine, <5> pH 8.5, 50°C, wild-type enzyme [10]) [10]

8706 <5> (D-alanine, <5> pH 8.5, 50°C, V33A mutant enzyme [10]) [10]

15600 <5> (2-oxoglutarate, <5> pH 8.5, 50°C [11]) [11]

16800 <5> (D-alanine, <5> pH 8.0, 37°C, wild-type enzyme [8]) [8]

20400 <5> (D-alanine, <5> pH 8.5, 50°C [11]) [11]

Specific activity (U/mg)

0.088 <3> [3]

90 <2> [1]

K$_m$-Value (mM)

0.06 <4> (2-oxoglutarate, <4> pH 7.9, 30°C [6]) [6]

0.63 <5> (2-oxoglutarate, <5> pH 8.0, 37°C, wild-type enzyme [8]) [8]

0.67 <5> (2-oxoglutarate, <5> pH 8.0, 37°C, L201W mutant enzyme [8]) [8]

1.2 <5> (D-alanine, <5> pH 8.0, 37°C, L201W mutant enzyme [8]; pH 8.5, 50°C, E32D mtant enzyme [10]) [8, 10]

1.3 <1> (D-alanine, <1> pH 8.0, 37°C, cosubstrate 2-oxobutanoate [7]) [7]

1.4 <5> (2-oxoglutarate, <5> pH 8.5, 50°C, E32Q mutant enzyme [10]) [10]

1.5 <5> (D-alanine, <5> pH 8.0, 37°C, L201A mutant enzyme [8]) [8]

1.8 <5> (2-oxoglutarate, <5> pH 8.5, 50°C, E32A mutant enzyme [10]) [10]

1.8 <5> (D-alanine, <5> pH 8.5, 50°C [11]) [11]

2 <5> (2-oxoglutarate, <5> pH 8.5, 50°C [11]) [11]

2.1 <5> (2-oxoglutarate, <5> pH 8.5, 50°C, E32D mutant enzyme [10]) [10]

2.1 <5> (D-alanine, <5> pH 8.5, 50°C, wild-type and E32Q mutant enzyme [10]) [10]

2.4 <5> (2-oxoglutarate, <5> pH 8.5, 50°C, Y33A mutant enzyme [10]) [10]

2.4 <4> (D-alanine, <4> pH 7.9, 30°C [6]) [6]

2.5 <5> (2-oxoglutarate, <5> pH 8.5, 50°C, wild-type enzyme [10]) [10]

2.7 <3> (D-alanine, <3> pH 8.3, 37°C [3]) [3]

3.4 <1, 2> (2-oxoglutarate, <2> cosubstrate D-alanine [2]; <1> pH 8.0, 37°C, cosubstrate D-alanine [7]) [2, 7]

4.2 <1> (D-alanine, <1> pH 8.0, 37°C, cosubstrate 2-oxoglutarate [7]) [4, 7]

4.8 <5> (2-oxoglutarate, <5> pH 8.0, 37°C, L201A mutant enzyme [8]) [8]

4.9 <3> (2-oxoglutarate, <3> pH 8.3, 37°C [3]) [3]

10.1 <5> (2-oxoglutarate, <5> pH 8.5, 50°C, Y31A mutant enzyme [10]) [10]

14 <1> (2-oxobutanoate, <1> cosubstrate D-alanine [7]) [7]

15.5 <5> (2-oxoisovalerate, <5> pH 8.5, 50°C, V33A mutant enzyme [10]) [10]

20 <5> (D-alanine, <5> pH 8.0, 37°C, wild-type enzyme [8]) [8]

28 <5> (2-oxoisovalerate, <5> pH 8.5, 50°C, wild-type enzyme [10]) [10]

32.2 <5> (2-oxoisovalerate, <5> pH 8.5, 50°C, K35A mutant enzyme [10])
[10]
35.6 <5> (2-oxoglutarate, <5> pH 8.5, 50°C, K35A mutant enzyme [10]) [10]

K_i-Value (mM)
0.000011 <1> (D-cycloserine, <1> pH 8.0, 37°C [7]) [7]
0.00225 <1> (β-chloro-D-alanine, <1> pH 8.0, 37°C [7]) [7]
0.01 <1> (β-chloro-D-alanine) [4]
0.057 <1> (D-penicillamine, <1> pH 8.0, 37°C [7]) [7]
0.8 <1> (L-penicillamine, <1> pH 8.0, 37°C [7]) [7]

pH-Optimum
7.8-7.9 <4> [6]
8.2-8.8 <2, 3> (<3> with D-alanine and 2-oxoglutarate [3]) [1, 3]

Temperature optimum (°C)
30 <4> (<4> assay at [6]) [6]
37 <1, 3> (<1,3> assay at [3,4,7]) [3, 4, 7]

4 Enzyme Structure

Molecular weight
53000 <2> (<2> sedimentation equilibrium centrifugation [1]) [1]
58000 <1> (<1> sedimentation equilibrium centrifugation [7]) [7]
60000 <1, 4> (<4> gel filtration [6]; <1> gel filtration [7]) [6, 7]
65000 <3> (<3> gel filtration [3]) [3]

Subunits
dimer <1, 5> (<1> 2 * 30000, SDS-PAGE [4]; <5> 2 * 32500 [9]) [4, 7, 9]

5 Isolation/Preparation/Mutation/Application

Source/tissue
seedling <3> [3]

Purification
<1> [9]
<2> (protamine sulfate, ammonium sulfate, DEAE-Sephadex, hydroxylapatite
[1]) [1]
<3> (ammonium sulfate, DEAE-Sephadex, partial purification [3]) [3]
<4> (DEAE-cellulose, Sephadex G-200, partial purification [6]) [6]
<5> (wild-type, L201A and L201W mutant enzyme, DEAE-toyopearl [8]; re-
combinant wild-type and mutant enzymes, Resource Q, phenyl-superose
[10]) [8, 10]

Crystallization

<1> [7]

<5> (crystallized from solutions containing 0.05 mM pyridoxal 5'-phosphate, 1-2% dioxane, 5 mM sodium azide, 18-22% poly(ethylene glycol) 4000, and 25 mM 2-(N-morpholino)ethanesulfonic acid, pH 5.5-6.5 by the hanging drop method, 1.94 A resolution [9]; pyridoxal 5'-phosphate and pyridoxamine 5'-phosphate form of L201A mutant enzyme, crystallization from solutions containing 100 mM Tris-HCl, pH 8.5-9.9, 29-33% polyethylenglycol 3350, 300-700 mM sodium acetate, and 5 mM sodium azide by the hanging drop method, crystal structures at 2.0 A resolution [13]; pyridoxal 5'-phosphate form: hanging drop method, 25% polyethylene glycol 4000, 200 mM ammonium sulfate, 100 mM sodium acetate, pH 4.6 and 1 mM 2-oxoglutarate, single yellow crystals after 2 weeks, enzyme complexed with N-(5'-phosphopyridoxyl)-D-alanine: hanging drop method, 26-28% polyethylene glycol 4000, 300 mM sodium acetate, 100 mM Tris-chloride, pH 8.5, crystal structure at 2.4 and 1.9 A resolution, respectively [14]) [9, 13, 14]

Cloning

<5> (expression of wild-type, L201A and L201W mutant enzymes in Escherichia coli [8]) [8, 9, 10, 16]

Engineering

E32A <5> (<5> 1% of wild-type activity [10]) [10]
E32D <5> (<5> 65% of wild-type activity [10]) [10]
E32Q <5> (<5> 10% of wild-type activity [10]) [10]
K33A <5> (<5> 64% of wild-type activity [10]) [10]
L201A <5> (<5> 2% of wild-type k_{cat} [8]) [8, 13]
L201W <5> (<5> 0.043% of wild-type k_{cat} [8]) [8]
P119G/R120G/P121G <5> (<5> higher activity than wild-type with both pyruvate and 2-oxoglutarate as amino acceptors and a variety of D-amino acids except for D-alanine and D-aspartate, reduced thermostability [12]) [12]
V33A <5> (<5> 110% of wild-type activity [10]) [10]
Y31A <5> (<5> 0.4% of wild-type activity [10]) [10]
Additional information <5> (<5> replacement of the loop core P119-R120-P121 with glycine chains of different lengths: 1, 3, or 5 glycines, mutant forms are much more active than the wild type enzyme in the overall reactions with various amino acids and pyruvate [15]) [15]

6 Stability

Temperature stability

50 <5> (<5> no loss of activity after 55 min [11]; <5> no loss of activity after 1.5 h [12]) [11, 12]

Storage stability

<2>, -10°C, 20 mM potassium phosphate, pH 7.0, 2 months, no loss of activity [1]

References

[1] Martinez-Carrion, M.; Jenkins, W.T.: D-Alanine-D-glutamate transaminase. I. Purification and characterization. J. Biol. Chem., **240**, 3538-3546 (1965)

[2] Martinez-Carrion, M.; Jenkins, W.T.: D-Alanine-D-glutamate transaminase. II. Inhibitors and the mechanism of transamination of D-amino acids. J. Biol. Chem., **240**, 3547-3552 (1965)

[3] Ogawa, T.; Fukuda, M.: Occurrence of D-amino acid aminotransferase in pea seedlings. Biochem. Biophys. Res. Commun., **52**, 998-1002 (1973)

[4] Soper, T.S.; Jones, W.M.; Lerner, B.; Trop, M.; Manning, J.M.: Inactivation of bacterial D-amino acid transaminase by β-chloro-D-alanine. J. Biol. Chem., **252**, 3170-3175 (1977)

[5] Soper, T.S.; Manning, J.M.; Marcotte, P.A.; Walsh, C.T.: Inactivation of bacterial D-amino acid transaminases by the olefinic amino acid D-vinylglycine. J. Biol. Chem., **252**, 1571-1575 (1977)

[6] Gosling, J.P.; Fottrell, P.F.: The partial purification and some properties of D-alanine aminotransferase from Rhizobium japonicum. Biochem. Soc. Trans., **1**, 252-254 (1973)

[7] Yonaha, K.; Misono, H.; Yamamoto, T.; Soda, K.: D-amino acid aminotransferase of Bacillus sphaericus. Enzymologic and spectrometric properties. J. Biol. Chem., **250**, 6983-6989 (1975)

[8] Kishimoto, K.; Yoshimura, T.; Esaki, N.; Sugio, S.; Manning, J.M.; Soda, K.: Role of leucine 201 of thermostable D-amino acid aminotransferase from a thermophile, Bacillus sp. YM-1. J. Biochem., **117**, 691-696 (1995)

[9] Sugio, S.; Petsko, G.A.; Manning, J.M.; Soda, K.; Ringe, D.: Crystal structure of a D-amino acid aminotransferase: how the protein controls stereoselectivity. Biochemistry, **34**, 9661-9669 (1995)

[10] Ro, H.S.; Hong, S.P.; Seo, H.J.; Yoshimura, T.; Esaki, N.; Soda, K.; Kim, H.S.; Sung, M.H.: Site-directed mutagenesis of the amino acid residues in β-strand III [Val30-Val36] of D-amino acid aminotransferase of Bacillus sp. YM-1. FEBS Lett., **398**, 141-145 (1996)

[11] Fuchikami, Y.; Yoshimura, T.; Gutierrez, A.; Soda, K.; Esaki, N.: Construction and properties of a fragmentary D-amino acid aminotransferase. J. Biochem., **124**, 905-910 (1998)

[12] Gutierrez, A.; Yoshimura, T.; Fuchikami, Y.; Soda, K.; Esaki, N.: A mutant D-amino acid aminotransferase with broad substrate specificity: construction by replacement of the interdomain loop Pro119-Arg120-Pro121 by Gly-Gly-Gly. Protein Eng., **11**, 53-58 (1998)

[13] Sugio, S.; Kashima, A.; Kishimoto, K.; Peisach, D.; Petsko, G.A.; Ringe, D.; Yoshimura, T.; Esaki, N.: Crystal structures of L201A mutant of D-amino acid aminotransferase at 2.0 Å resolution: implication of the structural role of Leu201 in transamination. Protein Eng., **11**, 613-619 (1998)

[14] Peisach, D.; Chipman, D.M.; Van Ophem, P.W.; Manning, J.M.; Ringe, D.: Crystallographic study of steps along the reaction pathway of D-amino acid aminotransferase. Biochemistry, **37**, 4958-4967 (1998)

[15] Gutierrez, A.; Yoshimura, T.; Fuchikami, Y.; Esaki, N.: Modulation of activity and substrate specificity by modifying the backbone length of the distant interdomain loop of D-amino acid aminotransferase. Eur. J. Biochem., **267**, 7218-7223 (2000)

[16] Ro, H.S.: Effects of salts on the conformation and catalytic properties of D-amino acid aminotransferase. J. Biochem. Mol. Biol., **35**, 306-312 (2002)

(S)-3-Amino-2-methylpropionate transaminase

2.6.1.22

1 Nomenclature

EC number
2.6.1.22

Systematic name
(S)-3-amino-2-methylpropanoate:2-oxoglutarate aminotransferase

Recommended name
(S)-3-amino-2-methylpropionate transaminase

Synonyms
L-3-aminoisobutyrate transaminase
L-3-aminoisobutyric aminotransferase
aminotransferase, L-3-aminoisobutyrate
β-aminobutyric transaminase
Additional information (<2> not identical with EC 2.6.1.61, may be identical with EC 2.6.1.19 [2])

CAS registry number
9031-95-2

2 Source Organism

<1> *Sus scrofa* [1]
<2> *Rattus norvegicus* (male Sprague Dawley [2]) [2, 3]

3 Reaction and Specificity

Catalyzed reaction
(S)-3-amino-2-methylpropanoate + 2-oxoglutarate = 2-methyl-3-oxopropanoate + L-glutamate (also acts on β-alanine and other ω-amino acids having carbon chains between 2 and 5, not identical with EC2.6.1.61 (R)-3-amino-2-methylproprionate transaminase)

Reaction type
amino group transfer

Natural substrates and products
  S (S)-3-amino-2-methylpropanoate + 2-oxoglutarate <2> (<2> involved in ω-amino acid metabolism in mammals [2]) [2]
  P 2-methyl-3-oxopropanoate + L-glutamate

Substrates and products

S (S)-3-amino-2-methylpropanoate + 2-oxoglutarate <1, 2> (<1, 2> i.e. L-β-aminoisobutyrate, not D-isomer [1, 2]) (Reversibility: r <1, 2> [1, 3]; ? <2> [2]) [1, 2]

P 2-methyl-3-oxopropanoate + L-glutamate <1, 2> [1, 2]

S (S)-3-amino-2-methylpropanoate + pyruvate <2> (Reversibility: r <2> [3]) [3]

P D-methylmalonate semialdehyde + L-alanine <2> [3]

S 4-aminobutanoate + 2-oxoglutarate <1, 2> (<1> transamination at 92% the rate of (S)-3-amino-2-methylpropanoate [1]; <2> transamination at 150% the rate of (S)-3-amino-2-methylpropanoate [3]) (Reversibility: r <2> [3]; ? <1> [1]) [1, 3]

P succinic semialdehyde + L-glutamate <1> [1]

S 5-aminopentanoate + 2-oxoglutarate <1, 2> (<1> transamination at 33% the rate of (S)-3-amino-2-methylpropanoate [1]; <2> transamination at 145% the rate of (S)-3-amino-2-methylpropanoate [3]) (Reversibility: r <2> [3]; ? <1> [1]) [1, 3]

P 5-oxopentanoate + L-glutamate

S β-alanine + 2-oxoglutarate <1, 2> (<1, 2> as good as (S)-3-amino-2-methylpropanoate [1-3]) (Reversibility: r <2> [3]; ? <1, 2> [1, 2]) [1, 2]

P malonic semialdehyde + L-glutamate <2> [2]

S Additional information <1, 2> (<1> no substrates are 3-aminobutanoate, 2-ethyl-β-alanine, glycine, L-α-alanine, 2-aminoisobutanoate, lysine, taurine, O-phosphoethanolamine, homocarnosine, glyoxylate, poor substrates are 6-aminohexanoate or ornithine, no substrates are pyruvate, oxaloacetate [1]; <2> no substrates are glyoxylate, pyruvate, oxaloacetate, phenylpyruvate [3]) [1, 3]

P ?

Inhibitors

(S)-3-amino-2-methylpropanoate <2> (<2> β-alanine as substrate, not (R)-isomer [2]) [2]

2-oxoglutarate <1> (<1> in excess, substrate inhibition [1]) [1]

5-fluorouracil <2> [3]

5-iodouracil <2> [3]

6-azathymine <2> [3]

6-azauracil <2> [3]

antiserum against β-alanine:2-oxoglutarate aminotransferase <2> [2]

gabaculine <2> [3]

pyridoxal 5'-phosphate <1> (<1> inhibits at high concentrations, activates at low concentrations [1]) [1]

Additional information <2> (<2> not inhibitory: 6-azauridine, 6-azauridine 5'-phosphate [3]) [3]

Cofactors/prosthetic groups

pyridoxal 5'-phosphate <1, 2> (<1> activates at low concentrations, inhibits at high concentrations [1]) [1, 2]

Specific activity (U/mg)

0.047 <2> (<2> kidney enzyme, pH 8.8, 37°C [3]) [3]
0.096 <2> (<2> brain enzyme, pH 8.8, 37°C [3]) [3]
0.133 <2> (<2> liver enzyme, pH 8.8, 37°C [3]) [3]
0.185 <1> (<1> pH 7.5, 37°C [1]) [1]

K_m-Value (mM)

1.1 <2> (β-alanine, <2> pH 8.8, 37°C [3]) [3]
2.7 <2> ((S)-3-amino-2-methylpropanoate, <2> pH 8.8, 37°C [2,3]) [2, 3]

K_i-Value (mM)

0.0071 <2> (gabaculine, <2> pH 8.8, 37°C [3]) [3]
0.7 <2> (6-azauracil, <2> pH 8.8, 37°C [3]) [3]
1.8 <2> (5-fluorouracil, <2> uncompetitive against 2-oxoglutarate, pH 8.8, 37°C [3]) [3]
1.9 <2> (5-fluorouracil, <2> competitive to β-alanine, pH 8.8, 37°C [3]) [3]
2.7 <2> ((S)-3-amino-2-methylpropanoate, <2> pH 8.8, 37°C [2]) [2]

pH-Optimum

9.1 <1> [1]
9.2 <2> [3]

pH-Range

8.5-9.5 <1> (<1> about half-maximal activity at pH 8.5 and 9.5 [1]) [1]

Temperature optimum (°C)

37 <1, 2> (<1,2> assay at [1,2]) [1, 2]

4 Enzyme Structure

Molecular weight

105000 <2> (<2> gel filtration [3]) [3]

Subunits

dimer <2> (<2> 2 x 56000, SDS-PAGE [3]) [3]

5 Isolation/Preparation/Mutation/Application

Source/tissue

brain <2> [3]
kidney <1> [1, 3]
liver <2> [2, 3]

Localization

cytoplasm <2> [2]
mitochondrion <2> [3]

Purification

<1> [1]
<2> (partial [2]) [2, 3]

6 Stability

Storage stability

<1>, -20°C, in 0.05 M phosphate buffer, pH 7.5, at least 3 weeks [1]
<2>, -25°C, in 0.1 M phosphate buffer, pH 7.0, 1mM EDTA, 2 mM 2-mercap-toethanol, 0.04 mM pyridoxal-5'-phosphate, few days, 50% loss of activity [3]
<2>, 4°C, in 0.1 M phosphate buffer, pH 7.0, 1mM EDTA, 2 mM 2-mercap-toethanol, 0.04 mM pyridoxal-5'-phosphate, 1 week, no loss of activity [3]

References

[1] Kakimoto, Y.; Kanazawa, A.; Taniguchi, K.; Sano, I.: β-Aminoisobutyrate-α-ketoglutarate transaminase in relation to β-aminoisobutyric aciduria. Biochim. Biophys. Acta, **156**, 374-380 (1968)
[2] Tamaki, N.; Fujimoto, S.; Mizota, C.; Kikugawa, M.: Identity of β-alanine-oxo-glutarate aminotransferase and L-β-aminoisobutyrate aminotransferase in rat liver. Biochim. Biophys. Acta, **925**, 238-240 (1987)
[3] Tamaki, N.; Sakata, S.F.; Matsuda, K.: Purification, properties, and sequencing of aminoisobutyrate aminotransferases from rat liver. Methods Enzymol., **324**, 376-389 (2000)

4-Hydroxyglutamate transaminase 2.6.1.23

1 Nomenclature

EC number
2.6.1.23

Systematic name
4-hydroxy-L-glutamate:2-oxoglutarate aminotransferase

Recommended name
4-hydroxyglutamate transaminase

Synonyms
4-hydroxyglutamate aminotransferase
HGA
aminotransferase, 4-hydroxyglutamate
Additional information <1> (<1> may be identical with EC 2.6.1.1 aspartate transaminase [1]) [1]

CAS registry number
37277-86-4

2 Source Organism

<1> *Rattus norvegicus* [1, 2]

3 Reaction and Specificity

Catalyzed reaction
4-hydroxy-L-glutamate + 2-oxoglutarate = 4-hydroxy-2-oxoglutarate + L-glutamate (oxaloacetate can replace 2-oxoglutarate)

Reaction type
amino group transfer

Natural substrates and products
S   4-hydroxy-L-glutamate + oxo acid <1> (<1> metabolism of 4-hydroxy-L-glutamate in rat liver [1, 2]) (Reversibility: r <1> [1, 2]) [1, 2]
P   4-hydroxy-2-oxoglutarate + L-glutamate <1> [1, 2]

Substrates and products

S 4-hydroxy-L-glutamate + 2-oxoglutarate <1> (<1> enzyme utilises ery-
    thro-L-4-hydroxyglutamate and threo-L-4-hydroxyglutamate to the same
    extent, but no D-isomers [1]) (Reversibility: r <1> [1,2]) [1, 2]

P 4-hydroxy-2-oxoglutarate + L-glutamate <1> [1, 2]

S 4-hydroxy-L-glutamate + oxaloacetate <1> (Reversibility: r <1> [1,2]) [1,
    2]

P 4-hydroxy-2-oxoglutarate + L-aspartate

S Additional information <1> (<1> substrate stereospecificity [1]; <1> pyr-
    uvate is a poor substrate [1]; <1> no activity with pyruvate, glyoxylate
    [2]) [1, 2]

P ?

Inhibitors

L-glutarate <1> (<1> 63% inhibition at 0.5 mM [1]) [1]
N-ethylmaleimide <1> (<1> 50% inhibition at 5 mM [1]) [1]
hydroxylamine <1> (<1> 70% inhibition at 1 mM [1]) [1]
maleate <1> (<1> 65% inhibition at 5 mM [1]) [1]
p-chloromercuribenzoate <1> (<1> 77% inhibition at 0.5 mM [1]) [1]
phenylhydrazine <1> (<1> 84% inhibition at 1 mM [1]) [1]

Cofactors/prosthetic groups

Additional information <1> (<1> no cofactor requirement [1]) [1]

Specific activity (U/mg)

0.49 <1> (<1> partially purified enzyme [1]) [1]

K_m-Value (mM)

0.41 <1> (2-oxoglutarate, <1> pH 8.1, 37°C [1]) [1]
5.7 <1> (4-hydroxyglutamate, <1> pH 8.0, 37°C [2]) [2]
33 <1> (L-glutamate, <1> pH 8.1, 37°C [1]) [1]
240 <1> (erythro-L-hydroxyglutamic acid, <1> pH 8.1, 37°C [1]) [1]

pH-Optimum

8 <1> [2]
8.2 <1> (<1> with erythro-L-hydroxyglutamic acid [1]) [1]

pH-Range

6.7-9.5 <1> (<1> pH 6.7: about 45% of activity maximum, pH 9.5: about 55%
of activity maximum [1]) [1]

Temperature optimum (°C)

37 <1> (<1> assay at [1,2]) [1, 2]

5 Isolation/Preparation/Mutation/Application

Source/tissue

liver <1> [1, 2]

Localization
cytosol <1> [1]

Purification
<1> (partial [1,2]; 15fold [2]) [1, 2]

References

[1] Goldstone, A.; Adams, E.: Metabolism of γ-hydroxyglutamic acid. I. Conversion to α-hydroxy-γ-ketoglutarate by purified glutamic-aspartic transaminase of rat liver. J. Biol. Chem., **237**, 3476-3485 (1962)
[2] Kuratomi, K.; Fukunaga, K.; Kobayashi, Y.: The metabolism of γ hydroxyglutamate in rat liver. II. A transaminase concerned in γ-hydroxyglutamate metabolism. Biochim. Biophys. Acta, **78**, 629-636 (1963)

Diiodotyrosine transaminase

2.6.1.24

1 Nomenclature

EC number
2.6.1.24

Systematic name
3,5-diiodo-L-tyrosine:2-oxoglutarate aminotransferase

Recommended name
diiodotyrosine transaminase

Synonyms
aminotransferase, diiodotyrosine
diiodotyrosine aminotransferase
halogenated tyrosine aminotransferase
halogenated tyrosine transaminase

CAS registry number
9033-18-5

2 Source Organism

<1> *Rattus norvegicus* [1]

3 Reaction and Specificity

Catalyzed reaction
3,5-diiodo-L-tyrosine + 2-oxoglutarate = 3,5-diiodo-4-hydroxyphenylpyruvate + L-glutamate (also acts on 3,5-dichloro-, 3,5-dibromo- and 3-iodo-L-tyrosine, thyroxine and triiodothyronine)

Reaction type
amino group transfer

Natural substrates and products
S 3,5-diiodo-L-tyrosine + 2-oxoglutarate <1> (Reversibility: ? <1> [1]) [1]
P 3,5-diiodo-4-hydroxyphenylpyruvate + L-glutamate <1> [1]

Substrates and products
S 3,5-dibromo-L-tyrosine + 2-oxoglutarate <1> (<1> at 127% the rate of 3,5-diiodo-L-tyrosine [1]) (Reversibility: ? <1> [1]) [1]
P 3,5-dibromo-4-hydroxyphenylpyruvate + L-glutamate

S   3,5-dichloro-L-tyrosine + 2-oxoglutarate <1> (<1> at 110% the rate of 3,5-diiodo-L-tyrosine [1]) (Reversibility: ? <1> [1]) [1]

P   3,5-dichloro-4-hydroxyphenylpyruvate + L-glutamate

S   3,5-diiodo-L-tyrosine + 2-oxoglutarate <1> (<1> specific for 2-oxogluta-rate as amino acceptor [1]) (Reversibility: ? <1> [1]) [1]

P   3,5-diiodo-4-hydroxyphenylpyruvate + L-glutamate <1> [1]

S   3,5-diiodo-L-tyrosine + oxaloacetate <1> (<1> 30% substitution of 2-ox-oglutarate by oxaloacetate as amino acceptor [1]) (Reversibility: ? <1> [1]) [1]

P   3,5-diiodo-4-hydroxyphenylpyruvate + L-aspartate <1> [1]

S   3-(4-fluorophenyl)alanine + 2-oxoglutarate <1> (<1> i.e. *p*-fluoro-DL-phenylalanine, at 3.1% the rate of 3,5-diiodo-L-tyrosine [1]) (Reversibil-ity: ? <1> [1]) [1]

P   (4-fluorophenyl)pyruvate + L-glutamate

S   3-iodo-L-tyrosine + 2-oxoglutarate <1> (<1> at 76% the rate of 3,5-diio-do-L-tyrosine [1]) (Reversibility: ? <1> [1]) [1]

P   3-iodo-4-hydroxyphenylpyruvate + L-glutamate

S   L-phenylalanine + 2-oxoglutarate <1> (<1> at 3.9% the rate of 3,5-diiodo-L-tyrosine [1]) (Reversibility: ? <1> [1]) [1]

P   phenylpyruvate + L-glutamate

S   L-tryptophan + 2-oxoglutarate <1> (<1> at 3.6% the rate of 3,5-diiodo-L-tyrosine [1]) (Reversibility: ? <1> [1]) [1]

P   3-indole-2-oxopropanoate + L-glutamate

S   L-tyrosine + 2-oxoglutarate <1> (<1> at 3.9% the rate of 3,5-diiodo-L-tyrosine [1]) (Reversibility: ? <1> [1]) [1]

P   (4-hydroxyphenyl)pyruvate + L-glutamate

S   Additional information <1> (<1> substrate specificity [1]; <1> no activ-ity with pyruvate as amino acceptor [1]; <1> no activity with: D-diiodo-tyrosine, tyrosine analogues with nitro, amino or hydroxy substituents in position 3 or 3 and 5 [1]) [1]

P   ?

Inhibitors
3,5-diiodo-4-hydroxyphenyl-DL-lactic acid <1> [1]
3,5-diiodo-4-hydroxyphenylacetate <1> (<1> strong inhibition, mixed type [1]) [1]
4-chloromercuribenzoate <1> [1]
4-hydroxyphenyl-DL-lactic acid <1> (<1> weak [1]) [1]
4-hydroxyphenylacetic acid <1> [1]
o-iodosobenzoate <1> [1]

Cofactors/prosthetic groups
pyridoxal 5'-phosphate <1> (<1> required [1]) [1]

Specific activity (U/mg)
0.53 <1> (<1> partially purified enzyme [1]) [1]

K$_m$-Value (mM)
1.1-1.3 <1> (3,5-diiodo-L-tyrosine, <1> pH 6.2, 37°C [1]) [1]
3.4-3.7 <1> (3,5-dibromo-L-tyrosine, <1> pH 6.5, 37°C [1]) [1]
6.1 <1> (3,5-dichloro-L-tyrosine, <1> pH 6.-L-7, 37°C [1]) [1]
6.5 <1> (3-iodo-L-tyrosine, <1> pH 7.4, 37°C [1]) [1]

pH-Optimum
6 <1> (<1> with substrate diiodo-L-tyrosine [1]) [1]
6.5 <1> (<1> with substrate dibromo-L-tyrosine, dichloro-L-tyrosine [1]) [1]
7-7.5 <1> (<1> with substrate monoiodo-L-tyrosine [1]) [1]

Temperature optimum (°C)
37 <1> (<1> assay at [1]) [1]

4 Enzyme Structure

Molecular weight
80000 <1> (<1> gel filtration [1]) [1]

5 Isolation/Preparation/Mutation/Application

Source/tissue
kidney <1> [1]

Localization
mitochondrion <1> [1]

Purification
<1> (30fold [1]) [1]

6 Stability

Storage stability
<1>, 0°C, 50 mM sodium phosphate buffer, 1 mM EDTA, 1 mM 2-mercap-
toethanol, 0.15 M NaCl, approximately 4 mg protein/ml, 40% loss of activity
after 1 week [1]

References

[1] Nakano, M.: Purification and properties of halogenated tyrosine and thyroid
hormone transaminase from rat kidney mitochondria. J. Biol. Chem., **242**,
73-81 (1967)

Thyroxine transaminase

1 Nomenclature

EC number
2.6.1.25 (deleted, included in EC 2.6.1.24)

Recommended name
thyroxine transaminase

Thyroid-hormone transaminase

1 Nomenclature

EC number
2.6.1.26

Systematic name
L-3,5,3'-triiodothyronine:2-oxoglutarate aminotransferase

Recommended name
thyroid-hormone transaminase

Synonyms
3,5-dinitrotyrosine transaminase
aminotransferase, thyroid hormone

CAS registry number
51004-29-6

2 Source Organism

<1> *Oryctolagus cuniculus* [1]

3 Reaction and Specificity

Catalyzed reaction
L-3,5,3'-triiodothyronine + 2-oxoglutarate = 3,5,3'-triiodothyropyruvate + L-glutamate

Reaction type
amino group transfer

Natural substrates and products
S L-3,5,3'-triiodothyronine + 2-oxoglutarate <1> (Reversibility: r <1> [1])
   [1]
P 3-[4-(4-hydroxy-3-iodophenoxy)-3,5-diiodophenyl]-2-oxopropanoate + L-glutamate <1> (<1> i.e. 3,5,3'-triiodothyropyruvate + L-Glu [1]) [1]

Substrates and products
S 3,5-dibromo-L-tyrosine + 2-oxoglutarate <1> (Reversibility: r <1> [1])
   [1]
P 3-(3,5-dibromo-4-hydroxyphenyl)-2-oxopropanoate + L-glutamate

S 3,5-diiodo-D-tyrosine + 2-oxoglutarate <1> (<1> very low activity [1]) (Reversibility: r <1> [1]) [1]

P 3-(3,5-diiodo-4-hydroxyphenyl)-2-oxopropanoate + L-glutamate

S 3,5-diiodo-L-tyrosine + 2-oxoglutarate <1> (Reversibility: r <1> [1]) [1]

P 3-(3,5-diiodo-4-hydroxyphenyl)-2-oxopropanoate + L-glutamate

S 3,5-dinitro-L-tyrosine + 2-oxoglutarate <1> (Reversibility: r <1> [1]) [1]

P 3-(3,5-dinitro-4-hydroxyphenyl)-2-oxopropanoate + L-glutamate <1> (i.e. 3,5-dinitro-*p*-hydroxyphenylpyruvic acid + Glu) [1]

S 3-amino-L-tyrosine + 2-oxoglutarate <1> (<1> very low activity [1]) (Reversibility: r <1> [1]) [1]

P 3-(3-amino-4-hydroxyphenyl)-2-oxopropanoate + L-glutamate

S 3-iodo-L-tyrosine + 2-oxoglutarate <1> (Reversibility: r <1> [1]) [1]

P 3-(3-iodo-4-hydroxyphenyl)-2-oxopropanoate + L-glutamate

S 3-nitro-L-tyrosine + 2-oxoglutarate <1> (Reversibility: r <1> [1]) [1]

P 3-(3-nitro-4-hydroxyphenyl)-2-oxopropanoate + L-glutamate

S L-3,5,3'-triiodothyronine + 2-oxoglutarate <1> (Reversibility: r <1> [1]) [1]

P 3-[4-(4-hydroxy-3-iodophenoxy)-3,5-diiodophenyl]-2-oxopropanoate + L-glutamate <1> (<1> i.e. 3,5,3'-triiodothyropyruvate + L-Glu [1]) [1]

S L-3,5,3'-triiodothyronine + oxaloacetate <1> (Reversibility: r <1> [1]) [1]

P 3-[4-(4-hydroxy-3-iodophenoxy)-3,5-diiodophenyl]-2-oxopropanoate + glycine <1> (<1> i.e. 3,5,3'-triiodothyrophenylpyruvate + Gly [1]) [1]

S L-3,5,3'-triiodothyronine + pyruvate <1> (Reversibility: r <1> [1]) [1]

P 3-[4-(4-hydroxy-3-iodophenoxy)-3,5-diiodophenyl]-2-oxopropanoate + L-alanine <1> (<1> i.e. 3,5,3'-triiodothyrophenylpyruvate + L-Ala [1]) [1]

S L-phenylalanine + 2-oxoglutarate <1> (<1> low activtiy [1]) (Reversibility: r <1> [1]) [1]

P 3-(4-hydroxyphenyl)-2-oxopropanoate + L-glutamate

S L-thyroxine + 2-oxoglutarate <1> (Reversibility: r <1> [1]) [1]

P 3-[4-(4-hydroxy-3,5-diiodophenoxy)-3,5-diiodophenyl]-2-oxopropanoate + L-glutamate

S L-tryptophan + 2-oxoglutarate <1> (<1> low activtiy [1]) (Reversibility: r <1> [1]) [1]

P 3-indole-2-oxopropanoate + L-glutamate

S L-tyrosine + 2-oxoglutarate <1> (<1> low activtiy [1]) (Reversibility: r <1> [1]) [1]

P 2-oxo-3-phenylpropanoate + L-glutamate

S Additional information <1> (<1> specific for 2-oxoglutarate [1]; <1> pyruvate and oxalacetate are ineffective with 3,5-dinitro-L-tyrosine and only partially effective with iodinated tyrosines [1]) [1]

P ?

Inhibitors

p-hydroxymercuribenzoate <1> (<1> complete inhibition at 0.1 mM [1]) [1]

Cofactors/prosthetic groups

pyridoxal 5'-phosphate <1> (<1> partial requirement [1]) [1]

Activating compounds

2-mercaptoethanol <1> (<1> absolutely required [1]) [1]

Turnover number (min^{-1})

1.4 <1> (L-thyroxine, <1> pH 7.8, 37°C [1]) [1]
1.9 <1> (L-triiodothyronine, <1> pH 7.8, 37°C [1]) [1]
66 <1> (3,5-diiodo-L-tyrosine, <1> pH 7.8, 37°C [1]) [1]
221 <1> (3-iodo-L-tyrosine, <1> pH 7.8, 37°C [1]) [1]
396 <1> (3,5-diiodo-L-tyrosine, <1> pH 7.8, 37°C [1]) [1]
2391 <1> (3,5-dinitro-L-tyrosine, <1> pH 7.8, 37°C [1]) [1]

Specific activity (U/mg)

0.013 <1> (<1> purified enzyme, substrates L-3,5,3'-triiodothyronine + 2-oxo-glutarate [1]) [1]
15 <1> (<1> purified enzyme, substrates 3,5-dinitro-L-tyrosine + 2-oxoglu-tarate [1]) [1]

K$_m$-Value (mM)

0.034 <1> (L-triiodothyronine, <1> pH 7.8, 37°C [1]) [1]
0.2 <1> (2-oxoglutarate, <1> cosubstrate L-thyroxine, pH 7.8, 37°C [1]) [1]
0.2 <1> (2-oxoglutarate, <1> cosubstrate triiodothyronine, pH 7.8, 37°C [1]) [1]
0.9 <1> (3,5-diiodo-L-tyrosine, <1> pH 7.8, 37°C [1]) [1]
2 <1> (2-oxoglutarate, <1> cosubstrate 3,5-dinitro-L-tyrosine, pH 7.8, 37°C [1]) [1]
2 <1> (pyruvate, <1> cosubstrate L-triiodothyronine, pH 7.8, 37°C [1]) [1]
2.12 <1> (L-thyronine, <1> pH 7.8, 37°C [1]) [1]
3.8 <1> (3-iodo-L-tyrosine, <1> pH 7.8, 37°C [1]) [1]
4.4 <1> (3,5-dinitro-L-tyrosine, <1> pH 7.8, 37°C [1]) [1]

pH-Optimum

7.4-8.6 <1> (<1> broad optimum [1]) [1]

pH-Range

7-8.6 <1> (<1> pH 7.0: 30% of maximal activity with L-triiodothyronine, 15% of maximal activity with 3,5-dinitro-L-tyrosine, pH 7.4-8.6: maximal activity with triiodothyronine and 3,5-dinitro-L-tyrosine [1]) [1]

4 Enzyme Structure

Molecular weight

95000 <1> (<1> glycerol density gradient centrifugation [1]) [1]

Subunits

dimer <1> (<1> 2 * 48000, SDS-PAGE [1]) [1]

5 Isolation/Preparation/Mutation/Application

Source/tissue
  brain <1> (<1> low activity [1]) [1]
  heart <1> (<1> low activity [1]) [1]
  kidney <1> [1]
  liver <1> [1]
  lung <1> (<1> low activity [1]) [1]
  skeletal muscle <1> (<1> low activity [1]) [1]
  Additional information <1> (<1> not in serum [1]) [1]

Localization
  cytosol <1> [1]

Purification
  <1> [1]

6 Stability

Storage stability
  <1>, -20°C, 30% loss of activity per month [1]

References

[1] Soffer, R.L.; Hechtman, P.; Savage, M.: L-Triiodothyronine aminotransferase. J. Biol. Chem., **248**, 1224-1230 (1973)

Tryptophan transaminase

2.6.1.27

1 Nomenclature

EC number
2.6.1.27

Systematic name
L-tryptophan:2-oxoglutarate aminotransferase

Recommended name
tryptophan transaminase

Synonyms
5-hydroxytryptophan-ketoglutaric transaminase
L-phenylalanine-2-oxoglutarate aminotransferase
L-tryptophan aminotransferase
L-tryptophan transaminase
aminotransferase, tryptophan
hydroxytryptophan aminotransferase
tryptophan aminotransferase

CAS registry number
9022-98-4

2 Source Organism

<1> *Sus scrofa* [1]
<2> *Zea mays* (2 isozymes: L-TAT-1 and L-TAT-2, 1 D-tryptophan aminotransferase D-TAT [4]) [4]
<3> *Phaseolus aureus* [10]
<4> *Candida maltosa* [6]
<5> *Rhodosporidium toruloides* [8]
<6> *Clostridium sporogenes* (strain 175 [2]) [2]
<7> *Streptomyces griseus* (strain ATCC 12648 [3]) [3]
<8> *bacterium* (isolated from the rhizosphere of Festuca octoflora [5]) [5]
<9> *Rattus norvegicus* [7, 9]

3 Reaction and Specificity

Catalyzed reaction
L-tryptophan + 2-oxoglutarate = indolepyruvate + L-glutamate (also acts on 5-hydroxytryptophan and, to a lesser extent, on the phenyl amino acids)

Reaction type
amino group transfer

Natural substrates and products
S   L-tryptophan + 2-oxoglutarate <6-8> (<8> role in microbial synthesis of auxins, influence on plant growth and development [5]; <6> first step in metabolic path for the conversion of L-tryptophan to indolepropionate [2]; <7> initial step of biosynthetic pathway of the antibiotic indolmycin [3]) (Reversibility: ? <6-8> [2, 3, 5]) [2, 3, 5]

P   L-glutamate + 3-indole-2-oxopropanoate

Substrates and products
S   3-methyltryptophan + 2-oxoglutarate <7> (<7> can also inhibit the enzyme, stereospecific for positions 2 and 3 [3]) (Reversibility: ? <7> [3]) [3]

P   3-(3-methylindole)-2-oxopropanoate + L-glutamate

S   5-hydroxytryptophan + 2-oxoglutarate <1, 9> (<1> DL-5-hydroxytryptophan, 62% as effective as L-phenylalanine [1]) (Reversibility: ? <1, 9> [1, 9]) [1, 9]

P   3-(5-hydroxyindole)-2-oxopropanoate + L-glutamate

S   5-hydroxytryptophan + oxaloacetate <9> (Reversibility: ? <9> [9]) [9]

P   3-(5-hydroxyindole)-2-oxopropanoate + L-aspartate

S   D-tryptophan + 2-oxoglutarate <5> (<5> activity can be due to a different enzyme [8]) (Reversibility: ? <5> [8]) [8]

P   L-glutamate + 3-indole-2-oxopropanoate

S   D-tryptophan + pyruvate <2> (<2> no activity with oxaloacetate and 2-oxoglutarate as amino group acceptors [4]; <2> D-TAT [4]) (Reversibility: ? <2> [4]) [4]

P   3-indole-2-oxopropanoate + L-alanine

S   DL-*p*-fluorophenylalanine + 2-oxoglutarate <1> (41% as effective as phenylalanine [1]) (Reversibility: ? <1> [1]) [1]

P   3-(4-fluorophenyl)-2-oxopropanoate + L-glutamate

S   L-3,4-dihydroxyphenylalanine + 2-oxoglutarate <1> (<1> 46% as effective as L-phenylalanine [1]) (Reversibility: ? <1> [1]) [1]

P   3-(3,4-dihydroxyphenyl)-2-oxopropanoate + L-glutamate

S   L-aspartate + 2-oxoglutarate <9> (Reversibility: ? <9> [9]) [9]

P   oxaloacetate + L-glutamate <9> [9]

S   L-aspartate + oxaloacetate <9> (Reversibility: ? <9> [9]) [9]

P   oxaloacetate + L-aspartate <9> [9]

S   L-aspartate + phenylpyruvate <1> (<1> 10% as effective as L-glutamate [1]) (Reversibility: r <1> [1]) [1]

P   2-oxosuccinic acid + L-phenylalanine

S L-histidine + 2-oxoglutarate <1> (<1> 35% as effective as L-phenylalanine [1]) (Reversibility: ? <1> [1]) [1]

P 3-(1H-imidazol-4-yl)-2-oxopropanoate + L-glutamate

S L-phenylalanine + 2-oxoglutarate <1, 6, 7, 9> (Reversibility: r <1> [1]; ? <6, 7, 9> [2, 3, 9]) [1-3, 9]

P L-glutamate + phenylpyruvate <1> [1]

S L-phenylalanine + 2-oxosuccinic acid <1> (<1> 70% as effective as 2-oxoglutarate [1]) (Reversibility: r <1> [1]) [1]

P phenylpyruvate + L-aspartate

S L-phenylalanine + oxaloacetate <9> (Reversibility: ? <9> [9]) [9]

P phenylpyruvate + L-aspartate

S L-tryptophan + 2-oxoglutarate <1-9> (<8> specific for the substrates [5]; <1> 52% of the activity with L-phenylalanine [1]; <2> 2-oxoglutarate is more effective than pyruvate, oxaloacetate and glyoxylate [4]) (Reversibility: ? <1-9> [1-10]) [1-10]

P L-glutamate + 3-indole-2-oxopropanoate <3, 6-8> (<3, 8> i.e. indole-3-pyruvate [5, 10]) [2, 3, 5, 10]

S L-tryptophan + glyoxylate <2> (<2> isozymes L-TAT-1 and L-TAT-2 [4]) (Reversibility: ? <2> [4]) [4]

P 3-indole-2-oxopropanoate + glycine

S L-tryptophan + oxaloacetate <9> (Reversibility: ? <9> [9]) [9]

P 3-indole-2-oxopropanoate + L-aspartate

S L-tryptophan + oxaloacetate <2> (<2> isozymes L-TAT-1 and L-TAT-2 [4]) (Reversibility: ? <2> [4]) [4]

P 3-indole-2-oxopropanoate + L-aspartate

S L-tyrosine + 2-oxoglutarate <1, 6, 7, 9> (<1> 49% of the activity with L-phenylalanine [1]) (Reversibility: ? <1, 6, 7, 9> [1-3, 9]) [1-3, 9]

P L-glutamate + 3-(4-hydroxyphenyl)-2-oxopropanoate

S L-tyrosine + oxaloacetate <9> (Reversibility: ? <9> [9]) [9]

P 3-(4-hydroxyphenyl)-2-oxopropanoate + L-aspartate

S Additional information <1, 3, 4, 7> (<1, 3, 4> substrate specificity [1, 6, 10]; <1> no activity with D-phenylalanine and D-glutamic acid [1]; <7> no activity with D-tryptophan [3]) [1, 3, 6, 10]

P ?

Inhibitors

3-methyltryptophan <7> (<7> can also act as substrate [3]) [3]

4-fluorophenylalanine <1> (<1> 32 mM, 50% inhibition [1]) [1]

L-glutamate <9> (<9> inhibits activity towards tryptophan [9]) [9]

p-chloromercuribenzoate <1> (<7> no inhibition [3]; <1> reduced by preincubation with L-phenylalanine and pyridoxal phosphate and reversed by a subsequent preincubation with 2-mercaptoethanol [1]) [1]

Additional information <1, 9> (<1> reduced activity in cacodylate, Tris and borate buffers [1]; <1> not affected by metal chelators and high concentration of ammonium sulfate [1]) [1, 7]

Cofactors/prosthetic groups

pyridoxal 5'-phosphate <1, 2, 6-8> (<2,6> required for full activation [2,4]; <6> K_m: 0.00218 mM [2]; <1> pyridoxamine 5'-phosphate and pyridoxal phosphate are effective in partially restoring the apoenzyme activity and in stimulating the holoenzyme pyridoxamine phosphate, pyridoxamine phosphate is somewhat more effective than pyridoxal phosphate [1]) [1-5]

pyridoxamine 5'-phosphate <1, 6> (<6> activates [2]; <1> pyridoxamine 5'-phosphate and pyridoxal phosphate are effective in partially restoring the apoenzyme activity and in stimulating the holoenzyme pyridoxamine phosphate, pyridoxamine phosphate is somewhat more effective than pyridoxal phosphate [1]) [1, 2]

Activating compounds

2-mercaptoethanol <1> (<1> preincubation, activates [1]) [1]

phosphate <1> (<1> maximal activity in presence of phosphate buffer, in absence reaction proceeds at 50% of that observed in optimal phosphate concentration [1]) [1]

Additional information <1, 5, 6> (<6> high salt concentrations activate [2]; <1,6> no activation by: pyridoxine, pyridoxal, pyridoxamine [1,2]) [1, 2, 8]

Specific activity (U/mg)

7.04 <1> (<1> reverse reaction, purified enzyme [1]) [1]

23.6 <6> (<6> partially purified enzyme [2]) [2]

K_m-Value (mM)

0.158 <6> (2-oxoglutarate, <6> cosubstrate phenylalanine , pH 8.4, 22°C approximately [2]) [2]

0.74 <1> (2-oxoglutarate, <1> cosubstrate phenylalanine, pH 8.0, 37°C [1]) [1]

1.22 <8> (L-tryptophan, <8> pH 8.0, 40°C [5]) [5]

2.68 <6> (L-tryptophan, <6> cosubstrate tryptophan, pH 8.4, 22°C approximately [2]) [2]

3.8 <1> (L-tyrosine, <1> pH 8.0, 37°C [1]) [1]

6 <1> (L-3,4-dihydroxyphenylalanine, <1> pH 8.0, 37°C [1]) [1]

7 <1> (4-fluorophenylalanine, <1> pH 8.0, 37°C [1]) [1]

15 <1> (L-tryptophan, <1> pH 8.0, 37°C [1]) [1]

50 <1> (L-phenylalanine, <1> pH 8.0, 37°C [1]) [1]

Additional information <3, 4, 9> [6, 7, 10]

pH-Optimum

8 <1, 8, 9> (<1> assay at [1]) [1, 5, 9]

8-9 <1, 2> (<1> phenylalanine [1]; <2> both L-TAT isozymes, D-TAT [4]) [1, 4]

8.4 <6> [2]

pH-Range

7-9.3 <6> (<6> pH 7.0: about 55% of activity maximum, pH 9.3: about 60% of activity maximum [2]) [2]

Temperature optimum (°C)

22 <6> (<6> assay at [2]) [2]
30 <2> (<2> D-TAT [4]) [4]
37 <1> (<1> assay at [1]) [1]
40 <8> [5]
50-60 <2> (<2> both L-TAT isozymes [4]) [4]

4 Enzyme Structure

Molecular weight

45000 <2> (<2> isozyme form L-TAT-2, gel filtration [4]) [4]
55000 <2, 9> (<2> D-TAT, gel filtration [4]) [4, 9]
80000 <2> (<2> isozyme form L-TAT-1, gel filtration [4]) [4]
97000 <6> (<6> sucrose density gradient centrifugation [2]) [2]

5 Isolation/Preparation/Mutation/Application

Source/tissue

brain <1, 9> (<1> cortex [1]) [1, 9]
coleoptile <2> [4]
cortex <1> [1]
leaf <3> [10]
liver <9> [7]
root <3> [10]
seedling <3> [10]

Localization

cytosol <3, 9> (<9> also synaptosomal cytosol [9]) [7, 9, 10]
particle-bound <5> (<5> no sucess in solubilization of the enzyme from the particles [8]) [8]
synaptosome <9> (<9> cytosol [9]) [9]
Additional information <9> (<9> no activity in mitochondria [9]) [9]

Purification

<1> (about 900fold [1]) [1]
<2> (partial, 3 isozymes: L-TAT-1 and L-TAT-2, D-TAT [4]) [4]
<3> (28fold [10]) [10]
<6> (partial, 200fold [2]) [2]
<7> (partial, 3fold [3]) [3]
<8> (35fold [5]) [5]
<9> [9]

6 Stability

Temperature stability
0-10 <7> (<7> 12 h, complete loss of activity [3]) [3]

General stability information
<6>, low buffer concentrations: loss of activity [2]
<6>, unstable upon freezing and thawing [2]

Storage stability
<1>, -20°C, 0.40 M potassium phosphate buffer, pH 8.0, highly purified enzyme stable for at least 2 weeks [1]
<1>, -20°C, 100fold purified enzyme, 0.4 M potassium phosphate, pH 8.0, stable for at least 9 months [1]
<6>, 0°C, partially purified enzyme, stable for at least 48 h [2]
<7>, -20°C, best stored by freezing the crude cell-free extract with 10% glycerol or by freezing the 45-60% ammonium sulfate precipitate in a phosphate buffer solution [3]

References

[1] George, H.; Gabay, S.: Brain aromatic aminotransferase. I. Purification and some properties of pig brain L-phenylalanine-2-oxoglutarate aminotransferase. Biochim. Biophys. Acta, **167**, 555-566 (1968)
[2] O'Neil, S.R.; DeMoss, R.D.: Tryptophan transaminase from Clostridium sporogenes. Arch. Biochem. Biophys., **127**, 361-369 (1968)
[3] Speedie, M.K.; Hornemann, U.; Floss, H.G.: Isolation and characterization of tryptophan transaminase and indolepyruvate C-methyltransferase. Enzymes involved in indolmycin biosynthesis in Streptomyces griseus. J. Biol. Chem., **250**, 7819-7825 (1975)
[4] Koshiba, T.; Mito, N.; Miyakado, M.: L- And D-tryptophan aminotransferases from maize coleoptiles. J. Plant Res., **106**, 25-29 (1993)
[5] Frankenberger, W.T.Jr.; Poth, M.: L-Tryptophan transaminase of a bacterium isolated from the rhizosphere of Festuca octoflora (Graminae). Soil Biol. Biochem., **20**, 299-304 (1988)
[6] Bode, R.; Birnbaum, D.: Characterization of three tryptophan aminotransferases from Candida maltosa. Prog. Tryptophan Serotonin Res. Proc.-Meet. Int. Study Group Tryptophan Res. ISTRY (Schlossberger, H.G., ed.) 4th, Meeting Date1983, de Gruyter, **Berlin**, 769-772 (1984)
[7] Stanley, J.; Nicholas, A.; Thompson, I.; Pogson, C.: Tryptophan aminotransferase activity in rat liver. Prog. Tryptophan Serotonin Res., Proc.-Meet. Int. Study Group Tryptophan Res. ISTRY (Schlossberger, H.G., ed.) 4th, Meeting Date1983, de Gruyter, **Berlin**, 665-668 (1984)
[8] Lesch, T.; Bode, R.; Birnbaum, D.: Transamination of L- and D-tryptophan by a soluble and a particle-bound enzyme fraction of Rhodosporidium toruloides. Biochem. Physiol. Pflanz., **174**, 546-554 (1979)

[9] Minatogawa, Y.; Noguchi, T.; Kido, R.: Purification, characteriaztion and identification of tryptophan aminotransferase from rat brain. J. Neurochem., **27**, 1097-1101 (1976)

[10] Truelsen, T.A.: Indole-3-pyruvic acid as an intermediate in the conversion of tryptophan to indole-3-acetic acid. I. Characterization of tryptophan transaminase from mung bean seedlings. Physiol. Plant., **26**, 289-295 (1972)

Tryptophan-phenylpyruvate transaminase

1 Nomenclature

EC number
2.6.1.28

Systematic name
L-tryptophan:phenylpyruvate aminotransferase

Recommended name
tryptophan-phenylpyruvate transaminase

Synonyms
L-tryptophan-α-ketoisocaproate aminotransferase
aminotransferase, tryptophan-phenylpyruvate

CAS registry number
37277-87-5

2 Source Organism

<1> *Agrobacterium tumefaciens* [1]
<2> *Pseudomonas sp.* [2]

3 Reaction and Specificity

Catalyzed reaction
L-tryptophan + phenylpyruvate = indolepyruvate + L-phenylalanine

Reaction type
amino group transfer

Natural substrates and products
 S L-tryptophan + phenylpyruvate <1> (<1> conversion of tryptophan into
    the plant indolylacetic acid [1]) (Reversibility: ? <1> [1]) [1]
 P 3-indole-2-oxopropanoate + L-phenylalanine

Substrates and products
 S L-tryptophan + 2-oxobutanoate <2> (<2> weak activity [2]) (Reversibil-
    ity: r <2> [2]) [2]
 P 3-indole-2-oxopropanoate + 2-aminobutanoate
 S L-tryptophan + 2-oxoglutarate <1> (<2> no activity [2]) (Reversibility: r
    <1> [1]) [1]

P 3-indole-2-oxopropanoate + L-glutamate
S L-tryptophan + 2-oxohexanoate <2> (Reversibility: r <2> [2]) [2]
P 3-indole-2-oxopropanoate + L-norleucine
S L-tryptophan + 2-oxoisohexanoate <2> (Reversibility: r <2> [2]) [2]
P 3-indole-2-oxopropanoate + L-leucine
S L-tryptophan + 2-oxoisopentanoate <1, 2> (<1,2> i.e. L-Trp + dimethyl-pyruvate [1,2]) (Reversibility: r <1,2> [1,2]) [1, 2]
P 3-indole-2-oxopropanoate + L-valine
S L-tryptophan + 2-oxopentanoate <2> (Reversibility: r <2> [2]) [2]
P 3-indole-2-oxopropanoate + L-norvaline
S L-tryptophan + p-hydroxyphenylpyruvate <1> (Reversibility: r <1> [1]) [1]
P 3-indole-2-oxopropanoate + L-tyrosine
S L-tryptophan + phenylpyruvate <1, 2> (<1> phenylpyruvate is the most effective amino acceptor, in the reverse reaction phenylalanine is the most effective amino donor [1]) (Reversibility: r <1,2> [1,2]) [1, 2]
P 3-indole-2-oxopropanoate + L-phenylalanine <1, 2> (<1,2> i.e. indolepyruvate + L-Phe [1,2]) [1, 2]
S L-tryptophan + pyruvate <2> (<2> 2% of the activity with α-ketoisocaproate [2]) (Reversibility: r <2> [2]) [2]
P 3-indole-2-oxopropanoate + L-alanine
S phenylalanine + 2-oxobutanoate <2> (<2> low activity [2]) (Reversibility: r <2> [2]) [2]
P L-phenylpyruvate + 2-aminobutanoate
S phenylalanine + 2-oxohexanoate <2> (Reversibility: r <2> [2]) [2]
P L-phenylpyruvate + L-norleucine
S phenylalanine + 2-oxoisobutanoate <2> (Reversibility: r <2> [2]) [2]
P L-phenylpyruvate + L-valine
S phenylalanine + 2-oxoisohexanoate <2> (<2> i.e α-ketoisocaproate, highest activity as amino acceptor [2]) (Reversibility: r <2> [2]) [2]
P L-phenylpyruvate + L-leucine
S phenylalanine + pyruvate <2> (<2> 2% of the activity with α-ketoisocaproate [2]) (Reversibility: r <2> [2]) [2]
P L-phenylpyruvate + L-alanine
S phenylpyruvate + DL-5-hydroxytryptophan <2> (Reversibility: r <2> [2]) [2]
P L-phenylalanine + 3-(5-hydroxyindole)-2-oxopropanoate
S phenylpyruvate + DL-histidine <1> (<1> low activity [1]) (Reversibility: r <1> [1]) [1]
P L-phenylalanine + 3-(1H-imidazol-4-yl)-2-oxopropanoate
S phenylpyruvate + DL-isoleucine <1, 2> (Reversibility: r <1,2> [1,2]) [1, 2]
P L-phenylalanine + 3-methyl-2-oxopentanoate
S phenylpyruvate + DL-leucine <1, 2> (Reversibility: r <1,2> [1,2]) [1, 2]
P L-phenylalanine + 4-methyl-2-oxopentanoate
S phenylpyruvate + DL-methionine <1, 2> (<2> low activity [2]) (Reversibility: r <1,2> [1,2]) [1, 2]
P L-phenylalanine + 4-methylsulfanyl-2-oxobutanoate

S phenylpyruvate + DL-valine <1, 2> (Reversibility: r <1,2> [1,2]) [1, 2]
P L-phenylalanine + 3-methyl-2-oxobutanoate
S phenylpyruvate + L-2-iminobutanoate <2> (Reversibility: r <2> [2]) [2]
P L-phenylalanine + ?
S phenylpyruvate + L-citrulline <2> (<2> low activity [2]) (Reversibility: r <2> [2]) [2]
P L-phenylalanine + 2-oxo-5-ureido-pentanoate
S phenylpyruvate + L-norleucine <2> (Reversibility: r <2> [2]) [2]
P L-phenylalanine + 2-oxohexanoate
S phenylpyruvate + L-tyrosine <1, 2> (Reversibility: r <1,2> [1,2]) [1, 2]
P L-phenylalanine + 3-(4-hydroxyphenyl)-2-oxopropanoate
S phenylpyruvate + L-norvaline <2> (Reversibility: r <2> [2]) [2]
P L-phenylalanine + 2-oxopentanoate
S phenylpyruvate + phenylglycine <2> (Reversibility: r <2> [2]) [2]
P L-phenylalanine + phenylglyoxylate
S Additional information <1, 2> (<1> glutamic acid, aspartic acid, alanine, threonine, arginine, glutamine and serine are practically inactive [1]; <2> no activity with glutamate, 2-oxoglutarate [2]) [1, 2]
P ?

Specific activity (U/mg)

76 <2> (<2> purified enzyme [2]) [2]
Additional information <1> [1]

K_m-Value (mM)

0.03 <2> (2-oxoisohexanoate, <2> with L-tryptophan, pH 8.0, 30°C [2]) [2]
0.08 <2> (2-oxohexanoate, <2> with tryptophan, pH 8.0, 30°C [2]) [2]
0.11 <2> (2-oxoisopentanoate, <2> with L-tryptophan, pH 8.0, 30°C [2]) [2]
0.12 <2> (2-oxopentanoate, <2> with L-tryptophan, pH 8.0, 30°C [2]) [2]
0.2 <2> (indolepyruvate, <2> with L-leucine, pH 8.0, 30°C [2]) [2]
0.4 <2> (L-leucine, <2> with phenylpyruvate, pH 8.0, 30°C [2]) [2]
0.47 <2> (L-isoleucine, <2> with phenylpyruvate, pH 8.0, 30°C [2]) [2]
0.8 <2> (L-phenylalanine, <2> with 2-oxoisohexanoate, pH 8.0, 30°C [2]) [2]
1 <2> (L-tyrosine, <2> with 2-oxoisohexanoate, pH 8.0, 30°C [2]) [2]
1.4 <2> (2-oxobutanoate, <2> with L-tryptophan, pH 8.0, 30°C [2]) [2]
1.4 <2> (L-tryptophan, <2> with 2-oxoisohexanoate, pH 8.0, 30°C [2]) [2]
3.4 <2> (L-norleucine, <2> with phenylpyruvate, pH 8.0, 30°C [2]) [2]
4.5 <2> (L-norvaline, <2> with phenylpyruvate, pH 8.0, 30°C [2]) [2]
4.5 <2> (L-valine, <2> with phenylpyruvate, pH 8.0, 30°C [2]) [2]
12 <2> (pyruvate, <2> with L-tryptophan, pH 8.0, 30°C [2]) [2]
23.8 <2> (L-histidine, <2> with 2-oxoisohexanoate, pH 8.0, 30°C [2]) [2]
30 <2> (L-2-aminobutanoate, <2> with phenylpyruvate, pH 8.0, 30°C [2]) [2]
250 <2> (L-alanine, <2> with 2-oxoisohexanoate, pH 8.0, 30°C [2]) [2]

pH-Optimum

8.5 <2> [2]
9.6 <1> [1]

pH-Range

8.6-10.6 <1> (<1> pH 8.6: about 70% of activity maximum, pH 10.6: about 80% of activity maximum [1]) [1]

Temperature optimum (°C)

70 <1> (<1> increase in activity from 20°C to 70°C [1]) [1]

4 Enzyme Structure

Molecular weight

90000 <2> (<2> sedimentation equilibrium analysis [2]) [2]
110000 <2> (<2> gel filtration [2]) [2]

5 Isolation/Preparation/Mutation/Application

Purification

<1> (67.5fold [1]) [1]
<2> (to homogeneity [2]) [2]

6 Stability

Temperature stability

45 <1> (<1> 30 min, with pyridoxal 5'-phosphate, no loss of activity [1]) [1]
70 <1> (<1> 5 min, without substrate or pyridoxal 5'-phosphate, complete loss of activity [1]) [1]
Additional information <1> (<1> markedly heat-stable in presence of tryptophan, phenylpyruvate or pyridoxal 5'-phosphate [1]) [1]

General stability information

<1>, freezing causes loss of activity [1]

Storage stability

<1>, -30°C, crude enzyme, stable for more than 3 months [1]
<1>, 0-4°C, more than 72 h, partially purified enzyme loses activity [1]

References

[1] Sukanya, N.K.; Vaidyanathan, C.S.: Aminotransferases of Agrobacterium tumefaciens. Transamination between tryptophan and phenylpyruvate. Biochem. J., **92**, 594-598 (1964)
[2] Koide, Y.; Honma, M.; Shimomura, T.: L-tryptophan-α-ketoisocaproate aminotransferase from Pseudomonas sp.. Agric. Biol. Chem., **44**, 2013-2019 (1980)

Diamine transaminase

1 Nomenclature

EC number
2.6.1.29

Systematic name
diamine:2-oxoglutarate aminotransferase

Recommended name
diamine transaminase

Synonyms
amine transaminase
amine-ketoacid transaminase
aminotransferase, diamine
diamine aminotransferase
diamine-ketoglutaric transaminase

CAS registry number
9031-83-8

2 Source Organism

<1> *Escherichia coli* [1, 2]
<2> *Streptomyces clavuligerus* (strain NRRL 3585 [4]) [3, 4]
<3> *Streptomyces griseus* (strain NRRL 3851 [4]) [4]
<4> *Nocardia lactamdurans* (strain NRRL 3802 [4]) [4]
<5> *Streptomyces lividans* (strain TK24 [4]) [4]
<6> *Streptomyces phaeochromogenes* (strain B2196 [4]) [4]
<7> *Streptomyces viridochromogenes* (strain CUB416 [4]) [4]
<8> *Streptomyces glaucescens* (strain GLAO [4]) [4]
<9> *Streptomyces venezuelae* (strain ISP5230 [4]) [4]
<10> *Streptomyces parvulus* (strain ISP5048 [4]) [4]
<11> *Streptomyces rimosus* (strain NRRL 2234 [4]) [4]

3 Reaction and Specificity

Catalyzed reaction
an α,ω-diamine + 2-oxoglutarate = an ω-aminoaldehyde + L-glutamate

Reaction type
amino group transfer

Natural substrates and products

S cadaverine + 2-oxoglutarate <2-11> (<2-11> enzyme catalyzes the second step in lysine catabolism [4]) (Reversibility: ? <2-11> [4]) [4]

P 1-piperidine + L-glutamate <2-11> [4]

S putrescine + 2-oxoglutarate <1> (Reversibility: ? <1> [1, 2]) [1, 2]

P 4-aminobutanal + L-glutamate <1> [1, 2]

Substrates and products

S (+)-1,4-diaminobutan-2-ol + 2-oxoglutarate <1> (<1> 30% of the activity with putrescine [2]) [2]

P ?

S 1,6-diaminohexane + 2-oxoglutarate <1> (Reversibility: ? <1> [2]) [2]

P 6-aminohexanal + L-glutamate <1> [2]

S 1,7-diaminoheptane + 2-oxoglutarate <1> (<1> 30% of the activity with putrescine [1,2]) (Reversibility: ? <1> [1,2]) [1, 2]

P 7-aminoheptanal + L-glutamate <1> [1, 2]

S 4-aminobutanoate + 2-oxoglutarate <1> (<1> 11% of the activity with putrescine [1]) (Reversibility: ? <1> [1,2]) [1, 2]

P 4-oxobutanoate + L-glutamate <1> [1, 2]

S cadaverine + 2-oxoglutarate <1-11> (<1> equally active as putrescine [1,2]) (Reversibility: ? <1-11> [1,2,4]) [1, 2, 4]

P 1-piperidine + L-glutamate <1-11> [1, 2, 4]

S putrescine + 2-oxoglutarate <1> (Reversibility: ? <1> [1,2]) [1, 2]

P 4-aminobutanal + L-glutamate <1> [1, 2]

S putrescine + pyruvate <1> (Reversibility: ? <1> [1,2]) [1, 2]

P 4-aminobutanal + L-alanine <1> [1, 2]

S Additional information <1> (<1> inactive with oxaloacetate [1]; <1> inactive with 1,3-diaminopropane, lysine, ornithine, spermidine [1,2]; <1> inactive with histamine, tyramine, 4-aminobutanol, β-alanine [2]) [1, 2]

P ?

Inhibitors

cyanide <1> (<1> complete inhibition at 0.5 mM [2]) [2]

hydroxylamine <1> (<1> complete inhibition at 0.5 mM [2]) [2]

semicarbazide <1> (<1> complete inhibition at 0.5 mM [2]) [2]

Cofactors/prosthetic groups

pyridoxal 5'-phosphate <1> (<1> required [1]; <1> purified enzyme normally active without added pyridoxal phosphate, absolute requirement when irradiated with UV light or treated with 2 mM phenylhydrazine and then with Norite [2]) [1, 2]

Specific activity (U/mg)

0.00064 <10> [4]

0.00064 <4> [4]

0.0011 <11> [4]

0.00116 <8> [4]

0.00156 <9> [4]

0.00167 <7> [4]

0.00192 <3> [4]
0.0021 <5> [4]
0.00238 <2> [4]
0.00248 <6> [4]
1.63 <1> [1, 2]
Additional information <2> (<2> values with different nitrogen sources [3])
[3]

K$_m$-Value (mM)

0.88 <1> (2-oxoglutarate, <1> pH 7.6, room temperature [1,2]) [1, 2]
2.7 <1> (pyruvate, <1> pH 7.6, room temperature [1,2]) [1, 2]

pH-Optimum

9-10 <1> [1, 2]

pH-Range

8-10.5 <1> (<1> almost completely inactive at pH 7, pH 8 about 40% of maximum activity, pH 10.5 about 20% of maximum activity [1]) [1]

Temperature optimum (°C)

37 <2-11> (<2-11> assay at [4]) [4]

5 Isolation/Preparation/Mutation/Application

Purification

<1> (precipitation with ammonium sulphate followed by column chromatography [1,2]) [1, 2]

6 Stability

Storage stability

<1>, -15°C, 0.03 M Tris-HCl buffer, pH 7.6, 1 month [2]

References

[1] Kim, K. H.: Purification and properties of a diamine α-ketoglutarate transaminase from Escherichia coli. J. Biol. Chem., **239**, 783-786 (1964)
[2] Kim, K. H.; Tchen, T.T.: Diamine-α-ketoglutarate aminotransferase (Escherichia coli). Methods Enzymol., **17B**, 812-815 (1971)
[3] Bascarán, V.; Sánchez, L.; Hardisson, C.; Braña, A.F.: Stringent response and initiation of secondary metabolism in Streptomyces clavuligerus. J. Gen. Microbiol., **137**, 1625-1634 (1991)
[4] Madduri, K.; Stuttard, C.; Vining, L.C.: Lysine catabolism in Streptomyces spp. is primarily through cadaverine: β-lactam producers also make α-aminoadipate. J. Bacteriol., **171**, 299-302 (1989)

Pyridoxamine-pyruvate transaminase 2.6.1.30

1 Nomenclature

EC number
2.6.1.30

Systematic name
pyridoxamine:pyruvate aminotransferase

Recommended name
pyridoxamine-pyruvate transaminase

Synonyms
aminotransferase, pyridoxamine-pyruvate
pyridoxamine-pyruvate aminotransferase
pyridoxamine-pyruvic transaminase

CAS registry number
9023-38-5

2 Source Organism

<1> *Pseudomonas sp.* (strain MA-1 [1, 2, 4, 5]) [1, 2, 4, 5]
<2> *soil bacterium* [3]

3 Reaction and Specificity

Catalyzed reaction
pyridoxamine + pyruvate = pyridoxal + L-alanine (<1> mechanism [5])

Reaction type
amino group transfer

Natural substrates and products
S pyridoxamine + pyruvate <1, 2> (<1> oxaloacetate and several other keto acids tested cannot replace pyruvate as amino group acceptor [1]) (Reversibility: r <1, 2> [1-5]) [1-5]
P pyridoxal + L-alanine <1, 2> [1-5]
S Additional information <1> (<1> enzyme is part of the degradative pathway for vitamin B_6 compounds utilized by Pseudomonas sp. MA-1 [4]) [4]
P ?

Substrates and products

S 2-norpyridoxal + L-alanine <1> (Reversibility: r <1> [2]) [2]
P 2-norpyridoxamine + pyruvate <1> [2]
S 3-hydroxypyridine-4-aldehyde + L-alanine <1> (Reversibility: r <1> [2]) [2]
P 3-hydroxy-4-aminomethylpyridine + pyruvate <1> [2]
S 5-deoxypyridoxal + L-alanine <1, 2> (Reversibility: r <1> [2]; ? <2> [3]) [2, 3]
P 5-deoxypyridoxamine + pyruvate <1> [2]
S ω-methylpyridoxal + L-alanine <1> (<1> i.e. 3-hydroxy-5-hydroxy-methyl-2-ethylpyridine-4-carboxaldehyde [2]) (Reversibility: r <1> [2]) [2]
P ω-methylpyridoxamine + pyruvate <1> [2]
S pyridoxamine + pyruvate <1, 2> (<1> oxaloacetate and several other keto acids tested cannot replace pyruvate as amino group acceptor [1]) (Reversibility: r <1,2> [1-5]) [1-5]
P pyridoxal + L-alanine <1, 2> [1-5]
S Additional information <1> (<1> not active with: 3-deoxypyridoxal, pyridine-4-aldehyde, O-methylpyridoxal, 4-nitrosalicylaldehyde [2]; <1> determination of the active site stoichiometry and the pH dependence of the dissociation constant for 5'-deoxypyridoxal [4]; <1> temperature-jump and stopped-flow kinetic investigation of the rate and mechanism of the reaction of 5'-deoxy-pyridoxal with the enzyme [5]) [2, 4, 5]
P ?

Inhibitors

3-deoxypyridoxal <1> (<1> inhibits pyridoxamine formation from pyridoxal + alanine [2]) [2]

4-nitrosalicylaldehyde <1> (<1> inhibits pyridoxamine formation from pyridoxal + alanine [2]) [2]

N-methylpyridoxal <1> (<1> inhibits pyridoxamine formation from pyridoxal + alanine [2]) [2]

O-methylpyridoxal <1> (<1> inhibits pyridoxamine formation from pyridoxal + alanine [2]) [2]

pyridine-4-aldehyde <1> (<1> inhibits pyridoxamine formation from pyridoxal + alanine [2]) [2]

Additional information <1> (<1> phosphate is neither essential nor inhibitory [1]) [1]

Cofactors/prosthetic groups

Additional information <1> (<1> pyridoxal 5'-phosphate is not required for activity [1]) [1]

Activating compounds

Additional information <1> (<1> phosphate is neither essential nor inhibitory [1]) [1]

Metals, ions

K$^+$ <1> (<1> activates [4]) [4]

Turnover number (min⁻¹)

Additional information <1> [1]

Specific activity (U/mg)

7.3 <1> (<1> purified enzyme [1]) [1]

K_m-Value (mM)

0.009 <1> (5-deoxypyridoxal, <1> pH 8.9, 25°C [2]) [2]

0.012 <1> (pyridoxal, <1> pH 8.9, 25°C [2]) [2]

0.013 <1> (pyridoxamine, <1> pH 8.9, 25°C [2]) [2]

0.014 <1> (5-deoxypyridoxamine, <1> pH 8.9, 25°C [2]) [2]

0.043 <1> (pyridoxamine, <1> pH 8.5, 37°C [1]) [1]

0.1 <1> (pyruvate, <1> pH 8.5, 37°C [1]) [1]

0.16 <1> (2-norpyridoxamine, <1> pH 8.9, 25°C [2]) [2]

0.35 <1> (pyridoxamine, <1> with 3-hydroxy-4-aminomethylpyridine, pH 8.9, 25°C [2]) [2]

0.35 <1> (pyruvate, <1> with 3-hydroxy-4-aminomethylpyridine, pH 8.9, 25°C [2]) [2]

0.4 <1> (pyruvate, <1> with ω-methylpyridoxamine, pH 8.9, 25°C [2]) [2]

0.42 <1> (pyruvate, <1> with 5-deoxypyridoxamine, pH 8.9, 25°C [2]) [2]

0.52 <1> (pyruvate, <1> with 2-norpyridoxamine, pH 8.9, 25°C [2]) [2]

0.58 <1> (ʟ-alanine, <1> with ω-methylpyridoxal, pH 8.9, 25°C [2]) [2]

0.59 <1> (2-norpyridoxal, <1> pH 8.9, 25°C [2]) [2]

0.89 <1> (3-hydroxypyridine-4-aldehyde, <1> pH 8.9, 25°C [2]) [2]

1.3 <1> (ω-methylpyridoxal, <1> pH 8.9, 25°C [2]) [2]

1.6 <1> (ʟ-alanine, <1> with pyridoxal, pH 8.9, 25°C [2]) [2]

1.9 <1> (ʟ-alanine, <1> with 5-deoxypyridoxal, pH 8.9, 25°C [2]) [2]

2.7 <1> (ʟ-alanine, <1> with 2-norpyridoxal, pH 8.9, 25°C [2]) [2]

7.4 <1> (pyruvate, <1> with 3-hydroxy-4-aminomethylpyridine, pH 8.9, 25°C [2]) [2]

64.5 <1> (ʟ-alanine, <1> with 3-hydroxypyridine-4-aldehyde, pH 8.9, 25°C [2]) [2]

Additional information <1, 2> (<1> kinetics [2,4,5]; <2> effect of pH on kinetic parameters [3]) [2-5]

pH-Optimum

8.5 <1> [1]

pH-Range

7-10 <2> [3]

Temperature optimum (°C)

25 <1> (<1> assay at [5]) [5]

37 <1> (<1> assay at [1,4]) [1, 4]

5 Isolation/Preparation/Mutation/Application

Purification

<1> [1]

Crystallization

<1> (addition of saturated ammonium sulfate solution to the concentrated protein solution until the first permanent turbidity appears, several days at 5°C [1]) [1]

6 Stability

Temperature stability

70 <1> (<1> 10 min, stable [1]) [1]

General stability information

<1>, K$^+$ stabilizes [4]

Storage stability

<1>, loss of activity within 12 h at 25°C in sodium phosphate buffer, but is stable in potassium buffer under the same conditions [4]

References

[1] Wada, H.; Snell, E.E.: Enzymatic transamination pf pyridoxamine. II. Crystalline pyridoxamine-pyruvate transaminase. J. Biol. Chem., **237**, 133-137 (1962)

[2] Ayling, J.E.; Snell, E.E.: Relation of structure to activity of pyridoxal analogs as substrates for pyridoxamine pyruvate transaminase. Biochemistry, **7**, 1626-1636 (1968)

[3] Ford, J.B.: Effect of pH and temperature on kinetic parameters of pyridoxamine-pyruvate transaminase. J. Ala. Acad. Sci., **49**, 16-30 (1978)

[4] Gilmer, P.J.; McIntire, W.S.; Kirsch, J.F.: Pyridoxamine-pyruvate transaminase. 1. Determination of the active site stoichiometry and the pH dependence of the dissociation constant for 5-deoxypyridoxal. Biochemistry, **16**, 5241-5246 (1977)

[5] Gilmer, P.J.; Kirsch, J.F.: Pyridoxamine-pyruvate transaminase. 2. Temperature-jump and stopped-flow kinetic investigation of the rates and mechanism of the reaction of 5-deoxypyridoxal with the enzyme. Biochemistry, **16**, 5246-5253 (1977)

Pyridoxamine-oxaloacetate transaminase 2.6.1.31

1 Nomenclature

EC number
2.6.1.31

Systematic name
pyridoxamine:oxaloacetate aminotransferase

Recommended name
pyridoxamine-oxaloacetate transaminase

Synonyms
aminotransferase, pyridoxamine-oxalacetate
pyridoxamine-oxalacetate aminotransferase

CAS registry number
37277-88-6

2 Source Organism

<1> *Escherichia coli* [1]
<2> *Oryctolagus cuniculus* [1]
<3> *Rattus norvegicus* [2]

3 Reaction and Specificity

Catalyzed reaction
pyridoxamine + oxaloacetate = pyridoxal + L-aspartate

Reaction type
amino group transfer

Natural substrates and products
S pyridoxamine + oxaloacetate <1-3> (Reversibility: r <1-3> [1, 2]) [1, 2]
P pyridoxal + L-aspartate <1, 2> [1]

Substrates and products
S pyridoxamine + oxaloacetate <1-3> (Reversibility: r <1-3> [1, 2]) [1, 2]
P pyridoxal + L-aspartate <1, 2> [1]
S Additional information <1, 2> (<1, 2> no substrates are: pyruvate, 2-oxo-glutarate, 3-oxoglutarate, 2-oxisovalerate, 2-oxoisocaproate, 2-oxocapro-ate, phenylpyruvate, mesoxalate, ketopantoate [1]) [1]
P ?

Inhibitors

2-oxoglutarate <3> (<3> inhibition at pH 7.0 is higher than at pH 8.5, 3.33 mM in presence of equimolar concentrations of oxaloacetate [2]) [2]

2-oxomalonate <3> (<3> inhibition at pH 7.0 is higher than at pH 8.5, 0.4 mM in presence of equimolar concentrations of oxaloacetate [2]) [2]

3,3-dimethylglutamic acid <3> (<3> not at pH 7.0, weak at pH 8.0 [2]) [2]

3-oxoadipate <3> (<3> inhibition at pH 7.0 is higher than at pH 8.5, 3.33 mM in presence of equimolar concentrations of oxaloacetate [2]) [2]

L-aspartate <2, 3> (<2> no inhibition by D-aspartate [1]) [1, 2]

L-glutamate <2, 3> (<2> no inhibition by D-glutamate [1]) [1, 2]

adipic acid <3> (<3> not at pH 7.0, weak at pH 8.0 [2]) [2]

α-glycerophosphate <1, 2> [1]

benzylmalonic acid <3> (<3> not at pH 7.0, weak at pH 8.0 [2]) [2]

β-glycerophosphate <1, 2> [1]

diethylstilbestrol diphosphate <3> [2]

pyridoxal 5'-phosphate <1-3> [1, 2]

pyridoxamine phosphate <1-3> [1, 2]

Activating compounds

D-aspartate <3> (<3> increases activity [2]) [2]

arsenate <3> (<3> activation, less effective than phosphate [2]) [2]

diphosphate <3> (<3> activation, less effective than phosphate [2]) [2]

phosphate <3> (<3> essential, activation in absence of diethylstilbestrol and pyridoxal 5'-phosphate [2]) [2]

Specific activity (U/mg)

0.066 <3> (<3> purified enzyme [2]) [2]

0.076 <1> (<1> partially purified enzyme [1]) [1]

0.081 <2> (<2> partially purified enzyme [1]) [1]

Additional information <1, 2> [1]

K_m-Value (mM)

0.012 <3> (oxaloacetate, <3> somewhat higher at very high levels of pyridoxamine, 14 mM [2]) [2]

0.034 <1> (oxaloacetate, <1> pH 8.5, 37°C [1]) [1]

0.57 <2> (oxaloacetate, <2> pH 8.5, 37°C [1]) [1]

0.7 <2> (pyridoxamine, <2> pH 8.5, 37°C [1]) [1]

2.3 <1> (pyridoxamine, <1> pH 8.5, 37°C [1]) [1]

pH-Optimum

8.5 <1-3> [1, 2]

pH-Range

7.5-9.5 <3> (<3> about 50% of activity maximum at pH 7.5 and pH 9.5 [2]) [2]

Temperature optimum (°C)

37 <1, 2> (<1,2> assay at [1]) [1]

5 Isolation/Preparation/Mutation/Application

Source/tissue
  kidney <3> [2]
  liver <2> [1]

Purification
  <1> [1]
  <2> [1]
  <3> (979fold [2]) [2]

6 Stability

Storage stability
  <3>, frozen, 20% loss of activity after 2 months [2]

References

[1] Wada, H.; Snell, E.E.: Enzymatic transamination of pyridoxamine. I. With oxaloacetate and α-ketoglutarate. J. Biol. Chem., **237**, 127-132 (1962)
[2] Wu, H.L.C.; Mason, M.: Pyridoxamine-oxaloacetic transaminase of rat kidney. J. Biol. Chem., **239**, 1492-1497 (1964)

Valine-3-methyl-2-oxovalerate transaminase 2.6.1.32

1 Nomenclature

EC number
2.6.1.32

Systematic name
L-valine:(S)-3-methyl-2-oxopentanoate aminotransferase

Recommended name
valine-3-methyl-2-oxovalerate transaminase

Synonyms
alanine-valine transaminase
aminotransferase, valine-3-methyl-2-oxovalerate
valine-2-keto-methylvalerate aminotransferase
valine-isoleucine aminotransferase
valine-isoleucine transaminase

CAS registry number
9023-14-7

2 Source Organism

<1> *Pisum sativum* [1]

3 Reaction and Specificity

Catalyzed reaction
L-valine + (S)-3-methyl-2-oxopentanoate = 3-methyl-2-oxobutanoate + L-isoleucine

Reaction type
amino group transfer

Substrates and products
  S L-valine + (S)-3-methyl-2-oxopentanoate <1> (Reversibility: ? <1> [1])
  [1]
  P 3-methyl-2-oxobutanoate + L-isoleucine <1> [1]

5 Isolation/Preparation/Mutation/Application

Source/tissue
sprout <1> [1]

Purification
<1> (partial [1]) [1]

References

[1] Kagan, Z.S.; Dronov, A.S.; Kretovich, V.L.: Separation of valine-glutamic aminotransferase from valine-isoleucine aminotransferase in pea sprouts. Dokl. Akad. Nauk SSSR, **175**, 1171-1174 (1967)

dTDP-4-Amino-4,6-dideoxy-D-glucose transaminase

1 Nomenclature

EC number

2.6.1.33

Systematic name

dTDP-4-amino-4,6-dideoxy-D-glucose:2-oxoglutarate aminotransferase

Recommended name

dTDP-4-amino-4,6-dideoxy-D-glucose transaminase

Synonyms

TDP-4-keto-6-deoxy-D-glucose transaminase
aminotransferase, thymidine diphospho-4-amino-4,6-dideoxyglucose
thymidine diphospho-4-amino-4,6-dideoxyglucose aminotransferase
thymidine diphospho-4-amino-6-deoxyglucose aminotransferase
thymidine diphospho-4-keto-6-deoxy-D-glucose transaminase
thymidine diphospho-4-keto-6-deoxy-D-glucose-glutamic transaminase

CAS registry number

9023-19-2

2 Source Organism

<1> *Escherichia coli* (strain B [1]) [1]

3 Reaction and Specificity

Catalyzed reaction

dTDP-4-amino-4,6-dideoxy-D-glucose + 2-oxoglutarate = dTDP-4-dehydro-6-deoxy-D-glucose + L-glutamate

Reaction type

amino group transfer

Natural substrates and products

S TDP-4-keto-6-deoxy-D-glucose + L-glutamate <1> (Reversibility: r <1> [1]) [1]

P TDP-4-amino-4,6-dideoxy-D-glucose + 2-oxoglutarate <1> [1]

Substrates and products

S   TDP-4-keto-6-deoxy-D-glucose + L-glutamate <1> (<1> specific for the substrates [1]) (Reversibility: r <1> [1]) [1]

P   TDP-4-amino-4,6-dideoxy-D-glucose + 2-oxoglutarate <1> [1]

S   Additional information <1> (<1> DL-glutamine, L-glutamine, D-glutamate, L-asparagine, DL-aspartate and L-alanine are inactive as amino donor, TDP-4-amino-4,6-dideoxy-D-galactose is inactive in the reverse reaction, neither pyruvate nor oxaloacetate can replace 2-oxoglutarate [1]) [1]

P   ?

Cofactors/prosthetic groups

pyridoxal 5'-phosphate <1> (<1> absolute requirement, K_m is 0.00003 mM in reverse reaction, 0.00012 mM in forward reaction [1]) [1]

Specific activity (U/mg)

0.047 <1> [1]

K_m-Value (mM)

0.032 <1> (TDP-4-amino-4,6-dideoxy-D-glucose, <1> 38°C, pH 7.0-7.6 [1]) [1]

0.14 <1> (TDP-4-keto-6-deoxy-D-glucose, <1> 38°C, pH 7.0-7.6 [1]) [1]

0.22 <1> (2-oxoglutarate, <1> 38°C, pH 7.0-7.6 [1]) [1]

5.1 <1> (L-glutamate, <1> 38°C, pH 7.0-7.6 [1]) [1]

pH-Optimum

6-8 <1> [1]

Temperature optimum (°C)

38 <1> (<1> assay at [1]) [1]

5 Isolation/Preparation/Mutation/Application

Purification

<1> (partial from strain B, 28fold [1]) [1]

6 Stability

Storage stability

<1>, 0°C or -15°C, partially purified enzyme, stable for several weeks [1]

<1>, 0°C, crude cell extract, stable for several days [1]

References

[1] Matsuhashi, M.; Strominger, J.L.: Thymidine diphosphate 4-acetamido-2,6-dideoxyhexoses. 3. Purification and properties of thymidine diphosphate 4-keto-6-deoxy-D-glucose transaminase from Escherichia coli strain B. J. Biol. Chem., **241**, 4738-4744 (1966)

UDP-2-Acetamido-4-amino-2,4,6-trideoxyglucose transaminase

2.6.1.34

1 Nomenclature

EC number
2.6.1.34

Systematic name
UDP-2-acetamido-4-amino-2,4,6-trideoxyglucose:2-oxoglutarate aminotransferase

Recommended name
UDP-2-acetamido-4-amino-2,4,6-trideoxyglucose transaminase

Synonyms
aminotransferase, uridine diphospho-4-amino-2-acetamido-2,4,6-trideoxyglucose
uridine diphospho-4-amino-2-acetamido-2,4,6-trideoxyglucose aminotransferase

CAS registry number
37277-89-7

2 Source Organism

<1> *Diplococcus pneumoniae* (type XIV, ATCC 6314 [1]) [1]

3 Reaction and Specificity

Catalyzed reaction
UDP-2-acetamido-4-amino-2,4,6-trideoxyglucose + 2-oxoglutarate = UDP-2-acetamido-4-dehydro-2,6-dideoxyglucose + L-glutamate (A pyridoxal-phosphate protein.)

Reaction type
amino group transfer

Natural substrates and products
S UDP-2-acetamido-4-keto-2,6-dideoxyhexose + L-glutamate <1> (Reversibility: r <1> [1]) [1]
P UDP-2-acetamido-4-amino-2,4,6-trideoxyhexose + 2-oxoglutarate <1> [1]

Substrates and products

S   UDP-2-acetamido-4-keto-2,6-dideoxyhexose + L-glutamate <1> (<1> glu-
     tamate can only partially be replaced by glutamine and cannot be re-
     placed by asparagine and aspartate [1]) (Reversibility: r <1> [1]) [1]
P   UDP-2-acetamido-4-amino-2,4,6-trideoxyhexose + 2-oxoglutarate <1> [1]

Activating compounds

pyridoxal 5'-phosphate <1> (<1> absolute requirement [1]) [1]
Additional information <1> (<1> not affected by EDTA [1]) [1]

Metals, ions

Additional information <1> (<1> not affected by $MgCl_2$ at 0.01 M [1]) [1]

pH-Optimum

7.9 <1> [1]

Temperature optimum (°C)

37 <1> (<1> assay at [1]) [1]

5 Isolation/Preparation/Mutation/Application

Purification

<1> (type XIV, ATCC 6314, partial [1]) [1]

6 Stability

Storage stability

<1>, -15°C, stable for at least 3 weeks [1]

References

[1] Distler, J.; Kaufman, B.; Roseman, S.: Enzymic synthesis of a diamino sugar
    nucleotide by extracts of type XIV Diplococcus pneumoniae. Arch. Biochem.
    Biophys., **116**, 466-478 (1966)

Glycine-oxaloacetate transaminase

1 Nomenclature

EC number
2.6.1.35

Systematic name
glycine:oxaloacetate aminotransferase

Recommended name
glycine-oxaloacetate transaminase

Synonyms
GOAT <2, 3> [2, 3]
aminotransferase, glycine-oxalacetate
glycine-oxalacetate aminotransferase

CAS registry number
37277-90-0

2 Source Organism

<1> *Micrococcus denitrificans* (inducible [1]) [1]
<2> *Rhodobacter capsulatus* (inducible [3]; strain E1F1 [3]) [3]
<3> *Rhodopseudomonas acidophila* (inducible [2]; strain 7050, DSM 137 [2])
[2]

3 Reaction and Specificity

Catalyzed reaction
glycine + oxaloacetate = glyoxylate + L-aspartate

Reaction type
amino group transfer

Natural substrates and products
S glyoxylate + L-aspartate <1-3> (<3> biosynthetic pathway, overview [2];
<1> key reaction in 3-hydroxy-aspartate pathway of glyoxylate metabo-
lism [1]) (Reversibility: r <1, 2> [1, 3]; ? <3> [2]) [1-3]
P glycine + oxaloacetate <1-3> [1-3]

464

Substrates and products

S glyoxylate + L-aspartate <1-3> (<1> reverse reaction: specific for glyoxylate as amino acceptor [1]; <1> favoured reaction, equilibrium in direction of glycine formation [1]; <1> poor substrates are asparagine or threo-L-β-hydroxyaspartate [1]) (Reversibility: r <1, 2> [1, 3]; ? <3> [2]) [1-3]

P glycine + oxaloacetate <1-3> [1-3]

S glyoxylate + L-serine <1> (<1> transamination at about 30% the rate of L-aspartate [1]) (Reversibility: r <1> [1]) [1]

P hydroxypyruvate + glycine <1> [1]

Inhibitors

L-serine <1> (<1> competitive versus L-aspartate [1]) [1]

erythro-L-hydroxyaspartate <1> (<1> competitive versus L-aspartate and L-serine [1]) [1]

Additional information <3> (<3> no inhibition by methionine sulfoximine [2]) [2]

Cofactors/prosthetic groups

pyridoxal 5'-phosphate <1> (<1> requirement [1]) [1]

Activating compounds

L-asparagine <3> (<3> enzyme induction when used as sole nitrogen source during growth [2]) [2]

L-glutamine <3> (<3> enzyme induction when used as sole nitrogen source during growth [2]) [2]

Metals, ions

Additional information <1> (<1> no metal ion requirement [1]) [1]

Specific activity (U/mg)

0.043 <1> (<1> partially purified enzyme [1]) [1]

0.23 <2> (<2> activity depends on growth substrates [3]) [3]

K$_m$-Value (mM)

0.43 <1> (glyoxylate, <1> pH 7.1, 25°C, with substrate L-aspartate [1]) [1]

1.9 <1> (L-aspartate, <1> pH 7.1, 25°C [1]) [1]

pH-Optimum

7.1 <1, 3> (<3> assay at [2]) [1, 2]

7.8 <2> (<2> assay at [3]) [3]

5 Isolation/Preparation/Mutation/Application

Purification

<1> (partial, 40fold [1]) [1]

6 Stability

pH-Stability
7 <1> (<1> quite stable at 2°C [1]) [1]

Temperature stability
2 <1> (<1> quite stable at pH 7.0 [1]) [1]

Storage stability
<1>, 2°C, partially purified enzyme, pH 7.0, quite stable [1]

References

[1] Gibbs, R.G.; Morris, J.G.: Formation of glycine from glyoxylate in Micrococcus denitrificans. Biochem. J., **99**, 27P-28P (1966)
[2] Herbert, R.A.; Macfarlane, G.T.: Asparagine and glutamine metabolism in Rhodopseudomonas acidophila. Arch. Microbiol., **128**, 233-238 (1980)
[3] Caballero, F.J.; Igeño, I.; Cárdenas, J.; Castillo, F.: Regulation of reduced nitrogen assimilation in Rhodobacter capsulatus E1F1. Arch. Microbiol., **152**, 508-511 (1989)

L-Lysine 6-transaminase

1 Nomenclature

EC number
2.6.1.36

Systematic name
L-lysine:2-oxoglutarate 6-aminotransferase

Recommended name
L-lysine 6-transaminase

Synonyms
L-lysine-α-ketoglutarate 6-aminotransferase
L-lysine-α-ketoglutarate aminotransferase
LAT
aminotransferase, lysine 6-
lysine 6-aminotransferase
lysine ε-aminotransferase
lysine ε-transaminase
lysine:2-ketoglutarate 6-aminotransferase

CAS registry number
9054-68-6

2 Source Organism

<1> *Flavobacterium fuscum* [1]
<2> *Flavobacterium lutescens* (IFO 3084 [4, 5, 10, 14]; i.e. Achromobacter liquidum [2, 3]) [2-7, 9, 10, 14]
<3> *Candida utilis* [8]
<4> *Streptomyces clavuligerus* (NRRL 3585 [12]) [11-13, 15, 16]

3 Reaction and Specificity

Catalyzed reaction
L-lysine + 2-oxoglutarate = 2-aminoadipate 6-semialdehyde + L-glutamate (a pyridoxal phosphate protein, the product (allysine) is converted into the intramolecularly dehydrated form, 1-piperideine 6-carboxylate; <2> characterization of the half and overall reactions [9]; <2> stereochemistry [4])

467

Reaction type

amino group transfer

Natural substrates and products

S L-lysine + 2-oxoglutarate <1-4> (<1> first step of metabolism of lysine [1]; <2> involved in production of L-pipecolic acid [10]; <4> first step in biosynthesis of cephalosporins, response of activity to carbon source [11]; <4> time profile of enzyme activity [12]; <4> study of temporal and spatial distribution of enzyme [15]) [1-12, 15]

P 2-aminoadipate 6-semialdehyde + L-glutamate

Substrates and products

S 5-N-acetyl-L-ornithine + 2-oxoglutarate <2> (<2> 4.0% of the activity with L-lysine [3, 9]) (Reversibility: ? <2> [3, 9]) [3, 9]

P 5-(N-acetylamino)-2-oxopentanoate + L-glutamate

S 6-N-acetyl-L-lysine + 2-oxoglutarate <2> (<2> 5.7% of the activity with L-lysine [3]; <2> 5.7% of the activity with L-lysine [9]) (Reversibility: ? <2> [3,9]) [3, 9]

P 6-(N-acetylamino)-2oxohexanoate + L-glutamate

S L-alanine + 2-oxoglutarate <2> (<2> very slow [5]) (Reversibility: ? <2> [5]) [5]

P pyruvate + L-glutamate

S L-lysine + 2-oxoadipate <3> (<3> 7% of the activity with 2-oxoglutarate [8]) (Reversibility: ? <3> [8]) [8]

P 2-aminoadipate 6-semialdehyde + L-homoglutarate

S L-lysine + 2-oxobutanoate <2> (<2> 5.0% of the activity with 2-oxoglutarate [3]; <2> 0.46% of the activity with 2-oxoglutarate [9]) (Reversibility: ? <2> [3, 9]) [3, 9]

P 2-aminoadipate 6-semialdehyde + 2-aminobutanoate

S L-lysine + 2-oxoglutarate <1-4> (<2> 2-oxoglutarate is the only active amino acceptor [2]) (Reversibility: ? <1-4> [1-11, 16]) [1-11, 16]

P 2-aminoadipate 6-semialdehyde + L-glutamate <1, 2> (<1, 2> the product allysine is converted into the intramolecularly dehydrated form 1-piperideine 6-carboxylate [1, 3, 4]) [1-4]

S L-lysine + 2-oxohexanoate <2> (<2> 8.7% of the activity with 2-oxoglutarate [3]) (Reversibility: ? <2> [3]) [3]

P 2-aminoadipate 6-semialdehyde + 2-aminohexanoate

S L-lysine + 2-oxopentanoate <2> (<2> 5.3% of the activity with 2-oxoglutarate [3]; <2> 0.46% of the activity with 2-oxoglutarate [9]) (Reversibility: ? <2> [3, 9]) [3, 9]

P 2-aminoadipate 6-semialdehyde + 2-aminopentanoate

S L-lysine + oxaloacetate <2-4> (<2> 5.3% of the activity with 2-oxoglutarate [3]; <3> 38% of the activity with 2-oxoglutarate [8]; <2> 0.46% of the activity with 2-oxoglutarate [9]; <4> with very low efficiency [16]) (Reversibility: ? <2-4> [3, 8, 9, 16]) [3, 8, 9, 16]

P 2-aminoadipate 6-semialdehyde + L-aspartate

S L-lysine + pyruvate <2-4> (<3> 19% of the activity with 2-oxoglutarate [8]; <2> 4.7% of the activity with 2-oxoglutarate [9]; <4> with very low efficiency [16]) (Reversibility: ? <2-4> [8, 9, 16]) [8, 9, 16]
P 2-aminoadipate 6-semialdehyde + L-alanine
S L-lysine hydroxamate + 2-oxoglutarate <2> (<2> 5.1% of the activity with L-lysine [9]) (Reversibility: ? <2> [9]) [9]
P ? + L-glutamate
S L-ornithine + 2-oxoglutarate <2, 4> (<2> 55% of the activity with L-lysine [3, 9]) (Reversibility: ? <2,4> [2, 3, 9, 16]) [2, 3, 9, 16]
P ? + L-glutamate
S L-thialysine + 2-oxoglutarate <3> (<3> 13% of the activity with L-lysine [8]) (Reversibility: ? <3> [8]) [8]
P ? + L-glutamate
S N-(2'-aminoethyl)-2,3-diaminopropionate + 2-oxoglutarate <2> (<2> 3.6% of the activity with L-lysine [3]) (Reversibility: ? <2> [3]) [3]
P 2-aminoadipate 6-semialdehyde-hydroxamate + L-glutamate
S S-(2-N-acetylaminoethyl)-L-cysteine + 2-oxoglutarate <2> (<2> 3.3% of the activity with L-lysine [3]) (Reversibility: ? <2> [3]) [3]
P S-(2-N-acetylaminoethyl)-2-oxopropanoic acid + L-glutamate
S S-(2-aminoethyl)-L-cysteine + 2-oxoglutarate <2, 3> (<3> 16% of the activity with L-lysine [8]) [3, 8]
P ? + L-glutamate
S Additional information <2, 4> (<2> no substrate: D-lysine, 2-N-acetyl-L-lysine, 2-N-acetyl-L-ornithine, S-(2-aminoethyl)α-L-cysteine [3]; <4> lysine is substrate and inducer of the enzyme [13]) [3, 13]
P ?

Inhibitors
5-aminopentanoate <2> [3]
5-hydroxylysine <2> (<2> DL- [3]) [3]
Br⁻ <2> (<2> stimulates alanine transamination, inhibits lysine transamination [5]) [5]
Cl⁻ <2> (<2> stimulates alanine transamination, inhibits lysine transamination [5]) [5]
F⁻ <2> (<2> stimulates alanine transamination, inhibits lysine transamination [5]) [5]
I⁻ <2> (<2> stimulates alanine transamination, inhibits lysine transamination [5]) [5]
O-(2-aminoethyl)-DL serine <2> [3]
acetate <2> (<2> stimulates alanine transamination, inhibits lysine transamination [5]) [5]
formate <2> (<2> stimulates alanine transamination, inhibits lysine transamination [5]) [5]
phenylglyoxal <2> (loss of activity with lysine, no effect on activity with L-alanine) [5]
propionate <2> (<2> inhibits alanine transamination and lysine transamination [5]) [5]

Additional information <2> (<2> incubation of the enzyme with L-lysine in the presence of high concentrations of phosphate gives an inactive form of the enzyme (semiapoenzyme), reactivation with pyridoxal 5'-phosphate [2]) [2]

Cofactors/prosthetic groups

pyridoxal 5'-phosphate <2, 3> (<2> a pyridoxal phosphate protein [2, 7]; <2, 3> 2 mol of pyridoxal 5'-phosphate per mol of holoenzyme [2, 4, 8]; <2> K_m: 0.000357 mM [2]; <3> K_m 0.015 mM [8]; <2> 2 mol per mol of enzyme, one bound to subunit B, participating in catalytic reaction [6]) [2, 4, 6-8]
pyridoxamine 5'-phosphate <2> (<2> can replace pyridoxal 5'-phosphate, K_m: 0.00715 mM [2]) [2]

Activating compounds

Br^- <2> (<2> stimulates alanine transamination, inhibits lysine transamination [5]) [5]
Cl^- <2> (<2> stimulates alanine transamination, inhibits lysine transamination [5]) [5]
F^- <2> (<2> stimulates alanine transamination, inhibits lysine transamination [5]) [5]
I^- <2> (<2> stimulates alanine transamination, inhibits lysine transamination [5]) [5]
acetate <2> (<2> stimulates alanine transamination, inhibits lysine transamination [5]) [5]
formate <2> (<2> stimulates alanine transamination, inhibits lysine transamination [5]) [5]

Specific activity (U/mg)

18.67 <2> (<2> pH 8.0, 37°C [2]) [2]
Additional information <2, 3> [4, 7, 8]

K_m-Value (mM)

0.13 <2> (2-oxoglutarate, <2> plus L-ornithine, pH 8.0, 37°C [2,3]) [2, 3]
0.5 <2> (2-oxoglutarate, <2> L-lysine, pH 8.0, 37°C [2,3]) [2, 3]
2 <2> (L-ornithine, <2> pH 8.0, 37°C [2,3]) [2, 3]
2 <2> (S-(2-aminoethyl)-L-cysteine, <2> pH 8.0, 37°C [3]) [3]
2.5 <3> (L-lysine, <3> pH 8.5, 37°C [8]) [8]
2.8 <2> (L-lysine, <2> pH 8.0, 37°C [2,3]) [2, 3]
3.8 <3> (2-oxoglutarate, <2> pH 8.0, 37°C [3]) [8]

pH-Optimum

7-7.5 <4> [16]
7.5 <2> (<2> substrates L-ornithine + 2-oxoglutarate [2,3]) [2, 3]
8.3-8.5 <2> (<2> substrates lysine + 2-oxoglutarate [2,3]) [2, 3]
8.5 <3> [8]
9 <2> (<2> substrate S-(2-aminoethyl)-L-cysteine [3]) [3]

pH-Range

6-9.5 <2> (<2> about 40% of maximum activity at pH 6 and pH 9.5, substrates ornithine + 2-oxoglutarate [2]) [2]
7-10 <2> (<2> pH 7: about 50% of maximum activity, pH 10: about 30% of maximum activity [2]) [2]
7.2-10.2 <3> (<3> about 50% of maximum activity at pH 7.2 and pH 10.2 [8]) [8]

Temperature optimum (°C)

30 <4> [16]
37 <1-3> (<3> assay at [8]) [1, 3, 8]
40 <3> [8]

4 Enzyme Structure

Molecular weight

51300 <4> (<4> gel filtration [16]) [16]
83000 <3> (<3> gel filtration [8]) [8]
110000 <2> (<2> native PAGE [14]) [14]
116000 <2> (<2> sedimentation equilibrium method [2,3]) [2, 3]

Subunits

dimer <3> (<3> 2 * 40000, SDS-PAGE [8]; <2> 2 * 53000, SDS-PAGE, 2 * 53300, deduced from gene sequence [14]) [8, 14]
tetramer <2> (<2> 1 * 24000 (A) + 1 * 28000 (B1) + 1 * 28000 (B2) + 1 * 45000-46000 (C), SDS-disc gel electrophoresis, gel filtration in presence of 6 M guanidine-HCl, equilibrium centrifugation in presence of 6 M urea [4,6]) [4, 6]

5 Isolation/Preparation/Mutation/Application

Purification

<1> (partial [1]) [1]
<2> [2, 3, 7, 14]
<3> [8]
<4> (partial [16]) [16]

Crystallization

<2> [2, 3]

Cloning

<2> [14]

Engineering

Additional information <4> (<4> fusion protein of enzyme to green fluorescent protein, study of temporal and spatial distribution of enzyme [15]) [15]

6 Stability

pH-Stability

6-7.5 <2> (<2> 55°C, 5 min, stable, partially purified enzyme [2]) [2]

Temperature stability

55 <2> (<2> pH 6.0-7.5, 5 min, stable, partially purified enzyme [2]) [2]

65 <2> (<2> 5 min, activated by heat treatment [7]) [7]

75 <2> (<2> 5 min, inactivation [7]) [7]

Additional information <2> (<2> pyridoxal 5'-phosphate, 0.1 mM, or 2-mercaptoethanol, 0.02%, protects against heat inactivation [2]) [2]

General stability information

<2>, pyridoxal 5'-phosphate, 0.1 mM, or 2-mercaptoethanol, 0.02%, protects against heat inactivation [2]

Storage stability

<2>, 0-5°C, crystalline enzyme, 0.2 M potassium phosphate buffer, pH 7.2, 50% saturation of ammonium sulfate, 0.1 mM pyridoxal 5'-phosphate, 0.01% 2-mercaptoethanol, stable for several weeks [3]

References

[1] Soda, K.; Misono, H.; Yamamoto, T.: L-Lysine:α-ketoglutarate aminotransferase. I. Identification of a product, Δ^1-piperideine-6-carboxylic acid. Biochemistry, **7**, 4102-4109 (1968)

[2] Soda, K.; Misono, H.: L-Lysine:α-ketoglutarate aminotransferase. II. Purification, crystallization, and properties. Biochemistry, **7**, 4110-4119 (1968)

[3] Soda, K.; Misono, H.: L-Lysine-ketoglutarate aminotransferase (Acetobacter liquidum). Methods Enzymol., **17B**, 222-228 (1971)

[4] Tanizawa, K.; Yoshimura, T.; Soda, K.: L-Lysine transaminase from Flavobacterium lutescens. Methods Enzymol., **113**, 96-102 (1985)

[5] Yoshimura, T.; Tanizawa, K.; Tanaka, H.; Soda, K.: The effect of carboxylates and halides on L-lysine 6-aminotransferase-catalyzed reactions. J. Biochem., **95**, 559-565 (1984)

[6] Yagi, T.; Misono, H.; Kurihara, N.; Yamamoto, T.; Soda, K.: L-Lysine: 2-oxoglutarate 6-aminotransferase. Subunit structure composed of non-identical polypeptides and pyridoxal 5-phosphate-binding subunit. J. Biochem., **87**, 1395-1402 (1980)

[7] Yagi, T.; Yamamoto, T.; Soda, K.: A novel purification procedure of L-lysine 6-aminotransferase from Flavobacterium lutescence. Biochim. Biophys. Acta, **614**, 63-70 (1980)

[8] Hammer, T.; Bode, R.: Purification and characterization of an inducible L-lysine: 2-oxoglutarate 6-aminotransferase from Candida utilis. J. Basic Microbiol., **32**, 21-27 (1992)

[9] Yagi, T.; Misono, H.; Tanizawa, K.; Yoshimura, T.; Soda, K.: Characterization of the half and overall reactions catalyzed by L-lysine:2-oxoglutarate 6-aminotransferase. J. Biochem., **109**, 61-65 (1991)

[10] Fujii, T.; Mukaihara, M.; Agematu, H.; Tsunekawa, H.: Biotransformation of L-lysine to L-pipecolic acid catalyzed by L-lysine 6-aminotransferase and pyrroline-5-carboxylate reductase. Biosci. Biotechnol. Biochem., **66**, 622-627 (2002)

[11] Rius, N.; Demain, A.L.: Regulation of lysine ε-aminotransferase by carbon source and lack of control by phosphate in Streptomyces clavuligerus. Appl. Microbiol. Biotechnol., **48**, 735-737 (1997)

[12] Malmberg, L.H.; Hu, W.S.; Sherman, D.H.: Effects of enhanced lysine ε-aminotransferase activity on cephamycin biosynthesis in Streptomyces clavuligerus. Appl. Microbiol. Biotechnol., **44**, 198-205 (1995)

[13] Rius, N.; Maeda, K.; Demain, A.L.: Induction of L-lysine ε-aminotransferase by L-lysine in Streptomyces clavuligerus, producer of cephalosporins. [Erratum to document cited in CA125:326465]. FEMS Microbiol. Lett., **146**, 319 (1997)

[14] Fujii, T.; Narita, T.; Agematu, H.; Agata, N.; Isshiki, K.: Characterization of L-lysine 6-aminotransferase and its structural gene from Flavobacterium lutescens IFO3084. J. Biochem., **128**, 391-397 (2000)

[15] Khetan, A.; Hu, W.S.; Sherman, D.H.: Heterogeneous distribution of lysine 6-aminotransferase during cephamycin C biosynthesis in Streptomyces clavuligerus demonstrated using green fluorescent protein as a reporter. Microbiology, **146 (Pt 8)**, 1869-1880 (2000)

[16] Romero, J.; Martin, J.F.; Liras, P.; Demain, A.L.; Rius, N.: Partial purification, characterization and nitrogen regulation of the lysine ε-aminotransferase of Streptomyces clavuligerus. J. Ind. Microbiol. Biotechnol., **18**, 241-246 (1997)

2-Aminoethylphosphonate-pyruvate transaminase

2.6.1.37

1 Nomenclature

EC number

2.6.1.37

Systematic name

(2-aminoethyl)phosphonate:pyruvate aminotransferase

Recommended name

2-aminoethylphosphonate-pyruvate transaminase

Synonyms

(2-aminoethyl)phosphonate aminotransferase

(2-aminoethyl)phosphonate transaminase

(2-aminoethyl)phosphonic acid aminotransferase

2-aminoethylphosphonate aminotransferase

2-aminoethylphosphonate-pyruvate aminotransferase

aminotransferase, (2-aminoethyl)phosphonate

CAS registry number

37277-91-1

2 Source Organism

<1> *Bacillus cereus* [1]

<2> *Pseudomonas aeruginosa* [2-4]

<3> *Salmonella enterica* (serovar Thyphimurium [5]) [5, 6]

<4> *Pseudomonas aeruginosa* (encoded by phnW gene [5]) [5]

3 Reaction and Specificity

Catalyzed reaction

(2-aminoethyl)phosphonate + pyruvate = 2-phosphonoacetaldehyde + L-alanine (<2> ping-pong bi-bi mechanism [2]; <2> stereochemistry, 2-aminoethylphosphonate-pyruvate transaminase catalyses the abstraction of the pro-S hydrogen atom at the prochiral C2 carbon of 2-aminoethylphosphonate [3]; <3> bi-bi ping-pong mechanism [5])

Reaction type

amino group transfer

Natural substrates and products

S   (2-aminoethyl)phosphonate + pyruvate <1> [1]
P   2-phosphonoacetaldehyde + L-alanine

Substrates and products

S   (2-aminoethyl)phosphonate + pyruvate <1, 2, 3> (<2> trivial name cilia-
tine [2]; <2> highly specific for pyruvate and (2-aminoethyl)phospho-
nate, no activity with D-penicillamine, N-ethylmaleimide, iodoacetamide,
3-aminopropylphosphonate and taurine [2]; <3> higly specific for (2-ami-
noethyl)phosphonate, no activity with oxaloacetate as NH_3 donor) (Reversibility: ? <1, 2> [1-3]; r <3> [5]) [1-3, 5]
P   2-phosphonoacetaldehyde + L-alanine <1, 2, 3> [1-3, 5]
S   2-aminoethylarsonic acid + pyruvate <2> (Reversibility: ? <2> [4]) [4]
P   alanine + 2-arsonoacetaldehyde <2> [4]
S   2-phosphonoacetaldehyde + D-alanine <3> (<3> L-alanine is preferred
but not absolutely required [5]) (Reversibility: ? <3> [5]) [5]
P   (2-aminoethyl)phosphonate + pyruvate <3> [5]
S α-ketoglutarate + L-alanine <3> (<3> 0.25% of the k_{cat} observed with 2-
phosphonoacetaldehyde [5]) (Reversibility: ? <3> [5]) [5]
P   L-glutamate + pyruvate <3> [5]

Inhibitors

4-aminobutyrate <2> (<2> inhibitor/substrate ratio = 3, 8% inhibition [2]) [2]
D-cycloserine <2> (<2> 1 mM, 40% inhibition [2]) [2]
DL-1-aminobutylphosphonate <2> (<2> inhibitor/substrate ratio = 3, 22%
inhibition [2]) [2]
DL-1-aminoethylphosphonate <1> (weak inhibition [1]) [1]
DL-1-aminopentylphosphonate <2> (<2> inhibitor/substrate ratio = 3, 22%
inhibition [2]) [2]
DL-1-aminopropylphosphonate <1> (<1> weak inhibition [1]) [1]
$HgCl_2$ <2> (<2> 1 mM, 100% inhibition [2]) [2]
L-cysteine <2> (<2> 10 mM, 35% inhibition [2]) [2]
acetate <2> (<2> inhibitor/substrate ratio = 3, 20% inhibition [2]) [2]
aminomethylphosphonate <2> (<2> inhibitor/substrate ratio = 3, 50% inhibition [2]) [2]
aminomethylsulfonate <2> (<2> inhibitior/substrate ratio = 3, 20% inhibition [2]) [2]
aminooxyacetate <2> (<2> 1 mM, 100% inhibition [2]) [2]
β-alanine <2> (<2> inhibitor/substrate ratio = 3, 20% inhibition [2]) [2]
ethylphosphonate <2> (<2> inhibitor/substrate ratio = 3, 94% inhibition [2])
[2]
ethylsulfonate <2> (<2> inhibitor/substrate ratio = 3, 12% inhibition [2]) [2]
hydroxylamine <2> (<2> 1 mM, 50% inhibition [2]) [2]
methylphosphonate <2> (<2> inhibitor/substrate ratio = 3, 90% inhibition
[2]) [2]
phenylhydrazine <2> (<2> 1 mM, 20% inhibition [2]) [2]
propylphosphonate <2> (<2> inhibitor/substrate ratio = 3, 65% inhibition
[2]) [2]

Cofactors/prosthetic groups

pyridoxal 5'-phosphate <2> (<2> a pyridoxal phosphate protein, 4 mol of pyridoxal phosphate per mol of enzyme [2]) [2]

Turnover number (min^{-1})

1.2 <3> ((2-aminoethyl)phosphonate, <3> pH 8.5, 25°C, synthesis of 2-phosphonoacetaldehyde, R340K and R340A mutant enzymes [5]) [5]

1.2 <3> (2-phosphonoacetaldehyde, <3> pH 8.5, 25°C, synthesis of (2-aminoethyl)phosphonate, R340A mutant enzyme [5]) [5]

2.4 <3> (D-alanine, <3> pH 8.5, 25°C, synthesis of (2-aminoethyl)phosphonate [5]) [5]

36 <3> (2-phosphonoacetaldehyde, <3> pH 8.5, 25°C, synthesis of (2-aminoethyl)phosphonate, R340K mutant enzyme [5]) [5]

444 <3> ((2-aminoethyl)phosphonate, <3> pH 8.5, 25°C, synthesis of 2-phosphonoacetaldehyde [5]) [5]

558 <3> (2-phosphonoacetaldehyde, <3> pH 8.5, 25°C, synthesis of (2-aminoethyl)phosphonate [5]) [5]

558 <3> (L-alanine, <3> pH 8.5, 25°C, synthesis of (2-aminoethyl)phosphonate [5]) [5]

Specific activity (U/mg)

16.64 <2> [2]

K$_m$-Value (mM)

0.009 <3> (2-phosphonoacetaldehyde, <3> pH 8.5, 25°C, synthesis of (2-aminoethyl)phosphonate [5]) [5]

0.15 <3> (pyruvate, <3> pH 8.5, 25°C, synthesis of 2-phosphonoacetaldehyde [5]) [5]

0.19 <3> (2-phosphonoacetaldehyde, <3> pH 8.5, 25°C, R340A mutant enzyme [5]) [5]

0.5 <3> (pyruvate, <3> pH 8.5, 25°C, R340K mutant enzyme [5]) [5]

1.11 <3> ((2-aminoethyl)phosphonate, <3> pH 8.5, 25°C, synthesis of 2-phosphonoacetaldehyde [5]) [5]

1.4 <3> (L-alanine, <3> pH 8.5, 25°C, synthesis of (2-aminoethyl)phosphonate [5]) [5]

2.8 <3> (D-alanine, <3> pH 8.5, 25°C, R340K mutant enzyme [5]) [5]

2.9 <3> (2-phosphonoacetaldehyde, <3> pH 8.5, 25°C, R340K mutant enzyme [5]) [5]

3.5 <2> (pyruvate, <2> pH 8.5, 30°C [2]) [2]

3.7 <3> (D-alanine, <3> pH 8.5, 25°C, R340A mutant enzyme [5]) [5]

3.85 <2> (2-aminoethylphosphonate, <2> pH 8.5, 30°C [2]) [2]

4 <2> (2-aminoethylarsonic acid, <2> pH 8.5, 30°C [4]) [4]

6.1 <3> (pyruvate, <3> pH 8.5, 25°C, R340A mutant enzyme [5]) [5]

11 <3> (D-alanine, <3> pH 8.5, 25°C [5]) [5]

20 <3> (L-alanine, <3> pH 8.5, 25°C, R340K mutant enzyme [5]) [5]

26 <3> (2-aminoethylphosphonate, <3> pH 8.5, 25°C, R340A mutant enzyme [5]) [5]

140 <3> (L-alanine, <3> pH 8.5, 25°C, R340A mutant enzyme [5]) [5]

K$_i$-Value (mM)

1 <2> (ethylphosphonate) [2]
1.55 <2> (methylphosphonate) [2]
2.8 <2> (propylphosphonate) [2]
3.85 <2> (aminomethylphosphonate) [2]

pH-Optimum

6.5-9.5 <3> (<3> synthesis of 2-phosphonoacetaldehyde [5]) [5]
7.5-8.5 <3> (<3> synthesis of (2-aminoethyl)phosphonate [5]) [5]
8.4 <1> (<1> assay at [1]) [1]
8.5-9 <2> [2]

pH-Range

7.3-10.8 <2> (<2> approx. 20% of maximal activity at pH 7.3, approx. 60% of
maximal activity at pH 10.8, no activity above pH 11.5 [2]) [2]

Temperature optimum (°C)

25 <1> (<1> assay at [1]) [1]
50 <2> [2]

Temperature range (°C)

30-65 <1> (<1> approx. 70% of maximal activity at 30°C, approx. 35% of
maximal activity at 65°C [1]) [1]

4 Enzyme Structure

Molecular weight

65000 <2> (<2> gel filtration [2]) [2]
100000 <3> (<3> gel filtration [5]) [5]

Subunits

dimer <3> (<3> 2 * 42000, SDS-PAGE [5]) [5]
tetramer <2, 4> (<2> 4 * 16500, SDS-PAGE [2]; <4> 4 * 16500, SDS-PAGE
[5]) [2, 5]

5 Isolation/Preparation/Mutation/Application

Purification

<2> (poly(ethyleneimine), heat treatment, ammonium sulfate, DEAE-cellu-
lose, hydroxyapatite, Ultrogel AcA [2]) [2]
<3> (recombinant enzyme, amonium sulfate, DEAE-cellulose [5]; enzyme
containing Se-Met [6]) [5, 6]

Crystallization

<3> (crystals of Se-Met containing 2-aminoethylphosphonate-pyruvate
transaminase, hanging drop vapor diffusion at 4°C, enzyme solution containing 0.45 mM protein, 8 mM phosphate buffer, pH 7.5, 0.8 mM dithiothreitol
and 20 mM phosphonacetaldehyde is equilibrated against a reservoir solution

containing 200 mM ammonium acetate, 100 mM sodium citrate, pH 5.0, and 10-13% monomethyl polyethylene glycol 5000, yellow crystals appear within 2 weeks [6]) [6]

Cloning
<3> (expression in Escherichia coli [5]) [5]

Engineering
D168A <3> (<3> no activity [5]) [5]
K194L <3> (<3> no activity [5]) [5]
K194R <3> (<3> no activity [5]) [5]
R340A <3> (<3> partial activity [5]) [5]
R340K <3> (<3> partial activity [5]) [5]

6 Stability

Temperature stability
70 <2> (<2> inactivation [2]) [2]

Storage stability
<2>, -20°C, several months, no loss of activity [2]

References

[1] La Nauze, J.M.; Rosenberg, H.: The identification of 2-phosphonoacetalde-hyde as an intermediate in the degradation of 2-aminoethylphosphonate by Bacillus cereus. Biochim. Biophys. Acta, **165**, 438-447 (1968)

[2] Dumora, C.; Lacoste, A.M.; Cassaigne, A.: Purification and properties of 2-aminoethylphosphonate:pyruvate aminotransferase from Pseudomonas aer-uginosa. Eur. J. Biochem., **133**, 119-125 (1983)

[3] Lacoste, A.M.; Dumora, C.; Balas, L.; Hammerschmidt, F.; Vercauteren, J.: Stereochemistry of the reaction catalysed by 2-aminoethylphosphonate ami-notransferase. A 1H-NMR study. Eur. J. Biochem., **215**, 841-844 (1993)

[4] Lacoste, A.M.; Dumora, C.; Ali, B.R.S.; Neuzil, E.; Dixon, H.B.F.: Utilization of 2-aminoethylarsonic acid in Pseudomonas aeruginosa. J. Gen. Microbiol., **138**, 1283-1287 (1992)

[5] Kim, A.D.; Baker, A.S.; Dunaway-Mariano, D.; Metcalf, W.W.; Wanner, B.L.; Martin, B.M.: The 2-aminoethylphosphonate-specific transaminase of the 2-aminoethylphosphonate degradation pathway. J. Bacteriol., **184**, 4134-4140 (2002)

[6] Chen, C.C.; Zhang, H.; Kim, A.D.; Howard, A.; Sheldrick, G.M.; Mariano-Dunaway, D.; Herzberg, O.: Degradation pathway of the phosphonate cilia-tine: crystal structure of 2-aminoethylphosphonate transaminase. Biochem-istry, **41**, 13162-13169 (2002)

Histidine transaminase

1 Nomenclature

EC number

2.6.1.38

Systematic name

L-histidine:2-oxoglutarate aminotransferase

Recommended name

histidine transaminase

Synonyms

Hat-1 <1> [3]

HisAT <4> [5]

histidine aminotransferase

histidine-2-oxoglutarate aminotransferase

CAS registry number

37277-92-2

2 Source Organism

<1> *Mus musculus* (mouse, inbred strains C57BL, DBA, Peru, SM and SWR [3]) [3]

<2> *Pseudomonas acidovorans* [1]

<3> *Pseudomonas testosteroni* (N.C.I.B. 10808 [2]) [2]

<4> *Streptomyces tendae* (Tue901/AEC6, mutant hut-11 [4,5]) [4, 5]

3 Reaction and Specificity

Catalyzed reaction

L-histidine + 2-oxoglutarate = imidazol-5-yl-pyruvate + L-glutamate

Reaction type

amino group transfer

Natural substrates and products

S imidazol-5-yl-pyruvate + L-glutamate <2-4> (<3> essential for utilization of imidazolyl-L-lactate as sole carbon source [2]; <4> involved in nikko-mycin biosynthesis, essential in the utilization of imidazol-L-lactate and imidazole pyruvate as carbon sources [4,5]) (Reversibility: r <2-4> [1, 2, 4, 5]) [1, 2, 4, 5]

P histidine + 2-oxoglutarate <2, 3> [1, 2]

Substrates and products

S L-histidine + 2-oxobutyrate <4> (Reversibility: r <4> [5]) [5]
P imidazol-5-yl-pyruvate + 2-aminobutanoate
S L-histidine + glyoxylate <4> (Reversibility: r <4> [5]) [5]
P imidazol-5-yl-pyruvate + glycine
S L-histidine + pyruvate <4> (<4> ping pong reaction mechanism [5]) (Reversibility: r <4> [5]) [5]
P imidazol-5-yl-pyruvate + L-alanine
S histidine + 2-oxocaproate <4> (Reversibility: r <4> [5]) [5]
P imidazol-5-yl-pyruvate + L-norleucine
S histidine + 2-oxoglutarate <4> (Reversibility: r <4> [5]) [5]
P imidazol-5-yl-pyruvate + L-glutamate
S histidine + 2-oxoisovalerate <4> (Reversibility: r <4> [5]) [5]
P imidazol-5-yl-pyruvate + L-valine
S histidine + 2-oxovalerate <4> (Reversibility: r <4> [5]) [5]
P imidazol-5-yl-pyruvate + L-norvaline
S imidazol-5-yl-pyruvate + L-glutamate <2-4> (<3> reaction mechanism ping pong, great specificity for histidine as amino donor, highly specific for 2-oxoglutarate in the reverse reaction, very slight activity with 2-oxomalonate or 4-methyl-2-oxopentanoate, L-tyrosine, phenylalanine or L-tryptophan [2]) (Reversibility: r <2-4> [1, 2, 4]) [1, 2, 4]
P L-histidine + 2-oxoglutarate <2, 3> [1, 2]
S Additional information <2-4> (<2> no activity observed if L-alanine or L-aspartate replaces glutamate, no activity observed if 2-oxoglutarate is replaced by glyoxylate, pyruvate, 2-oxobutyrate or 2-oxoisocaproate [1]; <3> L-histidinol phosphate, D-histidine, 5'-N-methyl-L-histidine and pyruvate are no substrates [2]; <4> does not use aromatic amino acids as amino substrates, L-phenylalanine, L-tyrosine, L-tryptophan and D-histidine are no substrates, does not react with L-histidinol phosphate, no activity on gels towards L-alanine, L-leucine, L-lysine, L-glutamate and L-aspartate with pyruvate as amino acceptor, no activity towards L-aspartate with 2-oxoglutarate [5]) [1, 2, 5]
P ?

Inhibitors

bromopyruvate <4> [4]
p-chloromercuribenzoate <3> [2]

Cofactors/prosthetic groups

pyridoxal 5'-phosphate <3, 4> (<3> inactive in absence, K_m 0.000059 mM [2]; <4> activity is dependent on the coenzyme and a keto acceptor such as pyruvate or 2-oxoglutarate [4]) [2, 4]

Specific activity (U/mg)

0.0015 <4> (<4> substrate L-tryptophan [4]) [4]
0.0019 <4> (<4> substrate L-histidine [4]) [4]
0.002 <4> (<4> substrate L-tyrosine [4]) [4]

0.0021 <4> (<4> substrate L-phenylalanine [4]) [4]
0.866 <4> [5]
17.8 <3> [2]

K_m-Value (mM)

2.5 <3> (imidazolylpyruvate, <3> pH 8.0, 30°C [2]) [2]
4 <3> (L-histidine, <3> pH 8.0, 30°C [2]) [2]
5 <3> (L-glutamate, <3> pH 8.0, 30°C [2]) [2]
10 <4> (pyruvate, <4> pH 8.0, 37°C [5]) [5]
25 <4> (L-histidine, <4> pH 8.0, 37°C [5]) [5]

pH-Optimum

7.2 <4> [5]
8 <3> [2]

pH-Range

7.5-8.5 <3> (<3> optimal activity in this pH range, outside this range activity decreases sharply [2]) [2]

Temperature optimum (°C)

37 <4> [5]

4 Enzyme Structure

Molecular weight

70000 <3> (<3> gel filtration [2]) [2]
85000 <4> (<4> gel filtration [5]) [5]
110000 <4> (<4> gel filtration [4]) [4]

Subunits

homodimer <4> (<4> 2 * 45000, SDS-PAGE [5]) [5]

5 Isolation/Preparation/Mutation/Application

Source/tissue

liver <1> [3]

Localization

cytosol <1> [3]

Purification

<1> [3]
<3> [2]
<4> [5]

6 Stability

Temperature stability
50-70 <4> (<4> preincubation of the reaction mixture at 50, 60 and 70°C results in a increase of the reaction rate [4]) [4]

General stability information
<3>, purified preparations are unstable, activity is lost during purification, EDTA, DTT or 2-mercaptoethanol stabilizes during purification, pyridoxal 5'-phosphate restores activity [2]

References

[1] Coote, J.G.; Hassall, H.: The role of imidazol-5-yl-lactate-nicotinamide-adenine dinucleotide phosphate oxidoreductase and histidine-2-oxoglutarate aminotransferase in the degradation of imidazol-5-yl-lactate by Pseudomonas acidovorans. Biochem. J., 111, 237-239 (1969)

[2] Hacking, A.J.; Hassall, H.: The purification and properties of L-histidine–2-oxoglutarate aminotransferase from Pseudomonas testosteroni. Biochem. J., 147, 327-334 (1975)

[3] Bulfield, G.: Genetic variation in the activity of the histidine catabolic enzymes between inbred strains of mice: a structural locus for a cytosol histidine aminotransferase isozyme (Hat-1). Biochem. Genet., 16, 1233-1241 (1978)

[4] Roos, U.; Mattern, S.; Schrempf, H.; Bormann, C.: Histidine aminotransferase activity in Streptomyces tendae and its correlation with nikkomycin production. FEMS Microbiol. Lett., 97, 185-190 (1992)

[5] Roos, U.; Bormann, C.: Purification and characterization of L-histidine aminotransferase from nikkomycin-producing Streptomyces tendae Tu901. J. Gen. Microbiol., 139, 2773-2778 (1993)

2-Aminoadipate transaminase 2.6.1.39

1 Nomenclature

EC number

2.6.1.39

Systematic name

L-2-aminoadipate:2-oxoglutarate aminotransferase

Recommended name

2-aminoadipate transaminase

Synonyms

2-aminoadipate aminotransferase
2-aminoadipic aminotransferase
AADAT <3> [12]
AAT <4> [2]
AadAT <1, 3, 4> [6-11]
GKAT <5> [1]
α-aminoadipate aminotransferase
glutamate-α-ketoadipate transaminase
glutamic-ketoadipic transaminase
halogenated tyrosine aminotransferase <4> [5]
kynurenine/α-aminoadipate aminotransferase <4> [11]

CAS registry number

9033-00-5

2 Source Organism

<1> *Bos taurus* (bovine [6]) [6]
<2> *Homo sapiens* (human, UniProt-ID Q8N5Z0 [12]) [12]
<3> *Homo sapiens* (human [10,12]) [10, 12]
<4> *Rattus norvegicus* (Sprague-Dawley [6]; albino rat [3]; Wistar [4]) [2-9, 11]
<5> *Saccharomyces cerevisiae* (bakers' yeast, haploid strains MO-11-48A and X1049-2B, diploid strain F2, clonal isolate of commercial bakers' yeast [1]) [1]

3 Reaction and Specificity

Catalyzed reaction

L-2-aminoadipate + 2-oxoglutarate = 2-oxoadipate + L-glutamate (a pyridoxal-phosphate protein)

Reaction type

amino group transfer

Natural substrates and products

S L-2-aminoadipate + 2-oxoglutarate <1, 3-5> (<1, 3, 4> lysine catabolic pathway [2, 3, 6, 9, 11, 12]; <3> lysine and tryptophan metabolism [10]) (Reversibility: r <1, 3-5> [1-3, 5, 6, 9-12]) [1-3, 5, 6, 9-12]

P 2-oxoadipate + L-glutamate <5> [1]

Substrates and products

S 2-aminoadipate + pyruvate <1, 4> (Reversibility: r <1,4> [6]) [6]

P 2-oxoadipate + L-alanine

S 3,5-di-iodotyrosine + 2-oxoglutarate <1, 4> (Reversibility: r <1, 4> [5, 6]) [5, 6]

P 3,5-diiodophenylpyruvate + L-glutamate

S DL-2-aminopimelic acid + 2-oxoglutarate <1, 4> (Reversibility: r <1, 4> [5, 6]) [5, 6]

P 2-oxoadipate + L-glutamate

S L-2-aminoadipate + 2-oxoglutarate <1, 3-5> (Reversibility: r <1, 3-5> [1-3, 5, 6, 9-12]) [1-3, 5, 6, 9-12]

P 2-oxoadipate + L-glutamate <1, 3-5> [1, 6, 10, 12]

S L-histidine + 2-oxoglutarate <1, 4> (Reversibility: r <1,4> [6]) [6]

P 3-(1H-imidazol-4-yl)-2-oxopropanoate + L-glutamate

S L-histidine + pyruvate <1, 4> (Reversibility: r <1,4> [6]) [6]

P 3-(1H-imidazol-4-yl)-2-oxopropanoate + L-alanine

S L-norleucine + 2-oxoglutarate <1, 4> (Reversibility: r <1,4> [5,6]) [5, 6]

P 2-oxohexanoate + L-glutamate

S L-tryptophan + 2-oxoglutarate <3> (Reversibility: ? <3> [10]) [10]

P 3-indole-2-oxopropanoate + L-glutamate

S kynurenine + 2-oxoglutarate <3, 4> (Reversibility: ir <3,4> [6,10,11]) [6, 10, 11]

P kynurenic acid + L-glutamate

S Additional information <1, 3, 4> (<4> 2-aminoadipate aminotransferase and kynurenine aminotransferase, EC 2.6.1.7, activities are properties of a single protein [2-6, 11]; <4> KAT and AadAT activities are associated with 2 different proteins [7,9]; <1> no kynurenine aminotransferase activity [6]; <1, 4> only slight activity with phenylalanine, tryptophan, tyrosine, aspartate or alanine as amino group donors, glyoxylate is no amino group acceptor [6]; <4> less than 2% activity with L-aspartate, L-alanine, L-α-aminobutyrate, glycine, L-valine, L-leucine, L-isoleucine, L-norvaline, L-ε-aminocaproate, L-threonine, L-serine, L-homoserine, L-cysteine, L-methionine, L-lysine, L-arginine, L-citrulline, L-ornithine, L-histidine, L-tyrosine, L-tryptophan, L-phenylalanine, L-proline and L-glutamate [5]; <3> 2 isoenzymes, AadAT-I and AadAT-II, isoenzyme AadAT-II shows additional activity with tryptophan or kynurenine-2-oxoglutarate reaction, only slight activity with asparagine, alanine, arginine, ornithine, isoleucine, valine, leucine, lysine, serine, threonine, phenylalanine, tyrosine, histidine, glutamine, methionine or aspartate with 2-oxoglutarate [10]) [2-7, 9-11]

P ?

Inhibitors

3-methylglutaric acid <1, 4> (<4> 6 mM, 38% inhibition [6]; <1> 6 mM, 2% inhibition [6]) [6]

L-serine-O-sulfate <4> [3]

adipic acid <1, 4> (<4> 6 mM, 60% inhibition [6]; <1> 6 mM, 48% inhibition [6]; <4> competitive inhibition with respect to L-kynurenine, but not to 2-oxoglutarate [2]) [2, 6]

α-aminoadipate <4> (<4> competitive inhibition against kynurenine or 3-hydroxykynurenine [4]) [4]

azelaic acid <1, 4> (<4> 6 mM, 38% inhibition [6]; <1> 6 mM, 36% inhibition [6]) [6]

decanoic acid <1, 4> (<4> 6 mM, 45% inhibition [6]; <1> 6 mM, 50% inhibition [6]) [6]

diethylglutaric acid <1, 4> (<4> 6 mM, 20% inhibition [6]; <1> 6 mM, 20% inhibition [6]) [6]

dimethylglutaric acid <1, 4> (<4> 6 mM, 42% inhibition [6]; <1> 6 mM, 15% inhibition [6]) [6]

glutaric acid <4> (<4> 6 mM, 5% inhibition [6]) [6]

kynurenic acid <1, 4> (<4> 1 mM, 40% inhibition [6]; <1> 1 mM, 30% inhibition [6]) [6]

pimelic acid <1, 4> (<4> 6 mM, 32% inhibition [6]; <1> 6 mM, 28% inhibition [6]) [6]

Additional information <1, 4> (<1,4> oxaloacetic acid is no inhibitor [6]) [6]

Cofactors/prosthetic groups

pyridoxal 5'-phosphate <1, 3-5> [1, 3, 6, 12]

Specific activity (U/mg)

0.53 <1> [6]

0.77 <3> (<3> isoenzyme AadAT-I [10]) [10]

1.5 <4> (<4> kynurenine aminotransferase activity [4]) [4]

1.56 <4> [11]

2.65 <4> (<4> kynurenine aminotransferase activity [3,5]) [3, 5]

6.3 <4> (<4> α-aminoadipate aminotransferase activity [4]) [4]

10.68 <3> (<3> isoenzyme AadAT-II [10]) [10]

26.1 <4> (<4> α-aminoadipate aminotransferase activity [3,5]) [3, 5]

K_m-Value (mM)

0.01 <4> (2-oxoadipate, <4> pH 6.5, 37°C [4]) [4]

0.25 <3> (2-aminoadipate, <3> pH 8.0, 37°C isoenzyme AadAT-II [10]) [10]

0.5 <4> (2-oxoadipic acid, <4> pH 7.0, 37°C [6]) [6]
1 <3> (2-oxoadipate, <3> pH 8.0, 37°C isoenzyme AadAT-II [10]) [10]
1.1 <3> (2-oxoglutarate, <3> pH 8.0, 37°C isoenzyme AadAT-I [10]) [10]
1.4 <3> (glutamate, <3> pH 8.0, 37°C [10]) [10]
2.5 <3> (2-oxoadipate, <3> pH 8.0, 37°C isoenzyme AadAT-I [10]) [10]
3.2 <3> (2-oxoglutarate, <3> pH 8.0, 37°C isoenzyme AadAT-II [10]) [10]
4.6 <4> (L-glutamic acid, <4> pH 7.0, 37°C [6]) [6]
12.5 <3> (L-glutamate, <3> pH 8.0, 37°C, isoenzyme AadAT-II [10]) [10]
20 <3> (2-aminoadipate, <3> pH 8.0, 37°C [10]) [10]

pH-Optimum
6.5 <4> (<4> kynurenine aminotransferase activity [4]) [4]
7 <1, 4> [6]
8.5 <5> [1]
9-9.5 <3> (<3> isoenzyme AadAT-II [10]) [10]
9.5 <3> (<3> isoenzyme AadAT-I [10]) [10]

4 Enzyme Structure

Molecular weight
50800 <3> (<3> cDNA analysis [12]) [12]
82100 <4> (<4> gel filtration [3]) [3]
85000 <4> (<4> sucrose density gradient centrifugation [3]) [3, 5]
89000 <4> (<4> gel filtration [8]) [8]
98000 <3> (<3> isoenzyme AadAT-II, sucrose density gradient centrifugation [10]) [10]
100000 <4> (<4> gel filtration [11]) [11]
100000 <5> (<5> mitochondrial isoform glutamate-α-ketoadipate transaminase I, gel filtration [1]) [1]
104000 <3> (<3> isoenzyme AadAT-I, sucrose density gradient centrifugation [10]) [10]
140000 <5> (<5> cytoplasmic isoform glutamate-α-ketoadipate transaminase II, gel filtration [1]) [1]

Subunits
dimer <1, 4> (<4> 2 * 44500, SDS-PAGE [8]; <4> 2 * 41000, SDS-PAGE [6]; <4> 2 * 45500, SDS-PAGE [3]; <4> 2 * 45000, SDS-PAGE [5,11]; <1> 2 * 41000, SDS-PAGE [6]; <4> 2 * 49500, SDS-PAGE [4]; <3> 2 * 46000, SDS-PAGE [10]; <4> 2 * 47789, amino acid residues [11]) [3-6, 8, 10, 11]

5 Isolation/Preparation/Mutation/Application

Source/tissue
brain <3> [12]
heart <3> [12]
kidney <1, 3, 4> [2, 3, 5-9, 11, 12]
liver <3, 4> [2, 4, 6, 7, 10, 12]
ovary <3> [12]
pancreas <3> [12]
prostate <3> [12]
testis <3> [12]

Localization
cytoplasm <1, 3-5> (<5> glutamate-α-ketoadipate transaminase II [1]; <3> isoenzyme AadAT-I [10]) [1, 6, 10]
cytosol <3> [10]
mitochondrion <1, 3-5> (<5> glutamate-α-ketoadipate transaminase I [1]; <3> isoenzyme AadAT-II [10]) [1, 4-7, 10]

Purification
<1> [6]
<3> (2 isoenzymes, AadAT-I and AadAT-II [10]) [10]
<4> [2-9, 11]
<5> (partial [1]) [1]

Cloning
<2> [12]
<3> (gene AADAT sequenced [12]) [12]
<4> (kynurenine/α-aminoadipate aminotransferase cDNA cloned and expressed in HEK-293 cells [11]) [11]

Application
medicine <4> (<4> L-2-aminoadipate, the substrate for AadAT, is a well known astroglial-specific toxin, knowledge of the cerebral disposition of this compound is instrumental for the elucidation of its mechanism of toxicity and possible relevance in pathology [11]) [11]
pharmacology <3> (<3> L-α-aminoadipate is a component of the precursor to penicillin and cephalosporin [12]) [12]

6 Stability

pH-Stability
6.5-7.5 <4> (<4> apoenzyme is most stable in this range during purification [5]) [5]
9-9.5 <3> (<3> isoenzyme AadAT-I, sensitive to pH changes, activity is quickly lost at a higher pH, isoenzyme AadAT-II is less sensitive to pH changes [10]) [10]

Temperature stability

45-70 <5> (<5> stable against heating at 45°C, 50°C, 55°C, 60°C, 65°C and 70°C for 10 min [1]) [1]

50-70 <4> (<4> both activities are not changed by heating at 50°C for 30 min, activities shows a parallel decrease with time at 60°C and 70°C, both are completely lost by incubation at 70°C for 20 min [4]) [4]

60-70 <4> (<4> inactivation is initiated at 60°C and is near completion at 70°C [3]) [3]

General stability information

<1>, 0.1 mM dithiothreitol improves the stability of the enzyme by 50% [6]

Storage stability

<1>, 4°C, partially purified enzyme in 8 mM potassium phosphate buffer, pH 7.1, 100% activity is lost when stored for 14 days [6]

<4>, -15°C, crude enzyme, little activity is lost over 3 months [2]

<4>, -80°C, stable for more than 1 month [4]

<4>, 4°C, partially purified enzyme in 8 mM potassium phosphate buffer, pH 7.1, 15% activity is lost when stored for 14 days [6]

References

[1] Matsuda, M.; Ogur, M.: Separation and specificity of the yeast glutamate-α-ketoadipate transaminase. J. Biol. Chem., **244**, 3352-3358 (1969)

[2] Tobes, M.C.; Mason, M.: L-kynurenine aminotransferase and L-α-aminoadipate aminotransferase. I. Evidence for identity. Biochem. Biophys. Res. Commun., **62**, 390-397 (1975)

[3] Tobes, M.C.; Mason, M.: α-Aminoadipate aminotransferase and kynurenine aminotransferase. Purification, characterization, and further evidence for identity. J. Biol. Chem., **252**, 4591-4599 (1977)

[4] Takeuchi, F.; Otsuka, H.; Shibata, Y.: Purification, characterization and identification of rat liver mitochondrial kynurenine aminotransferase with α-aminoadipate aminotransferase. Biochim. Biophys. Acta, **743**, 323-330 (1983)

[5] Hartline, R.A.: Kynurenine aminotransferase from kidney supernatant. Methods Enzymol., **113**, 664-672 (1985)

[6] Deshmukh, D.R.; Mungre, S.M.: Purification and properties of 2-aminoadipate: 2-oxoglutarate aminotransferase from bovine kidney. Biochem. J., **261**, 761-768 (1989)

[7] Mawal, M.R.; Deshmukh, D.R.: α-Aminoadipate and kynurenine aminotransferase activities from rat kidney. Evidence for separate identity. J. Biol. Chem., **266**, 2573-2575 (1991)

[8] Mawal, M.R.; Deshmukh, D.R.: Purification and properties of α-aminoadipate aminotransferase from rat kidney. Prep. Biochem., **21**, 63-73 (1991)

[9] Mawal, M.R.; Mukhopadhyay, A.; Deshmukh, D.R.: Purification and properties of kynurenine aminotransferase from rat kidney. Biochem. J., **279**, 595-599 (1991)

[10] Okuno, E.; Tsujimoto, M.; Nakamura, M.; Kido, R.: 2-Aminoadipate-2-oxo-glutarate aminotransferase isoenzymes in human liver: a plausible physiological role in lysine and tryptophan metabolism. Enzyme Protein, **47**, 136-148 (1993)

[11] Buchli, R.; Alberati-Giani, D.; Malherbe, P.; Kohler, C.; Broger, C.; Cesura, A.M.: Cloning and functional expression of a soluble form of kynurenine/α-aminoadipate aminotransferase from rat kidney. J. Biol. Chem., **270**, 29330-29335 (1995)

[12] Goh, D.L.; Patel, A.; Thomas, G.H.; Salomons, G.S.; Schor, D.S.; Jakobs, C.; Geraghty, M.T.: Characterization of the human gene encoding α-aminoadipate aminotransferase (AADAT). Mol. Genet. Metab., **76**, 172-180 (2002)

(R)-3-Amino-2-methylpropionate-pyruvate transaminase

2.6.1.40

1 Nomenclature

EC number
2.6.1.40

Systematic name
(R)-3-amino-2-methylpropanoate:pyruvate aminotransferase

Recommended name
(R)-3-amino-2-methylpropionate-pyruvate transaminase

Synonyms
D-3-aminoisobutyrate-pyruvate aminotransferase
D-3-aminoisobutyrate-pyruvate transaminase
D-BAIB aminotransferase
EC 2.6.1.61 (identical with deleted entry)
aminotransferase, D-3-aminoisobutyrate-pyruvate
β-aminoisobutyrate-pyruvate aminotransferase
Additional information (<1> may be identical with EC 2.6.1.44 [6])

CAS registry number
37279-00-8

2 Source Organism

<1> *Rattus norvegicus* [1-7]
<2> *Cavia porcellus* [1]
<3> *Homo sapiens* [1]
<4> *Sus scrofa* [1]

3 Reaction and Specificity

Catalyzed reaction
(R)-3-amino-2-methylpropanoate + pyruvate = 2-methyl-3-oxopropanoate + L-alanine (<1> may be identical with alanine-glyoxylate aminotransferase, EC 2.6.1.44 [6])

Reaction type
amino group transfer

Natural substrates and products

S (R)-3-amino-2-methylpropanoate + pyruvate <1-4> (<1-4> first step in aminoisobutyrate metabolism [1]) [1]

P 2-methyl-3-oxopropanoate + L-alanine <1-4> [1]

Substrates and products

S (R)-3-amino-2-methylpropanoate + 2-oxo-n-butanoate <1> (Reversibility: ? <1> [2]) [2]

P 2-methyl-3-oxopropanoate + 2-amino-n-butanoate

S (R)-3-amino-2-methylpropanoate + 2-oxo-n-pentanoate <1> (Reversibility: ? <1> [2]) [2]

P 2-methyl-3-oxopropanoate + norvaline

S (R)-3-amino-2-methylpropanoate + glyoxylate <1, 4> (<4> with 50% efficiency compared to pyruvate [1]; <1> with 90% efficiency compared to pyruvate [7]) (Reversibility: ? <1,4> [1-3, 6]; r <1> [7]) [1-3, 6, 7]

P 2-methyl-3-oxopropanoate + glycine

S (R)-3-amino-2-methylpropanoate + oxaloacetate <1> (<4> no substrate [1]; <1> with 63% efficiency compared to pyruvate [7]) (Reversibility: ? <1> [2,3]; r <1> [7]) [2, 3, 7]

P 2-methyl-3-oxopropanoate + L-aspartate

S (R)-3-amino-2-methylpropanoate + pyruvate <1-4> (<2> best substrate among ω-amino acids, highly specific for D-isomer [1]) (Reversibility: ? <1, 3, 4> [2, 3]; r <2> [1]) [1-3]

P 2-methyl-3-oxopropanoate + L-alanine <1-4> (<1> i.e. methylmalonyl semialdehyde [1]) [1-4]

S (S)-2-aminobutyrate + pyruvate <1> (Reversibility: ? <1> [6]) [6]

P 2-oxobutanoate + L-alanine

S 4-aminobutanoate + pyruvate <4> (<4> transamination at 19% the rate of (R)-3-amino-2-methylpropanoate [1]) (Reversibility: ? <4> [1]) [1]

P succinic semialdehyde + L-alanine

S 5-aminolevulinic acid + pyruvate <1> (Reversibility: ? <1> [2, 3]) [2, 3]

P 5-oxolevulinic acid + L-alanine

S DL-3-amino-n-butanoate + pyruvate <1> (<1> not [3]) (Reversibility: ? <1> [2]) [2]

P 3-oxobutanoate + L-alanine

S DL-α-ethyl-β-alanine + pyruvate <4> (<4> transamination at 33% the rate of (R)-3-amino-2-methylpropanoate [1]) (Reversibility: ? <4> [1]) [1]

P 2-ethyl-3-oxopropanoate + L-alanine

S NG,NG'-dimethyl-(S)-arginine + pyruvate <1> (Reversibility: ? <1> [2,3]) [2, 3]

P ? + L-alanine

S arginine + pyruvate <1> (<1> transamination at 40% the rate of (R)-3-amino-2-methylpropanoate [2]) (Reversibility: ? <1> [2]) [2]

P 2-oxo-5-guanidinovaleric acid + L-alanine

S β-alanine + pyruvate <1, 4> (<4> transamination at 66% the rate of (R)-3-amino-2-methylpropanoate [1]; <1> transamination at 66% the rate of

(R)-3-amino-2-methylpropanoate [7]) (Reversibility: ? <1,4> [1-3,7]) [1-3, 7]

P malonic semialdehyde + L-alanine

S glycine + pyruvate <1> (<1> not [3]) (Reversibility: ? <1> [2]) [2]

P glyoxylate + L-alanine

S Additional information <1-4> (<1,4> no substrate: 2-oxoglutarate [1, 3, 7]; <4> no substrate: 5-amino-n-pentanoate, 6-aminohexanoate, DL-3-aminobutyrate, oxaloacetate [1]; <1> no substrate: phenylpyruvate [3, 7]; <1> γ-aminobutanoate is neither substrate nor inhibitor [7]; <1> overview [2, 3, 7]) [1-3, 7]

P ?

Inhibitors

(R)-3-amino-2-methylpropanoate <1> (<1> β-alanine as substrate [6]) [6]

(S)-2-aminobutyrate <1> (<1> β-alanine as substrate [6]) [6]

(S)-alanine <1> (<1> with β-alanine + glyoxylate as substrate, not (R)-isomer [6]) [6]

5-aminolevulinic acid <1> (<1> β-alanine as substrate [6]) [6]

5-bromouracil <1> (<1> 1 M [5]) [5]

5-chlorouracil <1> (<1> 0.1 mM [5]) [5]

5-diazouracil <1> (<1> 1 M [5]) [5]

5-fluorouracil <1> (<1> strong, kinetics [5]) [5, 6]

5-fluorouridine <1> (<1> weak [5]) [5]

5-iodouracil <1> (<1> 1 M [5]) [5]

5-nitrouracil <1> (<1> 0.1 M [5]) [5]

5-thiouracil <1> (<1> 1 M, not 2-thiouracil [5]) [5]

6-azathymine <1> (<1> weak [6]) [6]

6-azauracil <1> (<1> weak, kinetics, not reversible by pyridoxal phosphate [3]) [3, 6]

6-azauridine <1> (<1> weak [3]) [3]

D-cycloserine <1> [3]

GSH <1> (<1> weak [2]) [2]

L-cysteine <1> [2]

N^γ,N^γ-dimethyl-(S)-arginine <1> (<1> β-alanine as substrate [6]) [6]

pyruvate <1> (<1> at high concentrations [6]) [5, 6]

α-fluoro-β-alanine <1> (<1> kinetics [5]) [5, 7]

amiooxyacetate <1> (<1> strong [3]) [3]

β-alanine <1> (<1> with (S)-alanine + glyoxylate as substrate [6]) [6]

gabaculine <1> (<1> i.e. 5-amino-1,3-cyclohexadienylcarboxylate, strong, irreversible, β-alanine or pyruvate protects, 4-aminobutyrate to a lesser extent, not 2-oxoglutarate [4]) [3, 4, 7]

glyoxylate <1> (<1> at high concentrations [6]) [6]

hydroxylamine <4> [1]

isoorotic acid <1> (<1> 1 M [5]) [5]

Additional information <1> (<1> no inhibitor: precursors of β-alanine or (R)-3-amino-2-methylpropanoate, i.e. orotic acid, uracil, thymine [3, 5]; <1>

no inhibitor: dihydrothymine, dihydrouracil, N-carbamoyl-β-alanine, N-carb-amoyl-DL-3-aminoisobutyrate [3]; <1> no inhibitor: aminouracil [5]) [3, 5]

Cofactors/prosthetic groups
pyridoxal 5'-phosphate <1, 4> (<1> requirement, tightly bound [2]) [1-6]

Activating compounds
2-mercaptoethanol <1, 4> (<1,4> activation only in the presence of pyridoxal phosphate [1,2]) [1, 2]
DTT <1> (<1> activation [2]) [2]

Specific activity (U/mg)
0.093 <4> (<4> pH 8.8, 37°C [1]) [1]
1.14 <1> (<1> pH 8.8, 37°C [2]) [2]
11.3 <1> (<1> pH 8.8, 37°C [7]) [7]
Additional information <1> (<1> assay principle [7]) [7]

K$_m$-Value (mM)
0.45 <1> (pyruvate, <1> pH 8.8, 37°C [3]) [3]
0.81 <1> (β-alanine, <1> pH 8.8, 37°C [1]) [3]

K$_i$-Value (mM)
Additional information <1> (<1> γ-aminobutanoate is neither substrate nor inhibitor [7]) [7]

pH-Optimum
9 <4> [1, 2]
9.5 <1> [3, 7]
Additional information <1> (<1> pI: 6.7 [3]) [3]

pH-Range
7.5-10.6 <4> (<4> about half-maximal activity at pH 7.5 and pH 10.6 [1]) [1]

Temperature optimum (°C)
37 <1-4> (<1-4> assay at [1-6]) [1-6]

4 Enzyme Structure

Molecular weight
180000 <1> (<1> gel filtration [2,6]) [2, 6]
220000 <1> (<1> gel filtration [3,7]) [3, 7]

Subunits
tetramer <1> (<1> 4 * 50000, reductive SDS-PAGE [2]; <1> 4 * 52000, SDS-PAGE [3,7]; <1> cross-linking experiments: tetramer composed of two di-mers [3]) [2, 3, 7]

5 Isolation/Preparation/Mutation/Application

Source/tissue
kidney <2> [1]
liver <1-4> [1-7]
Additional information <1, 2> (<2> distribution in guinea pig tissue [1]; <1> distribution in rat tissue [3]) [1, 3]

Localization
cytoplasm <1> [2, 3]

Purification
<1> [2, 3, 7]
<4> [1]

6 Stability

Storage stability
<1>, -20°C, concentrated enzyme solution, in 5% polyethylene glycol plus 1 mg/ml pyridoxal 5'-phosphate, several months [2]
<1>, 4°C, 0.05 mg protein/ml, in 0.1 M potassium phosphate buffer, pH 7, 1 mM EDTA, 2 mM 2-mercaptoethanol, about 50% loss of activity within a few days, 80% loss within 1 week [3, 7]
<1>, 4°C, concentrated enzyme solution, several days [2]
<4>, -20°C, at least 1 month [1]

References

[1] Kakimoto, Y.; Taniguchi, K.; Sano, I.: D-β-Aminoisobutyrate:pyruvate aminotransferase in mammalian liver and excretion of β-aminoisobutyrate by man. J. Biol. Chem., **244**, 335-340 (1969)
[2] Ueno, S.I.; Morino, H.; Sano, A.; Kakimoto, Y.: Purification and characterization of D-3-aminoisobutyrate-pyruvate aminotransferase from rat liver. Biochim. Biophys. Acta, **1033**, 169-175 (1990)
[3] Tamaki, N.; Kaneko, M.; Mizota, C.; Kikugawa, M.; Fujimoto, S.: Purification, characterization and inhibition of D-3-aminoisobutyrate aminotransferase from the rat liver. Eur. J. Biochem., **189**, 39-45 (1990)
[4] Kaneko, M.; Fujimoto, S.; Kikugawa, M.; Tamaki, N.: Irreversible inhibition of D-3-aminoisobutyrate-pyruvate aminotransferase by gabaculine. FEBS Lett., **276**, 115-118 (1990)
[5] Kaneko, M.; Kontani, M.; Kikugawa, M.; Tamaki, N.: Inhibition of D-3-aminoisobutyrate-pyruvate aminotransferase by 5-fluorouracil and α-fluoro-β-alanine. Biochim. Biophys. Acta, **1122**, 45-49 (1992)

[6] Kontani, M.; Kaneko, M.; Kikugawa, M.; Fujimoto, S.; Tamaki, N.: Identity of D-3-aminoisobutyrate-pyruvate aminotransferase with alanine-glyoxylate aminotransferase 2. Biochim. Biophys. Acta, **1156**, 161-166 (1993)

[7] Tamaki, N.; Sakata, S.F.; Matsuda, K.: Purification, properties, and sequencing of aminoisobutyrate aminotransferases from rat liver. Methods Enzymol., **324**, 376-389 (2000)

D-Methionine-pyruvate transaminase

2.6.1.41

1 Nomenclature

EC number
2.6.1.41

Systematic name
D-methionine:pyruvate aminotransferase

Recommended name
D-methionine-pyruvate transaminase

Synonyms
D-methionine aminotransferase
D-methionine transaminase
aminotransferase, D-methionine

CAS registry number
37277-93-3

2 Source Organism

<1> *Arachis hypogaea* (peanut cv. Starr [1]) [1]
<2> *Brassica sp.* (cauliflower [2]) [2]

3 Reaction and Specificity

Catalyzed reaction
D-methionine + pyruvate = 4-methylthio-2-oxobutanoate + L-alanine

Reaction type
amino group transfer

Natural substrates and products
S D-methionine + pyruvate <2> (<2> involved in ethylene synthesis from methionine [2]) [2]
P 4-methylthio-2-oxobutanoate + L-alanine

Substrates and products
S D-methionine + 2-oxoglutarate <1, 2> (<1, 2> less effective than pyruvate [1, 2]) (Reversibility: ? <1,2> [1, 2]) [1, 2]
P 4-methylthio-2-oxobutanoate + L-glutamate <1, 2> [1, 2]

S   D-methionine + oxaloacetate <1, 2> (<1,2> less effective than pyruvate [1,2]) (Reversibility: ? <1,2> [1,2]) [1, 2]

P   4-methylthio-2-oxobutanoate + L-aspartate <1, 2> [1, 2]

S   D-methionine + pyruvate <1, 2> (Reversibility: ? <1,2> [1,2]) [1, 2]

P   4-methylthio-2-oxobutanoate + L-alanine <1, 2> [1, 2]

S   L-methionine + pyruvate <2> (<2> one third of activity compared to D-methionine [2]) (Reversibility: ? <2> [2]) [2]

P   4-methylthio-2-oxobutanoate + L-alanine <2> [2]

Cofactors/prosthetic groups

pyridoxal 5'-phosphate <1> (<1> required for full enzyme activity [1]) [1]

Specific activity (U/mg)

Additional information <1> (<1> 10061 units/mg, 1 unit is defined as the quantity of enzyme that produces 1 nanoliter of ethylene per h of dark reaction [1]) [1]

pH-Optimum

7.5 <2> [2]

7.8 <1> (<1> assay at [1]) [1]

pH-Range

6-8 <2> (<2> 15% of maximal activity at pH 6.0, 66% of maximal activity at pH 8.0 [2]) [2]

Temperature optimum (°C)

25 <2> [2]

31 <1> (<1> assay at [1]) [1]

5 Isolation/Preparation/Mutation/Application

Source/tissue

embryo <1> (<1> germinating [1]) [1]

floret <2> [2]

Localization

mitochondrion <2> [2]

Purification

<1> (ammonium sulfate, cold precipitation, acetone precipitation, partial purification [1]) [1]

<2> (partial [2]) [2]

6 Stability

General stability information

<1>, pyridoxal 5'-phosphate stabilizes [1]

<1>, reducing agents, such as cysteine, 2-mercaptoethanol or dithiothreitol, are essential for maintaining activity during purification [1]

References

[1] Durham, J.I.; Morgan, P.W.; Prescott, J.M.; Lyman, C.M.: An aminotransferase specific for the D-enantiomorph of methionine. Phytochemistry, 12, 2123-2126 (1973)
[2] Mapson, L.W.; March, J.F.; Wardale, D.A.: Biosynthesis of ethylene. 4-methylmercapto-2-oxobutyric acid: an intermediate in the formation from methionine. Biochem. J., 115, 653-661 (1969)

Branched-chain-amino-acid transaminase 2.6.1.42

1 Nomenclature

EC number
2.6.1.42

Systematic name
branched-chain-amino-acid:2-oxoglutarate aminotransferase

Recommended name
branched-chain-amino-acid transaminase

Synonyms
BCAA aminotransferase <17> [30]
BCAA-AT <17, 33> [30, 32]
BCAT <15, 17, 22, 33, 38, 40> [15, 21, 23, 25, 27, 33, 39, 46, 48, 51, 53]
BCATc <17, 31, 33> [28, 34, 38, 43, 52]
BCATm <17, 31, 33> [28, 34, 35, 38, 43, 46, 52]
BcaT <20> [41]
EC 2.6.1.6 <37, 40> (<40> formerly [4,6]) [4, 6, 9]
L-branched chain amino acid aminotransferase
L-leucine-α-ketoglutarate transaminase <1> [8, 12]
TA-B <11> [16]
α-KGA <1> [12]
branched-chain amino acid aminotransferase
branched-chain amino acid-glutamate transaminase
branched-chain aminotransferase
eBCAT <15> [45, 53]
glutamate-branched-chain amino acid transaminase
hBCATm <17> [51]
leucine aminotranferase <40> [3]
leucine transaminase <40> [4]
transaminase B

CAS registry number
9054-65-3

2 Source Organism

<1> *Acetobacter suboxydans* (IFO 3172, synonym Gluconobacter suboxydans [8,12]) [8, 12]

<2> *Arabidopsis sp.* [49]

<3> *Arabidopsis sp.* (AtBCAT-2 [49]) [49]

<4> *Arabidopsis sp.* (AtBCAT-4 [49]) [49]

<5> *Arabidopsis sp.* (AtBCAT-1 [49]) [49]

<6> *Arabidopsis sp.* (AtBCAT-3 [49]) [49]

<7> *Arabidopsis sp.* (AtBCAT-5 [49]) [49]

<8> *Arabidopsis sp.* (AtBCAT-6 [49]) [49]

<9> *Bacillus brevis* [29]

<10> *Bos taurus* (bovine [32]) [32]

<11> *Brevibacterium flavum* (No.2247, ATCC 14067 [16]) [16]

<12> *Candida maltosa* [20]

<13> *Canis familiaris* (dog [22]) [22]

<14> *Endotinium sp.* [10]

<15> *Escherichia coli* (strain W [1]; strain W3110 [21, 39]; K-12 [14]) [1, 11, 14, 21, 25, 39, 45, 53]

<16> *Homo sapiens* (human, BCATm UniProt-ID O15382 [44]) [44]

<17> *Homo sapiens* (human [22, 30, 34, 38, 40, 44, 47, 48, 51, 52]) [22, 30, 34, 38, 40, 44, 47, 48, 51, 52]

<18> *Hordeum vulgare* (barley, L. cv. Ingrid [15]) [15]

<19> *Lactococcus lactis* (subsp. cremoris NCDO 763 and TIL 46, same nucleotide sequence accession number in DDBJ and EMBL [41]]) [41]

<20> *Lactococcus lactis* (LM0230 [36]) [36]

<21> *Macaca sp.* (macaque monkey [17]) [17]

<22> *Methanococcus aeolicus* [27]

<23> *Methanococcus maripaludis* (JJ, DSM 2067 [27]) [27]

<24> *Methanococcus voltae* (PS, DSM 1537 [27]) [27]

<25> *Neurospora crassa* [7]

<26> *Ovis aries* (sheep, BCATm UniProt-ID O97932 [35,37]) [35, 37]

<27> *Ovis aries* (sheep, BCATc UniProt-ID P24288 [37]) [37]

<28> *Ovis aries* (sheep, BCATm UniProt-ID P54687 [37]) [37]

<29> *Ovis aries* (sheep, BCATm UniProt-ID P54960 [37]) [37]

<30> *Ovis aries* (sheep, BCATm [37]) [37]

<31> *Ovis aries* (sheep [35,37]) [35, 37, 43]

<32> *Pseudomonas sp.* [13]

<33> *Rattus norvegicus* (rat, strain Wistar [5]; Sprague-Dawley [26]) [2, 5, 18, 22, 23, 26, 28, 32, 34, 38, 40, 46, 48, 53]

<34> *Saccharomyces cerevisiae* (YR6a, YPH501, Bat1p [31]) [31]

<35> *Saccharomyces cerevisiae* (YR6a, YPH501, Bat2p [31]) [31, 42, 44]

<36> *Saccharomyces cerevisiae* (strains MD101, 22, 133, 124, 138 and AE55, 16, 25 and 89 [42]) [42]
<37> *Salmonella typhimurium* [9, 24]
<38> *Schizosaccharomyces pombe* [33]
<39> *Staphylococcus carnosus* (strains BioCarna Ferment S1, PSM213, PSM215 [50]) [50]
<40> *Sus scrofa* (pig, hog [2-6, 18, 19, 23]) [2-6, 18, 19, 23]

3 Reaction and Specificity

Catalyzed reaction

L-leucine + 2-oxoglutarate = 4-methyl-2-oxopentanoate + L-glutamate (branched-chain amino acid + 2-oxoglutarate = corresponding keto acid + L-glutamate, also acts on L-isoleucine and L-valine. Different from EC 2.6.1.66 valine-pyruvate transaminase)

Reaction type

amino group transfer

Natural substrates and products

S   L-isoleucine + 2-oxoglutarate <1, 2, 9, 12, 14, 15, 17, 18, 20, 22-25, 31, 33, 36, 37, 39, 40> (<37> biosynthesis of L-isoleucine [24]) (Reversibility: r <1, 2, 9, 12, 14, 15, 17, 18, 20, 22-25, 31, 33, 36, 37, 39, 40> [2, 4-8, 10, 12, 14, 15, 18, 20-22, 24-29, 34, 35, 37-39, 41-43, 46, 47, 49-51, 53]) [2, 4-8, 10, 12, 14, 15, 18, 20-22, 24-29, 34-39, 41-43, 46, 47, 49-51, 53]

P   3-methyl-2-oxopentanoate + L-glutamate <33, 40> [5, 6]

S   L-leucine + 2-oxoglutarate <1, 2, 9, 12, 14, 15, 17, 18, 20, 22-25, 31-33, 36, 37, 40> (<33> sole transaminase in fetal rat liver [5]; <33> first step in the metabolism of branched-chain amino acids [34]; <33> key enzyme on the biosynthetic pathway of hydrophobic amino acids [53]; <12> biosynthesis of branched-chain amino acids [20]; <37> biosynthesis of L-leucine [24]; <25> mitochondrial isoenzyme functions in the biosynthesis of L-isoleucine, L-leucine and L-valine, cytoplasmic isoenzyme functions in amino acid catabolism [7]; <9, 15, 22> last step in the biosynthesis of the branched-chain amino acids L-isoleucine, L-valine and L-leucine [15, 27, 29, 39]; <22-24> biosynthesis of the branched-chain amino acids L-valine, L-isoleucine and L-leucine [27]; <36> involved in production of fusel alcohols during fermentation [42]; <2> last step of the synthesis or initial step of the degradation of branched chain amino acids [49]) (Reversibility: r <1, 2, 9, 12, 14, 15, 18, 20, 22-25, 31-33, 36, 37, 40> [1, 2, 4-8, 10, 12-15, 20-22, 24-30, 32, 34-39, 41-43, 46-51, 53]) [1, 2, 4-8, 10, 12-15, 18, 20, 21, 24-30, 32, 34-39, 41-43, 46-51, 53]

P   4-methyl-2-oxopentanoate + L-glutamate <33, 40> [5, 6]

S   L-valine + 2-oxoglutarate <1, 2, 9, 11, 14, 15, 18, 20, 22-25, 31-33, 36, 37, 39, 40> (<37> biosynthesis of L-valine [24]) (Reversibility: r <1, 2, 9, 11, 14, 15, 18, 20, 22-25, 31-33, 36, 37, 39, 40> [2, 4-8, 10-16, 21, 22, 24-29, 34-

39, 41-43, 46, 47, 49-51, 53]) [2, 4-8, 10-16, 21, 22, 24-29, 34-39, 41-43, 46, 47, 49-51, 53]

P 3-methyl-2-oxobutanoate + L-glutamate <33, 40> [5, 6]

S Additional information <33> (<33> specific for L-leucine, L-isoleucine or L-valine, no other amino acid would serve as an amino donor [5]) [5]

P ?

Substrates and products

S DL-2-aminoadipate + 2-oxoglutarate <40> (<40> heart enzyme [4]) (Reversibility: r <40> [4]) [4]

P 1,4-dicarboxy-2-oxobutanoate + L-glutamate

S DL-2-aminopimelate + 2-oxoglutarate <40> (<40> heart enzyme [4]) (Reversibility: r <40> [4]) [4]

P 2-oxoheptanedioate + L-glutamate

S DL-allo-isoleucine + 2-oxoglutarate <40> (<40> heart enzyme [4,18]) (Reversibility: r <40> [4,18]) [4, 18]

P 3-methyl-2-oxopentanoate + L-glutamate

S L-2-aminobutyrate + 2-oxo-isopentanoate <32> (Reversibility: r <32> [13]) [13]

P 2-oxobutanoate + L-valine

S L-2-aminobutyrate + 2-oxoglutarate <1, 17, 32, 40> (<40> heart enzyme [4]) (Reversibility: r <1, 17, 32, 40> [4, 12, 13, 30]) [4, 12, 13, 30]

P 2-oxobutanoate + L-glutamate

S L-2-aminobutyrate + 4-methyl-2-oxopentanoate <17> (Reversibility: r <17> [30]) [30]

P 2-oxobutanoate + L-leucine

S L-2-aminobutyrate + pyruvate <32> (Reversibility: r <32> [13]) [13]

P 2-oxobutanoate + L-alanine

S L-alanine + 2-oxoglutarate <9, 17, 18, 32> (Reversibility: r <9, 17, 18, 32> [13, 15, 29, 30]) [13, 15, 29, 30]

P 2-oxopropanoate + L-glutamate

S L-alanine + 2-oxoisohexanoate <18> (Reversibility: r <18> [15]) [15]

P pyruvate + L-leucine

S L-alanine + 2-oxoisopentanoate <18> (Reversibility: r <18> [15]) [15]

P pyruvate + L-valine

S L-alanine + 4-methyl-2-oxopentanoate <17> (Reversibility: r <17> [30]) [30]

P pyruvate + L-leucine

S L-alanine + pyruvate <20> (Reversibility: r <20> [41]) [41]

P pyruvate + L-alanine

S L-allo-isoleucine + 2-oxobutyrate <17> (Reversibility: r <17> [30]) [30]

P 3-methyl-2-oxopentanoate + 2-aminobutyrate

S L-allo-isoleucine + 2-oxoglutarate <33> (Reversibility: r <33> [28]) [28]

P 3-methyl-2-oxopentanoate + L-glutamate

S L-allo-isoleucine + 2-oxohexanoate <17> (Reversibility: r <17> [30]) [30]

P 3-methyl-2-oxopentanoate + 2-aminohexanoate

S L-allo-isoleucine + 2-oxoisohexanoate <17> (Reversibility: r <17> [30]) [30]

P 3-methyl-2-oxopentanoate + L-leucine

S L-allo-isoleucine + 2-oxoisopentanoate <17> (Reversibility: r <17> [30]) [30]

P 3-methyl-2-oxopentanoate + L-valine

S L-allo-isoleucine + 2-oxooctanoate <17> (Reversibility: r <17> [30]) [30]

P 3-methyl-2-oxopentanoate + 2-aminooctanoate

S L-arginine + 2-oxoglutarate <32> (Reversibility: r <32> [13]) [13]

P 2-oxo-5-guanidinopentanoate + L-glutamate

S L-asparagine + 2-oxoglutarate <1> (Reversibility: r <1> [12]) [12]

P 2,4-dioxo-4-aminobutanoate + L-glutamate

S L-aspartate + 2-oxoglutarate <9, 17, 18, 32> (Reversibility: r <9, 17, 18, 32> [13, 15, 29, 30]) [13, 15, 29, 30]

P oxaloacetate + L-glutamate

S L-aspartate + 2-oxoisohexanoate <18> (Reversibility: r <18> [15]) [15]

P oxaloacetate + L-leucine

S L-aspartate + 2-oxoisopentanoate <18> (Reversibility: r <18> [15]) [15]

P oxaloacetate + L-valine

S L-cysteine + 2-oxoglutarate <18, 20> (Reversibility: r <18, 20> [15,41]) [15, 41]

P 2-oxo-3-thiobutyrate + L-glutamate

S L-glutamate + 2-oxo-3-methylpentanoate <18> (Reversibility: r <18> [15]) [15]

P 2-oxoglutarate + L-isoleucine

S L-glutamate + 2-oxo-isohexanoate <18, 32> (Reversibility: r <18, 32> [13, 15]) [13, 15]

P 2-oxoglutarate + L-leucine

S L-glutamate + 2-oxo-isopentanoate <18, 32> (Reversibility: r <18, 32> [13, 15]) [13, 15]

P 2-oxoglutarate + L-valine

S L-glutamate + 2-oxobutyrate <32> (Reversibility: r <32> [13]) [13]

P 2-oxoglutarate + 2-aminobutyrate

S L-glutamate + 2-oxoglutarate <17, 25, 32, 33> (<25> mitochondrial enzyme [7]) (Reversibility: r <17, 25, 32, 33> [7, 13, 28, 30]) [7, 13, 28, 30]

P 2-oxoglutarate + L-glutamate

S L-glutamate + 4-methyl-2-oxopentanoate <17> (Reversibility: r <17> [30]) [30]

P 2-oxoglutarate + L-leucine

S L-glutamate + pyruvate <32> (Reversibility: r <32> [13]) [13]

P 2-oxoglutarate + L-alanine

S L-glutamine + 2-oxoglutarate <1> (Reversibility: r <1> [12]) [12, 28]

P 2,5-dioxo-5-aminopentanoate + L-glutamate

S L-histidine + 2-oxoglutarate <32> (Reversibility: r <32> [13]) [13]

P 2-oxo-3-imidazolpropanoate + L-glutamate

S L-isoleucine + 2-oxo-3-methylpentanoate <22> (Reversibility: r <22> [15, 27]) [15, 27]

P 2-oxoisohexanoate + L-isoleucine

S L-isoleucine + 2-oxo-isopentanoate <17, 20, 22, 32> (Reversibility: r <17, 20, 22, 32> [13, 15, 27, 30, 41]) [13, 15, 27, 30, 41]

P 2-oxoisohexanoate + L-valine

S L-isoleucine + 2-oxobutanoate <15> (Reversibility: r <15> [1]) [1]

P 3-methyl-2-oxopentanoate + 2-aminobutanoate

S L-isoleucine + 2-oxobutyrate <17, 32> (Reversibility: r <17, 32> [13, 30]) [13, 30]

P 2-oxoisohexanoate + 2-aminobutyrate

S L-isoleucine + 2-oxoglutarate <1, 2, 9, 12, 14, 15, 17, 18, 20, 22-25, 31-33, 36, 39, 40> (<32> lower reactivity than other origins [13]; <25> mitochondrial enzyme [7]) (Reversibility: r <1, 2, 9, 12, 14, 15, 17, 18, 20, 22-25, 31-33, 36, 39, 40> [2, 4, 5, 7, 8, 10, 12-15, 18, 20-22, 25-29, 34-36, 38, 39, 41-43, 46, 47, 49-51, 53]) [2, 4, 5, 7, 8, 10, 12-15, 18, 20-22, 25-29, 34-36, 38, 39, 41-43, 46, 47, 49-51, 53]

P 3-methyl-2-oxopentanoate + L-glutamate <33, 40> [5, 18]

S L-isoleucine + 2-oxohexanoate <17> (Reversibility: r <17> [30]) [30]

P L-norleucine + 2-oxoisohexanoate

S L-isoleucine + 2-oxoisohexanoate <17, 20, 22, 32> (Reversibility: r <17, 20, 22, 32> [13, 15, 27, 30, 41]) [13, 15, 27, 30, 41]

P 3-methyl-2-oxopentanoate + L-leucine

S L-isoleucine + 2-oxooctanoate <17> (Reversibility: r <17> [30]) [30]

P 2-oxoisohexanoate + 2-aminooctanoate

S L-isoleucine + 3-methyl-2-oxobutanoate <15> (Reversibility: r <15> [1]) [1]

P 3-methyl-2-oxopentanoate + L-valine

S L-isoleucine + 3-methylthio-2-oxobutanoate <15> (Reversibility: r <15> [1]) [1]

P 3-methyl-2-oxopentanoate + 2-amino-3-methylthiobutanoate

S L-isoleucine + 3-phenylpyruvate <20> (Reversibility: r <20> [41]) [41]

P 2-oxoisohexanoate + L-phenylalanine

S L-isoleucine + 4-methylthio-2-oxobutyrate <20> (Reversibility: r <20> [41]) [41]

P 3-methyl-2-oxopentanoate + L-methionine

S L-isoleucine + glyoxylate <18> (Reversibility: r <18> [15]) [15]

P 2-oxoisohexanoate + glycine

S L-isoleucine + oxaloacetate <18> (Reversibility: r <18> [15]) [15]

P 2-oxoisohexanoate + L-aspartate

S L-isoleucine + oxaloacetate <40> (<40> brain enzyme [6]) (Reversibility: r <40> [6]) [6]

P 3-methyl-2-oxopentanoate + L-aspartate

S L-isoleucine + pyruvate <18, 20, 32> (Reversibility: r <18, 20, 32> [13, 15, 41]) [13, 15, 41]

P 2-oxoisohexanoate + L-alanine

S L-leucine + 2-oxo-3-methiobutyrate <33> (Reversibility: r <33> [28]) [28]

P 2-oxoisohexanoate + L-methionine

S L-leucine + 2-oxo-3-methylpentanoate <18, 34> (Reversibility: r <18, 34> [15, 31]) [15, 31]
P 2-oxoisohexanoate + L-isoleucine
S L-leucine + 2-oxo-butyrate <17, 32, 33> (Reversibility: r <17, 32, 33> [13, 28, 30]) [13, 28, 30]
P 2-oxoisohexanoate + 2-aminobutyrate
S L-leucine + 2-oxo-hexanoate <33> (Reversibility: r <33> [28]) [28]
P 2-oxoisohexanoate + 2-aminohexanoate
S L-leucine + 2-oxo-isohexanoate <11, 20> (Reversibility: r <11,20> [16,41]) [16, 41]
P 2-oxoisohexanoate + L-leucine
S L-leucine + 2-oxo-pentanoate <33> (Reversibility: r <33> [28]) [28]
P 2-oxoisohexanoate + 2-aminopentanoate
S L-leucine + 2-oxoglutarate <1, 2, 9, 12, 14, 15, 17, 18, 20, 22-25, 31-33, 36, 39, 40> (Reversibility: r <1, 2, 9, 12, 14, 15, 17, 18, 20, 22-25, 31-33, 36, 39, 40> [1, 2, 4-8, 10, 12-15, 18, 20-22, 25-30, 32, 34-36, 38-43, 46-51, 53]) [1, 2, 4-8, 10, 12-15, 18, 20-22, 25-30, 32, 34-36, 38-43, 46-51, 53]
P 4-methyl-2-oxopentanoate + L-glutamate <33> [5]
S L-leucine + 2-oxohexanoate <17> (Reversibility: r <17> [30]) [30]
P 2-oxoisohexanoate + 2-aminohexanoate
S L-leucine + 2-oxoisohexanoate <11, 17, 18, 32-34> (Reversibility: r <11, 17, 18, 32-34> [13, 15, 16, 28, 30, 31]) [13, 16, 15, 28, 30, 31]
P 2-oxoisohexanoate + L-leucine
S L-leucine + 2-oxoisopentanoate <18, 33> (Reversibility: r <18, 33> [15, 28]) [15, 28]
P 2-oxoisohexanoate + L-valine
S L-leucine + 2-oxooctanoate <17> (Reversibility: r <17> [30]) [30]
P 2-oxoisohexanoate + 2-aminooctanoate
S L-leucine + 3-methyl-2-oxobutanoate <15, 17, 33, 34> (Reversibility: r <15, 17, 34> [1, 30, 31, 40]) [1, 30, 31, 40]
P 4-methyl-2-oxopentanoate + L-valine
S L-leucine + 3-methyl-2-oxopentanoate <15, 17, 33> (Reversibility: r <15, 17, 33> [1, 40]) [1, 40]
P 4-methyl-2-oxopentanoate + L-isoleucine
S L-leucine + 3-methyl-2-oxopentanoate <17> (Reversibility: r <17> [30]) [30]
P 2-oxoisohexanoate + L-amino-3-methyl-pentanoate
S L-leucine + 4-methyl-2-oxopentanoate <17, 33> (Reversibility: r <17, 33> [8, 32, 40]) [8, 32, 40]
P L-glutamate + 2-oxoglutarate
S L-leucine + DL-2-oxo-3-methylpentanoate <11, 33> (Reversibility: r <11, 33> [16, 28]) [16, 28]
P 2-oxoisohexanoate + L-isoleucine
S L-leucine + glyoxylate <18> (Reversibility: r <18> [15]) [15]
P 2-oxoisohexanoate + glycine
S L-leucine + oxaloacetate <18> (Reversibility: r <18> [15]) [15]
P 2-oxoisohexanoate + L-asparagine

S L-leucine + *p*-hydroxyphenylpyruvate <11, 17> (Reversibility: r <11, 17> [16, 30]) [16, 30]

P 2-oxoisohexanoate + L-tyrosine

S L-leucine + phenylpyruvate <11, 17, 33> (<33> 2-oxo-isohexanoic acid 100%, BCATm relative rate 4%, BCATc 6% [28]) (Reversibility: r <11, 17, 33> [16, 28, 30]) [16, 28, 30]

P 2-oxoisohexanoate + L-phenylalanine

S L-leucine + prephenate <11> (Reversibility: r <11> [16]) [16]

P 2-oxoisohexanoate + 2-amino-4-hydroxyphenyl-1,5-dicarboxypentanoate

S L-leucine + pyruvate <1, 17, 18, 32-34> (<33> 2-oxo-isohexanoic acid 100%, BCATm and BCATc, relative rate 6% [28]; <17> slight activity [30]; <1> 2-oxoglutarate 100%, relative rate 1% [12]) (Reversibility: r <1, 17, 18, 32-34> [12, 13, 15, 28, 30, 31]) [12, 13, 15, 28, 30, 31]

P 2-oxoisohexanoate + L-alanine

S L-lysine + 2-oxoglutarate <32> (Reversibility: r <32> [13]) [13]

P 2-oxo-6-aminohexanoate + L-glutamate

S L-methionine + 2-oxo-isohexanoate <32> (Reversibility: r <32> [13]) [13]

P 4-methylsulfanyl-2-oxobutanoate + L-leucine

S L-methionine + 2-oxobutyrate <32> (Reversibility: r <32> [13]) [13]

P 4-methylsulfanyl-2-oxobutanoate + 2-aminobutyrate

S L-methionine + 2-oxoglutarate <1, 9, 15, 18, 20, 25, 32, 33, 39, 40> (<18, 33> transaminated extremely poorly [15, 28]; <32> higher reactivity than other origins [13]; <25> cytoplasmic isoenzyme [7]) (Reversibility: r <1, 9, 15, 18, 20, 25, 32, 33, 39, 40> [1, 4, 7, 12, 13, 15, 18, 28, 29, 36, 39, 41, 50]) [1, 4, 7, 12, 13, 15, 18, 28, 29, 36, 39, 41, 50]

P 4-methylsulfanyl-2-oxobutanoate + L-glutamate

S L-methionine + pyruvate <32> (Reversibility: r <32> [13]) [13]

P 4-methylsulfanyl-2-oxobutanoate + L-alanine

S L-norleucine + 2-oxoglutarate <1, 9, 15, 33, 40> (<40> brain enzyme [6]) (Reversibility: r <1, 9, 15, 33, 40> [1, 2, 4, 6, 12, 28, 29]) [1, 2, 4, 6, 12, 28, 29]

P 2-oxohexanoate + L-glutamate <15, 40> [1, 2, 4, 6]

S L-norleucine + 4-methyl-2-oxopentanoate <17> (Reversibility: r <17> [30]) [30]

P 2-oxohexanoate + L-leucine

S L-norvaline + 2-oxoglutarate <1, 9, 15, 33, 40> (<40> brain enzyme [6]; <40> heart enzyme [2]) (Reversibility: r <1, 9, 15, 33, 40> [1, 2, 4, 6, 12, 28, 29]) [1, 4, 6, 12, 28, 29]

P 2-oxopentanoate + L-glutamate

S L-norvaline + 4-methyl-2-oxopentanoate <17> (Reversibility: r <17> [30]) [30]

P 2-oxopentanoate + L-leucine

S L-phenylalanine + 2-oxoglutarate <1, 9, 11, 15, 20, 25, 32> (<25> cytoplasmic isoenzyme [7]) (Reversibility: r <1, 9, 11, 15, 20, 25, 32> [7, 11-14, 16, 21, 29, 36, 39, 41]) [7, 11-14, 16, 21, 29, 36, 39, 41]

P phenylpyruvate + L-glutamate

S L-serine + 2-oxoglutarate <32> (Reversibility: r <32> [13]) [13]

P 3-hydroxy-2-oxopropanoate + L-glutamate

S L-threo-isoleucine + 2-oxoglutarate <33> (Reversibility: r <33> [28]) [28]

P 3-methyl-2-oxopentanoate + L-glutamate

S L-threonine + 2-oxoglutarate <32> (Reversibility: r <32> [13]) [13]

P 2-oxo-3-hydroxybutyrate + L-glutamate

S L-tryptophan + 2-oxoglutarate <1, 9, 15, 20, 32> (Reversibility: r <1, 9, 15, 20, 32> [12, 13, 21, 29, 39, 41]) [12, 13, 21, 29, 39, 41]

P 2-oxo-3-indolylpropanoate + L-glutamate

S L-tyrosine + 2-oxoglutarate <1, 9, 15, 20, 25> (<25> cytoplasmic isoenzyme [7]) (Reversibility: r <1, 9, 15, 20, 25> [7, 12, 14, 21, 29, 39, 41]) [7, 12, 14, 21, 29, 41]

P *p*-hydroxyphenylpyruvate + L-glutamate

S L-valine + 2-oxo-3-methylpentanoate <18> (Reversibility: r <18> [15]) [15]

P 3-methyl-2-oxobutanoate + L-isoleucine

S L-valine + 2-oxo-isohexanoate <32> (Reversibility: r <32> [13]) [13]

P 2-oxoisopentanoate + L-leucine

S L-valine + 2-oxobutyrate <17, 32> (Reversibility: r <17, 32> [13, 30]) [13, 30]

P 2-oxoisopentanoate + 2-aminobutanoate

S L-valine + 2-oxoglutarate <33> (Reversibility: r <33> [22]) [22]

P 2-oxoisopentanoate + L-glutamate

S L-valine + 2-oxoglutarate <1, 2, 9, 12, 14, 15, 17, 18, 20, 22-25, 31, 33, 36, 39, 40> (<25> mitochondrial enzyme [7]) (Reversibility: r <1, 2, 9, 12, 14, 15, 17, 18, 20, 22-25, 31, 33, 36, 39, 40> [1, 2, 4, 5, 7, 8, 10-12, 14, 15, 18, 20-22, 25-29, 34-36, 38, 39, 41-43, 46, 47, 49-51, 53]) [1, 2, 4, 5, 7, 8, 10-12, 14, 15, 18, 20-22, 25-29, 34-36, 38, 39, 41-43, 46, 47, 49-51, 53]

P 3-methyl-2-oxobutanoate + L-glutamate

S L-valine + 2-oxohexanoate <17> (Reversibility: r <17> [30]) [30]

P 2-oxoisopentanoate + 2-aminohexanoate

S L-valine + 2-oxoisopentanoate <18, 20> (Reversibility: r <18,20> [15,41]) [15, 41]

P 2-oxoisopentanoate + L-valine

S L-valine + 2-oxooctanoate <17> (Reversibility: r <17> [30]) [30]

P 2-oxoisopentanoate + 2-aminooctanoate

S L-valine + 3-methyl-2-oxopentanoate <15> (Reversibility: r <15> [1]) [1]

P 3-methyl-2-oxobutanoate + L-isoleucine

S L-valine + 4-methyl-2-oxopentanoate <17, 18> (Reversibility: r <17,18> [15,30]) [15, 30]

P 2-oxoisopentanoate + L-leucine

S L-valine + glyoxylate <18> (Reversibility: r <18> [15]) [15]

P 2-oxoisopentanoate + glycine

S L-valine + oxaloacetate <18> (Reversibility: r <18> [15]) [15]

P 2-oxoisopentanoate + L-aspartate

S L-valine + pyruvate <18, 32> (Reversibility: r <18,32> [13,15]) [13, 15]

P 2-oxoisopentanoate + L-alanine

S S-(1,1,2,2-tetrafluoroethyl)-L-cysteine + ? <17> (Reversibility: ? <17> [52]) [52]

P S-(1,1,2,2-tetrafluoroethyl)-2-oxo-3-thiobutyrate + ?

S S-(1,2-dichlorovinyl)-L-cysteine + ? <17> (Reversibility: ? <17> [52]) [52]

P S-(1,2-dichlorovinyl)-2-oxo-3-thiobutyrate + ?

S S-(2-chloro-1,1,2-trifluoroethyl)-L-cysteine + ? <17> (Reversibility: ? <17> [52]) [52]

P S-(2-chloro-1,1,2-trifluoroethyl)-2-oxo-3-thiobutyrate + ?

S S-methyl-L-cysteine + 2-oxoglutarate <40> (<40> heart enzyme [4]) (Reversibility: r <40> [4]) [4]

P S-methyl-2-oxopropanoate + L-glutamate

S β-chloro-L-alanine + ? <17> (Reversibility: ? <17> [52]) [52]

P 3-chloropyruvate + ?

S Additional information <1, 9, 15, 17, 18, 20, 25, 32, 33, 40> (<33> L-leucine is the best substrate for the heart enzyme [2]; <40> L-isoleucine is the best substrate for the heart enzyme, L-alanine, aspartic, α-amino-butyric acid and γ-aminobutyric acid, ε-aminocaproic acid, L-ornithine, L-methionine and L-phenylalanine are no substrates, pyruvate is not a good acceptor [2]; <40> activity less than 3% with glycine, L-alanine, L-lysine, L-phenylalanine, L-tryptophan, D-alloisoleucine, D-valine, D-leucine, L-threonine, L-histidine, L-arginine, L-cysteine, DL-homocyteine, β-amino-isobutyrate, γ-aminobutyrate, α-aminoisobutyrate, DL-α-aminocaprylate, L-serine, DL-homoserine, DL-β-aminobutyrate, DL-N-hydroxyleucine, DL-N-hydroxyvaline, L-aspartate, L-tyrosine, L-kynurenine, L-ornithine, DL-proline, L-glutamine, L-peniclliamine, DL-methionine sulfone, DL-methionine sulfoxide and DL-β-hydroxy-leucine [4]; <33> only little or no activity with L-tryptophan, L-phenylalanine, L-glutamine, L-alanine and L-aspartate, KIC, KIV, DL-2oxo-3-methylpentanoate, 2-oxoglutarate, 2-oxohexanoate, 2-oxopentanoate, 2-oxobutyrate, 2-oxo-3-methiobutyrate, pyruvate or phenylpyruvate are acceptors [28]; <40> B_6-vitamin-dependent enzyme [19]; <15> purified enzyme has no measurable L-aspartate-2-oxoglutarate activity [14]; <15> no activity towards L-aspartate [39]; <18> no significant activity with glycine [15]; <32> L-glutamine is no substrate [13]; <25> purified mitochondrial enzyme is not active with L-phenylalanine [7]; <17> recombinant human BCATm and BCATc have β-lyase activity towards 3 toxic L-cysteine S-conjugates, S-(1,1,2,2-tetrafluoroethyl)-L-cysteine, S-(1,2-dichlorovinyl)-L-cysteine and S-(2-chloro-1,1,2-trifluoroethyl)-L-cysteine and toward 3-chloro-L-alanine, BCATm is also active toward benzothiazolyl-L-cysteine, pyruvate formed from β-lyase substrates [52]; <17> novel co-repressor for thyroid hormone nuclear receptors [44]; <17> transamination activity with oxaloacetate is too low to be evaluated, DL-aminoadipate, glycine, L-phenylalanine, L-tyrosine and L-methionine are no substrates [30]; <1> L-aspartic acid, L-arginine, L-citrulline, L-lysine, L-ornithine, L-alanine, β-alanine and γ-aminobutyrate are not amino donors, oxalaceteate and glyoxylate are not amino acceptors, pyruvate is a weak amino acceptor, relative activity 1% [12]; <1> L-citrulline and L-methionine are poor amino donors [8]; <9> glycine and

D-leucine are no substrates, oxaloacetate, pyruvate and glyoxylate are not amino acceptors [29]; <20> with 2-oxoglutarate as acceptor L-alanine is no substrate, with L-aspartate as amino donor, oxaloacetate is no amino acceptor [41]) [2, 4, 7, 8, 12-15, 19, 28-30, 39, 41, 52]

P ?

Inhibitors

2-oxo-3-methylvalerate <18> (<18> measured in the forward reaction, concentrations higher than 1.0 mM inhibits the enzyme [15]) [15]

2-oxoisocaproate <17, 18, 32> (<18> measured in the forward reaction, concentrations higher than 1.0 mM inhibits the enzyme [15]) [13, 15, 52]

2-oxoisovalerate <18> (<18> measured in the forward reaction, concentrations higher than 1.0 mM inhibits the enzyme [15]) [15]

Co^{2+} <20> [41]

$CoCl_2$ <20> [41]

Cu^{2+} <20> [41]

$CuCl_2$ <20> [41]

$HgCl_2$ <1, 14, 40> [3, 10, 12]

KCN <17> [22]

N-ethylmaleimide <33, 40> [3, 28]

S-(1,1,2,2-tetrafluoroethyl)-L-cysteine <17> (<17> rapidly inactivated by the β-lyase substrate [52]) [52]

S-(1,2-dichlorovinyl)-L-cysteine <17> (<17> rapidly inactivated by the β-lyase substrate [52]) [52]

S-(2-chloro-1,1,2-trifluoroethyl)-L-cysteine <17> (<17> rapidly inactivated by the β-lyase substrate [52]) [52]

Tris <17> [47]

$ZnCl_2$ <20> [41]

aminooxyacetate <33> [28]

benzothiazolyl-L-cysteine <17> (<17> inhibits the L-leucine-2-oxoglutarate tranamination reaction of both isoenzymes [52]) [52]

β-chloro-L-alanine <17> (<17> rapidly inactivated by the β-lyase substrate [52]) [52]

cupric acetate <40> [3]

diethyldicarbonate <33> [28]

gabapentin <17, 33> (<17,33> structural analogue of leucine, competitive inhibitor of BCATc, does not inhibit BCATm [34]) [34]

hydrazine <9, 13, 17, 33> [22, 29]

hydroxylamine <1, 9, 13, 17, 20, 33, 40> [4, 12, 22, 29, 41]

iodoacetamide <20> [41]

iodoacetic acid <20> [41]

isocaproic acid <32> (<32> competitive inhibitor [13]) [13]

isoleucine <40> (<40> competitive inhibition of enzyme activity with leucine [2]) [2]

lead acetate <40> [3]

mersalyl <33> [28]

p-chloromercuribenzoate <1, 13, 14, 17, 33, 40> (<33> 5 mM, complete in-hibtition [22]; <40> 0.1 mM, complete inhibition [2]; <13,17> isoenzyme I, complete inhibition, isoenzyme III only partially inhibited [22]) [2, 3, 10, 12, 22, 28]
p-mercuribenzoate <37> [9]
phenylhydrazine <1, 9, 20, 40> [4, 12, 29, 41]
phenylpyruvate <11> (<11> shows substrate inhibition at 4 mM [16]) [16]
potassium cyanide <40> [4]
semicarbazide <9, 13> (<13> isoenzyme II is somewhat sensitive, isoenzyme I is not [22]) [22, 29]
silver acetate <40> [3]
sodium mersalyl <40> [3]
thiosemicarbazide <40> [4]
valine <40> (<40> competitive inhibition of enzyme activity with leucine [2]) [2]
zinc acetate <40> [3]
Additional information <13, 17, 33, 40> (<13,33> no inhibition with isonico-tinic acid hydrazide or KCN [22]; <40> less than 10% inhibition with iodoa-cetamide, iodoacetate, sodium arsenite, oxidized glutathione, potassium fer-ricyanide, stannous chloride, ceric sulfate, aluminium chloride, cobalt chlor-ide, ferric chloride, manganese chloride, magnesium chloride, calcium chlor-ide or ferrous ammonium sulfate [3]; <17> not inhibited by isonicotinic acid and semicarbazide [22]) [3, 22]

Cofactors/prosthetic groups
pyridoxal 5'-phosphate <1, 9, 14, 15, 17, 20, 31, 33, 37, 39, 40> (<33> K_m 0.025 mM [5]; <40> K_m 0.0065 mM [5]; <40> K_m 0.067 mM [2]; <1> K_m 0.00053 mM [12]; <9> K_m 0.0065 [29]; <31> BCATc [37]) [1, 2, 4-6, 10, 12, 21, 23, 24, 29, 30, 37, 39, 41, 45, 47, 50-53]
pyridoxamine 5'-phosphate <1> (<1> K_m 0.00062 mM [12]) [12]
Additional information <1> (<1> pyridoxine 5'-phosphate, pyridoxal, pyri-doxamine and pyridoxine are not effective as coenzymes [12]) [12]

Activating compounds
2-mercaptoethanol <1, 32, 33, 40> (<1> 1.0 mM, relative activity 101% [12]) [3-6, 12, 13, 19]
EDTA <1> (<1> 1.0 mM, relative activity 108% [12]) [12]
α,α'-dipyridyl <1> (<1> 1.0 mM, relative activity 103% [12]) [12]
glutathione <1> (<1> 1.0 mM, relative activity 109% [12]) [12]
neocuproin <1> (<1> 1.0 mM, relative activity 106% [12]) [12]
o-phenanthroline <1> (<1> 1.0 mM, relative activity 108% [12]) [12]
thioglycolate <1> (<1> 1.0 mM, relative activity 101% [12]) [12]

Turnover number (min^{-1})
132 <15> (L-tyrosine, <15> pH 8.0, 25°C [39]) [39]
174 <15> (L-phenylalanine, <15> pH 8.0, 25°C [39]) [39]
222 <15> (L-tryptophan, <15> pH 8.0, 25°C [39]) [39]
1020 <15> (L-methionine, <15> pH 8.0, 25°C [39]) [39]

1140 <15> (L-valine, <15> pH 8.0, 25°C [39]) [39]
2880 <15> (L-isoleucine, <15> pH 8.0, 25°C [39]) [39]
4680 <15> (L-leucine, <15> pH 8.0, 25°C [39]) [39]

Specific activity (U/mg)

0.006 <22> [27]
0.01855 <33> [5]
0.02 <18> [15]
0.65 <33> [46]
0.7 <20> (<20> substrate L-phenylalanine [36]) [36]
0.7273 <40> (<40> brain enzyme, substrate L-leucine [6]) [6]
1.82 <40> (<40> heart muscle enzyme, substrate L-leucine [2]) [2]
2.4 <20> (<20> substrate L-methionine [36]) [36]
2.5 <32> [13]
13.8 <9> [29]
15.9 <15> (<15> substrate L-valine + 2-oxoglutarate [14]) [14]
21.8 <33> [22]
23.9 <15> [21]
25 <20> (<20> substrate L-valine [36]) [36]
25 <25> (<25> cytoplasmic enzyme [7]) [7]
25.3 <40> (<40> in presence of 2-mercaptoethanol [3]) [3]
27 <20> (<20> substrate L-leucine [36]) [36]
27.3 <15> (<15> substrate L-isoleucine + 2-oxoglutarate [14]) [14]
42.8 <1> [12]
45.2 <40> [23]
47 <20> (<20> substrate L-isoleucine [36]) [36]
48.8 <25> (<25> mitochondrial enzyme [7]) [7]
49.7 <1> [8]
50-53 <40> [4]
66 <33> [26]
74 <13> [22]
85 <17> (<17> recombinant cytosolic enzyme [38]) [38]
89.81 <31> [35]
94 <20> (<20> substrate L-isoleucine + 2-oxoglutarate [41]) [41]
98.33 <33> [28]
125 <17> (<17> recombinant mitochondrial enzyme [38]) [38]
Additional information <15, 40> (<40> specific activity 4660 units, 1 unit is
defined as the change of absorbance units at 348 nm and 37° within 30 min
[19]; <15> final preparation, specific activity 23.9 U/mg [39]) [19, 39]

K$_m$-Value (mM)

0.06 <17> ((R)-3-methyl-2-oxopentanoate, <17> pH 8.0, 25°C, substrate L-
glutamate [40]) [40]
0.07 <33> ((R,S)-3-methyl-2-oxopentanoate, <33> pH 8.3, 25°C [32,40]) [32,
40]
0.09 <17> ((S)-3-methyl-2-oxopentanoate, <17> pH 8.3, 37°C, leucine as amino group donor [40]) [40]

0.1 <18> (2-oxoisohexanoate, <18> pH 8.4, 30°C, glutamate as amino group donor [15]) [15]

0.11 <33> (3-methyl-2-oxobutanoate, <33> pH 8.3, 25°C [32,40]) [32, 40]

0.14 <33> (4-methyl-2-oxopentanoate, <33> pH 8.3, 25°C [32,40]) [32, 40]

0.17 <17> ((S)-3-methyl-2-oxopentanoate, <17> pH 8.0, 25°C, substrate L-glutamate [40]) [40]

0.2 <18> (2-oxo-3-methylpentanoate, <18> pH 8.4, 30°C [15]) [15]

0.2 <15> (DL-2-oxo-3-methyl-n-pentanoate, <15> pH 8.0, 37°C [14]) [14]

0.21 <17> (2-oxoisohexanoate, <17> pH 8.3, 37°C, leucine as amino group donor [40]) [40]

0.21 <11> (DL-2-oxo-3-methylpentanoate, <11> pH 7.5, 25°C [16]) [16]

0.22 <9> (L-valine, <9> pH 8.0, 37°C [29]) [29]

0.24 <15> (2-oxoglutarate, <15> pH 8.0, 25°C, substrate L-tyrosine [21,39]) [21, 39]

0.25 <9> (L-leucine, <9> pH 8.0, 37°C [29]) [29]

0.26 <15> (2-oxoglutarate, <15> pH 8.0, 25°C, substrate L-phenylalanine [21,39]) [21, 39]

0.29 <14> (L-valine) [10]

0.3 <32> (2-oxoglutarate, <32> pH 8.0, 30°C [13]) [13]

0.3 <32> (L-leucine, <32> pH 8.0, 30°C [13]) [13]

0.3 <33> (L-leucine, <33> pH 8.4, 25°C, 2-oxoglutarate as amino group acceptor [40]) [40]

0.33 <1> (2-oxoisopentanoate, <1> pH 8.0, 37°C [12]) [12]

0.34 <11> (2-oxoisopentanoate, <11> pH 7.5, 25°C [16]) [16]

0.37 <11> (2-oxoisohexanoate, <11> pH 7.5, 25°C [16]) [16]

0.37 <17> (2-oxoisopentanoate, <17> pH 8.3, 37°C, leucine as amino group donor [40]) [40]

0.4 <18> (2-oxoisopentanoate, <18> pH 8.4, 30°C [15]) [15]

0.41 <17> (2-oxohexanoate, <17> pH 8.3, 37°C, leucine as amino group donor [40]) [40]

0.42 <15> (L-isoleucine, <15> pH 8.0, 25°C, 2-oxoglutarate as amino group acceptor [21,39]) [21, 39]

0.43 <9> (L-isoleucine, <9> pH 8.0, 37°C [29]) [29]

0.52 <15> (L-isoleucine, <15> pH 8.0, 37°C [14]) [14]

0.53 <17> (4-methyl-2-oxopentanoate, <17> pH 8.0, 25°C, substrate L-isoleucine [40]) [40]

0.56 <15> (2-oxoglutarate, <15> pH 8.0, 25°C, substrate L-tryptophan [21,39]) [21, 39]

0.56 <15> (2-oxoisohexanoate, <15> pH 8.0, 37°C [14]) [14]

0.56 <15> (2-oxoisopentanoate, <15> pH 8.0, 37°C [14]) [14]

0.56 <9> (L-glutamate, <9> pH 8.0, 37°C [29]) [29]

0.56 <40> (L-leucine, <40> pH 8.2, 37°C, brain enzyme [6]) [6]

0.56 <17> (isoleucine, <17> pH 8.4, 25°C, 2-oxoglutarate as amino group acceptor [40]) [40]

0.57 <40> (2-oxoglutarate, <40> pH 8.2, 37°C, brain enzyme [6]) [6]

0.58 <15> (L-leucine, <15> pH 8.0, 37°C [14]) [14]

0.6 <17> (β-chloro-L-alanine, <17> pH 7.4, 23°C, recombinant BCATm [52])
[52]

0.61 <17> (3-methyl-2-oxobutanoate, <17> pH 8.0, 25°C, substrate L-isoleucine [40]) [40]

0.62 <17> (L-leucine, <17> pH 8.4, 25°C, 2-oxoglutarate as amino group acceptor [40]) [40]

0.63 <40> (2-oxoglutarate, <40> pH 8.6, 25°C, heart enzyme [2]) [2]

0.64 <17> (2-oxovalerate, <17> pH 8.3, 37°C, L-leucine as amino group donor [40]) [40]

0.67 <40> (L-isoleucine, <40> pH 8.2, 37°C, brain enzyme [6]) [6]

0.68 <33> (2-oxoglutarate, <33> pH 8.4, 25°C, L-leucine as amino group donor [40]) [40]

0.75 <33> (L-leucine, <33> pH 8.2, 37°C [5]) [5]

0.77 <9> (2-oxoglutarate, <9> pH 8.0, 37°C, leucine as amino group donor
[29]) [29]

0.79 <14> (L-leucine) [10]

0.8 <18> (2-oxoglutarate, <18> pH 8.4, 30°C, L-valine as amino group donor
[15]) [15]

0.8 <18> (L-leucine, <18> pH 8.4, 30°C [15]) [15]

0.84 <33> (L-isoleucine, <33> pH 8.2, 37°C [5]) [5]

0.89 <15> (L-phenylalanine, <15> pH 8.0, 25°C [21]) [21]

0.89 <15> (L-phenylalanine, <15> pH 8.0, 25°C, 2-oxoglutarate as amino
group acceptor [39]) [39]

0.9 <14> (L-isoleucine) [10]

0.91 <1> (2-oxoisohexanoate, <1> pH 8.0, 37°C [12]) [12]

1 <13> (2-oxoglutarate, <13> pH 8.0, 37°C, isoenzyme III, isoleucine as amino group donor [22]) [22]

1 <15> (2-oxoglutarate, <15> pH 8.0, 25°C, substrate L-methionine [21,39])
[21, 39]

1 <17> (2-oxoglutarate, <17> pH 8.0, 37°C, isoenzyme III, leucine as amino
group donor [22]) [22]

1 <33> (2-oxoglutarate, <33> pH 8.2, 37°C [5]) [5]

1.1 <17> (2-oxoglutarate, <17> pH 8.0, 37°C, isoenzyme III, isoleucine as
amino group donor [22]) [22]

1.1 <32> (2-oxoisohexanoate, <32> pH 8.0, 30°C [13]) [13]

1.2 <32> (L-leucine, <32> pH 8.0, 30°C, activated with β-mercaptoethanol
[13]) [13]

1.2 <11> (phenylpyruvate, <11> pH 7.5, 25°C [16]) [16]

1.28 <15> (2-oxoglutarate, <15> pH 8.0, 37°C [14]) [14]

1.3 <18> (L-glutamate, <18> pH 8.4, 30°C, 2-oxoisocaproate as amino group
acceptor [15]) [15]

1.4 <13> (2-oxoglutarate, <13> pH 8.0, 37°C, isoenzyme III, leucine as amino
group donor [22]) [22]

1.4 <40> (L-valine, <40> pH8.2, 37°C, brain enzyme [6]) [6]

1.4 <11> (prephenate, <11> pH 7.5, 25°C [16]) [16]

1.5 <18> (2-oxoglutarate, <18> pH 8.4, 30°C, leucine as amino group donor
[15]) [15]

1.52 <17> (2-oxoglutarate, <17> pH 8.3, 37°C, leucine as amino group donor [40]) [40]

1.54 <17> (2-oxooctanoate, <17> pH 8.3, 37°C, leucine as amino group donor [40]) [40]

1.54 <17> (L-alloisoleucine, <17> pH 8.4, 25°C, 2-oxoglutarate as amino group acceptor [40]) [40]

1.66 <14> (L-leucine, <14> [10]) [10]

1.7 <13> (2-oxoglutarate, <13> pH 8.0, 37°C, isoenzyme III, valine as amino group donor [22]) [22]

1.7 <15> (2-oxoglutarate, <15> pH 8.0, 25°C, substrate L-valine [21,39]) [21, 39]

1.7 <17> (2-oxoglutarate, <17> pH 8.0, 37°C, isoenzyme III, valine as amino group donor [22]) [22]

1.7 <1> (L-norleucine, <1> pH 8.3, 37°C, 4-methyl-2-oxopentanoate as amino group acceptor [12]) [12]

1.8 <18> (2-oxoglutarate, <18> pH 8.4, 30°C, isoleucine as amino group donor [15]) [15]

1.82 <1> (L-leucine, <1> pH 8.0, 37°C [12]) [12]

2 <9> (L-phenylalanine, <9> pH 8.0, 37°C [29]) [29]

2.1 <13> (2-oxoglutarate, <13> pH 8.0, 37°C, isoenzyme I, leucine as amino group donor [22]) [22]

2.2 <15> (L-leucine, <15> pH 8.0, 25°C [21]) [21]

2.2 <15> (L-leucine, <15> pH 8.0, 25°C, 2-oxoglutarate as amino group acceptor [39]) [39]

2.3 <13> (L-leucine, <13> pH 8.0, 37°C, isoenzyme III, 2-oxoglutarate as amino group acceptor [22]) [22]

2.4 <15> (2-oxoglutarate, <15> pH 8.0, 25°C, substrate L-isoleucine [21,39]) [21, 39]

2.4 <17> (2-oxoglutarate, <17> pH 8.4, 25°C, substrate L-alloisoleucine [40]) [40]

2.45 <33> (L-glutamate, <33> pH 8.3, 25°C, (R,S)-3-methyl-2-oxopentanoate as amino group acceptor [32]) [32]

2.5 <1> (2-oxo-3-methylpentanoate, <1> pH 8.0, 37°C [12]) [12]

2.7 <13> (2-oxoglutarate, <13> pH 8.0, 37°C, isoenzyme I, valine as amino group donor [22]) [22]

2.7 <15> (L-isoleucine, <15> pH 8.0, 25°C, 2-oxoglutarate as amino group acceptor [39]) [39]

2.7 <15> (L-valine, <15> pH 8.0, 25°C [21]) [21]

2.9 <9> (L-tryptophan, <9> pH 8.0, 37°C [29]) [29]

2.96 <17> (L-valine, <17> pH 8.4, 25°C, 2-oxoglutarate as amino group acceptor [40]) [40]

3 <17> (L-isoleucine, <17> pH 8.0, 37°C, isoenzyme III, 2-oxoglutarate as amino group acceptor [22]) [22]

3 <13> (L-valine, <13> pH 8.0, 37°C, isoenzyme III, 2-oxoglutarate as amino group acceptor [22]) [22]

3.03 <1> (L-isoleucine, <1> pH 8.0, 37°C [12]) [12]

3.13 <15> (L-valine, <15> pH 8.0, 37°C [14]) [14]

3.2 <32> (L-glutamate, <32> pH 8.0, 30°C [13]) [13]

3.6 <33> (L-glutamate, <33> pH 8.3, 25°C, 3-methyl-2-oxobutanoate as amino group acceptor [32]) [32]

3.6 <13> (L-leucine, <13> pH 8.0, 37°C, isoenzyme I, 2-oxoglutarate as amino group acceptor [22]) [22]

3.62 <17> (2-oxobutyrate, <17> pH 8.3, 37°C, leucine as amino group donor [40]) [40]

3.8 <40> (L-isoleucine, <40> pH 8.6, 25°C [2]) [2]

3.8 <40> (L-leucine, <40> pH 8.6, 25°C [2]) [2]

4 <17> (2-oxoglutarate, <17> pH 8.4, 25°C, valine as amino group donor [40]) [40]

4 <33> (2-oxoglutarate, <33> pH 8.0, 37°C, leucine as amino group donor [22]) [22]

4.2 <13> (L-isoleucine, <13> pH 8.0, 37°C, isoenzyme III, 2-oxoglutarate as amino group acceptor [22]) [22]

4.3 <33> (L-valine, <33> pH 8.2, 37°C [5]) [5]

4.5 <17> (L-glutamate, <17> pH 8.0, 25°C, (R)-3-methyl-2-oxopentanoate as amino group acceptor [40]) [40]

4.57 <1> (2-oxoglutarate, <1> pH 8.0, 37°C, leucine as amino group donor [12]) [12]

4.7 <1> (L-norvaline, <1> pH 8.3, 37°C, 4-methyl-2-oxopentanoate as amino group acceptor [12]) [12]

4.8 <17> (2-oxoglutarate, <17> pH 8.0, 37°C, isoenzyme I, isoleucine as amino group donor [22]) [22]

5 <17> (2-oxoglutarate, <17> pH 8.0, 37°C, isoenzyme I, valine as amino group donor [22]) [22]

5.5 <17> (2-oxoglutarate, <17> pH 8.4, 25°C, isoleucine as amino group donor [40]) [40]

5.72 <1> (2-oxoglutarate, <1> pH 8.0, 37°C, valine as amino group donor [12]) [12]

6.45 <1> (L-valine, <1> pH 8.0, 37°C, 2-oxoglutarate as amino group acceptor [12]) [12]

6.6 <15> (2-oxoglutarate, <15> pH 8.0, 25°C, substrate L-leucine [21,39]) [21, 39]

6.65 <33> (L-glutamate, <33> pH 8.3, 25°C, 4-methyl-2-oxopentanoate as amino group acceptor [32]) [32]

6.67 <1> (2-oxoglutarate, <1> pH 8.0, 37°C, isoleucine as amino group donor [12]) [12]

6.8 <17> (L-leucine, <17> pH 8.0, 37°C, isoenzyme III, 2-oxoglutarate as amino group acceptor [22]) [22]

7 <15> (L-tyrosine, <15> pH 8.0, 25°C, 2-oxoglutarate as amino group acceptor [21,39]) [21, 39]

7.4 <13> (L-valine, <13> pH 8.0, 37°C, isoenzyme I, 2-oxoglutarate as amino group acceptor [22]) [22]

7.6 <17> (2-oxoglutarate, <17> pH 8.0, 37°C, isoenzyme I, leucine as amino group donor [22]) [22]

8.3 <17> (L-valine, <17> pH 8.0, 37°C, isoenzyme III, 2-oxoglutarate as amino group acceptor [22]) [22]

8.4 <17> (S-(1,1,2,2-tetrafluoroethyl)-L-cysteine, <17> pH 7.4, 23°C, recombinant BCATm [52]) [52]

10 <13> (2-oxoglutarate, <13> pH 8.0, 37°C, isoenzyme I, isoleucine as amino group donor [22]) [22]

10 <1> (L-aspartate, <1> pH 8.3, 37°C, 4-methyl-2-oxopentanoate as amino group acceptor [12]) [12]

10.3 <17> (L-isoleucine, <17> pH 8.0, 37°C, isoenzyme I, 2-oxoglutarate as amino group acceptor [22]) [22]

10.4 <17> (L-glutamate, <17> pH 8.0, 25°C, 3-methyl-2-oxobutanoate as amino group acceptor [40]) [40]

11 <40> (L-valine, <40> pH 8.6, 25°C [2]) [2]

11.1 <33> (L-leucine, <33> pH 8.0, 37°C [22]) [22]

12.5 <33> (2-oxoglutarate, <33> pH 8.0, 37°C, valine as amino group donor [22]) [22]

14.2 <33> (2-oxoglutarate, <33> pH 8.0, 37°C, isoleucine as amino group donor [22]) [22]

16.7 <1> (glutamate, <1> pH 8.0, 37°C, leucine as amino group donor [12]) [12]

18.1 <17> (L-glutamate, <17> pH 8.3, 37°C, 4-methyl-2-oxopentanoate as amino group acceptor [40]) [40]

18.2 <1> (L-glutamate, <1> pH 8.0, 37°C, isoleucine as amino group donor [12]) [12]

19 <15> (L-methionine, <15> pH 8.0, 25°C, 2-oxoglutarate as amino group acceptor [21,39]) [21, 39]

21.3 <17> (L-glutamate, <17> pH 8.0, 25°C, (S)-3-methyl-2-oxopentanoate as amino group acceptor [40]) [40]

22.2 <13> (L-isoleucine, <13> pH 8.0, 37°C, isoenzyme I, 2-oxoglutarate as amino group acceptor [22]) [22]

22.2 <33> (L-isoleucine, <33> pH 8.0, 37°C [22]) [22]

24 <11> (L-glutamate, <11> pH 7.5, 25°C [16]) [16]

25 <1> (L-glutamate, <1> pH 8.0, 37°C, valine as amino group donor [12]) [12]

25 <17> (L-leucine, <17> pH 8.0, 37°C, isoenzyme I, 2-oxoglutarate as amino group acceptor [22]) [22]

28.3 <17> (L-glutamate, <17> pH 8.0, 25°C, 4-methyl-2-oxopentanoate as amino group acceptor [40]) [40]

30.72 <17> (pyruvate, <17> pH 8.3, 37°C, leucine as amino group donor [40]) [40]

30.8 <17> (L-valine, <17> pH 8.0, 37°C, isoenzyme I, 2-oxoglutarate as amino group acceptor [22]) [22]

35.8 <17> (2-aminobutyrate, <17> pH 8.3, 37°C, 4-methyl-2-oxopentanoate as amino group acceptor [40]) [40]

72 <15> (L-tryptophan, <15> pH 8.0, 25°C, 2-oxoglutarate as amino group acceptor [21,39]) [21, 39]

143 <33> (L-valine, <33> pH 8.0, 37°C, 2-oxoglutarate as amino group acceptor [22]) [22]

K_i-Value (mM)

2.1 <33> (4-methyl-2-oxopentanoate, <33> pH 8.3, 25°C [32]) [32]
3.8 <32> (α-ketoisocaproate, <32> pH 8.0, 30°C [13]) [13]
6 <32> (isocaproic acid, <32> unactivated enzyme [13]) [13]
14 <32> (isocaproic acid, <32> activated enzyme [13]) [13]

pH-Optimum

6 <14> [10]
7.5 <20> [41]
7.5-8 <33> [22]
8 <17, 40> (<40> brain enzyme [6]; <17> isoenzyme III [22]) [6, 22]
8-8.7 <40> [23]
8.2 <33> [5, 32]
8.3 <9> [29]
8.3 <12, 17> (<12> isoenzyme AT-II [20]) [20, 30]
8.3-8.5 <40> [4]
8.4 <18> [15]
8.5 <25> (<25> cytoplasmic enzyme [7]) [7]
8.5-9 <1> [8]
8.5-9.5 <13> (<13> isoenzyme III [22]) [22]
8.6 <12, 40> (<12> isoenzyme AT-I [20]) [2, 20]
8.7 <18> (<18> using valine in the reverse reaction [15]) [15]
8.7 <40> (<40> heart enzyme [6]) [6]
8.8 <25> (<25> mitochondrial enzyme [7]) [7]
9 <1, 13, 17> (<13,17> isoenzyme I [22]) [12, 22]

pH-Range

4.6-9 <15> [11]
6-10 <40> [4]
6.5-10 <17> [30]
7-9 <33> [32]

Temperature optimum (°C)

35-40 <20> [41]
40-45 <1> [12]

Temperature range (°C)

30-60 <40> [2]

4 Enzyme Structure

Molecular weight

34000 <14> (<14> gel filtration [10]) [10]
39000 <40> (<40> calculated from S20 [6]) [6]
41000 <33> (<33> fetal liver nuclei, Western blot analysis [46]) [46]
41200 <33> (<33> predicted from amino acid sequence [46]) [46]
42330 <31> (<31> BCATm, calculated from amino acid sequence [35]) [35]
42500 <38> (<38> calculated from amino acid sequence [33]) [33]

43000 <33> (<33> fetal liver supernatant, Western blot analysis [46]) [46]
43070 <31> (<31> BCATc, calculated from amino acid sequence [37]) [37]
50000 <33> (<33> gel filtration [26]) [26]
64000 <20> (<20> gel filtration [41]) [41]
67500 <25> (<25> mitochondrial isoenzyme, gel filtration [7]) [7]
68000 <33> (<33> gel filtration [22]) [22]
75000 <40> (<40> sedimentation equilibrium [4]; <40> gel filtration [19,23]) [4, 19, 23]
79000 <33> (<33> sucrose density gradient centrifugation [22]) [22]
80000 <1, 13, 17> (<13,17> isoenzyme I, gel filtration [22]; <1> sedimentation equilibrium [12]) [12, 22]
82000 <1> (<1> gel filtration [12]) [12]
88000 <25> (<25> cytoplasmic isoenzyme, gel filtration [7]) [7]
90000 <17> (<17> isoenzyme III, gel filtration [22]) [22]
91000 <33> (<33> BCATc, gel filtration [28]) [28]
93000 <9, 12> (<12> both isoenzymes, gel filtration [20]; <9> gel filtration [29]) [20, 29]
95000 <18> (<18> gel filtration [15]) [15]
96000 <13> (<13> isoenzyme III, gel filtration [22]) [22]
150000 <22> (<22> gel filtration [27]) [27]
180000 <15> (<15> gel filtration [11]) [11]
182300 <15> (<15> sedimentation equilibrium centrifugation [14]) [14]
183000 <37> (<37> sedimentation equilibrium ultracentrifugation [9]) [9]
185000 <15> (<15> gel filtration [14]) [14]
195000 <15> (<15> gel filtration combined with angle laser light scattering technique [21]) [21]
197000 <15> (<15> sedimentation equilibrium [21]) [21]
204900 <37> (<37> calculated from amino acid sequence [24]) [24]

Subunits

dimer <9, 13, 17, 20, 31, 33, 40> (<33> 2 * 35000, SDS-PAGE [22]; <17> 2 * 41730, recombinant BCATm, homodimer [52]; <17> 2 * 43400, recombinant BCATc, homodimer [52]; <17> homodimer [47]; <13> 2 * 39000, isoenzyme I, SDS-PAGE [22]; <13> 2 * 43000, isoenzyme III, SDS-PAGE [22]; <9> 2 * 47000, SDS-PAGE [29]; <31> 2 * 41000, homodimer, SDS-PAGE [35]; <20> 2 * 38000, homodimer, SDS-PAGE [41]) [19, 22, 23, 29, 35, 41, 47, 52]

hexamer <15, 37> (<15> 6 * 34000, SDS-PAGE [21]; <15> 6 * 31500, SDS-PAGE [14]; <15> 6 * 33960 [39]; <15> 6 * 31500, trimer of dimers [45]; <37> 6 * 31500, SDS-PAGE [9]; <37> 6 * 33920, calculated from amino acid sequence [24]) [9, 11, 14, 21, 24, 39, 45]

monomer <33, 40> (<33> 1 * 43000, SDS-PAGE [26]; <33> 1 * 41000, BCATc, SDS-PAGE [28]; <33> 1 * 47000, BCATm, SDS-PAGE [28]) [6, 26, 28]

5   Isolation/Preparation/Mutation/Application

Source/tissue

aorta <10> (<10> endothelial cells [32]) [32]

astrocyte <33> [28, 34]

blood <31, 33> [43, 46]

brain <31, 33, 40> [6, 18, 28, 34, 37, 40, 46, 48]

fibroblast <17> [30, 40]

heart <33, 40> (<33,40> heart muscle [2,4]) [2-4, 6, 18, 19, 23, 26, 28, 32, 40, 46, 48]

hepatocyte <33> [28, 46]

kidney <33, 40> [2, 6, 28, 46, 48]

leaf <18> [15]

liver <33, 40> (<33,40> only slight activity [2,23]; <33> negligible activity in adult, high activity in fetal liver [28,46,48]) [2, 5, 18, 23, 28, 46, 48]

muscle <33> (<33> mixed [28]) [28]

neuron <33> [34]

ovary <33> [18, 28, 34, 40, 46]

pancreas <13, 17, 33, 40> [22, 23, 28, 46]

placenta <31, 33> [18, 28, 34, 35, 37, 40, 46]

root <18> [15]

seedling <18> [15]

shoot <18> [15]

skeletal muscle <31, 33, 40> [2, 18, 22, 23, 35, 37, 43, 46]

skin <17, 33> [30, 40, 46]

stomach <33, 40> [23, 46]

submandibular gland <21> [17]

Localization

cytoplasm <12, 33, 40> [2, 5, 6, 18, 20, 46]

cytosol <17, 25, 31, 33-35, 38, 40> [7, 23, 28, 30, 31, 33, 34, 37, 38, 43, 48, 52]

microsome <33> [2, 5]

mitochondrion <2, 17, 25, 31, 33-35, 38, 40> (<33> mainly located in the cytoplasm [5]; <25> inner membrane-matrix [7]; <31,34> mitochondrial matrix [31,35]) [2, 5-7, 23, 26, 28, 30, 31, 33-35, 38, 40, 43, 44, 46-49, 51, 52]

nucleus <17, 33> (<17> recombinant enzyme, expressed in transfected cells [44]) [5, 44, 46]

plastid <2> [49]

Purification

<1> (partial [8]) [8, 12]

<9> [29]

<11> (partially [16]) [16]

<12> (2 isoenzymes, AT-I and AT-II [20]) [20]

<13> [22]

<14> (partially [10]) [10]

<15> (recombinant enzyme [21,39]) [11, 14, 21, 25, 39]

<17> (recombinant Bcatm and BCATc, expressed in E. coli [38,48]) [22, 38, 47, 48]
<18> (partially [15]) [15]
<20> [41]
<21> (2 isoenzymes, partially [17]) [17]
<22> (partial [27]) [27]
<23> (partial [27]) [27]
<24> (partial [27]) [27]
<25> (partially [7]) [7]
<31> [35]
<32> (partially [13]) [13]
<33> (mitochondrial and cytosolic isoenzymes [28]; partially [32]) [5, 18, 22, 26, 28, 32, 34, 40]
<37> [9, 24]
<40> (hog heart, partial [2]) [2-4, 6, 19, 23]

Crystallization
<15> (crystallized in 2 crystal systems, monoclinic by sitting drop crystallization and tetragonal by hanging drop, monoclinic space group C2, a=93.9, b=143.6, c=143.9 and β=134.3°, tetragonal space group P422 or P4(1)22 and cell dimensions of a=b=101 A and c=249 A [25]; space group C222(1) [45]; space group C_2 [53]) [25, 39, 45, 53]
<17> (BCATm, orthorhombic space group P2(1)2(1)2(1), monoclinic P2(1), trigonal P3(2) [47]) [47, 51]
<37> (crystallized from recombinant E. coli [24]) [24]

Cloning
<2> (cDNA of 6 BCAT genes cloned, transient expression in Nicotiana tabacum protoplasts, complementation analysis in Saccharomyces cerevisiae [49]) [49]
<3> [49]
<4> [49]
<5> [49]
<6> [49]
<7> [49]
<8> [49]
<15> (ilvE gene of Escherichia coli is inserted into the region downstream of the tac-promotor, enzyme overproduced by about a 100fold in Escherichia coli W3110 [21]; K-12 strain carries the ilvE gene both on the host chromosome and on a plasmid [14]; cloning of ilv gene, Escherichia coli AB2227 transformed [39]) [14, 21, 39]
<16> [44]
<17> (isoform of BCATm, cloned using a yeast two-hybrid system, expressed in transfected monkey kidney CV1 cells [44]; vector pET-28a, overexpression of recombinant BCATm and BCATc in Escherichia coli [38]; recombinant BCATm and BCATc [52]) [38, 44, 48, 52]
<19> [41]

<20> (NCDO 763, gene cloned and sequenced with luciferase gene as the reporter [41]; LM0230 ilvE gene identified by complementation in Escherichia coli DL39 [36]) [36, 41]

<26> [35, 37]

<27> [37]

<28> [37]

<29> [37]

<30> [37]

<31> (BCATm cDNA [35]; cytosolic enzyme, cDNA [37]) [35, 37]

<33> (RT/PCR of BCATm cDNA [46]) [38, 46]

<34> (genes bat1 and bat2 [31]) [31]

<37> (ilvGEDAY gene cluster expressed in Escherichia coli JA199 [24]) [24]

<38> (isolation of eca39/BCAT gene [33]) [33]

<39> (ilvE gene cloned using degenerate oligonucleotides and PCR [50]) [50]

Application

nutrition <20, 39> (<20> widely used as starter in the cheese industry [41]; <39> used in the meat industry [50]) [41, 50]

6 Stability

pH-Stability

4.6-9 <15> (<15> almost insensitive to the pH of the medium [11]) [11]

5-10 <15> (<15> stable within 10 min [21]) [21]

5.5-6 <1> (<1> most stable in this range [12]) [12]

6-7 <40> (<40> most stable in this range [4]) [4]

6.5-9 <15> (<15> stable for 24 h at 25° [21]) [21]

8 <40> (<40> hog heart enzyme is rather unstable at pH 8.0 [2]) [2]

Temperature stability

0-40 <1> (<1> stable at 0-30°C, rapidly inactivated over 40°C at pH 6.0 [12]) [12]

30-60 <40> (<40> enzyme is inactivated by heat [2]) [2]

50-60 <20> (<20> partially inactivated by heating at 50°C, 40% inactivation after 30 min, more than 80% inactivation after 30 min at 60°C [41]) [41]

50-65 <33> (<33> labile to heat, activity is completely lost by heating at 65°C for 1 min, about 30% is lost within 1 min at 50°C and about 70% within 10 min [22]) [22]

65 <13, 17> (<13,17> not inactivated by heating for 10 min [22]) [22]

General stability information

<13>, 2-mercaptoethanol protects against inactivation by sulfhydryl reagents [22]

<17>, 2-mercaptoethanol protects against inactivation by sulfhydryl reagents [22]

<18>, enzyme is more stable in phosphate buffer than in Tris-Cl buffer, addition of glycerol increases stability [15]

<20>, pure enzyme is not stable [41]

<33>, 2-mercaptoethanol protects against inactivation by sulfhydryl reagents [22]

<33>, not stable to freezing without addition of dithiothreitol [26]

<40>, caproate protects the enzyme [19]

<40>, pyridoxal 5'-phosphate and mercaptoethanol stabilizes the heart enzyme, addition of 2-oxoglutarate does not increase its stability [2]

Storage stability

<9>, -20°C, 0.05 M potassium phosphate buffer, pH 7.5, 10% glycerol, 1 mN DTT, 0.1 M pyridoxal 5'-phosphate, purified enzyme can be stored at least for 3 months without loss of activity [29]

<18>, -25°C, partially purified enzyme is stable for several weeks [15]

<20>, -20°C, storage results in a very rapid loss of all activity [41]

<20>, 0°C, storage results in a very rapid loss of all activity [41]

<20>, 6°C, fairly stable for 1 week [41]

<33>, -70°C, when stored in presence of dithiothreitol activity is stable for at least 4 weeks [26]

<33>, 4°C, stable for several weeks without addition of dithiothreitol [28]

<40>, -10°C, concentrated solution frozen for 1month retains 90% of its original activity [4]

<40>, 2°C, preparation concentrated by ultracentrifugation, enzyme retains 75% of its activity for 4 months [19, 23]

<40>, 5°C, pH 7.0, purified enzyme loses activity, about 20% in 1 week [4]

<40>, 5°C, remarkably stable in crude extract when stored in 0.05 M sodium caproate, 0.5 mM EDTA, ph 6.1, no loss of activity after 1 month [3]

<40>, 5°C, stored in 0.05 M sodium diphosphate, 0.1 M β-mercaptoethanol, pH 8.5, loses about 50% activity over a period of several days [3]

References

[1] Rudman, D.; Meister, A.: Transamination in Escherichia coli. J. Biol. Chem., **200**, 591-604 (1953)

[2] Ichihara, A.; Koyama, E.: Transaminase of branched chain amino acids. I. Branched chain amino acids-α-ketoglutarate transaminase. J. Biochem., **59**, 160-169 (1966)

[3] Taylor, R.T.; Jenkins, W.T.: Leucine aminotransferase. 3. Activation by β-mercaptoethanol. J. Biol. Chem., **241**, 4406-4410 (1966)

[4] Taylor, R.T.; Jenkins, W.T.: Leucine aminotransferase. II. Purification and characterization. J. Biol. Chem., **241**, 4396-4405 (1966)

[5] Aki, K.; Ogawa, K.; Ichihara, A.: Transaminases of branched chain amino acids. IV. Purification and properties of two enzymes from rat liver. Biochim. Biophys. Acta, **159**, 276-284 (1968)

[6] Aki, K.; Yokojima, A.; Ichihara, A.: Transaminase of branched chain amino acids. VI. Purification and properties of the hog brain enzyme. J. Biochem., **65**, 539-544 (1969)

[7] Collins, M.; Wagner, R.P.: Branded chain amino acid aminotransferases of Neurospora crassa. Arch. Biochem. Biophys., **155**, 184-193 (1973)

[8] Tachiki, T.; Tochikura, T.: Separation of L-leucine-pyruvate and L-leucine-α-ketoglutarate transaminases in Acetobacter suboxydans and identification of their reaction products. Agric. Biol. Chem., **37**, 1439-1448 (1973)

[9] Lipscomb, E.L.; Horton, H.R.; Armstrong, F.B.: Molecular weight, subunit structure, and amino acid composition of the branched chain amino acid aminotransferase of Salmonella typhimurium. Biochemistry, **13**, 2071-2077 (1974)

[10] Wakita, M.; Hoshino, S.: A branched-chain amino acid aminotransferase from the rumen ciliate genus Entodinium. J. Protozool., **22**, 281-285 (1975)

[11] Monnier, N.; Montmitonnet, A.; Chesne, S.; Pelmont, J.: Transaminase B from Escherichia coli. I.-Purification and first properties. Biochimie, **58**, 663-675 (1976)

[12] Tachiki, T.; Tochikura, T.: Purification and characterization of L-leucine-α-ketoglutarate transaminase from Acetobacter suboxydans. Agric. Biol. Chem., **40**, 2187-2192 (1976)

[13] Koide, Y.; Honma, M.; Shimomura, T.: Branched chain amino acid aminotransferase of Pseudomonas sp.. Agric. Biol. Chem., **41**, 1171-1177 (1977)

[14] Lee-Peng, F.-C.; Hermodson, M.A.; Kohlhaw, G.B.: Transaminase B from Escherichia coli: quaternary structure, amino-terminal sequence, substrate specificity, and absence of a separate valine-α-ketoglutarate activity. J. Bacteriol., **139**, 339-345 (1979)

[15] Aarnes, H.: Branched chain amino acid aminotransferase in barley seedlings. Z. Pflanzenphysiol., **102**, 81-89 (1981)

[16] Shiio, I.; Mori, M.; Ozaki, H.: Amino acid aminotransferase in an amino acid-producing bacterium, Brevibacterium flavum. Agric. Biol. Chem., **46**, 2967-2977 (1982)

[17] Kawagishi, S.: Two isozymes of branched-chain amino acid aminotransferase in submandibular gland of the monkey. Shika Kiso Igakkai Zasshi, **26**, 947-952 (1984)

[18] Cooper, A.J.L.: Glutamate-branched-chain amino acid transaminase. Methods Enzymol., **113**, 71-73 (1985)

[19] Korpela, T.K.; Saarinen, R.: Affinity chromatography of B$_6$-vitamin-dependent enzymes: purification of pig-heart branched-chain amino acid transaminase. J. Chromatogr., **318**, 333-341 (1985)

[20] Bode, R.; Birnbaum, D.: Purification and properties of two branched-chain amino acid aminotransferases from the yeast Candida maltosa. Biochem. Physiol. Pflanz., **183**, 417-424 (1988)

[21] Inoue, K.; Kuramitsu, S.; Aki, K.; Watanabe, Y.; Takagi, T.; Nishigai, M.; Ikai, A.; Kagamiyama, H.: Branched-chain amino acid aminotransferase of Escherichia coli: overproduction and properties. J. Biochem., **104**, 777-784 (1988)

[22] Kido, R.: Pancreatic branched-chain-amino-acid aminotransferase. Methods Enzymol., **166**, 275-281 (1988)

[23] Korpela, T.K.: Purification of branched-chain-amino-acid aminotransferase from pig heart. Methods Enzymol., **166**, 269-274 (1988)

523

[24] Feild, M.J.; Nguyen, D.C.; Armstrong, F.B.: Amino acid sequence of Salmonella typhimurium branched-chain amino acid aminotransferase. Biochemistry, 28, 5306-5310 (1989)

[25] Kamitori, S.; Odagaki, Y.; Inoue, K.; Kuramitsu, S.; Kagamiyama, H.; Matsuura, Y.; Higuchi, T.: Crystallization and preliminary X-ray characterization of branched-chain amino acid aminotransferase from Escherichia coli. J. Biochem., 105, 671-672 (1989)

[26] Wallin, R.; Hall, T.R.; Hutson, S.M.: Purification of branched chain aminotransferase from rat heart mitochondria. J. Biol. Chem., 265, 6019-6024 (1990)

[27] Xing, R.; Whitman, W.B.: Characterization of enzymes of the branched-chain amino acid biosynthetic pathway in Methanococcus spp. J. Bacteriol., 173, 2086-2092 (1991)

[28] Hall, T.R.; Wallin, R.; Reinhart, G.D.; Hutson, S.M.: Branched chain aminotransferase isoenzymes. Purification and characterization of the rat brain isoenzyme. J. Biol. Chem., 268, 3092-3098 (1993)

[29] Kanda, M.; Hori, K.; Kurotsu, T.; Ohgishi, K.; Hanawa, T.; Saito, Y.: Purification and properties of branched chain amino acid aminotransferase from gramicidin S-producing Bacillus brevis. J. Nutr. Sci. Vitaminol., 41, 51-60 (1995)

[30] Schadewaldt, P.; Wendel, U.; Hammen, H.W.: Human branched-chain L-amino acid aminotransferase: activity and subcellular localization in cultured skin fibroblasts. Amino Acids, 9, 147-160 (1995)

[31] Kispal, G.; Steiner, H.; Court, D.A.; Rolinski, B.; Lill, R.: Mitochondrial and cytosolic branched-chain amino acid transaminases from yeast, homologs of the myc oncogene-regulated Eca39 protein. J. Biol. Chem., 271, 24458-24464 (1996)

[32] Schadewaldt, P.; Adelmeyer, F.: Coupled enzymatic assay for estimation of branched-chain L-amino acid aminotransferase activity with 2-oxo acid substrates. Anal. Biochem., 238, 65-71 (1996)

[33] Eden, A.; Benvenisty, N.: Characterization of a branched-chain amino-acid aminotransferase from Schizosaccharomyces pombe. Yeast, 14, 189-194 (1998)

[34] Hutson, S.M.; Berkich, D.; Drown, P.; Xu, B.; Aschner, M.; LaNoue, K.F.: Role of branched-chain aminotransferase isoenzymes and gabapentin in neurotransmitter metabolism. J. Neurochem., 71, 863-874 (1998)

[35] Faure, M.; Glomot, F.; Bledsoe, R.; Hutson, S.; Papet, I.: Purification and cloning of the mitochondrial branched-chain amino acid aminotransferase from sheep placenta. Eur. J. Biochem., 259, 104-111 (1999)

[36] Atiles, M.W.; Dudley, E.G.; Steele, J.L.: Gene cloning, sequencing, and inactivation of the branched-chain aminotransferase of Lactococcus lactis LM0230. Appl. Environ. Microbiol., 66, 2325-2329 (2000)

[37] Bonfils, J.; Faure, M.; Gibrat, J.F.; Glomot, F.; Papet, I.: Sheep cytosolic branched-chain amino acid aminotransferase: cDNA cloning, primary structure and molecular modelling and its unique expression in muscles. Biochim. Biophys. Acta, 1494, 129-136 (2000)

[38] Conway, M.E.; Hutson, S.M.: Mammalian branched-chain aminotrans-ferases. Methods Enzymol., **324**, 355-365 (2000)

[39] Kagamiyama, H.; Hayashi, H.: Branched-chain amino-acid aminotransfer-ase of Escherichia coli. Methods Enzymol., **324**, 103-113 (2000)

[40] Schadewaldt, P.: Determination of branched-chain L-amino-acid amino-transferase activity. Methods Enzymol., **324**, 23-32 (2000)

[41] Yvon, M.; Chambellon, E.; Bolotin, A.; Roudot-Algaron, F.: Characterization and role of the branched-chain aminotransferase (BcaT) isolated from Lac-tococcus lactis subsp. cremoris NCDO 763. Appl. Environ. Microbiol., **66**, 571-577 (2000)

[42] Eden, A.; Van Nedervelde, L.; Drukker, M.; Benvenisty, N.; Debourg, A.: Involvement of branched-chain amino acid aminotransferases in the pro-duction of fusel alcohols during fermentation in yeast. Appl. Microbiol. Biotechnol., **55**, 296-300 (2001)

[43] Faure, M.; Glomot, F.; Papet, I.: Branched-chain amino acid aminotransfer-ase activity decreases during development in skeletal muscles of sheep. J. Nutr., **131**, 1528-1534 (2001)

[44] Lin, H.M.; Kaneshige, M.; Zhao, L.; Zhang, X.; Hanover, J.A.; Cheng, S.Y.: An isoform of branched-chain aminotransferase is a novel co-repressor for thyroid hormone nuclear receptors. J. Biol. Chem., **276**, 48196-48205 (2001)

[45] Okada, K.; Hirotsu, K.; Hayashi, H.; Kagamiyama, H.: Structures of Escher-ichia coli branched-chain amino acid aminotransferase and its complexes with 4-methylvalerate and 2-methylleucine: Induced fit and substrate re-cognition of the enzyme. Biochemistry, **40**, 7453-7463 (2001)

[46] Torres, N.; Vargas, C.; Hernandez-Pando, R.; Orozco, H.; Hutson, S.M.; To-var, A.R.: Ontogeny and subcellular localization of rat liver mitochondrial branched chain amino-acid aminotransferase. Eur. J. Biochem., **268**, 6132-6139 (2001)

[47] Yennawar, N.; Dunbar, J.; Conway, M.; Hutson, S.; Farber, G.: The structure of human mitochondrial branched-chain aminotransferase. Acta Crystal-logr. Sect. D, **57**, 506-515 (2001)

[48] Cooper, A.J.; Conway, M.; Hutson, S.M.: A continuous 96-well plate spectro-photometric assay for branched-chain amino acid aminotransferases. Anal. Biochem., **308**, 100-105 (2002)

[49] Diebold, R.; Schuster, J.; Daeschner, K.; Binder, S.: The branched-chain ami-no acid transaminase gene family in Arabidopsis encodes plastid and mi-tochondrial proteins. Plant Physiol., **129**, 540-550 (2002)

[50] Madsen, S.M.; Beck, H.C.; Ravn, P.; Vrang, A.; Hansen, A.M.; Israelsen, H.: Cloning and inactivation of a branched-chain-amino-acid aminotransferase gene from Staphylococcus carnosus and characterization of the enzyme. Appl. Environ. Microbiol., **68**, 4007-4014 (2002)

[51] Conway, M.E.; Yennawar, N.; Wallin, R.; Poole, L.B.; Hutson, S.M.: Human mitochondrial branched chain aminotransferase: structural basis for sub-strate specificity and role of redox active cysteines. Biochim. Biophys. Acta, **1647**, 61-65 (2003)

[52] Cooper, A.J.; Bruschi, S.A.; Conway, M.; Hutson, S.M.: Human mitochon-drial and cytosolic branched-chain aminotransferases are cysteine S-conju-

gate β-lyases, but turnover leads to inactivation. Biochem. Pharmacol., **65**, 181-192 (2003)

[53] Goto, M.; Miyahara, I.; Hayashi, H.; Kagamiyama, H.; Hirotsu, K.: Crystal structures of branched-chain amino acid aminotransferase complexed with glutamate and glutarate: True reaction intermediate and double substrate recognition of the enzyme. Biochemistry, **42**, 3725-3733 (2003)

Aminolevulinate transaminase

1 Nomenclature

EC number

2.6.1.43

Systematic name

5-aminolevulinate:pyruvate aminotransferase

Recommended name

aminolevulinate transaminase

Synonyms

4,5-dioxovalerate aminotransferase
4,5-dioxovaleric acid aminotransferase
4,5-dioxovaleric acid transaminase
4,5-dioxovaleric transaminase
5-aminolevulinic acid transaminase
DOVA transaminase
DOVA-T <14> [20]
EC 2.6.1.19 <11> (<11> 2-aminobutyrate aminotransferase and aminolevulinate transaminase activity reside in the same protein [18]) [18]
EC 2.6.1.40 <6, 11> (<6,11> dimethylarginine-pyruvate aminotransferase and D-3-aminoisobutyrate pyruvate aminotransferase activities resides in the same protein [18,19]) [18, 19]
EC 2.6.1.44 <1, 3, 6, 10, 11> (<1,3,10> identical with alanine-glyoxylate aminotransferase [3,5,6]; <1,6,11> alanine-glyoxylate transaminase and aminolevulinate transaminase activities reside in the same protein [3,18,19]) [3, 5, 6, 18, 19]
L-alanine-4,5-dioxovalerate aminotransferase
L-alanine:4,5-dioxovaleric acid transaminase
L-alanine:dioxovalerate transaminase
alanine-dioxovalerate aminotransferase
alanine-γ,δ-dioxovalerate aminotransferase
alanine:4,5-dioxovalerate aminotransferase
aminolevulinate aminotransferase
aminolevulinic acid transaminase
dioxovalerate transaminase
γ,δ-dioxovalerate aminotransferase
γ,δ-dioxovaleric acid transaminase

CAS registry number

9012-46-8

2 Source Organism

<1>   *Bos taurus* (bovine [2,3]) [2, 3]
<2>   *Candida albicans* (3100 [13]) [13]
<3>   *Chlorella regularis* (S-50 [6]) [6]
<4>   *Clostridium tetanomorphum* (ATCC 15920 [4]) [4]
<5>   *Gallus gallus* [11]
<6>   *Leishmania donovani* (AG83, WHO nomenclature MHOM/IN/83/AG83, promastigote form [19]) [19]
<7>   *Mus musculus* (mouse [14,15]; strain C57BL/6J and DBA/2J [14]) [14, 15]
<8>   *Pennisetum typhoideum* (bajra [12]) [12]
<9>   *Pseudomonas riboflavina* (IFO 3140 [10]) [10]
<10> *Raphanus sativus* (radish [5]) [5]
<11> *Rattus norvegicus* (Wistar [7,9,16,18]) [7-9, 16-18]
<12> *Rhodopseudomonas spheroides* [1]
<13> *Saccharomyces cerevisiae* (3059, S288C [13]) [13]
<14> *Trypanosoma cruzi* (Tulahuen strain, Tul 2 stock [20]) [20]

3 Reaction and Specificity

Catalyzed reaction
5-aminolevulinate + pyruvate = 4,5-dioxopentanoate + L-alanine (a pyridoxal-phosphate protein)

Reaction type
amino group transfer

Natural substrates and products
S   L-alanine + 4,5-dioxopentanoate <1-7, 9, 11-14> (<9> δ-aminolevulinate pathway [10]; <11> heme biosynthetic pathway [7, 17, 18]; <11> alternate route for δ-aminolevulinate formation [18]; <1> biosynthesis of porphyrin precursors [3]; <6> first committed step in heme synthesis [19]) (Reversibility: ir <1-7, 9, 11-14> [1-4, 6-11, 13-15, 17-20]) [1-4, 6-11, 13-15, 17-20]
P   5-aminolevulinate + pyruvate <3-5, 7, 9, 11-13> [1-4, 6-11, 13-15, 17-20]

Substrates and products
S   D-alanine + 4,5-dioxopentanoate <1> (<1> 3% of the activity with L-alanine [2]) (Reversibility: ir <1> [2]) [2]
P   5-aminolevulinate + pyruvate
S   DL-alanine + 4,5-dioxopentanoate <1> (<1> 89% of the activity with L-alanine [2]) (Reversibility: ir <1> [2]) [2]
P   5-aminolevulinate + pyruvate
S   L-alanine + 4,5-dioxopentanoate <1-7, 9, 11-14> (Reversibility: ir <1-7, 9, 11-14> [1-4, 6-11, 13-15, 17-20]) [1-4, 6-11, 13-15, 17-20]
P   5-aminolevulinate + pyruvate <1-7, 9, 11-13> [1-4, 6-11, 13-15, 17-20]

S L-arginine + 4,5-dioxopentanoate <2> (<2> 6.8% of the activity with L-alanine [13]) (Reversibility: ir <2> [13]) [13]

P 5-aminolevulinate + ?

S L-arginine + 4,5-dioxopentanoate <2> (<2> 6.8% of the activity with L-alanine [13]) (Reversibility: ir <2> [13]) [13]

P 5-aminolevulinate + ?

S L-asparagine + 4,5-dioxopentanoate <2> (<2> 74% of the activity with L-alanine [13]) (Reversibility: ir <2> [13]) [13]

P 5-aminolevulinate + ?

S L-aspartate + 4,5-dioxopentanoate <2, 4> (<4> 67% of the activity with L-alanine [4]; <2> 39% of the activity with L-alanine [13]) (Reversibility: ir <2,4> [4,13]) [4, 13]

P 5-aminolevulinate + oxaloacetate

S L-glutamate + 4,5-dioxopentanoate <2-4, 6> (<3> 37% of the activity with L-alanine [6]; <4> 96% of the activity with L-alanine [4]; <2> 39% of the activity with L-alanine [13]; <6> 43.8% of the activity with L-alanine [19]) (Reversibility: ir <2-4,6> [4,6,13,19]) [4, 6, 13, 19]

P 5-aminolevulinate + 2-oxoglutarate

S L-glutamine + 4,5-dioxopentanoate <2, 3> (<3> 16% of the activity with L-alanine [6]; <2> 74% of the activity with L-alanine [13]) [6, 13]

P 5-aminolevulinate + ?

S L-histidine + 4,5-dioxopentanoate <4> (<4> 74% of the activity with L-alanine [4]) (Reversibility: ir <4> [4]) [4]

P 5-aminolevulinate + 3-(1H-imidazol-4-yl)-2-oxopropanoate

S L-lysine + 4,5-dioxopentanoate <1, 3> (<3> 15% of the activity with L-alanine [6]; <1> 9% of the activity with L-alanine [2]) (Reversibility: ir <1,3> [2,6]) [2, 6]

P 5-aminolevulinate + ?

S L-ornithine + 4,5-dioxopentanoate <1, 9, 12> (<12> less than 10% of the activity of α-alanine [1]; <9> 14% of the activity with L-alanine [10]; <1> 6% of the activity with L-alanine [2]) (Reversibility: ir <1, 9, 12> [1, 2, 10]) [1, 2, 10]

P 5-aminolevulinate + ?

S L-phenylalanine + 4,5-dioxopentanoate <4, 6> (<4> 71% of the activity with L-alanine [4]; <6> 90.6% of the activity with L-alanine [19]) (Reversibility: ir <4,6> [4,19]) [4, 19]

P phenylpyruvate + 5-aminolevulinate

S L-serine + 4,5-dioxopentanoate <3> (<3> 33% of the activity with L-alanine [6]) (Reversibility: ir <3> [6]) [6]

P 5-aminolevulinate + 3-hydroxy-2-oxopropanoate

S β-alanine + 4,5-dioxopentanoate <1, 6, 9> (<9> 39% of the activity with L-alanine [10]; <1> 23% of the activity with L-alanine [2]; <6> 65.6% of the activity with L-alanine [19]) (Reversibility: ir <1, 6, 9> [2, 10, 19]) [2, 10, 19]

P 5-aminolevulinate + malonic semialdehyde

S δ-amino-n-valerate + 4,5-dioxopentanoate <1, 12> (<12> transaminate at 95% of the rate of α-alanine [1]; <1> 10% of the activity with L-alanine [2]) (Reversibility: ir <1,12> [1,2]) [1, 2]

P 5-aminolevulinate + 5-oxo-n-pentanoate

S ε-amino-n-caproate + 4,5-dioxopentanoate <9, 12> (<12> transaminate at 103% of the rate of α-alanine [1]; <9> 34% of the activity with L-alanine [10]) (Reversibility: ir <9,12> [1,10]) [1, 10]

P 5-aminolevulinate + 6-oxo-n-hexanoate

S γ-amino-n-butyrate + 4,5-dioxopentanoate <1, 9, 12> (<12> transaminate at 122% of the rate of α-alanine [1]; <9> 55% of the activity with L-alanine [10]; <1> 10% of the activity with L-alanine [2]) (Reversibility: ir <1, 9, 12> [1, 2, 10]) [1, 2, 10]

P 5-aminolevulinate + succinic semialdehyde

S glycine + 4,5-dioxopentanoate <1-3, 6, 12> (<12> less than 10% of the activity of α-alanine [1]; <3> 21% of the activity with L-alanine [6]; <1> 9% of the activity with L-alanine [2]; <6> 81.2% of the activity with L-alanine [19]) (Reversibility: ir <1-3, 6, 12> [1, 2, 6, 13, 19]) [1, 2, 6, 13, 19]

P 5-aminolevulinate + glyoxylate

S Additional information <1-3, 6, 9, 11, 12> (<12> ability of β-alanine to replace α-alanine as amino group donor is lost during purification, α-amino-n-butyrate, α-aminoisobutyrate, β-amino-n-butyrate and β-aminoisobutyrate are inactive, serine, taurine, glutamine, asparagine, glutamic and aspartic acids, lysine and cysteine all have negligible activity [1]; <3> does not use 2-oxoglutarate as amino acceptor [6]; <9> L-lysine, L-arginine, L-glutamate, L-methionine, L-histidine, glycine, L-phenylalanine, L-threonine, L-asparagine, L-serine, L-proline, L-valine, L-isoleucine, L-leucine, L-tyrosine, L-cysteine and L-tryptophan are all inert as amino donors, methylglyoxal is a poor substrate for the enzyme [10]; <11> DL-alanine, D-alanine, ammonium sulfate, L-glutamate, glycine and phenylalanine cannot replace L-alanine [9]; <1> glutamate is no substrate, transamination of methylglyoxal yields aminoacetone, less than 1% efficient as amino acceptor as 4,5-dioxopentanoate [2]; <6> cysteine is no substrate [19]; <2> cannot utilize 2-oxoglutarate nor pyruvate as amino acceptors [13]) [1, 2, 6, 9, 10, 13, 19]

P ?

Inhibitors

2-oxoglutarate <1, 3, 4, 11> [2, 4, 6, 9]
4,5-dioxopentanoate <3, 11, 12> (<3,12> substrate inhibition [1,6]) [1, 6, 9]
5-aminolevulinate <1, 7> [2, 15]
DL-dithiothreitol <8> [12]
Hg^{2+} <8> [12]
Tris-HCl <1, 2> [2, 13]
acetyl acetone <12> [1]
amino-oxyacetic acid <3, 4, 12> [1, 4, 6]
benzoyl acetone <12> [1]
β-alanine <12> (<12> 8% inhibition [1]) [1]

β-chloroalanine <7> [15]
β-iodopropionate <12> (<12> 30% inhibition [1]) [1]
cadaverine <12> (<12> 14% inhibition [1]) [1]
cysteamine <12> [1]
cysteine <12> (<12> 90% inhibition [1]) [1]
diacetyl <4, 12> [1, 4]
ethyl acetoacetate <12> [1]
gabaculine <7> [15]
glyoxal <4> [4]
glyoxylic acid <1, 3, 4, 7, 10, 12> (<1> competitive inhibition [3]) [1, 3-6, 15, 16]
hemin <6-9, 11> (<11> uncompetitive inhibitor [9]; <11> mitochondrial iso-zyme [16]) [7-10, 12, 15-19]
homocysteine <12> [1]
hydroxylamine <1> [3]
imidazole <3> [6]
iodoacetate <11, 12> (<12> 30% inhibition [1]) [1, 9]
isonicotinic acid hydrazide <12> [1]
isonicotinic acid hydrazide <1> [3]
lysine <12> (<12> 20% inhibition [1]) [1]
methylglyoxal <1, 4, 6, 7, 9, 11> [2, 4, 9, 10, 15, 19]
ophthalaldehyde <12> [1]
ornithine <12> (<12> 20% inhibition [1]) [1]
oxaloacetic acid <4, 12> [1, 4]
p-chloromercuribenzoate <2, 12> (<12> 78% inhibition [1]) [1, 13]
p-chloromercuric benzene sulfonate <11> [9]
penicillamine <12> [1]
potassium cyanide <1> [3]
protoporphyrin <11> [8]
pyruvate <1, 3, 4, 7, 9, 11> [2, 4, 6, 9, 10, 15]
pyruvic acid <12> [1]
semicarbazide <1> [3]
succinate semialdehyde <1, 4> [2, 4]
succinic acid <12> [1]
thioglycollate <12> (<12> 74% inhibition [1]) [1]
Additional information <2, 12> (<12> iodoacetamide and chloroacetamide are not inhibitors [1]; <2> EDTA, N-ethylmaleimide, iodoacetic acid, cysteine and β-mercaptoethanol are no inhibitors [13]) [1, 13]

Cofactors/prosthetic groups

pyridoxal 5'-phosphate <1, 2, 4, 9, 11, 13> (<2> K_m 0.0005 mM [13]) [2, 4, 9, 10, 13]

Activating compounds

hemin <11> (<11> cytoplasmic isozyme, stimulates to 150% of the control [16]) [16]

Specific activity (U/mg)

  0.0266 <7> [14]

  0.16 <1> [2]

  0.186 <2> [13]

  0.22 <11> (<11> cytosolic isozyme [16]) [16]

  0.753 <1> [3]

  1.02 <9> [10]

  1.05 <11> [8]

  1.171 <11> [9]

  2.68 <3> [6]

  7.68 <4> [4]

  36.9 <11> [18]

K$_m$-Value (mM)

  0.00028 <2> (L-alanine, <2> pH 7.2, 60°C [13]) [13]

  0.0105 <2> (4,5-dioxopentanoate, <2> pH 7.2, 60°C [13]) [13]

  0.06 <5> (4,5-dioxopentanoate, <5> pH 7.0 [11]) [11]

  0.12 <3> (4,5-dioxopentanoate, <3> pH 8.0, 37°C [6]) [6]

  0.22 <7> (4,5-dioxopentanoate) [15]

  0.24 <1> (4,5-dioxopentanoate, <1> pH 6.9, 37°C [2]) [2]

  0.25 <11> (4,5-dioxopentanoate, <11> pH 7.6, 37°C [8]) [8]

  0.26 <4> (4,5-dioxopentanoate, <4> pH 6.9, 37°C [4]) [4]

  0.28 <11> (4,5-dioxopentanoate, <11> pH 6.9, 37°C [9]) [9]

  0.3 <1> (4,5-dioxopentanoate, <1> pH 7.0, 37°C, [3]) [3]

  0.4 <12> (4,5-dioxopentanoate, <12> pH 7.0, 37°C [1]) [1]

  0.4 <5> (L-alanine, <5> pH 7.0 [11]) [11]

  0.67 <1> (glyoxylate, <1> pH 7.0, 37°C [3]) [3]

  0.71 <10> (4,5-dioxopentanoate) [5]

  0.75 <9> (4,5-dioxopentanoate, <9> pH 8.0, 37°C [10]) [10]

  0.97 <7> (L-alanine) [15]

  1.3 <3> (glyoxylate, <3> pH 8.0, 37°C [6]) [6]

  1.44 <6> (4,5-dioxopentanoate, <6> pH 7.0, 65°C [19]) [19]

  1.7 <10> (L-alanine) [5]

  1.7 <9> (L-alanine, <9> pH 8.0, 37°C [10]) [10]

  1.8 <8> (4,5-dioxopentanoate) [12]

  1.96 <4> (L-alanine, <4> pH 6.9, 37°C [4]) [4]

  2.5 <1> (L-alanine, <1> pH 7.0, 37°C, 4,5-dioxopentanoate as amino acceptor [3]) [3]

  2.9 <11> (L-alanine, <11> pH 7.6, 37°C [8]) [8]

  3 <3> (L-alanine, <3> pH 8.0, 37°C, glyoxylate as amino acceptor [6]) [6]

  3.3 <11> (L-alanine, <11> pH 6.9, 37°C [9]) [9]

  3.44 <8> (L-alanine) [12]

  3.5 <3> (L-alanine, <3> pH 8.0, 37°C [6]) [6]

  3.7 <1> (L-alanine, <1> pH 6.9, 37°C [2]) [2]

  4 <1> (L-alanine, <1> pH 7.0, 37°C, glyoxylate as amino acceptor [3]) [3]

  8 <12> (L-alanine, <12> pH 7.0, 37°C [1]) [1]

K_i-Value (mM)

0.5 <3> (glyoxylate, <3> pH 8.0, 37°C [6]) [6]

30 <11> (hemin, <11> different concentrations of alanine at fixed 4,5-dioxopentanoate concentration [7]) [7]

42 <11> (hemin, <11> different concentrations of 4,5-dioxopentanoate concentration at fixed alanine concentration [7]) [7]

pH-Optimum

6.6 <11> (<11> cytoplasmic isozyme [16]) [8, 16]

6.9 <1> [2]

7 <12> [1]

7-8 <7> [15]

7.2 <2> [13]

8 <3, 10> [5, 6]

Temperature optimum (°C)

60 <2, 11, 13> [13, 18]

65 <6, 11> (<11> cytoplasmic isozyme [16]) [8, 16, 19]

Temperature range (°C)

20-80 <11> (<11> 3fold increase of activity from 37°C to 60°C [18]) [18]

4 Enzyme Structure

Molecular weight

59000 <2, 13> (<2,13> gel filtration [13]) [13]

60260 <2> (<2> gel filtration [13]) [13]

63000 <5> (<5> SDS-PAGE [11]) [11]

72000 <4> (<4> sedimentation equilibrium centrifugation, potassium phosphate buffer containing 4 M NaCl [4]) [4]

78000 <4> (<4> sedimentation equilibrium centrifugation, potassium phosphate buffer [4]) [4]

82000 <4> (<4> gel filtration [4]) [4]

111000 <4> (<4> gel filtration [4]) [4]

123000 <10> [5]

126000 <3> (<3> gel filtration [6]) [6]

146000 <7> (<7> SDS-PAGE [14]) [14]

168000 <8> (<8> gel filtration [12]) [12]

190000 <9> (<9> gel filtration [10]) [10]

210000 <11> (<11> gel filtration, gradient polyacrylamide gel electrophoresis [18]; <11> mitochondrial isozyme, gel filtration [16]) [16, 18]

225000 <11> (<11> gel filtration, Sepharose 6B [18]; <11> gel filtration [8]) [8, 18]

232000 <11> (<11> gel filtration [9]) [9]

238000 <1> (<1> gel filtration [3]) [3]

240000 <1> (<1> gel filtration [2]) [2, 3]

245000 <1> (<1> sucrose density gradient centrifugation [3]) [3]

260000 <11> (<11> cytoplasmic isozyme, gel filtration [16]) [16]

Subunits

dimer <3, 4, 10> (<3> 2 * 68000, SDS-PAGE [6]; <4> 2 * 55500, SDS-PAGE [4]) [4-6]

hexamer <11> (<11> 6 * 41000 [9]; <11> 6 * 37000, SDS-PAGE [8]) [8, 9]

monomer <2, 13> (<13> 1 * 59000, SDS-PAGE [13]; <2> 1 * 57544, SDS-PAGE [13]) [13]

tetramer <1, 8, 9, 11> (<8> 4 * 42000, SDS-PAGE [12]; <9> 4 * 48000, SDS-PAGE [10]; <11> 4 * 67000, cytoplasmic isozyme, homopolymer, SDS-PAGE [16]; <11> 4 * 50000, mitochondrial isozyme, SDS-PAGE [16,18]; <1> 4 * 58000, SDS-PAGE [3]) [3, 10, 12, 16, 18]

5 Isolation/Preparation/Mutation/Application

Source/tissue

embryo <5> [11]

epimastigote <14> [20]

hepatocyte <7> [15]

kidney <7, 11> [3, 8, 14, 16-18]

liver <1, 5, 7, 11> [2, 3, 7, 9, 11, 15, 16, 18]

seedling <8> [12]

spleen <7> [14]

Localization

cytosol <1, 2, 11> (<11> different isozymes in cytosol and mitochondrion [16]; <1> cytosol contains minimal enzymic activity [2]) [2, 13, 16]

mitochondrion <1, 2, 5, 7, 11> (<7, 11> mitochondrial matrix [14, 18]; <11> different isozymes in cytosol and mitochondrion [16]) [2, 3, 7, 9, 11, 13, 14, 16-18]

mitoplast <5, 7> [11, 14]

Purification

<1> [2, 3]

<2> [13]

<3> [6]

<4> [4]

<5> [11]

<7> [14]

<9> [10]

<11> [7-9, 16, 18]

<12> (partial [1]) [1]

<13> [13]

Crystallization

<1> [2]

<9> (crystallized as bright yellow needles [10]) [10]

Application

medicine <6, 14> (<6> immunological difference between the enzyme from mammalian source and this enzyme presents a potential target for chemotherapy, development of treatment for Leishmaniasis, a major parasitic disease in humans and opportunistic infections in patients with AIDS, an inhibitor specific for the leishmanial enzyme may be used as a therapeutic agent [19]; <14> target for chemotherapy against Chagas disease or American trypanosomiasis [20]) [19, 20]

6 Stability

Temperature stability

55 <3> (<3> retains 75% activity after 1 h incubation [6]) [6]
55-65 <11> (<11> retains more than 90% maximum activity through 55-65°C, beyond 65°C activity decreases sharply [18]) [18]
60 <1> (<1> 10 min, about 65% loss of activity [3]) [3]

General stability information

<1>, freezing and thawing does not affect activity [2]
<2>, repeated freezing and thawing inactivates the enzyme, stability of the enzyme is enhanced by presence of 5% glycerol [13]
<3>, stability is improved by 10-15% glycerol [6]
<11>, proteolytic hydrolysis of the cytosolic isozyme generates an enzymatically more active form [16]
<11>, purified enzyme is stable in 50 mM potassium buffer, pH 6.9, containing 10% glycerol, freezing and thawing does not affect activity significantly [9]
<11>, repeated freezing and thawing causes inactivation [18]
<12>, glutathione or β-mercaptoethanol restores activity of aged enzyme preparations, optimally at 1.0 mM [1]

Storage stability

<1>, -20°C, purified enzyme in 50 mM potassium phosphate buffer, pH 7.5, 10% glycerol, may be stored for at least 8 weeks without loss of either activity [3]
<1>, 0-5°C, 50 mM potassium phosphate buffer, pH 7.5, 10% glycerol, stored for 2 weeks, little or none of each activity is lost [3]
<2>, 4°C, stored in ice for 3 months without any appreciable loss of enzyme activity [13]
<3>, 4°C, 10-15% glycerol, 20% loss of activity after 5 days [6]
<3>, 4°C, unstable [6]
<9>, -15°C, stable for 3 months [10]
<11>, -20°C, 5 mM potassium phosphate buffer, pH 7.6, 10% glycerol [16]
<11>, -20°C, can be stored for nearly 1 year without significant loss of enzyme activity [18]
<11>, -20°C, remains stable for at least 1 month [9]

<11>, -70°C, stable for at least 2-3 months [9]
<11>, 4°C, can be stored for at least 1 week without any significant loss of activity [9]
<12>, -15°C, stable for several months [1]

References

[1] Neuberger, A.; Turner, J.M.: γ,δ-Dioxovalerate aminotransferase activity in Rhodopseudomonas spheroides. Biochim. Biophys. Acta, **67**, 342-345 (1963)

[2] Varticovski, L.; Kushner, J.P.; Burnham, B.F.: Biosynthesis of porphyrin precursors. Purification and characterization of mammalian L-alanine:γ,δ-dioxovaleric acid aminotransferase. J. Biol. Chem., **255**, 3742-3747 (1980)

[3] Noguchi, T.; Mori, R.: Biosynthesis of porphyrin precursors in mammals. Identity of alanine: γ,δ-dioxovalerate aminotransferase with alanine:glyoxylate aminotransferase. J. Biol. Chem., **256**, 10335-10339 (1981)

[4] Bajkowski, A.S.; Friedmann, H.C.: δ-Aminolevulinic acid formation. Purification and properties of alanine:4,5-dioxovalerate, aminotransferase and isolation of 4,5-dioxovalerate from Clostridium tetanomorphum. J. Biol. Chem., **257**, 2207-2211 (1982)

[5] Shioi, Y.; Doi, M.; Sasa, T.: Purification and characterization of L-alanine:4,5-dioxovalerate (glyoxylate) aminotransferase from radish (Raphanus sativus L.). Plant Cell Physiol., **25**, 1487-1493 (1984)

[6] Shioi, Y.; Nagamine, M.; Sasa, T.: Purification and properties of L-alanine:4,5-dioxovalerate aminotransferase from Chlorella regularis. Arch. Biochem. Biophys., **234**, 117-124 (1984)

[7] Shanker, J.; Datta, K.: Evidence of hemin as an end product inhibitor of L-alanine: 4,5-dioxovalerate transaminase in rat liver mitochondria. FEBS Lett., **189**, 129-132 (1985)

[8] Singh, N.K.; Datta, K.: Purification and characterization of rat kidney L-alanine: 4,5-dioxovalerate transaminase and inhibition by hemin. Biochim. Biophys. Acta, **827**, 305-309 (1985)

[9] Shanker, J.; Datta, K.: Affinity purification and properties of rat liver mitochondrial L-alanine:4,5-dioxovalerate transaminase and its inhibition by hemin. Arch. Biochem. Biophys., **248**, 652-657 (1986)

[10] Rhee, H.-i.; Murata, K.; Kimura, A.: L-alanine: 4,5-dioxovalerate aminotransferase from Pseudomonas riboflavina: purification and inactivation by methylglyoxal. J. Biochem., **103**, 1045-1049 (1988)

[11] Ades, I.Z.: Synthetic substrates for thrombin. Int. J. Biochem., **21**, 579-587 (1989)

[12] Prasad, D.D.K.; Prasad, A.R.K.: Characterization of L-alanine:4,5-dioxovaleric acid transaminasefrom bajra (Pennisetum typhoideum) seedlings. Biochem. Int., **18**, 149-161 (1989)

[13] Hoare, K.; Datta, K.: Characteristics of L-alanine:4,5-dioxovaleric acid transaminase: an alternate pathway of heme biosynthesis in yeast. Arch. Biochem. Biophys., **277**, 122-129 (1990)

[14] McKinney, C.E.; Ades, I.Z.: Production of δ-aminolevulinate: subcellular localization and purification of murine hepatic L-alanine: 4,5-dioxovaleric acid aminotransferase. Int. J. Biochem., 22, 347-357 (1990)

[15] McKinney, C.E.; Ades, I.Z.: Production of δ-aminolevulinic acid: characterization of murine liver 4,5-dioxovaleric acid: L-alanine aminotransferase. Int. J. Biochem., 23, 803-810 (1991)

[16] Singh, N.K.; Tyagi, R.K.; Datta, K.: Cytosolic L-alanine:4,5-dioxovalerate transaminase differs from the mitochondrial form. Eur. J. Biochem., 198, 581-587 (1991)

[17] Tyagi, R.K.; Datta, K.: In vitro translocation of L-alanine:4,5-dioxovalerate transaminase into rat kidney mitochondria. J. Biochem., 113, 557-562 (1993)

[18] Tyagi, R.K.; Datta, K.: A rapid, high-yield purification of L-alanine:4,5-dioxovalerate transaminase from rat kidney mitochondria using an improved enzyme assay method. Protein Expr. Purif., 5, 527-533 (1994)

[19] Sagar, R.; Salotra, P.; Bhatnagar, R.; Datta, K.: L-alanine: 4,5-dioxovalerate transaminase in Leishmania donovani that differs from mammalian enzyme. Microbiol. Res., 150, 419-423 (1995)

[20] Lombardo, M.E.; Araujo, L.S.; Batlle, A.: 5-Aminolevulinic acid synthesis in epimastigotes of Trypanosoma cruzi. Int. J. Biochem. Cell Biol., 35, 1263-1271 (2003)

Alanine-glyoxylate transaminase 2.6.1.44

1 Nomenclature

EC number

2.6.1.44

Systematic name

L-alanine:glyoxylate aminotransferase

Recommended name

alanine-glyoxylate transaminase

Synonyms

3-hydroxykynurenine transaminase/alanine glyoxylate transaminase <1> [21]

AGT

AGXT <18> [23]

EC 2.6.1.51 <36> (<36> serine-pyruvate aminotransferase identical with liver isoenzyme 1 [3,4,13]) [3, 4, 13]

L-alanine-glycine transaminase

L-alanine-glyoxylate aminotransferase

SPT/AGT <36> [13]

alanine-glyoxylate aminotransferase

alanine-glyoxylate aminotransferase isoenzyme 2 <36> [4]

alanine-glyoxylic aminotransferase

alanine:2-oxoglutarate aminotransferase <36> (<36> identical with [9]) [9]

cytosolic alanine aminotransferase <36> (<36> identical with [9]) [9]

serine pyruvate aminotransferase <14> [21]

serine:pyruvate/alanine:glyoxylate aminotransferase <36> [13, 18]

CAS registry number

9015-67-2

2 Source Organism

<1> *Aedes aegypti* [21]

<2> *Anas platyrhynchos domestica* (duck [10]) [10]

<3> *Anser anser domesticus* (goose [10]) [10]

<4> *Arabidopsis thaliana* (AGT1 nucleotide sequence [19]) [19]

<5> *Ateles paniscus* (red-faced black spider monkey [16]) [16]

<6> *Callimico goeldii* (Goeldis monkey [16]) [16]

<7> *Callithrix argentata* (silvery marmoset [16]) [16]

<8> *Callithrix jacchus* (common marmoset [16]) [16]
<9> *Canis familiaris* (dog [3]) [3]
<10> *Caranx chrysos* (yellow mackerel [6]) [6]
<11> *Ceropithecus diana* (Diana monkey [16]) [16]
<12> *Cheirogaleus medius* (fat tailed dwarf lemur [16]) [16]
<13> *Columba livia domestica* (pigeon [10]) [10]
<14> *Drosophila melanogaster* (fruit fly [21]) [21]
<15> *Felis catus* (cat [3]) [3]
<16> *Gallus gallus domesticus* (white leghorn, bantam [10]) [10]
<17> *Gorilla gorilla* (common gorilla [16]) [16]
<18> *Homo sapiens* (human [11, 12, 14-17, 20, 22-24]) [11, 12, 14-17, 20, 22-24]
<19> *Hylobates lar* (white-handed gibbon [16]) [16]
<20> *Lemur fulvis* (brown lemur [16]) [16]
<21> *Leontopithecus rosalia* (golden lion tamarin [16]) [16]
<22> *Loris tardigradus* (Slender loris [16]) [16]
<23> *Macaca fuscata* (japanese macaque [16]) [16]
<24> *Macaca nigra* (Celebes macaque [16]) [16]
<25> *Melopsittacus undulatus* (australian bugerigar [10]) [10]
<26> *Mus musculus* (mouse [3]) [3]
<27> *Nycticebus pygmaeus* (Lesser slow loris [16]) [16]
<28> *Pan troglodytes* (common chimpanzee, sequence of 5' region of the AGT gene [16]) [16]
<29> *Papio anubis* (anubis baboon [16]) [16]
<30> *Passer montanus saturatus* (sparrow [10]) [10]
<31> *Phasians versicolor* (pheasant [10]) [10]
<32> *Pithecia pithecia* (white-faced saki monkey [16]) [16]
<33> *Podda oryzivora* (Java sparrow [10]) [10]
<34> *Pongo pygmaeus* (orang-utan [16]) [16]
<35> *Pseudomonas sp.* [2]
<36> *Rattus norvegicus* [1, 3, 4, 9, 13, 18]
<37> *Saccharomyces cerevisiae* (101D, commercial baker's yeast [7]) [7]
<38> *Saimiri sciureus* (common squirrel monkey [16]) [16]
<39> *Sanguinus oedipus* (cotton-top tamarin [16]) [16]
<40> *Sardinops sp.* (sardine [6]) [6]
<41> *Scomberomorous sp.* (mackerel [6]) [6]
<42> *Sebastes carnatus* (gopher gray rock cod [6]) [6]
<43> *Serinus canarius* (canary [10]) [10]
<44> *Spinacia oleracea* (spinach [5]) [5]
<45> *Suncus murinus* [8]
<46> *Uroloncha striata domestica* (japanese mannikin [10]) [10]

3 Reaction and Specificity

Catalyzed reaction

L-alanine + glyoxylate = pyruvate + glycine (a pyridoxal-phosphate protein. With one component of the animal enzyme, 2-oxobutanoate can replace glyoxylate. A second component also catalyses the reaction of EC 2.6.1.51 serine -pyruvate transaminase)

Reaction type

amino group transfer

Natural substrates and products

S L-alanine + glyoxylate <1, 4, 9, 13-16, 18, 26, 35-37, 41, 44> (<37> regulatory enzyme in the glyoxylate pathway of glycine and serine biosynthesis from tricarboxylic acid-cycle intermediates [7]; <18> key role in the transamination/detoxification of glyoxylate [11, 23]; <4> photorespiratory enzyme [19]) (Reversibility: ir <1, 4, 9, 13-16, 18, 26, 35-37, 41, 44> [2-7, 10, 11, 15, 19-23]) [2-7, 10, 11, 15, 19-23]

P pyruvate + glycine <1, 4, 9, 13-16, 18, 26, 35-37, 41, 44> [2-7, 10, 11, 15, 19-23]

Substrates and products

S L-2-aminobutyrate + glyoxylate <1, 14, 36> (Reversibility: ir <1, 14, 36> [4, 21]) [4, 21]

P 2-oxobutanoate + glycine

S L-alanine + 2-oxobutyrate <36> (Reversibility: ir <36> [4]) [4]

P pyruvate + 2-aminobutanoate

S L-alanine + 4-methylthio-2-oxobutyrate <36> (<36> poor amino acceptor [4]) (Reversibility: ir <36> [4]) [4]

P pyruvate + methionine

S L-alanine + glyoxylate <1, 4, 9, 13-16, 18, 26, 35-37, 41, 44> (<26, 36> isoenzyme 1 [3]; <26,36> isoenzyme 2 highly specific, little or no activity with other amino donors/acceptors [3]) (Reversibility: ir <1, 4, 9, 13-16, 18, 26, 35-37, 41, 44> [2-7, 10, 11, 15, 19-23]) [2-7, 10, 11, 15, 19-23]

P pyruvate + glycine <1, 4, 9, 13-16, 18, 26, 35-37, 41, 44> [2-7, 10, 11, 15, 19-23]

S L-alanine + hydroxypyruvate <9, 15, 26, 36> (<26,36> isoenzyme 1 [3]) (Reversibility: ir <9, 15, 26, 36> [3]) [3]

P pyruvate + L-serine

S L-alanine + phenylpyruvate <9, 15, 26, 36> (<26, 36> isoenzyme 1 [3]; <9, 15> isoenzyme 1, less effectively acceptor [3]) (Reversibility: ir <9, 15, 26, 36> [3]) [3]

P pyruvate + L-phenylalanine

S L-alanine + pyruvate <36> (Reversibility: ir <36> [4]) [4]

P pyruvate + L-alanine

S L-arginine + pyruvate <1, 14> (Reversibility: ir <1, 14> [21]) [21]

P 5-guanidino-2-oxopentanoate + L-alanine

S L-asparagine + glyoxylate <26, 36> (<26, 36> isoenzyme 1 [3]) (Reversibility: ir <26, 36> [3]) [3]
P 4-amino-2,4-dioxobutanoate + glycine
S L-cysteine + pyruvate <1, 14> (Reversibility: ir <1,14> [21]) [21]
P 3-mercapto-2-oxopropanoate + L-alanine
S L-glutamate + glyoxylate <44> (Reversibility: ir <44> [5]) [5]
P 2-oxoglutaramate + glycine
S L-glutamine + glyoxylate <1, 26, 36> (<26,36> isoenzyme 1 [3]) (Reversibility: ir <1,26,36> [3,21]) [3, 21]
P 2-oxoglutaramate + glycine
S L-glutamine + pyruvate <1> (Reversibility: ir <1> [21]) [21]
P 2-oxoglutaramate + L-alanine
S L-histidine + glyoxylate <1, 14, 26, 36> (<26,36> isoenzyme 1 [3]) (Reversibility: ir <1,14,26,36> [3,21]) [3, 21]
P 3-(1H-imidazol-4-yl)-2-oxopropanoate + glycine
S L-histidine + pyruvate <14> (Reversibility: ir <14> [21]) [21]
P 3-(1H-imidazol-4-yl)-2-oxopropanoate + L-alanine
S L-isoleucine + glyoxylate <26> (<26> isoenzyme 1 [3]) (Reversibility: ir <26> [3]) [3]
P 3-methyl-2-oxopropanoate + glycine
S L-leucine + glyoxylate <26, 36> (<26,36> isoenzyme 1 [3]) (Reversibility: ir <26,36> [3]) [3]
P 4-methyl-2-oxopentanoate + glycine
S L-methionine + glyoxylate <1, 14, 26, 36> (<26,36> isoenzyme 1 [3]) (Reversibility: ir <1,14,26,36> [3,21]) [3, 21]
P 4-methylsulfanyl-2-oxobutanoate + glycine
S L-methionine + pyruvate <1, 14> (Reversibility: ir <1,14> [21]) [21]
P 4-methylsulfanyl-2-oxobutanoate + L-alanine
S L-phenylalanine + glyoxylate <1, 14, 26, 36> (<26, 36> isoenzyme 1 [3]) (Reversibility: ir <1,14,26,36> [3,21]) [3, 21]
P phenylpyruvate + glycine
S L-phenylalanine + pyruvate <1, 14> (Reversibility: ir <1,14> [21]) [21]
P phenylpyruvate + L-alanine
S L-serine + glyoxylate <1, 4, 9, 14, 15, 26, 36, 41> (<26, 36> isoenzyme 1 [3]) (Reversibility: ir <1, 4, 9, 14, 15, 26, 36, 41> [3, 4, 6, 19, 21]) [3, 4, 6, 19, 21]
P 3-hydroxy-2-oxopropanoate + glycine
S L-serine + pyruvate <1, 4, 9, 14, 15, 26, 36, 41> (<9, 15, 26, 36> isoenzyme 1 [3]) (Reversibility: ir <1, 4, 9, 14, 15, 26, 36, 41> [3, 6, 19, 21]) [3, 6, 19, 21]
P 3-hydroxy-2-oxopropanoate + L-alanine
S L-tryptophan + glyoxylate <1, 14> (Reversibility: ir <1,14> [21]) [21]
P 3-indole-2-oxopropanoate + glycine
S L-tryptophan + pyruvate <1, 14> (Reversibility: ir <1,14> [21]) [21]
P 3-indole-2-oxopropanoate + L-alanine
S L-tyrosine + glyoxylate <14, 26, 36> (<26, 36> isoenzyme 1 [3]) (Reversibility: ir <14,26,36> [3,21]) [3, 21]

P 3-(4-hydroxyphenyl)-2-oxopropanoate + glycine
S L-tyrosine + pyruvate <14> (Reversibility: ir <14> [21]) [21]
P 3-(4-hydroxyphenyl)-2-oxopropanoate + L-alanine
S L-valine + glyoxylate <26> (<26> isoenzyme 1 [3]) (Reversibility: ir <26> [3]) [3]
P 3-methyl-2-oxobutanoate + glycine
S L-aspartate + glyoxylate <36> (Reversibility: ir <36> [4]) [4]
P 2-oxosuccinate + glycine <36> [4]
S glycine + pyruvate <1, 14> (Reversibility: ir <1,14> [21]) [21]
P glyoxylate + L-alanine
S Additional information <1, 14, 36, 37> (<36> 2-oxoglutarate is not an amino acceptor, little or no activity with asparagine, glutamine, glutamate, threonine, cysteine, methionine, arginine, leucine, valine, isoleucine, phenylalanine, tryptophan, histidine or tyrosine as amino donors [4]; <9, 15, 26, 36> isoenzyme 1, little or no activity with 2-oxoglutarate as amino acceptor and alanine, serine, glutamic acid, isoleucine, methionine, glutamine, asparagine, valine, aspartic acid, leucine, phenylalanine, tyrosine, histidine, tryptophan or 5-hydroxytryptophan [3]; <37> utilizes only glyoxylate as amino acceptor, hydroxypyruvate, 2-oxoglutarate, phenylpyruvate and 2-oxo-4-methyl-thiobutyrate are inactive, only L-alanine as amino donor, L-serine, L-threonine, L-glutamic acid, L-glutamine, L-aspartic acid, L-asparagine, L-ornithine, L-leucine, L-valine, L-isoleucine, L-histidine, L-phenylalanine, L-tryptophan and L-tyrosine are no substrates [7]; <1,14> aminoadipate, asparagine, aspartate, glutamate, isoleucine, lysine, threonine and valine are no substrates [21]) [3, 4, 7, 21]
P ?

Inhibitors

L-glutamate <44, 45> (<45> GPT 1, alanine:2-oxoglutarate aminotransferase AGT activity [8]) [5, 8]
amino-oxyacetic acid <18, 37> (<18> competitive inhibitor [24]) [7, 10, 12, 23]
glycine <44> (<44> linear competitive inhibitor [5]) [5]
glyoxylate <36> (<36> above 5 mM [1]) [1]
hydroxylamine <9, 15, 26, 36, 37, 44> (<9, 15, 26, 36> isoenzyme 1 [3]; <26,36> isoenzyme [3]) [3, 5, 7]
isonicotinic acid hydrazide <26, 36, 37> (<26, 36> isoenzyme 1 and isoenzyme 2 [3]) [3, 7]
semicarbazide <9, 15, 26, 36, 37> (<9, 15, 26, 36> isoenzyme 1 [3]; <26,36> isoenzyme [3]) [3, 7]

Cofactors/prosthetic groups

pyridoxal 5'-phosphate <4, 13, 16, 18, 35-37> (<36> K_m 0.02 mM [4]; <35> activated by heat treatment with pyridoxal 5'-phosphate [2]) [2, 4, 7, 10, 13, 19, 20, 22]

Activating compounds

cyclic AMP <9, 15, 26, 36> (<9, 15, 26, 36> induced by [3]) [3]
glucagon <9, 15, 26, 36> (<9, 15, 26, 36> induced by, isoenzyme 1 is increased in activity [3]) [3]

Turnover number (min^{-1})

960 <1> (glyoxylate, <1> pH 7.0, 50°C, L-alanine as amino donor [21]) [21]
1200 <1> (L-alanine, <1> pH 7.0, 50°C, glyoxylate as amino acceptor [21]) [21]
6800 <14> (glyoxylate, <14> pH 7.0, 50°C, L-alanine as amino donor [21]) [21]
7880 <14> (L-alanine, <14> pH 7.0, 50°C, glyoxylate as amino acceptor [21]) [21]

Specific activity (U/mg)

2.64 <9> (<9> isoenzyme 2 [3]) [3]
2.93 <36> (<9> isoenzyme 2 [3]) [3]
4.62 <36> (<36> recombinant enzyme [13]) [13]
10.6 <13> [10]
21.4 <36> (isoenzyme 1 [3]) [3]
23.2 <26> (isoenzyme 1 [3]) [3]
51.8 <18> (<18> purified recombinant enzyme [17]) [17]
86.3 <15> (isoenzyme 1 [3]) [3]
91 <9> (isoenzyme 1 [3]) [3]
160 <36> (isoenzyme 1 [3]) [4]
180.2 <37> [7]

K$_m$-Value (mM)

0.07 <36> (glyoxylate, <36> pH 8.2, 37°C, isoenzyme 1, alanine as amino donor [3]) [3]
0.1 <15> (glyoxylate, <15> pH 8.2, 37°C, isoenzyme 1, alanine as amino donor [3]) [3]
0.13 <9> (glyoxylate, <9> pH 8.2, 37°C, isoenzyme 1, alanine as amino donor [3]) [3]
0.15 <44> (glyoxylate, <44> pH 7.0, 27°C, alanine as amino donor [5]) [5]
0.2 <26> (glyoxylate, <26> pH 8.2, 37°C, isoenzyme 1, alanine as amino donor [3]) [3]
0.23 <18> (glyoxylate, <18> pH 8.0, 37°C, recombinant His-AGT, L-alanine as amino donor [17]) [17]
0.24 <36> (L-alanine, <36> pH 7.5, 37°C, mitochondrial isozyme, glyoxylate as amino acceptor [1]) [1]
0.36 <18> (glyoxylate, <18> pH 7.4, 37°C, alanine as amino donor [23]) [23]
0.39 <4> (L-serine, <4> pH 7.0, 37°C, SGT activity, pyruvate as amino acceptor [19]) [19]
0.39 <18> (glyoxylate, <18> pH 8.0, 37°C, recombinant AGT-His, L-alanine as amino donor [17]) [17]
0.44 <36> (glyoxylate, <36> pH 8.0, 37°C, alanine as amino donor [4]) [4]
0.52 <45> (L-alanine, <45> isoenzyme AGT 1 [8]) [8]

0.67 <36> (glyoxylate, <36> pH 8.0, 37°C, 2-aminobutyrate as amino donor [4]) [4]

0.7 <37> (glyoxylate, <37> pH 8.2, 37°C, isoenzyme 2, alanine as amino donor [7]) [7]

0.72 <26> (glyoxylate, <26> pH 8.0, 37°C, alanine as amino donor [3]) [3]

0.88 <45> (L-alanine, <45> GPT 1, alanine:2-oxoglutarate aminotransferase AGT activity [8]) [8]

1 <36> (glyoxylate, <36> pH 8.2, 37°C, isoenzyme 2, alanine as amino donor [3]) [3]

1.11 <36> (L-alanine, <36> pH 7.5, 37°C, cytosolic isozyme, glyoxylate as amino acceptor [1]) [1]

1.4 <26> (L-alanine, <26> pH 8.2, 37°C, isoenzyme 1, glyoxylate as amino acceptor [3]) [3]

1.52 <4> (L-serine, <4> pH 7.0, 37°C, SGT activity, glyoxylate as amino acceptor [19]) [19]

1.6 <44> (L-alanine, <44> pH 7.0, 27°C, glyoxylate as amino acceptor [5]) [5]

1.6 <1> (glyoxylate, <1> pH 7.0, 50°C, alanine as amino donor [21]) [21]

1.7 <44> (L-glutamate, <44> pH 7.0, 27°C, glyoxylate as amino acceptor [5]) [5]

2.1 <36> (L-alanine, <36> pH 8.2, 37°C, isoenzyme 1, glyoxylate as amino acceptor [3]) [3]

2.1 <37> (L-alanine, <37> pH 8.2, 37°C, glyoxylate as amino acceptor [7]) [7]

2.5 <18> (glyoxylate, <18> pH 8.0, 37°C, L-alanine as amino donor [15]) [15]

2.5 <36> (pyruvate, <36> pH 8.0, 37°C, 2-aminobutyrate as amino donor [4]) [4]

2.7 <36> (2-oxobutyrate, <36> pH 8.0, 37°C, alanine as amino donor [4]) [4]

2.9 <36> (L-alanine, <36> pH 8.0, 37°C, 2-oxobutyrate as amino acceptor [4]) [4]

3 <9> (L-alanine, <9> pH 8.2, 37°C, isoenzyme 1, glyoxylate as amino acceptor [3]) [3]

3.3 <45> (L-alanine, <45> isoenzyme AGT 2 [8]) [8]

3.9 <9> (L-alanine, <9> pH 8.2, 37°C, isoenzyme 1, glyoxylate as amino acceptor [3]) [3]

9.1 <18> (L-alanine, <18> pH 8.0, 37°C, recombinant His-AGT, glyoxylate as amino acceptor [17]) [17]

9.4 <18> (L-alanine, <18> pH 8.0, 37°C, recombinant AGT-His, glyoxylate as amino acceptor [17]) [17]

11 <36> (L-alanine, <36> pH 8.0, 37°C, glyoxylate as amino acceptor [4]) [4]

13.5 <18> (L-alanine, <18> pH 8.0, 37°C, glyoxylate as amino acceptor [15]) [15]

14.9 <18> (L-alanine, <18> pH 7.4, 37°C, glyoxylate as amino acceptor [23]) [23]

18 <1> (L-alanine, <1> pH 7.0, 50°C, glyoxylate as amino acceptor [21]) [21]

25 <26> (L-alanine, <26> pH 8.2, 37°C, isoenzyme 2, glyoxylate as amino acceptor [3]) [3]

30 <36> (L-alanine, <36> pH 8.2, 37°C, isoenzyme 2, glyoxylate as amino acceptor [3]) [3]

51 <14> (pyruvate, <14> pH 7.0, 50°C, serine as amino donor [21]) [21]
60 <36> (2-aminobutyrate, <36> pH 8.0, 37°C, glyoxylate as amino acceptor [4]) [4]
64 <14> (glyoxylate, <14> pH 7.0, 50°C, alanine as amino donor [21]) [21]
100 <36> (2-aminobutyrate, <36> pH 8.0, 37°C, pyruvate as amino acceptor [4]) [4]
101.2 <4> (L-alanine, <4> pH 7.0, 37°C, glyoxylate as amino acceptor [19]) [19]
149 <14> (L-alanine, <14> pH 7.0, 50°C, glyoxylate as amino acceptor [21]) [21]
256 <14> (L-serine, <14> pH 7.0, 50°C, glyoxylate as amino acceptor [21]) [21]

K_i-Value (mM)

1.8 <44> (L-glutamate, <44> pH 7.0, 27°C [5]) [5]
1.8 <45> (L-glutamate, <45> GPT 1, alanine:2-oxoglutarate aminotransferase AGT activity [8]) [8]
8.7 <44> (glycine, <44> pH 7.0, 27°C [5]) [5]

pH-Optimum

7-9 <36> (<36> cytosolic isozyme, no distinct pH optimum [1]) [1]
7-10 <1, 14> [21]
7.1 <44> [5]
7.5-8.5 <18> (<18> recombinant enzyme [17]) [17]
7.8-8 <18> [15]
8 <16> [10]
8.1-8.4 <26> (<26> isoenzyme 2 [3]) [3]
8.3-8.6 <9, 15, 26, 36> [3]
8.5 <37> [7]
8.5-8.8 <36> (<36> isoenzyme 2 [3]) [3]
8.6 <36> (<36> mitochondrial isozyme [1]) [1]
9 <1, 13, 14> (<1,14> best pH for activity towards alanine [21]) [10, 21]

pH-Range

6-10 <1, 14> [21]
7-8.2 <18> [15]
7-9 <36> [1]

Temperature optimum (°C)

30-80 <14> (<14> alanine as substrate [21]) [21]
60 <1> (<1> alanine as substrate [21]) [21]

Temperature range (°C)

20-80 <1, 14> [21]

4 Enzyme Structure

Molecular weight

77000 <9, 15, 26, 36> (<36> isoenzyme 1, gel filtration [3]) [3]

80000 <9, 15, 26, 36, 37> (<36> peroxisomal and mitochondrial alanine-glyoxylate aminotransferase 1 [3, 9]; <9, 15, 26, 36, 37> sucrose-density-gradient centrifugation [3,7]) [3, 7, 9]

90000 <4, 13, 18> (<13> gel filtration, sucrose density gradient centrifugation [10]) [10, 14, 19]

92140 <4> (<4> AGT1, calculated from gel filtration [19]) [19]

110000 <36> [9]

170000 <36> (<36> isoenzyme 2, gel filtration [3]) [3]

175000 <9, 36> (<36> liver isoenzyme 2, gel filtration [3,4]) [3, 4]

200000 <13, 16, 41> (<41> alanine-glyoxylate aminotransferase I [6]; <13,16> mitochondrial holo enzyme, sucrose density gradient centrifugation [10]) [6, 10]

210000 <45> (<45> isoenzyme AGT 2 [8]) [8]

213000 <36> (<36> analytical ultracentrifugation [4]) [4]

220000 <36> (<36> gel filtration [4]) [4, 9]

236000 <36> (<36> sucrose density gradient centrifugation [4]) [4]

Subunits

dimer <4, 9, 13, 15, 18, 26, 36, 37> (<36> 2 * 44400, homodimer, recombinant enzyme, SDS-PAGE [13]; <36> 2 * 45000, SDS-PAGE [18]; <13> 2 * 45000, holoenzyme, SDS-PAGE [10]; <37> 2 * 40000, SDS-PAGE [7]; <18> 2 * 43000, homodimer [20]; <18> 2 * 43000, immunoreaction [11]; <18> 2 * 43000, SDS-PAGE; [14, 17]; <9, 15, 26, 36> 2 * 38000, isoenzyme 1, SDS-PAGE [3]; <4> 2 * 45000, SDS-PAGE [19]) [3, 7, 10, 11, 13, 14, 17-20, 22]

monomer <16> (<16> 1 * 45000, apoenzyme, SDS-PAGE [10]) [10]

tetramer <13, 16, 36, 41> (<36> 4 * 56000, SDS-PAGE [4]; <16> 4 * 50000, mitochondrial holoenzyme, antibody cross reaction [10]; <13> 4 * 50000, mitochondrial holoenzyme, antibody cross reaction [10]) [4, 6, 10]

5 Isolation/Preparation/Mutation/Application

Source/tissue

embryo <16> [10]

hepatocyte <18> [11, 22]

kidney <36> [4]

leaf <4, 44> [5, 19]

liver <2, 3, 9, 13, 15, 16, 18, 25, 26, 30, 31, 33, 36, 41, 43, 45, 46> [1, 3, 4, 6, 8-15, 17, 18, 20, 21, 23]

seedling <4> [19]

Localization

cytosol <18, 36> [1, 9, 11]

mitochondrion <7, 8, 12, 16, 20, 22, 27, 36, 39> (<16> apoenzyme in the peroxisomes, holoenzyme in the mitochondrion [10]) [1, 3, 9, 10, 14, 16, 18]

peroxisome <2-4, 7, 8, 12, 13, 16-18, 20, 22, 24, 25, 27-34, 36, 39, 43, 44, 46> (<16, 31, 46> apoenzyme in the peroxisomes, holoenzyme in the mitochondrion [10]; <2, 3, 13, 25, 30, 33, 43> holoenzyme in the peroxisomes [10]) [1, 5, 9-12, 14-19, 22-24]

Purification

<4> (recombinant enzyme [19]) [19]

<9> (isoenzyme 1 [3]) [3]

<10> (partially [6]) [6]

<13> [10]

<15> (isoenzyme 1 [3]) [3]

<16> [10]

<18> (recombinant enzyme [17, 20, 22, 23]) [11, 17, 20, 22, 23]

<26> (isoenzyme 1 and 2 [3]) [3]

<36> (co-purified with 2-aminobutyrate aminotranferase [4]; isoenzyme 1 and 2 [3]; recombinant enzyme SPT10 [13]) [3, 4, 13, 18]

<37> [7]

<40> (partially [6]) [6]

<41> [6]

<42> (partially [6]) [6]

Crystallization

<18> (crystals belong to space group P4(1)2(1)2 or its enantiomorph with uni-cell parameters a = b = 90.81, c = 142.62 A [20]) [20, 22, 24]

Cloning

<1> (cDNA isolation [21]) [21]

<4> (cDNA for AGT1 expressed in Escherichia coli BL-21 [19]) [19]

<5> (sequencing of the 5' region of the AGT gene [16]) [16]

<6> (sequencing of the 5' region of the AGT gene [16]) [16]

<8> (single AGT gene cloned [16]) [16]

<11> (sequencing of the 5' region of the AGT gene [16]) [16]

<14> [21]

<17> (sequencing of the 5' region of the AGT gene [16]) [16]

<18> (human AGT expressed in Escherichia coli B834(DE3) [20]; mammalian expression vector pHYK, expressed in COS-1 cells [14]; AGXT cDNA cloned, AGXT*LTM expressed as a GST-fusion protein in Escherichia coli BL21(RIL) and in Sf9 cells [23]; cloned and expressed in Escherichia coli JM109 [17]) [11, 14, 17, 20, 22, 23]

<19> (sequencing of the 5' region of the AGT gene [16]) [16]

<21> (sequencing of the 5' region of the AGT gene [16]) [16]

<23> (sequencing of the 5' region of the AGT gene [16]) [16]

<28> (sequencing of the 5' region of the AGT gene [16]) [16]

<29> (sequencing of the 5' region of the AGT gene [16]) [16]

<32> (sequencing of the 5' region of the AGT gene [16]) [16]

<36> (Escherichia coli DH1 transformed with pRspt10 [13]; mitochondrial and peroxisomal isozymes generated from a single gene by alternative transcription initiation, cloned and expressed in Escherichia coli DH5 and ER1458 [18]) [13, 18]

<38> (sequencing of the 5' region of the AGT gene [16]) [16]

Application

medicine <18> (<18> hereditary disease primary hyperoxaluria type 1 is caused by a deficiency of the liver-specific peroxisomal enzyme alanine:-glyoxylate aminotransferase, diagnosis with selective inhibitors and enzyme assays [12,15,20]) [11, 12, 14, 15, 17, 22-24]

6 Stability

General stability information

<16>, pyridoxal 5'-phosphate stabilizes during purification [10]

Storage stability

<9>, -20°C, 50 mM potassium phosphate buffer, pH 7.5, 0.1 mM pyridoxal 5'-phosphate, 1 mM 2-mercaptoethanol, may be stored for at least 4 weeks without loss of either activity [3]

<9>, 0-6°C, 50 mM potassium phosphate buffer, pH 7.5, 0.1 mM pyridoxal 5'-phosphate, 1 mM 2-mercaptoethanol, may be stored for at least 2 weeks with little loss of either activity [3]

<13>, -20°C, peroxisomal holoenzyme, may be stored for at least 4 weeks without loss of activity [10]

<13>, 4°C, peroxisomal holoenzyme, 50% activity is lost when stored for 6 days [10]

<15>, -20°C, 50 mM potassium phosphate buffer, pH 7.5, 0.1 mM pyridoxal 5'-phosphate, 1 mM 2-mercaptoethanol, may be stored for at least 4 weeks without loss of either activity [3]

<15>, 0-6°C, 50 mM potassium phosphate buffer, pH 7.5, 0.1 mM pyridoxal 5'-phosphate, 1 mM 2-mercaptoethanol, may be stored for at least 2 weeks with little loss of either activity [3]

<16>, -20°C, peroxisomal apoenzyme, may be stored for at least 4 weeks without loss of activity [10]

<16>, 4°C, peroxisomal apoenzyme, 50% activity is lost when stored for 6 days [10]

<26>, -20°C, 50 mM potassium phosphate buffer, pH 7.5, 0.1 mM pyridoxal 5'-phosphate, 1 mM 2-mercaptoethanol, may be stored for at least 4 weeks without loss of either activity [3]

<26>, 0-6°C, 50 mM potassium phosphate buffer, pH 7.5, 0.1 mM pyridoxal 5'-phosphate, 1 mM 2-mercaptoethanol, may be stored for at least 2 weeks with little loss of either activity [3]

<36>, -20°C, 50 mM potassium phosphate buffer, 0.1 mM pyridoxal 5'-phosphate, 1 mM 2-mercaptoethanol, may be stored for at least 4 weeks without loss of either activity [3]

<36>, -20°C, crude homogenate can be stored for 6-12 months with little loss of activity [4]

<36>, -20°C, purified enzyme, 50% glycerol, can be stored for at least 3 months with little loss in activity [4]

<36>, -20°C, purified enzyme, freezing results in formation of floculence and complete loss of activity [4]

<36>, 0-6°C, 50 mM potassium phosphate buffer, pH 7.5, 0.1 mM pyridoxal 5'-phosphate, 1 mM 2-mercaptoethanol, may be stored for at least 2 weeks with little loss of either activity [3]

<37>, -20°C, 25 mM-potassium phosphate buffer, pH 7.5, containing 100 mM NaCl, 0.1 mM pyridoxal 5'-phosphate and 10% glycerol, my be stored for at least 3 months without loss of activity [7]

<37>, 0-4°C, little or no activity is lost when stored for 2 weeks [7]

<44>, -18°C, stable for about 2 weeks [5]

References

[1] Rowsell, E.V.; Snell, K.; Carnie, J.A.; Rowsell, K.V.: The subcellular distribution of rat liver L-alanine-glyoxylate aminotransferase in relation to a pathway for glucose formation involving glyoxylate. Biochem. J., 127, 155-156 (1972)

[2] Koide, Y.; Honma, M.; Shimomura, T.: L-Alanine-α-keto acid aminotransferase of Pseudomonas sp.. Agric. Biol. Chem., 41, 781-784 (1977)

[3] Noguchi, T.; Okuno, E.; Takada, Y.; Minatogawa, Y.; Okai, K.; Kido, R.: Characteristics of hepatic alanine-glyoxylate aminotransferase in different mammalian species. Biochem. J., 169, 113-122 (1978)

[4] Okuno, E.; Minatogawa, Y.; Kido, R.: Co-purification of alanine-glyoxylate aminotransferase with 2-aminobutyrate aminotransferase in rat kidney. Biochim. Biophys. Acta, 715, 97-104 (1982)

[5] Nakamura, Y.; Tolbert, N.E.: Serine: glyoxylate, alanine:glyoxylate, and glutamate:glyoxylate aminotransferase reactions in peroxisomes from spinach leaves. J. Biol. Chem., 258, 7631-7638 (1983)

[6] Noguchi, T.; Fujiwara, S.; Takada, Y.; Mori, T.; Nagano, M.; Hanada, N.; Saeki, E.; Yasuo, O.: Comp. Biochem. Physiol. B Comp. Biochem., 77, 279-283 (1984)

[7] Takada, Y.; Noguchi, T.: Characteristics of alanine: glyoxylate aminotransferase from Saccharomyces cerevisiae, a regulatory enzyme in the glyoxylate pathway of glycine and serine biosynthesis from tricarboxylic acid-cycle intermediates. Biochem. J., 231, 157-163 (1985)

[8] Okuno, E.; Ishikawa, T.; Kawai, J.; Kido, R.: Alanine:glyoxylate aminotransferase activities in liver of Suncus murinus (Insectivora). Comp. Biochem. Physiol. B Comp. Biochem., 90, 773-778 (1988)

[9] Kobayashi, S.; Hayashi, S.; Fujiwara, S.; Noguchi, T.: Identity of alanine:-glyoxylate aminotransferase with alanine:2-oxoglutarate aminotransferase in rat liver cytosol. Biochimie, **71**, 471-475 (1989)

[10] Sakuraba, H.; Fujiwara, S.; Noguchi, T.: Purification and characterization of peroxisomal apo and holo alanine:glyoxylate aminotransferase from bird liver. Arch. Biochem. Biophys., **286**, 453-460 (1991)

[11] Danpure, C.J.; Fryer, P.; Griffiths, S.; Guttridge, K.M.; Jennings, P.R.; Allsop, J.; Moser, A.B.; Naidu, S.; Moser, H.W.; et al.: Cytosolic compartmentalization of hepatic alanine:glyoxylate aminotransferase in patients with aberrant peroxisomal biogenesis and its effect on oxalate metabolism. J. Inher. Metab. Dis., **17**, 27-40 (1994)

[12] Horvath, V.A.P.; Wanders, R.J.A.: Aminooxy acetic acid: a selective inhibitor of alanine:glyoxylate aminotransferase and its use in the diagnosis of primary hyperoxaluria type I. Clin. Chim. Acta, **243**, 105-114 (1995)

[13] Ishikawa, K.; Kaneko, E.; Ichiyama, A.: Pyridoxal 5'-phosphate binding of a recombinant rat serine: pyruvate/alanine:glyoxylate aminotransferase. J. Biochem., **119**, 970-978 (1996)

[14] Leiper, J.M.; Oatey, P.B.; Danpure, C.J.: Inhibition of alanine:glyoxylate aminotransferase 1 dimerization is a prerequisite for its peroxisome-to-mitochondrion mistargeting in primary hyperoxaluria type 1. J. Cell. Biol., **135**, 939-951 (1996)

[15] Rumsby, G.; Weir, T.; Samuell, C.T.: A semiautomated alanine:glyoxylate aminotransferase assay for the tissue diagnosis of primary hyperoxaluria type 1. Ann. Clin. Biochem., **34**, 400-404 (1997)

[16] Holbrook, J.D.; Birdsey, G.M.; Yang, Z.; Bruford, M.W.; Danpure, C.J.: Molecular adaptation of alanine: glyoxylate aminotransferase targeting in primates. Mol. Biol. Evol., **17**, 387-400 (2000)

[17] Lumb, M.J.; Danpure, C.J.: Functional synergism between the most common polymorphism in human alanine:glyoxylate aminotransferase and four of the most common disease-causing mutations. J. Biol. Chem., **275**, 36415-36422 (2000)

[18] Oda, T.; Uchida, C.; Miura, S.: Mitochondrial targeting signal-induced conformational change and repression of the peroxisomal targeting signal of the precursor for rat liver serine:pyruvate/alanine:glyoxylate aminotransferase. J. Biochem., **127**, 665-671 (2000)

[19] Liepman, A.H.; Olsen, L.J.: Peroxisomal alanine: glyoxylate aminotransferase (AGT1) is a photorespiratory enzyme with multiple substrates in Arabidopsis thaliana. Plant J., **25**, 487-498 (2001)

[20] Zhang, X.; Roe, S.M.; Pearl, L.H.; Danpure, C.J.: Crystallization and preliminary crystallographic analysis of human alanine:glyoxylate aminotransferase and its polymorphic variants. Acta Crystallogr. Sect. D, **57**, 1936-1937 (2001)

[21] Han, Q.; Li, J.: Comparative characterization of Aedes 3-hydroxykynurenine transaminase/alanine glyoxylate transaminase and Drosophila serine pyruvate aminotransferase. FEBS Lett., **527**, 199-204 (2002)

[22] Danpure, C.J.; Lumb, M.J.; Birdsey, G.M.; Zhang, X.: Alanine:glyoxylate aminotransferase peroxisome-to-mitochondrion mistargeting in human hereditary kidney stone disease. Biochim. Biophys. Acta, 1647, 70-75 (2003)

[23] Santana, A.; Salido, E.; Torres, A.; Shapiro, L.J.: Primary hyperoxaluria type 1 in the Canary Islands: A conformational disease due to I244T mutation in the P11L-containing alanine:glyoxylate aminotransferase. Proc. Natl. Acad. Sci. USA, 100, 7277-7282 (2003)

[24] Zhang, X.; Roe, S.M.; Hou, Y.; Bartlam, M.; Rao, Z.; Pearl, L.H.; Danpure, C.J.: Crystal structure of alanine:glyoxylate aminotransferase and the relationship between genotype and enzymatic phenotype in primary hyperoxaluria type 1. J. Mol. Biol., 331, 643-652 (2003)

Serine-glyoxylate transaminase

2.6.1.45

1 Nomenclature

EC number
2.6.1.45

Systematic name
L-serine:glyoxylate aminotransferase

Recommended name
serine-glyoxylate transaminase

Synonyms
L-serine glyoxylate aminotransferase
aminotransferase, serine-glyoxylate
serine-glyoxylate aminotransferase

CAS registry number
37259-57-7

2 Source Organism

<1> *Phaseolus vulgaris* (kidney bean, cv. Red Kidney [1,2]) [1, 2]
<2> *Pisum sativum* (pea, cv. Little Marvel [3]) [3]
<3> *Spinacia oleracea* (spinach [4]) [4]
<4> *Cucumis sativus* (cucumber, cv. Improved Long Green [5]) [5]
<5> *Nicotiana tabacum* (tobacco [6]) [6]
<6> *Secale cereale* (rye [8,9]) [8, 9, 10, 11]
<7> *Hyphomicrobium methylovorum* (GM2, glycine-resistant mutant derived from parent strain KM146 [7,12]) [7, 12, 13, 15, 16]
<8> *Hyphomicrobium sp.* (strains NCIB10099, X, JTS-811, ZV622 and 53-49 [13]) [13]
<9> *Methylobacterium extorquens* (strain AM1 [13]) [13]
<10> *Methylobacterium organophilum* (strain XX [13]) [13]
<11> *Hordeum vulgare* [14]

3 Reaction and Specificity

Catalyzed reaction
L-serine + glyoxylate = 3-hydroxypyruvate + glycine (<1, 3, 7> ping-pong reaction mechanism [1, 4, 7, 15, 16]; <6> ping pong bi bi mechanism [8];

<7> reaction is catalyzed via the initial formation of an external Schiff base intermediate [16])

Reaction type
amino group transfer

Natural substrates and products
S   L-serine + glyoxylate <1-4, 7, 11> (<3> physiologically irreversible [4]; <1> involved in glycine metabolism [1,2]; <4> reaction in photorespiratory glycolate pathway [5]; <7> plays essential role in methanol assimilation through serine pathway [7]; <11> reduced activity in vivo results in the accumulation of serine and to a smaller extend, of glycine, indicating that the flux through the photorespiratory pathway is restricted [14]; <7> part of the C-1 assimilation pathway [16]) [1-5, 7, 14, 16]
P   3-hydroxypyruvate + glycine

Substrates and products
S   L-alanine + 2-oxomalonate <7> (Reversibility: ? <7> [15]) [15]
P   pyruvate + 2-aminomalonate <7> [15]
S   L-alanine + glyoxylate <3, 4, 6, 7> (<6> 20% of L-serine transamination activity [9]; <3> poor substrate [4]) (Reversibility: ? <3, 4, 6, 7> [4, 5, 8, 9, 15]) [4, 5, 8, 9, 15]
P   pyruvate + glycine <3, 4, 6, 7> [4, 5, 8, 9, 15]
S   L-asparagine + glyoxylate <2, 5, 6> (<2> transamination at 38% the rate of L-serine [3]; <6> 20% of L-serine transamination actiivty [9]) (Reversibility: r <2,6> [3,9]; ? <5> [6,8]) [3, 6, 8, 9]
P   2-oxosuccinamate + glycine <2, 5, 6> [3, 6, 8, 9]
S   L-asparagine + hydroxypyruvate <6> (<6> 20% of L-serine transamination activity [9]) (Reversibility: ? <6> [9]) [9]
P   2-oxosuccinamate + L-serine <6> [9]
S   L-serine + 2-oxomalonate <7> (Reversibility: ? <7> [7]) [7, 15]
P   3-hydroxypyruvate + 2-aminomalonate <7> [7, 15]
S   L-serine + 2-oxosuccinamate <2> (<2> poor substrate [3]) (Reversibility: ? <2> [3]) [3]
P   3-hydroxypyruvate + L-asparagine <2> [3]
S   L-serine + glyoxylate <1-11> (<1,2> preferred substrates [1-3]; <7> nearly irreversible, trace amounts of L-serine [7]; <2> equilibrium towards glycine production [3]; <1,2,6> reverse reaction at 4-11% the rate of forward reaction [2,3,9]; <7> highly specific [7]; <7> no activity with D-serine [7]; <2,7> no or trace activity with oxaloacetate, 2-oxoglutarate [3,7]; <2,7> no activity with Phe, Arg, Val, Trp, Thr, Met [3,7]; <2> no activity with Tyr, Ile, Pro, Cys, Leu, Asp [3]; <7> no activity with Ala, Glu, Gln, His, oxamate, 2-oxo-n-butanoate, 3-methyl-2-oxo-butanoate, 2-methyl-DL-serine, L-serine-O-sulfate, DL-serine hydroxamate, O-phospho-L-serine [7]) (Reversibility: r <1-6> [1-6, 8, 9]; ir <7> [7, 12]; ? <8-11> [13,14]) [1-14]
P   3-hydroxypyruvate + glycine <1-11> [1-14]

S L-serine + hydroxypyruvate <2> (<2> poor substrate [3]) (Reversibility: ? <2> [3]) [3]

P 3-hydroxypyruvate + L-serine <2> [3]

S L-serine + pyruvate <1-3, 6> (<2> equilibrium towards alanine production [3]; <3> poor substrate, 10% of glyoxylate transamination activity [4]; <2,6> 8% of glyoxylate transamination activity [3,9]; <1> 6% of glyoxylate transamination activity [2]) (Reversibility: r <1,2> [2,3]; ? <3,6> [1,4,9]) [1-4, 9]

P 3-hydroxypyruvate + L-alanine <1, 2, 3, 6> [1-4, 9]

Inhibitors

$(NH_4)_2SO_4$ <1> (<1> 0.05 mM, 45% inhibition, 5 mM, complete inhibition [1]) [1]

3-chloro-L-alanine <7> [7]

$AgNO_3$ <7> (<7> strong inhibition [7]) [7]

D-serine <6> (<6> competitive inhibition [8]) [8]

$HgCl_2$ <7> (<7> strong inhibition [7]) [7]

KCN <7> (<7> weak inhibition [7]) [7]

L-alanine <3> (<3> competitive vs. L-serine [4]) [4]

L-asparagine <2> (<2> competitive vs. serine [3]) [3]

L-serine <2> (<2> competitive vs. asparagine [3]) [3]

$MgCl_2$ <7> (<7> weak inhibition [7]) [7]

N-ethylmaleimide <1, 6> (<1> 1 mM, 41% inhibition [1]; <6> weak, but significant inhibition [11]) [1, 2, 11]

NH_4^+ <5> (<5> reversible inhibition [6]) [6]

NH_4Cl <1, 6> (<1> glycine-hydroxypyruvate transamination is only weakly inhibited [2]; <1> 0.1 mM, 42% inhibition, 10 mM, complete inhibition [1]; <6> competitive vs. L-serine [10]) [2, 10]

NaF <7> (<7> weak inhibition [7]) [7]

SH-group inhibitor <6> [8]

aminooxyacetate <2, 6> (<2> 1 mM, complete inhibition [3]; <6> competitive inhibition [8]) [3, 8]

ammonium acetate <1> (<1> 0.1 mM, 42% inhibition, 10 mM, complete inhibition [1]) [1]

β-chloroalanine <7> (<7> competitive vs. L-alanine, uncompetitive vs. 2-oxomalonate [15]) [15]

cycloserine <7> (<7> weak inhibition [7]) [7]

formaldehyde <6> (<6> glyoxylate protects [8]) [8]

glycine <3> (<3> competitive vs. L-serine [4]) [4]

glyoxylate <2, 5, 6> (<5> irreversible inhibition only in the presence of NH_4^+, inactivation in the absence of amino acid substrates [6]; <6> amino acid substrates partially protect [8]; <2> competitive vs. pyruvate [3]; <6> presence of D-serine protects against inhibition [10]) [3, 6, 8, 10]

hydroxylamine <1-3, 6, 7> (<2> 0.1 mM, 97% inhibition, 85% inhibition in the presence of 0.1 mM pyridoxal phosphate [3]; <3> 0.002 mM, 92% inhibition, serine, L-alanine and glycine protect [4]; <6> 0.1 mM, 70% inhibition [8]; <1> 0.01 mM, 98% inhibition of serine-glyoxylate transamination [1];

<1> 0.01 mM, 85% inhibition of glycine-hydroxypyruvate transamination [2])
[1-4, 7, 8]
iodoacetamide <7> [7]
isonicotinic acid hydrazide <6> (<6> 0.1 mM, 205 inhibition [8]) [8]
oxalate <7> (<7> competitive vs. 2-oxomalonate, uncompetitive vs. L-serine
[15]) [15]
p-chloromercuribenzoate <1, 7> (<1> 1 mM, 37% inhibition [1]) [1, 7]
p-hydroxymercuribenzoate <1, 6> (<6> weak but significant inhibition [11])
[2, 11]
penicillamine <7> (<7> weak inhibition [7]) [7]
phenylhydrazine <7> (<7> weak inhibition [7]) [7]
pyruvate <2> (<2> competitive vs. glyoxylate [3]) [3]
semicarbazide <7> [7]
Additional information <5, 6, 7> (<5> not inhibited with or without NH_4^+ by
oxalate, formate, acetaldehyde, pyruvate, hydroxypyruvate, 2-oxoglutarate
[6]; <7> not inhibited by $NaBH_4$, 3-chloro-D-alanine, 1,10-phenanthroline,
2,2'-dipyridyl, 8-hydroxyquinoline, EDTA, iodoacetate, $FeCl_3$, $AlCl_3$, $CdCl_2$,
$CoCl_2$, $NiCl_2$, $CaCl_2$, NaN_3, $ZnCl_2$, $PbCl_2$, LiCl, $CuCl_2$, $BaCl_2$, ascorbate, cysteamine, N-ethylmaleimide [7]; <6> not inhibited by 5,5'-dithio-bis(2-nitro-
benzoate) [11]) [6, 7, 11]

Cofactors/prosthetic groups

pyridoxal 5'-phosphate <2, 3, 6, 7> (<2, 3, 6, 7> required for activity [3, 4, 7,
8]; <7> a pyridoxal phosphate protein, 4 mol per mol enzyme [7]; <6>
loosely bound [8]; <2> K_m-value: 0.0025 mM [3]) [3, 4, 7, 8, 16]

Specific activity (U/mg)

2.18 <3> [4]
2.41 <5> [6]
6.97 <1> [1]
34.4 <4> [5]
39.3 <2> [3]
53.2 <6> [9]
69.5 <7> [7]
72 <7> (<7> recombinant enzyme [12]) [12]

K_m-Value (mM)

0.15 <3> (glyoxylate, <3> pH 7.0, 27°C [4]) [4]
0.23 <7> (glyoxylate, <7> pH 7.0, 30°C [7]) [7]
0.28 <7> (L-serine, <7> pH 8.0 [15]) [15]
0.39 <1> (L-serine, <1> cosubstrate pyruvate [1]) [1]
0.5 <2> (L-serine, <2> pH 8.1, 30°C, cosubstrate pyruvate [3]) [3]
0.6 <2> (L-serine, <2> pH 8.1, 30°C, cosubstrate glyoxylate [3]) [3]
0.6 <1> (glyoxylate, <1> pH 8.0, 30°C, cosubstrate L-serine [1]) [1]
0.63 <1> (3-hydroxypyruvate, <1> pH 8.0, 30°C, cosubstrate L-alanine [2])
[2]
0.68 <7> (glyoxylate, <7> recombinant enzyme [12]) [12]
0.71 <1> (L-serine, <1> pH 8.0, 30°C, cosubstrate glyoxylate [1]) [1]

1.1 <1> (3-hydroxypyruvate, <1> pH 8.0, 30°C, cosubstrate glycine [2]) [2]
1.13 <7> (2-oxomalonate, <7> pH 8.0 [15]) [15]
2 <2> (pyruvate, <2> pH 8.1, 30°C, cosubstrate L-serine [3]) [3]
2.4 <6> (L-serine) [10]
2.5 <2> (pyruvate, <2> pH 8.1, 30°C, cosubstrates: L-asparagine, glyoxylate, L-asparagine [3]) [3]
2.7 <3> (L-serine, <3> pH 7.0, 27°C [4]) [4]
3.86 <7> (L-serine, <7> recombinant enzyme [12]) [12]
3.9 <2> (L-asparagine, <2> pH 8.1, 30°C, cosubstrate glyoxylate [3]) [3]
4.5 <2> (L-asparagine, <2> pH 8.1, 30°C, cosubstrate pyruvate [3]) [3]
4.6 <2> (glyoxylate, <2> pH 8.1, 30°C, cosubstrate L-serine [3]) [3]
4.98 <7> (L-serine) [7]
11 <1> (glycine, <1> pH 8.0, 30°C [2]) [2]
20 <1> (L-alanine, <1> pH 8.0, 30°C [2]) [2]
Additional information <1> (<1> kinetic study [1]) [1]

K$_i$-Value (mM)

0.00012 <6> (aminooxyacetate) [8]
0.0036 <7> (β-chloroalanine, <7> pH 8.0 [15]) [15]
0.12 <2> (L-serine, <2> pH 8.1, 30°C [3]) [3]
0.27 <7> (oxalate, <7> pH 8.0 [15]) [15]
0.4 <2> (glyoxylate, <2> pH 8.1, 30°C [3]) [3]
1 <6> (NH$_4^+$) [10]
1.6 <6> (D-serine) [8]
3 <2> (L-asparagine, <2> pH 8.1, 30°C [3]) [3]
25 <2> (pyruvate, <2> pH 8.1, 30°C [3]) [3]
33 <3> (glycine, <3> pH 7.0, 27°C [4]) [4]
45 <3> (L-alanine, <3> pH 7.0, 27°C [4]) [4]

pH-Optimum

7 <3> [4]
8 <7> [7]
8.1 <2> [3]
8.2 <1> [1]
8.5 <1> (<1> substrates L-alanine and 3-hydroxypyruvate [2]) [2]
9 <1> (<1> substrates glycine and 3-hydroxypyruvate [2]) [2]
Additional information <7> (<7> pI: 6.9 [7]) [7]

pH-Range

6-9 <3> (<3> approx. 60% of maximal activity at pH 6, approx. half-maximal activity at pH 9 [4]) [4]
7.5-9.2 <2> (<2> approx. 50% activity at pH 7.5 and 9.2, no activity below pH 6.7 and above pH 9.9 [3]) [3]

Temperature optimum (°C)

27 <3> (<3> assay at [4]) [4]
30 <1, 2, 6> (<1,2,6> assay at [1-3,9]) [1-3, 9]
40 <7> [7]

4 Enzyme Structure

Molecular weight

85000-91200 <6> (<6> gel filtration [9]) [9]
105000 <2> (<2> gel filtration [3]) [3]
140000 <7> (<7> HPLC, gel filtration [7]) [7]
150000 <7> (<7> Sephacryl S-200, gel filtration [7]) [7]
170000 <4> (<4> gel filtration, purification reduces molecular weight to 62000 Da [5]) [5]

Subunits

dimer <6> (<6> 2 * 43000, SDS-PAGE [9]) [9]
oligomer <4> (<4> x * 45000 + x * 47000, SDS-PAGE [5]) [5]
tetramer <7> (<7> 4 * 40000, SDS-PAGE [7]; <7> 4 * 40000, recombinant enzyme, SDS-PAGE [12]) [7, 12]

5 Isolation/Preparation/Mutation/Application

Source/tissue

cotyledon <4> [5]
leaf <1-3, 5> [1-4, 6]
seedling <6> (<6> 7 days old [8,9]) [8, 9, 10, 11]

Localization

peroxisome <2-4, 11> (<4> marker enzyme [5]) [3-5, 14]
soluble <1, 7> [1, 7]

Purification

<1> (DEAE-cellulose, agarose gel, Brushite, Sephadex g-200, partial purification [1]) [1, 2]
<2> (pH 5.4, polyethylene glycol 6000, DEAE-Sephadex A25, Sephacryl S-300, partial purification [3]) [3]
<3> (partial [4]) [4]
<4> (polyethyleneimine, ammonium sulfate, Ultrogel AcA 34, DEAE-cellulose, isoenzymes A and B [5]) [5]
<5> (ammonium sulfate, agarose A-15, chromatofocusing, partial purification [6]) [6]
<6> [9]
<7> (ammonium sulfate, DEAE-Sephacel, phenyl-Sepharose, CM-toyopearl, Cellulofine [7]; recombinant enzyme [12]) [7, 12]

Cloning

<7> (expression in Escherichia coli [12]) [12]

6 Stability

pH-Stability

6-11 <7> (<7> stable for 30 min at 30°C [7]) [7]

Temperature stability

35 <7> (<7> and below, 30 min stable [7]) [7]
40 <7> (<7> 30 min, 16% loss of activity [7]) [7]
45 <7> (<7> 52% loss of activity after 30 min [7]) [7]
50 <7> (<7> 30 min, 93% loss of activity [7]) [7]
55 <7> (<7> 30 min, inactivation [7]) [7]

General stability information

<5>, imidazole decreases stability [6]
<6>, sucrose, pyridoxal phosphate or 2-mercaptoethanol stabilize [9]

Storage stability

<2>, 5°C, stable overnight, 20% loss of activity per day on further storage, in 20% ethylene glycol or 15% glycerol, 20% loss of activity within 5 days [3]
<3>, -18°C, 2 weeks [4]
<5>, -10°C, 2 months [6]
<7>, -20°C, 10 mM phosphate buffer pH 7, 0.1 mM pyridoxal phosphate, 0.1 mM dithiothreitol, 45% v/v glycerol, at least 5 months [7]

References

[1] Smith, I.K.: Purification and characterization of serine:glyoxylate amino-transferase from kidney bean (Phaseolus vulgaris). Biochim. Biophys. Acta, **321**, 156-164 (1973)

[2] Carpe, A.I.; Smith, I.K.: Serine-glyoxylate aminotransferase from kidney bean (Phaseolus vulgaris). II. The reverse reactions. Biochim. Biophys. Acta, **370**, 96-101 (1974)

[3] Ireland, R.J.; Joy, K.W.: Purification and properties of an asparagine amino-transferase from Pisum sativum leaves. Arch. Biochem. Biophys., **223**, 291-296 (1983)

[4] Nakamura, Y.; Tolbert, N.E.: Serine:glyoxylate, alanine:glyoxylate, and glu-tamate:glyoxylate aminotransferase reactions in peroxisomes from spinach leaves. J. Biol. Chem., **258**, 2763-2768 (1983)

[5] Hondred, D.; Hunter, J.McC.; Keith, R.; Titus, D.E.; Becker, W.M.: Isolation of serine:glyoxylate aminotransferase from cucumber cotyledons. Plant Physiol., **79**, 95-102 (1985)

[6] Havir, E.A.: Inactivation of serine:glyoxylate and glutamate:glyoxylate ami-notransferases from tobaco leaves by glyoxylate in the presence of ammo-nium ion. Plant Physiol., **80**, 473-478 (1986)

[7] Izumi, Y.; Yoshida, T.; Yamada, H.: Purification and characterization of ser-ine-glyoxylate aminotransferase from a serine-producing methylotroph,

Hyphomicrobium methylovorum GM2. Eur. J. Biochem., **190**, 285-290 (1990)

[8] Paszkowski, A.: Some properties of serine: glyoxylate aminotransferase from rye seedlings (Secale cereale L.). Acta Biochim. Pol., **38**, 437-448 (1991)

[9] Paszkowski, A.; Niedzielska, A.: Serine:glyoxylate aminotransferase from the seedlings of rye (Secale cereale L.). Acta Biochim. Pol., **37**, 277-282 (1990)

[10] Paszkowski, A.: Inhibition of glutamate: glyoxylate and serine: Glyoxylate aminotransferases from rye seedlings (Secale cereale L.) by glyoxylate or glyoxylate in the presence of ammonium ion. Acta Physiol. Plant., **16**, 217-223 (1994)

[11] Paszkowski, A.: The hydrosulfide groups of glutamate:glyoxylate and serine:glyoxylate aminotransferases from rye (Secale cereale L.) seedlings. Acta Physiol. Plant., **17**, 85-90 (1995)

[12] Hagishita, T.; Yoshida, T.; Izumi, Y.; Mitsunaga, T.: Cloning and expression of the gene for serine-glyoxylate aminotransferase from an obligate methylotroph Hyphomicrobium methylovorum GM2. Eur. J. Biochem., **241**, 1-5 (1996)

[13] Hagishita, T.; Yoshida, T.; Izumi, Y.; Mitsunaga, T.: Immunological characterization of serine-glyoxylate aminotransferase and hydroxypyruvate reductase from a methylotrophic bacterium, Hyphomicrobium methylovorum GM2. FEMS Microbiol. Lett., **142**, 49-52 (1996)

[14] Wingler, A.; Ann, V.J.; Lea, P.J.; Leegood, R.C.: Serine: glyoxylate aminotransferase exerts no control on photosynthesis. J. Exp. Bot., **50**, 719-722 (1999)

[15] Karsten, W.E.; Ohshiro, T.; Izumi, Y.; Cook, P.F.: Initial velocity, spectral, and pH studies of the serine-glyoxylate aminotransferase from Hyphomicrobiuim methylovorum. Arch. Biochem. Biophys., **388**, 267-275 (2001)

[16] Karsten, W.E.; Cook, P.F.: Detection of intermediates in reactions catalyzed by PLP-dependent enzymes: O-acetylserine sulfhydrylase and serine-glyoxalate aminotransferase. Methods Enzymol., **354**, 223-237 (2002)

Diaminobutyrate-pyruvate transaminase 2.6.1.46

1 Nomenclature

EC number
2.6.1.46

Systematic name
L-2,4-diaminobutanoate:pyruvate aminotransferase

Recommended name
diaminobutyrate-pyruvate transaminase

Synonyms
L-diaminobutyric acid transaminase
aminotransferase, diaminobutyrate-pyruvate
diaminobutyrate-pyruvate aminotransferase

CAS registry number
37277-95-5

2 Source Organism

<1> *Xanthomonas sp.* [1]
<2> *Ectothiorhodospira halochloris* (strain DSM 1059 [2]) [2]
<3> *Halomonas elongata* (strain DSM 2581 [2]; strain OUT300318 [3]) [2, 3]

3 Reaction and Specificity

Catalyzed reaction
L-2,4-diaminobutanoate + pyruvate = L-aspartate 4-semialdehyde + L-alanine

Reaction type
amino group transfer

Natural substrates and products
S L-2,4-diaminobutanoate + pyruvate <1> (Reversibility: r <1> [1]) [1]
P L-2-amino-4-oxobutanoate + L-alanine <1> [1]
S L-2-amino-4-oxobutanoate + L-glutamate <2-3> (<2,3> involved in the biosynthesis of the osmolyte ectoine in halophilic eubacteria [2]) (Reversibility: ? <2-3> [2]; r <3> [3]) [2, 3]
P L-2,4-diaminobutanoate + 2-oxoglutarate <2-3> [2, 3]

Substrates and products

S L-2,4-diaminobutanoate + pyruvate <1> (Reversibility: r <1> [1]) [1]
P L-2-amino-4-oxobutanoate + L-alanine <1> [1]
S L-2-amino-4-oxobutanoate + L-glutamate <2-3> (Reversibility: ? <2-3> [2]; r <3> [3]) [2, 3]
P L-2,4-diaminobutanoate + 2-oxoglutarate <2-3> [2, 3]

Inhibitors

hydroxylamine <1> (<1> 3.3 mM: 87% inhibition, 0.33 mM: 58.3% inhibition, 0.033 mM: 16.6% inhibition [1]) [1]
semicarbazide hydrochloride <1> (<1> 20 mM: 83% inhibition, 10 mM: 63% inhibition [1]) [1]
sodium borohydride <1> (<1> 5 mM: complete inhibition [1]) [1]
Additional information <1> (<1> EDTA has no effect [1]) [1]

Metals, ions

K^+ <3> (<3> required for activity and stability [3]) [3]
Additional information <1> (<1> probably no need for heavy-metal ions [1]) [1]

Specific activity (U/mg)

0.00092 <1> [1]
0.054 <2> [2]
0.171 <3> [2]
Additional information <3> [3]

K_m-Value (mM)

1.2 <2> (L-glutamate, <2> pH 9.1, 37°C [2]) [2]
1.85 <3> (L-glutamate, <2> pH 9.1, 37°C [2]) [2]
4.5 <3> (L-2-amino-4-oxobutanoate, <3> pH 8.5, 25°C [3]) [3]
9.1 <3> (L-glutamate, <3> pH 8.5, 25°C [3]) [3]

pH-Optimum

7.9 <3> [2]
8.2 <2> [2]
8.6-8.7 <3> [3]
9 <1> [1]

Temperature optimum (°C)

25 <3> (<3> in presence of 0.4 M NaCl [3]) [3]
37 <2-3> (<2-3> assay at [2]) [2]

4 Enzyme Structure

Molecular weight

44000 <3> (<3> gel filtration [3]) [3]

5 Isolation/Preparation/Mutation/Application

Purification
<1> (ammonium sulfate fractionation [1]) [1]
<3> (ammonium sulfate precipitation, Sepharose CL-6B [3]) [3]

6 Stability

Storage stability
<3>, 0°C, 100 mM 2,4-diaminobutyric acid [3]

References

[1] Rajagopal Rao, D.; Hariharan, K.; Vijayalakshmi, K.R.: A study of the metabolism of L-α γ-diaminobutyric acid in a Xanthomonas species. Biochem. J.,
   114, 107-115 (1969)
[2] Peters, P.; Galinski, E.A.; Trüper, H.G.: The biosynthesis of ectoine. FEMS
   Microbiol. Lett., **71**, 157-162 (1990)
[3] Ono, H.; Sawada, K.; Khunajakr, N.; Tao, T.; Yamamoto, M.; Hiramoto, M.;
   Shinmyo, A.; Takano, M.; Murooka, Y.: Characterization of biosynthetic enzymes for ectoine as a compatible solute in a moderately halophilic eubacter-
   ium, Halomonas elongata. J. Bacteriol., **181**, 91-99 (1999)

Alanine-oxomalonate transaminase   2.6.1.47

1 Nomenclature

EC number
2.6.1.47

Systematic name
L-alanine:oxomalonate aminotransferase

Recommended name
alanine-oxomalonate transaminase

Synonyms
L-alanine-ketomalonate transaminase
alanine-ketomalonate (mesoxalate) transaminase
alanine-oxomalonate aminotransferase
aminotransferase, alanine-oxomalonate

CAS registry number
37277-96-6

2 Source Organism

<1> *Bombyx mori* [1]
<2> *Rattus norvegicus* [1]
<3> *Oryctolagus cuniculus* [1]

3 Reaction and Specificity

Catalyzed reaction
L-alanine + oxomalonate = pyruvate + aminomalonate

Reaction type
amino group transfer

Natural substrates and products
  S DL-alanine + oxomalonate <1-3> (<1-3> glutamate, 2-aminobutyrate, as-
    partate can also act as amino donors in decreasing order of reactivity,
    probably involved in the biosynthesis of glycine [1]) [1]
  P pyruvate + aminomalonic acid

Substrates and products

S DL-alanine + oxomalonate <1-3> (<1-3> glutamate, 2-aminobutyrate and aspartate can also act as amino donors in decreasing order of reactivity [1]) (Reversibility: ? <1-3> [1]) [1]

P pyruvate + aminomalonic acid <1-3> [1]

Inhibitors

KCN <1> (<1> 1 mM, 38% inhibition [1]) [1]
benzoquinone <1> (<1> 1 mM, 84% inhibition [1]) [1]
hydroxylamine <1> (<1> 2 mM, 58% inhibition [1]) [1]
p-chloromercuribenzoate <1> (<1> 1 mM, 99% inhibition [1]) [1]

pH-Optimum

9-9.2 <1> [1]

Temperature optimum (°C)

37 <1> (<1> assay at [1]) [1]

5 Isolation/Preparation/Mutation/Application

Source/tissue

heart <3> [1]
liver <2, 3> [1]
silk gland <1> (<1> posterior [1]) [1]

Purification

<1> (partial [1]) [1]

References

[1] Nagayama, H.; Muramatsu, M.; Shimura, K.: Enzymatic formation of amino-malonic acid from ketomalonic acid. Nature, **181**, 417-418 (1958)

5-Aminovalerate transaminase 2.6.1.48

1 Nomenclature

EC number
2.6.1.48

Systematic name
5-aminopentanoate:2-oxoglutarate aminotransferase

Recommended name
5-aminovalerate transaminase

Synonyms
5-aminovalerate aminotransferase
δ-aminovalerate aminotransferase
δ-aminovalerate transaminase
δ-aminovaleric acid-glutamic acid transaminase <7> [1]

CAS registry number
37277-97-7

2 Source Organism

<-2> no activity in *Saccharomyces cerevisiae* [5]
<-1> no activity in *Candida boidinii* (CBS 5777 [5]) [5]
<1> *Candida famata* (CBS 8109 [5]) [5]
<2> *Candida guilliermondii* (var.membranaefaciens [2,4]) [2, 4]
<3> *Candida tropicalis* (ATCC 20366 [5]) [5]
<4> *Candida utilis* (CBS 621 [5]) [5]
<5> *Clostridium aminovalericum* (T2-7 [3]) [3]
<6> *Pseudomonas putida* (KT2440, nucleotide sequence accession number [6]) [6]
<7> *Pseudomonas sp.* [1]
<8> *Rhodotorula glutinis* (NCYC 61 [5]) [5]

3 Reaction and Specificity

Catalyzed reaction
5-aminopentanoate + 2-oxoglutarate = 5-oxopentanoate + L-glutamate (a pyridoxal-phosphate protein)

565

Reaction type
amino group transfer

Natural substrates and products
S   5-aminopentanoate + 2-oxoglutarate <1-8> (<7> metabolism of L-lysine
[1]; <2> enzyme is strongly induced in cells grown on δ-aminovalerate or
on lysine as only nitrogen source [2]; <2, 3, 6> catabolism of lysine [4-6])
(Reversibility: r <1-8> [1-5]) [1-6]

P   5-oxovalerate + L-glutamate <1-8> [1-6]

Substrates and products
S   4-aminobutyrate + 2-oxoglutarate <2, 3, 5> (<2> relative activity 40%
[2]) (Reversibility: r <2, 3, 5> [2, 3, 5]) [2, 3, 5]

P   succinic semialdehyde + L-glutamate

S   5-aminopentanoate + 2-oxoglutarate <1-8> (Reversibility: r <1-8> [1-6])
[1-6]

P   5-oxovalerate + L-glutamate <1-8> [1-6]

S   6-aminohexanoate + 2-oxoglutarate <5> (Reversibility: r <5> [3]) [3]

P   adipic semialdehyde + L-glutamate <5> [3]

S   L-leucine + 2-oxoglutarate <2> (<2> relative activity 10% [2]) (Reversibility: r <2> [2]) [2, 3]

P   4-methyl-2-oxopentanoate + L-glutamate

S   L-norleucine + 2-oxoglutarate <2, 3> (<2> relative activity 4.8% [2]) (Reversibility: r <2,3> [2,5]) [2, 3, 5]

P   2-oxohexanoate + L-glutamate

S   glycine + 2-oxoglutarate <2> (<2> relative activity 1% [2]) (Reversibility:
r <2> [2]) [2, 3]

P   glyoxlate + L-glutamate

Inhibitors
2,3-diaminopropionate <2> (<2> noncompetitive inhibition with respect to
δ-aminovalerate, competitive inhibition to 2-oxoglutarate [2]) [2]

2,4-diaminobutyrate <2> (<2> noncompetitive inhibition with respect to δ-
aminovalerate, competitive inhibition to 2-oxoglutarate [2]) [2]

L-lysine <2> (<2> noncompetitive inhibition with respect to δ-aminovalerate, competitive inhibition to 2-oxoglutarate [2]) [2]

L-ornithine <2> (<2> mixed inhibition with respect to δ-aminovalerate,
competitive inhibition to 2-oxoglutarate [2]) [2]

L-alanine <2> (<2> noncompetitive inhibition with respect to δ-aminovalerate, competitive inhibition to 2-oxoglutarate [2]) [2]

β-alanine <2> (<2> noncompetitive inhibition with respect to δ-aminovalerate, competitive inhibition to 2-oxoglutarate [2]) [2]

glycine <2> (<2> noncompetitive inhibition with respect to δ-aminovalerate,
competitive inhibition to 2-oxoglutarate [2]) [2]

hydroxylamine <7> [1]

p-chloromercuribenzoate <7> [1]

semicarbazide <7> [1]
Additional information <2> (<2> α-aminobutyrate, 4-aminobutyrate, α-aminovalerate and ε-aminocaproate does not cause any inhibition [2]) [2]

Cofactors/prosthetic groups
pyridoxal 5'-phosphate <2, 5> (<2> K_m 0.0227 mM [2]) [2, 3]

Activating compounds
2-mercaptoethanol <5> [3]

Specific activity (U/mg)
0.004 <4> [5]
0.005 <1> [5]
0.006 <8> [5]
0.026 <3> [5]
0.103 <7> [1]
15.5 <5> [3]
20.8 <2> [2]

K_m-Value (mM)
3.6 <2> (2-oxoglutarate, <2> pH 7.8, 35°C [2]) [2]
4 <5> (δ-aminovaleric acid, <5> pH 8.2, 37°C [3]) [3]
4.1 <7> (glutamic acid, <7> pH 8.0, 30°C [1]) [1]
4.7 <5> (2-oxoglutarate, <5> pH 8.2, 37°C [3]) [3]
4.9 <2> (δ-aminovaleric acid, <2> pH 7.8, 35°C [2]) [2]
8 <7> (2-oxoglutarate, <7> pH 8.0, 30°C [1]) [1]
10 <7> (δ-aminovaleric acid, <7> pH 8.0, 30°C [1]) [1]

K_i-Value (mM)
5 <2> (β-alanine, <2> pH 7.8, 35°C, competitive inhibition with respect to 2-oxoglutarate [2]) [2]
5.6 <2> (L-ornithine, <2> pH 7.8, 35°C, competitive inhibition with respect to 2-oxoglutarate [2]) [2]
7 <2> (2,3-diaminopropionate, <2> pH 7.8, 35°C, competitive inhibition with respect to 2-oxoglutarate [2]) [2]
9.5 <2> (ornithine, <2> pH 7.8, 35°C, mixed inhibition [2]) [2]
10 <2> (glycine, <2> pH 7.8, 35°C, competitive inhibition with respect to 2-oxoglutarate [2]) [2]
11.5 <2> (2,4-diaminobutyrate, <2> pH 7.8, 35°C, competitive inhibition with respect to 2-oxoglutarate [2]) [2]
12.5 <2> (L-lysine, <2> pH 7.8, 35°C, competitive inhibition with respect to 2-oxoglutarate [2]) [2]
15 <2> (alanine, <2> pH 7.8, 35°C, competitive inhibition with respect to 2-oxoglutarate [2]) [2]
16 <2> (β-alanine, <2> pH 7.8, 35°C, noncompetitive inhibition with respect to δ-aminovalerate [2]) [2]
50 <2> (2,3-diaminopropionate, <2> pH 7.8, 35°C, noncompetitive inhibition with respect to δ-aminovalerate [2]) [2]

90 <2> (glycine, <2> pH 7.8, 35°C, noncompetitive inhibition with respect to δ-aminovalerate [2]) [2]

120 <2> (alanine, <2> pH 7.8, 35°C, noncompetitive inhibition with respect to δ-aminovalerate [2]) [2]

120 <2> (lysine, <2> pH 7.8, 35°C, noncompetitive inhibition with respect to δ-aminovalerate [2]) [2]

190 <2> (2,4-diaminobutyrate, <2> pH 7.8, 35°C, noncompetitive inhibition with respect to δ-aminovalerate [2]) [2]

pH-Optimum

7.8-8.5 <2> [2, 4]

8.2 <5> (<5> half-maximal rates at pH 6.5 and 9.8 [3]) [3]

9 <7> [1]

pH-Range

6-9.5 <7> [1]

Temperature optimum (°C)

37 <5> [3]

40 <2> [2, 4]

40-55 <2> (<2> maximal activity at 55, 45 and 40°C [4]) [4]

4 Enzyme Structure

Molecular weight

118000 <2> (<2> gel filtration [2]) [2]

Subunits

dimer <2> (<2> 2 * 60000, SDS-PAGE [2]) [2]

5 Isolation/Preparation/Mutation/Application

Purification

<2> [2]

<3> (partially [5]) [5]

<5> (partially [3]) [3]

<7> [1]

Cloning

<6> (gene davT identified using a transposon to generate transcriptional fusions by insertional mutagenesis [6]) [6]

6 Stability

Temperature stability

50 <2> (<2> preserves 50% of its activity after being heated for 7 min, in presence of 2-oxoglutarate [2]) [2]

General stability information

<2>, 2-oxoglutarate and pyridoxal 5'-phosphate protects against thermal denaturation [2]

<3>, 2-oxoglutarate and protects against thermal denaturation [5]

References

[1] Ichihara, A.; Ichihara, E.A.; Suda, M.: Metabolism of L-lysine by bacterial enzymes. IV. δ-aminovaleric acid-glutamic acid transaminase. J. Biochem., **48**, 412-420 (1960)

[2] Der Garabedian, P.A.: Candida δ-aminovalerate: α-ketoglutarate aminotransferase: purification and enzymologic properties. Biochemistry, **25**, 5507-5512 (1986)

[3] Barker, H.A.; DÁri, L.; Kahn, J.: Enzymatic reactions in the degradation of 5-aminovalerate by Clostridium aminovalericum. J. Biol. Chem., **262**, 8994-9003 (1987)

[4] Der Garabedian, P.A.; Vermeersch, J.J.: Lysine degradation in Candida. Characterization and probable role of L-norleucine-leucine, 4-aminobutyrate and δ-aminovalerate:2-oxoglutarate aminotransferases. Biochimie, **71**, 497-503 (1989)

[5] Large, P.J.; Robertson, A.: The route of lysine breakdown in Candida tropicalis. FEMS Microbiol. Lett., **82**, 209-213 (1991)

[6] Espinosa-Urgel, M.; Ramos, J.L.: Expression of a Pseudomonas putida aminotransferase involved in lysine catabolism is induced in the rhizosphere. Appl. Environ. Microbiol., **67**, 5219-5224 (2001)

Dihydroxyphenylalanine transaminase

1 Nomenclature

EC number
2.6.1.49

Systematic name
3,4-dihydroxy-L-phenylalanine:2-oxoglutarate aminotransferase

Recommended name
dihydroxyphenylalanine transaminase

Synonyms
L-dopa transaminase
aminotransferase, dihydroxyphenylalanine
aspartate-DOPP transaminase, ADT
dihydroxyphenylalanine aminotransferase
dopa aminotransferase
dopa transaminase
glutamate-DOPP transaminase, GDT
phenylalanine-DOPP transaminase, PDT

CAS registry number
37277-98-8

2 Source Organism

<1> *Cavia porcellus* [1]
<2> *Homo sapiens* [2]
<3> *Rattus norvegicus* [2]
<4> *Enterobacter cloacae* (NB 320 [3]) [3]
<5> *Alcaligenes faecalis* (IAM 1015 [4]) [4]

3 Reaction and Specificity

Catalyzed reaction
3,4-dihydroxy-L-phenylalanine + 2-oxoglutarate = 3,4-dihydroxyphenylpyruvate + L-glutamate

Reaction type
amino group transfer

Substrates and products

S 3-(3,4-dihydroxyphenyl)-L-alanine + 2-oxoglutarate <1-3> (<1-3> i.e. L-Dopa [1,2]; <1> most effective amino acceptor [1]) (Reversibility: ? <1-3> [1,2]) [1, 2]

P 3-(3,4-dihydroxyphenyl)pyruvate + L-glutamate <1-3> (<1-3> i.e. DOPP [1,2]) [1, 2]

S 3-(3,4-dihydroxyphenyl)-L-alanine + L-phenylpyruvate <2, 3> (<2,3> i.e. L-Dopa [2]) (Reversibility: ? <2,3> [2]) [2]

P 3-(3,4-dihydroxyphenyl)pyruvate + L-phenylalanine <2, 3> [2]

S 3-(3,4-dihydroxyphenyl)-L-alanine + oxaloacetic acid <1> (<1> i.e. L-Dopa, 10% of activity with 2-oxoglutarate [1]) (Reversibility: ? <1> [1]) [1]

P 3-(3,4-dihydroxyphenyl)pyruvate + L-aspartate <1> [1]

S 3-(3,4-dihydroxyphenyl)pyruvate + L-aspartate <4, 5> (<4, 5> aspartate-3,4-dihydroxyphenylpyruvate transaminase, ADT [3, 4]) (Reversibility: ? <4, 5> [3,4]) [3, 4]

P 3,4-dihydroxy-L-phenylalanine + oxaloacetate <4, 5> (<4, 5> i.e. L-Dopa [3,4]) [3, 4]

S 3-(3,4-dihydroxyphenyl)pyruvate + L-glutamate <4, 5> (<4, 5> glutamate-3,4-dihydroxyphenylpyruvate transaminase, GDT [3, 4]) (Reversibility: ? <4,5> [3,4]) [3, 4]

P 3-(3,4-dihydroxyphenyl)-L-alanine + 2-oxoglutarate <4, 5> (<4, 5> i.e. L-Dopa [3,4]) [3, 4]

S 3-(3,4-dihydroxyphenyl)pyruvate + L-phenylalanine <4, 5> (<4, 5> phenylalanine-3,4-dihydroxyphenylpyruvate transaminase, PDT [3, 4]) (Reversibility: ? <4,5> [3,4]) [3, 4]

P 3,4-dihydroxy-L-phenylalanine + L-phenylpyruvate <4, 5> (<4, 5> i.e. L-Dopa [3,4]) [3, 4]

S L-tyrosine + 2-oxoglutarate <1, 2, 3> (<1, 2, 3> 8% of activity with L-Dopa [1,2]) (Reversibility: ? <1,2,3> [1,2]) [1, 2]

P 3-(4-hydroxyphenyl)-2-oxopropanoate + L-glutamate <1, 2, 3> [1, 2]

Inhibitors

3-hydroxy-4-benzyloxyamine dihydrogen phosphate <2, 3> (<2, 3> NSD 1055, up to 30 mM [2]) [2]

Al^{3+} <5> (<5> strong inhibition [4]) [4]

Co^{2+} <5> (<5> weak inhibition [4]) [4]

Cu^{2+} <5> (<5> strong inhibition [4]) [4]

Fe^{2+} <5> (<5> strong inhibition [4]) [4]

KCN <1> (<1> 1 mM, 93% inhbition, pyridoxal 5'-phosphate protects [1]) [1]

Ni^{2+} <5> (<5> inhibition of ADT and GDT, not PDT [4]) [4]

Zn^{2+} <5> (<5> weak inhibition [4]) [4]

carbidopa <3> (<3> MK 486, inhibition is more marked if 2-oxoglutarate is used as amino group acceptor rather than phenylpyruvate, pyridoxal 5'-phosphate protects [2]) [2]

iodoacetic acid <1> (<1> 1 mM, 75% inhibition [1]) [1]

p-chloromercuriphenyl sulfonic acid <1> (<1> 0.02 mM, 92% inhibition, addition of 1 mM glutathione protects to 100% [1]) [1]
Additional information <2, 3> (<2,3> not inhibited by benserazide [2]) [2]

Activating compounds
pyridoxal phosphate <1> (<1> a pyridoxal phosphate protein, K_m: 0.016 mM [1]) [1]

Specific activity (U/mg)
0.0366 <1> [1]

K_m-Value (mM)
0.48 <1> (2-oxoglutarate, <1> pH 8.2, 37°C [1]) [1]
41 <1> (L-Dopa, <1> pH 8.2, 37°C [1]) [1]

pH-Optimum
7.4 <4, 5> (<4> PDT [3]; <5> GDT [4]) [3, 4]
8.2 <1> [1]
8.5 <4> (<4> GDT, PDT [3]) [3]
9-10 <5> (<5> ADT [4]) [4]
9.5 <5> (<5> PDT [4]) [4]

pH-Range
7-9 <1> (<1> approx. 35% of maximal activity at pH 7.0, approx. 60% of activity at pH 9.0 [1]) [1]

Temperature optimum (°C)
37 <1> (<1> assay at [1]) [1]
45 <4> [3]

Temperature range (°C)
35-55 <4> (<4> approx. 40% of maximal activity at 35°C, approx. 90% of maximal activity at 55°C [3]) [3]

5 Isolation/Preparation/Mutation/Application

Source/tissue
brain <1> [1]
liver <2, 3> [2]

Purification
<1> (ammonium sulfate, 52°C, calcium phosphate gel, alumina C, Sephadex G-200 [1]) [1]

6 Stability

pH-Stability
8 <5> (<5> 45°C, 30 min, enzyme GDT stable [4]) [4]
8-8.5 <4> (<4> 45°C, 30 min, enzyme GDT stable [3]) [3]

8.5-9 <5> (<5> 45°C, 30 min, enzyme PDT stable [4]) [4]
8.5-9 <5> (<5> 45°C, 30 min, enzyme PDT stable [4]) [4]

Temperature stability
30-40 <4> (<4> 15 min, heat activation, enzyme GDT [3]) [3]
55 <1> (<1> 6 min, pH 7.0, stable [1]) [1]
60 <4, 5> (<4,5> 15 min, complete loss of activity, enzyme GDT [3,4]) [3, 4]
Additional information <1> (<1> pyridoxal phosphate, Ca^{2+}, Mg^{2+} or Mn^{2+} protects against heat inactivation [1]) [1]

General stability information
<1>, pyridoxal phosphate, Ca^{2+}, Mg^{2+} or Mn^{2+} protects against heat inactivation [1]

References

[1] Fonnum, F.; Larsen, K.: Purification and properties of dihydroxyphenylalanine transaminase from guinea pig brain. J. Neurochem., **12**, 589-598 (1965)
[2] Waterhouse, M.J.; Chia, Y.C.; Lees, G.J.: Inhibition of human and rat hepatic aminotransferase activity with L-3,4-dihydroxyphenylalanine by inhibitors of peripheral aromatic amino acid decarboxylase. Mol. Pharmacol., **15**, 108-114 (1979)
[3] Nagasaki, T.; Sugita, M.; Fukawa, H.; Lin, H.-T.: DOPA production with Enterobacter cloacae NB 320 by transaminase reaction. Agric. Biol. Chem., **39**, 363-369 (1975)
[4] Nagasaki, T.; Sugita, M.; Fukawa, H.: Studies on DOPA transaminase of Alcaligenes faecalis. Agric. Biol. Chem., **37**, 1701-1706 (1973)

Glutamine-scyllo-inositol transaminase 2.6.1.50

1 Nomenclature

EC number
2.6.1.50

Systematic name
L-glutamine:2,4,6/3,5-pentahydroxycyclohexanone aminotransferase

Recommended name
glutamine-scyllo-inositol transaminase

Synonyms
L-glutamine-keto-scyllo-inositol aminotransferase
L-glutamine-scyllo-inosose transaminase
L-glutamine:2-deoxy-scyllo-inosose <5> [5]
L-glutamine:inosose aminotransferase <4> [3]
aminocyclitol aminotransferase <2> [2]
glutamine scyllo-inosose aminotransferase
glutamine-scyllo-inosose aminotransferase

CAS registry number
9033-03-8

2 Source Organism

<-2> no activity in *Streptomyces lividans* (66 strain 1326; 66 strain TK23, host strains, free of enzymatic activity [4]) [4]
<-1> no activity in *Escherichia coli* (host strain, free of enzymatic activity [4]) [4]
<1> *Bacillus circulans* [5]
<2> *Micromonospora purpurea* (ATCC 15835 [2]) [2]
<3> *Streptomyces bikiniensis* (ATCC 11062 [1]; NRRL 1049b [4]) [1, 4]
<4> *Streptomyces flavopersicus* (ATCC 19756 [3]) [3]
<5> *Streptomyces fradiae* (IFO12773 [5]) [5]
<6> *Streptomyces fradiae* (IFO 13147 [5]) [5]
<7> *Streptomyces griseocarneus* (ATCC 12628 [3]) [3]
<8> *Streptomyces griseus* (ATCC 12475 [1]; N2-3-11, DSM 40236 [4]) [1, 4]
<9> *Streptomyces hygroscopicus* (subsp. glebosus ATCC 14607 [1,3]; subsp. glebosus DSM 40823 [4]) [1, 3, 4]
<10> *Streptomyces kanamyceticus* (JCM4433 [5]) [5]
<11> *Streptomyces ornatus* (ATCC 23265 [1]) [1]

<12> *Streptomyces ribosidificus* (JCM4923 [5]) [5]

3 Reaction and Specificity

Catalyzed reaction

L-glutamine + 2,4,6/3,5-pentahydroxycyclohexanone = 2-oxoglutaramate + 1-amino-1-deoxy-scyllo-inositol (a pyridoxal-phosphate protein)

Reaction type

amino group transfer

Natural substrates and products

S   L-glutamine + 2,4,6/3,5-pentahydroxycyclohexanone <2-4, 7-9, 11> (<2> 2-deoxystreptamine biosynthesis [2]; <8> streptidine subpathway of streptomycin biosynthesis [4]) (Reversibility: r <2-4, 7-9, 11> [1-5]) [1-5]

P   2-oxoglutaramate + 1-amino-1-deoxy-scyllo-inositol

Substrates and products

S   2-deoxystreptamine + 2-ketoglutaramate <2> (Reversibility: r <2> [2]) [2]

P   2-deoxy-keto-inosamine + L-glutamine <2> [2]

S   2-deoxystreptamine + 2-oxoglutaramate <2> (Reversibility: r <2> [2]) [2]

P   2-deoxy-scyllo-inositol + L-glutamine

S   2-deoxystreptamine + keto-scyllo-inositol <2> (Reversibility: r <2> [2]) [2]

P   2-deoxy-scyllo-inositol + amino-deoxy-keto-scyllo-inositol

S   L-glutamine + 1D-4-keto-myo-inositol <9> (Reversibility: r <9> [1]) [1]

P   2-oxoglutaramate + amino-deoxy-1D-4-keto-myo-inositol

S   L-glutamine + 2,4,6/3,5-pentahydroxycyclohexanone <2-4, 7-9, 11> (<2> keto-scyllo-inositol [2]) (Reversibility: r <2-4,7-9,11> [1-5]) [1-5]

P   2-oxoglutaramate + 1-amino-1-deoxy-scyllo-inositol

S   L-glutamine + 2-oxoadipate <2> (Reversibility: r <2> [2]) [2]

P   2-oxoglutaramate + L-2-amino-1,6-hexandioate

S   L-glutamine + 2-oxobutyrate <2> (Reversibility: r <2> [2]) [2]

P   2-oxoglutaramate + 2-aminobutyrate

S   L-glutamine + 2-oxocaproate <2> (Reversibility: r <2> [2]) [2]

P   2-oxoglutaramate + L-norleucine

S   L-glutamine + 2-oxoglutaramate <2> (Reversibility: r <2> [2]) [2]

P   2-oxoglutaramate + L-glutamine

S   L-glutamine + 2-oxoglutarate <2, 9> (Reversibility: r <2,9> [1,2]) [1, 2]

P   2-oxoglutaramate + L-glutamate

S   L-glutamine + 2-oxovalerate <2> (Reversibility: r <2> [2]) [2]

P   2-oxoglutaramate + L-norvaline

S   L-glutamine + DL-epi-inosose <2> (Reversibility: r <2> [2]) [2]

P   2-oxoglutaramate + amino-deoxy-inosose

S   L-glutamine + glyoxlate <2> (Reversibility: r <2> [2]) [2]

P   2-oxoglutaramate + glycine

S L-glutamine + oxaloacetate <2> (Reversibility: r <2> [2]) [2]
P 2-oxoglutaramate + L-aspartate
S L-glutamine + pyruvate <2, 9> (Reversibility: r <2,9> [1,2]) [1, 2]
P 2-oxoglutaramate + L-alanine
S L-glutamine + scyllo-inosose <8> (Reversibility: r <8> [4]) [4]
P scyllo-inosamine + 2-oxoglutaramate <8> (<8> spontaneously cyclization of 2-oxoglutaramate to 2-pyrrolidone-5-hydroxy-5-carboxylic acid [4]) [4]
S N^3-methyl-2-deoxy-streptamine + inosose <4, 7, 9> (Reversibility: r <4,7,9> [3]) [3]
P aminodeoxyinositol + 1-keto-2-deoxy-3-methylaminodeoxy-scyllo-inositol
S N^3-methyl-2-deoxy-streptamine + keto-scyllo-inositol <2> (Reversibility: r <2> [2]) [2]
P amino-2-deoxy-N^3-methyl-scyllo-inositol + amino-deoxy-keto-scyllo-inositol
S N^3-methyl-2-deoxystreptamine + 2-oxoglutaramate <2> (Reversibility: r <2> [2]) [2]
P 2-amino-2-deoxy-N^3-methyl-scyllo-inositol + L-glutamine
S aminodeoxy-scyllo-inositol + 2-oxoglutaramate <2> (Reversibility: r <2> [2]) [2]
P scyllo-inositol + L-glutamine
S aminodeoxy-scyllo-inositol + keto-scyllo-inositol <2> (Reversibility: r <2> [2]) [2]
P amino-scyllo-inositol + aminodeoxy-keto-scyllo-inositol
S myo-inositol + diaminocyclitol + oxalacetate <4, 9> (Reversibility: r <4,9> [3]) [3]
P amino-deoxyinositol + ketoaminocyclitol +malate
S myo-inositol + glutamine + oxalacetate <4, 9> (Reversibility: r <4,9> [3]) [3]
P amino-deoxyinositol + 2-oxoglutarate + malate + NH_4^+
S streptamine + 2-oxoglutaramate <2> (Reversibility: r <2> [2]) [2]
P amino-scyllo-inositol + L-glutamine
S streptamine + keto-scyllo-inositol <2> (Reversibility: r <2> [2]) [2]
P amino-scyllo-inositol + amino-deoxy-keto-scyllo-inositol

Inhibitors
5,5'-dithio-bis(2-nitrobenzoic acid) <2> (<2> reduces specific activity by 50% [2]) [2]
p-hydroxymercuribenzoate <2> (<2> reduces specific activity by 50% [2]) [2]

Cofactors/prosthetic groups
pyridoxal 5'-phosphate <2, 3, 8, 9, 11> [1, 2]

Specific activity (U/mg)
0.24 <2> [2]

pH-Optimum
7.5-8.5 <2> [2]

pH-Range

7.5-9 <2> [2]

4 Enzyme Structure

Molecular weight

45500 <8> (<8> SDS-PAGE [4]) [4]
64000-76000 <2> (<2> gel filtration [2]) [2]
95000-100000 <8> (<8> gel filtration [4]) [4]

Subunits

dimer <8> (<8> 2 * 45000, SDS-PAGE [4]) [4]

5 Isolation/Preparation/Mutation/Application

Source/tissue

mycelium <2-4, 7-9, 11> [1-3]

Purification

<2> [2]
<8> (recombinant enzyme, pJAW76-2, expressed in Streptomyces lividans [4]) [4]

Cloning

<1> (btrS gene identified and heterologously expressed [5]) [5]
<5> [5]
<6> (btrS homologous gene identified and sequenced [5]) [5]
<8> (gene StsC identified, cloned by PCR and expressed in Escherichia coli BL21 and Streptomyces lividans [4]) [4]
<10> (btrS homologous gene identified and sequenced [5]) [5]
<12> (btrS homologous gene identified and sequenced [5]) [5]

Application

pharmacology <2, 4, 5> (<2> 2-deoxystreptamine is a component of numerous clinically important antibiotics such as gentamicin, neomycin, tobramycin, amikacin and hygromycin [2]; <7> 5'-hydroxystreptomycin production [3]; <4> spectinomycin production [3]; <5> biosynthesis of aminoglycoside antibiotics [5]) [2, 3, 5]

6 Stability

Temperature stability

50 <2> (<2> heat labile, loses all activity upon incubation for 20 min [2]) [2]
55 <3> (<3> relatively resistant to heating for 5 min [1]) [1]

General stability information

<2>, activity rapidly lost when dialyzed [2]

<3>, pyridoxal 5'-phosphate or pyruvate stabilizes during dialysis [1]

Storage stability

<2>, -20°C, stored at concentrations greater than 1mg/ml, stable for several months with little loss of activity [2]

<2>, 4°C, inactivated within 1 month [2]

<8>, 6°C, purified enzyme stable for several weeks [4]

References

[1] Walker, J.B.: L-glutamine:keto-scyllo-inositol aminotransferase. Methods Enzymol., **43**, 439-443 (1975)

[2] Lucher, L.A.; Chen, Y.M.; Walker, J.B.: Reactions catalyzed by purified L-glutamine:keto-scyllo-inositol aminotransferase, an enzyme required for biosynthesis of aminocyclitol antibiotics. Antimicrob. Agents Chemother., **33**, 452-459 (1989)

[3] Walker, J.B.: Enzymic synthesis of aminocyclitol moieties of aminoglycoside antibiotics from inositol by Streptomyces sp.: detection of glutamine-aminocyclitol aminotransferase and diaminocyclitol aminotransferase activities in a spectinomycin producer. J. Bacteriol., **177**, 818-822 (1995)

[4] Ahlert, J.; Distler, J.; Mansouri, K.; Piepersberg, W.: Identification of stsC, the gene encoding the L-glutamine:scyllo-inosose aminotransferase from streptomycin-producing Streptomycetes. Arch. Microbiol., **168**, 102-113 (1997)

[5] Tamegai, H.; Eguchi, T.; Kakinuma, K.: First identification of Streptomyces genes involved in the biosynthesis of 2-deoxystreptamine-containing aminoglycoside antibiotics: genetic and evolutionary analysis of L-glutamine:2-deoxy-scyllo-inose aminotransferase genes. J. Antibiot., **55**, 1016-1018 (2002)

Serine-pyruvate transaminase 2.6.1.51

1 Nomenclature

EC number
2.6.1.51

Systematic name
L-serine:pyruvate aminotransferase

Recommended name
serine-pyruvate transaminase

Synonyms
Dm-Spat <2> [16]
SPT
SPT/AGT <1, 4, 7, 10> [9-11, 13, 15]
SPT10 <10> (<10> recombinant enzyme [9]) [9]
alanine:glyoxylate aminotransferase-1 <10> [18]
hydroxypyruvate:L-alanine transaminase
isoenzyme 1 of histidine-pyruvate aminotransferase <10> [4]
serine:pyruvate/alanine:glyoxylate aminotransferase <1, 4, 7, 10> [9-11, 13-15]

CAS registry number
9030-88-0

2 Source Organism

<1> *Canis familiaris* (dog [1,5,11,12]) [1, 5, 11, 12]
<2> *Drosophila melanogaster* [16]
<3> *Felis catus* [5]
<4> *Homo sapiens* (human [10-12]) [10-12]
<5> *Mus musculus* (mouse [5]) [5]
<6> *Neurospora crassa* [1]
<7> *Oryctolagus cuniculus* (rabbit, Japanese white [11]) [11, 12]
<8> *Phaseolus vulgaris* (kidney bean, L. cv. Red Kidney [3]) [3]
<9> *Pisum sativum* (var. Alaska [2]) [2]
<10> *Rattus norvegicus* (Donryu strain [4]; strain Wistar [11,14]) [4, 5, 7-11, 13-15, 17, 18]
<11> *Salmo gairdneri* (rainbow trout [6]) [6]

3 Reaction and Specificity

Catalyzed reaction

L-serine + pyruvate = 3-hydroxypyruvate + L-alanine (a pyridoxal-phosphate protein. The liver enzyme may be identical with EC 2.6.1.44 alanine-glyoxylate transaminase)

Reaction type

amino group transfer

Natural substrates and products

S L-alanine + 3-hydroxypyruvate <1> (Reversibility: r <1> [1]) [1]

P pyruvate + L-serine

S L-serine + pyruvate <1-5, 7-11> (<1, 3, 5, 7, 10> gluconeogenesis from serine [5,1 1]; <1, 4, 7, 9> L-serine metabolism [2, 11, 12, 15]; <4, 10> glyoxylate metabolism [11, 12]; <2> glyoxylate detoxification, physiological role in tryptophan catabolism [16]) (Reversibility: r <1-5, 7-11> [1-6, 8, 9, 11-16]) [1-6, 8, 9, 11-16]

P 3-hydroxypyruvate + L-alanine <1> [1]

Substrates and products

S 5-hydroxytryptophan + pyruvate <5, 10> (Reversibility: ? <5,10> [4, 5]) [4, 5]

P 3-(5-hydroxyindole)-2-oxopropanoate + alanine

S L-alanine + 3-hydroxypyruvate <1, 5, 6, 10> (Reversibility: r <1, 5, 6, 10> [1, 5, 8]) [1, 5, 8]

P pyruvate + L-serine +

S L-alanine + glyoxylate <2, 10> (Reversibility: ? <2,10> [9,16]) [9, 16]

P pyruvate + glycine <10> [9]

S L-alanine + phenylpyruvate <10> (Reversibility: ? <10> [4]) [4]

P pyruvate + L-phenylalanine <10> [4]

S L-asparagine + phenylpyruvate <10> [4]

P 2-oxosuccinamate + L-phenylalanine <10> [4]

S L-leucine + phenylpyruvate <5, 10> (Reversibility: ? <5,10> [4,5]) [4, 5]

P 4-methyl-2-oxopentanoate + L-phenylalanine <10> [4]

S L-leucine + pyruvate <2> (Reversibility: ? <2> [16]) [16]

P 4-methyl-2-oxopentanoate + L-alanine

S L-phenylalanine + glyoxylate <2> (Reversibility: ? <2> [16]) [16]

P phenylpyruvate + glycine

S L-phenylalanine + pyruvate <2, 5, 10> (Reversibility: ? <2, 5, 10> [4, 5, 16]) [4, 5, 16]

P phenylpyruvate + L-alanine <10> [4]

S L-serine + glyoxylate <1-5, 8, 10> (Reversibility: r <1-5, 8, 10> [3, 5, 11, 12, 16]) [3, 5, 11, 12, 16]

P glycine + 3-hydroxypyruvate <4, 10> [8, 11]

S L-serine + oxaloacetate <10> (Reversibility: ? <10> [4]) [4]

P 3-hydroxypyruvate + L-aspartate <10> [4]

S L-serine + phenylpyruvate <1, 3, 5, 10> (<1,3> little activity detected [5])
(Reversibility: r <1,3,5,10> [5]) [5]

P 3-hydroxypyruvate + L-phenylalanine <5, 10> [5]

S L-serine + pyruvate <1-5, 7-11> (Reversibility: r <1-5, 7-11> [1-6, 8, 9,
11-16]) [1-6, 8, 9, 11-16]

P 3-hydroxypyruvate + L-alanine <1> [1]

S L-tryptophan + glyoxylate <2> (Reversibility: ? <2> [16]) [16]

P 3-indole-2-oxopropanoate + glycine

S L-tryptophan + pyruvate <10> (Reversibility: ? <10> [4]) [4]

P 3-indole-2-oxopropanoate + L-alanine

S L-tyrosine + glyoxylate <2> (Reversibility: ? <2> [16]) [16]

P 3-(4-hydroxyphenyl)-2-oxopropanoate + glycine

S L-tyrosine + pyruvate <2, 5, 10> (Reversibility: ? <2, 5, 10> [4, 5, 16]) [4,
5, 16]

P 3-(4-hydroxyphenyl)-2-oxopropanoate + L-alanine

S 2-amino-butyrate + glyoxylate <2> (Reversibility: ? <2> [16]) [16]

P 2-oxobutanoate + glycine

S L-arginine + pyruvate <2> (Reversibility: ? <2> [16]) [16]

P 5-guanidino-2-oxopentanoate + L-alanine

S L-asparagine + pyruvate <5, 10> [4, 5]

P 2-oxosuccinamate + L-alanine <5, 10> [4, 5]

S L-cysteine + pyruvate <2> (Reversibility: ? <2> [16]) [16]

P 3-mercapto-2-oxo-propanoate + L-alanine

S L-glutamic acid + 3-hydroxypyruvate <1> (Reversibility: r <1> [1]) [1]

P 2-oxo-1,5-pentandioate + serine

S L-glutamine + phenylpyruvate <10> (Reversibility: ? <10> [4]) [4]

P 4-carbamoyl-2-oxobutanoate + L-phenylalanine <10> [4]

S L-glutamine + pyruvate <5, 10> (Reversibility: ? <5,10> [4,5]) [4, 5]

P 4-carbamoyl-2-oxobutanoate + L-alanine <5, 10> [4, 5]

S glycine + 3-hydroxypyruvate <1, 8> (Reversibility: r <1,8> [1,3]) [1, 3]

P glyoxylate + L-serine

S glycine + 3-hydroxypyruvate <1, 10> (Reversibility: r <1,10> [1,4]) [1, 4]

P L-serine + glyoxylate <1, 10> [1, 4]

S glycine + pyruvate <2> (Reversibility: ? <2> [16]) [16]

P glyoxylate + L-alanine

S L-histidine + glyoxylate <2> (Reversibility: ? <2> [16]) [16]

P 3-(1H-imidazol-4-yl)-2-oxopropanoate + glycine

S L-histidine + pyruvate <2, 5, 10> (Reversibility: ? <2, 5, 10> [4, 5, 16]) [4,
5, 16]

P 3-(1H-imidazol-4-yl)-2-oxopropanoate + L-alanine <10> [4]

S L-isoleucine + pyruvate <5> (Reversibility: ? <5> [5]) [5]

P 3-methyl-2-oxopentanoate + L-alanine <5> [5]

S L-lysine + pyruvate <5> (Reversibility: ? <5> [5]) [5]

P 6-amino-2-oxohexanoate + L-alanine

S L-methionine + glyoxylate <2> (Reversibility: ? <2> [16]) [16]

P 4-methylsulfanyl-2-oxobutanoate + glycine

S L-methionine + phenylpyruvate <10> (Reversibility: ? <10> [4]) [4]

P 4-methylsulfanyl-2-oxobutanoate + L-phenylalanine <10> [4]
S methionine + pyruvate <2, 5, 10> (Reversibility: ? <2, 5, 10> [4, 5, 16]) [4, 5, 16]
P 4-methylsulfanyl-2-oxobutanoate + L-alanine <5, 10> [4, 5]
S L-ornithine + phenylpyruvate <10> (Reversibility: ? <10> [4]) [4]
P 5-amino-2-oxopentanoate + L-phenylalanine <10> [4]
S L-ornithine + pyruvate <10> (Reversibility: ? <10> [4]) [4]
P 5-amino-2-oxopentanoate + L-alanine <10> [4]
S L-threonine + phenylpyruvate <10> (Reversibility: ? <10> [4]) [4]
P 3-hydroxy-2-oxobutanoate + L-phenylalanine
S L-valine + pyruvate <5> (Reversibility: ? <5> [5]) [5]
P 3-methyl-2-oxobutanoate + L-alanine
S Additional information <1-5, 10, 11> (<1> D-alanine is no substrate [1]; <10> identical with histidine-pyruvate aminotransferase isoenzyme 1, isoleucine and oxoglutarate are no substrates [4]; <1, 3, 11> highly specific for serine as the amino donor [5, 6]; <1, 3, 5, 10> 2-oxoglutarate, indol-3-ylpyruvate and p-hydroxyphenylpyruvate are no substrates [5]; <4, 10> class IV aminotransferase, identical with EC 2.6.1.44 [9, 10]; <2> amino-adipate, asparagine, aspartate, glutamate, isoleucine, lysine, threonine and valine are no substrates [16]) [1, 4-6, 9, 10, 16]
P ?

Inhibitors

KCN <10> [4]
N-ethylmaleimide <8> [3]
histidine <10> (<10> competitive inhibitor of serine transamination [4]) [4]
hydroxylamine <1, 3, 5, 8, 10> [3, 5]
isonicotinic acid hydrazide <5, 10> [4, 5]
semicarbazide <1, 3, 5, 10> [4, 5]
serine <10> (<10> competitive inhibitor of histidine transamination [4]) [4]
Additional information <1, 3, 8> (<1,3> not inhibited by isonicotinic acid hydrazide [5]; <8> less sensitive to N-ethylmaleimide and p-chloromercuribenzoate, not inhibited by NH_4^+ [3]) [3, 5]

Cofactors/prosthetic groups

pyridoxal 5'-phosphate <1, 10> [1, 4, 8, 9]

Activating compounds

glucagon <10> (<10> mitochondrial enzyme is markedly induced by [9,15]) [9, 15]

Turnover number (min^{-1})

6800 <2> (glyoxylate) [16]
7880 <2> (alanine) [16]

Specific activity (U/mg)

2.02 <10> (<10> alanine-hydroxypyruvate transamination [5]) [5]
2.23 <5> (<5> alanine-hydroxypyruvate transamination [5]) [5]
3.47 <10> (<10> serine-pyruvate aminotransferase activity [4]) [4]

3.58 <5> (<5> serine-pyruvate transamination [5]) [5]
4.08 <10> (<10> serine-pyruvate transamination [5]) [5]
4.62 <10> (<10> recombinant enzyme [9]) [9]
5.41 <10> (<10> recombinant enzyme [8]) [8]
8.2 <9> (<9> source leaf, age 10 days [2]) [2]
8.3 <9> (<9> source apical meristem, age 16 days [2]) [2]
8.5 <9> (<9> source leaf, age 11 days [2]; <9> source apical meristem, age 15 days [2]) [2]
8.6 <9> (<9> source apical meristem, age 13 days [2]) [2]
8.8 <9> (<9> source leaf, age 13 days [2]) [2]
9.4 <9> (<9> source apical meristem, age 10 days [2]) [2]
9.5 <9> (<9> source apical meristem, age 11 days [2]) [2]
11.1 <9> (<9> source leaf, age 16 days [2]) [2]
13 <1> (<1> serine-pyruvate transamination [5]) [5]
13.1 <9> (<9> source leaf, age 15 days [2]) [2]
13.12 <1> [5]
14 <3> (<3> serine-pyruvate transamination [5]) [5]
16.63 <3> (<3> alanine-hydroxypyruvate transamination [5]) [5]
25.7 <10> (<10> histidine-pyruvate aminotransferase activity [4]) [4]
31.38 <1> (<1> alanine-hydroxypyruvate transamination [5]) [5]

K_m-Value (mM)

0.01 <10> (glyoxylate, <10> pH 7.4, 37°C [11]) [11]
0.0246 <10> (glyoxylate, <10> pH 8.0, 30°C, native enzyme [8]) [8]
0.0336 <10> (glyoxylate, <10> pH 8.0, 30°C, recombinant enzyme [8]) [8]
0.21 <11> (pyruvate, <11> pH 8.4-9.2, 15°C [6]) [6]
0.39 <8> (L-serine, <8> pH 8.0, 30°C [3]) [3]
0.48 <10> (pyruvate, <10> pH 7.4, 37°C [11]) [11]
1 <10> (pyruvate, <10> pH 8.0, 37°C, serine as amino donor [5]) [5]
1.1 <10> (pyruvate, <10> pH 8.0, 30°C, native and recombinant enzyme [8]) [8]
1.2 <5> (pyruvate, <5> pH 8.0, 37°C, serine as amino donor [5]) [5]
1.6 <3> (pyruvate, <3> pH 8.0, 37°C, serine as amino donor [5]) [5]
1.7 <1> (pyruvate, <1> pH 8.0, 37°C, serine as amino donor [5]) [5]
2.2 <11> (L-alanine, <11> pH 8.4-9.2, 15°C [6]) [6]
4 <11> (hydroxypyruvate, <11> pH 8.4-9.2, 15°C [6]) [6]
8 <1> (L-serine, <1> pH 8.0, 37°C, pyruvate as amino acceptor [5]) [5]
8.1 <11> (L-serine, <11> pH 8.4-9.2, 15°C, K_m increases with rising temperature, remains constant over the range 5-30°C [6]) [6]
10 <3> (L-serine, <3> pH 8.0, 37°C, pyruvate as amino acceptor [5]) [5]
16 <10> (L-serine, <10> pH 8.0, 37°C, pyruvate as amino acceptor [5]) [5]
18 <5> (L-serine, <5> pH 8.0, 37°C, pyruvate as amino acceptor [5]) [5]
38 <8> (pyruvate, <8> pH 8.0, 30°C [3]) [3]
51 <2> (pyruvate, <2> pH 7.0, 50°C, recombinant enzyme [16]) [16]
64 <2> (glyoxylate, <2> pH 7.0, 50°C, recombinant enzyme [16]) [16]
149 <2> (L-alanine, <2> pH 7.0, 50°C, recombinant enzyme [16]) [16]
256 <2> (L-serine, <2> pH 7.0, 50°C, recombinant enzyme [16]) [16]

pH-Optimum
  8.4-9.2 <11> [6]
  8.5-9 <1, 3, 5, 10> [5]
  9 <10> [4]

pH-Range
  8-9.5 <10> [4]

4 Enzyme Structure

Molecular weight
  73000 <10> (<10> gel filtration [4]) [4]
  73000-79000 <1, 3, 5, 10> (<1, 3, 5, 10> sucrose density gradient centrifugation [5]) [5]
  75000 <10> (<10> sucrose density gradient centrifugation [4]) [4]
  75000-80000 <1, 3, 5, 10> (<1,3,5,10> gel filtration [5]) [5]
  80000 <10> (<10> gel filtration [9]) [9]

Subunits
  dimer <1, 3, 5, 10> (<1> 2 * 36000-40000, SDS-PAGE [5]; <10> 2 * 43000,
  recombinant enzyme, SDS-PAGE [8]; <10> 2 * 38000, SDS-PAGE [4]; <10> 2
  * 36000-40000, SDS-PAGE [5]; <5> 2 * 36000-40000, SDS-PAGE [5]; <10> 2 *
  43000, SDS-PAGE [10,17]; <3> 2 * 36000-40000, SDS-PAGE [5]) [4, 5, 8-10,
  17]
  homodimer <10> (<10> 2 * 44400, recombinant enzyme, SDS-PAGE [9]) [9]

5 Isolation/Preparation/Mutation/Application

Source/tissue
  apical meristem <9> [2]
  hepatocyte <10> [7]
  kidney <1> [1]
  leaf <8, 9> [2, 3]
  liver <1, 3-5, 7, 10, 11> [1, 4-15, 17, 18]
  seed <9> (<9> barely detectable [2]) [2]

Localization
  cytoplasm <1, 4, 7, 10> [7, 8, 11]
  cytosol <10> [4]
  mitochondrion <1, 10> [4, 5, 7-15, 17, 18]
  peroxisome <4, 7, 10> [7-15, 17]

Purification
  <1> [5]
  <2> (recombinant enzyme [16]) [16]
  <3> [5]
  <5> [5]

<8> (co-purification of serine:glyoxylate aminotransferase [3]) [3]
<10> (native and recombinant enzyme [8]; isoenzyme 1 of histidine-pyru-
vate aminotransferase [4]; recombinant enzyme [9]; mitochondrial and peroxisomal isoforms [14]) [4, 5, 8, 9, 14]
<11> [6]

Cloning

<2> (expressed in SF9 cells, using a insect/baculovirus expression system
[16]) [16]
<4> (plasmid containing cDNA of human SPT cloned [10]) [10]
<10> (cDNA cloned and expressed in Escherichia coli DH1 [8,9]; cloned and
expressed in Escherichia coli DH5 and Escherichia coli ER1458 [14]; C-terminal SPT mutants constructed and examined in transfected COS-1 cells
[13,17]; SPT/AGT gene is linked to the CAT gene and transiently expressed
in human hepatoma-derived HepG2 cells [15]; binding sites identified by
using a luciferase reporter assay with HepG2 cells, DNase I footprinting analysis and gel shift experiments [18]) [8, 9, 13-15, 17, 18]

Application

medicine <4> (<4> absence of human SPT/AGT or mistargeting of this enzyme causes primary hyperoxaluria type 1, an inborn error of glyoxylate metabolism characterized by increased oxalate production [10,11]) [10, 11]

6 Stability

Temperature stability

70 <10> (<10> heating of the purified enzyme for different lengths of time
produces equivalent losses of both activities [4]) [4]

General stability information

<4>, enzyme recovered in the supernatant is resistant to proteinase K in pres-
ence or absence of Triton X-100, enzyme recovered in the precipitate is sensitive to the protease [10]
<10>, precursor of mitochondrial enzyme is highly sensitive to proteinase K
digestion, peroxisomal enzyme is fairly resistant to the protease [13]

Storage stability

<1>, -20°C, 50 mM potassium phosphate buffer, pH 7.5, containing 0.1 M
pyridoxal 5'-phosphate, may be stored for at least 4 weeks without loss of
activity [5]
<1>, 0-6°C, 50 mM potassium phosphate buffer, pH 7.5, containing 0.1 M
pyridoxal 5'-phosphate, little loss with storage for at least 2 weeks [5]
<1>, 5°C, acetone powder of fresh liver stored in a vacuum desiccator over
aluminia retains its enzymatic activity for months [1]
<3>, -20°C, 50 mM potassium phosphate buffer, pH 7.5, containing 0.1 M
pyridoxal 5'-phosphate, may be stored for at least 4 weeks without loss of
activity [5]

<3>, 0-6°C, 50 mM potassium phosphate buffer, pH 7.5, containing 0.1 M pyridoxal 5'-phosphate, little loss with storage for at least 2 weeks [5]

<5>, -20°C, 50 mM potassium phosphate buffer, pH 7.5, containing 0.1 M pyridoxal 5'-phosphate, may be stored for at least 4 weeks without loss of activity [5]

<5>, 0-6°C, 50 mM potassium phosphate buffer, pH 7.5, containing 0.1 M pyridoxal 5'-phosphate, little loss with storage for at least 2 weeks [5]

<10>, -20°C, 50 mM potassium phosphate buffer, pH 7.5 may be stored for at least 6 weeks without loss of either activity [4]

<10>, 0-5°C, little or none of either activity is lost when the enzyme is stored for 2 weeks [4]

References

[1] Sallach, H.J.: Formation of serine from hydroxypyruvate and L-alanine. J. Biol. Chem., 223, 1101-1108 (1956)

[2] Cheung, G.P.; Rosenblum, I.Y.; Sallach, H.J.: Comparative studies of enzymes related to serine metabolism in hogher plants. Plant Physiol., 13, 1813-1820 (1968)

[3] Smith, I.K.: Purification and characterization of serine:glyoxylate aminotransferase from kidney bean (Phaseolus vulgaris). Biochim. Biophys. Acta, 321, 156-164 (1973)

[4] Noguchi, T.; Okuno, E.; Kido, R.: Identity of isoenzyme 1 of histidine-pyruvate aminotransferase with serine-pyruvate aminotransferase. Biochem. J., 159, 607-613 (1976)

[5] Noguchi, T.; Takada, Y.; Kido, R.: Characteristics of hepatic serine-pyruvate aminotransferase in different mammalian species. Biochem. J., 161, 609-614 (1977)

[6] Youngson, A.; Cowey, C.B.; Walton, M.J.: Some properties of serine pyruvate aminotransferase purified from liver of rainbow trout Salmo gairdneri. Comp. Biochem. Physiol. B, 73, 393-398 (1982)

[7] Yokota, S.; Oda, T.: Fine localization of serine:pyruvate aminotransferase in rat hepatocytes revealed by a post-embedding immunocytochemical technique. Histochemistry, 80, 591-595 (1984)

[8] Oda, T.; Miyajima, H.; Suzuki, Y.; Ito, T.; Yokota, S.; Hoshino, M.; Ichiyama, A.: Purification and characterization of the active serine: pyruvate aminotransferase of rat liver mitochondria expressed in Escherichia coli. J. Biochem., 106, 460-467 (1989)

[9] Ishikawa, K.; Kaneko, E.; Ichiyama, A.: Pyridoxal 5'-phosphate binding of a recombinant rat serine: pyruvate/alanine:glyoxylate aminotransferase. J. Biochem., 119, 970-978 (1996)

[10] Oda, T.; Funai, T.; Miura, S.: In vitro association with peroxisomes and conformational change of peroxisomal serine:pyruvate/alanine:glyoxylate aminotransferase in rat and human livers. Biochem. Biophys. Res. Commun., 228, 341-346 (1996)

[11] Xue, H.H.; Sakaguchi, T.; Fujie, M.; Ogawa, H.; Ichiyama, A.: Flux of the L-serine metabolism in rabbit, human, and dog livers. Substantial contributions of both mitochondrial and peroxisomal serine:pyruvate/alanine:-glyoxylate aminotransferase. J. Biol. Chem., 274, 16028-16033 (1999)

[12] Ichiyama, A.; Xue, H.-H.; Oda, T.; Uchida, C.; Sugiyama, T.; Maeda-Nakai, E.; Sato, K.; Nagai, E.; Watanabe, S.; Takayama, T.: Oxalate synthesis in mammals: properties and subcellular distribution of serine:pyruvate/alanine:glyoxylate aminotransferase in the liver. Mol. Urol., 4, 333-340 (2000)

[13] Oda, T.; Mizuno, T.; Ito, K.; Funai, T.; Ichiyama, A.; Miura, S.: Peroxisomal and mitochondrial targeting of serine:pyruvate/alanine:glyoxylate aminotransferase in rat liver. Cell Biochem. Biophys., 32, 277-281 (2000)

[14] Oda, T.; Uchida, C.; Miura, S.: Mitochondrial targeting signal-induced conformational change and repression of the peroxisomal targeting signal of the precursor for rat liver serine:pyruvate/alanine:glyoxylate aminotransferase. J. Biochem., 127, 665-671 (2000)

[15] Sugiyama, T.; Uchida, C.; Oda, T.; Kitagawa, M.; Hayashi, H.; Ichiyama, A.: Involvement of CCAAT/enhancer-binding protein in regulation of the rat serine:pyruvate/alanine:glyoxylate aminotransferase gene expression. FEBS Lett., 508, 16-22 (2001)

[16] Han, Q.; Li, J.: Comparative characterization of Aedes 3-hydroxykynurenine transaminase/alanine glyoxylate transaminase and Drosophila serine pyruvate aminotransferase. FEBS Lett., 527, 199-204 (2002)

[17] Mizuno, T.; Ito, K.; Uchida, C.; Kitagawa, M.; Ichiyama, A.; Miura, S.; Fujita, K.; Oda, T.: Analyses in transfected cells and in vitro of a putative peroxisomal targeting signal of rat liver serine:pyruvate aminotransferase. Histochem. Cell Biol., 118, 321-328 (2002)

[18] Uchida, C.; Oda, T.; Sugiyama, T.; Otani, S.; Kitagawa, M.; Ichiyama, A.: The role of Sp1 and AP-2 in basal and protein kinase A-induced expression of mitochondrial serine:pyruvate aminotransferase in hepatocytes. J. Biol. Chem., 277, 39082-39092 (2002)

Phosphoserine transaminase 2.6.1.52

1 Nomenclature

EC number
2.6.1.52

Systematic name
O-phospho-L-serine:2-oxoglutarate aminotransferase

Recommended name
phosphoserine transaminase

Synonyms
3-phosphoserine aminotransferase
AspAT <6> [6]
L-phosphoserine aminotransferase
PSAT
PSAT α <5> [13]
PSAT β <5> [13]
PSerAT <3> [11]
hydroxypyruvic phosphate-glutamic transaminase
phosphohydroxypyruvate transaminase
phosphohydroxypyruvic-glutamic transaminase
phosphoserine aminotransferase

CAS registry number
9030-90-4

2 Source Organism

<1> *Arabidopsis thaliana* (ecotype Columbia [10]) [10]
<2> *Bacillus circulans* (subsp. alkalophilus, ATCC 21783 [8]) [8]
<3> *Bos taurus* (beef [4]; bovine [3,11]) [3, 4, 11]
<4> *Escherichia coli* [12]
<5> *Homo sapiens* [13]
<6> *Methanobacterium thermoformicicum* (strain SF⁻4 [6]) [6]
<7> *Ovis aries* [1, 2]
<8> *Scenedesmus obliquus* (mutant C-2 A' [5,7]) [5, 7]
<9> *Spinacia oleracea* (spinach, cv. Parade [9]) [9, 10]

3 Reaction and Specificity

Catalyzed reaction
O-phospho-L-serine + 2-oxoglutarate = 3-phosphonooxypyruvate + L-glutamate (a pyridoxal-phosphate protein)

Reaction type
amino group transfer

Natural substrates and products
S L-glutamate + 3-phosphohydroxypyruvate <1, 4, 5, 7-9> (<5, 7-9> phos-
phorylated pathway for serine biosynthesis [1, 5, 9, 13]) (Reversibility: r
<1, 4, 5, 7-9> [1, 2, 5, 9-13]) [1, 2, 5, 9-13]
P O-phospho-L-serine + 2-oxoglutarate

Substrates and products
S 2-amino-4-phosphonobutyrate + 2-oxoglutarate <3> (Reversibility: r <3>
[4]) [4]
P 2-oxophosphonobutanoate + L-glutamate
S 2-amino-5-phosphonovalerate + 2-oxoglutarate <3> (Reversibility: r <3>
[4]) [4]
P 5-phosphono-2-oxopentanoate + L-glutamate
S L-alanine + 3-phosphonooxypyruvate <7> (<7> L-alanine as amino
group donor, 10% of the activity with L-glutamate [1]) (Reversibility: r
<7> [1, 2]) [1, 2]
P O-phospho-L-serine + pyruvate
S L-glutamate + 3-phosphohydroxypyruvate <1, 3-5, 7> (<7> forward reac-
tion [1]) (Reversibility: r <1, 3-5, 7> [1, 2, 4, 9-13]) [1, 2, 4, 9-13]
P O-phospho-L-serine + 2-oxoglutarate
S L-glutamate + 4,5-dioxopentanoate <8> (Reversibility: ? <8> [7]) [7]
P 5-aminolevulinate + 2-oxoglutarate
S O-phospho-L-serine + 2-oxoglutarate <3, 7, 8> (<7> reverse reaction [1])
(Reversibility: r <3, 7, 8> [1, 2, 4, 5, 7]) [1, 2, 4, 5, 7]
P 3-phosphonooxypyruvate + L-glutamate <3, 7, 8> [1, 2, 4, 5, 7]
S homocysteate + 2-oxoglutarate <3> (Reversibility: r <3> [4]) [4]
P 4-mercapto-2-oxobutanoate + L-glutamate
S Additional information <2, 3, 6-8> (<7> hydroxypyruvate is unreactive [1,
2]; <8> 4,5-dioxovalerate with glutamate as amino donor is effective as
competitive substrate to phosphohydroxypyruvate in the forward reaction
and yields 5-aminolevulinate, 4,5-dioxovalerate and glutamate-1-semialde-
hyde can both serve as competitive aminoacceptor in the reverse reaction
with phosphoserine and as substrate with 2-oxoglutarate as aminoacceptor
[7]; <3> 2-amino-3-phosphonopropionate, 3-amino-3-phosphonopropio-
nate 4-amino-4-phosphonobutyrate, 1-amino-1,3-phosphonopropane, 2-
amino-6-phosphohexanoate, serine, aspartate, cysteine sulphinate and cy-
steate are no substrates [4]; <6> aspartate, alanine and phosphoserine as
amino donors, 2-oxoglutarate and glyoxylate as amino acceptors [6]; <2>
high aspartate aminotransferase side activity [8]) [1, 2, 4, 6-8]
P ?

Inhibitors

2-amino-4-phosphonobutyrate <3> [4]
2-oxoglutarate <3> [11]
L-cysteine <1, 9> [10]
L-glutamate <3> [11]
L-phosphoserine <3> (<3> product inhibition [11]) [11]

Cofactors/prosthetic groups

pyridoxal 5'-phosphate <2-4, 7-9> (<3> K_m 0.005 mM [11]) [1, 2, 5, 8, 9, 11, 12]

Specific activity (U/mg)

1.29 <7> [2]
4.16 <2> [8]
13.2 <3> [3]
13.84 <8> [5]
15 <3> [11]
Additional information <7> (<7> 13 U/mg [1]) [1]

K_m-Value (mM)

0.005 <3> (3-phosphohydroxypyruvate, <3> pH 7.5, 30°C [11]) [4, 11]
0.035 <3> (L-phosphoserine, <3> pH 7.5, 30°C [11]) [11]
0.07 <1> (L-glutamate, <1> pH 8.2, 25°C [10]) [10]
0.083 <8> (L-phosphoserine, <8> pH 8.0, 35°C [7]) [7]
0.15 <9> (L-glutamate, <9> pH 8.2, 25°C [10]) [10]
0.18 <8> (2-oxoglutarate, <8> pH 8.0, 35°C [7]) [7]
0.233 <8> (2-oxoglutarate, <8> pH 8.0, 35°C [7]) [7]
0.25 <7> (3-phosphohydroxypyruvate, <7> pH 8.2, 25°C [1]) [1, 2]
0.7 <7> (L-glutamate, <7> pH 8.2, 25°C [1]) [1, 2]
0.8 <3> (2-oxoglutarate, <3> pH 7.5, 30°C [11]) [11]
0.84 <8> (4,5-dioxopentanoate, <8> pH 8.0, 35°C [7]) [7]
1.2 <3> (L-glutamate, <3> pH 7.5, 30°C [11]) [11]
3 <9> (3-phosphohydroxypyruvate, <9> pH 8.2, 25°C [10]) [10]
5 <1> (3-phosphohydroxypyruvate, <1> pH 8.2, 25°C [10]) [10]

K_i-Value (mM)

0.045 <3> (L-phosphoserine, <3> pH 7.5, 30°C [11]) [11]
0.24 <3> (2-amino-4-phosphonobutyrate) [4]
2.55 <3> (2-oxoglutarate, <3> pH 7.5, 30°C [11]) [11]
7.42 <3> (L-glutamate, <3> pH 7.5, 30°C [11]) [11]

pH-Optimum

6.8-7.2 <3> [4]
6.8-8.2 <8> [7]
7 <8> (<8> phosphoserine with 2-oxoglutarate as amino acceptor [7]) [7]
8.15 <7> [1, 2]

pH-Range

5-9.5 <8> [7]

7.8-8.8 <7> (<7> at pH 7.8 about 45% of maximum activity, at pH 8.8 about 35% of maximum activity [1]) [1]

4 Enzyme Structure

Molecular weight

35200 <5> (<5> PSATα, predicted from cDNA [13]) [13]

39790 <2> (<2> predicted from cDNA [8]) [8]

40000 <5> (<5> PSATβ, predicted from cDNA [13]) [13]

41000 <1> (<1> SDS-PAGE [10]) [10]

41680 <6> (<6> deduced from amino acid sequence [6]) [6]

70000 <2> (<2> gel filtration [8]) [8]

77000 <3> (<3> calculated from Stokes radius [3]) [3]

80000 <8> (<8> gel filtration [5]) [5]

90700 <3> (<3> sedimentation equilibrium ultracentrifugation [3]) [3]

96000 <7> (<7> Trautman modification of the Archibald approach to equilibrium method [1]) [1]

Subunits

dimer <2, 3, 8> (<8> 2 * 40000, SDS-PAGE [5]; <3> 2 * 43000, SDS-PAGE [3, 11]; <2> 2 * 37000, homodimer, SDS-PAGE [8]) [3, 5, 8, 11]

5 Isolation/Preparation/Mutation/Application

Source/tissue

brain <5, 7> (<5> mRNA expressed in [13]) [1, 2, 13]

kidney <5> (<5> mRNA expressed in [13]) [13]

leaf <9> [9]

liver <3, 5> (<5> mRNA expressed in [13]) [11, 13]

pancreas <5> (<5> mRNA expressed in [13]) [13]

root <1> [10]

seedling <1> [10]

shoot <1> [10]

Additional information <5> (<5> very weak mRNA expression in thymus, prostate, testis and colon [13]) [13]

Localization

chloroplast <9> [9]

plastid <1, 9> [9, 10]

proplastid <9> [9]

Purification

<1> (recombinant enzyme [10]) [10]

<2> [8]

<3> [3, 4, 11]
<7> [1, 2]
<8> [5, 7]

Crystallization
<4> (space group P2(1)2(1)2(1) [12]) [12]

Cloning
<1> (cDNA encoding PSAT, single gene mapped on the lower arm of chromo-some 4, overexpressed in Escherichia coli AD494 and BL21 [10]) [10]
<2> (cloning of the PSAT gene, expression in Escherichia coli [8]) [8]
<5> (2 forms results from alternative splicing, PSATα and PSATβ, cDNA cloned and expressed in Escherichia coli and in human cell lines, leukaemia Jurkat, colon adenocarcinoma COLO 320DM, heptacellular carcinoma HepG2 and weakly in leukaemia MOLT-3, complementation of Saccharomyces cere-visiae SER-1 deletion mutation [13]) [13]
<6> (cloned and expressed in Escherichia coli [6]) [6]
<9> (cloned and expressed in Escherichia coli KL282 and BL21 [9]) [9, 10]

Application
medicine <5> (<5> useful for creating a new concept or target for the diag-nosis and therapy of human serine metabolism-related disorders and cancer [13]) [13]

6 Stability

General stability information
<7>, enzyme can be inactivated by dialysis against cysteine [1]

Storage stability
<3>, -80°C, stored in individual aliquots or stored as pellet concentrated by ammonium sulfate precipitation, stable for at least 3 months [3]

References

[1] Hirsch, H.; Greenberg, D.M.: Studies on phosphoserine aminotransferase of sheep brain. J. Biol. Chem., **242**, 2283-2287 (1967)
[2] Hirsch-Kolb, H.; Greenberg, D.M.: Phosphoserine aminotransferase (sheep brain). Methods Enzymol., **17B**, 331-334 (1971)
[3] Lund, K.; Merrill, D.K.; Guynn, R.W.: Purification and properties of phos-phoserine aminotransferase from bovine liver. Arch. Biochem. Biophys., **254**, 319-328 (1987)
[4] Basurko, M.-J.; Marche, M.; Darriet, M.; Cassaigne, A.: Catalytic properties and specificity of phosphoserine aminotransferase from beef liver. Bio-chem. Soc. Trans., **17**, 787-788 (1989)

[5] Stolz, M.; Doernemann, D.: Purification, charcterization and N-terminal sequence of phosphoserine aminotransferase from the green alga Scenedesmus obliquus, mutant C-2 A'. Z. Naturforsch. C, **49c**, 63-69 (1993)

[6] Tanaka, T.; Yamamoto, S.; Moriya, T.; Taniguchi, M.; Hayashi, H.; Kagamiyama, H.; Oi, S.: Aspartate aminotransferase from a thermophilic for-
mate-utilizing methanogen, Methanobacterium thermoformicicum strain
SF-4: relation to serine and phosphoserine aminotransferases, but not to
the aspartate aminotransferase family. J. Biochem., **115**, 309-317 (1994)

[7] Stolz, M.; Doernemann, D.: Kinetic characteristics, substrate specificity and
catalytic properties of phosphoserine aminotransferase from the green alga
Scenedesmus obliquus, mutant C-2A'. Z. Naturforsch. C, **50c**, 630-637
(1995)

[8] Battchikova, N.; Himanen, J.P.; Ahjolahti, M.; Korpela, T.: Phosphoserine
aminotransferase from Bacillus circulans subsp. alkalophilus: purification,
gene cloning and sequencing. Biochim. Biophys. Acta, **1295**, 187-194 (1996)

[9] Saito, K.; Takagi, Y.; Ling, H.C.; Takahashi, H.; Noji, M.: Molecular cloning,
characterization and expression of cDNA encoding phosphoserine aminotransferase involved in phosphorylated pathway of serine biosynthesis from
spinach. Plant Mol. Biol., **33**, 359-366 (1997)

[10] Ho, C.L.; Noji, M.; Saito, M.; Yamazaki, M.; Saito, K.: Molecular characterization of plastidic phosphoserine aminotransferase in serine biosynthesis
from Arabidopsis. Plant J., **16**, 443-452 (1998)

[11] Basurko, M.J.; Marche, M.; Darriet, M.; Cassaigne, A.: Phosphoserine aminotransferase, the second step-catalyzing enzyme for serine biosynthesis.
IUBMB Life, **48**, 525-529 (1999)

[12] Hester, G.; Stark, W.; Moser, M.; Kallen, J.; Markovic-Housley, Z.; Jansonius,
J.N.: Crystal structure of phosphoserine aminotransferase from Escherichia
coli at 2.3 A resolution: comparison of the unligated enzyme and a complex
with α-methyl-l-glutamate. J. Mol. Biol., **286**, 829-850 (1999)

[13] Baek, J.Y.; Jun do, Y.; Taub, D.; Kim, Y.H.: Characterization of human phosphoserine aminotransferase involved in the phosphorylated pathway of L-
serine biosynthesis. Biochem. J., **373**, 191-200 (2003)

Glutamate synthase

1 Nomenclature

EC number
2.6.1.53 (transferred to EC 1.4.1.13)

Recommended name
glutamate synthase

Pyridoxamine-phosphate transaminase

1 Nomenclature

EC number

2.6.1.54

Systematic name

pyridoxamine-5'-phosphate:2-oxoglutarate aminotransferase (D-glutamate-forming)

Recommended name

pyridoxamine-phosphate transaminase

Synonyms

aminotransferase, pyridoxamine phosphate

pyridoxamine 5'-phosphate transaminase

pyridoxamine 5'-phosphate-α-ketoglutarate transaminase

CAS registry number

9074-84-4

2 Source Organism

<1> *Clostridium kainantoi* (IFO 3353 [1]) [1]

3 Reaction and Specificity

Catalyzed reaction

pyridoxamine 5'-phosphate + 2-oxoglutarate = pyridoxal 5'-phosphate + D-glutamate

Reaction type

amino group transfer

Substrates and products

S 4-aminobutanoate + 2-oxoglutarate <1> (<1> about 8% of activity compared to pyridoxamine 5'-phosphate [1]) (Reversibility: ? <1> [1]) [1]

P 4-oxobutanoate + D-glutamate <1> [1]

S pyridoxamine + 2-oxoglutarate <1> (<1> about 25% of activity compared to pyridoxamine 5'-phosphate [1]) (Reversibility: ? <1> [1]) [1]

P pyridoxal + D-glutamate <1> [1]

S pyridoxamine 5'-phosphate + 2-oxoglutarate <1> (Reversibility: r <1> [1]) [1]

P pyridoxal 5'-phosphate + D-glutamate <1> [1]

Inhibitors

D-glutamate <1> (<1> slight inhibition [1]) [1]
L-cysteine <1> (<1> about 35% inhibition [1]) [1]
diisopropyl phosphofluoride <1> (<1> 0.3 mM, complete inhibition [1]) [1]
pyridoxal 5'-phosphate <1> (<1> product inhibition [1]) [1]
Additional information <1> (<1> not inhibited by sulfhydryl compounds, e.g. PCMB, phenylmercuric acetate and mercuric chloride [1]) [1]

Activating compounds

4-aminobutyrate <1> (<1> stimulation, 177% of non-stimulated activity [1]) [1]
D-asparagine <1> (<1> stimulation, 112% of non-stimulated activity [1]) [1]
D-cysteine <1> (<1> stimulation, 135% of non-stimulated activity [1]) [1]
D-methionine <1> (<1> stimulation, 127% of non-stimulated activity [1]) [1]
DL-ornithine <1> (<1> stimulation, 112% of non-stimulated activity [1]) [1]
L-asparagine <1> (<1> stimulation, 112% of non-stimulated activity [1]) [1]
L-glutamate <1> (<1> stimulation, 166% of non-stimulated activity [1]) [1]
L-lysine <1> (<1> stimulation, 105% of non-stimulated activity [1]) [1]
L-methionine <1> (<1> stimulation, 114% of non-stimulated activity [1]) [1]
L-ornithine <1> (<1> stimulation, 112% of non-stimulated activity [1]) [1]
glycine <1> (<1> stimulation, 137% of non-stimulated activity [1]) [1]
Additional information <1> (<1> stimulating effect of amino compounds may be caused by disappearance of the inhibitory product of the reaction, pyridoxal 5'-phosphate, forming the corresponding Schiff base [1]) [1]

Metals, ions

Additional information <1> (<1> metal ions do not accelerate the reaction [1]) [1]

Specific activity (U/mg)

0.281 <1> [1]

K_m-Value (mM)

0.004 <1> (pyridoxamine 5'-phosphate, <1> pH 8.0, 37°C [1]) [1]
0.01 <1> (pyridoxal 5'-phosphate, <1> pH 8.0, 37°C [1]) [1]
0.015 <1> (2-oxoglutarate, <1> cosubstrate pyridoxamine 5'-phosphate, pH 8.0, 37°C [1]) [1]
9 <1> (D-glutamate, <1> pH 8.0, 37°C [1]) [1]

pH-Optimum

8 <1> [1]

Temperature optimum (°C)

37 <1> (<1> assay at [1]) [1]

5 Isolation/Preparation/Mutation/Application

Purification

<1> (ammonium sulfate, protamine treatment, DEAE-cellulose, hydroxylapa-
tite, DEAE-Sephadex, Sephadex G-200, partial purification [1]) [1]

References

[1] Tani, Y.; Ukita, M.; Ogata, K.: Studies on vitamin B$_6$ metabolism in microorganisms. Part X. Further purification and characterization of pyridoxamine
5'-phosphate-α-ketoglutarate transaminase from Clostridium kainantoi. Agric. Biol. Chem., **36**, 181-188 (1972)

Taurine-2-oxoglutarate transaminase 2.6.1.55

1 Nomenclature

EC number
2.6.1.55

Systematic name
taurine:2-oxoglutarate aminotransferase

Recommended name
taurine-2-oxoglutarate transaminase

Synonyms
aminotransferase, taurine
taurine transaminase
taurine-α-ketoglutarate aminotransferase
taurine-glutamate transaminase
taurine:α-ketoglutarate aminotransferase

CAS registry number
9076-52-2

2 Source Organism

<1> *Achromobacter superficialis* (ICR-B-89 [2]) [1-5]
<2> *Achromobacter polymorph* (KR B-88 [2,4]) [2, 4]

3 Reaction and Specificity

Catalyzed reaction
taurine + 2-oxoglutarate = sulfoacetaldehyde + L-glutamate

Reaction type
amino group transfer

Substrates and products
- **S** 3-aminopropanesulfonate + 2-oxoglutarate <1, 2> (<1> 43% of the activity with taurine [2,3]) (Reversibility: ? <1,2> [1-3]) [1-3]
- **P** 2-oxopropanesulfonate + L-glutamate <1, 2> [1-3]
- **S** 4-aminobutyrate + 2-oxoglutarate <1> (<1> 60% of the activity with taurine [2,3]) (Reversibility: ? <1> [2,3]) [2, 3]
- **P** L-glutamate + acetaldehyde + sulfite <1> [5]

S   DL-3-aminobutyrate + 2-oxoglutarate <1> (<1> 14% of the activity with taurine [2,3]) (Reversibility: ? <1> [2,3]) [2, 3]
P   3-oxobutyrate + L-glutamate <1> [2, 3]
S   DL-3-aminoisobutyrate + 2-oxoglutarate <1, 2> (<1> 208% of the activity with taurine [2]) (Reversibility: ? <1,2> [1-3]) [1-3]
P   2-methyl-3-oxopropanoate + L-glutamate <1, 2> [1-3]
S   β-alanine + 2-oxoglutarate <1, 2> (<1> 184% of the activity with taurine [2]) (Reversibility: ? <1,2> [1-3]) [1-3]
P   3-oxopropanoate + L-glutamate <1, 2> [1-3]
S   hypotaurine + 2-oxoglutarate <1, 2> (<1> 601% of the activity with taurine [2,3]; <1> 6 times the rate of taurin transamination [5]) (Reversibility: ? <1,2> [2,3,5]) [2, 3, 5]
P   2-oxoethanesulfinic acid + L-glutamate <1, 2> [2, 3, 5]
S   taurine + 2-oxoglutarate <1, 2> (<1,2> no activity with pyruvate, phenylpyruvate, oxaloacetate [1-3]; <1> no activity with aminomethanesulfonate, glycine, 5-aminopentanoate and amines [2]; <1,2> no activity with L- and D-amino acids [2]) (Reversibility: r <1,2> [2,4]) [1-4]
P   sulfoacetaldehyde + L-glutamate <1, 2> [1, 2, 4]

Inhibitors
3-methyl-2-benzothiazolone hydrazone hydrochloride <1> (<1> weak inhibition [2,3]) [2, 3]
D-cycloserine <1> (<1> 0.5 mM; 20% inhibition [2,3]) [2, 3]
HgCl₂ <1> (<1> 0.1 mM, 98% inhibition [3]) [2, 3]
L-cycloserine <1> (<1> 0.5 mM, 97% inhibition [3]) [2, 3]
hydroxylamine <1> (<1> 0.5 mM, complete inhibition [3]) [2, 3]
phenylhydrazine <1> (<1> 0.5 mM, 17% inhibition [2,3]) [2, 3]
semicarbazide <1> (<1> 0.5 mM, 21% inhibition [2,3]) [2, 3]
Additional information <1> (<1> not inhibited by p-chloromercuribenzoate, monoiodoacetic acid, D-penicillamine, L-penicillamine and EDTA [2,3]) [2, 3]

Cofactors/prosthetic groups
pyridoxal 5'-phosphate <1, 2> (<1,2> a pyridoxal phosphate protein [1,2]; <1> 4 mol of pyridoxal phosphate per mol of enzyme [1,3]; <1> enzyme inactivated by treatment with ammonium sulfate is activated by incubation with pyridoxal 5'-phosphate at 45-60°C for about 10 min [1,3]; <1> K_m: 0.005 mM [2,3]) [1-3]

Specific activity (U/mg)
0.519 <1> [2]
5.19 <1> [1]

K_m-Value (mM)
11 <1> (2-oxoglutarate, <1> pH 8.0, 37°C, cosubstrate DL-3-aminoisobutyrate [2,3]) [2, 3]
12 <1> (taurine, <1> pH 8.0, 37°C [2]) [2, 3]
16 <1> (hypotaurine, <1> pH 8.0, 37°C [2]) [2, 3]
17 <1> (β-alanine, <1> pH 8.0, 37°C [2]) [2, 3]

pH-Optimum
    7.8-8 <1> [2, 3]
    8 <1> (<1> assay at [1]) [1]

Temperature optimum (°C)
    30 <1> (<1> assay at [1,5]) [1, 5]
    60 <1> [2]

4 Enzyme Structure

Molecular weight
    156000 <1> (<1> sedimentation equilibrium [1,2]) [1, 2]

5 Isolation/Preparation/Mutation/Application

Purification
    <1> (DEAE-cellulose, Sephadex G-150, hydroxylapatite, ammonium sulfate, Sephadex g-150, crystalization [1]) [1, 2]

Crystallization
    <1> (ammonium sulfate at pH 7.0-7.6, crystals appear after 2-3 days [2]) [1, 2]

6 Stability

pH-Stability
    6-9 <1> (<1> 60°C, 10 min, stable [3]) [3]

Temperature stability
    60 <1> (<1> pH 6.0-9.0, 10 min, stable [3]) [3]

General stability information
    <1>, treatment with ammonium sulfate during purification leads to a significant loss of activity [2]

Storage stability
    <1>, 0-5°C, crystalline enzyme in 10 mM potassium phosphate buffer, pH 7.0, 60% saturation of ammonium sulfate, 100 mM pyridoxal 5'-phosphate, 0.01% 2-mercaptoethanol, several months, no loss of activity [2, 3]

References

[1] Toyama, S.; Misono, H.; Soda, K.: Crystalline taurine: α-ketoglutarate aminotransferase from Achromobacter superficialis. Biochem. Biophys. Res. Commun., **46**, 1374-1379 (1972)

[2] Yonaha, K.; Toyama, S.; Soda, K.: Taurine-glutamate transaminase. Methods Enzymol., **113**, 102-108 (1985)

[3] Toyama, S.; Misono, H.; Soda, K.: Properties of taurine: α-ketoglutarate aminotransferase of Achromobacter superficialis. Inactivation and reactivation of enzyme. Biochim. Biophys. Acta, **523**, 75-81 (1978)

[4] Toyama, S.; Soda, K.: Occurrence of taurine: α-ketoglutarate aminotransferase in bacterial extracts. J. Bacteriol., **109**, 533-538 (1972)

[5] Tanaka, H.; Toyama, S.; Tsukahara, H.; Soda, K.: Transamination of hypotaurine by taurine:α-ketoglutarate aminotransferase. FEBS Lett., **45**, 111-113 (1974)

1D-1-Guanidino-3-amino-1,3-dideoxy-scyllo-inositol transaminase **2.6.1.56**

1 Nomenclature

EC number
2.6.1.56

Systematic name
1D-1-guanidino-3-amino-1,3-dideoxy-scyllo-inositol:pyruvate aminotransferase

Recommended name
1D-1-guanidino-3-amino-1,3-dideoxy-scyllo-inositol transaminase

Synonyms
L-alanine-N-amidino-3-(or 5-)keto-scyllo-inosamine transaminase
guanidinoaminodideoxy-scyllo-inositol-pyruvate aminotransferase

CAS registry number
57127-19-2

2 Source Organism

<-1> no activity in *Streptomyces hygroscopius forma glebosus* (ATCC 14607 [2]) [2]
<1> *Streptomyces bikiniensis* (ATCC 11062 [1,2]) [1, 2]
<2> *Streptomyces griseus* (ATCC 12475 [2]) [2]
<3> *Streptomyces ornatus* (ATCC 23265 [2]) [2]

3 Reaction and Specificity

Catalyzed reaction
1D-1-guanidino-3-amino-1,3-dideoxy-scyllo-inositol + pyruvate = 1D-1-guanidino-1-deoxy-3-dehydro-scyllo-inositol + L-alanine

Reaction type
amino group transfer

Natural substrates and products
S L-alanine + 1D-1-guanidino-1-deoxy-3-dehydro-scyllo-inositol <1-3> (<1> streptomycin biosynthesis [1,2]) (Reversibility: r <1-3> [1]) [1, 2]
P pyruvate + 1D-1-guanidino-3-amino-1,3-dideoxy-scyllo-inositol <1> [1]

Substrates and products

S L-alanine + 1D-1-guanidino-1-deoxy-3-dehydro-scyllo-inositol <1-3> (Reversibility: r <1-3> [1,2]) [1, 2]

P pyruvate + 1D-1-guanidino-3-amino-1,3-dideoxy-scyllo-inositol <1> [1]

S L-glutamate + 1D-1-guanidino-1-deoxy-3-dehydro-scyllo-inositol <1> (Reversibility: r <1> [1,2]) [1, 2]

P 5-amino-2,5-dioxopentanoic acid + 1D-1-guanidino-3-amino-1,3-dideoxy-scyllo-inositol

S L-glutamine + 1D-1-guanidino-1-deoxy-3-dehydro-scyllo-inositol <1> (<1> reaction should be relatively irreversible because of cyclization or enzymic deamidation of the α-ketoglutaramate formed [2]) (Reversibility: ir <1> [1,2]) [1, 2]

P α-ketoglutaramate + 1D-1-guanidino-3-amino-1,3-dideoxy-scyllo-inositol

S Additional information <1> (<1> D-amino acids are no substrates, also catalyzes transamination between 1D-1-guanidino-3-amino-1,3-dideoxy-scyllo-inositol and 1D-1-guanidino-1-deoxy-3-keto-scyllo inositol [2]) [2]

P ?

Cofactors/prosthetic groups
pyridoxal 5'-phosphate <1> [1]

6 Stability

Temperature stability
55 <1> (<1> inactivated by heating for 5 min [2]) [2]

General stability information
<1>, unstable to dialysis in absence of pyridoxal 5'-phosphate, pyruvate spares the pyridoxal 5'-phosphate requirement during dialysis [2]

References

[1] Walker, J.B.; Walker, M.S.: Streptomycin biosynthesis. Transamination reactions involving inosamines and inosadiamines. Biochemistry, **8**, 763-770 (1969)

[2] Walker, J.B.: L-Alanine:1D-1-guanidino-1-deoxy-3-deto-scyllo-inositol aminotransferase. Methods Enzymol., **43**, 462-465 (1975)

Aromatic-amino-acid transaminase 2.6.1.57

1 Nomenclature

EC number
2.6.1.57

Systematic name
aromatic-amino-acid:2-oxoglutarate aminotransferase

Recommended name
aromatic-amino-acid transaminase

Synonyms
AAT I-III <12> (<12> i.e. aromatic aminotransferase I, II and III [15]) [15]
AAT1 <19> [23]
AT-IA <5> [7]
ArAT <1> [19]
ArAT-ITL <4> (<4> i.e. aromatic aminotransferase II from Thermococcus litoralis [6]) [6]
ArATPf <15> [20]
ArATPh <27> [32]
AraT <26> [30]
AroAT II <15> [33]
aminotransferase, aromatic amino acid
arom. amino acid transferase
arom.-amino-acid transaminase
aromatic amino acid aminotransferase
aromatic amino acid transaminase
aromatic amino transferase I <10> [12]
aromatic aminotransferase
pdArAT <21> [25]
pdAroAT <23> [27]
Additional information (<1> controlled proteolysis with subtilisin converts the enzyme into EC 2.6.1.1 [1]; <1> may be identical with EC 2.6.1.5 [1])

CAS registry number
37332-38-0

2 Source Organism

<1> *Escherichia coli* [1, 3, 5, 9, 10, 19, 21, 31]
<2> *Rhizobium leguminosarum* (biovar trifolii [2]) [2]

<3> *Rattus norvegicus* (vitamin B6-deficient strain [4]; aromatic amino acid-pyruvate transaminase and aromatic amino acid-2-oxoglutarate transaminase [14]) [4, 14]

<4> *Thermococcus litoralis* (aromatic aminotransferases I and II [6]) [6]

<5> *Brevibacterium linens* (strain 47 [7]) [7]

<6> *Bacillus brevis* (gramicidin S-producing [8]) [8]

<7> *Corynebacterium glutamicum* (isoenzymes AT-I and AT-II [11]) [11]

<8> *Brevibacterium flavum* (isoenzymes AT-I and AT-II [11]) [11]

<9> *Brevibacterium ammoniagenes* [11]

<10> *Klebsiella aerogenes* (aromatic aminotransferase I [12]) [12]

<11> *Flavobacterium sp.* (strains CB 60 and CB 6 [13]) [13]

<12> *Candida maltosa* [15]

<13> *Bacillus subtilis* (isoforms A and B2, major form is B2 [16]) [16, 17]

<14> *Methanococcus aeolicus* [6, 18]

<15> *Pyrococcus furiosus* (aromatic aminotransferase II [33]) [20, 33]

<16> *Trypanosoma brucei* [22]

<17> *Crithidia fasciculata* [22]

<18> *Mus musculus* [22]

<19> *Azospirillum brasilense* (rhizosphere microorganism, strain UAP 14, aromatic amino acid transaminase AAT1, predominant isoform [23]) [23]

<20> *Vigna radiata* (mung bean [24]) [24]

<21> *Paracoccus denitrificans* [25]

<22> *Lactococcus lactis* (lactis S3 [26]) [26]

<23> *Paracoccus denitrificans* [27, 29]

<24> *Saccharomyces cerevisiae* (ARO 8 gene encoding aromatic aminotransferase I [28]) [28]

<25> *Saccharomyces cerevisiae* (ARO 9 gene encoding aromatic aminotransferase II, expression is induced when aromaticc amino acids are present in the growth medium [28]) [28]

<26> *Lactococcus lactis* (strain cremoris [30]) [30]

<27> *Pyrococcus horikoshii* [32]

<28> *Enterobacter sp.* (BK2K-1 [34]) [34]

3 Reaction and Specificity

Catalyzed reaction

an aromatic amino acid + 2-oxoglutarate = an aromatic oxo acid + L-glutamate (<13> ping pong mechanism [17]; <4> two-step mechanism with a pyridoxamine intermediate [6]; <11> ping-pong bi-bi mechanism [13]; <1> ping-pong mechanism [5]; <1> proposed reaction mechanism with aspartate and phenylalanine [31])

Reaction type

amino group transfer

Natural substrates and products

S L-phenylalanine + 2-oxoglutarate <4, 5, 15, 26> (<4> may play a catabolic role in proteolysis, generation of glutamate [6]; <5> probably key enzyme for utilization of aromatic amino acids as sole nitrogen source in Brevibacterium linens 47 [7]; <26> enzyme may be involved in the degradation of amino acids to aroma compounds [30]; <15> AroAT I and II may be involved in amino acid degradation [33]) [6, 7, 30, 33]

P phenylpyruvate + L-glutamate

S α-ketomethiobutyrate + phenylalanine <16, 17> (<16, 17> best amino donor [22]) [22]

P methionine + phenylpyruvate

S α-ketomethiobutyrate + tryptophan <16, 17> [22]

P methionine + indolepyruvate

S α-ketomethiobutyrate + tyrosine <16, 17> (<16, 17> probably involved in the recycling of methionine after the synthesis of polyamines [22]) [22]

P methionine + hydroxyphenylpyruvate

Substrates and products

S 2-amino-3-phenylpropanol + 2-oxoglutarate <11> (<11> aromatic aminotransferase from strain CB 6 [13]) (Reversibility: ? <11> [13]) [13]

P 2-oxo-3-phenylpropanol + L-glutarate <11> [13]

S 2-oxo-4-phenylbutyric acid + L-aspartate <28> (Reversibility: ? <28> [34]) [34]

P L-homophenylalanine + oxaloacetate <28> (<28> 99% enantiomeric excess of L-homophenylalanine [34]) [34]

S 3-iodo-L-tyrosine + 2-oxoglutarate <1> (Reversibility: ? <1> [3]) [3]

P 3-(3-iodophenyl)-2-oxopropanoate + L-glutamate <1> [3]

S 3-iodo-L-tyrosine + oxaloacetate <1> (Reversibility: ? <1> [3]) [3]

P 3-(3-iodophenyl)-2-oxopropanoate + L-aspartate <1> [3]

S 4-hydroxyphenylpyruvate + L-tryptophan <7-9> (Reversibility: ? <7-9> [11]) [11]

P L-tyrosine + 3-indole-2-oxopropanoate <7-9> [11]

S 5-hydroxytryptophan + 2-oxoglutarate <3> (<3> 19.5% of the activity with phenylalanine [4]) (Reversibility: ? <3> [4]) [4]

P 3-(5-hydroxyindole)-2-oxopropanoate + L-glutamate <3> [4]

S L-5-fluorotryptophan + 2-oxoglutarate <19> (Reversibility: ? <19> [23]) [23]

P 3-indole-2-oxopropanoate + L-glutamate <19> [23]

S L-5-methyltryptophan + 2-oxoglutarate <19> (Reversibility: ? <19> [23]) [23]

P 3-indole-2-oxopropanoate + L-glutamate <19> [23]

S L-alanine + α-ketomethiobutyrate <16, 17> (Reversibility: ? <16,17> [22]) [22]

P pyruvate + L-methionine <16, 17> [22]

S L-arogenate + indole pyruvate <12> (Reversibility: ? <12> [15]) [15]

P L-tryptophan + 3-(1-carboxy-4-hydroxycyclohexa-2,5-dienyl)2-oxopropanoate <12> [15]

S L-aspartate + 2-oxoglutarate <21> (Reversibility: r <21> [25]) [25]
P oxaloacetate + L-glutamate <21> [25]
S L-aspartate + 3-(4-hydroxyphenyl)pyruvate <21> (Reversibility: r <21> [25]) [25]
P L-tyrosine + oxaloacetate <21> [25]
S L-aspartate + α-ketomethiobutyrate <16, 17> (Reversibility: ? <16,17> [22]) [22]
P 2-oxobutanoate + L-methionine <16, 17> [22]
S L-cysteine + α-ketomethiobutyrate <16, 17> (Reversibility: ? <16,17> [22]) [22]
P 2-oxo-3-thiopropanoate + L-methionine <16, 17> [22]
S L-glutamine + indole pyruvate <12> (<12> AAT I and II [15]) (Reversibility: ? <12> [15]) [15]
P 2-oxoglutamine + L-tryptophan <12> [15]
S L-histidine + 2-oxoglutarate <2, 6, 10, 19> (<6> 27.8% of activity with phenylalanine [8]; <19> 28% of activity with L-tyrosine [23]) (Reversibility: ? <2, 6, 10, 19> [2, 8, 12, 23]) [2, 8, 12, 23]
P 3-(1H-imidazol-4-yl)-2-oxopropanoate + L-glutamate <2, 6, 10, 19> [2, 8, 12, 23]
S L-histidine + α-ketomethiobutyrate <17> (Reversibility: ? <17> [22]) [22]
P 3-(1H-imidazol-4-yl)-2-oxopropanoate + L-methionine <17> [22]
S L-histidine + indole pyruvate <12> (Reversibility: ? <12> [15]) [15]
P 3-(1H-imidazol-4-yl)-2-oxopropanoate + L-tryptophan <12> [15]
S L-isoleucine + α-ketomethiobutyrate <17> (Reversibility: ? <17> [22]) [22]
P 3-methyl-2-oxo-pentanoate + L-methionine <17> [22]
S L-leucine + 2-oxoglutarate <1, 22> (<1> weak activity [3]) (Reversibility: ? <1,22> [3,26]) [3, 5, 26]
P 4-methyl-2-oxopentanoate + L-glutamate <1, 22> [3, 5, 26]
S L-leucine + indole pyruvate <12> (Reversibility: ? <12> [15]) [15]
P 4-methyl-2-oxopentanoate + L-tryptophan <12> [15]
S L-lysine + α-ketomethiobutyrate <17> (Reversibility: ? <17> [22]) [22]
P 2-oxo-6-amino-hexanoate + L-methionine <17> [22]
S L-methionine + 2-oxoglutarate <1, 5, 6, 22> (<1> weak activity [3]; <6> 0.6% of activity with phenylalanine [8]) (Reversibility: ? <1, 5, 6, 22> [3, 7, 8, 26]) [3, 7, 8, 26]
P 4-methylsulfanyl-2-oxobutanoate + L-glutamate <1, 5, 6, 22> [3, 7, 8, 26]
S L-methionine + indole pyruvate <12> (<12> AAT I and II [15]) (Reversibility: ? <12> [15]) [15]
P 4-methylsulfanyl-2-oxobutanoate + L-tryptophan <12> [15]
S L-phenylalanine + 2-oxoglutarate <1-15, 19, 20, 21, 22, 26, 27> (<3> best amino acid substrate, specific for 2-oxoglutarate, little or no activity with glyoxylate, pyruvate, 2-oxo-4-methylthiobutyrate and α-ketoadipate [4]; <12> specificities of isoenzymes: AAT I, AAT II, AAT III [15]; <7-9> specificities of isoenzymes AT-I and AT-II [11]; <7-9> various combinations of keto acids and amino acids [11]; <5,12> completely inactive with D-isomers of aromatic amino acids [7,15]; <19> very low activity with L-

aspartate, L-valine and L-isoleucine [23]; <21> enzyme also reacts with 2-aminooctanoate and 2-aminoheptanoate [25]; <15> best substrate of Ar-oAT II [33]) (Reversibility: r <1, 3, 7-9, 15, 21> [4, 5, 11, 20, 25]; ? <2-6, 10-14, 19, 20, 22, 26, 27> [1-3, 6-10, 12-18, 23, 24, 26, 30, 32]) [1-13, 15-18, 19, 20, 21, 23, 24, 25, 26, 30, 32, 33]

P phenylpyruvate + L-glutamate <1-15, 19, 20, 21, 22, 26, 27> [1-13, 15-18, 19, 20, 21, 23, 24, 25, 26, 30, 32, 33]

S L-phenylalanine + oxaloacetate <1, 2, 5, 6> (<6> 3.5% of activity with 2-oxoglutarate [8]) (Reversibility: ? <1, 2, 5, 6> [2, 3, 7, 8]) [2, 3, 7, 8]

P phenylpyruvate + L-aspartate <1, 2, 5, 6> [2, 3, 7, 8]

S L-phenylalanine + pyruvate <6> (<6> 2.1% of the activity with 2-oxoglu-tarate [8]) (Reversibility: ? <6> [8]) [8]

P phenylpyruvate + L-alanine <6> [8]

S L-phenylalanine-methylester + 2-oxoglutarate <11> (<11> aromatic ami-notransferase from strain CB 6 [13]) (Reversibility: ? <11> [13]) [13]

P 2-oxo-3-phenylpropanoate methylester + L-glutarate <11> [13]

S L-tryptophan + 2-oxoglutarate <1-6, 10-14, 15, 19, 20, 21, 22, 27> (<1> weak activity [3]; <3> 39.7% of activity with phenylalanine [4]; <20> best amino donor, reverse reaction at 30% of the forward reaction, enzyme is able to transaminate alanine, arginine, aspartate, leucine and lysine to a lesser extent, enzyme uses oxaloacetate oxaloacetate and pyruvate as ami-no acceptors [24]) (Reversibility: r <1, 15, 20, 21> [1, 3, 5, 9, 10, 19, 20, 24, 25]; ? <2-6, 10-14, 19, 22, 27> [2, 4, 6, 7, 8, 12, 13, 15-18, 23, 24, 26, 32]) [1-10, 12, 13, 15-19, 20, 21, 23, 24, 25, 26, 32]

P 3-indole-2-oxopropanoate + L-glutamate <1-6, 10-14, 19, 20, 21, 22, 27> [1-10, 12, 13, 15-19, 20, 21, 23, 24, 25, 26, 32]

S L-tryptophan + indole pyruvate <12> (Reversibility: ? <12> [15]) [15]

P 3-indole-2-oxopropanoate + L-tryptophan <12> [15]

S L-tryptophan + oxaloacetate <1, 2, 5, 12> (Reversibility: ? <1, 2, 5, 12> [2, 3, 7, 15]) [2, 3, 7, 15]

P 3-indole-2-oxopropanoate + L-aspartate <1, 2, 5, 12> [2, 3, 7, 15]

S L-tryptophan + phenylpyruvate <12> (Reversibility: ? <12> [15]) [15]

P 3-indole-2-oxopropanoate + L-phenylalanine <12> [15]

S L-tyrosine + 2-oxoglutarate <1-9, 12, 14, 15, 19, 20, 21, 22, 27> (<3> 23.7% of activity with phenylalanine [4]; <19> best amino donor [23]; <20> reverse reaction at 40% of the forward reaction [24]) (Reversibility: r <1, 15, 20, 21> [1, 3, 5, 9, 10, 19, 20, 24, 25]; ? <9, 12, 14, 19, 22, 27> [2, 4, 6, 7, 8, 11, 15, 18, 23, 26, 32]) [1-11, 15, 18, 19, 20, 21, 23, 24, 25, 26, 32]

P p-hydroxyphenylpyruvate + L-glutamate <1-9, 12, 14, 15, 19, 20, 21, 22, 27> (<1> i.e. 3-(4-hydroxyphenyl)-2-oxobutanoate + L-Glu [5]) [1-11, 15, 18, 19, 20, 21, 23, 24, 25, 26, 32]

S L-tyrosine + indole pyruvate <12> (Reversibility: ? <12> [15]) [15]

P 4-hydroxyphenylpyruvate + L-tryptophan <12> [15]

S L-valine + α-ketomethiobutyrate <17> (Reversibility: ? <17> [22]) [22]

P 3-methyl-2-oxo-butanoate + L-methionine <17> [22]

S α-ketomethiobutyrate + L-glutamine <16, 17, 18> (<18> best amino donor [22]; <17> 36% of activity with phenylalanine [22]; <16> 18% of activity with phenylalanine [22]) (Reversibility: ? <16,17,18> [22]) [22]

P   2-oxoglutaramide + L-methionine <18> [22]

S α-ketomethiobutyrate + L-phenylalanine <16, 17, 18> (<16, 17> best amino donor [22]) (Reversibility: ? <16,17,18> [22]) [22]

P   L-methionine + phenylpyruvate <16, 17, 18> [22]

S α-ketomethiobutyrate + L-tryptophan <16, 17, 18> (Reversibility: ? <16, 17, 18> [22]) [22]

P   L-methionine + indolepyruvate <16, 17, 18> [22]

S α-ketomethiobutyrate + L-tyrosine <16, 17, 18> (Reversibility: ? <16, 17, 18> [22]) [22]

P   L-methionine + hydroxyphenylpyruvate <16, 17, 18> [22]

S   arginine + α-ketomethiobutyrate <17> (Reversibility: ? <17> [22]) [22]

P   ? + methionine <17> [22]

S   aspartate + 2-oxoglutarate <1, 5, 6> (<1> weak activity [3]; <6> 3.1% of the activity with phenylalanine [8]) (Reversibility: r <1> [3, 5, 9, 19, 21]; ? <5, 6> [7,8]) [3, 5, 7-9, 19, 21]

P   oxaloacetate + L-glutamate <1, 5, 6> [3, 5, 7-9, 19, 21]

S   aspartate + 3-phenylpyruvate <21> (Reversibility: r <21> [25]) [25]

P   oxaloacetate + L-phenylalanine <21> [25]

S   p-aminophenylalanine + 2-oxoglutarate <11> (<11> aromatic aminotransferase from strain CB 60 and CB 6 [13]) (Reversibility: ? <11> [13]) [13]

P   p-aminophenylpyruvate + L-glutamate <11> [13]

S   p-aminophenylserine + 2-oxoglutarate <11> (<11> aromatic aminotransferase from strain CB 60 and CB 6 [13]) (Reversibility: ? <11> [13]) [13]

P   3-hydroxy-2-oxo-3-(p-aminophenyl)propanoic acid + L-glutarate <11> [13]

S   p-aminophenylserinol + 2-oxoglutarate <11> (<11> aromatic aminotransferase from strain CB 60 and CB 6 [13]) (Reversibility: ? <11> [13]) [13]

P   3-(p-aminophenyl)-1,3-propanediol + L-glutarate <11> [13]

S   p-nitrophenylserine + 2-oxoglutarate <11> (<11> aromatic aminotransferase from strain CB 60 and CB 6 [13]) (Reversibility: ? <11> [13]) [13]

P   3-hydroxy-2-oxo-3-(p-nitrophenyl)propanoic acid + L-glutarate <11> [13]

S   prephenate + L-glutamate <7-9, 12> (<7-9> highest activity [11]) (Reversibility: ? <7-9,12> [11,15]) [11, 15]

P   1-carboxy-4-hydroxy-2,5-cyclohexadien-1-alanine + 2-oxoglutarate <7-9, 12> [11, 15]

S   prephenate + L-tryptophan <12> (Reversibility: ? <12> [15]) [15]

P   1-carboxy-4-hydroxy-2,5-cyclohexadien-1-alanine + 3-indole-2-oxopropanoate <12> [15]

Inhibitors

2,4,6-trichlorophenoxyacetic acid <23> [29]

3-(2-hydroxyphenyl)propionic acid <23> [29]

3-(2-methoxyphenyl)propionic acid <23> [29]
3-(3,4,5-trimethoxyphenyl)propionic acid <23> [29]
3-(3,4-dimethoxyphenyl)propionic acid <23> [29]
3-(3-methoxyphenyl)propionic acid <23> [29]
3-(4-hydroxyphenyl)propionic acid <23> [29]
3-(4-methoxyphenyl)propionic acid <23> [29]
3-(p-tolyl)propionic acid <23> [29]
3-cyclohexylpropionic acid <23> [29]
3-cyclopentylpropionic acid <23> [29]
3-indoleacetic acid <23> [29]
3-indolebutyric acid <23> [29]
3-indolepropionic acid <23> [29]
3-phenylpropionic acid <23> [29]
4-(2-thienyl)butyric acid <23> [29]
4-(4-nitrophenyl)butyric acid <23> [29]
4-aminohydrocinnamic acid <23> [29]
4-cyclohexylbutyric acid <23> [29]
4-hydroxyphenylpyruvate <12> (<12> substrate inhibition [15]) [15]
4-phenylbutyric acid <23> [29]
5-phenylvaleric acid <23> [29]
aminooxyacetate <10> [12]
cycloserine <11> (<11> 0.15 mM, 89% inhibition of aromatic aminotransferase from strain CB 6 [13]) [13]
cysteine <14> [18]
indole pyruvate <12> (<12> substrate inhibition [15]) [15]
indole-3-pyruvic acid <19> (<19> competitive inhibition [23]) [23]
isonicotinohydrazide <11> (<11> 36% inhibition of aromatic aminotransferase from strain CB 60, enzyme from strain CB 6 is not inhibited [13]) [13]
methionine <14> (<14> 10 mM, inhibits activity of isoenzyme ArAT-I with 1 mM tyrosine by 23%, isoenzyme ArAT-II is not affected [18]) [18]
phenylhydrazine <7-9, 11> (<11> 0.15 mM, 78% inhibition of aromatic aminotransferase from strain CB 60, 56% inhibition of enzyme from strain CB 6 [13]) [11, 13]
phenylpyruvate <12> (<12> substrate inhibition [15]) [15]
prephenate <12> (<12> substrate inhibition [15]) [15]
Additional information <11, 20> (<11> enzyme from strain CB 60 is not inhibited by cycloserine [13]; <20> not inhibited by idoleacetic acid [24]) [13, 24]

Cofactors/prosthetic groups

pyridoxal 5'-phosphate <1, 4-10, 12, 13, 15, 21, 22, 23, 27> (<1, 4, 5, 7-10, 12> a pyridoxal phosphate protein [1, 3, 5-7, 9, 11, 12, 15]; <6> one mol of pyridoxal phosphate per subunit [8]; <1> K_m: 0.01 mM [1]; <12> K_m: 0.001 mM, isoenzyme AAT I, 0.0005 mM, isoenzyme AAT II, AAT III [15]; <13> K_m: 0.025 mM [17]; <1> apoenzyme can be reactivated with pyridoxal 5'-phosphate to a maximum of 60-70% activity in the presence of 2-oxoglutarate [5]; <5,11> tightly bound to enzyme [7,13]; <1> Shiff base is formed between

Lys258 of ArAT and pyridoxal 5'-phosphate [19]; <21> 1 pyridoxal 5'-phosphate per subunit [25]) [1, 3, 5-9, 11, 12, 15, 17, 19, 20, 25, 26, 29, 32] pyridoxamine 5'-phosphate <16, 17, 18> [22]

Additional information <20> (<20> addition of pyridoxal 5'-phosphate increases the activity of the purified enzyme only slightly [24]) [24]

Turnover number (min^{-1})

3720 <15> (L-tryptophan, <15> pH 7.6, 80°C, cosubstrate 2-oxoglutarate [20]) [20]

4320 <15> (L-tyrosine, <15> pH 7.6, 80°C, cosubstrate 2-oxoglutarate [20]) [20]

8400 <1> (L-aspartate, <1> pH 8.0, 25°C, cosubstrate 2-oxoglutarate [9]) [9]

9100 <1> (L-phenylalanine, <1> pH 7.3, 37°C [3]) [3]

9600 <1> (L-tryptophan, <1> pH 8.0, 25°C, cosubstrate 2-oxoglutarate [9]) [9]

12600 <1> (L-tyrosine, <1> pH 8.0, 25°C, cosubstrate 2-oxoglutarate [9]) [9]

13500 <15> (2-oxoglutarate, <15> pH 7.6, 80°C [20]) [20]

15000 <1> (L-phenylalanine, <1> pH 8.0, 25°C, cosubstrate 2-oxoglutarate [9]) [9]

15000 <1> (L-tryptophan, <1> pH 7.3, 37°C [3]) [3]

15180 <15> (L-phenylalanine, <15> pH 7.6, 80°C, cosubstrate 2-oxoglutarate [20]) [20]

16700 <1> (2-oxoglutarate, <1> pH 7.3, 37°C [3]) [3]

17400 <1> (L-aspartate, <1> pH 8.0, 25°C, single-turnover, wild-type enzyme [21]) [21]

20000 <1> (L-tyrosine, <1> pH 7.3, 37°C [3]) [3]

21000 <1> (L-tryptophan, <1> pH 8.0, 25°C, single-turnover, wild-type enzyme [21]) [21]

30000 <1> (L-tyrosine, <1> pH 8.0, 25°C, single-turnover, wild-type enzyme [21]) [21]

30000 <1> (oxaloacetate, <1> pH 7.3, 37°C [3]) [3]

72000 <1> (L-phenylalanine, <1> pH 8.0, 25°C, single-turnover, wild-type enzyme [21]) [21]

Specific activity (U/mg)

0.49 <11> (<11> strain CB 6 [13]) [13]

1.59 <11> (<11> strain CB 60 [13]) [13]

1.8 <5> (<5> with tyrosine [7]) [7]

2.96 <5> (<5> with aspartate [7]) [7]

3.8 <5> (<5> with tryptophan [7]) [7]

6.5 <5> (<5> with phenylalanine [7]) [7]

15.4 <3> [4]

26 <19> [23]

70.2 <1> [3]

95 <13> [17]

110 <4> (<4> isoenzyme ArAT-I [6]) [6]

170 <1> [9]

181.8 <15> [20]

353 <6> [8]
424 <4> (<4> isoenzyme ArAT-II [6]) [6]
950 <13> [17]

K$_m$-Value (mM)

0.001 <12> (pyridoxal 5'-phosphate, <12> pH 8.6, 37°C, AAT I [15]) [15]
0.02 <10> (phenylpyruvate, <10> cosubstrate glutamate [12]) [12]
0.025 <13> (pyridoxal 5'-phosphate, <13> pH 7.5, 37°C [17]) [17]
0.032 <1, 13> (4-hydroxyphenylpyruvate, <1> pH 7.6, 37°C [5]) [5, 16]
0.042 <1> (L-tyrosine, <1> pH 7.6, 37°C [5]) [5]
0.05 <12> (4-hydroxyphenylpyruvate, <12> pH 8.6, 37°C, AAT I [15]) [15]
0.056 <1> (phenylpyruvate, <1> pH 7.6, 37°C [5]) [5]
0.059 <21> (3-(4-hydroxyphenyl)pyruvate, <21> pH 8.0, 25°C, cosubstrate aspartate [25]) [25]
0.06 <1> (L-phenylalanine, <1> pH 7.6, 37°C [5]) [5]
0.06 <12> (indolepyruvate, <12> pH 8.6, 37°C, AAT I [15]) [15]
0.07 <20> (L-phenylalanine) [24]
0.08 <10> (L-phenylalanine, <10> cosubstrate 2-oxoglutarate [12]) [12]
0.08 <13> (L-phenylalanine, <13> pH 7.6, 37°C, isoenzyme B2 [16]) [16]
0.08 <20> (L-tyrosine) [24]
0.08 <12> (phenylpyruvate, <12> pH 8.6, 37°C, AAT I [15]) [15]
0.08 <12> (prephenate, <12> pH 8.6, 37°C, AAT I [15]) [15]
0.083 <21> (phenylpyruvate, <21> pH 8.0, 25°C, cosubstrate aspartate [25]) [25]
0.095 <20> (L-tryptophan) [24]
0.1 <20> (indolepyruvate) [24]
0.11 <6> (L-tyrosine, <6> pH 8.0 [8]) [8]
0.12 <10> (oxaloacetate, <10> cosubstrate tryptophan [12]) [12]
0.14 <21> (2-oxoglutarate, <21> pH 8.0, 25°C, cosubstrate tryptophan [25]) [25]
0.14 <13> (L-phenylalanine, <13> pH 7.6, 37°C, isoenzyme A [16]) [16]
0.15 <6> (2-oxoglutarate, <6> pH 8.0 [8]) [8]
0.17 <6> (L-phenylalanine, <6> pH 8.0 [8]) [8]
0.19 <19> (L-tyrosine) [23]
0.21 <6> (L-tryptophan, <6> pH 8.0 [8]) [8]
0.23 <4> (2-oxoglutarate) [6]
0.23 <1> (2-oxoglutarate, <1> pH 7.6, 37°C [5]) [5]
0.26 <1> (L-phenylalanine, <1> pH 8.0, 25°C [9]) [9]
0.28 <1> (L-glutamate, <1> pH 7.6, 37°C [5]) [5]
0.31 <13> (L-tyrosine, <13> pH 7.6, 37°C, isoenzyme B2 [16]) [16]
0.32 <1> (L-tyrosine, <1> pH 8.0, 25°C, cosubstrate 2-oxoglutarate [9]) [9]
0.33 <1> (L-phenylalanine, <1> pH 7.3, 37°C [3]) [1, 3]
0.33 <16, 17> (α-ketomethiobutyrate, <16,17> 37°C [22]) [22]
0.35 <19> (L-histidine) [23]
0.36 <19> (L-5-fluorotryptophan) [23]
0.4 <12> (L-tyrosine, <12> pH 8.6, 37°C, AAT I [15]) [15]
0.43 <19> (L-phenylalanine) [23]

0.44 <4> (2-oxoglutarate, <4> pH 7.6, 78°C, isoenzyme ArAT-I [6]) [6]

0.44 <12> (L-phenylalanine, <12> pH 8.6, 37°C, AAT I [15]) [15]

0.45 <12> (L-arogenat, <12> pH 8.6, 37°C, AAT I [15]) [15]

0.49 <4> (2-oxoglutarate, <4> pH 7.6, 78°C, isoenzyme ArAT-II [6]) [6]

0.5 <1> (L-tryptophan, <1> pH 7.6, 37°C [5]; <1> pH 8.0, 25°C, cosubstrate 2-oxoglutarate [9]) [5, 9]

0.54 <3> (2-oxoglutarate, <3> cosubstrate phenylalanine [4]) [4]

0.55 <4> (L-phenylalanine, <4> pH 7.6, 78°C, isoenzyme ArAT-I [6]) [6]

0.57 <16, 17> (L-tyrosine, <16,17> 37°C [22]) [22]

0.58 <19> (2-oxoglutarate, <19> cosubstrate L-tyrosine [23]) [23]

0.58 <13> (L-tyrosine, <13> pH 7.6, 37°C, isoenzyme A [16]) [16]

0.59 <1> (2-oxoglutarate, <1> pH 8.0, 25°C, cosubstrate tryptophan [9]) [9]

0.59 <21> (2-oxoglutarate, <21> pH 8.0, 25°C, cosubstrate aspartate [25]) [25]

0.6 <10> (L-tryptophan, <10> cosubstrate 2-oxoglutarate [12]) [12]

0.62 <1> (L-tyrosine, <1> pH 7.3, 37°C [3]) [1, 3]

0.68 <1> (L-tryptophan, <1> pH 8.0, 25°C, single-turnover, wild-type enzyme [21]) [21]

0.7 <5> (L-tryptophan, <5> pH 8.5, 37°C [7]) [7]

0.76 <15> (2-oxoglutarate, <15> pH 7.6, 80°C [20]) [20]

0.8 <1> (2-oxoglutarate, <1> pH 8.0, 25°C, cosubstrate aspartate [9]) [9]

0.8 <20> (hydroxyphenylpyruvate) [24]

0.83 <1> (L-tyrosine, <1> pH 8.0, 25°C, single-turnover, wild-type enzyme [21]) [21]

0.97 <11> (2-oxoglutarate, <11> aromatic amino acid transferase from strain CB 6 [13]) [13]

1 <1> (L-phenylalanine, <1> pH 8.0, 25°C, single-turnover, wild-type enzyme [21]) [21]

1.05 <19> (L-tryptophan) [23]

1.15 <15> (L-phenylalanine, <15> pH 7.6, 80°C, cosubstrate 2-oxoglutarate [20]) [20]

1.2 <19> (L-5-methyltryptophan) [23]

1.3 <1> (2-oxoglutarate, <1> pH 8.0, 25°C, cosubstrate tyrosine [9]) [9]

1.3 <5> (L-phenylalanine, <5> pH 8.5, 37°C [7]) [7]

1.31 <15> (L-tryptophan, <15> pH 7.6, 80°C, cosubstrate 2-oxoglutarate [20]) [20]

1.33 <11> (L-phenylalanine, <11> aromatic amino acid transferase from strain CB 6 [13]) [13]

1.37 <11> (L-tryptophan, <11> aromatic amino acid transferase from strain CB 6 [13]) [13]

1.4 <21> (L-tryptophan, <21> pH 8.0, 25°C, cosubstrate 2-oxoglutarate [25]) [25]

1.4 <11> (L-tyrosine, <11> aromatic amino acid transferase from strain CB 6 [13]) [13]

1.4 <5> (L-tyrosine, <5> pH 8.5, 37°C [7]) [7]

1.44 <19> (2-oxoglutarate, <19> cosubstrate L-tryptophan [23]) [23]

1.47 <4> (L-tryptophan, <4> pH 7.6, 78°C, isoenzyme ArAT-I [6]) [6]

1.5 <1> (L-methionine, <1> pH 7.6, 37°C [5]) [5]

1.55 <4> (L-phenylalanine, <4> pH 7.6, 78°C, isoenzyme ArAT-II [6]) [6]

1.64 <16, 17> (L-tryptophan, <16,17> 37°C [22]) [22]

1.7 <1> (2-oxoglutarate, <1> pH 8.0, 25°C, cosubstrate phenylalanine [9]) [9]

1.7 <11> (2-oxoglutarate, <11> aromatic amino acid transferase from strain CB 60 [13]) [13]

1.8 <21> (L-aspartate, <21> pH 8.0, 25°C, cosubstrate 2-oxoglutarate [25]) [25]

2 <1> (2-oxo-4-methylpentanoate, <1> pH 7.6, 37°C [5]) [5]

2.08 <10> (2-oxoglutarate, <10> cosubstrate tryptophan [12]) [12]

2.1 <15> (L-tyrosine, <15> pH 7.6, 80°C, cosubstrate 2-oxoglutarate [20]) [20]

2.18 <16, 17> (phenylalanine, <16,17> 37°C [22]) [22]

2.37 <4> (L-tyrosine, <4> pH 7.6, 78°C, isoenzyme ArAT-II [6]) [6]

2.39 <4> (L-tyrosine, <4> pH 7.6, 78°C, isoenzyme ArAT-I [6]) [6]

2.5 <1> (2-oxoglutarate, <1> pH 7.3, 37°C [3]) [1, 3]

2.5 <12> (L-tryptophan, <12> pH 8.6, 37°C, AAT I [15]) [15]

2.7 <13> (L-tyrosine, <13> pH 7.5, 37°C [17]) [17]

3.1 <13> (2-oxoglutarate, <13> pH 7.6, 37°C, isoenzyme A [16]) [16]

3.13 <1> (oxaloacetate, <1> pH 7.3, 37°C [3]) [1, 3]

3.3 <12> (oxalacetate, <12> pH 8.6, 37°C, AAT I [15]) [15]

3.7 <6> (L-aspartate, <6> pH 8.0 [8]) [8]

3.8 <1> (L-aspartate, <1> pH 8.0, 25°C, cosubstrate 2-oxoglutarate [9]) [9]

3.8 <1> (oxaloacetate, <1> pH 7.6, 37°C [5]) [5]

4 <12> (L-leucine, <12> pH 8.6, 37°C, AAT I [15]) [15]

4.62 <4> (L-tryptophan, <4> pH 7.6, 78°C, isoenzyme ArAT-II [6]) [6]

5 <12> (2-oxoglutarate, <12> pH 8.6, 37°C, AAT I [15]) [15]

5 <1> (L-aspartate, <1> pH 7.6, 37°C [5]; <1> pH 8.0, 25°C, single-turnover, wild-type enzyme [21]) [5, 21]

5 <12> (L-histidine, <12> pH 8.6, 37°C, AAT I [15]) [15]

5 <12> (L-methionine, <12> pH 8.6, 37°C, AAT I [15]) [15]

5.1 <6> (oxaloacetate, <6> pH 8.0 [8]) [8]

5.8 <1> (L-leucine, <1> pH 7.6, 37°C [5]) [5]

7 <12> (L-glutamine, <12> pH 8.6, 37°C, AAT I [15]) [15]

7.1 <6> (L-leucine, <6> pH 8.0 [8]) [8]

7.7 <3> (L-phenylalanine, <3> pH 8.0. 37°C [4]) [4]

7.8 <13> (2-oxoglutarate, <13> pH 7.6, 37°C, isoenzyme B2 [16]) [16]

8.4 <5> (L-aspartate, <5> pH 8.5, 37°C [7]) [7]

8.8 <6> (pyruvate, <6> pH 8.0 [8]) [8]

9.8 <13> (L-phenylalanine, <13> pH 7.5, 37°C [17]) [17]

10 <1, 3> (L-tryptophan, <1> pH 7.3, 37°C [3]) [1, 3, 4]

11.8 <3> (L-tyrosine, <3> pH 8.0. 37°C [4]) [4]

15 <11> (L-tyrosine, <11> aromatic amino acid transferase from strain CB 60 [13]) [13]

16.6 <11> (L-phenylalanine, <11> aromatic amino acid transferase from strain CB 60 [13]) [13]

16.6 <11> (L-tryptophan, <11> aromatic amino acid transferase from strain CB 60 [13]) [13]

38 <13> (2-oxoglutarate, <13> pH 7.5, 37°C [17]) [17]

70 <10> (L-histidine, <10> cosubstrate 2-oxoglutarate [12]) [12]

Additional information <1, 12> (<1> K_m values of half-reactions [9]; <12> K_m values of AAT I and II are very similar for a variety of substrates, K_m values for AAT II are higher [15]) [9, 15]

K_i-Value (mM)

0.17 <19> (indole-3-pyruvic acid) [23]

0.3 <12> (4-hydroxyphenylpyruvate, <12> AAT III, substrate inhibition [15]) [15]

0.3 <12> (indolepyruvate, <12> AAT I and III, substrate inhibition [15]) [15]

0.4 <12> (4-hydroxyphenylpyruvate, <12> AAT I, substrate inhibition [15]) [15]

0.4 <12> (prephenate, <12> AAT I and III, substrate inhibition [15]) [15]

0.5 <12> (phenylpyruvate, <12> AAT I and III, substrate inhibition [15]) [15]

1.5 <12> (indolepyruvate, <12> AAT II, substrate inhibition [15]) [15]

10 <12> (4-hydroxyphenylpyruvate, <12> AAT II, substrate inhibition [15]) [15]

15 <12> (phenylpyruvate, <12> AAT II, substrate inhibition [15]) [15]

15 <12> (prephenate, <12> AAT II, substrate inhibition [15]) [15]

pH-Optimum

7.2-7.6 <1> (<1> phenylalanine [5]) [5]

7.3 <2, 13> (<2> phenylalanine [2]) [2, 17]

7.5 <1> (<1> tryptophan [1,3]) [1, 3]

7.6 <6> [8]

7.8-9 <11> (<11> strain CB 6 [13]) [13]

7.9 <2> (<2> histidine [2]) [2]

8 <1, 3, 10, 19> (<1> with tyrosine [1,3]; <1,3> with phenylalanine [1,3,4]; <10> with tryptophan [12]; <19> with L-phenylalanine and L-tyrosine as substrates [23]) [1, 3, 4, 12, 23]

8.4-9 <11> (<11> strain CB 60 [13]) [13]

8.5 <19> (<19> with L-tryptophan and its derivatives as substrates [23]) [23]

8.5-9 <5> (<5> phenylalanine [7]) [7]

8.7 <2> (<2> tryptophan [2]) [2]

pH-Range

5-10 <19> (<19> 50% of maximal activity at pH 6.0 and pH 9.5 [23]) [23]

6-9.5 <1> (<1> approx. 45% of maximal activity at pH 6.0,: approx. 60% of maximal activity at pH 9.5, phenylalanine [1]) [1]

6-9.6 <10> (<10> approx. 50% of maximal activity at pH 6.0 and pH 9.6 [12]) [12]

6-10 <1, 11> (<11> approx. 50% of maximal activity at pH 6.0 [13]; <1> approx. 40% of maximal activity at pH 6.0, tryptophan [1]; <11> approx. 60% of maximal activity at pH 10.0 [13]; <1> approx. 25% of maximal activity at pH 10.0, tryptophan [1]) [1, 13]

6.5-10 <5> (<5> approx. 45% of maximal activity at pH 6.5, approx. 50% of maximal activity at pH 10 [7]) [7]

7-9.5 <11> (<11> approx. 30% of maximal activity at pH 7.0 and pH 9.5 [13]) [13]

Temperature optimum (°C)

35 <2> [2]

37 <1, 3> (<1,3> assay at [1,4,14]) [1, 4, 14]

37-40 <5> (<5> at pH 8.5 [7]) [7]

40 <11> (<11> strain CB 60 [13]) [13]

57.5 <11> (<11> strain CB 6 [13]) [13]

90 <27> [32]

95-100 <4> [6]

Additional information <15> (<15> temperature optimum seems to be above 95°C [20]) [20]

Temperature range (°C)

30-95 <15> (<15> extremely thermostable aromatic aminotransferase from hyperthermophilic archaeon, very low activity at 30°C, approx. 50% of maximal activity at 65°C [20]) [20]

30-105 <4> (<4> virtually inactive at 30°C, approx. 50% of maximal ArAT-I activity at 70°C, approx. 50% of ArAT-II activity at 80°C, isoenzymes ArAT-I and II [6]) [6]

4 Enzyme Structure

Molecular weight

53000 <2> (<2> gel filtration [2]) [2]

55000-59000 <20> [24]

56000 <27> (<27> recombinant enzyme, gel filtration [32]) [32]

63500 <13> (<13> gel filtration [17]) [17]

64000 <13> (<13> form B2, gel filtration [16]) [16]

66000 <19> (<19> gel filtration [23]) [23]

71000 <6> (<6> gel filtration [8]) [8]

73000 <12, 13> (<13> form A, gel filtration [16]; <12> isoenzyme AAT-I, gel filtration [15]) [15, 16]

83000 <1, 21> (<1,21> gel filtration [9,25]) [9, 25]

84000 <22> (<22> homodimeric form, gel filtration [26]) [26]

85000 <12> (<12> isoenzyme AAT-II, gel filtration [15]) [15]

88000 <1> (<1> gel filtration [3]) [1, 3]

90000 <1, 10, 14> (<1> gel filtration [5]; <10> gel filtration [12]; <14> ArAT-II, gel filtration [18]) [5, 12, 18]

92000 <4, 15> (<4> isoenzyme ArAT-II, gel filtration [6]; <15> gel filtration [20]) [6, 20]
100000 <3> (<3> gel filtration, sucrose density gradient centrifugation [4]) [4]
105000 <12> (<12> isoenzyme AAT-III, gel filtration [15]) [15]
110000 <4> (<4> isoenzyme ArAT-I, gel filtration [6]) [6]
120000 <11> (<11> enzymes from strain CB 60 and CB 6, gel filtration [13]) [13]
126000 <5> (<5> gel filtration [7]) [7]
150000 <14> (<14> ArAT-I, gel filtration [18]) [18]
152000 <8> (<8> AT-I, gel filtration [11]) [11]
160000 <7> (<7> AT-I, gel filtration [11]) [11]
170000 <22> (<22> tetrameric form, gel filtration [26]) [26]
260000 <7-9> (<7-9> AT-II, gel filtration [11]) [11]

Subunits

dimer <1, 3, 4, 6, 10, 15, 19, 21, 22, 27> (<1> 2 * 42000-45000, SDS-PAGE [1]; <1> 2 * 46000, SDS-PAGE [5]; <1> 2 * 43000, SDS-PAGE [9]; <1> 2 * 43537, deduced from nucleotide sequence [9,10]; <3> 2 * 52000, SDS-PAGE [4]; <4> 2 * 47000, ArAT-I, SDS-PAGE [6]; <4> 2 * 45000, ArAT-II, SDS-PAGE [6]; <6> 2 * 35000, SDS-PAGE [8]; <10> 2 * 42000, SDS-PAGE [12]; <15> 2 * 44000, SDS-PAGE [20]; <19> 2 * 33000, SDS-PAGE [23]; <21> 2 * 42653, mass spectrometry [25]; <21> 2 * 42731, deduced from nucleotide sequence [25]; <22> 2 * 42000, dimeric enzyme form, SDS-PAGE [26]; <27> 2 * 42000, recombinant enzyme, SDS-PAGE [32]) [1, 4-6, 8-10, 12, 20, 23, 25, 26, 32]
tetramer <22> (<22> 4 * 42000, tetrameric enzyme form, SDS-PAGE [26]) [26]

5 Isolation/Preparation/Mutation/Application

Source/tissue

leaf <20> [24]
liver <18> [22]
shoot <20> [24]
small intestine <3> (<3> mucosa [4]; <3> not in muscle phase [4]) [4, 14]

Localization

cytosol <3> [4]

Purification

<1> (ammonium sulfate, 55°C, 5 min, calcium phosphate gel, DEAE-cellulose, isoelectric focusing [3]; ammonium sulfate, SAH-Sepharose, hydroxyapatite, pyridoxamine 5'-phosphate-Sepharose [5]) [1, 3, 5]
<3> (vitamin B6-deficient strain, ammonium sulfate, DEAE-cellulose, isoelectric focusing, Sephacryl s-200, hydroxylapatite [4]) [4, 14]
<4> (isoenzymes ArAT-I and ArAT-II, Q-Sepharose, DEAE-Sepharose, phenyl-Sepharose, Superdex-200, Mono Q [6]) [6]

<5> (Streptomycin sulfate and ammonium sulfate, DEAE-Trisacryl, HA-ultrogel, Sephacryl S-200, partial purification [7]) [7]

<6> (gramicidin S-producing strain, ammonium sulfate, heat treatment, DEAE-Sephadex, hydroxylapatite, pyridoxamine 5'-phosphate-Sepharose, Ultrogel Aca-34 [8]) [8]

<7> (isoenzymes AT-I and AT-II [11]) [11]

<8> (isoenzymes AT-I and AT-II [11]) [11]

<9> [11]

<10> (protamine, ammonium sulfate, DEAE-Sephadex, ultrogel AcA44, hydroxylapatite, isoelectric focusing [12]) [12]

<11> (strain CB 60, ammonium sulfate, DEAE-cellulose, Ultrogel AcA 34, strain CB 6, DEAE-cellulose, Ultrogel AcA 34 [13]) [13]

<12> (DEAE-cellulose, isolation of isoenzymes AAT I, AAT II and AAT III [15]) [15]

<13> (isoenzymes A and B2 [16]; 55°C, ammonium sulfate, DE52, phenylalanine-Sepharose, Sephadex G-150 [17]) [16, 17]

<14> (partial purification of enzyme forms ArAT-I and ArAT-II [18]) [18]

<15> (Q-Sepharose, ammonium sulfate, phenyl-Sepharose, S-Sepharose [20]) [20]

<19> (ammonium sulfate, DEAE-Sephacel, hydroxylapatite, Sephadex G-100, phenyl-Sepharose [23]) [23]

<20> (ammonium sulfate, gel filtration, anion exchange chromatography, Mono Q, phenyl-Superose [24]) [24]

<21> (ammonium sulfate, phenyl-Toyopearl, hydroxyapatite, Sephacryl S-200, recombinant enzyme [25]) [25]

<27> (recombinant enzyme, 80°C, HiTrap Q, HiLoad superdex 200 [32]) [32]

Crystallization

<23> (crystal structure of unliganded pdAroAT, pdAroAT in a complex with maleate and 3-phenylpropionate at 2.33 A, 2.5 A and 2.3 A resolution [27]; tertiary structures of pdAroAT complexed with nine kind of inhibitors: 3-indolebutyric acid, 4-phenylbutyric acid, 5-phenylvaleric acid, 4-(2-thienyl)butyric acid, cyclohexanepropionic acid, 4-amminohydrocinnamic acid, 3-(p-tolyl)-propionic acid and 3-(3,4-dimethoxyphenyl)propionic acid, crystals of the maleate complex of pdAroAT are made by the micro-seeding method using 24% poly(ethylene)glycol 4000, 200 mM sodium maleate, pH 5.7 and 5 mM $MgCl_2$ as precipitating buffer, cocrystallized inhibitor maleate is replaced by soaking the crystals in 24% poly(ethylene)glycol 4000, 100 mM sodium citrate, pH 5.7 containing 100 mM inhibitor for 10 h [29]) [27, 29]

<27> (hanging drop vapor diffusion, an equi-volume of 3 M 1,6-hexanediol solution at pH 7.5, 100 mM HEPES buffer, containing 10 mM $MgCl_2$ is added to a protein solution containing 1.6% AeATPh and 0.02 mM pyridoxal 5'-phosphate, a droplet of the solution is equilibrated with 1 ml of 3 M 1,6-hexane-di-ol solution, crystals are grown at room temperature for 1 week, crystal structure of the native enzyme 2.1 a resolution, heavy atom derivatives diffract to 3.0 A resolution [32]) [32]

Cloning

<1> (overexpression in Escherichia coli [9]) [9, 10, 31]
<15> (expression of AroAT II in Escherichia coli [33]) [33]
<21> (expression in Escherichia coli [25]) [25]
<27> (expression in Escherichia coli [32]) [32]
<28> (expression in Escherichia coli [34]) [34]
<24, 25> [28]

Engineering

K258A <1> (<1> pKa of the Schiff base formed between the coenzyme pyridoxal 5'-phosphate and Lys258 increases by 3.6 units [31]) [31]
R292A <1> (<1> same pKa value as wild-type [19]; <1> very low activity with aspartate, 5-10fold increase in K_m for aromatic amino acids [21]) [19, 21]
R292K <1> (<1> very low activity with aspartate, 10-100fold increase in K_m for aromatic amino acids [21]) [21]
R292L <1> (<1> very low activity with aspartate, 5-10fold increase in K_m for aromatic amino acids [21]) [21, 31]
R386L <1> (<1> pKa of the Schiff base formed between the coenzyme pyridoxal 5'-phosphate and Lys258 increases by 0.7 units [31]) [31]

6 Stability

pH-Stability

4-11 <27> (<27> no loss of activity after 24 h at 25°C [32]) [32]

Temperature stability

55 <1, 3, 10, 11, 13> (<1> 10 min, less than 10% of original activity towards tyrosine and phenylalanine [1]; <10> 10 min, 35% loss of activity [12]; <13> 10 min, 90% loss of activity without stabilizing agent, 10% loss of activity in the presence of 0.6 mM pyridoxal phosphate and 1 mM 2-oxoglutarate [17]; <3> approx. 30% and 27% loss of activity after 10 min with L-tyrosine and L-phenylalanine, respectively, as substrates [4]; <11> strain CB 60, complete inactivation after 8 min, 28% and 8% activity remain after 8 min in the presence of 2-oxoglutarate, strain CB 6, 68% loss of activity after 32 min, 35% and 22% loss of activity after 32 min in the presence of pyridoxal 5'-phosphate and 2-oxoglutarate respectively [13]; <13> 7 min, stable, isoenzymes A and B2 [16]; <1> 5% loss of activity after 5 min in the presence of 0.2 mM pyridoxal 5'-phosphate and 2 mM 2-oxoglutarate, 60% loss of activity in the absence of cofactor and substrate after 5 min, more than 90% after 10 min [3]) [1, 3, 4, 12, 13, 16, 17]
80 <15> (<15> holoenzyme, i.e. enzyme in its pyridoxal form, approx. 20% loss of phenylalanine transaminase activity after 6 h, apoenzyme, 70% loss of activity, complete protection of holoenzyme in the presence of pyridoxamine 5'-phosphate and 2-oxoglutarate [20]) [20]
95 <15, 27> (<15> 59% and 27% loss of activity after 13 h in the presence of pyridoxamine 5'-phosphate or 2-oxoglutarate [20]; <27> no loss of activity after 30 min at pH 6.5 [32]) [20, 32]

110 <27> (<27> half-life at pH 6.5: 30 min [32]) [32]
Additional information <11, 13> (<11,13> 2-oxoglutarate or pyridoxal 5'-phosphate stabilize against heat inactivation [13,17]) [13, 17]

General stability information

<1>, inactivated by freezing [5]
<1>, stable to freezing and thawing in phosphate buffers supplemented with pyridoxal phosphate, 2-oxoglutarate, dithiothreitol and EDTA [3]
<13>, stable to freezing and thawing [17]
<19>, 10% glycerol and 0.0001 mM pyridoxal 5'-phosphate stabilize [23]
<11, 13>, 2-oxoglutarate or pyridoxal 5'-phosphate stabilizes against heat inactivation [13, 17]

Storage stability

<1>, -25°C, protein concentration: 0.03-0.2 mg/ml, potassium phosphate pH 7.0, 1 mM EDTA, 1 mM dithiothreitol, 2 mM 2-oxoglutarate and 0.2 mM pyridoxal 5'-phosphate, several months, no loss of activity [3]
<3>, -20°C, at least 3 weeks, no loss of activity [4]
<3>, 0-4°C, 2 weeks, little loss of activity [4]
<5>, 4°C, 100 mM potassium phosphate buffer, pH 7.5, 0.002% NaN_3, at least 6 months, no loss of activity [7]
<6>, -20°C, 50 mM potassium phosphate buffer, pH 7.5, 10% glycerol, 10 mM DTT, 0.1 mM pyridoxal phosphate, at least 3 months, no loss of activity [8]
<10>, 4°C, glycerol, 10% stabilizes during storage [12]
<13>, 0°C, 1 month, 15% loss of activity [17]

References

[1] Mavrides, C.; Orr, W.: Multispecific aspartate and aromatic amino acid aminotransferases in Escherichia coli. J. Biol. Chem., **250**, 4128-4133 (1975)
[2] Perez-Galdona, R.; Corzo, J.; Leon-Barrios, M.A.; Gutierrez-Navarro, A.M.: Characterization of an aromatic amino acid aminotransferase from Rhizobium leguminosarum biovar trifolii. Biochimie, **74**, 539-544 (1992)
[3] Mavrides, C.: Transamination of aromatic amino acids in Escherichia coli. Methods Enzymol., **142**, 253-267 (1987)
[4] Noguchi, T.: Aromatic-amino-acid aminotransferase from small intestine. Methods Enzymol., **142**, 267-273 (1987)
[5] Powell, J.T.; Morrison, J.F.: The purification and properties of the aspartate aminotransferase and aromatic-amino-acid aminotransferase from Escherichia coli. Eur. J. Biochem., **87**, 391-400 (1978)
[6] Andreotti, G.; Cubellis, M.V.; Nitti, G.; Sannia, G.; Mai, X.; Marino, G.; Adams, M.W.W.: Characterization of aromatic aminotransferases from the hyperthermophilic archaeon Thermococcus litoralis. Eur. J. Biochem., **220**, 543-549 (1994)
[7] Lee, C.W.; Desmazeaud, M.J.: Partial purification and some properties of an aromatic-amino-acid and an aspartate aminotransferase in Brevibacterium linens 47. J. Gen. Microbiol., **131**, 459-467 (1985)

[8] Kanda, M.; Hori, K.; Kurotsu, T.; Miura, S.; Saito, Y.: Purification and properties of the aromatic amino acid aminotransferase from gramicidin S-producing Bacillus brevis. J. Biochem., **101**, 871-878 (1987)

[9] Hayashi, H.; Inoue, K.; Nagata, T.; Kuramitsu, S.; Kagamiyama, H.: Escherichia coli aromatic amino acid aminotransferase: characterization and comparison with aspartate aminotransferase. Biochemistry, **32**, 12229-12239 (1993)

[10] Kuramitsu, S.; Inoue, K.; Ogawa, H.; Kagamiyama, H.: Aromatic amino acid aminotransferase of Escherichia coli: nucleotide sequence of the tyrB gene. Biochem. Biophys. Res. Commun., **133**, 134-139 (1985)

[11] Fazel, A.M.; Jensen, R.A.: Aromatic aminotransferases in coryneform bacteria. J. Bacteriol., **140**, 580-587 (1979)

[12] Paris, C.G.; Magasanik, B.: Purification and properties of aromatic amino acid aminotransferase from Klebsiella aerogenes. J. Bacteriol., **145**, 266-271 (1981)

[13] Beschle, H.G.; Süssmuth, R.; Lingens, F.: Properties of aromatic-amino-acid aminotransferases from two chloramphenicol-resistant Flavobacteria. Hoppe-Seyler's Z. Physiol. Chem., **363**, 1365-1375 (1982)

[14] Nakamura, J.; Noguchi, T.; Kido, R.: Aromatic amino acid transaminase in rat intestine. Biochem. J., **135**, 815-818 (1973)

[15] Bode, R.; Birnbaum, D.: Charakterisierung von drei aromatischen Aminotransferasen aus Candida maltosa. Z. Allg. Mikrobiol., **24**, 67-75 (1984)

[16] Mavrides, C.; Comberton, M.: Aminotransferases for aromatic amino acids and aspartate in Bacillus subtilis. Biochim. Biophys. Acta, **524**, 60-67 (1978)

[17] Weigent, D.A.; Nester, E.W.: Purification and properties of two aromatic aminotransferases in Bacillus subtilis. J. Biol. Chem., **251**, 6974-6980 (1976)

[18] Xing, R.; Whitman, W.B.: J. Bacteriol., **174**, 541-548 (1974)

[19] Iwasaki, M.; Hayashi, H.; Kagamiyama, H.: Protonation state of the active-site Schiff base of aromatic amino acid aminotransferase: modulation by binding of ligands and implications for its role in catalysis. J. Biochem., **115**, 156-161 (1994)

[20] Andreotti, G.; Cubellis, M.V.; Nitti, G.; Sannia, G.; Mai, X.; Marino, G.; Adams, M.W.W.: Characterization of aromatic aminotransferases from the hyperthermophilic archaeon Thermococcus litoralis. Eur. J. Biochem., **220**, 543-549 (1994)

[21] Hayashi, H.; Inoue, K.; Mizuguchi, H.; Kagamiyama, H.: Analysis of the substrate-recognition mode of aromatic amino acid aminotransferase by combined use of quasisubstrates and site-directed mutagenesis: systematic hydroxy-group addition/deletion studies to probe the enzyme-substrate interactions. Biochemistry, **35**, 6754-6761 (1996)

[22] Berger, B.J.; Dai, W.W.; Wang, H.; Stark, R.E.; Cerami, A.: Aromatic amino acid transamination and methionine recycling in trypanosomatids. Proc. Natl. Acad. Sci. USA, **93**, 4126-4130 (1996)

[23] Soto-Urzua, L.; Xochinua-Corona, Y.G.; Flores-Encarnacion, M.; Baca, B.E.: Purification and properties of aromatic amino acid aminotransferases from Azospirillum brasilense UAP 14 strain. Can. J. Microbiol., **42**, 294-298 (1996)

[24] Simpson, R.M.; Nonhebel, H.M.; Christie, D.L.: Partial purification and characterization of an aromatic amino acid aminotransferase from mung bean. Planta, **201**, 71-77 (1997)

[25] Oue, S.; Okamoto, A.; Nakai, Y.; Nakahira, M.; Shibatani, T.; Hayashi, H.; Kagamiyama, H.: Paracoccus denitrificans aromatic amino acid aminotransferase: a model enzyme for the study of dual substrate recognition mechanism. J. Biochem., **121**, 161-171 (1997)

[26] Gao, S.; Steele, J.L.: Purification and characterization of oligomeric species of an aromatic amino acid aminotransferase from Lactococcus lactis subsp. lactis S3. J. Food Biochem., **22**, 197-211 (1998)

[27] Okamoto, A.; Nakai, Y.; Hayashi, H.; Hirotsu, K.; Kagamiyama, H.: Crystal structures of Paracoccus denitrificans aromatic amino acid aminotransferase: a substrate recognition site constructed by rearrangement of hydrogen bond network. J. Mol. Biol., **280**, 443-461 (1998)

[28] Iraqui, I.; Vissers, S.; Cartiaux, M.; Urrestarazu, A.: Characterisation of Saccharomyces cerevisiae ARO8 and ARO9 genes encoding aromatic aminotransferases I and II reveals a new aminotransferase subfamily. Mol. Gen. Genet., **257**, 238-248 (1998)

[29] Okamoto, A.; Ishii, S.; Hirotsu, K.; Kagamiyama, H.: The active site of Paracoccus denitrificans aromatic amino acid aminotransferase has contrary properties: flexibility and rigidity. Biochemistry, **38**, 1176-1184 (1999)

[30] Rijnen, L.; Bonneau, S.; Yvon, M.: Genetic characterization of the major lactococcal aromatic aminotransferase and its involvement in conversion of amino acids to aroma compounds. Appl. Environ. Microbiol., **65**, 4873-4880 (1999)

[31] Islam, M.M.; Hayashi, H.; Mizuguchi, H.; Kagamiyama, H.: The substrate activation process in the catalytic reaction of Escherichia coli aromatic amino acid aminotransferase. Biochemistry, **39**, 15418-15428 (2000)

[32] Matsui, I.; Matsui, E.; Sakai, Y.; Kikuchi, H.; Kawarabayasi, Y.; Ura, H.; Kawaguchi, S.I.; Kuramitsu, S.; Harata, K.: The molecular structure of hyperthermostable aromatic aminotransferase with novel substrate specificity from Pyrococcus horikoshii. J. Biol. Chem., **275**, 4871-4879 (2000)

[33] Ward, D.E.; De Vos, W.M.; Van der Oost, J.: Molecular analysis of the role of two aromatic aminotransferases and a broad-specificity aspartate aminotransferase in the aromatic amino acid metabolism of Pyrococcus furiosus. Archaea, **1**, 133-141 (2002)

[34] Cho, B.K.; Seo, J.H.; Kang, T.W.; Kim, B.G.: Asymmetric synthesis of L-homophenylalanine by equilibrium-shift using recombinant aromatic L-amino acid transaminase. Biotechnol. Bioeng., **83**, 226-234 (2003)